T0329316

Solar Energy Engineering
Processes and Systems

Solar Energy Engineering
Processes and Systems
Third Edition

Soteris A. Kalogirou

Department of Mechanical Engineering and
Materials Sciences and Engineering,
Cyprus University of Technology,
Limassol, Cyprus

ELSEVIER

ACADEMIC PRESS

An imprint of Elsevier

ISBN: 978-0-323-99350-0

For information on all Academic Press publications visit our website at https://www.elsevier.com/books-and-journals

Publisher: Megan Ball
Acquisitions Editor: Edward Payne
Editorial Project Manager: Rupinder K. Heron
Production Project Manager: Kamesh R
Cover Designer: Matthew Limbert

Typeset by TNQ Technologies

Working together to grow libraries in developing countries
www.elsevier.com • www.bookaid.org

Contents

CHAPTER 7   Industrial Process Heat, Chemistry Applications, and Solar Dryers 405

Preface to the third edition

Throughout my whole scientific life, I have been trying to convince people to use renewables for their everyday energy needs. This has been done through many open public lectures and a great deal of research and papers published in scientific journals, as well as numerous books, book chapters, and presentations at conferences as either scientific presentations or lectures I have given as an invited keynote speaker. Nowadays, I am quite happy because nearly everybody is talking about renewables. Unfortunately, this is connected to the tragic effects of war and the devastating destruction that many people on the planet suffer because of climate change effects. In any case, let's view this as an advantage and make our energy systems independent of fossil fuels and our environment cleaner, better, and more friendly to humans.

This book is the third edition of *Solar Energy Engineering: Processes and Systems*. It includes only a few substantial changes from the previous edition. The most important ones are the addition of two new appendices to show details of the System Advisor Model (SAM) program now used in various chapters for student assignments and the description of Excel functions Goal Seek and Solver, which can help solve many of the examples and problems in various chapters. Perhaps the most important change is a companion site (https://www.elsevier.com/books-and-journals/book-companion/9780323993500) with extensive information about the book, problem answers, errata to the second edition, dedicated Excel spreadsheets that can be used as tools for various calculations, and many more resources. Instructors or teachers using the book for teaching can write to the author directly to receive supporting material (excel files for solving problems, PowerPoints, and solution manual) at Soteris.kalogirou@cut.ac.cy.

SAM is used to predict the long-term performance of many energy systems employed for electricity generation or using electricity as backup power. This is done by performing annual simulations. The program can be freely downloaded from https://sam.nrel.gov and is described in more detail in Appendix 10.

Excel Goal Seek is a handy tool for finding optimum solutions to problems based on one variable parameter. This is typically used in the book, for example, to guess the correct glass temperature of a single glass-covered flat plate collector that makes the heat flow rate from the absorber plate to the glass and from the glass to the ambient equal. For more variables, the Solver add-in can be used. A typical example is to guess the correct temperatures of the two glasses of a double-glazed flat plate collector. Both tools are described in more detail in Appendix 9, which also provides examples.

The changes to particular chapters are as follows:

Various figures in Chapter 1 have been updated with current values for various parameters, like proven fuel reserves and energy consumption. New additions include the European Green Deal and the Sustainable Development Goals. A section has also been added that shows the worldwide adoption of various renewable energy sources up to and including the year 2021. Information about the state of the climate is updated, as well as a section on wind energy systems that is updated with new data.

Chapter 2 has not been substantially changed, with only a few new equations added for specific parameters. Chapter 3 was modified by adding more details on the wind and infrared-sensitive collectors (also called unglazed collectors) and a section on the outlook for solar collectors for water heating. Moreover, a section has been added on the thermal analysis of parabolic trough collectors for considering gas in the annulus between the receiver and the glass cover. Finally, Major Assignment #1 has been added about the design of a flat plate collector using the analytic equations presented in Chapter 2 (optical analysis) and Chapter 3 (thermal analysis).

The main material on collector performance evaluation in Chapter 4 has remained unchanged, but the section on standardization has been changed to generally reflect changes to both ISO and EN standards. Finally, more details have been added to the data acquisition systems section, and a new section has been added on the Internet of Things in data acquisition.

Chapter 5 has also not been substantially modified except to correct a few mistakes in the previous edition. Meanwhile, in Chapter 6, the heat pumps section has been modified, and a new section on photovoltaic powered heating and cooling has been added. Chapter 7 has also been modified very little, with only some new data on the industrial process heat systems installed worldwide and details of one recently installed in Cyprus that employs concrete storage. This chapter also includes an assignment on industrial process heat based on SAM. The primary modification of Chapter 8 is the addition of a section on membrane distillation, including solar power membrane distillation.

Chapter 9 has been modified by adding a subsection on perovskite cells in the section describing the types of PV technologies and a new section on remote detection of PV system faults, diagnosis, and cleaning. The chapter now includes three assignments, two using the SAM program and one using the TRNSYS V.18 demo program.

Chapter 10 has been modified to include details on newly constructed concentrating solar power systems and an assignment using SAM to design 50 and 100 MW CSP systems.

Chapter 11 has been changed to clarify a few issues related to the f-chart method, and gives details how to download and use the TRNSYS V.18 demo program. It also discusses its limitations and introduces the SAM program. Furthermore, a section has been added that introduces machine and deep learning, as well as Major Assignment #2 for students on designing a solar thermal system with an f-chart. Other new assignments include one on using TRNSYS to simulate a solar domestic hot water system and another using TRNSYS to simulate a daylighting system.

Chapter 12's main modification is the addition of two assignments, one on constructing spreadsheet programs to solve the problems of this chapter and a second that continues Major Assignment #2 from Chapter 11 to include economic analysis and optimization of the system designed in the Chapter 11 assignment.

Finally, Chapter 13's only major modification is an assignment based on SAM to design a 50 MW wind energy system.

Soteris A. Kalogirou
Cyprus University of Technology
Founding Member of the Cyprus Academy of Sciences, Letters and Arts

Introduction

1.1 General introduction to renewable energy technologies

The sun is the only star of our solar system located at its center. The earth and the other planets of our solar system orbit the sun. Energy from the sun in the form of solar radiation supports almost all life on earth via photosynthesis and drives the earth's climate and weather.

About 74% of the sun's mass is hydrogen, 25% is helium, and the rest is made up of trace quantities of heavier elements. The sun has a surface temperature of approximately 5500 K, giving it a white color, which, because of atmospheric scattering, appears yellow. The sun generates its energy by nuclear fusion of hydrogen nuclei to helium. Sunlight is the main source of energy to the surface of the earth that can be harnessed via a variety of natural and synthetic processes. The most important is photosynthesis, used by plants to capture the energy of solar radiation and convert it to chemical form. Generally, photosynthesis is the synthesis of glucose from sunlight, carbon dioxide, and water, with oxygen as a waste product. It is arguably the most important known biochemical pathway, and nearly all life on earth depends on it.

Basically all the forms of energy in the world as we know it are solar in origin. Oil, coal, natural gas, and wood were originally produced by photosynthetic processes, followed by complex chemical reactions in which decaying vegetation was subjected to very high temperatures and pressures over a long period of time. Even the energy of the wind and tide has a solar origin, since they are caused by differences in temperature in various regions of the earth.

Since prehistory, the sun has dried and preserved humankind's food. It has also evaporated seawater to yield salt. Since humans began to reason, they have recognized the sun as a motive power behind every natural phenomenon. This is why many of the prehistoric tribes considered the sun as a god. Many scripts of ancient Egypt say that the Great Pyramid, one of humankind's greatest engineering achievements, was built as a stairway to the sun (Anderson, 1977).

From prehistoric times, people realized that a good use of solar energy is beneficial. The Greek historian Xenophon in his "memorabilia" records some of the teachings of the Greek philosopher Socrates (470−399 BC) regarding the correct orientation of dwellings to have houses that were cool in summer and warm in winter.

The greatest advantage of solar energy compared with other forms of energy is that it is clean and can be supplied without environmental pollution. Over the past century, fossil fuels provided most of our energy, because these were much cheaper and more convenient than energy from alternative energy sources, and until recently, environmental pollution has been of little concern.

Twelve autumn days of 1973, after the Egyptian army stormed across the Suez Canal on October 12, changed the economic relation of fuel and energy as, for the first time, an international crisis was created over the threat of the "oil weapon" being used as part of Arab strategy. Both the price and the political weapon issues were quickly materialized when the six Gulf members of the Organization of Petroleum Exporting Countries (OPEC) met in Kuwait and abandoned the idea of holding any more price consultations with the oil companies, announcing at the same time that they were raising the price of their crude oil by 70%.

The rapid increase in oil demand occurred mainly because increasing quantities of oil, produced at very low cost, became available during the 1950 and 1960s from the Middle East and North Africa. For the consuming countries, imported oil was cheap compared with indigenously produced energy from solid fuels.

The proven world oil reserves are equal to 1707 billion barrels (2019), the world coal reserves are 113.9 billion tons (2018), and the world natural gas reserves are 187 trillion m^3 (2017). Total consumption in 2019 was 136,761 TWh: the sum of 53,620 TWh or 88.4 million barrels per day for oil, 43,849 TWh or 27.4 million tons per day for coal, and 39,252 TWh or 10.4 billion m^3 per day for natural gas. Therefore, the main problem is that proven reserves of oil and gas, at current rates of consumption, would be adequate to meet demand for only another 50.7 and 52.8 years, respectively. The reserves for coal are in a better situation; they would be adequate for at least the next 114 years.

If we try to see the implications of these limited reserves, we are faced with a situation in which the price of fuels will accelerate as the reserves are decreased. Considering that the price of oil has become firmly established as the price leader for all fuel prices, the conclusion is that energy prices will increase continuously over the next decades. In addition, there is growing concern about the environmental pollution caused by burning fossil fuels. This issue is examined in Section 1.3.

The sun's energy has been used by both nature and humankind throughout time in thousands of ways, from growing food to drying clothes; it has also been deliberately harnessed to perform a number of other jobs. Solar energy is used to heat and cool buildings (both actively and passively), heat water for domestic and industrial uses, heat swimming pools, power refrigerators, operate engines and pumps, desalinate water for drinking purposes, generate electricity, for chemistry applications, and many more operations. The objective of this book is to present various types of systems used to harness solar energy, their engineering details, and ways to design them, together with some examples and case studies.

1.2 Energy demand and renewable energy

Many alternative energy sources can be used instead of fossil fuels. The decision as to what type of energy source should be utilized in each case must be made on the basis of economic, environmental, and safety considerations. Because of the desirable environmental and safety aspects it is widely believed that solar energy should be utilized instead of other alternative energy forms because it can be provided sustainably without harming the environment.

If the world economy expands to meet the expectations of countries around the globe, energy demand is likely to increase, even if laborious efforts are made to increase the energy use efficiency. It is now generally believed that renewable energy technologies can meet much of the growing demand at prices that are equal to or lower than those usually forecast for conventional energy. By the middle of the 21st century, renewable sources of energy could account for a substantial portion of the world's electricity market and the market for fuels used directly.[1] Moreover, making a transition to a renewable energy-intensive economy would provide environmental and other benefits not measured in standard economic terms. An early analysis by Johanson et al. (1993) envisaged that by 2050, global carbon dioxide (CO_2) emissions would be reduced to 75% of their levels in 1985, provided that energy efficiency and renewables are widely adopted. In addition, such benefits could be achieved at no additional cost, because renewable energy is expected to be competitive with conventional energy (Johanson et al., 1993).

This promising outlook for renewables reflects impressive technical gains made during the past 3 decades as renewable energy systems benefited from developments in electronics, biotechnology, material sciences, and in other areas. For example, fuel cells developed originally for the space program opened the door to the use of hydrogen as a non-polluting fuel for transportation.

Moreover, because the size of most renewable energy equipment is small, renewable energy technologies can advance at a faster pace than conventional technologies. While large energy facilities require extensive construction in the field, most renewable energy equipment can be constructed in factories, like mirrors, trackers, and PV panels, where it is easier to apply modern manufacturing techniques that facilitate cost reduction. This is a decisive parameter that the renewable energy industry must consider in an attempt to reduce cost and increase the reliability of manufactured goods. The small scale of the equipment also makes the time required from initial design to operation short; therefore, any improvements can be easily identified and incorporated quickly into modified designs or processes.

The renewable energy-intensive scenario foreseen by many studies in the early 1990s (Johanson et al., 1993) has materialized today with good prospects for further development. The current (2021) situation for installed renewables worldwide is presented in Section 1.3.6. Johanson et al. (1993) proposed a scenario with a high rate of penetration by intermittent renewables without energy storage would be facilitated by emphasis on advanced natural gas-fired turbine power-generating systems. Such power-generating systems—characterized by low capital cost, high thermodynamic efficiency, and the flexibility to vary electrical output quickly in response to changes in the output of intermittent power-generating systems—would make it possible to back up the intermittent renewables at low cost, with little, if any, need for energy storage.

The key elements of a renewable energy-intensive future are likely to have the following key characteristics as foreseen by Johanson et al. (1993):

1. There would be a diversity of energy sources, the relative abundance of which would vary from region to region. For example, electricity could be provided by various combinations of hydroelectric power, intermittent renewable power sources (wind, solar thermal electric, and photovoltaic (PV)), biomass,[2] and geothermal sources. Fuels could be provided by methanol,

---

[1]This is according to a renewable energy-intensive scenario that would satisfy energy demands associated with an eightfold increase in economic output for the world by the middle of the 21st century. In the scenario considered, world energy demand continues to grow in spite of a rapid increase in energy efficiency.

[2]The term *biomass* refers to any plant matter used directly as fuel or converted into fluid fuel or electricity. Biomass can be produced from a wide variety of sources such as wastes of agricultural and forest product operations as well as wood, sugarcane, and other plants grown specifically as energy crops.

ethanol, hydrogen, and methane (biogas) derived from biomass, supplemented with hydrogen derived electrolytically from intermittent renewables.

2. Emphasis would be given to the efficient mixing of renewable and conventional energy supplies. This can be achieved with the introduction of energy carriers such as methanol and hydrogen. It is also possible to extract more useful energy from such renewable resources as hydropower and biomass, which are limited by environmental or land-use constraints. Most methanol exports could originate in sub-Saharan Africa and Latin America, where vast degraded areas are suitable for revegetation that will not be needed for cropland. Growing biomass on such lands for methanol or hydrogen production could provide a powerful economic driver for restoring these lands. Solar-electric hydrogen exports could come from the regions in North Africa and the Middle East that have good insolation.

3. Biomass would be widely used. Biomass would be grown sustainably and converted efficiently to electricity and liquid and gaseous fuels using modern technology without contributing to deforestation.

4. Intermittent renewables would provide a large quantity of the total electricity requirements cost-effectively, without the need for new electrical storage technologies.

5. Natural gas would play a major role in supporting the growth of a renewable energy industry. Natural gas-fired turbines, which have low capital costs and can quickly adjust their electrical output, can provide excellent backup for intermittent renewables on electric power grids. Natural gas would also help launch a biomass-based methanol industry.

6. A renewables-intensive energy future would introduce new choices and competition in energy markets. Growing trade in renewable fuels and natural gas would diversify the mix of suppliers and the products traded, which would increase competition and reduce the possibility of rapid price fluctuations and supply disruptions. This could also lead eventually to a stabilization of world energy prices with the creation of new opportunities for energy suppliers.

7. Most electricity produced from renewable sources would be fed into large electrical grids and marketed by electric utilities, without the need for electrical storage.

A renewable energy-intensive future is technically feasible, as proven today (2021), and the prospects are very good that a wide range of renewable energy technologies which are already competitive with conventional energy sources will probably be cheaper in a few years' time. However, to achieve such penetration of renewables, existing market conditions must change.

Fortunately, the policies needed to achieve the goals of increasing efficiency and expanding renewable energy markets are fully consistent with programs needed to encourage innovation and productivity growth throughout the economy. Given the right policy environment, energy industries will adopt innovations, driven by the same competitive pressures that revitalized other major manufacturing businesses around the world. Electric utilities have already shifted from being protected monopolies, enjoying economies of scale in large generating plants, to being competitive managers of investment portfolios that combine a diverse set of technologies, ranging from advanced generation, transmission, distribution, and storage equipment to efficient energy-using devices on customers' premises.

Capturing the potential for renewables requires new policy initiatives. Most of the following policy initiatives proposed by Johanson et al. (1993) to encourage innovation and investment in renewable technologies have already materialized:

1. Subsidies that artificially reduce the price of fuels that compete with renewables should be removed, or renewable energy technologies should be given equivalent incentives.
2. Taxes, regulations, and other policy instruments should ensure that consumer decisions are based on the full cost of energy, including environmental and other external costs not reflected in market prices.
3. Government support for research, development, and demonstration of renewable energy technologies should be increased to reflect the critical roles renewable energy technologies can play in meeting energy and environmental objectives.
4. Government regulations of electric utilities should be carefully reviewed to ensure that investments in new generating equipment are consistent with a renewables-intensive future and that utilities are involved in programs to demonstrate new renewable energy technologies.
5. Policies designed to encourage the development of the biofuels industry must be closely coordinated with both national agricultural development programs and efforts to restore degraded lands.
6. National institutions should be created or strengthened to implement renewable energy programs.
7. International development funds available for the energy sector should be increasingly directed to renewables. Today the majority of the investment for power goes to renewables, as shown in Section 1.3.6.

The integrating theme for all such initiatives, however, should be an energy policy aimed at promoting sustainable development. An example of this is the European Green Deal (EGD), which targets a reduction in GHG emissions of 55% by 2030 and deliver net zero by 2050 (European Commission, 2019). EU decarbonization goals will require the phasing out of most fossil fuels, notably coal, oil, and natural gas, by 2050, together with carbon capture and storage (CCS) and some extra removals of carbon from the atmosphere to compensate for those carbon emissions which cannot be completely controlled and phased out. Reducing CO_2 emissions to zero means banning sales of new vehicles with internal combustion engines (ICEs) by 2035, leading to a zero-emission EU fleet by 2050, when all existing ICE engines reach their end of life. Furthermore, the Renewable Energy Directive targets specify increased use of renewables in the energy mix from 32% to 40% by 2030, to have a renewable energy share for buildings at least 49% and others for heating and cooling, industry, and transport. It will not be possible to provide the energy needed to bring a decent standard of living to the world's poor or sustain the economic well-being of the industrialized countries in environmentally acceptable ways if the use of present energy sources continues. The path to a sustainable society requires more efficient energy use and a shift to a variety of renewable energy sources. Generally, the central challenge to policymakers in the next few decades is to develop economic policies that simultaneously satisfy both socioeconomic developmental and environmental challenges.

Such policies could be implemented in many ways. The preferred policy instruments will vary with the level of the initiative (local, national, or international) and the region. On a regional level, the preferred options will reflect differences in endowments of renewable resources, stages of economic development, and cultural characteristics. Here the region can be an entire continent. One example of this is the declaration of the European Union (EU) for the promotion of renewable energies as a key measure to ensure that Europe meets its climate change targets under the Kyoto Protocol.

According to the decision, central to the European Commission's (EC) action to ensure that the EU and member states meet their Kyoto targets is the European Climate Change Program launched in

2000 and the newer Green Deal as outlined above. Under this umbrella, the Commission, member states, and stakeholders identified and developed a range of cost-effective measures to reduce emissions.

To date, 35 measures have been implemented, including the EU Emissions Trading Scheme and legislative initiatives to promote renewable energy sources for electricity production, to expand the use of biofuels in road transport, and to improve the energy performance of buildings. Previously, the EC proposed an integrated package of measures to establish a new energy policy for Europe that would increase actions to fight climate change and boost energy security and competitiveness in Europe, and the proposals put the EU on course toward becoming a low-carbon economy. The new package sets a range of ambitious targets to be met originally by 2020, including improvement of energy efficiency by 20%, increasing the market share of renewables to 20%, and increasing the share of biofuels in transport fuels to 10%.

As a concrete first step toward this reduction, the EU has made a firm independent commitment to cut its emissions by at least 20% even before a global agreement is reached and irrespective of what others do.

Many scenarios describe how renewable energy will develop in the coming years. In a renewable energy-intensive scenario described before, global consumption of renewable resources reaches a level equivalent to 318 EJ (exa, $E = 10^{18}$) per annum (a) of fossil fuels by 2050—a rate comparable with the 1985 total world energy consumption, which was equal to 323 EJ. Although this figure seems to be very large, it is less than 0.01% of the 3.8 million EJ of solar energy reaching the earth's surface each year. The total electric energy produced from intermittent renewable sources (~ 34 EJ/a) would be less than 0.003% of the sunlight that falls on land and less than 0.1% of the energy available from wind. The amount of energy targeted for recovery from biomass could reach 206 EJ/a by 2050, which is also small compared with the rate (3800 EJ/a) at which plants convert solar energy to biomass. The production levels considered are therefore not likely to be constrained by resource availability. A number of other practical considerations, however, do limit the renewable resources that can be used. The renewable energy-intensive scenario considers that biomass would be produced sustainably, not harvested in virgin forests. About 60% of the biomass supply would come from plantations established on degraded land or excess agricultural land and the rest from residues of agricultural or forestry operations. Finally, the amounts of wind, solar thermal, and PV power that can be economically integrated into electric generating systems are very sensitive to patterns of electricity demand and weather conditions. The marginal value of these intermittent electricity sources typically declines as their share of the total electric market increases.

By making efficient use of energy and expanding the use of renewable technologies, the world can expect to have adequate supplies of fossil fuels well into the 21st century. However, in some instances regional declines in fossil fuel production can be expected because of resource constraints. Oil production outside the Middle East would decline slowly under the renewables-intensive scenario, so that one-third of the estimated ultimately recoverable conventional resources will remain in the ground in 2050. Although remaining conventional natural gas resources are comparable with those for conventional oil, with an adequate investment in pipelines and other infrastructure components, natural gas could be a major energy source for many years.

The next section reviews some of the most important environmental consequences of using conventional forms of energy. This is followed by a review of renewable energy technologies not included in this book.

1.3 Energy-related environmental problems and the role of renewables

Energy is considered a prime agent in the generation of wealth and a significant factor in sustainable development. Under the 17 Sustainable Development Goals (SDGs) of the United Nations, Energy is directly linked with SDG 7, Affordable and Clean Energy; and indirectly with SDG 12, Responsible Consumption and Production, and SDG 13, Climate Action. The importance of energy in sustainable development is recognized universally, and historical data verify that there is a strong relationship between the availability of energy and economic activity. Although in the early 1970s, after the oil crisis, the concern was on the cost of energy, during the past 2 decades the risk and reality of environmental degradation have become more apparent. The growing evidence of environmental problems is due to a combination of several factors since the environmental impact of human activities has grown dramatically. This is due to the increase of the world population, energy consumption, and industrial activity. Achieving solutions to the environmental problems that humanity faces today requires long-term potential actions for sustainable development. In this respect, renewable energy resources appear to be one of the most efficient and effective solutions hence its inclusion in the SDGs.

A few years ago, most environmental analysis and legal control instruments concentrated on conventional pollutants such as sulfur dioxide (SO_2), nitrogen oxides (NO_x), particulates, and carbon monoxide (CO). Recently, however, environmental concern has extended to the control of hazardous air pollutants, which are usually toxic chemical substances harmful even in small doses, as well as to other globally significant pollutants such as carbon dioxide (CO_2). Additionally, developments in industrial processes and structures have led to new environmental problems. Carbon dioxide as a GHG plays a vital role in global warming. Studies show that it is responsible for about two-thirds of the enhanced greenhouse effect. A significant contribution to the CO_2 emitted to the atmosphere is attributed to fossil fuel combustion (EPA, 2007).

The United Nations Conference on Environment and Development (UNCED), held in Rio de Janeiro, Brazil, in June 1992, addressed the challenges of achieving worldwide sustainable development. The goal of sustainable development cannot be realized without major changes in the world's energy system. Accordingly, Agenda 21, which was adopted by UNCED, called for "new policies or programs, as appropriate, to increase the contribution of environmentally safe and sound and cost-effective energy systems, particularly new and renewable ones, through less polluting and more efficient energy production, transmission, distribution, and use". Twenty-three years later and after the UNCED, the Kyoto Protocol, the Millennium Development Goals (MDGs), and the process from the MDGs until 2015, the United Nations has divided sustainability into 17 SDGs concluding with the 2030 Agenda for Sustainable Development and the Paris Agreement on Climate Change. Energy, as mentioned above, plays a major factor in the 2030 Agenda.

The division for sustainable development of the United Nations Department of Economics and Social Affairs defined sustainable development in 1992 as "development that meets the needs of the present without compromising the ability of future generations to meet their own needs" (United Nations, 1992). In 2015 and with the evolution of sustainability to the 17 SDGs, this clearly supports the definition. In 2015, 193 countries officially adopted Agenda 2030, titled "Transforming Our World: The 2030 Agenda for Sustainable Development"—also known as the 17 SDGs. A major role in this transformation is obviously energy, since SDG 7, Affordable and Clean Energy, focuses solely on that. Additionally, energy is behind the actions of other SDGs. Clearly, one of the main factors that will make a change to our future generations is energy, and one of the most important issues is the requirement for a supply of energy that is fully sustainable (Rosen, 1996; Dincer and Rosen, 1998), clean, and affordable. A secure supply of energy is generally agreed to be a necessary but not a sufficient requirement for development within a society. Furthermore, for a sustainable development within a society, it is required that a sustainable supply of energy and an effective and efficient utilization of energy resources are secure. Such a supply in the long term should be readily available at reasonable cost, sustainable, and able to be utilized for all the required tasks without causing negative societal impacts. This is the reason why there is a close connection between renewable sources of energy and sustainable development.

Sustainable development is a serious policy concept. In addition to the definition just given, it can be considered as a development that must not carry the seeds of destruction, because such a development is unsustainable. The concept of sustainability has its origin in fisheries and forest management in which prevailing management practices, such as overfishing or single-species cultivation, work for limited time, then yield diminishing results and eventually endanger the resource. Therefore, sustainable management practices should not aim for maximum yield in the short run but for smaller yields that can be sustained over time.

Pollution depends on energy consumption. In 2017, the world's daily oil consumption was 88.4 million barrels. Despite the well-known consequences of fossil fuel combustion on the environment, no decrease has been seen during the last decade. A large number of factors are significant in the determination of the future level of energy consumption and production. Such factors include population growth, economic performance, consumer tastes, and technological developments. Furthermore, government policies concerning energy and developments in the world energy markets certainly play a key role in the future level and pattern of energy production and consumption (Dincer, 1999).

Currently, 25% of the world population consumed 70% of the total energy supply, while the remaining 75% of the population was left with 30%. If the total population were to have the same consumption per inhabitant as the Organization for Economic Cooperation and Development member countries have on average, it would result in an increase in the 1984 world energy demand from 10 TW (tera, $T = 10^{12}$) to approximately 30 TW. An expected increase in the population from 4.7 billion in 1984 to 8.2 billion in 2020 would raise the figure to 50 TW.

The total primary energy demand in the world increased from 40,553 TWh in 1965 to 136,762 TWh in 2019, representing an average annual increase of about 4.4%.

The major sectors using primary energy sources include electrical power, transportation, heating, and industry. The International Energy Agency data shows that the electricity demand almost tripled from 1971 to 2002. This is because electricity is a very convenient form of energy to transport and use. Although primary energy use in all sectors has increased, their relative shares have decreased, except for transportation and electricity. The relative share of primary energy for electricity production in the world increased from about 20% in 1971 to about 30% in 2002 as electricity became the preferred form of energy for all applications.

Fueled by high increases in China and India, worldwide energy consumption may continue to increase at rates between 3% and 5% for at least a few more years, although at least China is leading the world with respect to installed renewables. However, such high rates of increase cannot continue for too long. Even at a 2% increase per year, the primary energy demand of 2002 would double by 2037 and triple by 2057. With such high energy demand expected 36 years from now, it is important to look at all the available strategies to fulfill the future demand, especially for electricity and transportation.

At present, 95% of all energy for transportation comes from oil. Therefore, the available oil resources and their production rates and prices greatly influence the future changes in transportation. An obvious replacement for oil would be biofuels such as ethanol, methanol, biodiesel, and biogases. It is believed that hydrogen is another alternative because, if it could be produced economically from renewable energy sources, it could provide a clean transportation alternative for the future.

Natural gas will be used at rapidly increasing rates to make up for the shortfall in oil production; however, it may not last much longer than oil itself at higher rates of consumption. Coal is the largest fossil resource available and the most problematic due to environmental concerns. All indications show that coal use will continue to grow for power production around the world because of expected increases in China, India, Australia, and other countries. This, however, would be unsustainable, from the environmental point of view, unless advanced clean coal technologies with carbon sequestration are deployed.

Another parameter to be considered is the world population. In the year 2021, this equaled 7.9 billion people. This is expected to be 9 billion by 2037 and 10 billion by 2057 (www.worldmeters.info/world.population/), and as economic development will certainly continue to grow, the global demand for energy is expected to increase. For example, the most populous country, China, increased its primary energy consumption by 15% from 2003 to 2004. Today, much evidence exists to suggest that the future of our planet and the generations to come will be negatively affected if humans keep degrading the environment. Currently, three environmental problems are internationally known: acid precipitation, stratospheric ozone depletion, and global climate change. These issues are analyzed in more detail in the following subsections.

1.3.1 Acid rain

Acid rain is a form of pollution depletion in which SO_2 and NO_x produced by the combustion of fossil fuels are transported over great distances through the atmosphere, where they react with water molecules to produce acids deposited via precipitation on the earth, causing damage to ecosystems that are exceedingly vulnerable to excessive acidity. Therefore, it is obvious that the solution to the issue of acid rain deposition requires an appropriate control of SO_2 and NO_x pollutants. These pollutants cause both regional and transboundary problems of acid precipitation.

Recently, attention also has been given to other substances, such as volatile organic compounds (VOCs), chlorides, ozone, and trace metals that may participate in a complex set of chemical transformations in the atmosphere, resulting in acid precipitation and the formation of other regional air pollutants.

It is well known that some energy-related activities are the major sources of acid precipitation. Additionally, VOCs are generated by a variety of sources and comprise a large number of diverse compounds. Obviously, the more energy we expend, the more we contribute to acid precipitation; therefore, the easiest way to reduce acid precipitation is by reducing energy consumption.

1.3.2 Ozone layer depletion

The ozone present in the stratosphere, at altitudes between 12 and 25 km, plays a natural equilibrium-maintaining role for the earth through absorption of ultraviolet (UV) radiation (240−320 nm) and absorption of infrared radiation (Dincer, 1998). A global environmental problem is the depletion of the stratospheric ozone layer, which is caused by the emissions of chlorofluorocarbons (CFCs), halons (chlorinated and brominated organic compounds), and NO_x. Ozone depletion can lead to increased levels of damaging UV radiation reaching the ground, causing increased rates of skin cancer and eye damage to humans, and is harmful to many biological species. It should be noted that energy-related activities are only partially (directly or indirectly) responsible for the emissions that lead to stratospheric ozone depletion. The most significant role in ozone depletion is played by the CFCs, which are mainly used in air-conditioning and refrigerating equipment as refrigerants, and NO_x emissions, which are produced by the fossil fuel and biomass combustion processes, natural denitrification, and nitrogen fertilizers.

In 1998, the size of the ozone hole over Antarctica was 25 million km^2 whereas in 2012 it was 18 million $km.^2$ It was about three million km^2 in 1993 (Worldwatch, 2007). Researchers expect the Antarctic ozone hole to remain severe in the next 10−20 years, followed by a period of slow healing. Models predict full recovery in 2040.

1.3.3 Global climate change

The term *greenhouse effect* has generally been used for the role of the whole atmosphere (mainly water vapor and clouds) in keeping the surface of the earth warm. Recently, however, it has been increasingly associated with the contribution of CO_2, which is estimated to contribute about 50% to the anthropogenic greenhouse effect. Additionally, several other gases, such as CH_4, CFCs, halons, N_2O, ozone, and peroxyacetylnitrate (also called *GHGs*), produced by the industrial and domestic activities can contribute to this effect, resulting in a rise of the earth's temperature. Increasing atmospheric concentrations of GHGs increase the amount of heat trapped (or decrease the heat radiated from the earth's surface), thereby raising the surface temperature of the earth. According to Colonbo (1992), the earth's surface temperature has increased by about 0.6°C over the past century, and as a consequence the sea level is estimated to have risen by perhaps 20 cm. These changes can have a wide range of effects on human activities all over the world. The role of various GHGs is summarized by Dincer and Rosen (1998).

According to the EU, climate change is happening. There is an overwhelming consensus among the world's leading climate scientists that global warming is being caused mainly by carbon dioxide and other GHGs emitted by human activities, chiefly the combustion of fossil fuels and deforestation.

A reproduction of the climate over the past 420,000 years was made recently using data from the Vostok ice core in Antarctica. An ice core is a core sample from the accumulation of snow and ice over many years that has recrystallized and trapped air bubbles from previous time periods. The composition of these ice cores, especially the presence of hydrogen and oxygen isotopes, provides a picture of the climate at the time. The data extracted from this ice core provide a continuous record of temperature and atmospheric composition. Two parameters of interest are the concentration of CO_2 in the atmosphere and the temperature. These are shown in Fig. 1.1, considering 1950 as the reference year. As can be seen, the two parameters follow a similar trend and have a periodicity of about 100,000 years. If one considers, however, the present CO_2 level, which is 420.99 (May 2022) ppm (www.co2now.org), the highest ever recorded, one can understand the implication that this would have on the temperature of the planet.

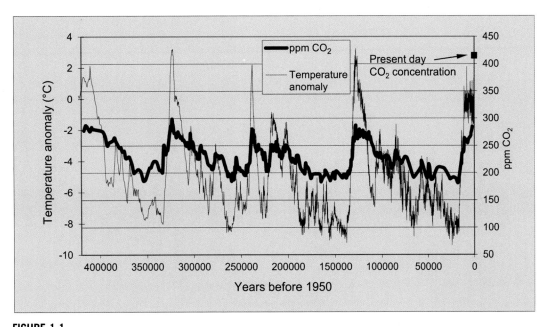

FIGURE 1.1

Temperature and CO_2 concentration from the vostok ice core.

Humans, through many of their economic and other activities, contribute to the increase of the atmospheric concentrations of various GHGs. For example, CO_2 releases from fossil fuel combustion, methane emissions from increased human activities, and CFC releases contribute to the greenhouse effect. Predictions show that if atmospheric concentrations of GHGs, mainly due to fossil fuel combustion, continue to increase at the present rates, the earth's temperature may increase by another $2-4°C$ in the next century. If this prediction is realized, the sea level could rise by $30-60$ cm before the end of this century (Colonbo, 1992). The impacts of such sea level increase can easily be understood and include flooding of coastal settlements, displacement of fertile zones for agriculture to higher latitudes, and decrease in availability of freshwater for irrigation and other essential uses. Thus, such consequences could put in danger the survival of entire populations.

1.3.4 Nuclear energy

Nuclear energy, although non-polluting, presents a number of potential hazards during the production stage and mainly for the disposal of radioactive waste. Nuclear power environmental effects include the effects on air, water, ground, and the biosphere (people, plants, and animals). Nowadays, in many countries, laws govern any radioactive releases from nuclear power plants. In this section some of the most serious environmental problems associated with electricity produced from nuclear energy are described. These include only the effects related to nuclear energy and not the emissions of other substances due to the normal thermodynamic cycle.

The first item to consider is radioactive gases that may be removed from the systems supporting the reactor cooling system. The removed gases are compressed and stored. The gases are periodically sampled and can be released only when the radioactivity is less than an acceptable level, according to certain standards. Releases of this nature are done very infrequently. Usually, all potential paths where radioactive materials could be released to the environment are monitored by radiation monitors (Virtual Nuclear Tourist, 2007).

Nuclear plant liquid releases are slightly radioactive. Very low levels of leakage may be allowed from the reactor cooling system to the secondary cooling system of the steam generator. However, in any case where radioactive water may be released to the environment, it must be stored and radioactivity levels reduced, through ion exchange processes, to levels below those allowed by the regulations.

Within the nuclear plant, a number of systems may contain radioactive fluids. Those liquids must be stored, cleaned, sampled, and verified to be below acceptable levels before release. As in the gaseous release case, radiation detectors monitor release paths and isolate them (close valves) if radiation levels exceed a preset set point (Virtual Nuclear Tourist, 2007).

Nuclear-related mining effects are similar to those of other industries and include generation of tailings and water pollution. Uranium milling plants process naturally radioactive materials. Radioactive airborne emissions and local land contamination were evidenced until stricter environmental rules aided in forcing cleanup of these sites.

As with other industries, operations at nuclear plants result in waste; some of it, however, is radioactive. Solid radioactive materials leave the plant by only two paths:

- Radioactive waste (e.g., clothes, rags, wood) is compacted and placed in drums. These drums must be thoroughly dewatered. The drums are often checked at the receiving location by regulatory agencies. Special landfills must be used.
- Spent resin may be very radioactive and is shipped in specially designed containers.

Generally, waste is distinguished into two categories: low-level waste (LLW) and high-level waste (HLW). LLW is shipped from nuclear plants and includes such solid waste as contaminated clothing, exhausted resins, or other materials that cannot be reused or recycled. Most anti-contamination clothing is washed and reused; however, eventually, as with regular clothing, it wears out. In some cases, incineration or super-compaction may be used to reduce the amount of waste that has to be stored in the special landfills.

HLW is considered to include the fuel assemblies, rods, and waste separated from the spent fuel after removal from the reactor. Currently the spent fuel is stored at the nuclear power plant sites in storage pools or in large metal casks. To ship the spent fuel, special transport casks have been developed and tested.

Originally, the intent had been that the spent fuel would be reprocessed. The limited amount of highly radioactive waste (also called HLW) was to be placed in glass rods surrounded by metal with low long-term corrosion or degradation properties. The intent was to store those rods in specially designed vaults where the rods could be recovered for the first 50–100 years and then made irretrievable for up to 10,000 years. Various underground locations can be used for this purpose, such as salt domes, granite formations, and basalt formations. The objective is to have a geologically stable location with minimal chance for groundwater intrusion. The intent had been to recover the plutonium and unused uranium fuel and then reuse it in either breeder or thermal reactors as mixed oxide fuel. Currently, France, Great Britain, and Japan are using this process (Virtual Nuclear Tourist, 2007).

1.3.5 Renewable energy technologies

Renewable energy technologies produce marketable energy by converting natural phenomena into useful forms of energy. These technologies use the sun's energy and its direct and indirect effects on the earth (solar radiation, wind, falling water, and various plants; i.e., biomass), gravitational forces (tides), and the heat of the earth's core (geothermal) as the resources from which energy is produced. These resources have massive energy potential; however, they are generally diffused and not fully accessible, and most of them are intermittent and have distinct regional variabilities. These characteristics give rise to difficult, but solvable, technical and economical challenges. Nowadays, significant progress is made by improving the collection and conversion efficiencies, lowering the initial and maintenance costs, and increasing the reliability and applicability of renewable energy systems.

Worldwide research and development in the field of renewable energy resources and systems has been carried out during the past 2 decades. Energy conversion systems that are based on renewable energy technologies appeared to be cost-effective compared with the projected high cost of oil. Furthermore, renewable energy systems can have a beneficial impact on the environmental, economic, and political issues of the world. At the end of 2001 the total installed capacity of renewable energy systems was equivalent to 9% of the total electricity generation (Sayigh, 2001), whereas in 2021, the figure reached 28.3% (Renewable, 2022: Global Status Report, REN21). As was seen before, by applying the renewable energy-intensive scenario, the global consumption of renewable sources by 2050 would reach 318 EJ (Johanson et al., 1993).

The benefits arising from the installation and operation of renewable energy systems can be distinguished into three categories: energy saving, generation of new working posts, and decrease in environmental pollution.

The energy-saving benefit derives from the reduction in consumption of the electricity and diesel used conventionally to provide energy. This benefit can be directly translated into monetary units according to the corresponding production or avoiding capital expenditure for the purchase of imported fossil fuels.

Another factor of considerable importance in many countries is the ability of renewable energy technologies to generate jobs. The penetration of a new technology leads to the development of new production activities, contributing to the production, market distribution, and operation of the pertinent equipment. Specifically for the case of solar energy collectors, job creation is mainly related to the construction and installation of the collectors. The latter is a decentralized process, since it requires the installation of equipment in every building or for every individual consumer.

The most important benefit of renewable energy systems is the decrease in environmental pollution. This is achieved by the reduction of air emissions due to the substitution of electricity and conventional fuels. The most important effects of air pollutants on the human and natural environment are their impact on the public health, agriculture, and on ecosystems. It is relatively simple to measure the financial impact of these effects when they relate to tradable goods, such as the agricultural crops; however, when it comes to non-tradable goods, such as human health and ecosystems, things become more complicated. It should be noted that the level of the environmental impact and therefore the social pollution cost largely depend on the geographical location of the emission sources. Contrary to the conventional air pollutants, the social cost of CO_2 does not vary with the geographical characteristics of the source, as each unit of CO_2 contributes equally to the climate change thread and the resulting cost.

In 2021 all renewable energy sources combined account for 28.3% share of electricity production in the world (Renewable, 2022: Global Status Report, REN21), with hydroelectric power providing almost 38.2% of this amount. However, as renewable energy technologies mature and become even more cost-competitive in the future, they will be in a position to replace a major fraction of fossil fuels for electricity generation. Therefore, substituting fossil fuels with renewable energy for electricity generation must be an important part of any strategy for reducing CO_2 emissions into the atmosphere and combating global climate change.

In this book, emphasis is given to solar thermal systems. Solar thermal systems are nonpolluting and offer significant protection for the environment. The reduction of GHG pollution is the main advantage of utilizing solar energy. Therefore, solar thermal systems should be employed whenever possible to achieve a sustainable future.

The benefits of renewable energy systems can be summarized as follows (Johanson et al., 1993):

- *Social and economic development.* Production of renewable energy, particularly biomass, can provide economic development and employment opportunities, especially in rural areas, that otherwise have limited opportunities for economic growth. Renewable energy can thus help reduce poverty in rural areas and reduce pressure for urban migration.
- *Land restoration.* Growing biomass for energy on degraded lands can provide the incentive and financing needed to restore lands rendered nearly useless by previous agricultural or forestry practices. Although lands farmed for energy would not be restored to their original condition, the recovery of these lands for biomass plantations would support rural development, prevent erosion, and provide a better habitat for wildlife than at present.
- *Reduced air pollution.* Renewable energy technologies, such as methanol or hydrogen for fuel cell vehicles, produce virtually none of the emissions associated with urban air pollution and acid deposition, without the need for costly additional controls.
- *Abatement of global warming.* Renewable energy use does not produce carbon dioxide or other greenhouse emissions that contribute to global warming. Even the use of biomass fuels does not contribute to global warming, since the carbon dioxide released when biomass is burned equals the amount absorbed from the atmosphere by plants as they are grown for biomass fuel.
- *Fuel supply diversity.* There would be substantial interregional energy trade in a renewable energy-intensive future, involving a diversity of energy carriers and suppliers. Energy importers would be able to choose from among more producers and fuel types than they do today and thus would be less vulnerable to monopoly price manipulation or unexpected disruptions of supply. Such competition would make wide swings in energy prices less likely, leading eventually to stabilization of the world oil price. The growth in world energy trade would also provide new opportunities for energy suppliers. Especially promising are the prospects for trade in alcohol fuels, such as methanol, derived from biomass and hydrogen.
- *Reducing the risks of nuclear weapons proliferation.* Competitive renewable resources could reduce incentives to build a large world infrastructure in support of nuclear energy, thus avoiding major increases in the production, transportation, and storage of plutonium and other radioactive materials that could be diverted to nuclear weapons production.

Solar systems, including solar thermal and PVs, offer environmental advantages over electricity generation using conventional energy sources. The benefits arising from the installation and operation of solar energy systems fall into two main categories: environmental and socioeconomical issues.

From an environmental viewpoint, the use of solar energy technologies has several positive implications that include (Abu-Zour and Riffat, 2006):

- Reduction of the emission of the GHGs (mainly CO_2 and NO_x) and of toxic gas emissions (SO_2, particulates),
- Reclamation of degraded land,
- Reduced requirement for transmission lines within the electricity grid, and
- Improvement in the quality of water resources.

The socioeconomic benefits of solar technologies include:

- Increased regional and national energy independence,
- Creation of employment opportunities,
- Restructuring of energy markets due to penetration of a new technology and the growth of new production activities,
- Diversification and security (stability) of energy supply,
- Acceleration of electrification of rural communities in isolated areas, and
- Saving foreign currency.

It is worth noting that no artificial project can completely avoid some impact to the environment. The negative environmental aspects of solar energy systems include:

- Pollution stemming from production, installation, maintenance, and demolition of the systems,
- Noise during construction,
- Land displacement, and
- Visual intrusion.

These adverse impacts present difficult but solvable technical challenges.

The amount of sunlight striking the earth's atmosphere continuously is 1.75×10^5 TW. Considering a 60% transmittance through the atmospheric cloud cover, 1.05×10^5 TW reaches the earth's surface continuously. If the irradiance on only 1% of the earth's surface could be converted into electric energy with a 10% efficiency, it would provide a resource base of 105 TW, while the total global energy needs for 2050 are projected to be about 25–30 TW. The present state of solar energy technologies is such that single solar cell efficiencies have reached more than 20%, with concentrating PVs at about 40%, and solar thermal systems provide efficiencies of 40%–60%.

Solar PV panels have come down in cost from about $30/W to about $1.4/W in the past 3 decades. At $1.4/W panel cost, the overall system cost is around $2.5/W (depending on the size of the installation), which is now a very competitive cost. However, solar PV is already cost-effective in many off-grid applications. With net metering and governmental incentives, such as feed-in laws and other policies, even grid-connected applications such as building-integrated PV have become cost-effective. As a result, the worldwide growth in PV production has averaged more than 30% per year during the past 5 years.

Solar thermal power using concentrating solar collectors was the first solar technology that demonstrated its grid power potential. A total of 354 MWe solar thermal power plants have been operating continuously in California since 1985. Progress in solar thermal power stalled after that time because of poor policy and lack of R&D. However, the past 5 years have seen a resurgence of interest in this area, and a number of solar thermal power plants around the world are constructed and more are under construction. The cost of power from these plants (which so far is in the range of $0.12−$0.16/kWh) has the potential to go down to $0.05/kWh with scale-up and creation of a mass market. An advantage of solar thermal power is that thermal energy can be stored efficiently and fuels such as natural gas or biogas may be used as backup to ensure continuous operation.

1.3.6 Current state of renewable energy technologies in the world

Nowadays, a large number of renewable energy systems have been installed in various parts of the world. The purpose of this section is to show the amount of renewable energy systems capacities installed worldwide and to compare the energy produced by these systems to conventional fuel-based energy production. The total amount of renewable power capacity by the end of 2021 is 3146 GW, whereas the figure if hydro is excluded is down to 1945 GW (REN, 2022). From these, solar PVs constitute 942 GW, a 175 GW increase from the 2020 figure, wind power equals 845 GW compared with 743 in 2020, and solar thermal equals 522 GW_{th} compared with 501 GW_{th} for 2020 (REN, 2022). An outline of the general status of renewables in the world and a comparison of the figures in the last 6 years is shown in Table 1.1. The estimated renewable energy share for the global final energy consumption is 76.4% non-renewable and 12.6% renewable energy, whereas the figure for renewable electricity production is about 28.3% (REN, 2022). Another important characteristic is that it is estimated that worldwide more than 13 million people are working in jobs related to the renewable energy industry. Concerning the investment related to renewables, there is also an increasing trend through the years, and for the year 2020, 303.5 billion USD represents 69% of the total investment

Table 1.1 Renewable energy global indicators from 2016 to 2021.

Key figures	Unit	Year					
		2021	2020	2019	2018	2017	2016
Investment	Billion USD	365.9	303.5	298.4	296	326	274
Renewable including hydropower	GW	3146	2838	2581	2387	2197	2017
Renewable excluding hydropower	GW	1945	1668	1430	1252	1081	992
Hydropower	GW	1195	1170	1150	1135	1112	1095
Wind power capacity	GW	845	743	650	591	540	488
Solar photovoltaic capacity	GW	942	760	621	512	407	305
Bio-power capacity	GW	143	133	137	131	121	114
Geothermal power capacity	GW	14.5	14.1	14	13.2	12.8	12.1
Concentrating solar power	GW	6.0	6.2	6.1	5.5	4.8	4.7
Ocean power capacity	GW	0.5	0.5	0.5	0.5	0.5	0.5
Solar water heating capacity	GW_{th}	522	501	478	482	472	456

related to the construction of power plants, which shows that the attention worldwide now is on renewable energy systems. The same percentage applies for the year 2021.

As can be seen from Table 1.1, all indicators increase every year except the ocean power capacity, which has been steady over the last 6 years.

Solar thermal power includes systems using either thermal radiation or the light of solar irradiance as input. The former category includes solar thermal systems, which comprise both low-temperature systems, mostly for water heating and low-temperature industrial processes; and high-temperature systems, mostly for concentrating solar power (CSP) and high-temperature industrial processes; whereas the latter refers to solar PVs. The most common application of solar thermal power is solar water heating. This is used successfully in various parts of the world. The capacity of systems installed in the last decade is shown in Table 1.2. As can be seen, except for a small drop during 2019 because of some changes in China, all the other years show an increasing trend which is rather stabilized in the last few years. For low-temperature solar thermal systems, the most used type is the glazed collector, whereas unglazed collectors constitute about 5% of the total. Worldwide, because of the many applications in China, the most used type is the evacuated tube, whereas in Europe it is the flat plate collector (Weiss and Spörk-Dür, 2022). Another important application in this area is CSP. The installed capacity for CSP is not as big as the other renewables but is increasing year-by-year, as shown in Table 1.3. A slight drop in 2021 is due to decommissioning of nearly 300 MW of older CSP plants in the United States. Many more CSP plants are under development in the Middle East and North Africa region, China, and the USA. All these new plants will employ some form of thermal storage to allow operation during the night or on cloudy days.

The installed capacity for PV systems is shown in Table 1.4. As shown, PV technology is breaking a new threshold every year. In 2015, for the first time, installed capacity exceeded 200 GW (actual: 228 GW); 2016 exceeded 300 GW (actual: 305 GW); 2017 exceeded 400 GW (actual: 407 GW); 2018 exceeded 500 GW (actual: 512 GW); 2019 exceeded 600 GW (actual: 621 WG); and 2020 exceeded 700 GW (actual: 760 GW). A giant increase was observed in 2021, with a total installed capacity reaching 942 GW. It is interesting to note that 2021 alone saw an additional 175 GW of installed capacity, equivalent to installing more than 75,000 solar panels every hour worldwide. From the

Table 1.2 Solar water heating global capacity and annual additions from 2010 to 2021 in GW_{th}.

Year	Total	Added	%
2010	242	39	16.1
2011	285	43	15.1
2012	330	45	13.6
2013	374	44	11.8
2014	409	35	8.6
2015	435	26	6.0
2016	456	21	4.6
2017	472	16	3.4
2018	482	10	2.1
2019	478	−4	−0.8
2020	501	22	4.4
2021	522	21	4.0

Table 1.3 Concentrating solar thermal power global capacity from 2009 to 2021 in GW.

Year	Total	Added	%
2009	0.8	0.4	50.0
2010	1.2	0.4	33.3
2011	1.7	0.5	29.4
2012	2.5	0.8	32.0
2013	3.4	0.9	26.5
2014	4.3	0.9	20.9
2015	4.6	0.3	6.5
2016	4.7	0.1	2.1
2017	4.8	0.1	2.1
2018	5.5	0.7	12.7
2019	6.1	0.6	9.8
2020	6.2	0.1	1.6
2021	6.0	−0.2	−3.3

Table 1.4 Solar PV global capacity and annual additions from 2009 to 2021 in GW.

Year	Total	Added	%
2009	23	8	34.8
2010	39	17	43.6
2011	70	31	44.3
2012	100	30	30.0
2013	138	38	27.5
2014	178	40	22.5
2015	228	50	21.9
2016	305	77	25.2
2017	407	103	25.3
2018	512	104	20.3
2019	621	110	17.7
2020	760	139	18.3
2021	942	175	18.6

175 GW installed in 2021, a total of 54.9 GW were installed in China, with 26.9 GW in the USA, and 26.8 GW in the EU (IEA-PVPS, 2022). China also leads the world in total installed capacity with 308.5 GW.

Hydro energy comes in a wide variety of sizes ranging from the very large units which are greater than 10 MW to the pico size of less than 5 kW. Today the worldwide largest application is the 18 GW scheme at the Three Gorges in China. The four top countries in this area are China with 30%, Brazil with 9%, Canada and USA both with 7% (REN, 2022).

Table 1.5 Wind power global capacity, annual additions, and offshore capacity from 2009 to 2021 in GW.

Year	Total	Added	Offshore	%
2009	159	38	1.8	23.9
2010	198	39	2.9	19.7
2011	238	41	3.9	17.2
2012	283	45	5.2	15.9
2013	319	36	6.8	11.3
2014	370	52	8.5	14.1
2015	433	64	11.9	14.8
2016	488	55	14.2	11.3
2017	540	54	18.7	10.0
2018	591	51	23	8.6
2019	650	61	29	9.4
2020	743	93	35	12.5
2021	845	102	53.7	12.1

Another area of renewables is wind power, both onshore and offshore. As shown in Table 1.5, every year there is an increase of installed capacity, and the worldwide current capacity is 845 GW with 102 GW installed in 2021 (REN, 2022). Generally, wind has become the least-cost option for new power generating capacity in an increasing number of markets worldwide, provided that there is a good wind potential. The increase, year-by-year of wind energy for the last 13 years is shown in Table 1.5.

Finally, biomass is a broad area that includes biogas and biofuels (biodiesel). Biomass is mostly produced from crops. The challenge here is to use land for this purpose that is not currently cultivated for food production. The most important sector is the transportation sector, as nowadays more than 90% of transport depends on oil. At the end of 2021, the share of biomass in total final energy consumption and by end-use sector was 87.7% based on non-biomass and only 12.3% on biomass. The majority of the biofuels produced are ethanol and biodiesel (REN, 2022). Some scientists also consider waste, like MSW and landfills, renewable energy sources.

1.4 State of the climate

A good source of information on the state of climate in the year 2020 is the report published by the American Meteorological Society (Blunden and Boyer, 2020), which summarizes global and regional climate conditions and places them in the context of historical records (Blunden and Arndt, 2012) and US National Climatic Data Center (NCDC). The parameters examined here are global temperature and various gases found in the atmosphere.

1.4.1 Global temperature

Based on the National Oceanic and Atmospheric Administration and the US NCDC records, the global temperature has been rising gradually at a rate between 0.71 and 0.77°C per century since 1901 and between 0.14 and 0.17°C per decade since 1971. Data show that 2011 was the ninth warmest year since records began in 1979; 0.13°C above the 1981−2010 average whereas the upward trend for 1979−2011 was 0.12°C per decade (Blunden and Arndt, 2012).

The year 2020 was one of the three warmest years since global records began in the mid-to-late 1800s, with global land and ocean surface temperature 0.54−0.62°C above the 1981−2010 average.

The global average surface temperature has increased at an average rate of 0.08°C per decade since 1880, with a rate more than twice as high since 1981 (0.19−0.20°C per decade).

Along with the above-average surface temperatures, an unprecedented (since instrumental records began) geographic spread of heat waves and warm spells has occurred. Antarctica observed its highest temperature on record (18.3°C) at Esperanza in February. In August, Death Valley, California, reported the highest temperature observed anywhere on Earth since 1931 at 54.4°C.

The year 2020 began in El Nino−Southern Oscillation (ENSO)-neutral conditions, transitioning to La Nina by August. The global monthly temperature anomalies were high throughout 2020, with each month from January through November ranking among the five warmest for each respective month. December had the smallest temperature anomaly of the year. Despite the slightly cooler end to the year, this was the warmest non-El Nino year on record, surpassing 2017 by 0.07−0.11°C. Furthermore, the last 7 years (2014−20) were the seven warmest on record.

The majority of the top 10 warmest years on record have occurred in the past decade. The global temperature from 1850 until 2006 is shown in Fig. 1.2, together with the 5-year average values. As can be seen there is an upward trend that is more serious from the 1970s onward.

FIGURE 1.2

Global temperature since 1850.

1.4.2 Carbon dioxide

Carbon dioxide emitted from natural and anthropogenic (i.e., fossil fuel combustion) sources is partitioned into three reservoirs: atmosphere, oceans, and the terrestrial biosphere. The result of increased fossil fuel combustion has been that atmospheric CO_2 has increased from about 280 ppm (parts per million by dry air mole fraction) at the start of the industrial revolution to about 421 ppm in May 2022 (see Fig. 1.3). Carbon dioxide in fact has increased by 28.9 ppm since 2012 and exceeded 400 ppm for the first time since instrumental records began. Roughly half of the emitted CO_2 remains in the atmosphere and the remainder goes into the other two sinks: oceans and the land biosphere (which includes plants and soil carbon).

While the economic disruption associated with COVID-19 led to modest estimated reductions of 6%–7% in global anthropogenic carbon dioxide (CO_2) emissions, atmospheric CO_2 levels continued to grow rapidly—a reminder of its very long residence time in the atmosphere and the challenge of reducing atmospheric CO_2. As seen above, the climate has continued to respond to the resulting warming from these increases in CO_2 and other greenhouse gases such as methane and nitrous oxide, which also experienced record increases in 2020.

In 2010, anthropogenic carbon emissions to the atmosphere have increased globally to more than 9.1 ± 0.5 Pg/a (piga, $P = 10^{15}$). Most of this increase resulted from a 10% increase in emissions from China, the world's largest fossil fuel CO_2 emitter. During the 1990s, net uptake by the oceans was estimated at 1.7 ± 0.5 Pg/a, and by the land biosphere at 1.4 ± 0.7 Pg/a. The gross atmosphere–ocean and atmosphere–terrestrial biosphere (i.e., photosynthesis and respiration) fluxes are on the order of 100 Pg/a. Inter-annual variations in the atmospheric increase of CO_2 are not attributed to variations in

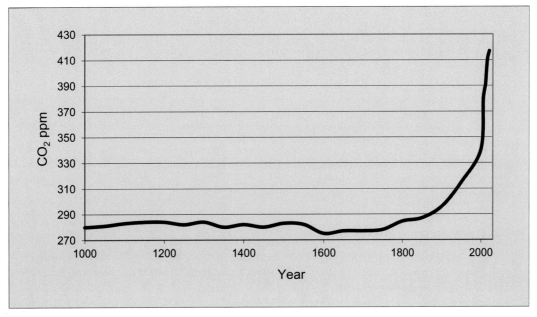

FIGURE 1.3

CO_2 levels in the past 1000 years.

fossil fuel emissions but rather to small changes in these net fluxes. Most attempts to explain the interannual variability of the atmospheric CO_2 increase have focused on short-term climate fluctuations (e.g., the El Niño/Southern Oscillation and post-mountain Pinatubo cooling), but the mechanisms, especially the role of the terrestrial biosphere, are poorly understood. To date, about 5% of conventional fossil fuels have been combusted. If combustion is stopped today, it is estimated that after a few hundred years, 15% of the total carbon emitted would remain in the atmosphere, and the remainder would be in the oceans.

In 2011, the globally averaged atmospheric CO_2 mole fraction was 390.4 ppm. This was slightly larger than the average increase from 2000 to 2010 of 1.96 ± 0.36 ppm/a. The record CO_2 concentration in 2021 (416.96 ppm) continues a trend toward increased atmospheric CO_2 since before the industrial era values of around 280 ppm. This continues the steady upward trend in this abundant and long-lasting GHG. Since 1900, atmospheric CO_2 has increased by 121 ppm (141%), with an average annual increase of 3.67 ppm since 2000.

1.4.3 Methane

The contribution of methane (CH_4) to anthropogenic radiative forcing, including direct ($\approx 70\%$) and indirect ($\approx 30\%$) effects, is about 0.7 W/m^2, or roughly half that of CO_2. Also, changes in the load of CH_4 feed back into atmospheric chemistry, affecting the concentrations of hydroxyl (OH) and ozone (O_3). The increase in CH_4 since the pre-industrial era is responsible for about half of the estimated increase in background tropospheric O_3 during that time. It should be noted that changes in OH concentration affect the lifetimes of other GHGs such as hydrochlorofluorocarbons (HCFCs) and hydrofluorocarbons (HFCs). Methane has a global warming potential (GWP) of 25; this means that, integrated over a 100-year timescale, the radiative forcing from a given pulse of CH_4 emissions is estimated to be 25 times greater than a pulse of the same mass of CO_2.

In 2020, CH_4 increased from 722 ppb (parts per billion, 10^9, by dry air mole fraction in pre-industrial times to 1879 ppb globally averaged methane (CH_4) concentration).

Stratospheric ozone over Antarctica in October 2017 reached a value of 131 Dobson units (DU) and the world average is about 300 DU. A DU is the most basic measure used in ozone research. The unit is named after G. M. B. Dobson, one of the first scientists to investigate atmospheric ozone. He designed the Dobson spectrometer, which is the standard instrument used to measure ozone from the ground. The Dobson spectrometer measures the intensity of solar UV radiation at four wavelengths, two of which are absorbed by ozone and two of which are not. One Dobson unit is defined to be 0.01 mm thickness at STP (standard temperature and pressure = 0°C and 1 atm pressure). For example, when in an area all the ozone in a column is compressed to STP and spread out evenly over the area and forms a slab of 3 mm thick, then the ozone layer over that area is 300 DU.

1.4.4 Carbon monoxide

Unlike CO_2 and CH_4, carbon monoxide (CO) does not strongly absorb terrestrial infrared radiation but affects climate through its chemistry. The chemistry of CO affects OH (which influences the lifetimes of CH_4 and HFCs) and tropospheric O_3 (which is by itself a GHG); so emissions of CO can be considered equivalent to emissions of CH_4. Current emissions of CO may contribute more to radiative forcing over decade timescales than emissions of anthropogenic nitrous oxide.

Because the lifetime of CO is relatively short (a few months), the anomaly of increased levels of CO in the atmosphere quickly disappeared and CO quickly returned to pre-1997 levels. Generally, however, carbon monoxide levels increase at about 6% per year on average.

1.4.5 Nitrous oxide and sulfur hexafluoride

Atmospheric nitrous oxide (N_2O) and sulfur hexafluoride (SF_6) are present in lower concentrations than CO_2, but the radiative forcing of each is far greater. Nitrous oxide is the third strongest GHG, while each SF_6 molecule is 23,900 times more effective as an infrared absorber than one CO_2 molecule and has an atmospheric lifetime of between 500 and 3200 years.

The concentration of both species has grown at a linear rate, N_2O at 0.76 ppb/a (0.25% per year) since 1978 and SF_6 at a rate of 0.22 ppt (parts per trillion, 10^{12}, by dry air mole fraction) per year ($\sim5\%$/a) since 1996. The concentration of 334.1 ppb N_2O in 2011 has added a radiative forcing of around 0.175 W/m^2 over the pre-industrial N_2O concentration of around 270 ppb. The 2021 value represents an increase of 10.5 ppb over the 2010 value and is higher than the average growth rate of 0.76 ppb/a shown above. Atmospheric N_2O is also a major source of stratospheric nitric oxide (NO), a compound that helps to catalytically destroy stratospheric O_3. The atmospheric concentration of SF_6 has grown due to its use as an electrical insulator for power transmission throughout the world. Its global mean concentration was over 10 ppt at the middle of 2020, an increase of 3 ppt over the 2010 value. While total radiative forcing of SF_6 from pre-industrial times to the present is relatively small, its long atmospheric lifetime, high atmospheric growth rate, and high GWP are a concern for the future.

1.4.6 Halocarbons

Concern over stratospheric ozone depletion has restricted or eliminated production of many halocarbons. The phase-out of human-produced halocarbons was the result of the 1987 Montreal Protocol on Substances that Deplete the Ozone Layer. As a result of these efforts, mixing ratios of many ozone-depleting gases have been declining at the earth's surface in recent years. Reports from many laboratories around the world that perform measurements of halocarbons show that tropospheric mixing ratios of CFC-12, the longest lived and most abundant human-made ozone-depleting gas in the atmosphere, peaked within the past few years. These measurements also show that mixing ratios of some halogenated gases continue to increase globally. The most rapid increases are in HCFCs and HFCs, which are chemicals commonly used as replacements for CFCs, halons, and other ozone-depleting gases. Although HCFCs contain chlorine (Cl) and deplete O_3 with a reduced efficiency compared with CFCs, HFCs do not participate in O_3 destroying reactions.

Changes in the direct radiative influence of long-lived halocarbons can be estimated from observed changes in atmospheric mixing ratios with knowledge of trace-gas radiative efficiencies. Such an analysis suggests that the direct radiative forcing of these gases was still increasing in 2021, though at a much slower rate than observed from 1970 to 1990.

1.4.7 Sea level

Global sea level has been rising over the past century, and the rate has increased in recent decades. In 2014, the global sea level was 66 mm above the 1993 average—the highest annual average in the satellite record (1993−present).

The two major causes of global sea level rise are thermal expansion caused by the warming of the ocean (since water expands as it warms) and increased melting of land-based ice, such as glaciers and ice sheets. The oceans are absorbing more than 90% of the increased atmospheric heat associated with emissions from human activity.

The sea level has risen 21–24 cm since 1880. In 2020, the global sea level set a new high record of 91.3 mm above 1993 levels. The rate of sea level rise is accelerating; it has more than doubled from 1.4 mm per year throughout most of the 20th century to 3.6 mm per year from 2006 to 2015. Even if the world follows a low greenhouse gas pathway, the global sea level will likely rise at least 0.3 m above 2000 levels by 2100. If a pathway with high emissions is followed, which is a worst-case scenario, the rise will be as much as 2.5 m above 2000 levels by 2100.

1.5 A brief history of solar energy

Solar energy is the oldest energy source ever used. The sun was adored by many ancient civilizations as a powerful god. The first known practical application was in drying for preserving food (Kalogirou, 2004).

Probably the oldest large-scale application known to us is the burning of the Roman fleet in the bay of Syracuse by Archimedes, the Greek mathematician and philosopher (287–212 BC). Scientists discussed this event for centuries. From 100 BC to 1100 AD, authors made reference to this event, although later it was criticized as a myth because no technology existed at that time to manufacture mirrors (Delyannis, 1967). The basic question was whether Archimedes knew enough about the science of optics to devise a simple way to concentrate sunlight to a point at which ships could be burned from a distance. Nevertheless, Archimedes had written a book, *On Burning Mirrors* (Meinel and Meinel, 1976), which is known only from references, since no copy survived.

The Greek historian Plutarch (46–120 AD) referred to the incident, saying that the Romans, seeing that indefinite mischief overwhelmed them from no visible means, began to think they were fighting with the gods.

In his book, *Optics*, Vitelio, a Polish mathematician, described the burning of the Roman fleet in detail (Delyannis and Belessiotis, 2000; Delyannis, 1967): "The burning glass of Archimedes composed of 24 mirrors, which conveyed the rays of the sun into a common focus and produced an extra degree of heat."

Proclus repeated Archimedes' experiment during the Byzantine period and burned the war fleet of enemies besieging Byzance in Constantinople (Delyannis, 1967).

Eighteen hundred years after Archimedes, Athanasius Kircher (1601–80) carried out some experiments to set fire to a woodpile at a distance to see whether the story of Archimedes had any scientific validity, but no report of his findings survives (Meinel and Meinel, 1976).

Many historians, however, believe that Archimedes did not use mirrors but the shields of soldiers, arranged in a large parabola, for focusing the sun's rays to a common point on a ship. This fact proved that solar radiation could be a powerful source of energy. Many centuries later, scientists again considered solar radiation as a source of energy, trying to convert it into a useable form for direct utilization.

Amazingly, the very first applications of solar energy refer to the use of concentrating collectors, which are, by their nature (accurate shape construction) and the requirement to follow the sun, more "difficult" to apply. During the eighteenth century, solar furnaces capable of melting iron, copper, and other metals were being constructed of polished iron, glass lenses, and mirrors. The furnaces were in use throughout Europe and the Middle East. One of the first large-scale applications was the solar furnace built by the well-known French chemist Lavoisier, who, around 1774, constructed powerful lenses to concentrate solar radiation (see Fig. 1.4). This attained the remarkable temperature of 1750°C. The furnace used a 1.32 m lens plus a secondary 0.2 m lens to obtain such temperature, which turned out to be the maximum achieved for 100 years. Another application of solar energy utilization in this century was carried out by the French naturalist Boufon (1747–48), who experimented with various devices that he described as "hot mirrors burning at long distance" (Delyannis, 2003).

During the nineteenth century, attempts were made to convert solar energy into other forms based upon the generation of low-pressure steam to operate steam engines. August Mouchot pioneered this field by constructing and operating several solar-powered steam engines between the years 1864 and 1878 in Europe and North Africa. One of them was presented at the 1878 International Exhibition in Paris (see Fig. 1.5). The solar energy gained was used to produce steam to drive a printing machine (Mouchot, 1878; 1880). Evaluation of one built at Tours by the French government showed that it was too expensive to be considered feasible. Another one was set up in Algeria. In 1875, Mouchot made a notable advance in solar collector design by making one in the form of a truncated cone reflector.

FIGURE 1.4

Solar furnace used by Lavoisier in 1774.

FIGURE 1.5

Parabolic collector powering a printing press at the 1878 Paris exposition.

Mouchot's collector consisted of silver-plated metal plates and had a diameter of 5.4 m and a collecting area of 18.6 m^2. The moving parts weighed 1400 kg.

Abel Pifre, a contemporary of Mouchot, also made solar engines (Meinel and Meinel, 1976; Kreider and Kreith, 1977). Pifre's solar collectors were parabolic reflectors made of very small mirrors. In shape they looked rather similar to Mouchot's truncated cones.

The efforts were continued in the United States, where John Ericsson, an American engineer, developed the first steam engine driven directly by solar energy. Ericsson built eight systems that had parabolic troughs by using either water or air as the working medium (Jordan and Ibele, 1956).

In 1901 A G. Eneas installed a 10 m diameter focusing collector that powered a water-pumping apparatus at a California farm. The device consisted of a large umbrella-like structure opened and inverted at an angle to receive the full effect of the sun's rays on the 1788 mirrors that lined the inside surface. The sun's rays were concentrated at a focal point where the boiler was located. Water within the boiler was heated to produce steam, which in turn powered a conventional compound engine and centrifugal pump (Kreith and Kreider, 1978).

In 1904, a Portuguese priest, Father Himalaya, constructed a large solar furnace. This was exhibited at the St. Louis World's Fair. This furnace appeared quite modern in structure, being a large, off-axis, parabolic horn collector (Meinel and Meinel, 1976).

In 1912, Frank Shuman, in collaboration with C.V. Boys, undertook to build the world's largest pumping plant in Meadi, Egypt. The system was placed in operation in 1913, using long parabolic cylinders to focus sunlight onto a long absorbing tube. Each cylinder was 62 m long, and the total area of the several banks of cylinders was 1200 m^2. The solar engine developed as much as 37–45 kW continuously for a 5-h period (Kreith and Kreider, 1978). Despite the plant's success, it was completely shut down in 1915 due to the onset of World War I and cheaper fuel prices.

During the past 50 years, many variations were designed and constructed using focusing collectors as a means of heating the heat-transfer or working fluid that powered mechanical equipment. The two primary solar technologies used are central receivers and distributed receivers employing various point and line focus optics to concentrate sunlight. Central receiver systems use fields of heliostats (two-axis tracking mirrors) to focus the sun's radiant energy onto a single tower-mounted receiver (SERI, 1987). Distributed receiver technology includes parabolic dishes, Fresnel lenses, parabolic troughs, and special bowls. Parabolic dishes track the sun in two axes and use mirrors to focus radiant energy onto a point focus receiver. Troughs and bowls are line focus tracking reflectors that concentrate sunlight onto receiver tubes along their focal lines. Receiver temperatures range from 100 °C in low-temperature troughs to close to 1500°C in dish and central receiver systems (SERI, 1987).

Today, many large solar plants have output in the megawatt range to produce electricity or process heat. The first commercial solar plant was installed in Albuquerque, New Mexico, in 1979. It consisted of 220 heliostats and had an output of 5 MW. The second was erected at Barstow, California, with a total thermal output of 35 MW. Most of the solar plants produce electricity or process heat for industrial use and they provide superheated steam at 673 K. Thus, they can provide electricity or steam to drive small-capacity conventional desalination plants driven by thermal or electrical energy.

Another area of interest, hot water and house heating, appeared in the mid-1930s but gained interest in the last half of the 1940s. Until then, millions of houses were heated by coal-burning boilers. The idea was to heat water and feed it to the radiator system that was already installed.

The manufacture of solar water heaters began in the early 1960s. The industry of solar water heater manufacturing expanded very quickly in many countries of the world. Typical solar water heaters in many cases are of the thermosiphon type and consist of two flat plate solar collectors having an absorber area between 3 and 4 m^2 and a storage tank with capacity between 150 and 180 L, all installed on a suitable frame. An auxiliary electric immersion heater or a heat exchanger, for central heating-assisted hot water production, is used in winter during periods of low solar insolation. Another important type of solar water heater is the forced circulation type. In this system, only the solar panels are visible on the roof, the hot water storage tank is located indoors in a plant room, and the system is completed with piping, a pump, and a differential thermostat. Obviously, this type is more appealing, mainly for architectural and aesthetic reasons, but it is also more expensive, especially for small installations (Kalogirou, 1997). More details on these systems are given in Chapter 5.

1.5.1 Photovoltaics

Becquerel discovered the PV effect in selenium in 1839. The conversion efficiency of the "new" silicon cells, developed in 1958, was 11%, although the cost was prohibitively high ($1000/W). The first practical application of solar cells was in space, where cost was not a barrier, since no other source of power is available. Research in the 1960s resulted in the discovery of other PV materials such as gallium arsenide (GaAs). These could operate at higher temperatures than silicon but were much more expensive. The global installed capacity of PVs at the end of 2011 was 67 GWp (Photon, 2012) or 70 GW according to REN21, whereas in 2021, this increased to 942 GW (Renewable, 2022: Global Status Report, REN21). PV cells are made of various semiconductors, which are materials that are only moderately good conductors of electricity. The materials most commonly used are silicon (Si) and compounds of cadmium sulfide (CdS), cuprous sulfide (Cu$_2$S), and gallium arsenide (GaAs).

Amorphous silicon cells are composed of silicon atoms in a thin homogenous layer rather than a crystal structure. Amorphous silicon absorbs light more effectively than crystalline silicon; so the cells can be thinner. For this reason, amorphous silicon is also known as a *thin-film* PV technology. Amorphous silicon can be deposited on a wide range of substrates, both rigid and flexible, which makes it ideal for curved surfaces and "foldaway" modules. Amorphous cells are, however, less efficient than crystalline-based cells, with typical efficiencies of around 6%, but they are easier and therefore cheaper to produce. Their low cost makes them ideally suited for many applications where high efficiency is not required and low cost is important.

Amorphous silicon (a-Si) is a glassy alloy of silicon and hydrogen (about 10%). Several properties make it an attractive material for thin-film solar cells:

1. Silicon is abundant and environmentally safe.
2. Amorphous silicon absorbs sunlight extremely well, so that only a very thin active solar cell layer is required (about 1 μm compared with 100 μm or so for crystalline solar cells), thus greatly reducing solar cell material requirements.
3. Thin films of a-Si can be deposited directly on inexpensive support materials such as glass, sheet steel, or plastic foil.

A number of other promising materials, such as cadmium telluride (CdTe) and copper indium diselenide (CIS), are now being used for PV modules. The attraction of these technologies is that they can be manufactured by relatively inexpensive industrial processes, in comparison to crystalline silicon technologies, yet they typically offer higher module efficiencies than amorphous silicon.

The PV cells are packed into modules that produce a specific voltage and current when illuminated. PV modules can be connected in series or in parallel to produce larger voltages or currents. PV systems can be used independently or in conjunction with other electrical power sources. Applications powered by PV systems include communications (both on earth and in space), remote power, remote monitoring, lighting, water pumping, and battery charging.

The two basic types of PV applications are the stand-alone and the grid-connected systems. Stand-alone PV systems are used in areas that are not easily accessible or have no access to mains electricity grids. A stand-alone system is independent of the electricity grid, with the energy produced normally being stored in batteries. A typical stand-alone system would consist of PV module or modules, batteries, and a charge controller. An inverter may also be included in the system to convert the direct current (DC) generated by the PV modules to the alternating current (AC) form required by normal appliances.

In the grid-connected applications, the PV system is connected to the local electricity network. This means that during the day, the electricity generated by the PV system can either be used immediately (which is normal for systems installed in offices and other commercial buildings) or sold to an electricity supply company (which is more common for domestic systems, where the occupier may be out during the day). In the evening, when the solar system is unable to provide the electricity required, power can be bought back from the network. In effect, the grid acts as an energy storage system, which means the PV system does not need to include battery storage.

When PVs started to be used for large-scale commercial applications about 20 years ago, their efficiency was well below 10%. Nowadays, their efficiency has increased to about 15%. Laboratory or experimental units can give efficiencies of more than 30%, but these have not been commercialized yet. Although 20 years ago PVs were considered a very expensive solar system, the present cost is around $2500, and there are good prospects for further reduction in the coming years. More details on PVs are included in Chapter 9 of this book.

1.5.2 Solar desalination

The lack of water was always a problem to humanity. Therefore, among the first attempts to harness solar energy was the development of equipment suitable for the desalination of seawater. Solar distillation has been in practice for a long time (Kalogirou, 2005).

As early as in the fourth century BC, Aristotle described a method to evaporate impure water and then condense it to obtain potable water. However, historically, probably one of the first applications of seawater desalination by distillation is depicted in the drawing shown in Fig. 1.6. The need to produce freshwater onboard emerged by the time the long-distance trips were possible. The drawing illustrates an account by Alexander of Aphrodisias in 200 AD, who said that sailors at sea boiled seawater and suspended large sponges from the mouth of a brass vessel to absorb what evaporated. In drawing this liquid off the sponges, they found that it was sweet water (Kalogirou, 2005).

Solar distillation has been in practice for a long time. According to Malik et al. (1985), the earliest documented work is that of an Arab alchemist in the fifteenth century, reported by Mouchot in 1869. Mouchot reported that the Arab alchemist had used polished Damascus mirrors for solar distillation.

Until medieval times, no important applications of desalination by solar energy existed. During this period, solar energy was used to fire alembics to concentrate dilute alcoholic solutions or herbal extracts for medical applications and to produce wine and various perfume oils. The stills, or alembics,

FIGURE 1.6

Sailors producing freshwater with seawater distillation.

were discovered in Alexandria, Egypt, during the Hellenistic period. Cleopatra the Wise, a Greek alchemist, developed many distillers of this type (Bittel, 1959). One of them is shown in Fig. 1.7 (Kalogirou, 2005). The head of the pot was called the *ambix*, which in Greek means the "head of the still", but this word was applied very often to the whole still. The Arabs, who overtook science and especially alchemy about the seventh century, named the distillers *Al-Ambiq*, from which came the name *alembic* (Delyannis, 2003).

Mouchot (1879), the well-known French scientist who experimented with solar energy, in one of his numerous books mentions that, in the fifteenth century, Arab alchemists used polished Damascus concave mirrors to focus solar radiation onto glass vessels containing saltwater to produce freshwater. He also reports on his own solar energy experiments to distill alcohol and an apparatus he developed with a metal mirror having a linear focus in which a boiler was located along its focal line.

Later on, during the Renaissance, Giovani Batista Della Porta (1535−1615), one of the most important scientists of his time, wrote many books, which were translated into French, Italian, and German. In one of them, *Magiae Naturalis*, which appeared in 1558, he mentions three desalination systems (Delyannis, 2003). In 1589, he issued a second edition in which, in the volume on distillation, he mentions seven methods of desalination. The most important of them is a solar distillation apparatus that converted brackish water into freshwater. In this, wide earthen pots were used, exposed to the intense heat of the solar rays to evaporate water, and the condensate collected into vases placed underneath (Nebbia and Nebbia-Menozzi, 1966). He also describes a method to obtain freshwater from the air (what is known today as the *humidification−dehumidification method*).

Around 1774, the great French chemist Lavoisier used large glass lenses, mounted on elaborate supporting structures, to concentrate solar energy on the contents of distillation flasks. The use of silver- or aluminum-coated glass reflectors to concentrate solar energy for distillation has also been described by Mouchot.

FIGURE 1.7

Cleopatra's alembic.

In 1870, the first American patent on solar distillation was granted to the experimental work of Wheeler and Evans. Almost everything we know about the basic operation of the solar stills and the corresponding corrosion problems is described in that patent. The inventors described the greenhouse effect, analyzed in detail the cover condensation and re-evaporation, and discussed the dark surface absorption and the possibility of corrosion problems. High operating temperatures were claimed as well as means of rotating the still to follow the solar incident radiation (Wheeler and Evans, 1870).

Two years later, in 1872, an engineer from Sweden, Carlos Wilson, designed and built the first large solar distillation plant, in Las Salinas, Chile (Harding, 1883); thus, solar stills were the first to be used on large-scale distilled water production. The plant was constructed to provide freshwater to the workers and their families at a saltpeter mine and a nearby silver mine. They used the saltpeter mine effluents, of very high salinity (140,000 ppm), as feed-water to the stills. The plant was constructed of wood and timber framework covered with one sheet of glass. It consisted of 64 bays having a total surface area of 4450 m^2 and a total land surface area of 7896 m^2. It produced 22.70 m^3 of freshwater per day (about 4.9 L/m^2). The still was operated for 40 years and was abandoned only after a freshwater pipe was installed, supplying water to the area from the mountains.

In the First World Symposium on Applied Solar Energy, which took place in November 1955, Maria Telkes described the Las Salinas solar distillation plant and reported that it was in operation for about 36 continuous years (Telkes, 1956a).

The use of solar concentrators in solar distillation was reported by Louis Pasteur, in 1928, who used a concentrator to focus solar rays onto a copper boiler containing water. The steam generated from the boiler was piped to a conventional water-cooled condenser in which distilled water was accumulated.

A renewal of interest in solar distillation occurred after the First World War, at which time several new devices had been developed, such as the roof-type, tilted wick, inclined tray, and inflated stills.

Before the Second World War only a few solar distillation systems existed. One of them, designed by C.G. Abbot, is a solar distillation device, similar to that of Mouchot (Abbot, 1930; 1938). At the same time some research on solar distillation was undertaken in the USSR (Trofimov, 1930; Tekutchev, 1938). During the years 1930–40, the dryness in California initiated the interest in desalination of saline water. Some projects were started, but the depressed economy at that time did not permit any research or applications. Interest grew stronger during the Second World War, when hundreds of Allied troops suffered from lack of drinking water while stationed in North Africa, the Pacific islands, and other isolated places. Then a team from MIT, led by Maria Telkes, began experiments with solar stills (Telkes, 1943). At the same time, the U.S. National Research Defense Committee sponsored research to develop solar desalters for military use at sea. Many patents were granted (Delano, 1946a;b; Delano and Meisner, 1946) for individual small plastic solar distillation apparatuses that were developed to be used on lifeboats or rafts. These were designed to float on seawater when inflated and were used extensively by the U.S. Navy during the war (Telkes, 1945). Telkes continued to investigate various configurations of solar stills, including glass-covered and multiple-effect solar stills (Telkes, 1951; 1953; 1956b).

The explosion of urban population and the tremendous expansion of industry after the Second World War again brought the problem of good-quality water into focus. In July 1952, the Office of Saline Water (OSW) was established in the United States, the main purpose of which was to finance basic research on desalination. The OSW promoted desalination application through research. Five demonstration plants were built, and among them was a solar distillation plant in Daytona Beach, Florida, where many types and configurations of solar stills (American and foreign) were tested (Talbert et al., 1970). G.O.G. Loef, as a consultant to the OSW in the 1950s, also experimented with solar stills, such as basin-type stills, solar evaporation with external condensers, and multiple-effect stills, at the OSW experimental station in Daytona Beach (Loef, 1954).

In the following years, many small-capacity solar distillation plants were erected on Caribbean islands by McGill University of Canada. Everett D. Howe, from the Sea Water Conversion Laboratory of the University of California, Berkeley, was another pioneer in solar stills who carried out many studies on solar distillation (Kalogirou, 2005).

Experimental work on solar distillation was also performed at the National Physical Laboratory, New Delhi, India, and the Central Salt and Marine Chemical Research Institute, Bhavnagar, India. In Australia, the Commonwealth Scientific and Industrial Research Organization (CSIRO) in Melbourne carried out a number of studies on solar distillation. In 1963, a prototype bay-type still was developed, covered with glass and lined with black polyethylene sheet (CSIRO, 1960). Solar distillation plants were constructed using this prototype still in the Australian desert, providing freshwater from saline well water for people and livestock. At the same time, V. A. Baum in the USSR was experimenting with solar stills (Baum, 1960; 1961; Baum and Bairamov, 1966).

Between 1965 and 1970, solar distillation plants were constructed on four Greek islands to provide small communities with freshwater (Delyannis, 1968). The design of the stills, done at the Technical University of Athens, was of the asymmetric glass-covered greenhouse type with aluminum frames. The stills used seawater as feed and were covered with single glass. Their capacity ranged from 2044 to 8640 m^3/day. In fact, the installation in the island of Patmos is the largest solar distillation plant ever built. On three more Greek islands, another three solar distillation plants were erected. These were plastic-covered stills (Tedlar) with capacities of 2886, 388, and 377 m^3/day, which met the summer freshwater needs of the Young Men's Christian Association campus.

Solar distillation plants were also constructed on the islands of Porto Santo and Madeira, Portugal, and in India, for which no detailed information exists. Today, most of these plants are not in operation. Although a lot of research is being carried out on solar stills, no large-capacity solar distillation plants have been constructed in recent years.

A number of solar desalination plants coupled with conventional desalination systems were installed in various locations in the Middle East. The majority of these plants are experimental or demonstration scale. A survey of these simple methods of distilled water production, together with some other, more complicated ones, is presented in Chapter 8.

1.5.3 Solar drying

Another application of solar energy is solar drying. Solar dryers have been used primarily by the agricultural industry. The objective in drying an agricultural product is to reduce its moisture contents to a level that prevents deterioration within a period of time regarded as the safe storage period. Drying is a dual process of heat transfer to the product from a heating source and mass transfer of moisture from the interior of the product to its surface and from the surface to the surrounding air. For many centuries farmers were using open sun drying. Recently, however, solar dryers have been used, which are more effective and efficient.

Drying by exposure to the sun is one of the oldest applications of solar energy, used for food preservation, such as vegetables, fruits, and fish and meat products. From the prehistoric times mankind used solar radiation as the only available thermal energy source to dry and preserve all necessary foodstuffs, to dry soil bricks for their homes and to dry animal skins for dressing.

The first known drying installation is in South of France and is dated at about 8000 BC. This is in fact a stone paved surface used for drying crops. Breeze or natural moderate wind velocities were combined with solar radiation to accelerate drying (Kroll and Kast, 1989).

Various other installations have been found around the world, dated between the years 7000 and 3000 BC. There are various combined installations, utilizing solar radiation combined with natural air circulation, used mainly for drying food. In Mesopotamia various sites have been found, for solar air drying of colored textile material and written clay plates. The first solely air drying installation for crops was found in Hindu river valley and is dated at about 2600 BC (Kroll and Kast, 1989).

The well-known Greek philosopher and physician, Aristotle (384−322 BC), described in detail the drying phenomena, and gave for first time, theoretical explanations of drying. Later on, biomass and wood were used to fire primitive furnaces to dry construction material, such as bricks and roof tiles, but food was exposed only to direct solar radiation (Belessiotis and Delyannis, 2011). The industry of conventional drying started about the eighteenth century but despite any modern methods developed, drying by exposure to the sun continues to be the main method for drying small amounts of agricultural products worldwide.

The objective of a dryer is to supply the product with more heat than is available under ambient conditions, increasing sufficiently the vapor pressure of the moisture held within the crop, thus enhancing moisture migration from within the crop and decreasing significantly the relative humidity of the drying air, hence increasing its moisture-carrying capability and ensuring a sufficiently low equilibrium moisture content.

In solar drying, solar energy is used as either the sole heating source or a supplemental source, and the air flow can be generated by either forced or natural convection. The heating procedure could involve the passage of the preheated air through the product or by directly exposing the product to solar radiation, or a combination of both. The major requirement is the transfer of heat to the moist product by convection and conduction from surrounding air mass at temperatures above that of the product, by radiation mainly from the sun and to a little extent from surrounding hot surfaces, or by conduction from heated surfaces in conduct with the product. More information on solar dryers can be found in Chapter 7.

1.5.4 Passive solar buildings

Finally, another area of solar energy is related to passive solar buildings. The term *passive system* is applied to buildings that include, as integral parts of the building, elements that admit, absorb, store, and release solar energy and thus reduce the need for auxiliary energy for comfort heating. These elements have to do with the correct orientation of buildings, the correct sizing of openings, the use of overhangs and other shading devices, and the use of insulation and thermal mass.

Before the advent of mechanical heating and cooling, passive solar building design was practiced for thousands of years as a means to provide comfortable indoor conditions and protect inhabitants from extreme weather conditions. People at those times considered factors such as solar orientation, thermal mass, and ventilation in the construction of residential dwellings, mostly by experience and the transfer of knowledge from generation to generation. The first solar architecture and urban planning methods were developed by both the Greeks and the Chinese. These methods specified that by orienting buildings toward the south, light and warmth can be provided. According to the "memorabilia" of Xenophon, mentioned in Section 1.1, Socrates said: "Now, supposing a house to have a southern aspect, sunshine during winter will steal in under the verandah, but in summer, when the sun traverses a path right over our heads, the roof will afford an agreeable shade, will it not?". These concepts, together with the others mentioned above, are nowadays considered by bioclimatic architecture. Most of these concepts are investigated in Chapter 6 of this book.

1.6 Other renewable energy systems

This section briefly reviews other renewable energy systems. Most of these, except wind energy, are not covered in this book. More details on these systems can be found in other publications.

1.6.1 Wind energy

Wind is generated by atmospheric pressure differences, driven by solar power. Of the total of 175,000 TW of solar power reaching the earth, about 1200 TW (0.7%) are used to drive the atmospheric pressure system. This power generates a kinetic energy reservoir of 750 EJ with a turnover time of 7.4 days (Soerensen, 1979). This conversion process takes place mainly in the upper layers of the atmosphere, at around 12 km height (where the "jet streams" occur). If it is assumed that about 4.6% of the kinetic power is available in the lowest strata of the atmosphere, the world wind potential is on the order of 55 TW. Therefore it can be concluded that, purely on a theoretical basis and disregarding the mismatch between supply and demand, the wind could supply an amount of electrical energy equal to the present world electricity demand.

As a consequence of the cubic (to the third power) relationship between wind speed and wind power (and hence energy), one should be careful in using average wind speed data (m/s) to derive wind power data (W/m^2). Local geographical circumstances may lead to mesoscale wind structures, which have a much higher energy content than one would calculate from the most commonly used wind speed frequency distribution (Rayleigh). Making allowances for the increase of wind speed with height, it follows that the energy available at, say, 25 m varies from around 1.2 $MWh/m^2/a$ to around 5 $MWh/m^2/a$ in the windiest regions. Higher energy levels are possible if hilly sites are used or local topography funnels a prevailing wind through valleys.

A brief historical introduction into wind energy

In terms of capacity, wind energy is the most widely used renewable energy source. Today there are many wind farms that produce electricity. Wind energy is, in fact, an indirect activity of the sun. Its use as energy goes as far back as 4000 years, during the dawn of historical times. It was adored, like the sun, as a god. For the Greeks, wind was the god Aeolos, the "flying man". After this god's name, wind energy is sometimes referred to as Aeolian energy (Delyannis, 2003).

In the beginning, about 4000 years ago, wind energy was used for the propulsion of sailing ships. In antiquity, this was the only energy available to drive ships sailing in the Mediterranean Basin and other seas, and even today, it is used for sailing small leisure boats. At about the same period, windmills, which were used mainly to grind various crops, appeared (Kalogirou, 2005).

It is believed that the genesis of windmills, though not proven, lay in the prayer mills of Tibet. The oldest very primitive windmills have been found at Neh, eastern Iran, and on the Afghanistan border (Major, 1990). Many windmills have been found in Persia, India, Sumatra, and Bactria. It is believed, in general, that many of the windmills were constructed by the Greeks, who immigrated to Asia with the troops of Alexander the Great (Delyannis, 2003). The earliest written document about windmills is a Hindu book of about 400 BC, called *Arthasastra of Kantilys*, in which there is a suggestion for the use of windmills to pump water (Soerensen, 1995). The next known record is from the Hero of Alexandria who described it in the first century AD. In Western Europe, windmills came later, during the twelfth century, with the first written reference in the 1040−1180 AD time frame (Merriam, 1980). Originally in the twelfth century, these were of the post-mill type in which the whole apparatus was mounted on a post so that it could be rotated to face the wind, and later in the fourteenth century, these were of the tower-mill type where only the top part of the windmill carrying the sails could be rotated (Sorensen, 2009a). The industrial revolution and the advent of steam power brought an end to the use of windmills.

A new use of the wind power was connected to the invention of the water pump and used extensively originally by farmers in the United States and subsequently in many parts of the world. This is of the classic lattice metal tower carrying a rotor made from galvanized steel vanes known as the California-type wind pumps (Sorensen, 2009a).

The famous Swiss mathematician, Leonhard Euler, developed the wind wheel theory and related equations, which are, even today, the most important principles for turbogenerators. The ancestor of today's vertical axis wind turbines was developed by Darrieus (1931), but it took about 50 years to be commercialized, in the 1970s. Scientists in Denmark first installed wind turbines during the Second World War to increase the electrical capacity of their grid. They installed 200 kW Gedser mill turbines, which were in operation until the 1960s (Dodge and Thresler, 1989).

Wind energy systems technology

The exploitation of wind energy today uses a wide range of machine sizes and types, giving a range of different economic performances. Today there are small machines up to about 300 kW and large-capacity ones that are in the megawatt range. A photograph of a wind park is shown in Fig. 1.8.

The technology of the wind turbine generators currently in use is only 30 years old, and investment in it so far has been rather modest, compared with other energy sources. Nearly all the wind turbines manufactured by industry are of the horizontal axis type, and most of them have a three-bladed rotor. However, for some years now, machines have been constructed with two blades to reduce the costs and prolong the life of machines by making them lighter and more flexible by reducing the number of high-technology components.

Worldwide by the end of 2021, 845 GW were installed, an increase of 102 GW from 2020 — including 83.4 GW onshore and nearly 18.6 GW offshore. Wind power provides a substantial share of electricity in a growing number of countries. For 2020, wind energy generated enough energy to provide an estimated 15% of the annual electricity consumption in the EU-27. Much higher shares are recorded in five individual Member States. Thus in 2020, wind energy covered an estimated 48% of Denmark's electricity demand, 38% for Ireland, 24.2% for the United Kingdom, 24% for Portugal, 23.2% for Germany, and 21.9% for Spain. In total, Europe added 13.8 GW of wind power capacity in 2020, of which 21% is operating offshore, bringing the region's total to nearly 210.4 GW. In fact, the year 2020 was Europe's third biggest year for new installations, after 2017 and 2019. Most new capacity in Europe was installed in the EU-27, with a total of 10.8 GW (8.4 GW onshore and 2.4 GW offshore), or net additions of 10.4 GW by accounting for decommissioning. The EU total wind capacity is equal to 179.3 GW, which corresponds to 164.7 GW onshore and 14.6 offshore. The top five countries are the Netherlands which added 2 GW, Spain with 1.7 GW, Germany with nearly 1.7 GW, France with 1.3 GW and Sweden with 1 GW — accounting for 70% of the total. The Netherlands was the top country in Europe and ranked fourth globally, adding nearly 2 GW, which is 0.3 GW more than the 2019 figure and has a cumulative total of 6.8 GW. Spain ranked second in the region and fifth globally for new capacity, adding more than 1.7 GW and the total capacity of the country is 27.4 GW. Although this is below the government's target set in 2020 for additions of 2.2 GW per year-to-year

FIGURE 1.8

A photograph of a wind park.

2030, it is a significant increase over annual installations during the years 2009 through 2018. Wind energy accounted for 21.9% of Spain's electricity generation in 2020. Germany is placed third in the EU and sixth globally for new capacity; however, total additions were the country's lowest in a decade. Germany added almost 1.7 GW and has a total of 62.6 GW (54.9 GW onshore and 7.7 GW offshore).

For the EU and the United Kingdom combined, wind energy generated an estimated 458 TWh in 2020, i.e., 10% more than the 417 TWh generated in 2019, which accounts for about 16.4% of the total electricity demand. Of this 13.4% corresponds to onshore and 3% to offshore wind. The increase relative to 2019 resulted from additional capacity, windy conditions early in the year and a drop in electricity demand due to COVID-related restrictions.

The investments made to achieve this level of development have led to a steady accumulation of field experience and organizational learning. Taken together, many small engineering improvements, better operation and maintenance practices, improved wind prospects, and a variety of other incremental improvements have led to steady cost reductions.

Technological advances promise continued cost reductions. For example, the falling cost of electronic controls has made it possible to replace mechanical frequency controls with electronic systems. In addition, modern computer technology has made it possible to substantially improve the design of blades and other components.

The value of wind electricity depends on the characteristics of the utility system into which it is integrated, as well as on regional wind conditions. Some areas, particularly warm coastal areas, have winds with seasonal and daily patterns that correlate with demand, whereas others have winds that do not. Analyses conducted in the United Kingdom, Denmark, and the Netherlands make it clear that wind systems have greater value if numerous generating sites are connected, because it is likely that wind power fluctuations from a system of turbines installed at many widely separated sites will be less than at any individual site.

More details on wind energy systems can be found in Chapter 13.

1.6.2 Biomass

Biomass energy is a generic term applied to energy production achieved from organic material broken down into two broad categories:

- *Woody biomass.* Forestry timber, residues and co-products, other woody material including thinning and cleaning from woodlands (known as *forestry arisings*), untreated wood products, energy crops such as willow, short rotation coppice, and miscanthus (elephant grass).
- *Non-woody biomass.* Animal wastes, industrial and biodegradable municipal products from food processing, and high-energy crops such as rape, sugarcane, and corn.

Biomass, mainly in the form of industrial and agricultural residues, provided electricity for many years with conventional steam turbine power generators. The United States currently has more than 8000 MWe of generating capacity fueled from biomass. Existing steam turbine conversion technologies are cost competitive in regions where low-cost biomass fuels are available, even though these technologies are comparatively inefficient at the small sizes required for biomass electricity production.

The performance of biomass electric systems can be improved dramatically by adapting to biomass advanced gasification technologies developed originally for coal. Biomass is a more attractive feedstock for gasification than coal because it is easier to gasify and has very low sulfur content; therefore, expensive sulfur removal equipment is not required. Biomass integrated gasifier—gas turbine power systems with efficiencies of more than 40% have been commercially available since the early 1990s. These systems offer high efficiency and low unit capital costs for base load power generation at relatively modest scales of 100 MWe or less and can compete with coal-fired power plants, even when fueled with relatively costly biomass feed stocks.

Another form of energy related to agriculture is biogas. Animal waste is usually used for the generation of electricity from biogas. In these systems, the manure from animals is collected and processed to produce methane, which can be used directly in a diesel engine driving a generator to produce electricity. This can be achieved with two processes; aerobic and anaerobic digestion. Aerobic digestion is the process that takes place in the presence of oxygen, whereas the term anaerobic means without air and hence anaerobic digestion refers to the special type of digestion, which takes place without oxygen. All animal manures are valuable sources of bioenergy. These are usually processed with anaerobic digestion. Anaerobic digestion offers solutions designed to control and accelerate the natural degradation process that occurs in stored manure. An anaerobic digester is a completely closed system, which allows more complete digestion of the odorous organic intermediates found in stored manure to less offensive compounds (Wilkie, 2005). Similar benefits can be obtained also from the aerobic digestion but its operational costs and complexity are greater than the anaerobic systems. Additionally, aerobic methods consume energy and produce large amounts of by-product sludge, which requires disposal compared with significantly less sludge produced in the anaerobic process. From the process engineering point of view, anaerobic digestion is relatively simple, even though the biochemical processes involved are very complex (Wilkie, 2005). Anaerobic digestion applications can be at ambient temperature ($15-25°C$), mesophilic temperature ($30-40°C$), or thermophilic ($50-60°C$) temperature, while farm digesters usually operate at mesophilic temperatures. For these systems to be feasible, large farms or consortiums of farms are required. This method also solves the problem of disposing of the manure, and as a by-product, we have the creation of a very good fertilizer. In the following subsections only biomass and biofuels are examined.

Sustainable biomass production for energy

The renewable energy-intensive global scenario described in Section 1.2 calls for some 400 million hectares of biomass plantations by the second quarter of the 21st century. If this magnitude of biomass is used, the questions raised are whether the net energy balances are sufficiently favorable to justify the effort, whether high biomass yields can be sustained over wide areas and long periods, and whether such plantations are environmentally acceptable (Johanson et al., 1993).

Achieving high plantation yields requires energy inputs, especially for fertilizers and harvesting and hauling the biomass. The energy content of harvested biomass, however, is typically $10-15$ times greater than the energy inputs.

However, whether such high yields can be achieved year after year is questionable. The question is critical because essential nutrients are removed from a site at harvest; if these nutrients are not replenished, soil fertility and yields will decline over time. Fortunately, replenishment is feasible with good management. Twigs and leaves, the parts of the plant in which nutrients tend to concentrate, should be left at the plantation site at harvest, and the mineral nutrients recovered as ash at energy conversion facilities should be returned to the plantation soils. Nitrogen losses can be restored through the application of chemical fertilizers; make-up requirements can be kept low by choosing species that are especially efficient in the use of nutrients. Alternatively, plantations can be made nitrogen self-sufficient by growing nitrogen-fixing species, perhaps intermixed with other species. In the future, it will be possible to reduce nutrient inputs by matching nutrient applications to a plant's cyclic needs.

Intensive planting and harvesting activities can also increase erosion, leading to productivity declines. Erosion risks for annual energy crops would be similar to those for annual food crops, and so the cultivation of such crops should be avoided on erodible lands. For crops such as trees and perennial grasses, average erosion rates are low because planting is so infrequent, typically once in every 10–20 years.

An environmental drawback of plantations is that they support far fewer species than natural forests. Accordingly, it is proposed here that plantations be established not on areas now occupied by natural forests but instead on deforested and otherwise degraded lands in developing countries and on excess agricultural lands in industrialized countries. Moreover, a certain percentage of land should be maintained in a natural state as sanctuary for birds and other fauna, to help control pest populations. In short, plantations would actually improve the status quo with regard to biological diversity.

Biofuels

Recent advancements in distillation and blending technologies are being widely recognized as influencing the global proliferation of biofuels. The idea of biofuels is not new; in fact, Rudolf Diesel envisaged the significance of biofuels back in the nineteenth century, stating, "The use of vegetable oils for engine fuels may seem insignificant today. But such oils may become in the course of time, as important as petroleum and the coal tar products of the present time" (Cowman, 2007).

Rudolf Diesel's first compression ignition engines ran on peanut oil at the World Exposition in Paris. The current drive toward greater use of biofuels is being pushed by the diversification of energy sources using renewable products, as reliance on carbon-based fuels becomes an issue, and the need to replace the methyl tertiary butyl ether (MTBE) component used in many of the world's petroleum products. The change from fuels with an MTBE component started as an environmental issue in various parts of the world.

Ethanol has been recognized as the natural choice for replacing MTBE, and the need for blending ethanol into petroleum products is now a global requirement. Brazil has long been the world's leader when it comes to fuel ethanol capacity, but the United States is trying to exceed this and other countries in the Western Hemisphere by rapidly growing its production. European legislation has set substantial targets for the coming years, and the recently published European Green Deal (EGD) supports a cost-effective transition to climate neutrality by 2050 by facilitating the phasing out of fossil fuels, in particular those that are most polluting. Under Directive 2009/28/EC on the promotion of the use of energy from renewable sources, the share of biofuels increased to a minimum of 10% in every Member State in 2020. Regarding the expansion of biofuels' use in the EU, the Directive aims to ensure the use of sustainable biofuels only, which generate a clear and net GHG saving without negative impact on biodiversity and land use. Standards for biofuels have already been established, with the undiluted base products being defined as B100 (100% biodiesel) and E100 (100% ethanol). Subsequent blending will modify this number, such as a blend of 80% petrol and 20% ethanol, defined as E20, or a blend of 95% diesel and 5% biodiesel, defined as B5 (Cowman, 2007).

Biodiesel can be used in any concentration with petroleum-based diesel fuel, and little or no modification is required for existing diesel engines. Biodiesel is a domestic renewable fuel for diesel engines and is derived from vegetable oils and animal fats, including used oils and fats. Soybean oil is the leading vegetable oil produced in the United States and the leading feedstock for biodiesel production. Biodiesel is not the same as a raw vegetable oil; rather, it is produced by a chemical process that removes the glycerin and converts the oil into methyl esters.

Utilizing the current petroleum distribution infrastructure, blending is typically carried out at the storage or loading terminal. The most common locations for blending are the storage tank, the load rack headers, or most effectively, at the load arm. The most important requirement for this process is the accurate volume measurement of each product. This can be done through sequential blending or ratio blending but most beneficially utilizing the side stream blending technique.

Although petroleum products containing MTBE could be blended at the refinery and transported to the truck or tanker loading terminals via a pipeline or railcar, ethanol-blended fuel contains properties that make this difficult. Ethanol, by nature, attracts any H_2O encountered on route or found in storage tanks. If this happens in a 10% blend and the concentration of H_2O in the blended fuel reaches 0.4%, the combined ethanol and H_2O drop out of the blend. The exact point of dropout depends on the ethanol percentage, make-up quantity, and temperature. If this dropout occurs, the ethanol combines with the H_2O and separates from the fuel, dropping to the bottom of the storage tank. The resulting blend goes out of specification, and getting back to the correct specification requires sending the contaminated ethanol back to the production plant.

The solution to this problem is to keep the ethanol in a clean, dry environment and blend the ethanol with the petroleum products when loading the transport trucks and tankers. Moving the blend point to the loading point minimizes the risk of fuels being contaminated by H_2O.

In general biodiesel processing, the fat or oil is degummed and then reacted with alcohol, such as methanol, in the presence of a catalyst to produce glycerin and methyl esters (biodiesel). Methanol is supplied in excess to assist in quick conversion, and the unused portion is recovered and reused. The catalyst employed is typically sodium or potassium hydroxide, which has already been mixed with the methanol (Cowman, 2007).

Although fuel produced from agriculture has had only marginal use in today's climate, there are political, environmental, legislative, and financial benefits for using biofuels. With oil prices remaining high and very unlikely to reduce, demand for biofuel will continue to rise and provide exciting growth prospects for both investors and equipment manufacturers.

1.6.3 Geothermal energy

Measurements show that the ground temperature below a certain depth remains relatively constant throughout the year. This is because the temperature fluctuations at the surface of the ground are diminished as the depth of the ground increases due to the high thermal inertia of the soil.

There are different geothermal energy sources. They may be classified in terms of the measured temperature as low ($<100°C$), medium ($100-150°C$), and high temperature ($>150°C$). The thermal gradient in the earth varies between 15 and 75°C per km depth; nevertheless, the heat flux is anomalous in different continental areas. The cost of electrical energy is generally competitive, 0.6–2.8 U S. cents/MJ (2–10 U S. cents/kWh), and 0.3%, or 177.5 billion MJ/a (49.3 billion kWh/a), of the world total electrical energy was generated in the year 2000 from geothermal resources (Baldacci et al., 1998).

Geothermal power based on current hydrothermal technology can be locally significant in those parts of the world where there are favorable resources. About 6 GWe of geothermal power were produced in the early 1990s and 15 GWe may be added during the next decade. If hot dry rock geothermal technology is successfully developed, the global geothermal potential will be much larger.

Deep geothermal heat plants operate with one- or two-hole systems. The high expenditure incurred in drilling holes discourages one from using this method in gaining thermal energy. The one-hole injection system or the use of existing single holes, made during crude oil or natural gas exploration, reduces the capital cost. In one-hole systems, the hole is adapted to locate in it a vertical exchanger with a double-pipe heat exchanger, in which the geothermal water is extracted via the inside pipe. Published characteristics allow the estimation of the gained geothermal heat energy flux as a function of the difference between the temperatures of extracted and injected water at different volume fluxes of the geothermal water. In general, the two-layer systems and two-hole systems are more advantageous than one-hole systems. More details of geothermal systems related to desalination are given in Chapter 8.

Ground coupled heat pumps

In these systems ground heat exchangers (GHE) are employed to exchange heat with the ground (see Chapter 8, Section 8.5.6). The ground can be used as an energy source, an energy sink, or for energy storage (Eckert, 1976). For the efficient use of the ground in energy systems, its temperature and other thermal characteristics must be known. Studies show that the ground temperature varies with depth. At the surface, the ground is affected by short-term weather variations, changing to seasonal variations as the depth increases. At deeper layers, the ground temperature remains almost constant throughout the seasons and years and is usually higher than that of the ambient air during the cold months of the year and lower during the warm months (Florides and Kalogirou, 2008). The ground therefore is divided into three zones:

1. The surface zone where hourly variations of temperature occur,
2. The shallow zone, with monthly variations, and
3. The deep zone, where the temperature is almost constant year round.

The structure and physical properties of the ground are factors affecting temperature, in all zones. The temperature of the ground is a function of the thermal conductivity, specific heat, density, geothermal gradient, water content, and water flow rate through the ground. Studies carried out in several locations in Cyprus (typical Mediterranean climate) show that according to the formation of the ground the surface zone reaches a depth of 0.5 m. The shallow zone penetrates to 7−8 m and there after the deep zone follows. Furthermore, the temperature of the ground of the island at the deep zone has a range between 18 and 23°C (Florides and Kalogirou, 2008).

GHE or earth heat exchangers (EHE), are devices used for the exploitation of the ground thermal capacity and the difference in temperature between ambient air and ground. A GHE is usually an array of buried pipes installed either horizontally or vertically into the ground. They use the ground as a heat source when operating in the heating mode and as a heat sink when operating in the cooling mode, with a fluid, usually air, water or a water−antifreeze mixture, to transfer the heat from or to the ground. They can contribute to the air-conditioning of a space, for water heating purposes and also for improving the efficiency of a heat pump.

Ground-coupled heat pumps (GCHPs) or geothermal heat pumps are systems combining a heat pump with a GHE for the heat exchange process, which improves the heat pump efficiency. Mainly, they are of two types; namely, the ground-coupled (closed-loop) system or the groundwater (open-loop) system. The type to be used is chosen according to the ground thermal characteristics, the available land for installation, and the groundwater availability and temperature.

Common heat pumps use an electrically driven compressor that compresses a refrigerant and raises its pressure and thus its temperature. The refrigerant has the ability to change state, from liquid to gas when heated and usually it boils at low temperatures. In the heating mode, the common heat pump's refrigerant absorbs heat from the environment and becomes gas. Then the gas is compressed mechanically and has its temperature increased. The refrigerant at this stage is of high pressure and temperature and exchanges heat with a lower temperature medium (gas or liquid) as it passes through a condenser heating the conditioned space. Having its temperature dropped it returns to the liquid stage and after passing through an expansion valve it becomes liquid at low temperature and pressure. Then the process starts all over again with the refrigerant absorbing heat from the environment cooling it as it passes through an evaporator.

GCHPs exchange heat with the ground instead of the atmosphere. An EHE and a liquid circulating pump are parts that are not included in common heat pumps and are used in the heat exchange process. The heat pumps can be used for both heating and cooling of buildings. This is achieved by reversing the cycle, which means that the condenser and evaporator reverse their roles. The efficiency of heat pumps is described by the coefficient of performance (COP) in the heating mode and the energy efficiency ratio (EER) in the cooling mode. The COP or EER is the ratio of the rate of net heat output to the total energy input expressed in consistent units and under designated rating conditions or is the ratio of the refrigerating capacity to the work absorbed by the compressor per unit time. Sometimes the efficiency is described by the seasonal performance factor, which is the average efficiency of the pump over the heating and cooling period or the seasonal energy efficiency ratio for cooling, which is the total cooling output of an air conditioner during its normal annual usage period for cooling divided by the total electric energy input during the same period. The COP or EER of GCHPs usually is higher than those of the common heat pumps and especially their seasonal performance factor, due to the fact that the ground temperature is more stable during the year, cooler in summer, and hotter in winter than the ambient air (Florides et al., 2011).

Both types of GHEs, open or closed loop, are pollutant free since the only effect they have on the ground is the small increase or decrease of its temperature in a certain distance around the borehole. The efficiency of GCHP depends on the temperature of the cold reservoir (T_C) and the temperature of the hot reservoir (T_H). For the same value of T_C, the efficiency of the refrigerator becomes greater when T_H is lower, that is the smaller the difference between T_H and T_C, the greater the COP. In equation form the ideal reversible Carnot cycle COP is given by:

$$\text{COP} = \frac{T_C}{T_H - T_C} \tag{1.1}$$

For example, calculating the thermodynamic COP of an ideal air-conditioning unit, working at an environmental temperature of 35°C and a room temperature of 20°C, with 5°C expansion temperature and 60°C compression temperature, we get a thermodynamic COP of 5.05 [=278/(333 − 278)]. It should be noted that this number will be around 3.7 for an actual unit. By keeping the same variables as above, but reverting into a ground-coupled refrigerator exchanging heat with the ground at 22°C and with the compression temperature reduced to 35°C instead of 60°C, the thermodynamic COP increases to 9.26 [=278/(308 − 278)] with obvious advantages in electricity consumption. Again in this case, the COP for an actual unit is around 7.7. GCHPs are considered to be improvement on the common water-cooled heat pumps.

1.6.4 Hydrogen

Hydrogen, though the most common element in the universe, is not found in its pure form on earth and must be either electrolyzed from water or stripped out from natural gas, both of which are energy-intensive processes that result in GHG emissions. Hydrogen is an energy carrier and not a fuel, as is usually wrongly asserted. Hydrogen produced electrolytically from wind or direct solar power sources and used in fuel cell vehicles can provide zero-emission transportation. As for any fuel, appropriate safety procedures must be followed. Although the hazards of hydrogen are different from those of the various hydrocarbon fuels now in use, they are not greater.

The basic question is how to produce hydrogen in a clean efficient way. Using natural gas, coal, or even nuclear power to produce hydrogen in many ways defeats the purpose of moving toward a future powered by hydrogen. In the first two instances, GHGs are emitted in the process of producing the hydrogen, whereas in the last case, nuclear waste is generated.

As a nearly ideal energy carrier, hydrogen will play a critical role in a new decentralized energy infrastructure that can provide power to vehicles, homes, and industries. However, the process of making hydrogen with fossil-based power can involve the emission of significant levels of GHGs.

Although the element of hydrogen is the most abundant one in the universe, it must be extracted from biomass, water, or fossil fuels before it can take the form of an energy carrier. A key issue in the future is to promote the generation of electricity from wind and then use that electricity to produce hydrogen.

Extracting hydrogen from water involves a process called *electrolysis*, defined as splitting elements apart using an electric current. Energy supplied from an external source, such as wind or the burning of a fossil fuel, is needed to drive the electrochemical reaction. An electrolyzer uses DC to separate water into its component parts, hydrogen and oxygen. Supplementary components in the electrolyzer, such as pumps, valves, and controls, are generally supplied with AC from a utility connection. Water is "disassociated" and ions are transported through the electrolyte. Hydrogen is collected at the cathode and oxygen at the anode. The process requires pure water.

Electrolysis of water is the decomposition of water (H_2O) into oxygen (O_2) and hydrogen (H_2) gas. This is achieved by passing an electric current through the water.

As shown schematically in Fig. 1.9, an electric DC power source is connected to two electrodes, which usually are in the form of plates to increase the surface area, typically made from inert metals (platinum or stainless steel). By doing so hydrogen appears at the cathode, the negatively charged electrode, and oxygen will appear at the anode, the positively charged electrode. Under ideal conditions, the number of moles of hydrogen generated is twice the number of moles of oxygen.

Electrolysis of pure water requires excess energy otherwise the process is very slow. The efficiency of electrolysis is increased with the addition of an electrolyte, such as a salt, an acid, or a base. If electrolysis is applied in pure water, H^+ cations will accumulate at the anode and hydroxyl OH− anions will accumulate at the cathode. Unless a very large potential is applied electrolysis of pure water is very slow, limited by the overall conductivity.

If a water-soluble electrolyte is added, the conductivity of the water increases considerably. The electrolyte disassociates into cations and anions; the anions move toward the anode and neutralize the buildup of positively charged H^+, whereas the cations move toward the cathode and neutralize the buildup of negatively charged OH−.

FIGURE 1.9

Schematic diagram of the process of electrolysis.

Industrial electrolysis cells are very similar to basic unit shown in Fig. 1.9, using platinum plates or honeycombs as electrodes in an attempt to increase the electrode's surface area.

Two variations of the basic process are the high-pressure and high-temperature electrolysis. In high-pressure electrolysis hydrogen is compressed at around 120–200 bar. By pressurizing the hydrogen in the electrolyzer, the need for an external hydrogen compressor is eliminated whereas the average energy consumption for internal compression is very small of the order of 3%. High-temperature electrolysis, also called *steam electrolysis*, can combine electrolysis with a heat engine. High temperature increases the efficiency of electrolysis reaction and the process is very effective as generally heat energy is cheaper than electricity.

Despite considerable interest in hydrogen, however, there is a significant downside to producing it by means of fossil fuel-generated electricity due to the emissions related to the electrolysis process. Hydrogen fuel promises little GHG mitigation if a developing hydrogen economy increases demand for fossil fuel electricity. On the other hand, using cleanly produced hydrogen can fundamentally change our relationship with the natural environment.

Electrolytic hydrogen may be attractive in regions such as Europe, South and East Asia, North Africa, and the southwestern United States, where prospects for biomass-derived fuels are limited because of either high population density or lack of water. Land requirements are small for both wind and direct solar sources, compared with those for biomass fuels. Moreover, as with wind electricity, producing hydrogen from wind would be compatible with the simultaneous use of the land for other purposes such as ranching or farming. Siting in desert regions, where land is cheap and insolation is good, may be favored for PV−hydrogen systems because little water is needed for electrolysis. The equivalent of 2−3 cm of rain per year on the collectors—representing a small fraction of total precipitation, even for arid regions—would be enough.

Electrolytically produced hydrogen will probably not be cheap. If hydrogen is produced from wind and PV electricity, the corresponding cost of pressurized electrolytic hydrogen to the consumer would be about twice that for methanol derived from biomass; moreover, a hydrogen fuel cell car would cost more than a methanol fuel cell car because of the added cost for the hydrogen storage system. Despite these extra expenses, the life-cycle cost for a hydrogen fuel cell car would be marginally higher than for a gasoline internal combustion engine car, which is about the same as for a battery-powered electric vehicle.

The transition to an energy economy in which hydrogen plays a major role could be launched with hydrogen derived from biomass. Hydrogen can be produced thermochemically from biomass using the same gasifier technology that would be used for methanol production. Although the downstream gas processing technologies would differ from those used for methanol production, in each case the process technologies are well established. Therefore, from a technological perspective, making hydrogen from biomass is no more difficult than making methanol. Biomass-derived hydrogen delivered to users in the transport sector would typically cost only half as much as hydrogen produced electrolytically from wind or PV sources.

Probably the best way to utilize hydrogen is with a fuel cell. A *fuel cell* is an electrochemical energy conversion device in which hydrogen is converted into electricity. Generally, fuel cells produce electricity from external supplies of fuel (on the anode side) and oxidant (on the cathode side). These react in the presence of an electrolyte. Generally, the reactants flow in and reaction products flow out while the electrolyte remains in the cell. Fuel cells can operate continuously as long as the necessary flows are maintained. A hydrogen cell uses hydrogen as fuel and oxygen as an oxidant. Fuel cells differ from batteries in that they consume reactants, which must be replenished, whereas batteries store electrical energy chemically in a closed system. Additionally, while the electrodes within a battery react and change as a battery is charged or discharged, a fuel cell's electrodes are catalytic and relatively stable. More details on fuel cells are given in Chapter 7.

1.6.5 Ocean energy

The various forms of ocean energy are abundant but often available far away from the consumer sites. The world's oceans have the capacity to provide cheap energy. Right now, there are very few ocean energy power plants, and most are fairly small.

The energy of the ocean can be used in three basic ways (Energy Quest, 2007):

- Use the ocean's waves (wave energy conversion).
- Use the ocean's high and low tides and tide currents (tidal energy conversion).
- Use temperature differences in the water (ocean thermal energy conversion (OTEC)).

Unlike other renewable energy sources that rely on sophisticated technologies and advanced materials, such as PVs, most ocean renewable energy systems are inherently simple, since they are made from concrete and steel. Additionally, most of the ocean systems rely on proven technologies, such as hydraulic rams and low-head hydropower turbines and impellers. The ocean's energy resource is large and well understood. It is superior to wind and solar energy, since ocean waves and currents traveling in deep water maintain their characteristics over long distances and the state of the sea can easily be predicted accurately more than 48 h in advance. Therefore, although wave energy is variable, like all renewable energy sources, it is more predictable than solar or wind energy. Similarly, tidal currents are created because of the interaction of the tides and the ocean floor and are thus very predictable and generally more constant than wind and solar energy. Additionally, the high density of water makes the resource concentrated, so moving water carries a lot of energy (Katofsky, 2008). The disadvantage of ocean systems is the need to apply mechanical systems that must be robust and withstand the harsh marine environment.

Wave energy conversion systems convert the kinetic energy of the waves into mechanical energy directly to drive a generator to produce electricity in a special construction, where the oscillating movement of waves is converted into air pressure. This in sequence drives a special wind turbine to produce electricity. Obviously the larger the relative height of the waves, the better. The waves must also be present for many hours of the year.

Tidal energy systems again utilize special turbines or propellers located underwater, which convert the movement of the water due to the tide into mechanical energy to drive an electric generator to produce electricity. Tide is a consequence of the rotation of the moon around the earth and local geography of the seafloor and coastlines. These systems are feasible in places where the tide distance is extended to hundreds of meters. The greatest advantage of these systems is that tides are easily predictable.

Finally, the OTEC systems use the temperature difference of surface and deep water to produce energy. Rankine cycle using a low-pressure turbine is the most employed heat cycle for OTEC. Both closed-cycle and open-cycle engines can be employed. Closed-cycle engines use refrigerants (ammonia or R-134a) as the working fluids, whereas open-cycle engines use vapor produced from the seawater itself as the working fluid. OTEC can also supply cold water as a by-product, which can be used for air-conditioning and refrigeration.

The various ocean energy systems are described briefly in the following sections.

Wave energy

Wave power is the transport of energy by ocean surface waves and the harnessing of that energy to produce useful work such as electricity generation and seawater desalination. The equipment used to exploit wave power is called a wave energy converter (WEC).

Kinetic energy (movement) exists in the moving waves of the ocean and can be used to power a turbine. These systems fundamentally convert the kinetic energy of the waves into electricity by capturing either the vertical oscillation or the linear motion of the waves. Individual devices range in sizes of about 100 kW to about 2 MW (Katofsky, 2008). In the simple example shown in Fig. 1.10, the wave rises into a chamber. The rising water forces the air out of the chamber. The moving air spins a

FIGURE 1.10

Principle of operation of a wave energy converter.

turbine that can turn a generator. When the wave goes down, air flows through the turbine and back into the chamber through doors that are normally closed.

This is only one type of wave energy system. Others actually use the up-and-down motion of the wave to power a piston that moves up and down inside a cylinder. That piston can also turn a generator. Most wave energy systems are very small and are applied mainly to power a warning buoy or small lighthouses.

Although a massive potential exists, wave power generation is not currently a widely employed commercial technology. The first attempts to use this power date back to 1890. Only quite recently, in 2008, the first experimental wave farm was opened in Portugal, at the Aguçadoura Wave Park.

Waves are generated by wind blowing over the surface of the seawater. When the waves propagate at a slower speed than the wind speed, at the boundary of the two media an energy transfer exists from the wind to the waves. The wave height is determined by wind speed, the depth, and topography of the seafloor and by the duration of time the wind is blowing. Generally, the larger the waves the more powerful they are whereas the wave power depends also on the water density and wave speed and wavelength. It should be noted that a matching practical limit exists and variations in time or distance will not produce larger waves and thus when this limit is reached, the waves are fully developed.

One of the first applications of wave power was constructed around 1910 by Bochaux-Praceique, used to power his house at Royan, a small seaside city near Bordeaux in France. This was the first oscillating water column-type WEC. Subsequently, the research on wave energy carried out by Yoshio Masuda in the 1940s was very important as he tested various concepts of wave energy devices at sea to power navigation lights.

The interest in wave energy was renewed in the 1970s, motivated by the first oil crisis. At this time a number of researchers re-investigated the potential of generating useful energy from ocean waves and some important inventions were produced, like the Salter's or Edinburgh Duck developed by Stephen Salters in 1974. This unit attained a remarkable efficiency of 81% as the Duck's curved cam-like body can stop 90% of wave motion and convert 90% of that energy to useful electricity. More recently, the interest in wave energy as a renewable energy system has grown due to climate change issues.

Wave power devices are mainly classified according to the method used to capture the wave energy (point absorber or buoy, oscillating water column, tapered channels, oscillating flaps), location (shoreline, nearshore and offshore), and the power delivery system (elastomeric hose pump, pump to shore, hydroelectric turbine, hydraulic ram, and air turbine). Parabolic or tapered channel reflectors are used to amplify the height of the wave and create a head of water, which can be used to drive a conventional low-head hydro turbine (Sorensen, 2009b). Another design uses a flap that oscillates because of the action of the moving waves like a pendulum and this motion can be converted into electricity using hydraulic rams (Sorensen, 2009b). Once this energy is captured and converted into electricity, power must be transported to the point of use or connected to the grid. Some of the important applications of WECs are the following:

- *Protean WEC.* When deployed this system sits on the ocean surface and converts the relative movement between the static ocean floor and the floating buoy into energy.
- *Pelamis WEC.* This consists of a series of semi-submerged cylindrical sections linked by hinged joints. As waves pass along the length of the converter the sections move relative to one another and the wave-induced motion of the sections is resisted by hydraulic cylinders, which pump

high-pressure oil through hydraulic motors. Finally, the hydraulic motors drive the electrical generator to produce electricity.

- *Wave Dragon energy converter.* In the Wave Dragon large wing reflectors focus the waves up a ramp into an offshore reservoir. The water returns to the ocean by the force of gravity via hydroelectric generators.

Although wave energy conversion is still in the early development stage and a realistic picture of costs is impossible to be made, initial estimates give values of about €0.06–€0.12 per kWh (Sorensen, 2009b).

Tidal energy

Another form of ocean energy system is called *tidal energy*. This is a form of hydropower system that converts the energy of tides into electricity. When tides come into the shore, they can be trapped in reservoirs behind dams. Then when the tide drops, the water behind the dam can be allowed to flow, just like in a regular hydroelectric power plant. Tidal technologies can also employ underwater turbines or propellers driven by the flowing water. Such technologies can be deployed in streams and rivers as well.

Tidal energy has been used since about the eleventh century, when small dams were built along ocean estuaries and small streams. The tidal water behind these dams was used to turn water wheels to mill grains. Tidal barrage systems are in commercial operation in a few locations, but their further development is questionable because of their environmental impact in blocking off large estuaries (Katofsky, 2008).

Tidal energy works well when there is a large increase in tides. An increase of at least 5 m between low tide and high tide is needed. There are only a few places on earth where this tide change occurs.

Some power plants are already operating using this idea. One plant, the La Rance Station, in France, makes enough energy from tides (240 MW) to power 240,000 homes. It began making electricity in 1966. It produces about one-fifth of a regular nuclear or coal-fired power plant. It generates more than 10 times the power of the next largest tidal station in the world, the 17 MW Canadian Annapolis station.

Although tidal power is not widely adopted yet, it has the potential to supply large quantities of electricity in the near future. The greatest advantage of these systems compared with other renewable energy systems is that tides are more predictable than wind energy and solar power. Despite this, the costs involved are still relatively high and there is a limited availability of sites with high tidal potential. Many recent technological achievements in system and turbine designs, however, with the adoption of new axial and cross-flow turbines, give promise for a much lower cost of electricity produced.

Tidal forces are created because of the periodic variations in gravitational attraction exerted by celestial bodies, and they create motions or currents in the oceans of the earth. The intensity of this motion depends mainly on the earth rotation, the position of the moon and sun relative to the earth, and the local topography of the seafloor and coastline. The greatest effect is due to the orbital characteristics of the earth–moon system, and to a lesser extent in the earth–sun system. Because the earth's tides and currents are due to the rotation of the earth and the gravitational interaction with the moon and sun, this form of energy resource is renewable and practically inexhaustible. Other currents are caused by geothermal gradients.

The conversion of tidal currents is similar to the conversion of kinetic energy from the wind. Therefore, many of the proposed designs to harness this potential have resemblances to the wind turbines. As the water density is much bigger than that of air much higher energy densities are present, which leads to lower exploitable current velocities and smaller diameter turbines.

A tidal generator converts the energy of tides and currents into useful electricity. As can be understood, the greater the tidal variation and the higher the tidal current velocities, the greater is the potential of a site for electricity generation. Tidal power can be classified into three generating methods:

- *Tidal stream generator.* Tidal stream generators use the kinetic energy of moving water to power turbines, similar to the way wind turbines use wind.
- *Tidal barrage.* Tidal barrages use the potential energy created because of the difference in hydrostatic height between high and low tides. Barrages are in fact dams constructed across the opening of a tidal estuary (see Fig. 1.11).
- *Dynamic tidal power.* Dynamic tidal power is not applied yet. In principle it is a technology that could convert the kinetic energy in tidal flows or currents into useful electricity. Systems can be built into the sea or ocean, without enclosing an area. For this purpose wind-type turbines can be used underwater.

A horizontal axis turbine comprises a propeller with two or more blades. The turbine can be mounted on a tower fixed to the seafloor, which is more suitable for shallow waters, or can be deployed below a floating support, for deep waters. To increase the efficiency of a horizontal axis turbine, the water flow around the turbine can be controlled with a shroud. These designs however are large underwater constructions and to avoid problems, sites of major shipping lanes and local fisheries must be avoided (Sorensen, 2009b).

Ocean thermal energy conversion (OTEC)

OTEC systems use the temperature difference of surface and deep water to make energy. This idea is not new but actually dates back to 1881, when a French engineer by the name of Jacques D'Arsonval first thought of OTEC. Ocean water gets colder the deeper you go from the surface, and at a great depth the ocean gets very cold. It is warmer on the surface because sunlight warms the water.

Power plants can be built that use this difference in temperature to make energy. A difference of at least 21°C is needed between the warmer surface water and the colder deep ocean water. Using this type of energy source, an OTEC system is being demonstrated in Hawaii.

FIGURE 1.11

Principle of operation of tidal barrage.

OTEC relies on the principle of thermodynamics that a source of heat and a source of cold can be used to drive a heat engine. It is well known from the laws of thermodynamics that a heat engine gives greater efficiency and power when a large temperature difference exists. In the oceans the temperature difference between surface and deep water is quite low in the order of 20–25°C. So the main technical challenge of OTEC is to generate significant amounts of power efficiently from small temperature difference. The greatest temperature differences can be found in the tropics, and these offer the greatest possibilities for OTEC. Tropical oceans have surface water temperatures between 24 and 33°C, whereas the temperature 500 m below the surface temperature drops between 9 and 5°C (Sorensen, 2009b). Maps of the word showing the magnitude of the resource are presented by Rajagopalan and Nihous (2013). A temperature difference of about 20°C gives a thermodynamic efficiency of 6.7% but when pumping energy is considered, this drops to about 3% as OTEC cycles must compensate with several cubic meters per second seawater flow rates per MW of net electricity produced. For example to generate 1 MW of electricity an OTEC plant requires 4 m³/s of warm seawater and 2 m³/s of cold seawater. To supply a 100 MW plant a pipe 11 m in diameter would be required (Sorensen, 2009b). These systems have the potential to offer great amounts of energy although a variation of 1°C in the seawater thermal resource corresponds to a change in net power output of 15% (Rajagopalan and Nihous, 2013). OTEC plants, however, can operate continuously providing a base load supply of electrical power.

The first operational OTEC system was built in Cuba in 1930 and generated 22 kW. The most commonly used heat cycle for OTEC is the Rankine employing a low-pressure turbine and systems can be either closed cycle or open cycle. The former cycles use volatile working fluids (refrigerants) such as ammonia or R-134a, whereas open-cycle engines use vapor produced from the seawater. Useful by-products of OTEC system are the supply of large quantities of cold water, which can be used for air-conditioning or refrigeration and freshwater distilled from the sea, which can be used as a freshwater supply. Additionally the fertile deep ocean water can feed various biological processes. A schematic diagram of possible applications of OTEC is shown in Fig. 1.12.

Three types of thermodynamic cycles can be used for OTEC; open, closed, and hybrid. One of the main problems is to pump the cold seawater from the deep ocean to the surface. This can be achieved by ordinary pumping and desalination. In the latter, desalinating seawater at the ocean floor lowers its density and this is the driving force to raise it to the surface. An alternative to high-cost pipes, which bring cold water to the surface, is to pump a vaporized low-boiling-point fluid into the required depth,

FIGURE 1.12

OTEC applications.

which reduces pumping volume with consequent lowering of the costs. In any case OTEC plants require a long large diameter intake pipe, which is usually 1 km or more in length, to bring cold water to the surface.

Open-cycle OTEC uses warm surface water to produce electricity directly by pumping it in a low-pressure container, which causes its boiling. The expanding steam thus produced drives a low-pressure turbine/generator set. The steam produced in this way is pure freshwater, which is finally condensed by exposing it to cold temperatures from deep ocean water. This method produces also desalinated freshwater, which is an added advantage.

Closed-cycle systems use a fluid with low boiling point, such as ammonia or other refrigerants, expanding in a turbine so as to generate electricity. To achieve this, warm surface seawater is pumped through a heat exchanger to vaporize the volatile fluid. Cold water, which is pumped through a different heat exchanger, condenses the vapor, which is then recirculated through the system.

A hybrid cycle, as its name implies, combines the characteristics of both closed- and open-cycle systems. In these systems, warm seawater is fed in a vacuum chamber where it is flash evaporated, as in the open-cycle process. The steam thus produced vaporizes the working fluid, which is usually ammonia, of a closed-cycle loop on the other side of the evaporator. The vaporized working fluid then drives a turbine/generator unit while the steam condensed within the heat exchanger is used to produce desalinated water.

OTEC has the potential to produce large quantities of electrical power. For this purpose a number of systems have been proposed generally falling into three categories; land based, shelf based, and floating.

Land-based or near-shore facilities offer a number of advantages over those located in water; they do not require sophisticated mooring and lengthy power cables and compared with the open-ocean systems they require simple maintenance. These systems can be installed in sheltered areas to protect them from storms and the weather. Additionally, all products such as electricity, desalinated water, and cold water can be transmitted easily to the grid and water network.

Shelf-based OTEC plants can be mounted on a shelf at depths up to about 100 m to avoid the turbulent wave zone and to be closer to the cold water resource. This type of plant can be fixed to the sea bottom, similar to the way used for offshore oil rigs. Because of the conditions encountered by operating an OTEC plant in deeper water and open ocean, the expenditure involved is bigger compared with land-based systems. There are also problems associated with the difficulties related to the delivery of the produced electricity and freshwater. In fact, depending on the distance of the plant from the shore, power delivery can require long underwater cables, which makes shelf-mounted plants less attractive.

As their name implies, floating OTEC plants are located off-shore on special platforms and although they can be used for large power systems, they present a number of difficulties, for example related to the fixing of the platform with cables to the seabed and the power delivery. For the latter the problems are similar to those presented by the shelf-mounted system, whereas cables attached to floating platforms are more prone to damage, especially during heavy storms.

Exercise

Perform a review of the current status of energy consumption, by sector and type of fuel, and the current status of the use of renewables in your country. It is highly recommended that you use data from the statistical service of your country and the Internet. Suggest various measures to increase the contribution of renewables.

References

Abbot, C.G., 1930. Smithsonian Inst. Mish. Coll. 98 (5). Publ. No. 3530.

Abbot, C.G., December 27, 1938. Solar Distilling Apparatus. U.S. Patent No. 2,141.330.

Abu-Zour, A., Riffat, S., 2006. Environmental and economic impact of a new type of solar louver thermal collector. Int. J. Low Carbon Technol. 1 (3), 217–227.

Anderson, B., 1977. Solar Energy: Fundamentals in Building Design. McGraw-Hill, New York.

Baldacci, A., Burgassi, P.D., Dickson, M.H., Fanelli, M., 1998. Non-electric utilization of geothermal energy in Italy. In: Proceedings of World Renewable Energy Congress V, Part I, September 20–25, Florence, Italy. Pergamon, UK, p. 2795.

Baum, V.A., 1960. Technical characteristics of solar stills of the greenhouse type (in Russian). In: Thermal Power Engineering, Utilization of Solar Energy, vol. 2, pp. 122–132. Academy of Science, USSR Moscow.

Baum, V.A., 1961. Solar Distillers, UN Conference on New Sources of Energy. Paper 35/S/119: 43. United Nations, New York.

Baum, V.A., Bairamov, R., 1966. Prospects of solar stills in Turkmenia. Sol. Energy 10 (1), 38–40.

Belessiotis, V., Delyannis, E., 2011. Solar drying. Sol. Energy 85 (8), 1665–1691.

Bittel, A., 1959. Zur geschichte multiplikativer trennverfahren. Chem. Ing. Tech. 31 (6), 365–424.

Blunden, J., Arndt, D.S. (Eds.), 2012. State of the climate in 2011. Bull. Am. Meteorol. Soc. 93 (7), S1–S264.

Blunden, J., Boyer, T. (Eds.), 2020. State of the climate in 2020. Bull. Amer. Meteor. Soc. 102 (8), Si–S475.

Colonbo, U., 1992. Development and the global environment. In: Hollander, J.M. (Ed.), The Energy–Environment Connection. Island Press, Washington, DC, pp. 3–14.

Cowman, T., January–February 2007. Biofuels focus. Refocus 48–53.

CSIRO, 1960. An Improved Diffusion Still. Australian Patent No. 65.270/60.

Darrieus, G.J.M., 1931. Turbine having its Rotating Shaft Transverse to the Flow of the Current. U.S. Patent 1.850.018.

Delano, W.R.P., 1946a. Process and Apparatus for Distilling Liquids. U.S. Patent 2.402.737.

Delano, W.R.P., 1946b. Solar Still with No Fogging Window. U.S. Patent 2.413.101.

Delano, W.R.P., Meisner, W.E., August 5, 1946. Solar Distillation Apparatus. U.S. Patent 2.405.118.

Delyannis, A., 1967. Solar stills provide island inhabitants with water. Sun Work 10 (1), 6–8.

Delyannis, A., 1968. The Patmos solar distillation plant. Sol. Energy 11, 113–115.

Delyannis, E., 2003. Historic background of desalination and renewable energies. Sol. Energy 75 (5), 357–366.

Delyannis, E., Belessiotis, V., 2000. The history of renewable energies for water desalination. Desalination 128, 147–159.

Dincer, I., 1998. Energy and environmental impacts: present and future perspectives. Energy Sources 20 (4–5), 427–453.

Dincer, I., 1999. Environmental impacts of energy. Energy Pol. 27 (14), 845–854.

Dincer, I., Rosen, M.A., 1998. A worldwide perspective on energy, environment and sustainable development. Int. J. Energy Res. 22 (15), 1305–1321.

Dodge, D.M., Thresler, R.W., 1989. Wind technology today. Adv. Sol. Energy 5, 306–395.

Eckert, E., 1976. The ground used as energy source, energy sink, or for energy storage. Energy 1, 315–323.

Energy Quest, 2007. Energy Quest. Available at: www.energyquest.ca.gov/story/chapter14.html.

EPA, 2007. Environmental Protection Agency. Available at: www.epa.gov/globalwarming/index.html.

European Commission, 2019. The European Green Deal. Communication from the Commission to the European Parliament, the European Council. European Commission, Brussels, Belgium. https://ec.europa.eu/info/sites/info/files/european-green-deal-communication_en.pdf.

Florides, G., Kalogirou, S.A., 2008. First in situ determination of the thermal performance of a U-pipe borehole heat exchanger in Cyprus. Appl. Therm. Eng. 28, 157–163.

Florides, G.A., Pouloupatis, P.D., Kalogirou, S.A., Messaritis, V., Panayides, I., Zomeni, Z., Partasides, G., Lizides, A., Sophocleous, A., Koutsoumpas, K., 2011. The geothermal characteristics of the ground and the potential of using ground coupled heat pumps in Cyprus. Energy 36 (8), 5027−5036.

Harding, J., 1883. Apparatus for solar distillation. In: Proceedings of the Institution of Civil Engineers, London., Vol. 73, pp. 284−288.

IEA-PVPS, 2022. Snapshot of Global PV Markets 2022. IEA-PVPS − Annual Report 2022, Report IEA-PVPS T1-42: 2022.

Johanson, T.B., Kelly, H., Reddy, A.K.N., Williams, R.H., 1993. Renewable fuels and electricity for a growing world economy: defining and achieving the potential. In: Johanson, T.B., Kelly, H., Reddy, A.K.N., Williams, R.H. (Eds.), Renewable Energy: Sources for Fuels and Electricity. Earthscan. Island Press, Washington, DC, pp. 1−71.

Jordan, R.C., Ibele, W.E., 1956. Mechanical energy from solar energy. In: Proceedings of the World Symposium on Applied Solar Energy, pp. 81−101.

Kalogirou, S.A., 1997. Solar water heating in Cyprus—current status of technology and problems. Renew. Energy 10, 107−112.

Kalogirou, S.A., 2004. Solar thermal collectors and applications. Prog. Energy Combust. Sci. 30 (3), 231−295.

Kalogirou, S.A., 2005. Seawater desalination using renewable energy sources. Prog. Energy Combust. Sci. 31 (3), 242−281.

Katofsky, R., 2008. Ocean energy: technology basics. Renew. Energy Focus, 34−36.

Kreider, J.F., Kreith, F., 1977. Solar Heating and Cooling. McGraw-Hill, New York.

Kreith, F., Kreider, J.F., 1978. Principles of Solar Engineering. McGraw-Hill, New York.

Kroll, K., Kast, W., 1989. Trocknen und trockner in der produktion. In: Geschichtliche Entwicklung der Trocknungstechnik, vol. 3. Springer Verlag, Berlin, p. 574.

Loef, G.O.G., 1954. Demineralization of Saline Water with Solar Energy. OSW Report No. 4, PB 161379, p. 80.

Major, J.K., 1990. Water, wind and animal power. In: McNeil, J. (Ed.), An Encyclopaedia of the History of Technology. Rutledge, R. Clay Ltd, Bungay, UK, pp. 229−270.

Malik, M.A.S., Tiwari, G.N., Kumar, A., Sodha, M.S., 1985. Solar Distillation. Pergamon Press, Oxford, UK.

Meinel, A.B., Meinel, M.P., 1976. Applied Solar Energy—an Introduction. Addison-Wesley Publishing Company, Reading, MA.

Merriam, M.F., 1980. Characteristics and uses of wind machines. In: Dicknson, W.C., Cheremisinoff, P.N. (Eds.), Solar Energy Technology Handbook, Part A, Engineering Fundamentals. Marcel Dekker, New York, pp. 665−718.

Mouchot, A., 1878. Resultat des experiences faites en divers points de l'Algerie, pour l'emploi industrielle de la chaleur solaire [Results of some experiments in Algeria, for industrial use of solar energy]. C. R. Acad. Sci. 86, 1019−1021.

Mouchot, A., 1879. La Chaleur Solaire et ses Applications Industrielles [Solar Heat and Its Industrial Applications]. Gauthier-Villars, Paris, pp. 238 and 233.

Mouchot, A., 1880. Utilization industrielle de la chaleur solaire [Industrial utilization of solar energy]. C. R. Acad. Sci. 90, 1212−1213.

Nebbia, G., Nebbia-Menozzi, G., April 18−19, 1966. A Short History of Water Desalination. Acqua Dolce dal Mare, 11 Inchiesta Inter., Milano 129−172.

Photon, 2012. 27.7 GW of Grid-Connected PV Added in 2011, Photon, January 2012.

Rajagopalan, K., Nihous, G.C., 2013. Estimates of global thermal energy conversion (OTEC) resources using an ocean general model. Renew. Energy 50, 532−540.

REN, 2022. Renewables 2022 − Global Status Report. REN 21.

Rosen, M.A., 1996. The role of energy efficiency in sustainable development. Technol. Soc. 15 (4), 21−26.

Sayigh, A.A.W., 2001. Renewable Energy: Global Progress and Examples. Renewable Energy, WREN 15−17.

SERI, 1987. Power from the Sun: Principles of High Temperature Solar Thermal Technology.

Sorensen, B. (Ed.), 2009a. Renewable Energy Focus Handbook. Academic Press, Elsevier, ISBN 978-0-12-374705-1, pp. 435−444. Wind power, (Chapter 9.1).

Sorensen, B. (Ed.), 2009b. Renewable Energy Focus Handbook. Academic Press, Elsevier, ISBN 978-0-12-374705-1, pp. 403−409. Ocean power, (Chapter 7.1).

Sorensen, B., 1979. Renewable Energy. Academic Press, London.

Sorensen, B., 1995. History of, and recent progress in, wind energy utilization. Annu. Rev. Energy Environ. 20, 387−424.

Talbert, S.G., Eibling, J.A., Loef, G.O.G., 1970. Manual on Solar Distillation of Saline Water. R&D Progress Report No. 546. US Department of the Interior, Battelle Memorial Institute, Columbus, OH, p. 270.

Telkes, M., January 1943. Distilling Water with Solar Energy. Report to Solar Energy Conversion Committee, MIT.

Telkes, M., 1945. Solar Distiller for Life Rafts. U.S. OfficeTtechnical Service, Report No. 5225 MIT, OSRD. In: Final Report, to National Defense Research Communication, 11.2. vol., p. 24

Telkes, M., 1951. Solar Distillation Produces Fresh Water from Seawater. MIT Solar Energy Conversion Project, No. 22, p. 24.

Telkes, M., 1953. Fresh water from seawater by solar distillation. Ind. Eng. Chem. 45 (5), 1080−1114.

Telkes, M., 1956a. Solar stills. In: Proceedings of World Symposium on Applied Solar Energy, pp. 73−79.

Telkes, M., 1956b. Research for Methods on Solar Distillation. OSW Report No. 13, PB 161388, p. 66.

Tekutchev, A.N., 1938. Physical Basis for the Construction and Calculation of a Solar Still with Fluted Surface, vol. 2. Transactions of the Uzbekistan State University, Samarkand.

Trofimov, K.G., 1930. The Use of Solar Energy in the National Economy. Uzbek SSR State Press, Tashkent.

United Nations, 1992. Rio Declaration on Environment and Development. Available from: www.un.org/documents/ga/conf151/aconf15126-1annex1.htm.

Virtual Nuclear Tourist, 2007. Virtual Nuclear Tourist. Available from: www.nucleartourist.com.

Wheeler, N.W., Evans, W.W., 1870. Evaporating and Distilling with Solar Heat. U.S. Patent No. 102.633.

Weiss, W., Spörk-Dür, M., 2022. Solar heat worldwide, global market development and trends in 2021. In: Solar Heating and Cooling Program, IEA, 2022.

Wilkie, A.C., 2005. Anaerobic digestion of dairy manure: design and process considerations. In: Dairy Manure Management Conference, NRAES-176, pp. 301−312.

Worldwatch, 2007. Worldwatch. Available from: www.worldwatch.org.

Environmental Characteristics

The sun is a sphere of intensely hot gaseous matter with a diameter of 1.39×10^9 m (see Fig. 2.1). The sun is about 1.5×10^8 km away from earth, so, because thermal radiation travels with the speed of light in vacuum (about 300,000 km/s), after leaving the sun solar energy reaches our planet in 8 min and 20 s. As observed from the earth, the sun disk forms an angle of 32 min of a degree. This is important in many applications, especially in concentrator optics, where the sun cannot be considered as a point source and even this small angle is significant in the analysis of the optical behavior of the collector. The sun has an effective blackbody temperature of 5760 K. The temperature in the central region is much higher. In effect, the sun is a continuous fusion reactor in which hydrogen is turned into helium. The sun's total energy output is 3.8×10^{20} MW, which is equal to 63 MW/m^2 of the sun's surface. This energy radiates outward in all directions. The earth receives only a tiny fraction of the total radiation emitted, equal to 1.7×10^{14} kW; however, even with this small fraction, it is estimated that 84 min of solar radiation falling on earth is equal to the world energy demand for 1 year (about 900 EJ). As seen from the earth, the sun rotates around its axis about once every 4 weeks.

As observed from earth, the path of the sun across the sky varies throughout the year. The shape described by the sun's position, considered at the same time each day for a complete year, is called the *analemma* and resembles a figure 8 aligned along a north–south axis. The most obvious variation in the sun's apparent position through the year is a north–south swing over 47° of angle (because of the 23.5° tilt of the earth axis with respect to the sun), called *declination* (see Section 2.2). The north–south swing in apparent angle is the main cause for the existence of seasons on earth.

Knowledge of the sun's path through the sky is necessary to calculate the solar radiation falling on a surface, the solar heat gain, the proper orientation of solar collectors, the placement of collectors to avoid shading, and many more factors that are not of direct interest in this book. The objective of this chapter is to describe the movements of the sun relative to the earth that give to the sun its east–west trajectory across the sky. The variation of solar incidence angle and the amount of solar energy received are analyzed for a number of fixed and tracking surfaces. The environment in which a solar system works depends mostly on the solar energy availability. Therefore, this is analyzed in some detail. The general weather of a location is required in many energy calculations. This is usually presented as a typical meteorological year (TMY) file, which is described in the last section of this chapter.

2.1 Reckoning of time

In solar energy calculations, apparent solar time (AST) must be used to express the time of day. AST is based on the apparent angular motion of the sun across the sky. The time when the sun crosses the

FIGURE 2.1

Sun–earth relationship.

meridian of the observer is the local solar noon. It usually does not coincide with the 12:00 o'clock time of a locality. To convert the local standard time (LST) to AST, two corrections are applied; the equation of time (ET) and longitude correction. These are analyzed next.

2.1.1 Equation of time

Due to factors associated with the earth's orbit around the sun, the earth's orbital velocity varies throughout the year, so the AST varies slightly from the mean time kept by a clock running at a uniform rate. The variation is called the *equation of time* (ET). The ET arises because the length of a day, that is, the time required by the earth to complete one revolution about its own axis with respect to the sun, is not uniform throughout the year. Over the year, the average length of a day is 24 h; however, the length of a day varies due to the eccentricity of the earth's orbit and the tilt of the earth's axis from the normal plane of its orbit. Due to the ellipticity of the orbit, the earth is closer to the sun on January 3 and furthest from the sun on July 4. Therefore the earth's orbiting speed is faster than its average speed for half the year (from about October through March) and slower than its average speed for the remaining half of the year (from about April through September).

The values of the ET as a function of the day of the year (N) can be obtained approximately from the following equations:

$$ET = 9.87 \sin(2B) - 7.53 \cos(B) - 1.5 \ \sin(B) \ [\text{min}] \tag{2.1}$$

and

$$B = (N - 81)\frac{360}{364} \tag{2.2}$$

A graphical representation of Eq. (2.1) is shown in Fig. 2.2, from which the ET can be obtained directly.

2.1.2 Longitude correction

The standard clock time is reckoned from a selected meridian near the center of a time zone or from the standard meridian, the Greenwich, which is at longitude of $0°$. Since the sun takes 4 min to transverse $1°$ of longitude, a longitude correction term of $4\times$ (Standard longitude [SL] – Local longitude [LL])

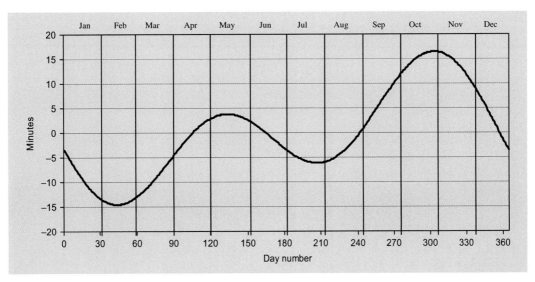

FIGURE 2.2

Equation of time.

should be either added or subtracted to the standard clock time of the locality. This correction is constant for a particular longitude, and the following rule must be followed with respect to sign convention. If the location is east of the standard meridian, the correction is added to the clock time. If the location is west, it is subtracted. The general equation for calculating the AST is:

$$AST = LST + ET \pm 4(SL - LL) - DS \tag{2.3}$$

where

LST = local standard time;
ET = equation of time;
SL = standard longitude;
LL = local longitude; and
DS = daylight saving (it is either 0 or 60 min).

If a location is east of Greenwich, the sign of Eq. (2.3) is minus ($-$), and if it is west, the sign is plus ($+$). If a daylight saving time is used, this must be subtracted from the LST. The term DS depends on whether daylight saving time is in operation (usually from end of March to end of October) or not. This term is usually ignored from this equation and considered only if the estimation is within the DS period.

EXAMPLE 2.1

Find the equation of AST for the city of Nicosia, Cyprus.

Solution

For the locality of Cyprus, the SL is 30°E. The city of Nicosia is at an LL of 33.33° east of Greenwich. Therefore, the longitude correction is $-4 \times (30-33.33) = +13.32$ min. Thus, Eq. (2.3) can be written as:

$$AST = LST + ET + 13.32 \text{ (min)}$$

2.2 Solar angles

The earth makes one rotation about its axis every 24 h and completes a revolution about the sun in a period of approximately 365.25 days. This revolution is not circular but follows an ellipse with the sun at one of the foci, as shown in Fig. 2.3. The eccentricity, e, of the earth's orbit is very small, equal to 0.01673. Therefore, the orbit of the earth round the sun is almost circular. The sun−earth distance, R, at perihelion (shortest distance, at January 3) and aphelion (longest distance, at July 4) is given by Garg (1982):

$$R = a(1 \pm e) \tag{2.4}$$

where a = mean sun−earth distance = 149.5985×10^6 km.

The plus sign in Eq. (2.4) is for the sun−earth distance when the earth is at the aphelion position and the minus sign for the perihelion position. The solution of Eq. (2.4) gives values for the longest distance equal to 152.1×10^6 km and for the shortest distance equal to 147.1×10^6 km, as shown in Fig. 2.3. The difference between the two distances is only 3.3%. The mean sun−earth distance, a, is defined as half the sum of the perihelion and aphelion distances.

The sun's position in the sky changes from day to day and from hour to hour. It is common knowledge that the sun is higher in the sky in the summer than in winter. The relative motions of the sun and earth are not simple, but they are systematic and thus predictable. Once a year, the earth moves around the sun in an orbit that is elliptical in shape. As the earth makes its yearly revolution around the sun, it rotates every 24 h about its axis, which is tilted at an angle of 23° 27.14 min (23.45°) to the plane of the elliptic, which contains the earth's orbital plane and the sun's equator, as shown in Fig. 2.3.

The most obvious apparent motion of the sun is that it moves daily in an arc across the sky, reaching its highest point at midday. As winter becomes spring and then summer, the sunrise and sunset points

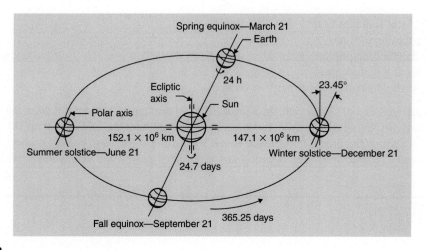

FIGURE 2.3

Annual motion of the earth about the sun.

move gradually northward along the horizon. In the Northern Hemisphere, the days get longer as the sun rises earlier and sets later each day and the sun's path gets higher in the sky. On June 21 the sun is at its most northerly position with respect to the earth. This is called the *summer solstice* and during this day the daytime is at a maximum. Six months later, on December 21, the *winter solstice*, the reverse is true and the sun is at its most southerly position (see Fig. 2.4). In the middle of the 6-month range, on March 21 and September 21, the length of the day is equal to the length of the night. These are called *spring* and *fall equinoxes*, respectively. The summer and winter solstices are the opposite in the Southern Hemisphere; that is, summer solstice is on December 21 and winter solstice is on June 21. It should be noted that all these dates are approximate and that there are small variations (difference of a few days) from year to year.

For the purposes of this book, the Ptolemaic view of the sun's motion is used in the analysis that follows, for simplicity; that is, since all motion is relative, it is convenient to consider the earth fixed and to describe the sun's virtual motion in a coordinate system fixed to the earth with its origin at the site of interest.

For most solar energy applications, one needs reasonably accurate predictions of where the sun will be in the sky at a given time of day and year. In the Ptolemaic sense, the sun is constrained to move with 2 degrees of freedom on the celestial sphere; therefore, its position with respect to an observer on earth can be fully described by means of two astronomical angles, the solar altitude (α) and the solar azimuth (z). The following is a description of each angle, together with the associated formulation. An approximate method for calculating these angles is by means of sun path diagrams (see Section 2.2.2).

Before giving the equations of solar altitude and azimuth angles, the solar declination and hour angle need to be defined. These are required in all other solar angle formulations.

Declination, δ

As shown in Fig. 2.3 the earth axis of rotation (the polar axis) is always inclined at an angle of 23.45° from the ecliptic axis, which is normal to the ecliptic plane. The ecliptic plane is the plane of orbit of

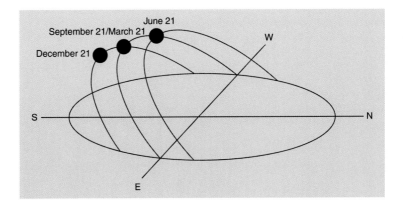

FIGURE 2.4

Annual changes in the sun's position in the sky (northern hemisphere).

the earth around the sun. As the earth rotates around the sun it is as if the polar axis is moving with respect to the sun. The solar declination is the angular distance of the sun's rays north (or south) of the equator, north declination designated as positive. As shown in Fig. 2.5 it is the angle between the sun—earth centerline and the projection of this line on the equatorial plane. Declinations north of the equator (summer in the Northern Hemisphere) are positive, and those south are negative. Figure 2.6 shows the declination during the equinoxes and the solstices. As can be seen, the declination ranges from 0° at the spring equinox to +23.45° at the summer solstice, 0° at the fall equinox, and −23.45° at the winter solstice.

The variation of the solar declination throughout the year is shown in Fig. 2.7. The declination, δ, in degrees for any day of the year (N) can be calculated approximately by the equation (ASHRAE, 2007):

$$\delta = 23.45 \sin\left[\frac{360}{365}(284 + N)\right] \tag{2.5}$$

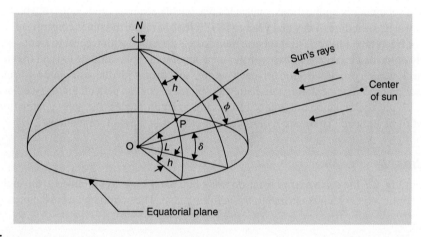

FIGURE 2.5

Definition of latitude, hour angle, and solar declination.

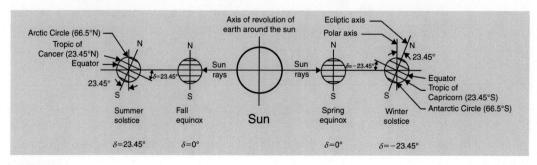

FIGURE 2.6

Yearly variation of solar declination.

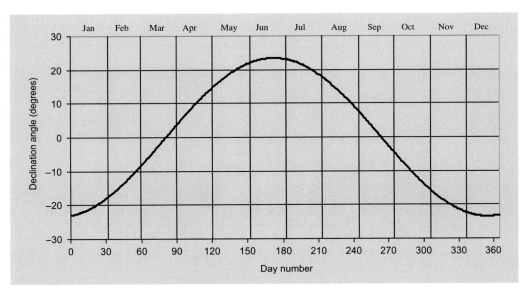

FIGURE 2.7

Declination of the sun.

Declination can also be given in radians[1] by the Spencer formula (Spencer, 1971):

$$\delta = 0.006918 - 0.399912\cos(\Gamma) + 0.070257\sin(\Gamma)$$
$$- 0.006758\cos(2\Gamma) + 0.000907\sin(2\Gamma) \tag{2.6}$$
$$- 0.002697\cos(3\Gamma) + 0.00148\sin(3\Gamma)$$

where Γ is called the *day angle*, given (in radians) by:

$$\Gamma = \frac{2\pi(N-1)}{365} \tag{2.7}$$

The solar declination during any given day can be considered constant in engineering calculations (Kreith and Kreider, 1978; Duffie and Beckman, 1991).

As shown in Fig. 2.6, the Tropics of Cancer (23.45°N) and Capricorn (23.45°S) are the latitudes where the sun is overhead during summer and winter solstices, respectively. Another two latitudes of interest are the Arctic (66.5°N) and Antarctic (66.5°S) Circles. As shown in Fig. 2.6, at winter solstice all points north of the Arctic Circle are in complete darkness, whereas all points south of the Antarctic Circle receive continuous sunlight. The opposite is true for the summer solstice. During spring and fall equinoxes, the North and South Poles are equidistant from the sun and daytime is equal to nighttime, both of which equal 12 h.

Because the day number, the hour of the month, and the average day of each month are frequently required in solar geometry calculations, Table 2.1 is given for easy reference.

---

[1] Radians can be converted to degrees by multiplying by 180 and dividing by π.

Table 2.1 Day number and recommended average day for each month.

Month	Day number	Hour of the month	Average day of the month		
			Date	N	δ (degrees)
January	i	k	17	17	−20.92
February	$31 + i$	$744 + k$	16	47	−12.95
March	$59 + i$	$1416 + k$	16	75	−2.42
April	$90 + i$	$2160 + k$	15	105	9.41
May	$120 + i$	$2880 + k$	15	135	18.79
June	$151 + i$	$3624 + k$	11	162	23.09
July	$181 + i$	$4344 + k$	17	198	21.18
August	$212 + i$	$5088 + k$	16	228	13.45
September	$243 + i$	$5832 + k$	15	258	2.22
October	$273 + i$	$6552 + k$	15	288	−9.60
November	$304 + i$	$7296 + k$	14	318	−18.91
December	$334 + i$	$8016 + k$	10	344	−23.05

Hour angle, *h*

The hour angle, h, of a point on the earth's surface is defined as the angle through which the earth would turn to bring the meridian of the point directly under the sun. Figure 2.5 shows the hour angle of point P as the angle measured on the earth's equatorial plane between the projection of OP and the projection of the sun–earth center to center line. The hour angle at local solar noon is zero, with each 360/24 or 15° of longitude equivalent to 1 h, afternoon hours being designated as positive. Expressed symbolically, the hour angle in degrees is:

$$h = \pm 0.25 \text{ (Number of minutes from local solar noon)} \tag{2.8}$$

where the plus sign applies to afternoon hours and the minus sign to morning hours.

The hour angle can also be obtained from the AST; that is, the corrected local solar time:

$$h = (\text{AST} - 12)15 \tag{2.9}$$

At local solar noon, AST = 12 and $h = 0°$. Therefore, from Eq. (2.3), the LST (the time shown by our clocks at local solar noon) is:

$$\text{LST} = 12 - \text{ET} \mp 4(\text{SL} - \text{LL}) \tag{2.10}$$

EXAMPLE 2.2

Find the equation for LST at local solar noon for Nicosia, Cyprus.

Solution

For the location of Nicosia, Cyprus, from Example 2.1,

$$\text{LST} = 12 - \text{ET} - 13.32 \text{ (min)}$$

EXAMPLE 2.3

Calculate the apparent solar time on March 10 at 2:30 p.m. for the city of Athens, Greece (23°40′E longitude).

Solution

The ET for March 10 ($N = 69$) is calculated from Eq. (2.1), in which the factor B is obtained from Eq. (2.2) as:

$$B = 360/364(N - 81) = 360/364(69 - 81) = -11.87$$

$$\begin{aligned}\text{ET} &= 9.87 \sin(2B) - 7.53 \cos(B) - 1.5 \sin(B) \\ &= 9.87 \sin(-2 \times 11.87) - 7.53 \cos(-11.87) - 1.5 \sin(-11.87)\end{aligned}$$

Therefore,

$$\text{ET} = -11.04 \text{ min} \sim -11 \text{ min}$$

The standard meridian for Athens is 30°E longitude. Therefore, the AST at 2:30 p.m., from Eq. (2.3), is:

$$\begin{aligned}\text{AST} &= 14{:}30 - 4(30 - 23.66) - 0{:}11 = 14{:}30 - 0{:}25 - 0{:}11 \\ &= 13{:}54, \quad \text{or} \quad 1{:}54 \text{ pm}\end{aligned}$$

Solar altitude angle, α

The solar altitude angle is the angle between the sun's rays and a horizontal plane, as shown in Fig. 2.8. It is related to the solar zenith angle, Φ, which is the angle between the sun's rays and the vertical. Therefore,

$$\Phi + \alpha = \pi/2 = 90° \tag{2.11}$$

The mathematical expression for the solar altitude angle is:

$$\sin(\alpha) = \cos(\Phi) = \sin(L)\sin(\delta) + \cos(L)\cos(\delta)\cos(h) \tag{2.12}$$

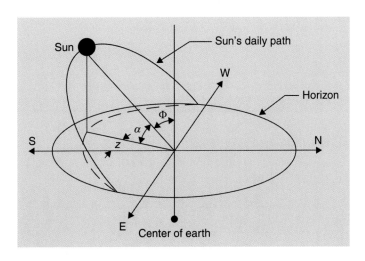

FIGURE 2.8

Apparent daily path of the sun across the sky from sunrise to sunset.

where L = local latitude, defined as the angle between a line from the center of the earth to the site of interest and the equatorial plane. Values north of the equator are positive and those south are negative.

Solar azimuth angle, z

The solar azimuth angle, z, is the angle of the sun's rays measured in the horizontal plane from due south (true south) for the Northern Hemisphere or due north for the Southern Hemisphere; westward is designated as positive. The mathematical expression for the solar azimuth angle is:

$$\sin(z) = \frac{\cos(\delta)\sin(h)}{\cos(\alpha)} \tag{2.13a}$$

This equation is correct, provided that $\cos(h) > \tan(\delta)/\tan(L)$ (ASHRAE, 1975). If not, it means that the sun is behind the E–W line, as shown in Fig. 2.4, and the azimuth angle for the morning hours is $-\pi + |z|$ and for the afternoon hours is $\pi - z$.

Another equation that calculates solar azimuth angle and does not require correction is:

$$z = \pm \cos^{-1}\left(\frac{\cos(\Phi)\sin(L) - \sin(\delta)}{\sin(\Phi)\cos(L)}\right) \tag{2.13b}$$

The $\pm$ sign is according to the sign of the hour angle.

At solar noon, by definition, the sun is exactly on the meridian, which contains the north–south line, and consequently, the solar azimuth is $0°$. Therefore the noon altitude α_n is:

$$\alpha_n = 90° - L + \delta \tag{2.14}$$

EXAMPLE 2.4

What are the maximum and minimum noon altitude angles for a location at $40°$ latitude?

Solution

The maximum angle is at summer solstice, where δ is maximum, that is, $23.5°$. Therefore, the maximum noon altitude angle is $90° - 40° + 23.5° = 73.5°$.

The minimum noon altitude angle is at winter solstice, where δ is minimum, that is, $-23.5°$. Therefore, the minimum noon altitude angle is $90° - 40° - 23.5° = 26.5°$.

Sunrise and sunset times and day length

The sun is said to rise and set when the solar altitude angle is 0. So, the hour angle at sunset, h_{ss}, can be found by solving Eq. (2.12) for h when $\alpha = 0°$:

$$\sin(\alpha) = \sin(0) = 0 = \sin(L)\sin(\delta) + \cos(L)\cos(\delta)\cos(h_{ss})$$

or

$$\cos(h_{ss}) = -\frac{\sin(L)\sin(\delta)}{\cos(L)\cos(\delta)}$$

which reduces to:

$$\cos(h_{ss}) = -\tan(L)\tan(\delta) \tag{2.15}$$

where h_{ss} is taken as positive at sunset.

Since the hour angle at local solar noon is $0°$, with each $15°$ of longitude equivalent to 1 h, the sunrise and sunset time in hours from local solar noon is then:

$$H_{ss} = -H_{sr} = 1/15 \cos^{-1}[-\tan(L)\tan(\delta)] \tag{2.16}$$

The sunrise and sunset hour angles for various latitudes are shown in Figure A3.1 in Appendix 3.

The day length is twice the sunset hour, since the solar noon is at the middle of the sunrise and sunset hours. Therefore, the length of the day in hours is:

$$\text{Day length} = 2/15 \cos^{-1}[-\tan(L)\tan(\delta)] \tag{2.17}$$

EXAMPLE 2.5

Find the equation for sunset standard time for Nicosia, Cyprus.

Solution

The LST at sunset for the location of Nicosia, Cyprus, from Example 2.1 is:

$$\text{Sunset standard time} = H_{ss} - \text{ET} - 13.32 \text{ (min)}$$

EXAMPLE 2.6

Find the solar altitude and azimuth angles at 2 h after local noon on June 16 for a city located at $40°$N latitude. Also find the sunrise and sunset hours and the day length.

Solution

From Eq. (2.5), the declination on June 16 ($N = 167$) is:

$$\delta = 23.45 \sin\left[\frac{360}{365}(284 + 167)\right] = 23.35°$$

From Eq. (2.8), the hour angle, 2 h after local solar noon is:

$$h = +0.25(120) = 30°$$

From Eq. (2.12), the solar altitude angle is:

$$\sin(\alpha) = \sin(40)\sin(23.35) + \cos(40)\cos(23.35)\cos(30) = 0.864$$

Therefore,

$$\alpha = 59.75°$$

From Eq. (2.13), the solar azimuth angle is:

$$\sin(z) = \cos(23.35)\frac{\sin(30)}{\cos(59.75)} = 0.911$$

Continued

EXAMPLE 2.6–CONT'D

Therefore,

$$z = 65.67°$$

From Eq. (2.17), the day length is:

$$\text{Day length} = 2/15\cos^{-1}[-\tan(40)\tan(23.35)] = 14.83\ h$$

This means that the sun rises at $12 - 7.4 = 4.6 = 4{:}36$ a.m. solar time and sets at $7.4 = 7{:}24$ p.m. solar time. As no longitude is given in the problem, we cannot estimate the actual time the clock of a person at that location will show. This can be accomplished with Eq. (2.10).

Incidence angle, θ

The solar incidence angle, θ, is the angle between the sun's rays and the normal on a surface. For a horizontal plane, the incidence angle, θ, and the zenith angle, Φ, are the same. The angles shown in Fig. 2.9 are related to the basic angles, shown in Fig. 2.5, with the following general expression for the angle of incidence (Kreith and Kreider, 1978; Duffie and Beckman, 1991):

$$\begin{aligned}
\cos(\theta) = {} & \sin(L)\sin(\delta)\cos(\beta) - \cos(L)\sin(\delta)\sin(\beta)\cos(Z_s) \\
& + \cos(L)\cos(\delta)\cos(h)\cos(\beta) \\
& + \sin(L)\cos(\delta)\cos(h)\sin(\beta)\cos(Z_s) \\
& + \cos(\delta)\sin(h)\sin(\beta)\sin(Z_s)
\end{aligned} \tag{2.18}$$

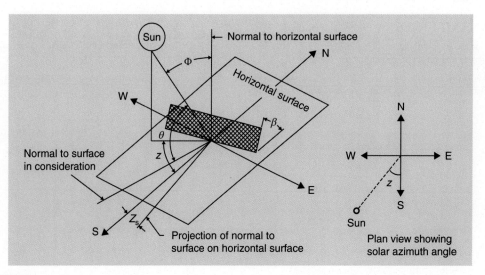

FIGURE 2.9

Solar angles diagram.

where

β = surface tilt angle from the horizontal;
Z_s = surface azimuth angle, the angle between the normal to the surface projected to a horizontal plane from true south, westward is designated as positive.

For certain cases Eq. (2.18) reduces to much simpler forms

- *For horizontal surfaces, $\beta = 0°$ and $\theta = \Phi$, and Eq. (2.18) reduces to Eq. (2.12).*
- *For vertical surfaces, $\beta = 90°$ and Eq. (2.18) becomes:*

$$\cos(\theta) = -\cos(L)\sin(\delta)\cos(Z_s) + \sin(L)\cos(\delta)\cos(h)\cos(Z_s) + \cos(\delta)\sin(h)\sin(Z_s) \qquad (2.19)$$

- *For a south-facing tilted surface in the Northern Hemisphere, $Z_s = 0°$ and Eq. (2.18) reduces to:*

$$\cos(\theta) = \sin(L)\sin(\delta)\cos(\beta) - \cos(L)\sin(\delta)\sin(\beta)$$
$$+ \cos(L)\cos(\delta)\cos(h)\cos(\beta) + \sin(L)\cos(\delta)\cos(h)\sin(\beta)$$

which can be further reduced to:

$$\cos(\theta) = \sin(L-\beta)\sin(\delta) + \cos(L-\beta)\cos(\delta)\cos(h) \qquad (2.20)$$

At solar noon, $\theta_{noon} = |L-\delta-\beta|$ and $z_{noon} = |L-\delta|$

- *For a north-facing tilted surface in the Southern Hemisphere, $Z_s = 180°$ and Eq. (2.18) reduces to:*

$$\cos(\theta) = \sin(L+\beta)\sin(\delta) + \cos(L+\beta)\cos(\delta)\cos(h) \qquad (2.21)$$

Similarly, at solar noon: $\theta_{noon} = |-L+\delta-\beta|$ and $z_{noon} = |-L+\delta|$

Equation (2.18) is a general relationship for the angle of incidence on a surface of any orientation. As shown in Eqs. (2.19)–(2.21), it can be reduced to much simpler forms for specific cases.

EXAMPLE 2.7

A surface tilted 45° from horizontal and pointed 10° west of due south is located at 35°N latitude. Calculate the incident angle at 2 h after local noon on June 16.

Solution

From Example 2.6 we have $\delta = 23.35°$ and the hour angle $= 30°$. The solar incidence angle θ is calculated from Eq. (2.18):

$$\cos(\theta) = \sin(35)\sin(23.35)\cos(45) - \cos(35)\sin(23.35)\sin(45)\cos(10)$$
$$+ \cos(35)\cos(23.35)\cos(30)\cos(45) + \sin(35)\cos(23.35)\cos(30)\sin(45)\cos(10)$$
$$+ \cos(23.35)\sin(30)\sin(45)\sin(10) = 0.769$$

Therefore,

$$\theta = 39.72°$$

2.2.1 The incidence angle for moving surfaces

For the case of solar-concentrating collectors, some form of tracking mechanism is usually employed to enable the collector to follow the sun. This is done with varying degrees of accuracy and modes of tracking, as indicated in Fig. 2.10.

Tracking systems can be classified by the mode of their motion. This can be about a single axis or about two axes (Fig. 2.10a). In the case of a single-axis mode, the motion can be in various ways: parallel to the earth's axis (Fig. 2.10b), north–south (Fig. 2.10c), or east–west (Fig. 2.10d). The following equations are derived from the general Eq. (2.18) and apply to planes moved, as indicated in each case. The amount of energy falling on a surface per unit area for the summer and winter solstices and the equinoxes for latitude of 35°N is investigated for each mode. This analysis has been performed with a radiation model. This is affected by the incidence angle, which is different for each mode. The type of model used here is not important, since it is used for comparison purposes only.

Full tracking

For a two-axis tracking mechanism, keeping the surface in question continuously oriented to face the sun (see Fig. 2.10a) at all times has an angle of incidence, θ, equal to:

$$\cos(\theta) = 1 \tag{2.22}$$

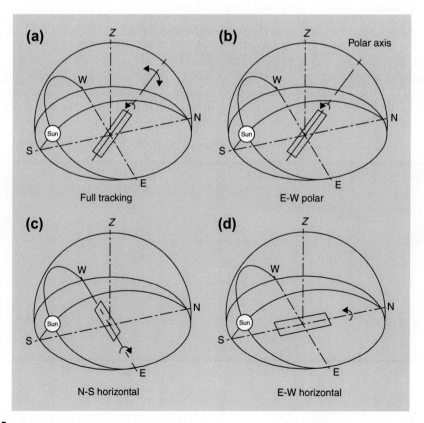

FIGURE 2.10

Collector geometry for various modes of tracking.

or $\theta = 0°$. This, of course, depends on the accuracy of the mechanism. The full tracking configuration collects the maximum possible sunshine. The performance of this mode of tracking with respect to the amount of radiation collected during 1 day under standard conditions is shown in Fig. 2.11.

The slope of this surface (β) is equal to the solar zenith angle (Φ), and the surface azimuth angle (Z_s) is equal to the solar azimuth angle (z).

Tilted N–S axis with tilt adjusted daily

For a plane moved about a north–south axis with a single daily adjustment so that its surface normal coincides with the solar beam at noon each day, θ is equal to (Meinel and Meinel, 1976; Duffie and Beckman, 1991):

$$\cos(\theta) = \sin^2(\delta) + \cos^2(\delta)\cos(h) \tag{2.23}$$

For this mode of tracking, we can accept that, when the sun is at noon, the angle of the sun's rays and the normal to the collector can be up to a 4° declination, since for small angles $\cos(4°) = 0.998 \sim 1$. Figure 2.12 shows the number of consecutive days that the sun remains within this 4° "declination window" at noon. As can be seen in Fig. 2.12, most of the time the sun remains close to either the summer solstice or the winter solstice, moving rapidly between the two extremes. For nearly 70 consecutive days, the sun is within 4° of an extreme position, spending only 9 days in the 4° window, at the equinox. This means that a seasonally tilted collector needs to be adjusted only occasionally.

The problem encountered with this and all tilted collectors, when more than one collector is used, is that the front collectors cast shadows on adjacent ones. Therefore, in terms of land utilization, these collectors lose some of their benefits when the cost of land is taken into account. The performance of this mode of tracking (see Fig. 2.13) shows the peaked curves typical for this assembly.

Polar N–S axis with E–W tracking

For a plane rotated about a north–south axis parallel to the earth's axis, with continuous adjustment, θ is equal to:

$$\cos(\theta) = \cos(\delta) \tag{2.24}$$

FIGURE 2.11

Daily variation of solar flux, full tracking.

FIGURE 2.12

Number of consecutive days the sun remains within 4° declination.

FIGURE 2.13

Daily variation of solar flux: tilted N–S axis with tilt adjusted daily.

This configuration is shown in Fig. 2.10b. As can be seen, the collector axis is tilted at the polar axis, which is equal to the local latitude. For this arrangement, the sun is normal to the collector at equinoxes ($\delta = 0°$) and the cosine effect is maximum at the solstices. The same comments about the tilting of the collector and shadowing effects apply here as in the previous configuration. The performance of this mount is shown in Fig. 2.14.

The equinox and summer solstice performance, in terms of solar radiation collected, are essentially equal; that is, the smaller air mass for summer solstice offsets the small cosine projection effect. The winter noon value, however, is reduced because these two effects combine. If it is desired to increase the winter performance, an inclination higher than the local latitude would be required; but the

FIGURE 2.14

Daily variation of solar flux: polar N—S axis with E—W tracking.

physical height of such configuration would be a potential penalty to be traded off in cost-effectiveness with the structure of the polar mount. Another side effect of increased inclination is shadowing by the adjacent collectors, for multi-row installations.

The slope of the surface varies continuously and is given by:

$$\tan(\beta) = \frac{\tan(L)}{\cos(Z_s)} \tag{2.25a}$$

The surface azimuth angle is given by:

$$Z_s = \tan^{-1}\frac{\sin(\Phi)\sin(z)}{\cos(\theta')\sin(L)} + 180 C_1 C_2 \tag{2.25b}$$

where

$$\cos(\theta') = \cos(\Phi)\cos(L) + \sin(\Phi)\sin(L)\cos(z) \tag{2.25c}$$

$$C_1 = \begin{cases} 0 & \text{if} \quad \left(\tan^{-1}\frac{\sin(\Phi)\sin(z)}{\cos(\theta')\sin(L)}\right) z \ge 0 \\ 1 & \text{otherwise} \end{cases} \tag{2.25d}$$

$$C_2 = \begin{cases} 1 & \text{if} \quad z \ge 0° \\ -1 & \text{if} \quad z < 0° \end{cases} \tag{2.25e}$$

Horizontal E—W axis with N—S tracking

For a plane rotated about a horizontal east—west axis with continuous adjustment to minimize the angle of incidence, θ can be obtained from (Kreith and Kreider, 1978; Duffie and Beckman, 1991):

$$\cos(\theta) = \sqrt{1 - \cos^2(\delta)\sin^2(h)} \tag{2.26a}$$

or from this equation (Meinel and Meinel, 1976):

$$\cos(\theta) = \sqrt{\sin^2(\delta) + \cos^2(\delta)\cos^2(h)} \tag{2.26b}$$

The basic geometry of this configuration is shown in Fig. 2.10c. The shadowing effects of this arrangement are minimal. The principal shadowing is caused when the collector is tipped to a maximum degree south ($\delta = -23.5°$) at winter solstice. In this case, the sun casts a shadow toward the collector at the north. This assembly has an advantage in that it approximates the full tracking collector in summer (see Fig. 2.15), but the cosine effect in winter greatly reduces its effectiveness. This mount yields a rather "square" profile of solar radiation, ideal for leveling the variation during the day. The winter performance, however, is seriously depressed relative to the summer one.

The slope of this surface is given by:

$$\tan(\beta) = \tan(\Phi)|\cos(z)| \tag{2.27a}$$

The surface orientation for this mode of tracking changes between 0° and 180°, if the solar azimuth angle passes through ±90°. For either hemisphere,

$$\begin{aligned} \text{If } |z| < 90°, \quad &Z_s = 0° \\ \text{If } |z| > 90°, \quad &Z_s = 180° \end{aligned} \tag{2.27b}$$

Horizontal N−S axis with E−W tracking

For a plane rotated about a horizontal north−south axis with continuous adjustment to minimize the angle of incidence, θ can be obtained from (Kreith and Kreider, 1978; Duffie and Beckman, 1991),

$$\cos(\theta) = \sqrt{\sin^2(\alpha) + \cos^2(\delta)\sin^2(h)} \tag{2.28a}$$

FIGURE 2.15

Daily variation of solar flux: horizontal E−W axis with N−S tracking.

or from this equation (Meinel and Meinel, 1976):

$$\cos(\theta) = \cos(\Phi)\cos(h) + \cos(\delta)\sin^2(h) \tag{2.28b}$$

The basic geometry of this configuration is shown in Fig. 2.10d. The greatest advantage of this arrangement is that very small shadowing effects are encountered when more than one collector is used. These are present only at the first and last hours of the day. In this case the curve of the solar energy collected during the day is closer to a cosine curve function (see Fig. 2.16).

The slope of this surface is given by:

$$\tan(\beta) = \tan(\Phi)|\cos(Z_s - z)| \tag{2.29a}$$

The surface azimuth angle (Z_s) is 90° or −90°, depending on the solar azimuth angle:

$$
\begin{aligned}
\text{If } z > 0°, \quad & Z_s = 90° \\
\text{If } z < 0°, \quad & Z_s = -90°
\end{aligned}
\tag{2.29b}
$$

Comparison

The mode of tracking affects the amount of incident radiation falling on the collector surface in proportion to the cosine of the incidence angle. The amount of energy falling on a surface per unit area for the four modes of tracking for the summer and winter solstices and the equinoxes are shown in Table 2.2. This analysis has been performed with the same radiation model used to plot the solar flux figures in this section. Again, the type of the model used here is not important, because it is used for comparison purposes only. The performance of the various modes of tracking is compared with the full tracking, which collects the maximum amount of solar energy, shown as 100% in Table 2.2. From this table it is obvious that the polar and the N−S horizontal modes are the most suitable for one-axis tracking, since their performance is very close to the full tracking, provided that the shadow effects of the polar mode and the low winter performance of the N−S horizontal mode is not a problem.

FIGURE 2.16

Daily variation of solar flux: horizontal N−S axis and E−W tracking.

Table 2.2 Comparison of energy received for various modes of tracking.

Tracking mode	Solar energy received (kWh/m^2)			Percentage to full tracking		
	E	SS	WS	E	SS	WS
Full tracking	8.43	10.60	5.70	100	100	100
E—W polar	8.43	9.73	5.23	100	91.7	91.7
N—S horizontal	7.51	10.36	4.47	89.1	97.7	60.9
E—W horizontal	6.22	7.85	4.91	73.8	74.0	86.2

E, *equinoxes;* SS, *summer solstice;* WS, *winter solstice.*

2.2.2 Sun path diagrams

For practical purposes, instead of using the preceding equations, it is convenient to have the sun's path plotted on a horizontal plane, called a *sun path diagram,* and to use the diagram to find the position of the sun in the sky at any time of the year. As can be seen from Eqs. (2.12) and (2.13), the solar altitude angle, α, and the solar azimuth angle, z, are functions of latitude, L, hour angle, h, and declination, δ. In a two-dimensional plot, only two independent parameters can be used to correlate the other parameters; therefore, it is usual to plot different sun path diagrams for different latitudes. Such diagrams show the complete variations of hour angle and declination for a full year. Figure 2.17 shows the sun path diagram for 35°N latitude. Lines of constant declination are labeled by the value of the angles. Points of constant hour angles are clearly indicated. This figure is used in combination with Fig. 2.7 or Eqs. (2.5)–(2.7); that is, for a day in a year, Fig. 2.7 or the equations can be used to estimate declination, which is then entered together with the time of day and converted to solar time using Eq. (2.3) in Fig. 2.17 to estimate solar altitude and azimuth angles. It should be noted that Fig. 2.17 applies for the Northern Hemisphere. For the Southern Hemisphere, the sign of the declination should be reversed. Figures A3.2 through A3.4 in Appendix 3 show the sun path diagrams for 30, 40, and 50°N latitudes.

2.2.3 Shadow determination

In the design of many solar energy systems, it is often required to estimate the possibility of the shading of solar collectors or the windows of a building by surrounding structures. To determine the shading, it is necessary to know the shadow cast as a function of time during every day of the year. Although mathematical models can be used for this purpose, a simpler graphical method is presented here, which is suitable for quick practical applications. This method is usually sufficient, since the objective is usually not to estimate exactly the amount of shading but to determine whether a position suggested for the placement of collectors is suitable or not.

Shadow determination is facilitated by the determination of a surface-oriented solar angle, called the *solar profile angle.* As shown in Fig. 2.18, the solar profile angle, p, is the angle between the normal to a surface and the projection of the sun's rays on a plane normal to the surface. In terms of the solar altitude angle, α, solar azimuth angle, z, and the surface azimuth angle, Z_s, the solar profile angle p is given by the equation:

$$\tan(p) = \frac{\tan(\alpha)}{\cos(z - Z_s)} \tag{2.30a}$$

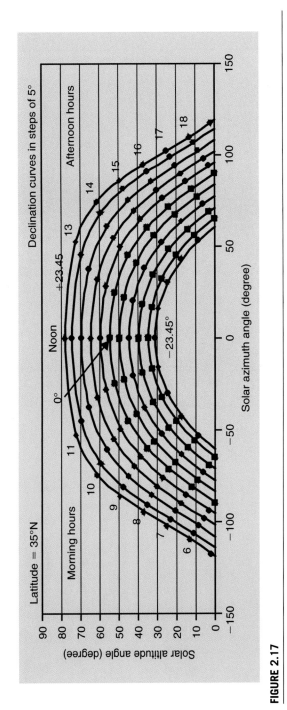

FIGURE 2.17

Sun path diagram for 35°N latitude.

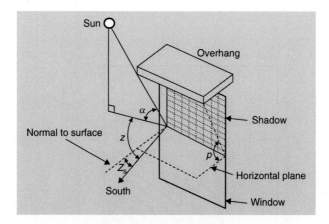

FIGURE 2.18

Geometry of solar profile angle, p, in a window overhang arrangement.

A simplified equation is obtained when the surface faces due south, that is, $Z_s = 0°$, given by:

$$\tan(p) = \frac{\tan(\alpha)}{\cos(z)} \tag{2.30b}$$

The sun path diagram is often very useful in determining the period of the year and hours of day when shading will take place at a particular location. This is illustrated in the following example.

EXAMPLE 2.8

A building is located at 35°N latitude and its side of interest is located 15° east of south. We want to investigate the time of the year that point x on the building will be shaded, as shown in Fig. 2.19.

Note: Distances marked with * can also be obtained from simple trigonometry

FIGURE 2.19

Shading of building in Example 2.8.

EXAMPLE 2.8—CONT'D

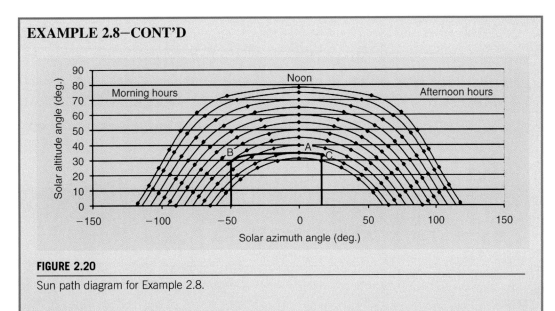

FIGURE 2.20

Sun path diagram for Example 2.8.

Solution

The upper limit of profile angle for shading point x is 35 and 15° west of true south. This is point A drawn on the sun path diagram, as shown in Fig. 2.20. In this case, the solar profile angle is the solar altitude angle. Distance x–B is $(8.4^2 + 12^2)^{1/2} = 14.6$ m. For the point B, the altitude angle is $\tan(\alpha) = 8.4/14.6 \rightarrow \alpha = 29.9°$. Similarly, distance x–C is $(6.9^2 + 12^2)^{1/2} = 13.8$ m, and for point C, the altitude angle is $\tan(\alpha) = 8.4/13.8 \rightarrow \alpha = 31.3°$. Both points are as indicated on the sun path diagram in Fig. 2.20.

Therefore, point x on the wall of interest is shaded during the period indicated by the curve BAC in Fig. 2.20. It is straightforward to determine the hours that shading occurs, whereas the time of year is determined by the declination.

Solar collectors are usually installed in multi-rows facing the true south. There is, hence, a need to estimate the possibility of shading by the front rows of the second and subsequent rows. The maximum shading, in this case, occurs at local solar noon, and this can easily be estimated by finding the noon altitude, α_n, as given by Eq. (2.14) and checking whether the shadow formed shades on the second or subsequent collector rows. Generally, shading will not occur as long as the profile angle is greater than the angle θ_s, formed by the top corner of the front collector to the bottom end of the second row and the horizontal (see Fig. 5.25). If the profile angle at any time is less than θ_s, then a portion of the collectors in the second and subsequent rows will be shaded from beam radiation.

EXAMPLE 2.9

Find the equation to estimate the shading caused by a fin on a window shown in Fig. 2.21.

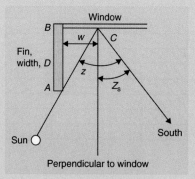

FIGURE 2.21

Fin and window assembly for Example 2.9.

Solution

The fin and window assembly are shown in Fig. 2.21.

From triangle ABC, the sides $AB = D$, $BC = w$, and angle A is $z - Z_s$.

Therefore, distance w is estimated by $w = D \tan(z - Z_s)$.

Shadow calculations for overhangs are examined in more detail in Chapter 6, Section 6.2.5.

2.3 Solar radiation

2.3.1 General

All substances, solid bodies as well as liquids and gases above the absolute zero temperature, emit energy in the form of electromagnetic waves.

The radiation that is important to solar energy applications is that emitted by the sun within the ultraviolet, visible, and infrared regions. Therefore, the radiation wavelength that is important to solar energy applications is between 0.15 and 3.0 μm. The wavelengths in the visible region lie between 0.38 and 0.72 μm.

This section initially examines issues related to thermal radiation, which includes basic concepts, radiation from real surfaces, and radiation exchanges between two surfaces. This is followed by the variation of extraterrestrial radiation, atmospheric attenuation, terrestrial irradiation, and total radiation received on sloped surfaces. Finally, it briefly describes radiation measuring equipment.

2.3.2 Thermal radiation

Thermal radiation is a form of energy emission and transmission that depends entirely on the temperature characteristics of the emissive surface. There is no intervening carrier, as in the other modes of heat transmission, that is, conduction and convection. Thermal radiation is in fact an electromagnetic wave that travels at the speed of light ($C \approx 300,000$ km/s in a vacuum). This speed is related to the wavelength (λ) and frequency (v) of the radiation as given by the equation:

$$C = \lambda v \tag{2.31}$$

When a beam of thermal radiation is incident on the surface of a body, part of it is reflected away from the surface, part is absorbed by the body, and part is transmitted through the body. The various properties associated with this phenomenon are the fraction of radiation reflected, called *reflectivity* (ρ); the fraction of radiation absorbed, called *absorptivity* (α); and the fraction of radiation transmitted, called *transmissivity* (τ). The three quantities are related by the following equation:

$$\rho + \alpha + \tau = 1 \tag{2.32}$$

It should be noted that the radiation properties just defined are not only functions of the surface itself but also functions of the direction and wavelength of the incident radiation. Therefore, Eq. (2.32) is valid for the average properties over the entire wavelength spectrum. The following equation is used to express the dependence of these properties on the wavelength:

$$\rho_\lambda + \alpha_\lambda + \tau_\lambda = 1 \tag{2.33}$$

where

ρ_λ = spectral reflectivity;
α_λ = spectral absorptivity; and
τ_λ = spectral transmissivity.

Similar expressions can be presented for the direction of radiation, in which case the subscript λ is replaced by θ. The angular variation of absorptance for black paint is illustrated in Table 2.3 for incidence angles of 0–90°. The absorptance for diffuse radiation is approximately 0.90 (Löf and Tybout, 1972).

Most solid bodies are opaque, so that $\tau = 0$ and $\rho + \alpha = 1$. If a body absorbs all the impinging thermal radiation such that $\tau = 0$, $\rho = 0$, and $\alpha = 1$, regardless of the spectral character or directional preference of the incident radiation, it is called a *blackbody*. This is a hypothetical idealization that does not exist in reality.

A blackbody is not only a perfect absorber but also characterized by an upper limit to the emission of thermal radiation. The energy emitted by a blackbody is a function of its temperature and is not evenly distributed over all wavelengths. The rate of energy emission per unit area at a particular wavelength is termed as the *monochromatic emissive power*. Max Planck was the first to derive a functional relation for the monochromatic emissive power of a blackbody in terms of temperature and

Table 2.3 Angular variation of absorptance for black paint.

Angle of incidence (°)	Absorptance
0–30	0.96
30–40	0.95
40–50	0.93
50–60	0.91
60–70	0.88
70–80	0.81
80–90	0.66

Reprinted from Löf and Tybout (1972) with permission from ASME.

wavelength. This was done by using the quantum theory, and the resulting equation, called *Planck's equation for blackbody radiation*, is given by:

$$E_{b\lambda} = \frac{C_1}{\lambda^5 \left(e^{C_2/\lambda T} - 1\right)} \tag{2.34}$$

where

$E_{b\lambda}$ = monochromatic emissive power of a blackbody (W/m^2 μm);
T = absolute temperature of the surface (K);
λ = wavelength (μm);
C_1 = constant = $2\pi h c_o^2$ = 3.74177×10^8 W μm^4/m^2;
C_2 = constant = $h c_o / k$ = 1.43878×10^4 μm K;
h = Planck's constant = 6.626069×10^{-34} Js;
c_o = speed of light in vacuum = 2.9979×10^8 m/s; and
k = Boltzmann's constant = 1.38065×10^{-23} J/K.

Equation (2.34) is valid for a surface in a vacuum or a gas. For other mediums it needs to be modified by replacing C_1 by C_1/n^2, where n is the index of refraction of the medium. By differentiating Eq. (2.34) and equating to 0, the wavelength corresponding to the maximum of the distribution can be obtained and is equal to $\lambda_{max} T = 2897.8$ μm K. This is known as *Wien's displacement law.* Figure 2.22 shows the spectral radiation distribution for blackbody radiation at three temperature sources. The curves have been obtained by using the Planck's equation.

The total emissive power, E_b, and the monochromatic emissive power, $E_{b\lambda}$, of a blackbody are related by:

$$E_b = \int_0^\infty E_{b\lambda} \, d\lambda \tag{2.35}$$

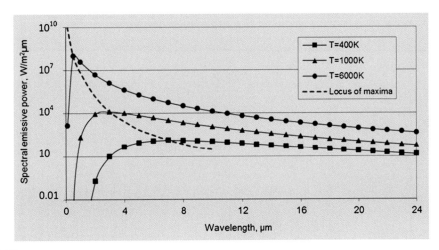

FIGURE 2.22

Spectral distribution of blackbody radiation.

Substituting Eq. (2.34) into Eq. (2.35) and performing the integration results in the Stefan–Boltzmann law:

$$E_b = \sigma T^4 \tag{2.36a}$$

where σ = the Stefan–Boltzmann constant = 5.6697×10^{-8} W/m²K⁴.

In many cases, it is necessary to know the amount of radiation emitted by a blackbody in a specific wavelength band $\lambda_1 \rightarrow \lambda_2$. This is done by modifying Eq. (2.35) as:

$$E_b(0 \rightarrow \lambda) = \int_0^\lambda E_{b\lambda}\, d\lambda \tag{2.36b}$$

Since the value of $E_{b\lambda}$ depends on both λ and T, it is better to use both variables as:

$$E_b(0 \rightarrow \lambda T) = \int_0^{\lambda T} \frac{E_{b\lambda}}{T}\, d\lambda T \tag{2.36c}$$

Therefore, for the wavelength band of $\lambda_1 \rightarrow \lambda_2$, we get:

$$E_b(\lambda_1 T \rightarrow \lambda_2 T) = \int_{\lambda_1 T}^{\lambda_2 T} \frac{E_{b\lambda}}{T}\, d\lambda T \tag{2.36d}$$

which results in $E_b(0 \rightarrow \lambda_1 T) - E_b(0 \rightarrow \lambda_2 T)$. Table 2.4 presents a tabulation of $E_b(0 \rightarrow \lambda T)$ as a fraction of the total emissive power, $E_b = \sigma T^4$, for various values of λT, also called *fraction of radiation emitted* from a blackbody at temperature T in the wavelength band from $\lambda = 0$ to λ, $f_{0-\lambda T}$ or for a particular temperature, $f\lambda$. The values are not rounded, because the original table, suggested by Dunkle

Table 2.4 Fraction of blackbody radiation as a function of λT.

λT (µm K)	$E_b(0 \rightarrow \lambda T)/\sigma T^4$	λT (µm K)	$E_b(0 \rightarrow \lambda T)/\sigma T^4$	λT (µm K)	$E_b(0 \rightarrow \lambda T)/\sigma T^4$
555.6	1.70E−08	4000.0	0.48085	7,444.4	0.83166
666.7	7.56E−07	4111.1	0.50066	7,555.6	0.83698
777.8	1.06E−05	4222.2	0.51974	7,666.7	0.84209
888.9	7.38E−05	4333.3	0.53809	7,777.8	0.84699
1,000.0	3.21E−04	4444.4	0.55573	7,888.9	0.85171
1,111.1	0.00101	4555.6	0.57267	8,000.0	0.85624
1,222.2	0.00252	4666.7	0.58891	8,111.1	0.86059
1,333.3	0.00531	4777.8	0.60449	8,222.2	0.86477
1,444.4	0.00983	4888.9	0.61941	8,333.3	0.86880
1,555.6	0.01643	5000.0	0.63371	8,888.9	0.88677
1,666.7	0.02537	5111.1	0.64740	9,444.4	0.90168
1,777.8	0.03677	5222.2	0.66051	10,000.0	0.91414
1,888.9	0.05059	5333.3	0.67305	10,555.6	0.92462
2,000.0	0.06672	5444.4	0.68506	11,111.1	0.93349
2,111.1	0.08496	5555.6	0.69655	11,666.7	0.94104
2,222.2	0.10503	5666.7	0.70754	12,222.2	0.94751
2,333.3	0.12665	5777.8	0.71806	12,777.8	0.95307
2,444.4	0.14953	5888.9	0.72813	13,333.3	0.95788
2,555.5	0.17337	6000.0	0.73777	13,888.9	0.96207
2,666.7	0.19789	6111.1	0.74700	14,444.4	0.96572
2,777.8	0.22285	6222.1	0.75583	15,000.0	0.96892
2,888.9	0.24803	6333.3	0.76429	15,555.6	0.97174
3,000.0	0.27322	6444.4	0.77238	16,111.1	0.97423
3,111.1	0.29825	6555.6	0.78014	16,666.7	0.97644
3,222.2	0.32300	6666.7	0.78757	22,222.2	0.98915
3,333.3	0.34734	6777.8	0.79469	22,777.8	0.99414
3,444.4	0.37118	6888.9	0.80152	33,333.3	0.99649
3,555.6	0.39445	7000.0	0.80806	33,888.9	0.99773
3,666.7	0.41708	7111.1	0.81433	44,444.4	0.99845
3,777.8	0.43905	7222.2	0.82035	50,000.0	0.99889
3,888.9	0.46031	7333.3	0.82612	55,555.6	0.99918

(1954), recorded λT in micrometer-degrees Rankine (µm °R), which were converted to micrometer-Kelvins (µm K) in Table 2.4.

The fraction of radiation emitted from a blackbody at temperature T in the wavelength band from $\lambda = 0$ to λ can be solved easily on a computer using the polynomial form, with about 10 summation terms for good accuracy, suggested by Siegel and Howell (2002):

$$f_{0-\lambda T} = \frac{15}{\pi^4} \sum_{n=1}^{\infty} \left[\frac{e^{-n\omega}}{n} \left(\omega^3 + \frac{3\omega^2}{n} + \frac{6\omega}{n^2} + \frac{6}{n^3} \right) \right] \tag{2.36e}$$

where $\omega = C_2/\lambda T$.

A blackbody is also a perfect diffuse emitter, so its intensity of radiation, I_b, is a constant in all directions, given by:

$$E_b = \pi I_b \qquad (2.37)$$

Of course, real surfaces emit less energy than corresponding blackbodies. The ratio of the total emissive power, E, of a real surface to the total emissive power, E_b, of a blackbody, both at the same temperature, is called the *emissivity* (ε) of a real surface; that is,

$$\varepsilon = \frac{E}{E_b} \qquad (2.38)$$

The emissivity of a surface not only is a function of surface temperature but also depends on wavelength and direction. In fact, the emissivity given by Eq. (2.38) is the average value over the entire wavelength range in all directions, and it is often referred to as the *total* or *hemispherical emissivity*. Similar to Eq. (2.38), to express the dependence on wavelength, the monochromatic or spectral emissivity, $\varepsilon\lambda$, is defined as the ratio of the monochromatic emissive power, $E\lambda$, of a real surface to the monochromatic emissive power, $E_b\lambda$, of a blackbody, both at the same wavelength and temperature:

$$\varepsilon_\lambda = \frac{E_\lambda}{E_{b\lambda}} \qquad (2.39)$$

Kirchoff's law of radiation states that, for any surface in thermal equilibrium, monochromatic emissivity is equal to monochromatic absorptivity:

$$\varepsilon_\lambda(T) = \alpha_\lambda(T) \qquad (2.40)$$

The temperature (T) is used in Eq. (2.40) to emphasize that this equation applies only when the temperatures of the source of the incident radiation and the body itself are the same. It should be noted, therefore, that the emissivity of a body on earth (at normal temperature) cannot be equal to solar radiation (emitted from the sun at $T = 5760$ K). Equation (2.40) can be generalized as:

$$\varepsilon(T) = \alpha(T) \qquad (2.41)$$

Equation (2.41) relates the total emissivity and absorptivity over the entire wavelength. This generalization, however, is strictly valid only if the incident and emitted radiation have, in addition to the temperature equilibrium at the surfaces, the same spectral distribution. Such conditions are rarely met in real life; to simplify the analysis of radiation problems, however, the assumption that monochromatic properties are constant over all wavelengths is often made. Such a body with these characteristics is called a *gray body*.

Similar to Eq. (2.37) for a real surface, the radiant energy leaving the surface includes its original emission and any reflected rays. The rate of total radiant energy leaving a surface per unit surface area is called the *radiosity* (J), given by:

$$J = \varepsilon E_b + \rho H \qquad (2.42)$$

where

E_b = blackbody emissive power per unit surface area (W/m^2);
H = irradiation incident on the surface per unit surface area (W/m^2);

ε = emissivity of the surface; and

ρ = reflectivity of the surface.

There are two idealized limiting cases of radiation reflection: the reflection is called *specular* if the reflected ray leaves at an angle with the normal to the surface equal to the angle made by the incident ray, and it is called *diffuse* if the incident ray is reflected uniformly in all directions. Real surfaces are neither perfectly specular nor perfectly diffuse. Rough industrial surfaces, however, are often considered as diffuse reflectors in engineering calculations.

A real surface is both a diffuse emitter and a diffuse reflector and hence, it has diffuse radiosity; that is, the intensity of radiation from this surface (I) is constant in all directions. Therefore, the following equation is used for a real surface:

$$J = \pi \times I \tag{2.43}$$

EXAMPLE 2.10

A glass with transmissivity of 0.92 is used in a certain application for wavelengths 0.3 and 3.0 μm. The glass is opaque to all other wavelengths. Assuming that the sun is a blackbody at 5760 K and neglecting atmospheric attenuation, determine the percent of incident solar energy transmitted through the glass. If the interior of the application is assumed to be a blackbody at 373 K, determine the percent of radiation emitted from the interior and transmitted out through the glass.

Solution

For the incoming solar radiation at 5760 K, we have:

$$\lambda_1 T = 0.3 \times 5760 = 1728 \ \text{μm K}$$

$$\lambda_2 T = 3 \times 5760 = 17280 \ \text{μm K}$$

From Table 2.4 by interpolation, we get:

$$\frac{E_b(0 \to \lambda_1 T)}{\sigma T^4} = 0.0317 = 3.17\%$$

$$\frac{E_b(0 \to \lambda_2 T)}{\sigma T^4} = 0.9778 = 97.78\%$$

Therefore, the percent of solar radiation incident on the glass in the wavelength range 0.3–3 μm is:

$$\frac{E_b(\lambda_1 T \to \lambda_2 T)}{\sigma T^4} = 97.78 - 3.17 = 94.61\%$$

In addition, the percentage of radiation transmitted through the glass is 0.92 × 94.61 = 87.04%. For the outgoing infrared radiation at 373 K, we have:

$$\lambda_1 T = 0.3 \times 373 = 111.9 \ \text{μm K}$$

$$\lambda_2 T = 3 \times 373 = 1119.0 \ \text{μm K}$$

From Table 2.4, we get:

$$\frac{E_b(0 \to \lambda_1 T)}{\sigma T^4} = 0.0 = 0\%$$

$$\frac{E_b(0 \to \lambda_2 T)}{\sigma T^4} = 0.00101 = 0.1\%$$

The percent of outgoing infrared radiation incident on the glass in the wavelength 0.3–3 μm is 0.1%, and the percent of this radiation transmitted out through the glass is only 0.92 × 0.1 = 0.092%. This example, in fact, demonstrates the principle of the greenhouse effect; that is, once the solar energy is absorbed by the interior objects, it is effectively trapped.

EXAMPLE 2.11

A surface has a spectral emissivity of 0.87 at wavelengths less than 1.5 μm, 0.65 at wavelengths between 1.5 and 2.5 μm, and 0.4 at wavelengths longer than 2.5 μm. If the surface is at 1000 K, determine the average emissivity over the entire wavelength and the total emissive power of the surface.

Solution

From the data given, we have:

$$\lambda_1 T = 1.5 \times 1000 = 1500 \ \mu m \ K$$

$$\lambda_2 T = 2.5 \times 1000 = 2500 \ \mu m \ K$$

From Table 2.4 by interpolation, we get:

$$\frac{E_b(0 \rightarrow \lambda_1 T)}{\sigma T^4} = 0.01313$$

and

$$\frac{E_b(0 \rightarrow \lambda_2 T)}{\sigma T^4} = 0.16144$$

Therefore,

$$\frac{E_b(\lambda_1 T \rightarrow \lambda_2 T)}{\sigma T^4} = 0.16144 - 0.01313 = 0.14831$$

and

$$\frac{E_b(\lambda_2 T \rightarrow \infty)}{\sigma T^4} = 1 - 0.16144 = 0.83856$$

The average emissive power over the entire wavelength is given by:

$$\varepsilon = 0.87 \times 0.01313 + 0.65 \times 0.14831 + 0.4 \times 0.83856 = 0.4432$$

and the total emissive power of the surface is:

$$E = \varepsilon \sigma T^4 = 0.4432 \times 5.67 \times 10^{-8} \times 1000^4 = 25129.4 \ W/m^2$$

The other properties of the materials can be obtained using the Kirchhoff's law given by Eq. (2.40) or Eq. (2.41) as demonstrated by the following example.

EXAMPLE 2.12

The variation of the spectral absorptivity of an opaque surface is 0.2 up to the wavelength of 2 μm and 0.7 for bigger wavelengths. Estimate the average absorptivity and reflectivity of the surface from radiation emitted from a source at 2500 K. Determine also the average emissivity of the surface at 3000 K.

Solution

At a temperature of 2500 K:

$$\lambda_1 T = (2 \ \mu m)(2500 \ K) = 5000 \ \mu m \ K.$$

Therefore, from Table 2.4: $\dfrac{E_b(0 \rightarrow \lambda_1 T)}{\sigma T^4} = f_{\lambda 1} = 0.63371$

Continued

EXAMPLE 2.12–CONT'D

The average absorptivity of the surface is:

$$\alpha(T) = \alpha_1 f_{\lambda 1} + \alpha_2(1 - f_{\lambda 1}) = (0.2)(0.63371) + (0.7)(1 - 0.63371) = 0.383$$

As the surface is opaque from Eq. (2.32): $\alpha + \rho = 1$. So, $\rho = 1 - \alpha = 1 - 0.383 = 0.617$.

Using Kirchhoff's law, from Eq. (2.41) $\varepsilon(T) = \alpha(T)$. So the average emissivity of this surface at $T = 3000$ K is:

$$\lambda_1 T = (2 \ \mu m)(3000 \ K) = 6000 \ \mu m \ K.$$

Therefore, from Table 2.4: $\dfrac{E_b(0 \rightarrow \lambda_1 T)}{\sigma T^4} = f_{\lambda 1} = 0.73777$

And $\varepsilon(T) = \varepsilon_1 f_{\lambda 1} + \varepsilon_2(1 - f_{\lambda 1}) = (0.2)(0.73777) + (0.7)(1 - 0.73777) = 0.331.$

2.3.3 Transparent plates

When a beam of radiation strikes the surface of a transparent plate at angle θ_1, called the *incidence angle*, as shown in Fig. 2.23, part of the incident radiation is reflected and the remainder is refracted, or bent, to angle θ_2, called the *refraction angle*, as it passes through the interface. Angle θ_1 is also equal to the angle at which the beam is specularly reflected from the surface. Angles θ_1 and θ_2 are not equal when the density of the plane is different from that of the medium through which the radiation travels. Additionally, refraction causes the transmitted beam to be bent toward the perpendicular to the surface of higher density. The two angles are related by the Snell's law:

$$n = \frac{n_2}{n_1} = \frac{\sin \theta_1}{\sin \theta_2} \tag{2.44}$$

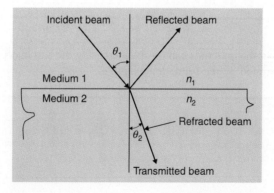

FIGURE 2.23

Incident and refraction angles for a beam passing from a medium with refraction index n_1 to a medium with refraction index n_2.

where n_1 and n_2 are the refraction indices and n is the ratio of refraction index for the two media forming the interface. The refraction index is the determining factor for the reflection losses at the interface. A typical value of the refraction index is 1.000 for air, 1.526 for glass, and 1.33 for water.

Expressions for perpendicular and parallel components of radiation for smooth surfaces were derived by Fresnel as:

$$r_\perp = \frac{\sin^2(\theta_2 - \theta_1)}{\sin^2(\theta_2 + \theta_1)} \tag{2.45}$$

$$r_\parallel = \frac{\tan^2(\theta_2 - \theta_1)}{\tan^2(\theta_2 + \theta_1)} \tag{2.46}$$

Equation (2.45) represents the perpendicular component of unpolarized radiation and Eq. (2.46) represents the parallel one. It should be noted that *parallel* and *perpendicular* refer to the plane defined by the incident beam and the surface normal.

Properties are evaluated by calculating the average of these two components as:

$$r = \frac{1}{2}\left(r_\perp + r_\parallel\right) \tag{2.47}$$

For normal incidence, both angles are 0 and Eq. (2.47) can be combined with Eq. (2.44) to yield:

$$r_{(0)} = \left(\frac{n_1 - n_2}{n_1 + n_2}\right)^2 \tag{2.48}$$

If one medium is air ($n = 1.0$), then Eq. (2.48) becomes:

$$r_{(0)} = \left(\frac{n - 1}{n + 1}\right)^2 \tag{2.49}$$

Similarly, the transmittance, τ_r (subscript r indicates that only reflection losses are considered), can be calculated from the average transmittance of the two components as follows:

$$\tau_r = \frac{1}{2}\left(\frac{1 - r_\parallel}{1 + r_\parallel} + \frac{1 - r_\perp}{1 + r_\perp}\right) \tag{2.50a}$$

For a glazing system of N covers of the same material, it can be proven that:

$$\tau_r = \frac{1}{2}\left(\frac{1 - r_\parallel}{1 + (2N - 1)r_\parallel} + \frac{1 - r_\perp}{1 + (2N - 1)r_\perp}\right) \tag{2.50b}$$

The transmittance, τ_α (subscript α indicates that only absorption losses are considered), can be calculated from:

$$\tau_\alpha = e^{\left(-\frac{KL}{\cos\theta_2}\right)} \tag{2.51}$$

where K is the extinction coefficient, which can vary from 4 m^{-1} (for high-quality glass) to 32 m^{-1} (for low-quality glass), and L is the thickness of the glass cover.

The transmittance, reflectance, and absorptance of a single cover (by considering both reflection and absorption losses) are given by the following expressions. These expressions are for the perpendicular components of polarization, although the same relations can be used for the parallel components:

$$\tau_\perp = \frac{\tau_\alpha(1-r_\perp)^2}{1-(r_\perp\tau_\alpha)^2} = \tau_\alpha\frac{1-r_\perp}{1+r_\perp}\left(\frac{1-r_\perp^2}{1-(r_\perp\tau_\alpha)^2}\right) \tag{2.52a}$$

$$\rho_\perp = r_\perp + \frac{(1-r_\perp)^2\tau_\alpha^2 r_\perp}{1-(r_\perp\tau_\alpha)^2} = r_\perp(1+\tau_\alpha\tau_\perp) \tag{2.52b}$$

$$\alpha_\perp = (1-\tau_\alpha)\left(\frac{1-r_\perp}{1-r_\perp\tau_\alpha}\right) \tag{2.52c}$$

Since, for practical collector covers, τ_α is seldom less than 0.9 and r is of the order of 0.1, the transmittance of a single cover becomes:

$$\tau \cong \tau_\alpha\tau_r \tag{2.53}$$

The absorptance of a cover can be approximated by neglecting the last term of Eq. (2.52c):

$$\alpha \cong 1-\tau_\alpha \tag{2.54}$$

and the reflectance of a single cover could be found (keeping in mind that $\rho = 1-\alpha-\tau$) as:

$$\rho \cong \tau_\alpha(1-\tau_r) = \tau_\alpha - \tau \tag{2.55}$$

For a two-cover system of not necessarily same materials, the following equation can be obtained (subscript 1 refers to the outer cover and 2 to the inner one):

$$\tau = \frac{1}{2}\left[\left(\frac{\tau_1\tau_2}{1-\rho_1\rho_2}\right)_\perp + \left(\frac{\tau_1\tau_2}{1-\rho_1\rho_2}\right)_{||}\right] = \frac{1}{2}\left(\tau_\perp + \tau_{||}\right) \tag{2.56}$$

$$\rho = \frac{1}{2}\left[\left(\rho_1 + \frac{\tau\rho_2\tau_1}{\tau_2}\right)_\perp + \left(\rho_1 + \frac{\tau\rho_2\tau_1}{\tau_2}\right)_{||}\right] = \frac{1}{2}\left(\rho_\perp + \rho_{||}\right) \tag{2.57}$$

EXAMPLE 2.13

A solar energy collector uses a single glass cover with a thickness of 4 mm. In the visible solar range, the refraction index of glass, n, is 1.526 and its extinction coefficient K is 32 m^{-1}. Calculate the reflectivity, transmissivity, and absorptivity of the glass sheet for the angle of incidence of 60° and at normal incidence (0°).

Solution
Angle of incidence = 60°
From Eq. (2.44), the refraction angle θ_2 is calculated as:

$$\theta_2 = \sin^{-1}\left(\frac{\sin\theta_1}{n}\right) = \sin^{-1}\left(\frac{\sin(60)}{1.526}\right) = 34.6°$$

EXAMPLE 2.13—CONT'D

From Eq. (2.51), the transmittance can be obtained as:

$$\tau_\alpha = e^{\left(-\dfrac{KL}{\cos(\theta_2)}\right)} = e^{\left(-\dfrac{32(0.004)}{\cos(34.6)}\right)} = 0.856$$

From Eqs. (2.45) and (2.46),

$$r_\perp = \frac{\sin^2(\theta_2 - \theta_1)}{\sin^2(\theta_2 + \theta_1)} = \frac{\sin^2(34.6 - 60)}{\sin^2(34.6 + 60)} = 0.185$$

$$r_\parallel = \frac{\tan^2(\theta_2 - \theta_1)}{\tan^2(\theta_2 + \theta_1)} = \frac{\tan^2(34.6 - 60)}{\tan^2(34.6 + 60)} = 0.001$$

From Eqs. (2.52a)–(2.52c), we have:

$$\tau = \frac{\tau_\alpha}{2}\left\{\frac{1 - r_\perp}{1 + r_\perp}\left[\frac{1 - r_\perp^2}{1 - (r_\perp \tau_\alpha)^2}\right] + \frac{1 - r_\parallel}{1 + r_\parallel}\left[\frac{1 - r_\parallel^2}{1 - \left(r_\parallel \tau_\alpha\right)^2}\right]\right\} = \frac{\tau_\alpha}{2}\left[\tau_\perp + \tau_\parallel\right]$$

$$= \frac{0.856}{2}\left\{\frac{1 - 0.185}{1 + 0.185}\left[\frac{1 - 0.185^2}{1 - (0.185 \times 0.856)^2}\right] + \frac{1 - 0.001}{1 + 0.001}\left[\frac{1 - 0.001^2}{1 - (0.001 \times 0.856)^2}\right]\right\}$$

$$= 0.428(0.681 + 0.998) = 0.719$$

$$\rho = \frac{1}{2}\left[r_\perp\left(1 + \tau_\alpha \tau_\perp\right) + r_\parallel\left(1 + \tau_\alpha \tau_\parallel\right)\right]$$

$$= 0.5[0.185(1 + 0.856 \times 0.681) + 0.001(1 + 0.856 \times 0.998)] = 0.147$$

$$\alpha = \frac{(1 - \tau_\alpha)}{2}\left[\left(\frac{1 - r_\perp}{1 - r_\perp \tau_\alpha}\right) + \left(\frac{1 - r_\parallel}{1 - r_\parallel \tau_\alpha}\right)\right]$$

$$= \frac{(1 - 0.856)}{2}\left(\frac{1 - 0.185}{1 - 0.185 \times 0.856} + \frac{1 - 0.001}{1 - 0.001 \times 0.856}\right) = 0.142$$

Normal incidence.
At normal incidence, $\theta_1 = 0°$ and $\theta_2 = 0°$. In this case, τ_α is equal to 0.880. There is no polarization at normal incidence; therefore, from Eq. (2.49),

$$r_{(0)} = r_\perp = r_\perp = \left(\frac{n - 1}{n + 1}\right)^2 = \left(\frac{1.526 - 1}{1.526 + 1}\right)^2 = 0.043$$

From Eqs. (2.52a)–(2.52c), we have:

$$\tau = \tau_\alpha \frac{1 - r_{(0)}}{1 + r_{(0)}}\left[\frac{1 - r_{(0)}^2}{1 - (r_{(0)}\tau_\alpha)^2}\right] = 0.880\left\{\frac{1 - 0.043}{1 + 0.043}\left[\frac{1 - 0.043^2}{1 - (0.043 \times 0.880)^2}\right]\right\} = 0.807$$

$$\rho = r_{(0)}\left(1 + \tau_\alpha \tau_{(0)}\right) = 0.043(1 + 0.880 \times 0.807) = 0.074$$

$$\alpha = (1 - \tau_\alpha)\left(\frac{1 - r_{(0)}}{1 - r_{(0)}\tau_\alpha}\right) = (1 - 0.880)\left(\frac{1 - 0.043}{1 - 0.043 \times 0.880}\right) = 0.119$$

2.3.4 Radiation exchange between surfaces

When studying the radiant energy exchanged between two surfaces separated by a non-absorbing medium, one should consider not only the temperature of the surfaces and their characteristics but also their geometric orientation with respect to each other. The effects of the geometry of radiant energy exchange can be analyzed conveniently by defining the term *view factor*, F_{12}, to be the fraction of radiation leaving surface A_1 that reaches surface A_2. If both surfaces are black, the radiation leaving surface A_1 and arriving at surface A_2 is $E_{b1}A_1F_{12}$, and the radiation leaving surface A_2 and arriving at surface A_1 is $E_{b2}A_2F_{21}$. If both surfaces are black and absorb all incident radiation, the net radiation exchange is given by:

$$Q_{12} = E_{b1}A_1F_{12} - E_{b2}A_2F_{21} \tag{2.58}$$

If both surfaces are of the same temperature, $E_{b1} = E_{b2}$ and $Q_{12} = 0$. Therefore,

$$A_1F_{12} = A_2F_{21} \tag{2.59}$$

It should be noted that Eq. (2.59) is strictly geometric in nature and valid for all diffuse emitters, irrespective of their temperatures. Therefore, the net radiation exchange between two black surfaces is given by:

$$Q_{12} = A_1F_{12}(E_{b1} - E_{b2}) = A_2F_{21}(E_{b1} - E_{b2}) \tag{2.60}$$

From Eq. (2.36a), $E_b = \sigma T^4$, Eq. (2.60) can be written as:

$$Q_{12} = A_1F_{12}\sigma\left(T_1^4 - T_2^4\right) = A_2F_{21}\sigma\left(T_1^4 - T_2^4\right) \tag{2.61}$$

where T_1 and T_2 are the temperatures of surfaces A_1 and A_2, respectively. As the term $(E_{b1} - E_{b2})$ in Eq. (2.60) is the energy potential difference that causes the transfer of heat, in a network of electric circuit analogy, the term $1/A_1F_{12} = 1/A_2F_{21}$ represents the resistance due to the geometric configuration of the two surfaces.

When surfaces other than black are involved in radiation exchange, the situation is much more complex, because multiple reflections from each surface must be taken into consideration. For the simple case of opaque gray surfaces, for which $\varepsilon = \alpha$, the reflectivity $\rho = 1 - \alpha = 1 - \varepsilon$. From Eq. (2.42), the radiosity of each surface is given by:

$$J = \varepsilon E_b + \rho H = \varepsilon E_b + (1 - \varepsilon)H \tag{2.62}$$

The net radiant energy leaving the surface is the difference between the radiosity, J, leaving the surface and the irradiation, H, incident on the surface; that is,

$$Q = A(J - H) \tag{2.63}$$

Combining Eqs. (2.62) and (2.63) and eliminating irradiation H results in:

$$Q = A\left(J - \frac{J - \varepsilon E_b}{1 - \varepsilon}\right) = \frac{A\varepsilon}{1 - \varepsilon}(E_b - J) = \frac{(E_b - J)}{R} \tag{2.64}$$

Therefore, the net radiant energy leaving a gray surface can be regarded as the current in an equivalent electrical network when a potential difference $(E_b - J)$ is overcome across a resistance $R = (1 - \varepsilon)/A\varepsilon$. This resistance, called *surface resistance*, is due to the imperfection of the surface as an emitter and absorber of radiation as compared with a black surface.

By considering the radiant energy exchange between 2 gray surfaces, A_1 and A_2, the radiation leaving surface A_1 and arriving at surface A_2 is $J_1 A_1 F_{12}$, where J is the radiosity, given by Eq. (2.42). Similarly, the radiation leaving surface A_2 and arriving surface A_1 is $J_2 A_2 F_{21}$. The net radiation exchange between the two surfaces is given by:

$$Q_{12} = J_1 A_1 F_{12} - J_2 A_2 F_{21} = A_1 F_{12}(J_1 - J_2) = A_2 F_{21}(J_1 - J_2) \tag{2.65}$$

Therefore, due to the geometric orientation that applies between the two potentials, J_1 and J_2, when 2 gray surfaces exchange radiant energy, there is a resistance, called *space resistance*, $R = 1/A_1 F_{12} = 1/A_2 F_{21}$.

An equivalent electric network for two the gray surfaces is illustrated in Fig. 2.24. By combining the surface resistance, $(1 - \varepsilon)/A\varepsilon$ for each surface and the space (or geometric) resistance, $1/A_1 F_{12} = 1/A_2 F_{21}$, between the surfaces, as shown in Fig. 2.24, the net rate of radiation exchange between the two surfaces is equal to the overall potential difference divided by the sum of resistances, given by:

$$Q_{12} = \frac{E_{b1} - E_{b2}}{\left[\dfrac{(1 - \varepsilon_1)}{A_1 \varepsilon_1}\right] + \dfrac{1}{A_1 F_{12}} + \left[\dfrac{(1 - \varepsilon_2)}{A_2 \varepsilon_2}\right]} = \frac{\sigma\left(T_1^4 - T_2^4\right)}{\left[\dfrac{(1 - \varepsilon_1)}{A_1 \varepsilon_1}\right] + \dfrac{1}{A_1 F_{12}} + \left[\dfrac{(1 - \varepsilon_2)}{A_2 \varepsilon_2}\right]} \tag{2.66}$$

In solar energy applications, the following geometric orientations between two surfaces are of particular interest.

A. For two infinite parallel surfaces, $A_1 = A_2 = A$ and $F_{12} = 1$, Eq. (2.66) becomes:

$$Q_{12} = \frac{A\sigma\left(T_1^4 - T_2^4\right)}{(1/\varepsilon_1) + (1/\varepsilon_2) - 1} \tag{2.67}$$

B. For two concentric cylinders, $F_{12} = 1$ and Eq. (2.66) becomes:

$$Q_{12} = \frac{A_1 \sigma\left(T_1^4 - T_2^4\right)}{(1/\varepsilon_1) + (A_1/A_2)[(1/\varepsilon_2) - 1]} \tag{2.68}$$

C. For a small convex surface, A_1, completely enclosed by a very large concave surface, A_2, $A_1 << A_2$ and $F_{12} = 1$, then Eq. (2.66) becomes:

$$Q_{12} = A_1 \varepsilon_1 \sigma\left(T_1^4 - T_2^4\right) \tag{2.69}$$

FIGURE 2.24

Equivalent electrical network for radiation exchange between 2 gray surfaces.

The last equation also applies for a flat-plate collector cover radiating to the surroundings, whereas case B applies in the analysis of a parabolic trough collector receiver where the receiver pipe is enclosed in a glass cylinder.

As can be seen from Eqs. (2.67)–(2.69), the rate of radiative heat transfer between surfaces depends on the difference of the fourth power of the surface temperatures. In many engineering calculations, however, the heat transfer equations are linearized in terms of the differences of temperatures to the first power. For this purpose, the following mathematical identity is used:

$$T_1^4 - T_2^4 = (T_1^2 - T_2^2)(T_1^2 + T_2^2) = (T_1 - T_2)(T_1 + T_2)(T_1^2 + T_2^2) \tag{2.70}$$

Therefore, Eq. (2.66) can be written as:

$$Q_{12} = A_1 h_r (T_1 - T_2) \tag{2.71}$$

with the radiation heat transfer coefficient, h_r, defined as:

$$h_r = \frac{\sigma(T_1 + T_2)(T_1^2 + T_2^2)}{\dfrac{1 - \varepsilon_1}{\varepsilon_1} + \dfrac{1}{F_{12}} + \dfrac{A_1}{A_2}\left(\dfrac{1 - \varepsilon_2}{\varepsilon_2}\right)} \tag{2.72}$$

For the special cases mentioned previously, the expressions for h_r are as follows:
Case A:

$$h_r = \frac{\sigma(T_1 + T_2)(T_1^2 + T_2^2)}{\dfrac{1}{\varepsilon_1} + \dfrac{1}{\varepsilon_2} - 1} \tag{2.73}$$

Case B:

$$h_r = \frac{\sigma(T_1 + T_2)(T_1^2 + T_2^2)}{\dfrac{1}{\varepsilon_1} + \dfrac{A_1}{A_2}\left(\dfrac{1}{\varepsilon_2} - 1\right)} \tag{2.74}$$

Case C:

$$h_r = \varepsilon_1 \sigma(T_1 + T_2)(T_1^2 + T_2^2) \tag{2.75}$$

It should be noted that the use of these linearized radiation equations in terms of h_r is very convenient when the equivalent network method is used to analyze problems involving conduction and/or convection in addition to radiation. The radiation heat transfer coefficient, h_r, is treated similarly to the convection heat transfer coefficient, h_c, in an electric equivalent circuit. In such a case, a combined heat transfer coefficient can be used, given by:

$$h_{cr} = h_c + h_r \tag{2.76}$$

In this equation, it is assumed that the linear temperature difference between the ambient fluid and the walls of the enclosure and the surface and the enclosure substances are at the same temperature.

EXAMPLE 2.14

The glass of a 1×2 m flat-plate solar collector is at a temperature of 80 °C and has an emissivity of 0.90. The environment is at a temperature of 15°C. Calculate the convection and radiation heat losses if the convection heat transfer coefficient is 5.1 W/m^{2}K.

Solution

In the following analysis, the glass cover is denoted by subscript 1 and the environment by 2. The radiation heat transfer coefficient is given by Eq. (2.75):

$$h_r = \varepsilon_1 \sigma (T_1 + T_2)(T_1^2 + T_2^2) = 0.90 \times 5.67 \times 10^{-8}(353 + 288)(353^2 + 288^2) = 6.789 \text{ W/m}^2 \text{ K}$$

Therefore, from Eq. (2.76),

$$h_{cr} = h_c + h_r = 5.1 + 6.789 = 11.889 \text{ W/m}^2 \text{ K}$$

Finally,

$$Q_{12} = A_1 h_{cr}(T_1 - T_2) = (1 \times 2)(11.889)(353 - 288) = 1545.6 \text{ W}$$

EXAMPLE 2.15

Two very large parallel plates are maintained at uniform temperatures of 900 and 400 K. The emissivities of the two surfaces are 0.3 and 0.8, respectively. What is the radiation heat transfer between the two surfaces?

Solution

As the areas of the two surfaces are not given the estimation is given per unit surface area of the plates. As the two plates are very large and parallel, Eq. (2.67) apply, so:

$$\frac{Q_{12}}{A} = \frac{\sigma(T_1^4 - T_2^4)}{\left(\dfrac{1}{\varepsilon_1}\right) + \left(\dfrac{1}{\varepsilon_2}\right) - 1} = \frac{5.67 \times 10^{-8}(900^4 - 400^4)}{\dfrac{1}{0.3} + \dfrac{1}{0.8} - 1} = 9976.6 \text{ W/m}^2$$

EXAMPLE 2.16

Two very long concentric cylinders of diameters 30 and 50 cm are maintained at uniform temperatures of 850 and 450 K. The emissivities of the two surfaces are 0.9 and 0.6, respectively. The space between the two cylinders is evacuated. What is the radiation heat transfer between the two cylinders per unit length of the cylinders?

Solution

For concentric cylinders Eq. (2.68) applies. Therefore,

$$Q_{12} = \frac{A_1 \sigma(T_1^4 - T_2^4)}{\left(\dfrac{1}{\varepsilon_1}\right) + \left(\dfrac{A_1}{A_2}\right)\left(\dfrac{1 - \varepsilon_2}{\varepsilon_2}\right)} = \frac{\pi \times 0.3 \times 1 \times 5.67 \times 10^{-8}(850^4 - 450^4)}{\dfrac{1}{0.9} + \left(\dfrac{30}{50}\right)\left(\dfrac{1 - 0.6}{0.6}\right)} = 17 \text{ kW}$$

2.3.5 Extraterrestrial solar radiation

The amount of solar energy per unit time, at the mean distance of the earth from the sun, received on a unit area of a surface normal to the sun (perpendicular to the direction of propagation of the radiation) outside the atmosphere is called the *solar constant*, G_{sc}. This quantity is difficult to measure from the surface of the earth because of the effect of the atmosphere. A method for the determination of the solar constant was first given in 1881 by Langley (Garg, 1982), who had given his name to the units of measurement as Langleys per minute (calories per square centimeter per minute). This was changed by the SI system to Watts per square meter (W/m^2).

When the sun is closest to the earth, on January 3, the solar heat on the outer edge of the earth's atmosphere is about 1400 W/m^2; and when the sun is farthest away, on July 4, it is about 1330 W/m^2.

Throughout the year, the extraterrestrial radiation measured on the plane normal to the radiation on the Nth day of the year, G_{on}, varies between these limits, as indicated in Fig. 2.25, in the range of 3.3% and can be calculated by (Duffie and Beckman, 1991; Hsieh, 1986):

$$G_{on} = G_{sc}\left[1 + 0.033 \cos\left(\frac{360N}{365}\right)\right] \tag{2.77a}$$

where

G_{on} = extraterrestrial radiation measured on the plane normal to the radiation on the Nth day of the year (W/m^2);
G_{sc} = solar constant (W/m^2).

A more accurate equation for the estimation of G_{on} is given by Spencer (1971). This equation gives an accuracy of $\pm 0.01\%$; the parameter Γ is given by Eq. (2.7) but is converted to degrees—i.e., 2π is replaced by 360:

$$G_{on} = G_{sc}[1.00011 + 0.034221 \cos(\Gamma) + 0.00128 \sin(\Gamma) \\ + 0.00719 \cos(2\Gamma) + 0.000077 \sin(2\Gamma)] \tag{2.77b}$$

FIGURE 2.25

Variation of extraterrestrial solar radiation with the time of year.

It should be noted that for engineering calculations Eq. (2.77a) is adequate.

The latest value of G_{sc} is 1366.1 W/m². This was adopted in 2000 by the American Society for Testing and Materials (ASTM), which developed an AM0 reference spectrum (ASTM E−490). The ASTM E−490 Air Mass Zero solar spectral irradiance is based on data from satellites, space shuttle missions, high-altitude aircraft, rocket soundings, ground-based solar telescopes, and modeled spectral irradiance. The spectral distribution of extraterrestrial solar radiation at the mean sun−earth distance is shown in Fig. 2.26. The spectrum curve of Fig. 2.26 is based on a set of data included in ASTM E−490 (Solar Spectra, 2007).

When a surface is placed parallel to the ground, the rate of solar radiation, G_{oH}, incident on this extraterrestrial horizontal surface at a given time of the year is given by:

$$G_{oH} = G_{on} \cos(\Phi) G_{sc} \left[1 + 0.033 \cos\left(\frac{360N}{365}\right)\right] [\cos(L)\cos(\delta)\cos(h) + \sin(L)\sin(\delta)] \qquad (2.78)$$

The total radiation, H_o, incident on an extraterrestrial horizontal surface during a day can be obtained by the integration of Eq. (2.78) over a period from sunrise to sunset. The resulting equation is:

$$H_o = \frac{24 \times 3600 G_{sc}}{\pi} \left[1 + 0.033 \cos\left(\frac{360N}{365}\right)\right]$$

$$\times \left[\cos(L)\cos(\delta)\sin(h_{ss}) + \left(\frac{\pi h_{ss}}{180}\right)\sin(L)\sin(\delta)\right] \qquad (2.79)$$

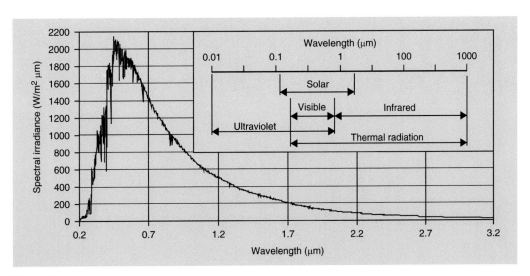

FIGURE 2.26

Standard curve giving a solar constant of 1366.1 W/m² and its position in the electromagnetic radiation spectrum.

where h_{ss} is the sunset hour in degrees, obtained from Eq. (2.15). The units of Eq. (2.79) are joules per square meter (J/m^2).

To calculate the extraterrestrial radiation on a horizontal surface for an hour period, Eq. (2.78) is integrated between hour angles, h_1 and h_2 (h_2 is larger). Therefore,

$$I_o = \frac{12 \times 3600 G_{sc}}{\pi}\left[1 + 0.033 \cos\left(\frac{360N}{365}\right)\right] \times \left\{\cos(L)\cos(\delta)[\sin(h_2) - \sin(h_1)]\right.$$

$$\left. + \left[\frac{\pi(h_2 - h_1)}{180}\right]\sin(L)\sin(\delta)\right\}$$

(2.80)

It should be noted that the limits h_1 and h_2 may define a time period other than 1 h.

EXAMPLE 2.17

Determine the extraterrestrial normal radiation and the extraterrestrial radiation on a horizontal surface on March 10 at 2:00 p.m. solar time for 35°N latitude. Determine also the total solar radiation on the extraterrestrial horizontal surface for the day.

Solution

The declination on March 10 ($N = 69$) is calculated from Eq. (2.5):

$$\delta = 23.45 \sin\left[\frac{360}{365}(284 + 69)\right] = -4.8°$$

The hour angle at 2:00 p.m. solar time is calculated from Eq. (2.8):

$$h = 0.25 \ (\text{number of minutes from local solar noon}) = 0.25(120) = 30°$$

The hour angle at sunset is calculated from Eq. (2.15):

$$h_{ss} = \cos^{-1}[-\tan(L)\tan(\delta)] = \cos^{-1}[-\tan(35)\tan(-4.8)] = 86.6°$$

The extraterrestrial normal radiation is calculated from Eq. (2.77):

$$G_{on} = G_{sc}\left[1 + 0.033 \cos\left(\frac{360N}{365}\right)\right] = 1366\left[1 + 0.033 \cos\left(\frac{360 \times 69}{365}\right)\right] = 1383 \ W/m^2$$

The extraterrestrial radiation on a horizontal surface is calculated from Eq. (2.78):

$$G_{oH} = G_{on} \cos(\Phi) = G_{on}[\sin(L)\sin(\delta) + \cos(L)\cos(\delta)\cos(h)]$$

$$= 1383[\sin(35)\sin(-4.8) + \cos(35)\cos(-4.8)\cos(30)] = 911 \ W/m^2$$

The total radiation on the extraterrestrial horizontal surface is calculated from Eq. (2.79):

$$H_o = \frac{24 \times 3600 G_{sc}}{\pi}\left[1 + 0.033 \cos\left(\frac{360N}{365}\right)\right]\left[\cos(L) \cos(\delta) \sin(h_{ss}) + \left(\frac{\pi h_{ss}}{180}\right) \sin(L) \sin(\delta)\right]$$

$$= \frac{24 \times 3600 \times 1366}{\pi}\begin{bmatrix}\cos(35) \cos(-4.8) \sin(86.6) \\ + \left(\frac{\pi \times 86.6}{180}\right) \sin(35) \sin(-4.8)\end{bmatrix} = 27.88 \ MJ/m^2$$

A list of definitions that includes those related to solar radiation is found in Appendix 2. The reader should familiarize himself or herself with the various terms and specifically with *irradiance*, which is the rate of radiant energy falling on a surface per unit area of the surface (units, watts per square meter [W/m²] symbol, *G*), whereas *irradiation* is incident energy per unit area on a surface (units, joules per square meter [J/m²]), obtained by integrating irradiance over a specified time interval. Specifically, for solar irradiance this is called *insolation*. The symbols used in this book are *H* for insolation for a day and *I* for insolation for an hour. The appropriate subscripts used for *G*, *H*, and *I* are beam (B), diffuse (D), and ground-reflected (G) radiation.

2.3.6 Atmospheric attenuation

The solar heat reaching the earth's surface is reduced below G_{on} because a large part of it is scattered, reflected back out into space, and absorbed by the atmosphere. As a result of the atmospheric interaction with the solar radiation, a portion of the originally collimated rays becomes scattered or nondirectional. Some of this scattered radiation reaches the earth's surface from the entire sky vault. This is called the *diffuse radiation*. The solar heat that comes directly through the atmosphere is termed *direct* or *beam radiation*. The insolation received by a surface on earth is the sum of diffuse radiation and the normal component of beam radiation. The solar heat at any point on earth depends on:

1. The ozone layer thickness
2. The distance traveled through the atmosphere to reach that point
3. The amount of haze in the air (dust particles, water vapor, etc.)
4. The extent of the cloud cover

The earth is surrounded by atmosphere that contains various gaseous constituents, suspended dust, and other minute solid and liquid particulate matter and clouds of various types. As the solar radiation travels through the earth's atmosphere, waves of very short length, such as X-rays and gamma rays, are absorbed in the ionosphere at extremely high altitude. The waves of relatively longer length, mostly in the ultraviolet range, are then absorbed by the layer of ozone (O_3), located about 15−40 km above the earth's surface. In the lower atmosphere, bands of solar radiation in the infrared range are absorbed by water vapor and carbon dioxide. In the long-wavelength region, since the extraterrestrial radiation is low and the H_2O and CO_2 absorption is strong, little solar energy reaches the ground.

Therefore, the solar radiation is depleted during its passage though the atmosphere before reaching the earth's surface. The reduction of intensity with increasing zenith angle of the sun is generally assumed to be directly proportional to the increase in air mass, an assumption that considers the atmosphere to be unstratified with regard to absorbing or scattering impurities.

The degree of attenuation of solar radiation traveling through the earth's atmosphere depends on the length of the path and the characteristics of the medium traversed. In solar radiation calculations, one standard *air mass* is defined as the length of the path traversed in reaching the sea level when the sun is at its zenith (the vertical at the point of observation). The air mass is related to the zenith angle, Φ (Fig. 2.27), without considering the earth's curvature, by the equation:

$$m = \frac{AB}{BC} = \frac{1}{\cos(\Phi)} \tag{2.81}$$

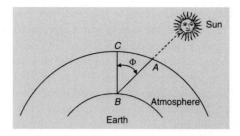

FIGURE 2.27

Air mass definition.

Therefore, at sea level when the sun is directly overhead, that is, when $\Phi = 0°$, $m = 1$ (air mass one); and when $\Phi = 60°$, we get $m = 2$ (air mass two). Similarly, the solar radiation outside the earth's atmosphere is at air mass zero. The graph of direct normal irradiance (solar spectrum) at ground level for air mass 1.5, which corresponds to $\Phi = 48°$, is shown in Appendix 4.

2.3.7 Terrestrial irradiation

A solar system frequently needs to be judged on its long-term performance. Therefore, knowledge of long-term monthly average daily insolation data for the locality under consideration is required. Daily mean total solar radiation (beam plus diffuse) incident on a horizontal surface for each month of the year is available from various sources, such as radiation maps or a country's meteorological service (see Section 2.4). In these sources, data, such as 24 h average temperature, monthly average daily radiation on a horizontal surface $\overline{H}$ (MJ/m^2 day), and monthly average clearness index, $\overline{K}_T$, are given together with other parameters, which are not of interest here.[2] The monthly average clearness index, $\overline{K}_T$, is defined as:

$$\overline{K}_T = \frac{\overline{H}}{\overline{H}_o} \tag{2.82a}$$

where

$\overline{H}$ = monthly average daily total radiation on a terrestrial horizontal surface (MJ/m^2 day);
$\overline{H}_o$ = monthly average daily total radiation on an extraterrestrial horizontal surface (MJ/m^2 day).

The bar over the symbols signifies a long-term average. The value of $\overline{H}_o$ can be calculated from Eq. (2.79) by choosing a particular day in the given month for which the daily total extraterrestrial insolation is estimated to be the same as the monthly mean value. Table 2.5 gives the values of $\overline{H}_o$ for each month as a function of latitude, together with the recommended dates of each month that would give the mean daily values of $\overline{H}_o$. The day number and the declination of the day for the recommended dates are shown in Table 2.1. For the same days, the monthly average daily extraterrestrial insolation on a horizontal surface for various months in kilowatt hours per square meter (kWh/m^2 day) for

---

[2]Meteorological data for various locations are shown in Appendix 7.

Table 2.5 Monthly average daily extraterrestrial insolation on horizontal surface (MJ/m^2).

Latitude	Jan 17	Feb 16	Mar 16	Apr 15	May 15	June 11	July 17	Aug 16	Sept 15	Oct 15	Nov 14	Dec 10
60°S	41.1	31.9	21.2	10.9	4.4	2.1	3.1	7.8	16.7	28.1	38.4	43.6
55°S	41.7	33.7	23.8	13.8	7.1	4.5	5.6	10.7	19.5	30.2	39.4	43.9
50°S	42.4	35.3	26.3	16.8	10.0	7.2	8.4	13.6	22.2	32.1	40.3	44.2
45°S	42.9	36.8	28.6	19.6	12.9	10.0	11.2	16.5	24.7	33.8	41.1	44.4
40°S	43.1	37.9	30.7	22.3	15.8	12.9	14.1	19.3	27.1	35.3	41.6	44.4
35°S	43.2	38.8	32.5	24.8	18.6	15.8	17.0	22.0	29.2	36.5	41.9	44.2
30°S	43.0	39.5	34.1	27.2	21.4	18.7	19.8	24.5	31.1	37.5	41.9	43.7
25°S	42.5	39.9	35.4	29.4	24.1	21.5	22.5	26.9	32.8	38.1	41.6	43.0
20°S	41.5	39.9	36.5	31.3	26.6	24.2	25.1	29.1	34.2	38.5	41.1	42.0
15°S	40.8	39.7	37.2	33.1	28.9	26.8	27.6	31.1	35.4	38.7	40.3	40.8
10°S	39.5	39.3	37.7	34.6	31.1	29.2	29.9	32.8	36.3	38.5	39.3	39.3
5°S	38.0	38.5	38.0	35.8	33.0	31.4	32.0	34.4	36.9	38.1	37.9	37.6
0	36.2	37.4	37.9	36.8	34.8	33.5	33.9	35.7	37.2	37.3	36.4	35.6
5°N	34.2	36.1	37.5	37.5	36.3	35.3	35.6	36.7	37.3	36.3	34.5	33.5
10°N	32.0	34.6	36.9	37.9	37.5	37.0	37.1	37.5	37.0	35.1	32.5	31.1
15°N	29.5	32.7	35.9	38.0	38.5	38.4	38.3	38.0	36.5	33.5	30.2	28.5
20°N	26.9	30.7	34.7	37.9	39.3	39.5	39.3	38.2	35.7	31.8	27.7	25.7
25°N	24.1	28.4	33.3	37.5	39.8	40.4	40.0	38.2	34.7	29.8	25.1	22.9
30°N	21.3	26.0	31.6	36.8	40.0	41.1	40.4	37.9	33.4	27.5	22.3	19.9
35°N	18.3	23.3	29.6	35.8	39.9	41.5	40.6	37.3	31.8	25.1	19.4	16.8
40°N	15.2	20.5	27.4	34.6	39.7	41.7	40.6	36.5	30.0	22.5	16.4	13.7
45°N	12.1	17.6	25.0	33.1	39.2	41.7	40.4	35.4	27.9	19.8	13.4	10.7
50°N	9.1	14.6	22.5	31.4	38.4	41.5	40.0	34.1	25.7	16.9	10.4	7.7
55°N	6.1	11.6	19.7	29.5	37.6	41.3	39.4	32.7	23.2	13.9	7.4	4.8
60°N	3.4	8.5	16.8	27.4	36.6	41.0	38.8	31.0	20.6	10.9	4.5	2.3

latitudes $-60°$ to $+60°$ is also shown graphically in Figure A3.5 in Appendix 3, from which we can easily interpolate.

Further to Eq. (2.82a) the daily clearness index K_T, can be defined as the ratio of the radiation for a particular day to the extraterrestrial radiation for that day given by:

$$K_T = \frac{H}{H_o} \qquad (2.82b)$$

Similarly, an hourly clearness index k_T, can be defined given by:

$$k_T = \frac{I}{I_o} \qquad (2.82c)$$

In all these equations the values of $\overline{H}$, H, and I can be obtained from measurements of total solar radiation on horizontal using a pyranometer (see Section 2.3.9).

To predict the performance of a solar system, hourly values of radiation are required. Because in most cases these types of data are not available, long-term average daily radiation data can be utilized to estimate long-term average radiation distribution. For this purpose, empirical correlations are usually used. Two such frequently used correlations are the Liu and Jordan (1977) correlation for the diffuse radiation and the Collares-Pereira and Rabl (1979) correlation for the total radiation.

According to the Liu and Jordan (1977) correlation,

$$r_d = \left(\frac{\pi}{24}\right) \frac{\cos(h) - \cos(h_{ss})}{\sin(h_{ss}) - \left(\frac{2\pi h_{ss}}{360}\right)\cos(h_{ss})} \qquad (2.83)$$

where

r_d = ratio of hourly diffuse radiation to daily diffuse radiation ($=I_D/H_D$);
h_{ss} = sunset hour angle (degrees); and
h = hour angle in degrees at the midpoint of each hour.

According to the Collares-Pereira and Rabl (1979) correlation,

$$r = \frac{\pi}{24}[\alpha + \beta \cos(h)] \frac{\cos(h) - \cos(h_{ss})}{\sin(h_{ss}) - \left(\frac{2\pi h_{ss}}{360}\right)\cos(h_{ss})} \qquad (2.84a)$$

where

r = ratio of hourly total radiation to daily total radiation ($=I/H$).

$$\alpha = 0.409 + 0.5016 \ \sin \ (h_{ss} - 60) \qquad (2.84b)$$

$$\beta = 0.6609 - 0.4767 \ \sin \ (h_{ss} - 60) \qquad (2.84c)$$

EXAMPLE 2.18

Given the following empirical equation,

$$\frac{\overline{H}_D}{\overline{H}} = 1.390 - 4.027\overline{K}_T + 5.531\overline{K}_T^2 - 3.108\overline{K}_T^3$$

where $\overline{H}_D$ is the monthly average daily diffuse radiation on horizontal surface—see Eq. (2.105a)—estimate the average total radiation and the average diffuse radiation between 11:00 a.m. and 12:00 p.m. solar time in the month of July on a horizontal surface located at 35°N latitude. The monthly average daily total radiation on a horizontal surface, $\overline{H}$, in July at the surface location is 23.14 MJ/m² day.

Solution

From Table 2.5 at 35°N latitude for July, $\overline{H}_o = 40.6$ MJ/m². Therefore,

$$\overline{K}_T = \frac{\overline{H}}{\overline{H}_o} = \frac{23.14}{40.6} = 0.570$$

Therefore,

$$\frac{\overline{H}_D}{\overline{H}} = 1.390 - 4.027(0.57) + 5.531(0.57)^2 - 3.108(0.57)^3 = 0.316$$

and

$$\overline{H}_D = 0.316\overline{H} = 0.316(23.14) = 7.31 \text{MJ/m}^2 \text{ day}$$

From Table 2.5, the recommended average day for the month is July 17 ($N = 198$). The solar declination is calculated from Eq. (2.5) as:

$$\delta = 23.45 \sin\left[\frac{360}{365}(284 + N)\right] = 23.45 \sin\left[\frac{360}{365}(284 + 198)\right] = 21.2°$$

The sunset hour angle is calculated from Eq. (2.15) as:

$$\cos(h_{ss}) = -\tan(L)\tan(\delta) \rightarrow h_{ss} = \cos^{-1}\left[-\tan(35)\tan(21.2)\right] = 106°$$

The middle point of the hour from 11:00 a.m. to 12:00 p.m. is 0.5 h from solar noon, or hour angle is −7.5°. Therefore, from Eqs. (2.84b), (2.84c) and (2.84a), we have:

$$\alpha = 0.409 + 0.5016 \sin(h_{ss} - 60) = 0.409 + 0.5016 \sin(106 - 60) = 0.77$$

$$\beta = 0.6609 - 0.4767 \sin(h_{ss} - 60) = 0.6609 - 0.4767 \sin(106 - 60) = 0.318$$

$$r = \frac{\pi}{24}(\alpha + \beta \cos(h)) \frac{\cos(h) - \cos(h_{ss})}{\sin(h_{ss}) - \left(\frac{2\pi h_{ss}}{360}\right)\cos(h_{ss})}$$

$$= \frac{\pi}{24}(0.77 + 0.318 \cos(-7.5)) \frac{\cos(-7.5) - \cos(106)}{\sin(106) - \left[\frac{2\pi(106)}{360}\right]\cos(106)} = 0.123$$

From Eq. (2.83), we have:

$$r_d = \left(\frac{\pi}{24}\right) \frac{\cos(h) - \cos(h_{ss})}{\sin(h_{ss}) - \left(\frac{2\pi h_{ss}}{360}\right)\cos(h_{ss})} = \left(\frac{\pi}{24}\right) \frac{\cos(-7.5) - \cos(106)}{\sin(106) - \left[\frac{2\pi(106)}{360}\right]\cos(106)} = 0.113$$

Finally,

$$\text{Average hourly total radiation} = 0.123(23.14) = 2.85 \text{ MJ/m}^2 \quad \text{or} \quad 2850 \text{ kJ/m}^2$$

$$\text{Average hourly diffuse radiation} = 0.113(7.31) = 0.826 \text{ MJ/m}^2 \quad \text{or} \quad 826 \text{ kJ/m}^2$$

2.3.8 Total radiation on tilted surfaces

Usually, collectors are not installed horizontally but at an angle to increase the amount of radiation intercepted and reduce reflection and cosine losses. Therefore, system designers need data about solar radiation on such tilted surfaces; measured or estimated radiation data, however, are mostly available either for normal incidence or for horizontal surfaces. Therefore, there is a need to convert these data to radiation on tilted surfaces.

The amount of insolation on a terrestrial surface at a given location for a given time depends on the orientation and slope of the surface.

A flat surface absorbs beam (G_{Bt}), diffuse (G_{Dt}), and ground-reflected (G_{Gt}) solar radiation; that is,

$$G_t = G_{Bt} + G_{Dt} + G_{Gt} \tag{2.85}$$

As shown in Fig. 2.28, the beam radiation on a tilted surface is:

$$G_{Bt} = G_{Bn}\cos(\theta) \tag{2.86}$$

and on a horizontal surface,

$$G_B = G_{Bn}\cos(\Phi) \tag{2.87}$$

where

G_{Bt} = beam radiation on a tilted surface (W/m^2);
G_B = beam radiation on a horizontal surface (W/m^2).

It follows that,

$$R_B = \frac{G_{Bt}}{G_B} = \frac{\cos(\theta)}{\cos(\Phi)} \tag{2.88}$$

where R_B is called the *beam radiation tilt factor*. The term $\cos(\theta)$ can be calculated from Eq. (2.86) and $\cos(\Phi)$ from Eq. (2.87). So the beam radiation component for any surface is:

$$G_{Bt} = G_B R_B \tag{2.89}$$

In Eq. (2.88), the zenith angle can be calculated from Eq. (2.12) and the incident angle θ can be calculated from Eq. (2.18) or, for the specific case of a south-facing fixed surface, from Eq. (2.20). Therefore, for a fixed surface facing south with tilt angle β, Eq. (2.88) becomes:

$$R_B = \frac{\cos(\theta)}{\cos(\Phi)} = \frac{\sin(L-\beta)\sin(\delta) + \cos(L-\beta)\cos(\delta)\cos(h)}{\sin(L)\sin(\delta) + \cos(L)\cos(\delta)\cos(h)} \tag{2.90a}$$

FIGURE 2.28

Beam radiation on horizontal and tilted surfaces.

Similarly to what is indicated in the notes after Eq. (2.20) and Eq. (2.21), at solar noon:

- For the northern hemisphere:

$$R_{B,noon} = \frac{\cos|L - \delta - \beta|}{\cos|L - \delta|} \qquad (2.90b)$$

- For the southern hemisphere:

$$R_{B,noon} = \frac{\cos|-L + \delta - \beta|}{\cos|-L + \delta|} \qquad (2.90c)$$

Equation (2.88) also can be applied to surfaces other than fixed, in which case the appropriate equation for $\cos(\theta)$, as given in Section 2.2.1, can be used. For example, for a surface rotated continuously about a horizontal east–west axis, from Eq. (2.26a), the ratio of beam radiation on the surface to that on a horizontal surface at any time is given by:

$$R_B = \frac{\sqrt{1 - \cos^2(\delta)\sin^2(h)}}{\sin(L)\sin(\delta) + \cos(L)\cos(\delta)\cos(h)} \qquad (2.90d)$$

EXAMPLE 2.19

Estimate the beam radiation tilt factor for a surface located at 35°N latitude and tilted 45° at 2:00 p.m. solar time on March 10. If the beam radiation at normal incidence is 900 W/m², estimate the beam radiation on the tilted surface.

Solution

From Example 2.17, $\delta = -4.8°$ and $h = 30°$. The beam radiation tilt factor is calculated from Eq. (2.90a) as:

$$R_B = \frac{\sin(L - \beta)\sin(\delta) + \cos(L - \beta)\cos(\delta)\cos(h)}{\sin(L)\sin(\delta) + \cos(L)\cos(\delta)\cos(h)}$$

$$= \frac{\sin(35 - 45)\sin(-4.8) + \cos(35 - 45)\cos(-4.8)\cos(30)}{\sin(35)\sin(-4.8) + \cos(35)\cos(-4.8)\cos(30)} = 1.312$$

Therefore, the beam radiation on the tilted surface is calculated from Eq. (2.89) as:

$$G_{Bt} = G_B R_B = 900(1.312) = 1181 \text{ W/m}^2$$

Isotropic sky model

Many models give the solar radiation on a tilted surface. The first one is the isotropic sky model developed originally by Hottel and Woertz (1942) and refined by Liu and Jordan (1960). According to this model, radiation is calculated as follows.

Diffuse radiation on a horizontal surface,

$$G_D = 2 \int_0^{\pi/2} G_R \cos(\Phi)\, d\Phi = 2G_R \tag{2.91}$$

where G_R = diffuse sky radiance (W/m^2 rad).

Diffuse radiation on a tilted surface,

$$G_{Dt} = \int_0^{\pi/2-\beta} G_R \cos(\Phi)\, d\Phi + \int_0^{\pi/2} G_R \cos(\Phi)\, d\Phi \tag{2.92}$$

where β is the surface tilt angle as shown in Fig. 2.28.

From Eq. (2.91), the second term of Eq. (2.92) becomes $G_R = G_D/2$. Therefore, Eq. (2.92) becomes:

$$G_{Dt} = \frac{G_D}{2} \int_0^{\pi/2-\beta} \cos(\Phi)\, d\Phi + \frac{G_D}{2} = \frac{G_D}{2}\left[\sin\left(\frac{\pi}{2}-\beta\right)\right] + \frac{G_D}{2} = G_D\left[\frac{1+\cos(\beta)}{2}\right] \tag{2.93}$$

Similarly, the ground-reflected radiation is obtained by $\rho_G(G_B + G_D)$, where ρ_G is ground albedo. Therefore, G_{Gt} is obtained as follows.

Ground-reflected radiation

$$\rho_G(G_B + G_D) = 2 \int_0^{\pi/2} G_r \cos(\Phi)\, d\Phi = 2G_r \tag{2.94}$$

where G_r is the isotropic ground-reflected radiance (W/m^2 rad).

Ground-reflected radiation on tilted surfaces,

$$G_{Gt} = \int_{\pi/2-\beta}^{\pi/2} G_r \cos(\Phi)\, d\Phi \tag{2.95}$$

Combining Eqs. (2.94) and (2.95) as before,

$$G_{Gt} = \rho_G(G_B + G_D)\left[\frac{1-\cos(\beta)}{2}\right] \tag{2.96}$$

Therefore, inserting Eqs. (2.93) and (2.96) into Eq. (2.85), we get:

$$G_t = R_B G_B + G_D \left[\frac{1 + \cos(\beta)}{2} \right] + (G_B + G_D)\rho_G \left[\frac{1 - \cos(\beta)}{2} \right] \tag{2.97}$$

The total radiation on a horizontal surface, G, is the sum of horizontal beam and diffuse radiation; that is,

$$G = G_B + G_D \tag{2.98}$$

Therefore, Eq. (2.97) can also be written as:

$$R = \frac{G_t}{G} = \frac{G_B}{G} R_B + \frac{G_D}{G} \left[\frac{1 + \cos(\beta)}{2} \right] + \rho_G \left[\frac{1 - \cos(\beta)}{2} \right] \tag{2.99}$$

where R is called the *total radiation tilt factor*.

Other radiation models

The isotropic sky model is the simplest model that assumes that all diffuse radiation is uniformly distributed over the sky dome and that reflection on the ground is diffuse. A number of other models have been developed by a number of researchers. Three of these models are summarized in this section: the Klucher model, the Hay—Davies model, and the Reindl model. The latter proved to give very good results in the Mediterranean region.

Klucher model

Klucher (1979) found that the isotopic model gives good results for overcast skies but underestimates irradiance under clear and partly overcast conditions, when there is increased intensity near the horizon and in the circumsolar region of the sky. The model developed by Klucher gives the total irradiation on a tilted plane:

$$G_t = G_B R_B + G_D \left[\frac{1 + \cos(\beta)}{2} \right] \left[1 + F' \sin^3 \left(\frac{\beta}{2} \right) \right] \left[1 + F' \cos^2(\beta) \sin^3(\Phi) \right]$$

$$+ (G_B + G_D)\rho \left[\frac{1 - \cos(\beta)}{2} \right] \tag{2.100}$$

where F' is a clearness index given by:

$$F' = 1 - \left(\frac{G_D}{G_B + G_D} \right)^2 \tag{2.101}$$

The first of the modifying factors in the sky diffuse component takes into account horizon brightening; the second takes into account the effect of circumsolar radiation. Under overcast skies, the clearness index F' becomes 0 and the model reduces to the isotropic model.

Hay—Davies model

In the Hay—Davies model, diffuse radiation from the sky is composed of an isotropic and circumsolar component (Hay and Davies, 1980) and horizon brightening is not taken into account. The anisotropy index, A, defined in Eq. (2.102), represents the transmittance through atmosphere for beam radiation:

$$A = \frac{G_{Bn}}{G_{on}} \tag{2.102}$$

The anisotropy index is used to quantify the portion of the diffuse radiation treated as circumsolar, with the remaining portion of diffuse radiation assumed isotropic. The circumsolar component is assumed to be from the sun's position. The total irradiance is then computed by:

$$G_t = (G_B + G_D A)R_B + G_D(1-A)\left[\frac{1+\cos(\beta)}{2}\right] + (G_B + G_D)\rho\left[\frac{1-\cos(\beta)}{2}\right] \tag{2.103}$$

Reflection from the ground is dealt with as in the isotropic model.

Reindl model

In addition to isotropic diffuse and circumsolar radiation, the Reindl model also accounts for horizon brightening (Reindl et al., 1990a;b) and employs the same definition of the anisotropy index, A, as described in Eq. (2.102). The total irradiance on a tilted surface can then be calculated using:

$$G_t = (G_B + G_D A)R_B + G_D(1-A)\left[\frac{1-\cos(\beta)}{2}\right]\left[1+\sqrt{\frac{G_B}{G_B+G_D}}\sin^3\left(\frac{\beta}{2}\right)\right]$$

$$+ (G_B + G_D)\rho\left[\frac{1-\cos(\beta)}{2}\right] \tag{2.104}$$

Reflection on the ground is again dealt with as in the isotropic model. Due to the additional term in Eq. (2.104), representing horizon brightening, the Reindl model provides slightly higher diffuse irradiances than the Hay–Davies model.

Insolation on tilted surfaces

The amount of insolation on a terrestrial surface at a given location and time depends on the orientation and slope of the surface. In the case of flat-plate collectors installed at a certain fixed angle, system designers need to have data about the solar radiation on the surface of the collector. Most measured data, however, are for either normal incidence or horizontal. Therefore, it is often necessary to convert these data to radiation on tilted surfaces. Based on these data, a reasonable estimation of radiation on tilted surfaces can be made. An empirical method for the estimation of the monthly average daily total radiation incident on a tilted surface was developed by Liu and Jordan (1977). In their correlation, the diffuse to total radiation ratio for a horizontal surface is expressed in terms of the monthly clearness index, $\overline{K}_T$, with the following equation:

$$\frac{\overline{H}_D}{\overline{H}} = 1.390 - 4.027\overline{K}_T + 5.531\overline{K}_T^2 - 3.108\overline{K}_T^3 \tag{2.105a}$$

Collares-Pereira and Rabl (1979) expressed the same parameter by also considering the sunset hour angle:

$$\frac{\overline{H}_D}{\overline{H}} = 0.775 + 0.00653(h_{ss} - 90) - [0.505 + 0.00455(h_{ss} - 90)]\cos(115\overline{K}_T - 103) \tag{2.105b}$$

where h_{ss} = sunset hour angle (degrees).

Erbs et al. (1982) also expressed the monthly average daily diffuse correlations by taking into account the season, as follows.

For $h_{ss} \leq 81.4°$ and $0.3 \leq \overline{K}_T \leq 0.8$,

$$\frac{\overline{H}_D}{\overline{H}} = 1.391 - 3.560\overline{K}_T + 4.189\overline{K}_T^2 - 2.137\overline{K}_T^3 \tag{2.105c}$$

For $h_{ss} > 81.4°$ and $0.3 \leq \overline{K}_T \leq 0.8$,

$$\frac{\overline{H}_D}{\overline{H}} = 1.311 - 3.022\overline{K}_T + 3.427\overline{K}_T^2 - 1.821\overline{K}_T^3 \tag{2.105d}$$

With the monthly average daily total radiation $\overline{H}$ and the monthly average daily diffuse radiation $\overline{H}_D$ known, the monthly average beam radiation on a horizontal surface can be calculated by:

$$\overline{H}_B = \overline{H} - \overline{H}_D \tag{2.106}$$

Like Eq. (2.99), the following equation may be written for the monthly total radiation tilt factor $\overline{R}$:

$$\overline{R} = \frac{\overline{H}_t}{\overline{H}} = \left(1 - \frac{\overline{H}_D}{\overline{H}}\right)\overline{R}_B + \frac{\overline{H}_D}{\overline{H}}\left[\frac{1 + \cos(\beta)}{2}\right] + \rho_G\left[\frac{1 - \cos(\beta)}{2}\right] \tag{2.107}$$

where

$\overline{H}_t$ = monthly average daily total radiation on a tilted surface (MJ/m^2 day);
$\overline{R}_B$ = monthly mean beam radiation tilt factor.

The term $\overline{R}_B$ is the ratio of the monthly average beam radiation on a tilted surface to that on a horizontal surface. Actually, this is a complicated function of the atmospheric transmittance, but according to Liu and Jordan (1977), it can be estimated by the ratio of extraterrestrial radiation on the tilted surface to that on a horizontal surface for the month. For surfaces facing directly toward the equator, it is given by:

$$\overline{R}_B = \frac{\cos(L - \beta)\cos(\delta)\sin(h'_{ss}) + (\pi/180)h'_{ss}\sin(L - \beta)\sin(\delta)}{\cos(L)\cos(\delta)\sin(h_{ss}) + (\pi/180)h_{ss}\sin(L)\sin(\delta)} \tag{2.108}$$

where h'_{ss} is sunset hour angle on the tilted surface (degrees), given by:

$$h'_{ss} = \min\left\{h_{ss}, \cos^{-1}[-\tan(L - \beta)\tan(\delta)]\right\} \tag{2.109}$$

It should be noted that, for the Southern Hemisphere, the term $(L - \beta)$ of Eqs. (2.108) and (2.109) changes to $(L + \beta)$.

For the same days as those shown in Table 2.5, the monthly average terrestrial insolation on a tilted surface for various months for latitudes $-60°$ to $+60°$ and for a slope equal to latitude and latitude plus $10°$, which is the usual collector inclination for solar water-heating collectors, is shown in Appendix 3, Figures A3.6 and A3.7, respectively.

EXAMPLE 2.20

For July, estimate the monthly average daily total solar radiation on a surface facing south, tilted 45°, and located at 35°N latitude. The monthly average daily insolation on a horizontal surface is 23.14 MJ/m² day. Ground reflectance is equal to 0.2.

Solution

From Example 2.18, we have: $\overline{H}_D/\overline{H} = 0.316$, $\delta = 21.2°$, and $h_{ss} = 106°$. The sunset hour angle for a tilted surface is given by Eq. (2.109):

$$h'_{ss} = \min\left\{h_{ss}, \cos^{-1}\left[-\tan(L-\beta)\tan(\delta)\right]\right\}$$

Here, $\cos^{-1}[-\tan(35-45)\tan(21.2)] = 86°$. Therefore,

$$h'_{ss} = 86°$$

The factor $\overline{R}_B$ is calculated from Eq. (2.108) as:

$$\overline{R}_B = \frac{\cos(L-\beta)\cos(\delta)\sin(h'_{ss}) + (\pi/180)h'_{ss}\sin(L-\beta)\sin(\delta)}{\cos(L)\cos(\delta)\sin(h_{ss}) + (\pi/180)h_{ss}\sin(L)\sin(\delta)}$$

$$= \frac{\cos(35-45)\cos(21.2)\sin(86) + (\pi/180)(86)\sin(35-45)\sin(21.2)}{\cos(35)\cos(21.2)\sin(106) + (\pi/180)(106)\sin(35)\sin(21.2)} = 0.735$$

From Eq. (2.107),

$$\overline{R} = \left(1 - \frac{\overline{H}_D}{\overline{H}}\right)\overline{R}_B + \frac{\overline{H}_D}{\overline{H}}\left[\frac{1+\cos(\beta)}{2}\right] + \rho_G\left[\frac{1-\cos(\beta)}{2}\right]$$

$$= (1-0.316)(0.735) + 0.316\left[\frac{1+\cos(45)}{2}\right] + 0.2\left[\frac{1-\cos(45)}{2}\right] = 0.802$$

Finally, the average daily total radiation on the tilted surface for July is:

$$\overline{H}_t = \overline{R}\overline{H} = 0.802(23.14) = 18.6 \ \text{MJ/m}^2 \ \text{day}$$

2.3.9 Solar radiation measuring equipment

A number of radiation parameters are needed for the design, sizing, performance evaluation, and research of solar energy applications. These include total solar radiation, beam radiation, diffuse radiation, and sunshine duration. Various types of equipment measure the instantaneous and long-term integrated values of beam, diffuse, and total radiation incident on a surface. This equipment usually employs the thermoelectric and photovoltaic effects to measure the radiation. Detailed description of this equipment is not within the scope of this book; this section is added, however, so the reader might know the types of available equipment. More details of this equipment can easily be found from manufacturers' catalogs on the Internet.

There are basically two types of solar radiation measuring instruments: the pyranometer (see Fig. 2.29) and the pyrheliometer (see Fig. 2.30). The former is used to measure total (beam and diffuse) radiation within its hemispherical field of view, whereas the latter is an instrument used for measuring the direct solar irradiance. The pyranometer can also measure the diffuse solar radiation if the sensing element is shaded from the beam radiation (see Fig. 2.31). For this purpose a shadow band is mounted with its axis tilted at an angle equal to the latitude of the location plus the declination for the day of

FIGURE 2.29

Photograph of a pyranometer.

FIGURE 2.30

Photograph of a solar pyrheliometer.

measurement. Since the shadow band hides a considerable portion of the sky, the measurements require corrections for that part of diffuse radiation obstructed by the band. Pyrheliometers are used to measure direct solar irradiance, required primarily to predict the performance of concentrating solar collectors. Diffuse radiation is blocked by mounting the sensor element at the bottom of a tube pointing directly at the sun. Therefore, a two-axis sun-tracking system is required to measure the beam radiation.

Finally, sunshine duration is required to estimate the total solar irradiation. The duration of sunshine is defined as the time during which the sunshine is intense enough to cast a shadow. Also, the duration of sunshine has been defined by the World Meteorological Organization as the time during which the beam solar irradiance exceeds the level of 120 W/m^2. Two types of sunshine recorders are used: the focusing type and a type based on the photoelectric effect. The focusing type consists of a solid glass sphere, approximately 10 cm in diameter, mounted concentrically in a section of a spherical bowl whose diameter is such that the sun's rays can be focused on a special card with time marking, held in place by grooves in the bowl. The record card is burned whenever bright sunshine exists. Thus, the portion of the burned trace provides the duration of sunshine for the day. The sunshine recorder

FIGURE 2.31

Photograph of a pyranometer with shading ring for measuring diffuse solar radiation.

based on the photoelectric effect consists of two photovoltaic cells, with 1 cell exposed to the beam solar radiation and the other cell shaded from it by a shading ring. The radiation difference between the 2 cells is a measure of the duration of sunshine.

The International Standards Organization (ISO) published a series of international standards specifying methods and instruments for the measurement of solar radiation. These are:

- **ISO 9059** (1990). *Calibration of field pyrheliometers by comparison to a reference pyrheliometer.* This International Standard describes the calibration of field pyrheliometers using reference pyrheliometers and sets out the calibration procedures and the calibration hierarchy for the transfer of the calibration. This International Standard is mainly intended for use by calibration services and test laboratories to enable a uniform quality of accurate calibration factors to be achieved.

- **ISO 9060** (1990). *Specification and classification of instruments for measuring hemispherical solar and direct solar radiation.* This standard establishes a classification and specification of instruments for the measurement of hemispherical solar and direct solar radiation integrated over the spectral range from 0.3 to 3 μm. According to the standard, pyranometers are radiometers designed for measuring the irradiance on a plane receiver surface, which results from the radiant fluxes incident from the hemisphere above, within the required wavelength range. Pyrheliometers are radiometers designed for measuring the irradiance that results from the solar radiant flux from a well-defined solid angle, the axis of which is perpendicular to the plane receiver surface.

- **ISO 9846** (1993). *Calibration of a pyranometer using a pyrheliometer.* This standard also includes specifications for the shade ring used to block the beam radiation, the measurement of diffuse radiation, and support mechanisms of the ring.
- **IS0 9847** (1992). *Calibration of field pyranometers by comparison to a reference pyranometer.* According to the standard, accurate and precise measurements of the irradiance of the global (hemispheric) solar radiation are required in:
  1. The determination of the energy available to flat-plate solar collectors.
  2. The assessment of irradiance and radiant exposure in the testing of solar- and non-solar-related material technologies.
  3. The assessment of the direct versus diffuse solar components for energy budget analysis, for geographic mapping of solar energy, and as an aid in the determination of the concentration of aerosol and particulate pollution and the effects of water vapor.

Although meteorological and resource assessment measurements generally require pyranometers oriented with their axes vertical, applications associated with flat-plate collectors and the study of the solar exposure of related materials require calibrations of instruments tilted at a predetermined non-vertical orientation. Calibrations at fixed tilt angles have applications that seek state-of-the-art accuracy, requiring corrections for cosine, tilt, and azimuth.

Finally, the International Standards Organization published a technical report, "lSO/TR 9901: *1990—Field pyranometers—Recommended practice for use*," the scope of which is self-explanatory.

2.4 The solar resource

The operation of solar collectors and systems depends on the solar radiation input and the ambient air temperature and their sequences. One of the forms in which solar radiation data are available is on maps. These give the general impression of the availability of solar radiation without details on the local meteorological conditions and, for this reason, must be used with care. One valuable source of such information is the Meteonorm. Two maps showing the annual mean global solar radiation for the years 1981−2000 for Europe and North America are shown in Figs. 2.32 and 2.33, respectively (Meteonorm, 2009). These are based on numerous climatological databases and computational models. Maps for other regions of the world can be obtained from the Meteonorm web site (Meteonorm, 2009).

Another way of estimating the average solar radiation $\overline{H}$ is by the use of the following equation (Page, 1964):

$$\overline{K}_{\mathrm{T}} = \frac{\overline{H}}{\overline{H}_{\mathrm{o}}} = a + b\frac{\overline{n}}{\overline{N}} \tag{2.110}$$

where

a and *b* = empirical constants;
$\overline{n}$ = monthly average daily hours of sunshine; and
$\overline{N}$ = monthly average maximum possible daily hours of sunshine, given by the day length (Eq. 2.17) of the average day of each month shown in Table 2.1.

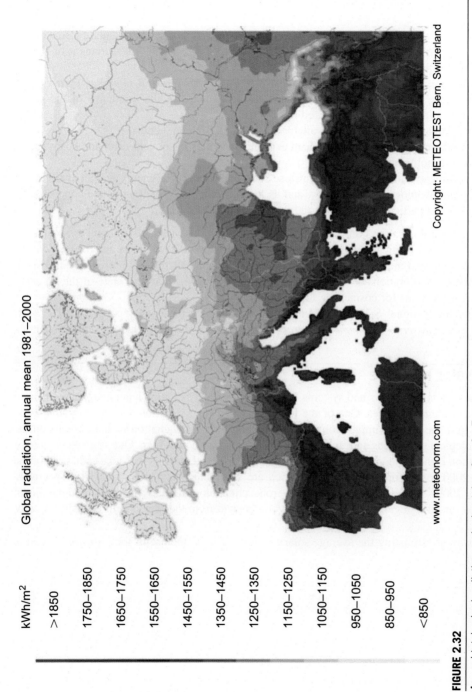

Global radiation, annual mean 1981–2000

kWh/m²

>1850

1750–1850

1650–1750

1550–1650

1450–1550

1350–1450

1250–1350

1150–1250

1050–1150

950–1050

850–950

<850

Copyright: METEOTEST Bern, Switzerland

www.meteonorm.com

Meteonorm database of Meteotest (www.Meteonorm.com).

FIGURE 2.32

Annual total solar irradiation on horizontal surface for Europe.

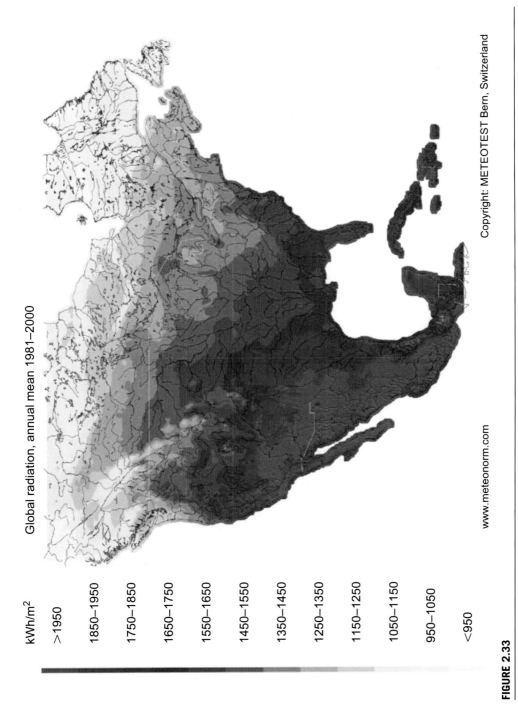

Global radiation, annual mean 1981–2000

kWh/m²

>1950

1850–1950

1750–1850

1650–1750

1550–1650

1450–1550

1350–1450

1250–1350

1150–1250

1050–1150

950–1050

<950

www.meteonorm.com

Copyright: METEOTEST Bern, Switzerland

Meteonorm database of Meteotest (www.Meteonorm.com).

FIGURE 2.33

Annual total solar irradiation on horizontal surface for North America.

As was seen before values of $\overline{H}_o$ can be obtained from Eq. (2.79) and for the average day of each month values can be obtained directly from Table 2.5. The constants a and b can be obtained from published tables or can be estimated from measured data in a location.

For the local climate, data in the form of a TMY are usually required. This is a typical year, which is defined as a year that sums up all the climatic information characterizing a period as long as the mean life of a solar system. In this way, the long-term performance of a collector or a system can be calculated by running a computer program over the reference year.

2.4.1 Typical meteorological year

A representative database of weather data for 1-year duration is known as the *test reference year* (TRY) or *typical meteorological year* (TMY). A TMY is a data set of hourly values of solar radiation and meteorological elements. It consists of months selected from individual years concatenated to form a complete year. The TMY contains values of solar radiation (global on horizontal and direct), ambient temperature, relative humidity, and wind speed and direction for all hours of the year. The selection of typical weather conditions for a given location is very crucial in computer simulations to predict the performance of solar systems and the thermal performance of buildings and has led various investigators to either run long periods of observational data or select a particular year that appears to be typical from several years of data. The intended use of a TMY file is for computer simulations of solar energy conversion systems and building systems (see Chapter 11, Section 11.5).

The adequacy of using an average or typical year of meteorological data with a simulation model to provide an estimate of the long-term system performance depends on the sensitivity of system performance to the hourly and daily weather sequences. Regardless of how it is selected, an "average" year cannot be expected to have the same weather sequences as those occurring over the long term. However, the simulated performance of a system for an "average year" may provide a good estimate of the long-term system performance, if the weather sequences occurring in the average year are representative of those occurring over the long term or the system performance is independent of the weather sequences (Klein et al., 1976). Using this approach, the long-term integrated system performance can be evaluated and the dynamic system's behavior can be obtained.

In the past, many attempts were made to generate such climatological databases for different areas around the world using various methodologies. One of the most common methodologies for generating a TMY is the one proposed by Hall et al. (1978) using the Filkenstein–Schafer (FS) statistical method (Filkenstein and Schafer, 1971).

The FS method algorithm is as follows: First, the cumulative distribution functions (CDFs) are calculated for each selected meteorological parameter and for each month, over the whole selected period as well as over each specific year of the period. To calculate the CDFs for each parameter, the data are grouped in a number of bins, and the CDFs are calculated by counting the cases in the same bin.

The next step is to compare the CDF of a meteorological parameter, such as global horizontal radiation, for each month for each specific year with the respective CDF of the long-term composite of all years in the selected period.

The FS is the mean difference of the long-term CDF, CDF_{LT}, and the specific month's CDF, CDF_{SM}, calculated in the bins used for the estimation of the CDFs, given by:

$$FS = \frac{1}{N} \sum_{i=1}^{N} |CDF_{LT}(z_i) - CDF_{SM}(z_i)| \tag{2.111}$$

where

N = number of bins (by default, $N = 31$);
z_i = value of the FS statistic for the particular month of the specific year and the meteorological parameter under consideration.

The next step is the application of the weighting factors, WF_j, to the FS statistics values, one for each of the considered meteorological parameters, FS_j, corresponding to each specific month in the selected period. In this way, a weighted sum, or average value, WS, is derived and this value is assigned to the respective month; that is,

$$WS = \frac{1}{M} \sum_{j=1}^{M} WF_j FS_j \tag{2.112}$$

with

$$\sum_{j=1}^{M} WF_j = 1 \tag{2.113}$$

where M = number of parameters in the database.

The user can change the WF values, thus examining the relative importance of each meteorological parameter in the final result. The smaller the WS, the better the approximation to a typical meteorological month (TMM).

Applying this procedure for all months of the available period, a composite year can be formed consisting of the selected months with the smallest WS values.

The root mean standard deviation (RMSD) of the total daily values of the global solar irradiance distribution for each month of each year can then be estimated with respect to the mean long-term hourly distribution and the FS statistics. The RMSD can be computed, and for each month, the year corresponding to the lowest value can be selected. The estimations are carried out according to the expression:

$$RMSD = \sqrt{\frac{\sum_{i=1}^{N} (x_i - \bar{x})}{N}} \tag{2.114}$$

where $\bar{x}$ = the average value of its parameter over the number of bins ($N = 31$).

A total of 8760 rows are included in a TMY file, each corresponding to an hour of the year. The format of TMY file suitable for earlier versions of the TRNSYS program is shown in Table 2.6.

Table 2.6 Format of TMY file suitable for the TRNSYS pogram up to version 14.

Month of year	Hour of month	I_B (kJ/m^2)^a	I (kJ/m^2)^b	Dry bulb temp^c	H_R^{d}	Wind velocity (m/s)	Wid direction^e
1	1	0	0	75	60.47	1	12
1	2	0	0	75	60.47	1	12
1	3	0	0	70	57.82	1	12
1	4	0	0	70	57.82	1	12
1	5	0	0	75	58.56	2	12
–	–	–	–	–	–	–	–
12	740	0	0	45	47.58	1	23
12	741	0	0	30	43.74	1	25
12	742	0	0	20	40.30	1	26
12	743	0	0	20	40.30	1	27
12	744	0	0	10	37.51	1	23

^{a}I_B = Direct (beam) normal solar radiation (integrated over previous hour) in kJ/m^2.
^{b}I = Global solar radiation on horizontal (integrated over previous hour) in kJ/m^2.
^{c}Degrees ×10 (°C).
^{d}Humidity ratio (H_R) in kg of water/kg of air × 10,000.
^{e}Degrees ÷ 10, expressed as 0 for wind from north, 9 for east, 18 for south, and so forth.

2.4.2 Typical meteorological year, second generation

A type 2 TMY format is completely different and consists of many more fields. Such a file can be used with detailed building analysis programs such as TRNSYS (version 16), DOE-2, BDA (Building Design Advisor), and Energy Plus. A TMY-2 file also contains a complete year (8760 items of data) of hourly meteorological data. Each hourly record in the file contains values for solar radiation, dry bulb temperature, and meteorological elements, such as illuminance, precipitation, visibility, and snowfall. Radiation and illumination data are becoming increasingly necessary in many simulation programs. A two-character source and an uncertainty flag are attached to each data value to indicate whether the data value was measured, modeled, or missing and provide an estimate of the uncertainty of the data value. By including the uncertainty flags, users can evaluate the potential impact of weather variability on the performance of solar systems or buildings.

The first record of each file is the file header that describes the station. The file header contains a five-digit meteorological station number, city, state (optional), time zone, latitude, longitude, and elevation. The field positions and definitions of these header elements, together with the values given for the TMY2 for Nicosia, Cyprus (Kalogirou, 2003), are shown in Table 2.7.

Following the file header, 8760 hourly data records provide a 1-year record of solar radiation, illuminance, and meteorological data, along with their source and uncertainty flags. Table 2.8 gives field

Table 2.7 Header elements in the TMY-2 format (first record only).

Field position	Element	Definition	Value used
002–006	5-Digit number	Weather station's number	17,609
008–029	City	City where the station is located (maximum 22 characters)	Nicosia
031–032	State	State where the station is located (2-letter abbreviation)	—
034–036	Time zone	Time zone: number of hours by which the local standard time is ahead of Greenwich (+ve E, −ve W)	2
038–044	Latitude	Latitude of the station:	
038		N = north of equator	N
040–041		Degrees	34
043–044		Minutes	53
046–053	Longitude	Longitude of the station:	
046		W = west, E = east	E
048–050		Degrees	33
052–053		Minutes	38
056–059	Elevation	Elevation of station in meters above sea level	162

positions and element definitions of each hourly record (Marion and Urban, 1995). Each hourly record begins with the year (field positions 2–3) from which the typical month was chosen, followed by the month, day, and hour information and the rest of the data as shown in Table 2.8 (Kalogirou, 2003).

For solar radiation and illuminance elements, the data values represent the energy received during the 60 min preceeding the hour indicated. For meteorological elements (with a few exceptions), observations or measurements were made at the hour indicated. A few of the meteorological elements have observations, measurements, or estimates made at daily, instead of hourly, intervals. Consequently, the data values for broadband aerosol optical depth, snow depth, and days since last snowfall represent the values available for the day indicated.

With the exception of extraterrestrial horizontal and extraterrestrial direct radiation, the two field positions immediately following the data value provide source and uncertainty flags both to indicate whether the data were measured, modeled, or missing and to provide an estimate of the uncertainty of the data. Source and uncertainty flags for extraterrestrial horizontal and extraterrestrial direct radiation are not provided, because these elements were calculated using equations considered to give exact values. Explanation of the uncertainty flags for the other quantities is given in Marion and Urban (1995).

Table 2.8 Data elements in the TMY-2 format (all except the first record).

Field position	Element	Value	Definition
002–009	Local standard time		
002–003	Year	2-Digit	Year
004–005	Month	1–12	Month
006–007	Day	1–31	Day of month
008–009	Hour	1–24	Hour of day in local standard time
010–013	Extraterrestrial horizontal irradiation	0–1415	Amount of solar radiation in Wh/m^2 received on a horizontal surface at the top of the atmosphere
014–017	Extraterrestrial direct normal irradiation	0–1415	Amount of solar radiation in Wh/m^2 received on a surface normal to the sun at the top of the atmosphere
018–023	Global horizontal irradiation		Total amount of direct and diffuse solar radiation in Wh/m^2 received on a horizontal surface
018–021	Data value	0–1200	
022	Flag for data source	A–H, ?	
023	Flag for data uncertainty	0–9	
024–029	Direct normal irradiation		Amount of solar radiation in Wh/m^2 received within a 5.7° field of view centered on the sun
024–027	Data value	0–1100	
028	Flag for data source	A–H, ?	
029	Flag for data uncertainty	0–9	
030–035	Diffuse horizontal irradiation		Amount of solar radiation in Wh/m^2 received from the sky (excluding the solar disk) on a horizontal surface
030–033	Data value	0–700	
034	Flag for data source	A–H, ?	
035	Flag for data uncertainty	0–9	
036–041	Global horizontal illuminance		Average total amount of direct and diffuse illuminance in hundreds of lux received on a horizontal surface
036–039	Data value	0–1300	0 to 1300 = 0 to 130,000 lux
040	Flag for data source	I, ?	
041	Flag for data uncertainty	0–9	
042–047	Direct normal illuminance		Average amount of direct normal illuminance in hundreds of lux received within a 5.7° field of view centered on the sun
042–045	Data value	0–1100	0 to 1100 = 0 to 110,000 lux
046	Flag for data source	I, ?	
047	Flag for data uncertainty	0–9	
048–053	Diffuse horizontal illuminance		Average amount of illuminance in hundreds of lux received from the sky (excluding the solar disk) on a horizontal surface

Table 2.8 Data elements in the TMY-2 format (all except the first record).—cont'd

Field position	Element	Value	Definition
048−051	Data value	0−800	0 to 800 = 0 to 80,000 lux
052	Flag for data source	I, ?	
053	Flag for data uncertainty	0−9	
054−059	Zenith luminance		Average amount of luminance at the sky's zenith in tens of Cd/m^2
054−057	Data value	0−7000	0 to 7000 = 0 to 70,000 Cd/m^2
058	Flag for data source	I, ?	
059	Flag for data uncertainty	0−9	
060−063	Total sky cover		Amount of sky dome in tenths covered by clouds or obscuring phenomena at the hour indicated
060−061	Data value	0−10	
062	Flag for data source	A−F	
063	Flag for data uncertainty	0−9	
064−067	Opaque sky cover		Amount of sky dome in tenths covered by clouds or obscuring phenomena that prevent observing the sky or higher cloud layers at the hour indicated
064−065	Data value	0−10	
066	Flag for data source	A−F	
067	Flag for data uncertainty	0−9	
068−073	Dry bulb temperature		Dry bulb temperature in tenths of a degree centigrade at the hour indicated.
068−071	Data value	−500 to 500	−500 to 500 = −50.0 to 50.0 °C
072	Flag for data source	A−F	
073	Flag for data uncertainty	0−9	
074−079	Dew point temperature		Dew point temperature in tenths of a degree centigrade at the hour indicated.
074−077	Data value	−600 to 300	−600 to 300 = −60.0 to 30.0 °C
078	Flag for data source	A−F	
079	Flag for data uncertainty	0−9	
080−084	Relative humidity		Relative humidity in percent at the hour indicated
080−082	Data value	0−100	
083	Flag for data source	A−F	
084	Flag for data uncertainty	0−9	
085−090	Atmospheric pressure		Atmospheric pressure at station in mbar at the hour indicated
085−088	Data value	700−1100	
089	Flag for data source	A−F	
090	Flag for data uncertainty	0−9	
091−095	Wind direction		Wind direction in degrees at the hour indicated. (N = 0 or 360, E = 90, S = 180, W = 270). For calm winds, wind direction equals zero.

Continued

Table 2.8 Data elements in the TMY-2 format (all except the first record).—cont'd

Field position	Element	Value	Definition
091–093	Data value	0–360	
094	Flag for data source	A–F	
095	Flag for data uncertainty	0–9	
096–100	Wind speed		Wind speed in tenths of meters per second at the hour indicated.
096–98	Data value	0–400	0 to 400 = 0–40.0 m/s
99	Flag for data source	A–F	
100	Flag for data uncertainty	0–9	
101–106	Visibility		Horizontal visibility in tenths of kilometers at the hour indicated.
101–104	Data value	0–1609	7777 = unlimited visibility
105	Flag for data source	A–F, ?	0 to 1609 = 0.0–160.9 km
106	Flag for data uncertainty	0–9	9999 = missing data
107–113	Ceiling height		Ceiling height in meters at the hour indicated.
107–111	Data value	0–30,450	77,777 = unlimited ceiling height
112	Flag for data source	A–F, ?	88,888 = cirroform
113	Flag for data uncertainty	0–9	99,999 = missing data
114–123	Present weather	–	Present weather conditions denoted by a 10-digit number.
124–128	Precipitable water		Precipitation water in millimeters at the hour indicated
124–126	Data value	0–100	
127	Flag for data source	A–F	
128	Flag for data uncertainty	0–9	
129–133	Aerosol optical depth		Broadband aerosol optical depth (broadband turbidity) in thousandths on the day indicated.
129–131	Data value	0–240	0 to 240 = 0.0 to 0.240
132	Flag for data source	A–F	
133	Flag for data uncertainty	0–9	
134–138	Snow depth		Snow depth in centimeters on the day indicated.
134–136	Data value	0–150	999 = missing data
137	Flag for data source	A–F, ?	
138	Flag for data uncertainty	0–9	
139–142	Days since last snowfall		Number of days since last snowfall.
139–140	Data value	0–88	88 = 88 or greater days
141	Flag for data source	A–F, ?	99 = missing data
142	Flag for data uncertainty	0–9	

From Marion, W., Urban, K., 1995. User's Manual for TMY2s Typical Meteorological Years. National Renewable Energy Laboratory, Colorado.

FIGURE 2.34

Format of TMY-2 file.

A sample of the Nicosia TMY-2 file, showing the data for the first days of January, including the header elements, can be seen in Fig. 2.34 (Kalogirou, 2003). It should be noted that the format of the TMY-2 for the Energy Plus program is a little different than the one shown in Fig. 2.34 since it includes after the header design conditions, extreme periods and holidays, and daylight saving data.

2.4.3 Typical meteorological year, third generation

The changes between TMY-2 and TMY-3 concern the format of the file and minor changes to the algorithm in selecting typical months. The latter concern changes to the persistence criteria implemented, which better accommodates the selection a TMM for periods or records with fewer years. Additionally, the code which prioritized the selection of months with measured solar data was removed. The effects of these changes between the TMY-2 and TMY-3 algorithm were evaluated as part of the TMY-3 production process (Wilcox and Marion, 2008). In particular, a few changes from the TMY-2 procedures were required to accommodate the use of data derived from only a 15-year period. For TMY-2 data, months with measured solar radiation data were preferred for selection as a typical month. TMY-3 procedures do not include this criterion because only modeled solar radiation data are included in the TMY-3 data to provide more consistent solar radiation values. For the TMY-3 data, using only 15 years instead of 30 years to select a candidate month, require that the persistence checks are relaxed to ensure that a candidate month could be selected. For the TMY-2 data, a candidate month is excluded from further consideration if it is the month with the most runs. For TMY-3, a candidate month is only excluded if it has more runs than every other candidate month. Consequently, if two candidate months tie for the most runs, neither is removed by the TMY-3 procedure, whereas the TMY-2 procedure would get rid of both candidate months. As an additional step, if the TMY-3 persistence procedure eliminates all candidate months, persistence was ignored and a month was selected from the candidate months that was closest to the long-term mean and median. This ensured the selection of a typical month for TMY-3 using 15-year or shorter data sets. No TMY for a site was produced however, if the pool of data was less than 10 years (Wilcox and Marion, 2008).

Generally, except for a few changes to the weighting criteria, which account for the relative importance of the solar radiation and meteorological elements, the TMY-2 and TMY-3 data sets were created using procedures similar to those developed by Sandia National Laboratories (Hall et al., 1978). The format for the TMY-3 data however is radically different from that of the TMY and TMY-2. The older TMY formats used columnar or positional layouts to optimize the data storage space. Such formats though are difficult to read, and also it is difficult to import specific data fields into many software packages. Therefore the comma-separated value (CSV) format is adopted in TMY-3 which is ubiquitous, and many existing programs and applications provide built-in functions to read or parse it.

For compatibility with existing software, the National Renewable Energy Laboratory (NREL) has produced an application to convert from TMY-3 to TMY-2 format. Despite the format differences however, the fields in the TMY-3 are very similar to those in the TMY-2 data set. Fundamental differences are measurement units, which are SI or equivalent in the TMY-3, the addition of new fields for surface albedo and liquid precipitation, and the removal of the fields for present weather, snow depth, and days since last snowfall that were present in the TMY-2 (Wilcox and Marion, 2008). These fields were removed because of incompatible changes in the nature of the source data or because the source data were not available for many stations. The TMY-3 data format has two header lines and 8760 lines of data, each with 68 data fields. The header line one contains the same data as the ones shown in Table 2.7 with the difference that the "city" is replaced by the "station name", whereas header line two contains data field name and units. The field format of the rest of the lines is similar to the TMY-2 format shown in Table 2.8 except for the local standard time data fields at the beginning which are changed to "Date" in MM/DD/YYYY format and "Time" in HH:MM format and the other changes at the end of the data fields as noted above.

The volcanic eruptions of El Chichón in Mexico in March 1982 and Mount Pinatubo in the Philippines in June 1991 injected large amounts of aerosols into the stratosphere. The aerosols spread northward and circulated around the earth. This phenomenon noticeably decreased the amount of solar radiation reaching the earth during May 1982 until December 1984 due to El Chichón and from June 1991 to December 1994 due to Pinatubo, after which the effects of the aerosols diminished. Consequently, these months were not used in any of the TMY procedures because they were considered as non-typical (Wilcox and Marion, 2008). The TMY-3 data files for many US locations are available for download from the NREL web site (NREL, 2012).

Exercises

2.1 As an assignment using a spreadsheet program and the relations presented in this chapter, try to create a program that estimates all solar angles according to the latitude, day of year, hour, and slope of surface.

2.2 As an assignment using a spreadsheet program and the relations presented in this chapter, try to create a program that estimates all solar angles according to the latitude, day of year, and slope of surface for all hours of a day.

2.3 Calculate the solar declination for the spring and fall equinoxes and the summer and winter solstices.

2.4 Calculate the sunrise and sunset times and day length for the spring and fall equinoxes and the summer and winter solstices at 45°N latitude and 35°E longitude.

2.5 Determine the solar altitude and azimuth angles at 10:00 a.m. local time for Rome, Italy, on June 10.

2.6 Calculate the solar zenith and azimuth angles, the sunrise and sunset times, and the day length for Cairo, Egypt, at 10:30 a.m. solar time on April 10.

2.7 Calculate the sunrise and sunset times and altitude and azimuth angles for London, England, on March 15 and September 15 at 10:00 a.m. and 3:30 p.m. solar times.

2.8 What is the apparent solar time (AST) in Denver, Colorado, on June 10 at 10:00 a.m. Mountain Standard Time?

2.9 A flat-plate collector in Nicosia, Cyprus, is tilted at 40° from horizontal and pointed 10° east of south. Calculate the solar incidence angle on the collector at 10:30 a.m. and 2:30 p.m. solar times on March 10 and September 10.

2.10 A vertical surface in Athens, Greece, faces 15° west of due south. Calculate the solar incidence angle at 10:00 a.m. and 3:00 p.m. solar times on January 15 and November 10.

2.11 By using the sun path diagram, find the solar altitude and azimuth angles for Athens, Greece, on January 20 at 10:00 a.m.

2.12 Two rows of 6 m wide by 2 m high flat-plate collector arrays tilted at 40° are facing due south. If these collectors are located in 35°N latitude, using the sun path diagram find the months of the year and the hours of day at which the front row will cast a shadow on the second row when the distance between the rows is 3 m. What should be the distance so there will be no shading?

2.13 Find the blackbody spectral emissive power at $\lambda = 8$ μm for a source at 400 K, 1000 K, and 6000 K.

2.14 Assuming that the sun is a blackbody at 5777 K, at what wavelength does the maximum monochromatic emissive power occur? What fraction of energy from this source is in the visible part of the spectrum in the range 0.38−0.78 μm?

2.15 What percentage of blackbody radiation for a source at 323 K is in the wavelength region 6−15 μm?

2.16 A 2-mm thick glass sheet has a refraction index of 1.526 and an extinction coefficient of 0.2 cm^{-1}. Calculate the reflectivity, transmissivity, and absorptivity of the glass sheet at 0°, 20°, 40°, and 60° incidence angles.

2.17 A flat-plate collector has an outer glass cover of 4 mm thick $K = 23$ m^{-1} and refractive index of 1.526, and a Tedlar inner cover with refractive index of 1.45. Calculate the reflectivity, transmissivity, and absorptivity of the complete cover at a 40° incidence angle by considering Tedlar to be of a very small thickness; that is, absorption within the material can be neglected.

2.18 The glass plate of a solar greenhouse has a transmissivity of 0.90 for wavelengths between 0.32 and 2.8 μm and is completely opaque at shorter and longer wavelengths. If the sun is a blackbody radiating energy to the earth's surface at an effective temperature of 5770 K and the interior of the greenhouse is at 300 K, calculate the percent of incident solar radiation transmitted through the glass and the percent of thermal radiation emitted by the interior objects that is transmitted out.

2.19 A 30-m^2 flat-plate solar collector is absorbing radiation at a rate of 900 W/m^2. The environment temperature is 25 °C and the collector emissivity is 0.85. Neglecting conduction and convection losses, calculate the equilibrium temperature of the collector and the net radiation exchange with the surroundings.

2.20 Two large parallel plates are maintained at 500 and 350 K, respectively. The hotter plate has an emissivity of 0.6 and the colder one 0.3. Calculate the net radiation heat transfer between the plates.

2.21 Find the direct normal and horizontal extraterrestrial radiation at 2:00 p.m. solar time on February 21 for 40°N latitude and the total solar radiation on an extraterrestrial horizontal surface for the day.

2.22 Estimate the average hourly diffuse and total solar radiation incident on a horizontal surface for Rome, Italy, on March 10 at 10:00 a.m. and 1:00 p.m. solar times if the monthly average daily total radiation is 18.1 MJ/m^2.

2.23 Calculate the beam and total radiation tilt factors and the beam and total radiation incident on a surface tilted at 45° toward the equator 1 h after local solar noon on April 15. The surface is located at 40°N latitude and the ground reflectance is 0.25. For that day, the beam radiation at normal incidence is $G_B = 710$ W/m^2 and diffuse radiation on the horizontal is $G_D = 250$ W/m^2.

2.24 For a south-facing surface located at 45°N latitude and tilted at 30° from the horizontal, calculate the hourly values of the beam radiation tilt factor on September 10.

2.25 A collector located in Berlin, Germany is tilted at 50° and receives a monthly average daily total radiation $\overline{H}$ equal to 17 MJ/m^2 day. Determine the monthly mean beam and total radiation tilt factors for October per unit area where the ground reflectance is 0.2. Also, estimate the monthly average daily total solar radiation on the surface.

References

ASHRAE, 1975. Procedure for Determining Heating and Cooling Loads for Computerizing Energy Calculations. ASHRAE, Atlanta.

ASHRAE, 2007. Handbook of HVAC Applications. ASHRAE, Atlanta. Note: The material quoted in this reference may also appear in subsequent ASHRAE handbooks, but here the year of the original version of the handbook used is shown.

Collares-Pereira, M., Rabl, A., 1979. The average distribution of solar radiation—correlations between diffuse and hemispherical and between daily and hourly insolation values. Sol. Energy 22 (2), 155−164.

Duffie, J.A., Beckman, W.A., 1991. Solar Engineering of Thermal Processes. John Wiley & Sons, New York.

Dunkle, R.V., 1954. Thermal radiation tables and application. ASME Trans 76, 549.

Erbs, D.G., Klein, S.A., Duffie, J.A., 1982. Estimation of the diffuse radiation fraction four hourly, daily and monthly-average global radiation. Sol. Energy 28 (4), 293−302.

Filkenstein, J.M., Schafer, R.E., 1971. Improved goodness of fit tests. Biometrica 58, 641−645.

Garg, H.P., 1982. Treatise on Solar Energy. Fundamentals of Solar Energy Research, vol. 1. John Wiley & Sons, New York.

Hall, I.J., Prairie, R.R., Anderson, H.E., Boes, E.C., 1978. Generation of typical meteorological years for 26 SOLMET stations. In: Sandia Laboratories Report, SAND 78-1601. Albuquerque, NM.

Hay, J.E., Davies, J.A., 1980. Calculations of the solar radiation incident on an inclined surface. In: Proceedings of the First Canadian Solar Radiation Data Workshop, vol. 59. Ministry of Supply and Services, Canada.

Hottel, H.C., Woertz, B.B., 1942. Evaluation of flat plate solar heat collector. ASME Trans 64, 91.

Hsieh, J.S., 1986. Solar Energy Engineering. Prentice-Hall, Englewood Cliffs, NJ.

Kalogirou, S.A., 2003. Generation of typical meteorological year (TMY-2) for Nicosia, Cyprus. Renew. Energy 28 (15), 2317−2334.

Klein, S.A., Beckman, W.A., Duffie, J.A., 1976. A design procedure for solar heating systems. Sol. Energy 18, 113−127.

Klucher, T.M., 1979. Evaluation of models to predict insolation on tilted surfaces. Sol. Energy 23 (2), 111−114.

Kreith, F., Kreider, J.F., 1978. Principles of Solar Engineering. McGraw-Hill, New York.

Liu, B.Y.H., Jordan, R.C., 1960. The interrelationship and characteristic distribution of direct, diffuse and total solar radiation. Sol. Energy 4 (3), 1−19.

Liu, B.Y.H., Jordan, R.C., 1977. Availability of solar energy for flat plate solar heat collectors. In: Liu, B.Y.H., Jordan, R.C. (Eds.), Application of Solar Energy for Heating and Cooling of Buildings. ASHRAE, Atlanta.

Löf, G.O.G., Tybout, R.A., 1972. Model for Optimizing Solar Heating Design. ASME Paper, 72-WA/SOL-8.

Marion, W., Urban, K., 1995. User's Manual for TMY2s Typical Meteorological Years. National Renewable Energy Laboratory, Colorado.

Meinel, A.B., Meinel, M.P., 1976. Applied Solar Energy—An Introduction. Addison-Wesley, Reading, MA.

Meteonorm, 2009. Maps. Available from: www.meteonorm.com.

NREL, 2012. TMY-3 Data Files. Available from: http://rredc.nrel.gov/solar/old_data/nsrdb/1991-2005/tmy3.

Page, J.K., 1964. The estimation of monthly mean values of daily total short-wave radiation of vertical and in-clined surfaces from sunshine records for latitudes 40°N-40°S. Proc. UN Conf. New Sources Energy 4, 378.

Reindl, D.T., Beckman, W.A., Duffie, J.A., 1990a. Diffuse fraction correlations. Sol. Energy 45 (1), 1—7.

Reindl, D.T., Beckman, W.A., Duffie, J.A., 1990b. Evaluation of hourly tilted surface radiation models. Sol. Energy 45 (1), 9—17.

Siegel, R., Howell, J.R., 2002. Thermal Radiation Heat Transfer, fourth ed. Taylor and Francis, New York.

Solar Spectra, 2007. Air Mass Zero. Available from: http://rredc.nrel.gov/solar/spectra/am0.

Spencer, J.W., 1971. Fourier series representation of the position of the sun. Search 2 (5), 172.

Wilcox, S., Marion, W., 2008. Users Manual for TMY3 Data Sets Technical Report NREL/TP-581-43156 (Revised May 2008).

Solar Energy Collectors

3

Solar energy collectors are special kinds of heat exchangers that transform solar radiation energy to internal energy of the transport medium. The major component of any solar system is the solar collector. This is a device that absorbs the incoming solar radiation, converts it into heat, and transfers the heat to a fluid (usually air, water, or oil) flowing through the collector. The solar energy collected is carried from the circulating fluid either directly to the hot water or space conditioning equipment or to a thermal energy storage tank, from which it can be drawn for use at night or on cloudy days.

There are basically two types of solar collectors: non-concentrating or stationary and concentrating. A non-concentrating collector has the same area for intercepting and absorbing solar radiation, whereas a sun-tracking concentrating solar collector usually has concave reflecting surfaces to intercept and focus the sun's beam radiation to a smaller receiving area, thereby increasing the radiation flux. Concentrating collectors are suitable for high-temperature applications. Solar collectors can also be distinguished by the type of heat transfer liquid used (water, non-freezing liquid, air, or heat transfer oil) and whether they are covered or uncovered. A large number of solar collectors are available on the market. A comprehensive list is shown in Table 3.1 (Kalogirou, 2003).

This chapter reviews the various types of collectors currently available. This is followed by the optical and thermal analysis of collectors.

3.1 Stationary collectors

Solar energy collectors are basically distinguished by their motion—stationary, single-axis tracking, and two-axis tracking—and the operating temperature. First, we'll examine the stationary solar collectors. These collectors are permanently fixed in position and do not track the sun. Three main types of collectors fall into this category:

1. Flat-plate collector (FPC).
2. Stationary compound parabolic collector (CPC).
3. Evacuated tube collector (ETC).

3.1.1 Flat-plate collector (FPC)

A typical flat-plate solar collector is shown in Fig. 3.1. When solar radiation passes through a transparent cover and impinges on the blackened absorber surface of high absorptivity, a large portion of this energy is absorbed by the plate and transferred to the transport medium in the fluid tubes, to be carried away for storage or use. The underside of the absorber plate and the two sides are well insulated

Table 3.1 Solar energy collectors.

Motion	Collector type	Absorber type	Concentration ratio	Indicative temperature range (°C)
Stationary	Flat-plate collector	Flat	1	30–80
	Evacuated tube collector	Flat	1	50–200
	Compound parabolic collector (CPC)	Tubular	1–5	60–240
Single-axis tracking	CPC	Tubular	5–15	60–300
	Linear fresnel reflector	Tubular	10–40	60–250
	Cylindrical trough collector	Tubular	15–50	60–300
	Parabolic trough collector	Tubular	10–85	60–400
Two-axis tracking	Parabolic dish reflector	Point	600–2000	100–1500
	Heliostat field collector	Point	300–1500	150–2000

Note: Concentration ratio is defined as the aperture area divided by the receiver/absorber area of the collector.

to reduce conduction losses. The liquid tubes can be welded to the absorbing plate or they can be an integral part of the plate. The liquid tubes are connected at both ends by large-diameter header tubes. The header and riser collector is the typical design for flat-plate collectors. An alternative is the serpentine design shown in Fig. 3.1a. This collector does not present the potential problem of uneven flow distribution in the various riser tubes of the header and riser design, but serpentine collectors cannot work effectively in thermosiphon mode (natural circulation) and need a pump to circulate the heat transfer fluid (see Chapter 5). The absorber plate can be a single sheet on which all risers are fixed, or each riser can be fixed on a separate fin (see Fig. 3.1b).

The transparent cover is used to reduce convection losses from the absorber plate through the restraint of the stagnant air layer between the absorber plate and the glass. It also reduces radiation losses from the collector because the glass is transparent to the shortwave radiation received by the sun, but it is nearly opaque to longwave thermal radiation emitted by the absorber plate (greenhouse effect).

The advantages of FPCs are that they are inexpensive to manufacture, they collect both beam and diffuse radiation, and they are permanently fixed in position, so no tracking of the sun is required. The collectors should be oriented directly toward the equator, facing south in the Northern Hemisphere and north in the Southern Hemisphere. The optimum tilt angle of the collector is equal to the latitude of the location, with angle variations of $10°-15°$ more or less, depending on the application (Kalogirou, 2003). If the application is solar cooling, then the optimum angle is latitude $-10°$ so that the sun will be perpendicular to the collector during summertime, when the energy will be mostly required. If the

FIGURE 3.1

Typical flat-plate collector. (a) Pictorial view of a flat-plate collector. (b) Photograph of a cut header-and-riser flat-plate collector.

application is space heating, then the optimal angle is latitude $+10°$; whereas for annual hot water production, it is latitude $+5°$, to have relatively better performance during wintertime, when hot water is mostly required.

The main components of an FPC, as shown in Fig. 3.2, are the following:

- *Cover.* One or more sheets of glass or other radiation-transmitting material.
- *Heat removal fluid passageways.* Tubes, fins, or passages that conduct or direct the heat transfer fluid from the inlet to the outlet.
- *Absorber plate.* Flat, corrugated, or grooved plates, to which the tubes, fins, or passages are attached. A typical attachment method is the embedded fixing shown in the detail of Fig. 3.2. The plate is usually coated with a high-absorptance low-emittance layer.
- *Headers or manifolds.* Pipes and ducts to admit and discharge the fluid.
- *Insulation.* Used to minimize the heat loss from the back and sides of the collector.

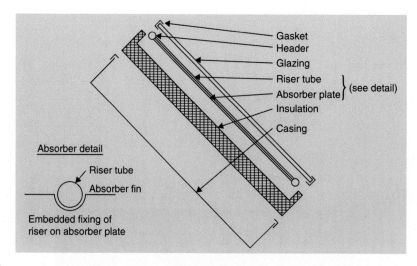

FIGURE 3.2

Exploded view of a flat-plate collector and absorber details.

- *Container.* The casing surrounds the aforementioned components and protects them from dust, moisture, and any other material.

Flat-plate collectors have been built in a wide variety of designs and from many different materials. They have been used to heat fluids such as water, water plus antifreeze additive, or air. Their major purpose is to collect as much solar energy as possible at the lowest possible total cost. The collector should also have a long effective life, despite the adverse effects of the sun's ultraviolet radiation and corrosion and clogging because of acidity, alkalinity, or hardness of the heat transfer fluid, freezing of water, or deposition of dust or moisture on the glazing and breakage of the glazing from thermal expansion, hail, vandalism, or other causes. These causes can be minimized by the use of tempered glass.

In the following two sections, more details are given about the glazing and absorber plate materials. Most of these details also apply to other types of collectors. A third section refers to the collector construction and types of absorber configurations used.

Glazing materials

Glass has been widely used to glaze solar collectors because it can transmit as much as 90% of the incoming shortwave solar irradiation while transmitting virtually none of the longwave radiation emitted outward by the absorber plate (see Example 2.10). Window glass usually has high iron content and is not suitable for use in solar collectors. Glass with low iron content has a relatively high transmittance for solar radiation (approximately 0.85–0.90 at normal incidence), but its transmittance is essentially zero for the longwave thermal radiation (5.0–50 μm) emitted by sun-heated surfaces.

Plastic films and sheets also possess high shortwave transmittance, but because most useable varieties also have transmission bands in the middle of the thermal radiation spectrum, they may have longwave transmittances as high as 0.40. Additionally, plastics are generally limited in the temperatures they can sustain without deteriorating or undergoing dimensional changes. Only a few types of plastics can withstand the sun's ultraviolet radiation for long periods. However, they are not broken by hail or stones, and in the form of thin films, they are completely flexible and have low mass.

The commercially available grades of window and greenhouse glass have normal incidence transmittances of about 0.87 and 0.85, respectively (ASHRAE, 2007). For direct radiation, the transmittance varies considerably with the angle of incidence.

Antireflective coatings and surface texture can improve transmission significantly. The effect of dirt and dust on collector glazing may be quite small, and the cleansing effect of an occasional rainfall is usually adequate to maintain the transmittance within 2%–4% of its maximum value. Dust is collected mostly during summertime, when rainfall is less frequent, but due to the high magnitude of solar irradiation during this period, the dust protects the collector from overheating.

The glazing should admit as much solar irradiation as possible and reduce the upward loss of heat as much as possible. Although glass is virtually opaque to the longwave radiation emitted by the collector plates, absorption of that radiation causes an increase in the glass temperature and a loss of heat to the surrounding atmosphere through radiation and convection. These effects are analyzed in Section 3.3.

Various prototypes of transparently insulated FPCs and CPCs (see Section 3.1.2) were built and tested in the 1990s (Spate et al., 1999). Low-cost, high-temperature resistant transparent insulating materials have been developed so that the commercialization of these collectors becomes feasible. A prototype FPC covered by transparent insulation was developed and tested by Benz et al. (1998), and the efficiency of the collector was proven comparable with that of ETCs (see Section 3.1.3). However, no commercial collectors of this type are available on the market yet.

Collector absorbing plates

The collector plate absorbs as much of the irradiation as possible through the glazing, while losing as little heat as possible upward to the atmosphere and downward through the back of the casing. The collector plates transfer the retained heat to the transport fluid. To maximize the energy collection, the absorber of a collector should have a coating that has high absorptance for solar radiation (short wavelength) and a low emittance for re-radiation (long wavelength). Such a surface is referred to as a *selective surface*. The absorptance of the collector surface for shortwave solar radiation depends on the nature and color of the coating and on the incident angle. Usually black color is used, but various color coatings have been proposed by Tripanagnostopoulos et al. (2000a;b), Wazwaz et al. (2002), and Orel et al. (2002), mainly for aesthetic reasons.

By suitable electrolytic or chemical treatment, surfaces can be produced with high values of solar radiation absorptance (α) and low values of longwave emittance (ε). Essentially, typical selective surfaces consist of a thin upper layer, which is highly absorbent to shortwave solar radiation but relatively transparent to longwave thermal radiation, deposited on a surface that has a high reflectance and low emittance for longwave radiation. Selective surfaces are particularly important when the collector surface temperature is much higher than the ambient air temperature. The cheapest absorber coating is matte black paint; however, this is not selective, and the performance of a collector produced in this way is low, suitable for operating temperatures not more than 40 °C above ambient.

Many methods and materials or material combinations have been used to obtain the desired property of spectral selectivity. The various selective absorbers can be divided into the following categories:

1. Intrinsic materials or mass absorbers.
2. Tandem stacks and inverse tandem stacks.
3. Multilayer stacks (interference stacks).
4. Metal particles in a dielectric or metal matrix (cermets).
5. Surface roughness.
6. Quantum size effects (QSEs).

A comprehensive review of these absorbers is given by Yianoulis et al. (2012).

An energy-efficient solar collector should absorb incident solar radiation, convert it to thermal energy, and deliver the thermal energy to a heat transfer medium with minimum losses at each step. It is possible to use several design principles and physical mechanisms to create a selective solar-absorbing surface. Solar absorbers, referred to as *tandem absorbers*, are based on two layers with different optical properties. A semiconducting or dielectric coating with high solar absorptance and high infrared transmittance on top of a non-selective highly reflecting material such as metal constitutes one type of tandem absorber. Another alternative is to coat a non-selective highly absorbing material with a heat mirror that has a high solar transmittance and high infrared reflectance (Wackelgard et al., 2001).

Today, commercial solar absorbers are made by electroplating, anodization, evaporation, sputtering, and applying solar selective paints. Of the many types of selective coatings developed, the most widely used is black chrome. Much of the progress in recent years has been based on the implementation of vacuum techniques for the production of fin-type absorbers used in low-temperature applications. The chemical and electrochemical processes used for their commercialization were readily taken over from the metal finishing industry. The requirements of solar absorbers used in high-temperature applications, however—namely, extremely low thermal emittance and high-temperature stability—were difficult to fulfill with conventional wet processes. Therefore, large-scale sputter deposition was developed in the late 1970s. Nowadays, the vacuum techniques are mature, are characterized by low cost, and have the advantage of being less environmentally polluting than the wet processes.

When the roughness of a surface is smaller than the wavelength of light impinging on the surface it behaves like a mirror, and when it is larger it strongly absorbs the light of smaller wavelengths. High solar absorptance is enhanced by multiple reflections between pyramidal, dendrite, or porous microstructure. Sometimes the materials with this property are called *wave front discriminating materials*. Some surfaces can be made to appear rough to obtain spectral selectivity by optical trapping of solar radiation. Such a structure can be produced by etching with a proper acid, a procedure called surface texturing. Properly textured surfaces appear rough and absorb solar energy while appearing highly reflective and mirror-like to thermal energy. The orientation of a textured surface affects its optical properties and can improve the absorption and emissivity of a spectrally selective material.

Quantum size effects (QSEs) can be utilized to achieve high absorption in the short wavelength region while maintaining a low thermal IR emittance. These occur in ultrathin films and dots. The critical thickness for the QSE in a metal film is 2–3 nm, and for a degenerate semiconductor, 10–50 nm.

Collector construction

For fluid-heating collectors, passages must be integral with or firmly bonded to the absorber plate. A major problem is obtaining a good thermal bond between tubes and absorber plates without incurring excessive costs for labor or materials. The materials most used for collector plates are copper, aluminum, and stainless steel. UV-resistant plastic extrusions are used for low-temperature applications. If the entire collector area is in contact with the heat transfer fluid, the thermal conductance of the material is not important. The convective heat loss in a collector is relatively insensitive to the spacing between the absorber and the cover in the range of 15—40 mm. The back insulation of a flat-plate collector is made from fiberglass or a mineral fiber mat that will not outgas at elevated temperatures. Building-grade fiberglass is unsatisfactory because the binders evaporate at high temperature and then condense on the collector cover, blocking incoming solar radiation.

Figure 3.3 shows a number of absorber plate designs for solar water and air heaters that have been used with varying degrees of success. Figure 3.3a shows a bonded sheet design, in which the fluid passages are integral to the plate to ensure good thermal conduct between the metal and the fluid. Figure 3.3b and c show fluid heaters with tubes soldered, brazed, or otherwise fastened to upper or lower surfaces of sheets or strips of copper (see also the detail in Fig. 3.2). Copper tubes are used most often because of their superior resistance to corrosion and high thermal conductivity.

Thermal cement, clips, clamps, or twisted wires have been tried in the search for low-cost bonding methods. Figure 3.3d shows the use of extruded rectangular tubing to obtain a larger heat transfer area between tube and plate. Mechanical pressure, thermal cement, or brazing may be used to make the assembly. Soft solder must be avoided because of the high plate temperature encountered at stagnation conditions, which could melt the solder.

The major difference between air- and water-based collectors is the need to design an absorber that overcomes the heat transfer penalty caused by lower heat transfer coefficients between air and the solar absorber. Air or other gases can be heated with flat-plate collectors, particularly if some type of extended surface (Fig. 3.3e) is used to counteract the low heat transfer coefficients between metal and air (Kreider, 1982). Metal or fabric matrices (Fig. 3.3f) (Kreider and Kreith, 1977; Kreider, 1982), thin corrugated metal sheets (Fig. 3.3g), or porous absorbers may be used, with selective surfaces applied to the latter when a high level of performance is required. The principal requirement of these designs is a large contact area between the absorbing surface and the air. The thermal capacity of air is much lower than water, hence larger volume flow rates of air are required, resulting in higher pumping power.

Another type of air collector, shown in Fig. 3.3h, is the transpired. Transpired air collectors are quite simple structures used for heating purposes in buildings. This collector consists of a perforated blackened metal sheet which is placed at close distance in front of a south facing building wall. A fan forces ambient air to pass through the perforation holes, which is thus heated and distributed inside the building, as shown in Fig. 3.3h.

Reduction of heat loss from the absorber can be accomplished either by a selective surface to reduce radiative heat transfer or by suppressing convection. Francia (1961) showed that a honeycomb made of transparent material, placed in the airspace between the glazing and the absorber, was beneficial.

Another category of collectors, not shown in Fig. 3.3, is the uncovered or unglazed solar collector, called a WISC—an abbreviation for "wind and infrared sensitive collector"—as the original names were misleading and not consequent in all technical cases. These are usually low-cost units that can offer cost-effective solar thermal energy in applications such as water preheating for domestic or industrial use, heating of swimming pools (Molineaux et al., 1994), space heating, and air heating for industrial or agricultural applications. Generally, since their heat losses increase strongly with the temperature difference between the absorber and the ambient, these collectors are used where the operating temperature of the collector is close to ambient, typically between 10 and 20 K. These collectors, usually called *panel collectors*, consist of a wide absorber sheet, made of plastic, containing

FIGURE 3.3

Various types of flat-plate solar collector absorber configurations for water and air.

closed-spaced fluid passages (see Fig. 3.4). Materials used for plastic panel collectors include polypropylene, polyethylene, acrylic, and polycarbonate.

The simplest solar thermal collector consists only of an absorber with channels for the heat transfer fluid integrated into the absorber. The absorber is made of a dark material for high absorption of solar radiation. Uncovered solar collectors are used mainly for swimming pool heating, preheating, and in combination with heat pumps only or as air-heating collectors for direct space-heating purposes.

Fluid passages

Header

FIGURE 3.4

Photograph of a plastic collector absorber plate.

Nowadays, the most common application of uncovered collectors is for pool heating and as a heat source for heat pumps. They are generally made of heavy-duty rubber or plastic treated with a UV light inhibitor to extend the life of the panels. The absorber consists either of many joined parallel polymer pipes or of extruded mats with channels made of a rubber compound. The joined pipes or mats are cut to length, and a manifold is added, which connects the ends into a supply and removal header pipe. Swimming pool water often flows directly through the channels (fluid passages in Fig. 3.4), picks up heat from the mat, and flows back into the pool.

In some cases, uncovered collectors are made of stainless steel where two sheets are bonded together, like a sandwich panel, with the heat transfer fluid flowing between them through expanded or impressed structures.

Another type of uncovered metal absorber is made from zinc plates, which can be used as a roofing material. In this case, the heat removal pipes are connected on the back side of the zinc plates. Uncovered metal collectors can also be used with the compression fluid of a heat pump. In some cases, the collector itself can function as the roof cover material at the same time, which means that no other roofing material is required with obvious economic benefits.

Flat-plate collectors are by far the most-used type of collector. Flat-plate collectors are usually employed for low-temperature applications, up to 80 °C, although some new types of collectors employing vacuum insulation or transparent insulation (TI) can achieve slightly higher values (Benz et al., 1998). Due to the introduction of highly selective coatings, actual standard flat-plate collectors can reach stagnation temperatures of more than 200 °C. With these collectors good efficiencies can be obtained up to temperatures of about 100 °C.

Lately some modern manufacturing techniques such as the use of laser and ultrasonic welding machines have been introduced by the industry that improve both the speed and the quality of welds. Both are used for welding of fins on risers, to improve heat conduction. The greatest advantage of the ultrasonic welding method is that the welding is performed at room temperature; therefore, deformation of the welded parts is avoided. However, this technique leaves a line across the absorber (at the welding point), which diminishes slightly the blackened collecting area. Laser welding provides a good seal between the absorber and the tubes without having the weak line associated with ultrasonic welding.

3.1.2 Compound parabolic collector (CPC)

Compound parabolic collectors (CPCs) are non-imaging concentrators. They have the capability of reflecting to the absorber all of the incident radiation within wide limits. Their potential as collectors of solar energy was pointed out by Winston (1974). The necessity of moving the concentrator to accommodate the changing solar orientation can be reduced by using a trough with two sections of a parabola facing each other, as shown in Fig. 3.5.

Compound parabolic concentrators can accept incoming radiation over a relatively wide range of angles. By using multiple internal reflections, any radiation entering the aperture within the collector acceptance angle finds its way to the absorber surface located at the bottom of the collector. The absorber can take a variety of configurations. It can be flat, bifacial, wedge, or cylindrical, as shown in Fig. 3.5. Details on the collector shape construction are presented in Section 3.6.1.

Two basic types of CPC collectors have been designed: symmetric and asymmetric. CPCs usually employ two main types of absorbers: the fin type with a pipe and tubular absorbers. The fin type can be flat, bifacial, or wedge, as shown in Fig. 3.5 for the symmetric type, and can be single channel or multichannel.

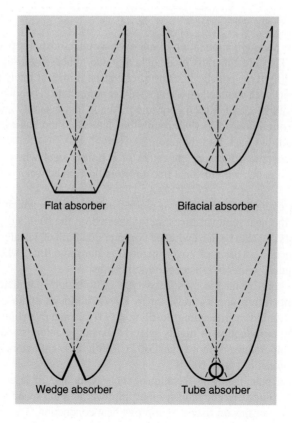

FIGURE 3.5

Various absorber types of compound parabolic collectors.

CPCs should have a gap between the receiver and the reflector to prevent the reflector from acting as a fin conducting heat away from the absorber. Because the gap results in a loss of reflector area and a corresponding loss of performance, it should be kept small. This is more important for flat receivers.

For higher temperature applications a tracking CPC can be used. When tracking is used, this is very rough or intermittent, since the concentration ratio is usually small and radiation can be collected and concentrated by one or more reflections on the parabolic surfaces.

CPCs can be manufactured either as one unit with one opening and one receiver (see Fig. 3.5) or as a panel (see Fig. 3.6a). When constructed as a panel, the collector looks like an FPC, as shown in Fig. 3.6b.

Another category of CPC collectors is the asymmetric type. This can be combined with a reverse or upside down absorber plate configuration. Following the initial investigations on this type of system presented by Kienzlen et al. (1988), in these configurations radiation is directed on the underside of the plate by a stationary concentrator of the CPC shape shown in Fig. 3.7a. In this way, heat losses from the absorber are significantly reduced because the upper side of the plate is well insulated and the convective losses are reduced because the convective current is blocked by the plate itself. Another configuration is the inclined design shown in Fig. 3.7b. Compared with a flat-plate collector, these designs have lower optical efficiency due to the scattering losses in the reflector but better efficiency at higher temperatures.

FIGURE 3.6

Panel CPC collector with cylindrical absorbers. (a) Schematic diagram. (b) Photo of a CPC panel collector installation.

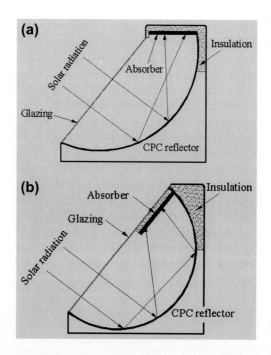

FIGURE 3.7

(a) Inverted flat-plate collector. (b) Inclined flat-plate collector.

A variation of this configuration is the double-sided CPC-type flat-plate collector investigated by Goetzberger et al. (1992) and Tripanagnostopoulos et al. (2000a;b). These are also called bifacial solar flat-plate collectors because the absorber is irradiated at both sides of the absorber. In the design presented by Goetzberger et al. (1992) the absorber is "insulated" in all sides with a transparent insulation whereas in the design presented by Tripanagnostopoulos et al. (2000a;b) a simple glazing was used either in one CPC mirror–absorber unit or in three CPC mirror–absorber units as shown in Fig. 3.8a and b, respectively.

FIGURE 3.8

Cross-section of (a) compound parabolic collector (CPC) with one mirror–absorber unit and (b) CPC with three mirror–absorber units.

3.1.3 Evacuated tube collector (ETC)

Conventional simple flat-plate solar collectors were developed for use in sunny warm climates. Their benefits, however, are greatly reduced when conditions become unfavorable during cold, cloudy, and windy days. Furthermore, weathering influences, such as condensation and moisture, cause early deterioration of internal materials, resulting in reduced performance and system failure. Evacuated heat pipe solar collectors (tubes) operate differently than the other collectors available on the market. These solar collectors consist of a heat pipe inside a vacuum-sealed tube, as shown in Fig. 3.9. In an actual installation, many tubes are connected to the same manifold as shown in Fig. 3.10.

Evacuated tube collectors have shown that the combination of a selective surface and an effective convection suppressor can result in good performance at high temperatures. The vacuum envelope reduces convection and conduction losses, so the collectors can operate at higher temperatures than FPCs. Like FPCs, they collect both direct and diffuse radiation, but their efficiency is higher at low incidence angles. This gives ETCs an advantage over FPCs in terms of daylong performance.

To guarantee tightness over 20 or so years is a huge challenge. However, circular glass tubes are resistant to pressure due to their edgeless shape, and for glass-glass-sealed tubes, the vacuum tightness is very high and reliable over time.

ETCs use liquid–vapor phase change materials to transfer heat at high efficiency. These collectors feature a heat pipe (a highly efficient thermal conductor) placed inside a vacuum-sealed tube. The pipe, which is a sealed copper pipe, is then attached to a black copper fin that fills the tube (absorber plate). Protruding from the top of each tube is a metal tip attached to the sealed pipe (condenser). The heat pipe contains a small amount of fluid (e.g., methanol or ethanol) that undergoes an evaporating–condensing cycle. In this cycle, solar heat evaporates the liquid and the vapor travels to the heat sink region, where it condenses and releases its latent heat. The condensed fluid returns to the solar

FIGURE 3.9

Schematic diagram of an evacuated tube collector.

FIGURE 3.10

Actual evacuated tube collector installation.

collector and the process is repeated. When these tubes are mounted, the metal tip projects into a heat exchanger (manifold), as shown in Fig. 3.9. Water or glycol flows through the manifold and picks up the heat from the tubes. The heated liquid circulates through another heat exchanger and gives off its heat to a process or water stored in a solar storage tank. Another possibility is to use the ETC connected directly to a hot water storage tank.

Because no evaporation or condensation above the phase change temperature is possible, the heat pipe offers inherent protection from freezing and overheating. This self-limiting temperature control is a unique feature of the evacuated heat pipe collector.

Evacuated tube collectors consist of a heat pipe inside a vacuum-sealed tube. The characteristics of a typical collector are shown in Table 3.2. ETCs on the market exhibit many variations in absorber shape. Evacuated tubes with CPC reflectors are also commercialized by several manufacturers. One such design, presented recently in an attempt to reduce cost and increase lifetime, shown in Fig. 3.11, consists of an all-glass Dewar-type ETC. This uses two concentric glass tubes, and the space in between the tubes is evacuated, creating a vacuum jacket. In this type of ETC, also called the Sidney type, the selective coating is deposited onto the outside surface of a glass tube domed at one end. This tube is then inserted into a second, larger diameter domed glass tube and the tubes are joined at the open end. The advantage of this design is that it is made entirely of glass and it is not necessary to penetrate the glass envelope to extract heat from the tube, eliminating leakage losses and keeping it cheaper than

Table 3.2 Characteristics of typical evacuated tube collector system.

Parameter	Value
Glass tube diameter	65 mm
Glass thickness	1.6 mm
Collector length	1965 mm
Absorber plate material	Copper
Coating	Selective
Absorber area	0.1 m^2

FIGURE 3.11

All-glass dewar-type evacuated tube collector.

the single-envelope system. However, these are suitable only for low-pressure systems and have the disadvantages that the tubes cannot be drained; if one tube breaks, all the working fluid may be lost (Morrison, 2001). This is also called a *wet tube ETC*. A variation of the wet tube ETC is a normal single-glass ETC in which water (or any other fluid) flows through the collector in either a U tube or a coaxial pipe.

Because a glass cylinder may break, for example, due to hail or its vacuum may fail over time, the vacuum glass tubes are usually removable and replaceable from the collector assembly. Often, evacuated tubes contain a barium sulfide-based "getter," as shown in Fig. 3.11, which helps to keep the vacuum stable through barium reacting with nitrogen gas chemically binding it in solid form. A resulting white area around the getter indicates that the vacuum is lost.

As evacuated tube collectors are relatively expensive, the cost-effectiveness of these collectors can be improved by reducing the number of tubes and using reflectors to concentrate the solar radiation onto the tubes. A diffuse reflector (reflectivity, $\rho = 0.6$) mounted behind the tubes, spaced one tube diameter apart, as shown in Fig. 3.12a, increases the absorbed energy in each tube by more than 25% for normal incidence. This system also presents a 10% increase in energy collection over a full day because of incidence angle effects. Greater enhancement per tube can be achieved by using CPC-type reflectors, as shown in Fig. 3.12b. Evacuated tube arrays with stationary concentrators may have stagnation temperatures exceeding 300 °C.

Another type of collector developed recently is the integrated compound parabolic collector (ICPC). This is an ETC in which, at the bottom part of the glass tube, a reflective material is fixed (Winston et al., 1999). In this case, either a CPC reflector, Fig. 3.13a, or a cylindrical reflector, Fig. 3.13b, is used. The latter does not achieve the concentration of the shaped reflector but has a very low manufacturing cost. In this way, the collector combines into a single unit the advantages of vacuum insulation and non-imaging stationary concentration. In another design, a tracking ICPC is developed that is suitable for high-temperature applications (Grass et al., 2000).

FIGURE 3.12

Evacuated tube collectors array with reflectors. (a) Flat diffuse reflector. (b) Compound parabolic collector reflector.

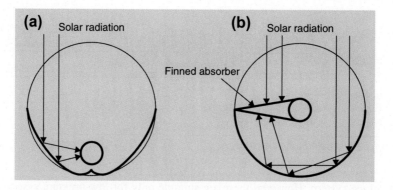

FIGURE 3.13

Integrated compound parabolic collector tubes. (a) Internal compound parabolic. (b) Circular reflector with finned absorber.

Evacuated tube collectors are produced in a variety of sizes, with outer diameters ranging from 30 to about 100 mm. The usual length of these collectors is about 2 m.

3.1.4 Outlook of the technology

Solar thermal collector technologies are already mature. Collectors for hot water heating, which are well-performing, affordable, durable, and of high quality, are on the market. However, there is still room for further improvements for specific application areas and new concepts through research and development in solar thermal technology. In the near future, the following developments are expected:

- Reduction of costs by introducing new materials and production methods.
- Extension of the collector technology to higher temperature ranges up to 250 °C for processes like industrial process heat.
- Integration of solar thermal collectors into the building envelope.
- Development of cheaper collectors for lower temperature applications.
- Development of multifunctional collectors like photovoltaic–thermal (PV/T) hybrid collectors to produce electricity and thermal energy simultaneously (see Section 9.8 in Chapter 9).
- Further improvement of air-heating collectors for the optimization and integration into air-heating and ventilation systems.

- Further development of large-area collectors to cover entire roofs of residential buildings, preferably at prefabricated designs.
- Development of multifunctional solar thermal façade elements which can provide adjusted daylighting, insulation, ventilation, and active solar thermal collection. Such systems can also be used for the refurbishment of existing buildings.

The last one is a new challenge to combine structural and functional building elements into one new construction as this often generates contradicting requirements. For this purpose, new, cheaper materials and constructions are required, which should be weather resistant over the required lifetime of the element, ideally equal to the life of the building, which is a huge design challenge.

3.2 Sun-tracking concentrating collectors

Energy delivery temperatures can be increased by decreasing the area from which the heat losses occur. Temperatures far above those attainable by FPCs can be reached if a large amount of solar radiation is concentrated on a relatively small collection area. This is done by interposing an optical device between the source of radiation and the energy-absorbing surface. Concentrating collectors exhibit certain advantages over the conventional flat-plate type (Kalogirou et al., 1994a). The main advantages are as follows:

1. The working fluid can achieve higher temperatures in a concentrator system than a flat-plate system of the same solar energy-collecting surface. This means that a higher thermodynamic efficiency can be achieved.
2. It is possible with a concentrator system to achieve a thermodynamic match between temperature level and task. The task may be to operate thermionic, thermodynamic, or other higher temperature devices.
3. The thermal efficiency is greater because of the small heat loss area relative to the receiver area.
4. Reflecting surfaces require less material and are structurally simpler than FPCs. For a concentrating collector, the cost per unit area of the solar-collecting surface is therefore less than that of a flat-plate collector.
5. Owing to the relatively small area of receiver per unit of collected solar energy, selective surface treatment and vacuum insulation to reduce heat losses and improve the collector efficiency are economically viable.

Their disadvantages are as follows:

1. Concentrator systems collect little or no diffuse radiation, depending on the concentration ratio.
2. Some form of tracking system is required to enable the collector to follow the sun.
3. Solar reflecting surfaces may lose their reflectance with time and may require periodic cleaning and refurbishing.

Many designs have been considered for concentrating collectors. Concentrators can be reflectors or refractors, can be cylindrical or parabolic, and can be continuous or segmented. Receivers can be convex, flat, cylindrical, or concave and can be covered with glazing or uncovered. Concentration ratios, that is, the ratio of aperture to absorber areas, can vary over several orders of magnitude, from as low as slightly above unity to high values on the order of 10,000. Increased ratios mean increased temperatures at which energy can be delivered, but consequently, these collectors have increased requirements for precision in optical quality and positioning of the optical system.

Because of the apparent movement of the sun across the sky, conventional concentrating collectors must follow the sun's daily motion. The sun's motion can be readily tracked by two methods. The first is the altazimuth method, which requires the tracking device to turn in both altitude and azimuth, that is, when performed properly, this method enables the concentrator to follow the sun exactly. Paraboloidal solar collectors generally use this system. The second one is one-axis tracking, in which the collector tracks the sun in only one direction, either from east to west or north to south. Parabolic trough collectors (PTCs) generally use this system. These systems require continuous and accurate adjustment to compensate for the changes in the sun's orientation. Relations on how to estimate the angle of incidence of solar radiation and the slope of the collector surface for these tracking modes are given in Chapter 2, Section 2.2.1.

The first type of solar concentrator, shown in Fig. 3.14, is effectively a flat-plate collector fitted with simple flat reflectors, which can markedly increase the amount of direct radiation reaching the collector. This is, in fact, a concentrator because the aperture is bigger than the absorber but the system is stationary. A comprehensive analysis and model of such a system is presented by Garg and Hrishikesan (1998). The model facilitates the prediction of the total energy absorbed by the collector at any hour of the day for any latitude for random tilt angles and azimuth angles of the collector and reflectors. This simple enhancement of FPCs was initially suggested by Tabor (1966).

FPCs can be equipped with flat reflectors, as shown in Fig. 3.14 or with the saw-toothed arrangement shown in Fig. 3.15, which is suitable for multi-row collector installations. In both cases, the simple flat diffuse reflectors can significantly increase the amount of direct radiation reaching the collector. The term *diffuse reflector* denotes a material that is not a mirror, avoiding the formation of an image of the sun on the absorber, which creates uneven radiation distribution and thermal stresses.

Another type of collector, the CPC, already covered under the stationary collectors, is also classified as concentrator. This can be stationary or tracking, depending on the acceptance angle. When tracking is used, this is very rough or intermittent, since the concentration ratio is usually small and radiation can be collected and concentrated by one or more reflections on the parabolic surfaces.

As was seen previously, one disadvantage of concentrating collectors is that, except at low concentration ratios, they can use only the direct component of solar radiation, because the diffuse component cannot be concentrated by most types. However, an additional advantage of concentrating

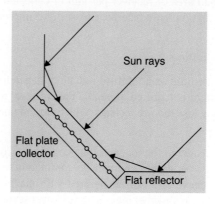

FIGURE 3.14

Flat-plate collector with flat reflectors.

FIGURE 3.15

Flat-plate collectors with saw-toothed reflectors.

collectors is that, in summer, when the sun rises well to the north of the east–west line, the sun follower, with its axis oriented north–south, can begin to accept radiation directly from the sun long before a fixed south-facing flat-plate collector can receive anything other than diffuse radiation from the portion of the sky that it faces. Thus, in relatively cloudless areas, the concentrating collector may capture more radiation per unit of aperture area than a flat-plate collector.

In concentrating collectors solar energy is optically concentrated before being transferred into heat. Concentration can be obtained by reflection or refraction of solar radiation by the use of mirrors or lenses. The reflected or refracted light is concentrated in a focal zone, thus increasing the energy flux in the receiving target. Concentrating collectors can also be classified into non-imaging and imaging, depending on whether the image of the sun is focused at the receiver. The concentrator belonging in the first category is the CPC, whereas all the other types of concentrators belong to the imaging type. The collectors falling into this category are:

1. Parabolic trough collector (PTC).
2. Linear Fresnel reflector (LFR).
3. Parabolic dish reflector (PDR).
4. Heliostat field collector (HFC).

3.2.1 Parabolic trough collector (PTC)

To deliver high temperatures with good efficiency a high-performance solar collector is required. Systems with light structures and low-cost technology for process heat applications up to 400 °C could be obtained with PTCs. PTCs can effectively produce heat at temperatures between 50 and 400 °C.

Parabolic trough collectors are made by bending a sheet of reflective material into a parabolic shape. A black metal tube, covered with a glass tube to reduce heat losses, is placed along the focal line of the receiver (see Fig. 3.16). When the parabola is pointed toward the sun, parallel rays incident on the reflector are reflected onto the receiver tube. The concentrated radiation reaching the receiver tube heats the fluid that circulates through it, transforming the solar radiation into useful heat. It suffices to use a single-axis tracking of the sun; therefore, long collector modules are produced. The collector can be oriented in an east–west direction, tracking the sun from north to south, or in a north–south

FIGURE 3.16

Schematic of a parabolic trough collector.

direction, tracking the sun from east to west. The advantages of the former tracking mode is that little collector adjustment is required during the day and the full aperture always faces the sun at noon but the collector performance during the early and late hours of the day is greatly reduced, due to large incidence angles (cosine loss). North–south oriented troughs have their highest cosine loss at noon and the lowest in the mornings and evenings, when the sun is due east or due west. Photographs of PTC collectors are shown in Fig. 3.17.

Over a period of 1 year, a horizontal north–south trough field usually collects slightly more energy than a horizontal east–west one. However, the north–south field collects a lot of energy in summer and much less in winter (see Chapter 2, Section 2.2.1). The east–west field collects more energy in winter than a north–south field and less in summer, providing a more constant annual output. Therefore, the choice of orientation usually depends on the application and whether more energy is needed during summer or winter.

Parabolic trough technology is the most advanced of the solar thermal technologies because of considerable experience with the systems and the development of a commercial industry to produce and market these systems. PTCs are built-in modules that are supported from the ground by simple pedestals at either end.

Parabolic trough collectors are the most mature solar technology to generate heat at temperatures up to 400 °C for solar thermal electricity generation or process heat applications. The biggest

FIGURE 3.17

Photos of actual parabolic trough collectors. (a) EuroTrough. (b) An industrial solar technology collector.

(A) From http://www.sbp.de/en#sun/show/1043-EuroTrough_Collector.

application of this type of system is the southern California power plants known as Solar Electric Generating Systems (SEGS), which have a total installed capacity of 354 MWe (Kearney and Price, 1992). SEGS I is 14 MWe, SEGS II−VII are 30 MWe each, and SEGS VIII and IX are 80 MWe each. Three collector designs have been used in these plants: LS-1 for SEGS I, LS-2 for SEGS II−VII, and LS-3 for part of SEGS VII, VIII, and IX. More details on this system are given in Chapter 10. Another important application of this type of collector is installed at Plataforma Solar de Almeria in southern Spain, mainly for experimental purposes, with a total installed capacity of the PTCs equal to 1.2 MW.

The receiver of a parabolic trough is linear. Usually, a tube is placed along the focal line to form an external surface receiver (see Fig. 3.16). The size of the tube, and therefore the concentration ratio, is determined by the size of the reflected sun image and the manufacturing tolerances of the trough. The surface of the receiver is typically plated with a selective coating that has a high absorptance for solar radiation but a low emittance for thermal radiation loss.

A glass cover tube is usually placed around the receiver tube to reduce the convective heat loss from the receiver, thereby further reducing the heat loss coefficient. A disadvantage of the glass cover tube is that the reflected light from the concentrator must pass through the glass to reach the absorber, adding a transmittance of about 0.9 (10% loss) when the glass is clean. The glass envelope usually has an antireflective coating to improve transmissivity. One way to further reduce convective heat loss from the receiver tube and thereby increase the performance of the collector, particularly for high-temperature applications, is to evacuate the space between the glass cover tube and the receiver. The total receiver tube length of PTCs is usually from 25 to 150 m.

New developments in the field of PTCs aim at cost reduction and improvements in technology. In one system, the collector mirrors can be washed automatically, drastically reducing the maintenance cost.

After a period of research and commercial development of the PTCs in the 1980s a number of companies entered the field, producing this type of collector for the temperature range between 50 and 300 °C, all of them with one-axis tracking. One such example is the solar collector produced by the Industrial Solar Technology (IST) Corporation. IST erected several process heat installations in the United States that by the end of the last century were up to 2700 m^2 of collector aperture area (Kruger et al., 2000).

The IST parabolic trough has been thoroughly tested and evaluated at the Sandia National Laboratory (Dudley, 1995) and the German Aerospace Center (DLR) (Kruger et al., 2000) for efficiency and durability.

The characteristics of the IST collector system are shown in Table 3.3.

Parabola construction

To achieve cost-effectiveness in mass production, the collector structure must feature not only a high stiffness-to-weight ratio, to keep the material content to a minimum, but also be amenable to low-labor manufacturing processes. A number of structural concepts have been proposed, such as steel framework structures with central torque tubes or double V trusses and fiberglass (Kalogirou et al., 1994b). A recent development in this type of collectors is the design and manufacture of the EuroTrough, a new PTC, in which an advanced lightweight structure is used to achieve cost-efficient solar power generation (Lupfert et al., 2000; Geyer et al., 2002). Based on environmental test data to date, mirrored glass appears to be the preferred mirror material, although self-adhesive reflective materials with lifetimes of 5−7 years exist in the market.

Table 3.3 Characteristics of industrial solar technology parabolic trough collector system.

Parameter	Value/Type
Collector rim angle	70°
Reflective surface	Silvered acrylic
Receiver material	Steel
Collector aperture	2.3 m
Receiver surface treatment	Highly selective blackened nickel
Absorptance	0.97
Emittance (80 °C)	0.18
Glass envelope transmittance	0.96
Absorber outside diameter	50.8 mm
Tracking mechanism accuracy	0.05°
Collector orientation	Axis in N—S direction
Mode of tracking	E—W horizontal

For the EuroTrough collector, a so-called torque box design has been selected, with less weight and fewer deformations of the collector structure due to dead weight and wind loading than the reference designs (LS-2 torque tube or the LS-3 V truss design, both applied in the Californian plants). This reduces torsion and bending of the structure during operation and results in increased optical performance and wind resistance. The weight of the steel structure has been reduced by about 14% as compared with the available design of the LS-3 collector. The central element of the box design is a 12 m long steel space-frame structure having a squared cross-section that holds the support arms for the parabolic mirror facets. The torque box is built out of only four steel parts. This leads to easy manufacturing and decreases the required effort and thus the cost for site assembling. The structural deformation of the new design is considerably less than that in the previous design (LS-3), which results in better performance of the collector.

Another method for producing lightweight troughs, developed by the author, is with fiberglass (Kalogirou et al., 1994b). For the production of the trough, a mold is required. The trough is in fact a negative copy of the mold. Initially, a layer of fiberglass is laid. Cavities produced with plastic conduits, covered with a second layer of fiberglass at the back of the collector surface, provide reinforcement in the longitudinal and transverse directions to increase rigidity, as shown in Fig. 3.18.

Tracking mechanisms

A tracking mechanism must be reliable and able to follow the sun with a certain degree of accuracy, return the collector to its original position at the end of the day or during the night, and track during periods of intermittent cloud cover. Additionally, tracking mechanisms are used for the protection of collectors, that is, they turn the collector out of focus to protect it from hazardous environmental and working conditions, such as wind gusts, overheating, and failure of the thermal fluid flow mechanism.

FIGURE 3.18

Fiberglass parabola details.

The required accuracy of the tracking mechanism depends on the collector acceptance angle. This is described in Section 3.6.3, and the method to determine it experimentally is given in Section 4.3.

Various forms of tracking mechanisms, ranging from complex to very simple, have been proposed. They can be divided into two broad categories: mechanical and electrical–electronic systems. The electronic systems generally exhibit improved reliability and tracking accuracy. These can be further subdivided into:

1. Mechanisms employing motors controlled electronically through sensors, which detect the magnitude of the solar illumination (Kalogirou, 1996).
2. Mechanisms using computer-controlled motors, with feedback control provided from sensors measuring the solar flux on the receiver (Briggs, 1980; Boultinghouse, 1982).

A tracking mechanism developed by the author (Kalogirou, 1996) uses three light-dependent resistors (LDRs), which detect the focus, sun–cloud, and day–night conditions and give instruction to a DC motor through a control system to focus the collector, follow approximately the sun path when cloudy conditions exist, and return the collector to the east during night.

The system, which was designed to operate with the required tracking accuracy, consists of a small direct current motor that rotates the collector via a speed reduction gearbox. A diagram of the system, together with a table showing the functions of the control system, is presented in Fig. 3.19. The system employs three sensors, of which A is installed on the east side of the collector shaded by the frame, whereas the other two (B and C) are installed on the collector frame. Sensor A acts as the "focus" sensor, that is, it receives direct sunlight only when the collector is focused. As the sun moves, sensor A becomes shaded and the motor turns "on". Sensor B is the "cloud" sensor, and cloud cover is assumed when illumination falls below a certain level. Sensor C is the "daylight" sensor. The condition when all three sensors receive sunlight is translated by the control system as daytime with no cloud passing over

FIGURE 3.19

Tracking mechanism, system diagram.

the sun and the collector in a focused position. The functions shown in the table of Fig. 3.19 are followed provided that sensor C is "on", that is, it is daytime.

The sensors used are light-dependent resistors (LDRs). The main disadvantage of LDRs is that they cannot distinguish between direct and diffuse sunlight. However, this can be overcome by adding an adjustable resistor to the system, which can be set for direct sunlight (i.e., a threshold value). This is achieved by setting the adjustable resistor so that for direct sunlight, the appropriate input logic level (i.e., 0) is set.

As mentioned previously, the motor of the system is switched on when any of the three LDRs is shaded. Which sensor is activated depends on the amount of shading determined by the value set on the adjustable resistor, that is, threshold value of radiation required to trigger the relays. Sensor A is always partially shaded. As the shading increases due to the movement of the sun, a value is reached that triggers the forward relay, which switches the motor on to turn the collector and therefore re-exposes sensor A.

The system also accommodates cloud cover, that is, when sensor B is not receiving direct sunlight, determined by the value of another adjustable resistor, a timer is automatically connected to the system and this powers the motor every 2 min for about 7 s. As a result, the collector follows approximately the sun's path and when the sun reappears the collector is refocused by the function of sensor A.

The system also incorporates two limit switches, the function of which is to stop the motor from going beyond the rotational limits. These are installed on two stops, which restrict the overall rotation of the collector in both directions, east and west. The collector tracks to the west as long as it is daytime. When the sun goes down and sensor C determines that it is night, power is connected to a reverse relay, which changes the motor's polarity and rotates the collector until its motion is restricted by the east limit switch. If there is no sun during the following morning, the timer is used to follow the sun's path as under normal cloudy conditions. The tracking system just described, comprising an electric motor and a gearbox, is for small collectors. For large collectors, powerful hydraulic units are required.

The tracking system developed for the EuroTrough collector is based on "virtual" tracking. The traditional sun-tracking unit with sensors that detect the position of the sun has been replaced by a system based on calculation of the sun position using a mathematical algorithm. The unit is implemented in the EuroTrough with a 13-bit optical angular encoder (resolution of 0.8 mrad) mechanically coupled to the rotation axis of the collector. By comparing both sun and collector axes positions by an electronic device, an order is sent to the drive system to induce tracking.

3.2.2 Fresnel collectors

Fresnel collectors have two variations: the Fresnel lens collector (FLC), shown in Fig. 3.20a, and the linear Fresnel reflector (LFR), shown in Fig. 3.20b. The former is made from a plastic material and shaped in the way shown to focus the solar rays to a point receiver, whereas the latter relies on an array of linear mirror strips that concentrate light onto a linear receiver. The LFR collector can be imagined as a broken-up parabolic trough reflector (see Fig. 3.20b), but unlike parabolic troughs, the individual strips need not be of parabolic shape. The strips can also be mounted on flat ground (field) and concentrate light on a linear fixed receiver mounted on a tower. A representation of an element of an LFR collector field is shown in Fig. 3.21. In this case, large absorbers can be constructed and the absorber does not have to move. The greatest advantage of this type of system is that it uses flat or elastically curved reflectors, which are cheaper than parabolic glass reflectors. Additionally, these are mounted close to the ground, thus minimizing structural requirements.

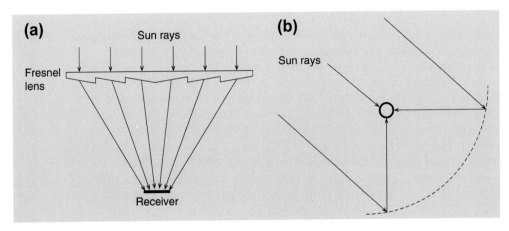

FIGURE 3.20

Fresnel collectors. (a) Fresnel lens collector. (b) Linear fresnel-type parabolic trough collector.

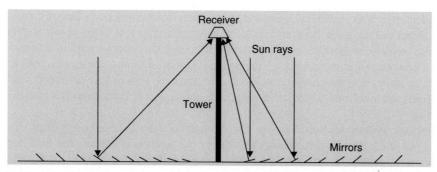

FIGURE 3.21

Schematic diagram of a downward-facing receiver illuminated from a linear fresnel reflector field.

The first to apply this principle was the great solar pioneer Giorgio Francia (1968), who developed both linear and two-axis tracking Fresnel reflector systems at Genoa, Italy, in the 1960s. These systems showed that elevated temperatures could be reached using such systems, but he moved on to two-axis tracking, possibly because advanced selective coatings and secondary optics were not available (Mills, 2001).

In 1979, the FMC Corporation produced a detailed project design study for 10 and 100 MWe LFR power plants for the U.S. Department of Energy (DOE). The larger plant would have used a 1.68 km linear cavity absorber mounted on 61 m height towers. The project, however, was never put into practice, because it ran out of DOE funding (Mills, 2001).

A later effort to produce a tracking LFR was made by the Israeli Paz company in the early 1990s by Feuermann and Gordon (1991). This used efficient secondary CPC-like optics and an evacuated tube absorber.

One difficulty with the LFR technology is that avoidance of shading and blocking between adjacent reflectors leads to increased spacing between reflectors. Blocking can be reduced by increasing the height of the absorber towers, but this increases cost. Compact linear Fresnel reflector (CLFR) technology has been recently developed at Sydney University in Australia. This is, in effect, a second type of solution for the Fresnel reflector field problem that has been overlooked until recently. In this design adjacent linear elements can be interleaved to avoid shading. The classical LFR system has only one receiver and there is no choice about the direction and orientation of a given reflector. However, if it is assumed that the size of the field will be large, as it must be in technology supplying electricity in the megawatt class, it is reasonable to assume that there will be many towers in the system. If they are close enough, then individual reflectors have the option of directing reflected solar radiation to at least two towers. This additional variable in the reflector orientation provides the means for much more densely packed arrays because patterns of alternating reflector orientation can be such that closely packed reflectors can be positioned without shading and blocking. The interleaving of mirrors between two receiving towers is shown in Fig. 3.22. The arrangement minimizes beam blocking by adjacent reflectors and allows high reflector densities and low tower heights to be used. Close spacing of reflectors reduces land usage, but in many cases, as in deserts, this is not a serious issue. The avoidance of large reflector spacing and tower heights is also an important cost issue when the cost of ground preparation, array substructure cost, tower structure cost, steam line thermal losses, and steam line cost are considered. If the technology is to be located in an area with limited land availability, such as in

FIGURE 3.22

Schematic diagram showing interleaving of mirrors in a compact linear fresnel reflector with reduced shading between mirrors.

urban areas or next to existing power plants, high array ground coverage can lead to maximum system output for a given ground area (Mills, 2001).

A good review of solar energy applications of Fresnel lenses is given by Xie et al. (2011). This includes both imaging and non-imaging systems.

3.2.3 Parabolic dish reflector (PDR)

A parabolic dish reflector (PDR), shown schematically in Fig. 3.23a, is a point-focus collector that tracks the sun in two axes, concentrating solar energy onto a receiver located at the focal point of the dish. The dish structure must fully track the sun to reflect the beam into the thermal receiver. For this purpose, tracking mechanisms similar to the ones described in the previous section are employed in double, so the collector is tracked in two axes. A photograph of a Eurodish collector is shown in Fig. 3.23b.

The receiver absorbs the radiant solar energy, converting it into thermal energy in a circulating fluid. The thermal energy can then be either converted into electricity using an engine–generator coupled directly to the receiver or transported through pipes to a central power conversion system. Parabolic dish systems can achieve temperatures in excess of 1500 °C. Because the receivers are distributed throughout a collector field, like parabolic troughs, parabolic dishes are often called *distributed receiver systems*. Parabolic dishes have several important advantages (De Laquil et al., 1993):

1. Because they are always pointing at the sun, they are the most efficient of all collector systems.
2. They typically have concentration ratios in the range of 600–2000 and thus are highly efficient at thermal energy absorption and power conversion systems.
3. They are modular collector and receiver units that can function either independently or as part of a larger system of dishes.

The main use of this type of concentrator is for parabolic dish engines. A parabolic dish engine system is an electric generator that uses sunlight instead of crude oil or coal to produce electricity. The major parts of a system are the solar dish concentrator and the power conversion unit. More details on this system are given in Chapter 10.

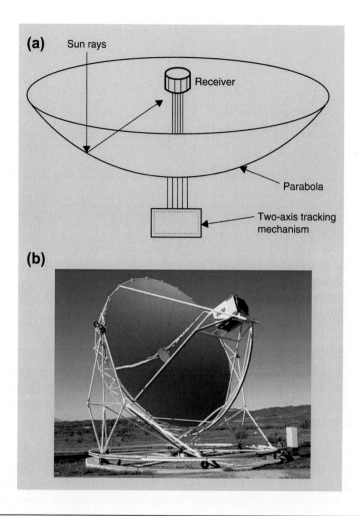

FIGURE 3.23

Parabolic dish collector. (a) Schematic diagram. (b) Photo of a eurodish collector.

From http://www.psa.es/webeng/instalaciones/discos.php.

Parabolic dish systems that generate electricity from a central power converter collect the absorbed sunlight from individual receivers and deliver it via a heat transfer fluid to the power conversion systems. The need to circulate heat transfer fluid throughout the collector field raises design issues such as piping layout, pumping requirements, and thermal losses.

3.2.4 Heliostat field collector (HFC)

For extremely high inputs of radiant energy, a multiplicity of flat mirrors, or heliostats, using altazimuth mounts can be used to reflect their incident direct solar radiation onto a common target, as shown in Fig. 3.24. This is called the *heliostat field* or *central receiver collector*. By using slightly concave mirror segments on the heliostats, large amounts of thermal energy can be directed into the cavity of a steam generator to produce steam at high temperature and pressure.

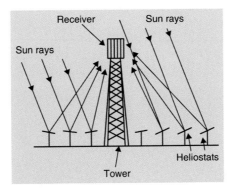

FIGURE 3.24

Schematic of central receiver system.

The concentrated heat energy absorbed by the receiver is transferred to a circulating fluid that can be stored and later used to produce power. Central receivers have several advantages (De Laquil et al., 1993):

1. They collect solar energy optically and transfer it to a single receiver, thus minimizing thermal energy transport requirements.
2. They typically achieve concentration ratios of 300−1500 and so are highly efficient, both in collecting energy and in converting it to electricity.
3. They can conveniently store thermal energy.
4. They are quite large (generally more than 10 MW) and thus benefit from economies of scale.

Each heliostat at a central receiver facility has from 50 to 150 m^2 of reflective surface, with four mirrors installed on a common pillar for economy, as shown in Fig. 3.25. The heliostats collect and concentrate sunlight onto the receiver, which absorbs the concentrated sunlight, transferring its energy to a heat transfer fluid. The heat transport system, which consists primarily of pipes, pumps, and valves, directs the transfer fluid in a closed loop among the receiver, storage, and power conversion systems. A thermal storage system typically stores the collected energy as sensible heat for later delivery to the power conversion system. The storage system also decouples the collection of solar energy from its conversion to electricity. The power conversion system consists of a steam generator, turbine generator, and support equipment, which convert the thermal energy into electricity and supply it to the utility grid.

In this case incident sunrays are reflected by large tracking mirrored collectors, which concentrate the energy flux toward radiative−convective heat exchangers, where energy is transferred to a working thermal fluid. After energy collection by the solar system, the conversion of thermal energy to electricity has many similarities with the conventional fossil-fueled thermal power plants (Romero et al., 2002).

The collector and receiver systems come in three general configurations. In the first, heliostats completely surround the receiver tower, and the receiver, which is cylindrical, has an exterior heat transfer surface. In the second, the heliostats are located north of the receiver tower (in the Northern Hemisphere), and the receiver has an enclosed heat transfer surface. In the third, the heliostats are located north of the receiver tower, and the receiver, which is a vertical plane, has a north-facing heat transfer surface. More details of these plants are given in Chapter 10.

FIGURE 3.25

Detail of a heliostat.

3.3 Thermal analysis of flat-plate collectors

In this section, the thermal analysis of the collectors is presented. The two major types of collectors, flat plate and concentrating, are examined separately. The basic parameter to consider is the collector thermal efficiency. This is defined as the ratio of the useful energy delivered to the energy incident on the collector aperture. The incident solar flux consists of direct and diffuse radiation. While flat-plate collectors can collect both, concentrating collectors can utilize direct radiation only if the concentration ratio is greater than 10 (Prapas et al., 1987).

In this section, the various relations required to determine the useful energy collected and the interaction of the various constructional parameters on the performance of a collector are presented.

3.3.1 Absorbed solar radiation

The prediction of collector performance requires information on the solar energy absorbed by the collector absorber plate. The solar energy incident on a tilted surface can be found by the methods presented in Chapter 2. As can be seen from Chapter 2, the incident radiation has three special components: beam, diffuse, and ground-reflected radiation. This calculation depends on the radiation model employed. Using the isotropic model on an hourly basis, Eq. (2.97) can be modified to give the absorbed radiation, S, by multiplying each term with the appropriate transmittance–absorptance product as follows:

$$S = I_B R_B (\tau\alpha)_B + I_D (\tau\alpha)_D \left[\frac{1 + \cos(\beta)}{2}\right] + \rho_G (I_B + I_D)(\tau\alpha)_G \left[\frac{1 - \cos(\beta)}{2}\right] \tag{3.1a}$$

where the terms $[1 + \cos(\beta)]/2$ and $[1 - \cos(\beta)]/2$ are the view factors from the collector to the sky and from the collector to the ground, respectively. The same equation can be used to estimate the monthly average absorbed solar radiation, $\overline{S}$, by replacing the hourly direct and diffuse radiation values with the appropriate monthly average values, $\overline{H}_B$ and $\overline{H}_D$, R_B with $\overline{R}_B$, and various $(\tau\alpha)$ values with monthly average values, $\overline{(\tau\alpha)}$ in Eq. (3.1a):

$$\overline{S} = \overline{H}_B\,\overline{R}_B\overline{(\tau\alpha)}_B + \overline{H}_D\overline{(\tau\alpha)}_D\left(\frac{1+\cos(\beta)}{2}\right) + \overline{H}\rho_G\overline{(\tau\alpha)}_G\left(\frac{1-\cos(\beta)}{2}\right) \tag{3.1b}$$

More details on this are given in Chapter 11.

The combination of cover with the absorber plate is shown in Fig. 3.26, together with a ray tracing of the radiation. As can be seen, of the incident energy falling on the collector, $\tau\alpha$ is absorbed by the absorber plate and $(1 - \alpha)\tau$ is reflected back to the glass cover. The reflection from the absorber plate is assumed to be diffuse, so the fraction $(1 - \alpha)\tau$ that strikes the glass cover is diffuse radiation and $(1 - \alpha)\tau\rho_D$ is reflected back to the absorber plate. The multiple reflection of diffuse radiation continues so that the fraction of the incident solar energy ultimately absorbed is:

$$(\tau\alpha) = \tau\alpha\sum_{n=0}^{\infty}[(1-\alpha)\rho_D]^n = \frac{\tau\alpha}{1-(1-\alpha)\rho_D} \tag{3.2}$$

Typical values of $(\tau\alpha)$ are 0.7–0.75 for window glass and 0.85–0.9 for low-iron glass. A reasonable approximation of Eq. (3.2) for most practical solar collectors is:

$$(\tau\alpha) \cong 1.01\tau\alpha \tag{3.3}$$

The reflectance of the glass cover for diffuse radiation incident from the absorber plate, ρ_D, can be estimated from Eq. (2.55) as the difference between τ_α and τ at an angle of 60°. For a collector with two covers, Eq. (2.57) can be used. For single covers, the following values can be used for ρ_D:

For $KL = 0.0125$, $\rho_D = 0.15$.
For $KL = 0.0370$, $\rho_D = 0.12$.
For $KL = 0.0524$, $\rho_D = 0.11$.

FIGURE 3.26

Radiation transfer between the glass cover and absorber plate.

For a given collector tilt angle, β, the following empirical relations, derived by Brandemuehl and Beckman (1980), can be used to find the effective incidence angle for diffuse radiation from sky, $\theta_{e,D}$, and ground-reflected radiation, $\theta_{e,G}$:

$$\theta_{e,D} = 59.68 - 0.1388\beta + 0.001497\beta^2 \tag{3.4a}$$

$$\theta_{e,G} = 90 - 0.5788\beta + 0.002693\beta^2 \tag{3.4b}$$

where

β = collector slope angle in degrees.

The proper transmittance can then be obtained from Eq. (2.53), whereas the angle-dependent absorptance from 0° to 80° can be obtained from Beckman et al. (1977):

$$\frac{a}{a_n} = 1 + 2.0345 \times 10^{-3}\theta_e - 1.99 \times 10^{-4}\theta_e^2 + 5.324 \times 10^{-6}\theta_e^3 - 4.799 \times 10^{-8}\theta_e^4 \tag{3.5a}$$

or from the polynomial fit for 0°–90° from (Duffie and Beckman, 2006):

$$\frac{a}{a_n} = 1 - 1.5879 \times 10^{-3}\theta_e + 2.7314 \times 10^{-4}\theta_e^2 - 2.3026 \times 10^{-5}\theta_e^3 + 9.0244 \times 10^{-7}\theta_e^4$$
$$- 1.80 \times 10^{-8}\theta_e^5 + 1.7734 \times 10^{-10}\theta_e^6 - 6.9937 \times 10^{-13}\theta_e^7 \tag{3.5b}$$

where

θ_e = effective incidence angle (degrees);
a_n = absorptance at normal incident angle, which can be found from the properties of the absorber.

Subsequently, Eq. (3.2) can be used to find $(\tau\alpha)_D$ and $(\tau\alpha)_G$. The incidence angle, θ, of the beam radiation required to estimate R_B can be used to find $(\tau\alpha)_B$.

Alternatively, $(\tau\alpha)_n$ can be found from the properties of the cover and absorber materials, and Fig. 3.27 can be used at the appropriate angle of incidence for each radiation component to find the three transmittance–absorptance products.

When measurements of incident solar radiation (I_t) are available, instead of Eq. (3.1a), the following relation can be used:

$$S = (\tau\alpha)_{av}I_t \tag{3.6}$$

where $(\tau\alpha)_{av}$ can be obtained from:

$$(\tau\alpha)_{av} \cong 0.96(\tau\alpha)_B \tag{3.7}$$

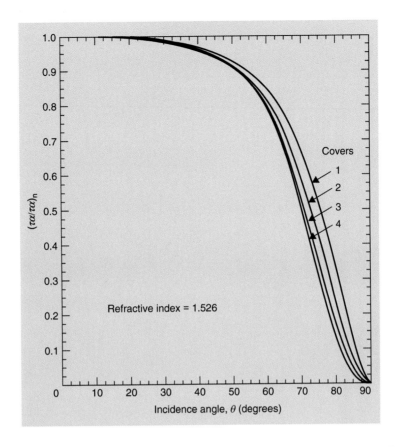

FIGURE 3.27

Typical $(\tau\alpha)/(\tau\alpha)_n$ curves for one to four glass covers.

Reprinted from Klein (1979), with permission from Elsevier.

EXAMPLE 3.1

For a clear winter day, $I_B = 1.42$ MJ/m^2 and $I_D = 0.39$ MJ/m^2. Ground reflectance is 0.5, incidence angle is 23°, and $R_B = 2.21$. Calculate the absorbed solar radiation by a collector having a glass with $KL = 0.037$, the absorptance of the plate at normal incidence, $a_n = 0.91$, and the refraction index of glass is 1.526. The collector slope is 60°.

Solution

Using Eq. (3.5a) for the beam radiation at $\theta = 23°$,

$$\frac{a}{a_n} = 1 + 2.0345 \times 10^{-3}\theta_e - 1.99 \times 10^{-4}\theta_e^2 + 5.324 \times 10^{-6}\theta_e^3 - 4.799 \times 10^{-8}\theta_e^4$$

$$= 1 + 2.0345 \times 10^{-3} \times 23 - 1.99 \times 10^{-4} \times 23^2 + 5.324 \times 10^{-6} \times 23^3 - 4.799 \times 10^{-8} \times 23^4 = 0.993$$

For the transmittance we need to calculate τ_α and τ_r. For the former, Eq. (2.51) can be used. From Eq. (2.44), $\theta_2 = 14.8°$. Therefore,

$$\tau_\alpha = e^{\left(-\dfrac{0.037}{\cos(14.8)}\right)} = 0.962$$

From Eqs. (2.45) and (2.46) $r_\perp = 0.054$ and $r_\parallel = 0.034$. Therefore, from Eq. (2.50a),

$$\tau_r = \frac{1}{2}\left(\frac{1-0.034}{1+0.034} + \frac{1-0.054}{1+0.054}\right) = 0.916$$

Continued

EXAMPLE 3.1—CONT'D

Finally, from Eq. (2.53),

$$\tau \cong \tau_\alpha \tau_r = 0.962 \times 0.916 = 0.881.$$

Alternatively, Eq. (2.52a) could be used with the above r values to obtain τ directly.
From Eq. (3.3),

$$(\tau\alpha)_B = 1.01\tau(a/a_n)a_n = 1.01 \times 0.881 \times 0.993 \times 0.91 = 0.804 \approx 0.80$$

From Eq. (3.4a), the effective incidence angle for diffuse radiation is:

$$\theta_{e,D} = 59.68 - 0.1388\beta + 0.001497\beta^2 = 59.68 - 0.1388 \times 60 + 0.001497 \times 60^2 = 57°$$

From Eq. (3.5a), for the diffuse radiation at $\theta = 57°$, $a/a_n = 0.949$.
From Eq. (2.44), for $\theta_1 = 57°$, $\theta_2 = 33°$. From Eqs. (2.45) and (2.46), $r_\perp = 0.165$ and $r_\parallel = 0$.
From Eq. (2.50a), $\tau_r = 0.858$, and from Eq. (2.51), $\tau_\alpha = 0.957$. From Eq. (2.53),

$$\tau = 0.957 \times 0.858 = 0.821$$

and from Eq. (3.3),

$$(\tau\alpha)_D = 1.01\tau(a/a_n)a_n = 1.01 \times 0.821 \times 0.949 \times 0.91 = 0.716 \approx 0.72$$

From Eq. (3.4b), the effective incidence angle for ground-reflected radiation is:

$$\theta_{e,G} = 90 - 0.5788\beta + 0.002693\beta^2 = 90 - 0.5788 \times 60 + 0.002693 \times 60^2 = 65°$$

From Eq. (3.5a), for the ground-reflected radiation at $\theta = 65°$, $a/a_n = 0.897$.
From Eq. (2.44), for $\theta_1 = 65°$, $\theta_2 = 36°$. From Eqs. (2.45) and (2.46), $r_\perp = 0.244$ and $r_\parallel = 0.012$.
From Eq. (2.50a), $\tau_r = 0.792$, and from Eq. (2.51), $\tau_\alpha = 0.955$. From Eq. (2.53),

$$\tau = 0.955 \times 0.792 = 0.756$$

And from Eq. (3.3),

$$(\tau\alpha)_G = 1.01\tau(a/a_n)a_n = 1.01 \times 0.756 \times 0.897 \times 0.91 = 0.623 \approx 0.62$$

In a different way, from Eq. (3.3),
$(\tau\alpha)_n = 1.01 \times 0.884 \times 0.91 = 0.812$ (note that for the transmittance the value for normal incidence is used, i.e., τ_n).
At $\theta = 0°$: $r_{(0)} = (0.526/2.526)^2 = 0.0434$. $\tau_{r(0)} = (1-0.0434)/(1 + 0.0434) = 0.917$
$\tau_\alpha = e^{-0.037} = 0.964$ and $\tau_n = \tau_\alpha \times \tau_{r(0)} = 0.884$.

From Fig. 3.27, for beam radiation at $\theta = 23°$, $(\tau\alpha)/(\tau\alpha)_n = 0.98$. Therefore,

$$(\tau\alpha)_B = 0.812 \times 0.98 = 0.796 \approx 0.80$$

From Fig. 3.27, for diffuse radiation at $\theta = 57°$, $(\tau\alpha)/(\tau\alpha)_n = 0.89$. Therefore,

$$(\tau\alpha)_D = 0.812 \times 0.89 = 0.722 \approx 0.72$$

From Fig. 3.27, for ground-reflected radiation at $\theta = 65°$, $(\tau\alpha)/(\tau\alpha)_n = 0.76$. Therefore,

$$(\tau\alpha)_G = 0.812 \times 0.76 = 0.617 \approx 0.62$$

All these values are very similar to the previously found values, but the effort required is much less.
Finally, the absorbed solar radiation is obtained from Eq. (3.1a):

$$S = I_B R_B (\tau\alpha)_B + I_D (\tau\alpha)_D \left[\frac{1 + \cos(\beta)}{2}\right] + \rho_G (I_B + I_D)(\tau\alpha)_G \left[\frac{1 - \cos(\beta)}{2}\right]$$

$$= 1.42 \times 2.21 \times 0.80 + 0.39 \times 0.72 \left[\frac{1 + \cos(60)}{2}\right] + 0.5 \times 1.81 \times 0.62 \left[\frac{1 - \cos(60)}{2}\right]$$

$$= 2.86 \text{ MJ/m}^2$$

3.3.2 Collector energy losses

When a certain amount of solar radiation falls on the surface of a collector, most of it is absorbed and delivered to the transport fluid, and it is carried away as useful energy. However, as in all thermal systems, heat losses to the environment by various modes of heat transfer are inevitable. The thermal network for a single-cover FPC in terms of conduction, convection, and radiation is shown in Fig. 3.28a and in terms of the resistance between plates in Fig. 3.28b. The temperature of the plate is T_p, the collector back temperature is T_b, and the absorbed solar radiation is S. In a simplified way, the various thermal losses from the collector can be combined into a simple resistance, R_L, as shown in Fig. 3.28c, so that the energy losses from the collector can be written as:

$$Q_{loss} = \frac{T_p - T_a}{R_L} = U_L A_c \left(T_p - T_a \right) \tag{3.8}$$

where

U_L = overall heat loss coefficient based on collector area A_c (W/m^2 K);
T_p = plate temperature (°C).

The overall heat loss coefficient is a complicated function of the collector construction and its operating conditions, given by the following expression:

$$U_L = U_t + U_b + U_e \tag{3.9}$$

where

U_t = top loss coefficient (W/m^2 K);
U_b = bottom heat loss coefficient (W/m^2 K); and
U_e = heat loss coefficient from the collector edges (W/m^2 K).

Therefore, U_L is the heat transfer resistance from the absorber plate to the ambient air. All these coefficients are examined separately. It should be noted that edge losses are not shown in Fig. 3.28.

In addition to serving as a heat trap by admitting shortwave solar radiation and retaining longwave thermal radiation, the glazing also reduces heat loss by convection. The insulating effect of the glazing is enhanced by the use of several sheets of glass or glass plus plastic.

Under steady-state conditions, the heat transfer from the absorber plate to the glass cover is the same as the energy lost from the glass cover to ambient. As shown in Fig. 3.28, the heat transfer upward from the absorber plate at temperature T_p to the glass cover at T_g and from the glass cover at T_g to ambient at T_a is by convection and infrared radiation. For the infrared radiation heat loss, Eq. (2.67) can be used. Therefore, the heat loss from absorber plate to glass is given by:

$$Q_{t,\text{absorber plate to glass cover}} = A_c h_{c,p-g} \left(T_p - T_g \right) + \frac{A_c \sigma \left(T_p^4 - T_g^4 \right)}{\left(1/\varepsilon_p \right) + \left(1/\varepsilon_g \right) - 1} \tag{3.10}$$

where

A_c = collector area (m^2);
$h_{c,p-g}$ = convection heat transfer coefficient between the absorber plate and glass cover (W/m^2 K);
ε_p = infrared emissivity of absorber plate; and
ε_g = infrared emissivity of glass cover.

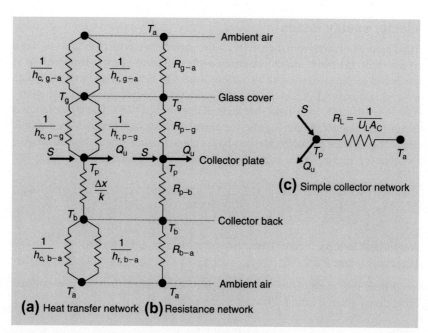

FIGURE 3.28

Thermal network for a single-cover collector in terms of (a) conduction, convection, and radiation; (b) resistance between plates; and (c) a simple collector network.

For tilt angles up to 60°, the convective heat transfer coefficient, $h_{c,p-g}$, is given by Hollands et al. (1976) for collector inclination (β) in degrees:

$$Nu = \frac{h_{c,p-g}L}{k}$$

$$= 1 + 1.446\left[1 - \frac{1708}{Ra \times \cos(\beta)}\right]^{+}\left\{1 - \frac{1708[\sin(1.8\beta)]^{1.6}}{Ra \times \cos(\beta)}\right\} + \left\{\left[\frac{Ra \times \cos(\beta)}{5830}\right]^{0.333} - 1\right\}^{+}$$

(3.11)

where the plus sign indicates that positive values should only be considered. Other more simplified relations are given in Appendix 5 (Table A5.8). The Rayleigh value, Ra, is given by:

$$Ra = \frac{g\beta'Pr}{\nu^2}(T_p - T_g)L^3$$

(3.12)

where

g = gravitational constant, = 9.81 m²/s;
β' = volumetric coefficient of expansion; for ideal gas, $\beta' = 1/T$;
Pr = Prandtl number;
L = absorber to glass cover distance (m); and
ν = kinetic viscosity (m²/s).

The fluid properties in Eq. (3.12) are evaluated at the mean gap temperature $(T_p + T_g)/2$. For vertical collectors, the convection correlation is given by Shewen et al. (1996) as:

$$\text{Nu} = \frac{h_{c,p-g}L}{k} = \left[1 + \left(\frac{0.0665\text{Ra}^{0.333}}{1 + (9600/\text{Ra})^{0.25}}\right)^2\right]^{0.5} \tag{3.13}$$

The radiation term in Eq. (3.10) can be linearized by the use of Eq. (2.73) as:

$$h_{r,p-g} = \frac{\sigma(T_p + T_g)\left(T_p^2 + T_g^2\right)}{(1/\varepsilon_p) + (1/\varepsilon_g) - 1} \tag{3.14}$$

Consequently, Eq. (3.10) becomes:

$$Q_{t,\text{absorber plate to glass cover}} = A_c\left(h_{c,p-g} + h_{r,p-g}\right)\left(T_p - T_g\right)\frac{T_p - T_g}{R_{p-g}} \tag{3.15}$$

in which:

$$R_{p-g} = \frac{1}{A_c\left(h_{c,p-g} + h_{r,p-g}\right)} \tag{3.16}$$

Similarly, the heat loss from glass cover to ambient is by convection to the ambient air (T_a) and radiation exchange with the sky (T_{sky}). For convenience, the combined convection–radiation heat transfer is usually given in terms of T_a only by:

$$Q_{t,\text{glass cover to ambient}} = A_c\left(h_{c,g-a} + h_{r,g-a}\right)\left(T_g - T_a\right) = \frac{T_g - T_a}{R_{g-a}} \tag{3.17}$$

where

$h_{c,g-a}$ = convection heat transfer coefficient between the glass cover and ambient due to wind $(\text{W/m}^2\,\text{K})$;

$h_{r,g-a}$ = radiation heat transfer coefficient between the glass cover and ambient $(\text{W/m}^2\,\text{K})$.

The radiation heat transfer coefficient is now given by Eq. (2.75), noting that, instead of T_{sky}, T_a is used for convenience, since the sky temperature does not affect the results much:

$$h_{r,g-a} = \varepsilon_g\sigma(T_g + T_a)\left(T_g^2 + T_a^2\right) \tag{3.18a}$$

If the sky temperature is considered:

$$h_{r,g-a} = \varepsilon_g\sigma(T_g + T_a)\left(T_g^2 + T_a^2\right)\frac{(T_g - T_{\text{sky}})}{(T_g - T_a)} \tag{3.18b}$$

The atmosphere does not have a uniform temperature. It radiates selectively at certain wavelengths and is essentially transparent in the wavelength range from 8 to 14 μm, while outside this range has absorbing bands covering much of the far infrared spectrum. Several relations have been proposed to associate T_{sky} (K) with measured meteorological variables. Two of them are given here. Additional relations are given in Appendix 5 (Table A5.9).

Swinbank (1963) correlation:

$$T_{sky} = 0.0552 T_a^{1.5} \tag{3.18c}$$

Berdahl and Martin (1984) correction:

$$T_{sky} = T_a \left(0.711 + 0.0056 T_{dp} + 0.000073 T_{dp}^2 + 0.013 \cos(15t) \right)^{0.25} \tag{3.18d}$$

where

T_a = ambient temperature (K);
T_{dp} = dew point temperature (°C); and
t = hour from midnight.

From Eq. (3.17),

$$R_{g-a} = \frac{1}{A_c \left(h_{c,g-a} + h_{r,g-a} \right)} \tag{3.19}$$

Since resistances R_{p-g} and R_{g-a} are in series, their resultant is given by:

$$R_t = R_{p-g} + R_{g-a} = \frac{1}{U_t A_c} \tag{3.20}$$

Therefore,

$$Q_t = \frac{T_p - T_a}{R_t} = U_t A_c \left(T_p - T_a \right) \tag{3.21a}$$

It should be noted that by applying Eq. (3.16) for R_{p-g} and Eq. (3.19) for R_{g-a} in Eq. (3.20) we obtain:

$$U_t = \left(\frac{1}{h_{c,p-g} + h_{r,p-g}} + \frac{1}{h_{c,g-a} + h_{r,g-a}} \right)^{-1} \tag{3.21b}$$

In some cases, collectors are constructed with two glass covers in an attempt to lower heat losses. In this case, another resistance is added to the system shown in Fig. 3.28 to account for the heat transfer from the lower to the upper glass covers. By following a similar analysis, the heat transfer from the lower glass at T_{g2} to the upper glass at T_{g1} is given by:

$$Q_{t,\text{lower cover to top cover}} = A_c \left(h_{c,g2-g1} + h_{r,g2-g1} \right) \left(T_{g2} - T_{g1} \right) = \frac{T_{g2} - T_{g1}}{R_{g2-g1}} \tag{3.22}$$

where

$h_{c,g2-g1}$ = convection heat transfer coefficient between the two glass covers (W/m^2 K);
$h_{r,g2-g1}$ = radiation heat transfer coefficient between the two glass covers (W/m^2 K).

The convection heat transfer coefficient can be obtained by Eqs. (3.11−3.13). The radiation heat transfer coefficient can be obtained again from Eq. (2.73) and is given by:

$$h_{r,g2-g1} = \frac{\sigma \left(T_{g2} + T_{g1} \right) \left(T_{g2}^2 + T_{g1}^2 \right)}{(1/\varepsilon_{g2}) + (1/\varepsilon_{g1}) - 1} \tag{3.23}$$

where ε_{g2} and ε_{g1} are the infrared emissivities of the bottom and top glass covers, respectively.

Finally, the resistance R_{g2-g1} is given by:

$$R_{g2-g1} = \frac{1}{A_c \left(h_{c,g2-g1} + h_{r,g2-g1} \right)} \tag{3.24}$$

In the case of collectors with two covers, Eq. (3.24) is added on the resistance values in Eq. (3.20). The analysis of a two-cover collector is given in Example 3.2.

In the preceding equations, solutions by iterations are required for the calculation of the top heat loss coefficient, U_t, since the air properties are functions of operating temperature. Because the iterations required are tedious and time consuming, especially for the case of multiple-cover systems, straightforward evaluation of U_t is given by the following empirical equation with sufficient accuracy for design purposes (Klein, 1975):

$$U_t = \frac{1}{\frac{N_g}{\frac{C}{T_p} \left[\frac{T_p - T_a}{N_g + f} \right]^{0.33}} + \frac{1}{h_w}} + \frac{\sigma \left(T_p^2 + T_a^2 \right) \left(T_p + T_a \right)}{\frac{1}{\varepsilon_p + 0.05N_g \left(1 - \varepsilon_p \right)} + \frac{2N_g + f - 1}{\varepsilon_g} - N_g} \tag{3.25}$$

where

$$f = \left(1 - 0.04h_w + 0.0005h_w^2 \right) \left(1 + 0.091N_g \right) \tag{3.26}$$

$$C = 365.9 \left(1 - 0.00883\beta + 0.0001298\beta^2 \right) \tag{3.27}$$

$$h_w = \frac{8.6V^{0.6}}{L^{0.4}} \tag{3.28}$$

It should be noted that, for the wind heat transfer coefficient, no well-established research has been undertaken yet, but until this is done, Eq. (3.28) can be used. The minimum value of h_w for still air conditions is 5 W/m² °C. Therefore, if Eq. (3.28) gives a lower value, this should be used as a minimum. Additional relations are given in Appendix 5 (Table A5.10).

The energy loss from the bottom of the collector is first conducted through the insulation and then by a combined convection and infrared radiation transfer to the surrounding ambient air. Because the temperature of the bottom part of the casing is low, the radiation term ($h_{r,b-a}$) can be neglected; thus the energy loss is given by:

$$U_b = \frac{1}{\frac{t_b}{k_b} + \frac{1}{h_{c,b-a}}} \tag{3.29}$$

where

t_b = thickness of back insulation (m);
k_b = conductivity of back insulation (W/m K); and
$h_{c,b-a}$ = convection heat loss coefficient from back to ambient (W/m² K).

The conduction resistance of the insulation behind the collector plate governs the heat loss from the collector plate through the back of the collector casing. The heat loss from the back of the plate rarely exceeds 10% of the upward loss. Typical values of the back surface heat loss coefficient are 0.3–0.6 W/m² K.

In a similar way, the heat transfer coefficient for the heat loss from the collector edges can be obtained from:

$$U_e = \frac{1}{\dfrac{t_e}{k_e} + \dfrac{1}{h_{c,e-a}}} \tag{3.30}$$

where

t_e = thickness of edge insulation (m);
k_e = conductivity of edge insulation (W/m K); and
$h_{c,e-a}$ = convection heat loss coefficient from edge to ambient (W/m² K).

As the U_L in Eq. (3.8) is multiplied by A_c the heat loss coefficient from the collector edges must be multiplied by A_e/A_c, where A_e is the total area of the four edges of the collector. The same applies for the bottom heat loss coefficient which must be multiplied by A_b/A_c if the two areas are not the same.

Typical values of the edge heat loss coefficient are 1.5–2.0 W/m² K.

EXAMPLE 3.2

Estimate the top heat loss coefficient of a collector that has the following specifications.

  Collector area = 2 m² (1 × 2 m).
  Collector slope = 35°.
  Number of glass covers = 2.
  Thickness of each glass cover = 4 mm.
  Thickness of absorbing plate = 0.5 mm.
  Distance between glass covers = 20 mm.
  Distance between inner glass cover and absorber = 40 mm.
  Mean absorber temperature, $T_p = 80\,°C = 353$ K.
  Ambient air temperature = 15 °C = 288 K.
  Absorber plate emissivity, $\varepsilon_p = 0.10$.
  Glass emissivity, $\varepsilon_g = 0.88$.
  Wind velocity = 2.5 m/s.

Solution
To solve this problem, the two glass cover temperatures are guessed and then by iteration are corrected until a satisfactory solution is reached by satisfying the following equations, obtained by combining Eqs. (3.15), (3.17), and (3.22):

$$(h_{c,p-g2} + h_{r,p-g2})(T_p - T_{g2}) = (h_{c,g2-g1} + h_{r,g2-g1})(T_{g2} - T_{g1})$$
$$= (h_{c,g1-a} + h_{r,g1-a})(T_{g1} - T_a)$$

However, to save time in this example, close to correct values are used. Assuming that $T_{g1} = 23.8\,°C$ (296.8 K) and $T_{g2} = 41.7\,°C$ (314.7 K), from Eq. (3.14),

$$h_{r,p-g2} = \frac{\sigma(T_p + T_{g2})(T_p^2 + T_{g2}^2)}{(1/\varepsilon_p) + (1/\varepsilon_{g2}) - 1} = \frac{(5.67 \times 10^{-8})(353 + 314.7)(353^2 + 314.7^2)}{(1/0.10) + (1/0.88) - 1}$$

$$= 0.835 \text{ W/m}^2 \text{ K}$$

EXAMPLE 3.2—CONT'D

Similarly, for the two covers, we have:

$$h_{r,g2-g1} = \frac{\sigma\left(T_{g2} + T_{g1}\right)\left(T_{g2}^2 + T_{g1}^2\right)}{(1/\varepsilon_{g2}) + (1/\varepsilon_{g1}) - 1} = \frac{(5.67 \times 10^{-8})(314.7 + 296.8)(314.7^2 + 296.8^2)}{(1/0.88) + (1/0.88) - 1}$$

$$= 5.098 \text{ W/m}^2 \text{ K}$$

From Eq. (3.18a) and noting that as no data are given $T_{sky} = T_a$, we have:

$$h_{r,g1-a} = \varepsilon_{g1}\sigma\left(T_{g1} + T_a\right)\left(T_{g1}^2 + T_a^2\right) = 0.88(5.67 \times 10^{-8})(296.8 + 288)(296.8^2 + 288^2)$$

$$= 4.991 \text{ W/m}^2 \text{ K}$$

From Table A5.1, in Appendix 5, the following properties of air can be obtained:
For $\frac{1}{2}(T_p + T_{g2}) = \frac{1}{2}(353 + 314.7) = 333.85$ K,

$$\nu = 19.51 \times 10^{-6} \text{ m}^2/\text{s}$$

$$\text{Pr} = 0.701$$

$$k = 0.0288 \text{ W/m K}$$

For $\frac{1}{2}(T_{g2} + T_{g1}) = \frac{1}{2}(314.7 + 296.8) = 305.75$ K,

$$\nu = 17.26 \times 10^{-6} \text{ m}^2/\text{s}$$

$$\text{Pr} = 0.707$$

$$k = 0.0267 \text{ W/m K}$$

By using these properties, the Rayleigh number, Ra, can be obtained from Eq. (3.12) and by noting that $\beta' = 1/T$.
For $h_{c,p-g2}$,

$$\text{Ra} = \frac{g\beta'\text{Pr}}{\nu^2}\left(T_p - T_{g2}\right)L^3 = \frac{9.81 \times 0.701 \times (353 - 314.7) \times 0.04^3}{333.85 \times \left(19.51 \times 10^{-6}\right)^2} = 132,648$$

For $h_{c,g2-g1}$,

$$\text{Ra} = \frac{g\beta'\text{Pr}}{\nu^2}\left(T_{g2} - T_{g1}\right)L^3 = \frac{9.81 \times 0.707 \times (314.7 - 296.8) \times 0.02^3}{305.75 \times \left(17.26 \times 10^{-6}\right)^2} = 10,904$$

Therefore, from Eq. (3.11), we have the following.
For $h_{c,p-g2}$,

$$h_{c,p-g2} = \frac{k}{L}\left(\begin{array}{c} 1 + 1.446\left[1 - \dfrac{1708}{\text{Ra} \times \cos(\beta)}\right]^+ \left\{1 - \dfrac{1708[\sin(1.8\beta)]^{1.6}}{\text{Ra} \times \cos(\beta)}\right\} \\[2em] + \left\{\left[\dfrac{\text{Ra} \times \cos(\beta)}{5830}\right]^{0.333} - 1\right\}^+ \end{array}\right)$$

$$= \frac{0.0288}{0.04}\left(\begin{array}{c} 1 + 1.446\left[1 - \dfrac{1708}{132,648 \times \cos(35)}\right]^+ \left\{1 - \dfrac{1708\,[\sin(1.8 \times 35)]^{1.6}}{132,648 \times \cos(35)}\right\} \\[2em] + \left\{\left[\dfrac{132,648 \times \cos(35)}{5830}\right]^{0.333} - 1\right\}^+ \end{array}\right)$$

$$= 2.918 \text{ W/m}^2 \text{ K}$$

Continued

EXAMPLE 3.2—CONT'D

For $h_{c,g2-g1}$,

$$h_{c,g2-g1} = \frac{k}{L} \left(\begin{array}{c} 1 + 1.446 \left[1 - \dfrac{1708}{\mathrm{Ra} \times \cos(\beta)} \right]^{+} \left\{ 1 - \dfrac{1708[\sin(1.8\beta)]^{1.6}}{\mathrm{Ra} \times \cos(\beta)} \right\} \\ + \left\{ \left[\dfrac{\mathrm{Ra} \times \cos(\beta)}{5830} \right]^{0.333} - 1 \right\}^{+} \end{array} \right)$$

$$= \frac{0.0267}{0.02} \left(\begin{array}{c} 1 + 1.446 \left[1 - \dfrac{1708}{10,904 \times \cos(35)} \right]^{+} \left\{ 1 - \dfrac{1708\,[\sin(1.8 \times 35)]^{1.6}}{10,904 \times \cos(35)} \right\} \\ + \left\{ \left[\dfrac{10,904 \times \cos(35)}{5830} \right]^{0.333} - 1 \right\}^{+} \end{array} \right)$$

$$= 2.852 \ \mathrm{W/m^2 \ K}$$

The convection heat transfer coefficient from glass to ambient is the wind loss coefficient given by Eq. (3.28). In this equation, the characteristic length is the length of the collector, equal to 2 m.
Therefore,

$$h_{c,g1-a} = h_w = 8.6(2.5)^{0.6}/2^{0.4} = 11.294 \ \mathrm{W/m^2 \ K}$$

To check whether the assumed values of T_{g1} and T_{g2} are correct, the heat transfer coefficients are substituted into Eqs. (3.15), (3.17), and (3.22):

$$Q_t/A_c = \left(h_{c,p-g2} + h_{r,p-g2} \right) \left(T_p - T_{g2} \right) = (2.918 + 0.835)(353 - 314.7) = 143.7 \ \mathrm{W/m^2}$$

$$Q_t/A_c = \left(h_{c,g2-g1} + h_{r,g2-g1} \right) \left(T_{g2} - T_{g1} \right) = (2.852 + 5.098)(314.7 - 296.8) = 142.3 \ \mathrm{W/m^2}$$

$$Q_t/A_c = \left(h_{c,g1-a} + h_{r,g1-a} \right) \left(T_{g1} - T_a \right) = (11.294 + 4.991)(296.8 - 288) = 143.3 \ \mathrm{W/m^2}$$

Since these three answers are not exactly equal, further trials should be made by assuming different values for T_{g1} and T_{g2}. This is a laborious process which, however, can be made easier by the use of a computer and artificial intelligence techniques, such as a genetic algorithm (see Chapter 11). Furthermore, the Excel function Solver or Goal Seek (for single-glazed collectors) can be used quite easily. Details on the use of these functions are given in Appendix 9. Following these techniques, the values that solve the problem are $T_{g1} = 296.80$ K and $T_{g2} = 314.81$ K. These two values give $Q_t/A_c = 143.3 \ \mathrm{W/m^2}$ for all cases. If we assume that the values $T_{g1} = 296.8$ K and $T_{g2} = 314.7$ K are correct (remember, they were chosen to be almost correct from the beginning), U_t can be calculated from:

$$U_t = \left(\frac{1}{h_{c,p-g2} + h_{r,p-g2}} + \frac{1}{h_{c,g2-g1} + h_{r,g2-g1}} + \frac{1}{h_{c,g1-a} + h_{r,g1-a}} \right)^{-1}$$

$$= \left(\frac{1}{2.918 + 0.835} + \frac{1}{2.852 + 5.098} + \frac{1}{11.294 + 4.991} \right)^{-1} = 2.204 \ \mathrm{W/m^2 \ K}$$

EXAMPLE 3.3

Repeat Example 3.2 using the empirical Eq. (3.25) and compare the results.

Solution

First, the constant parameters are estimated. The value of h_w is already estimated in Example 3.2 and is equal to 11.294 W/m^2 K.

From Eq. (3.26),

$$f = \left(1 - 0.04h_w + 0.0005h_w^2\right)\left(1 + 0.091N_g\right)$$

$$f = \left(1 - 0.04 \times 11.294 + 0.0005 \times 11.294^2\right)\left(1 + 0.091 \times 2\right) = 0.723$$

From Eq. (3.27),

$$C = 365.9\left(1 - 0.00883\beta + 0.0001298\beta^2\right)$$

$$C = 365.9\left(1 - 0.00883 \times 35 + 0.0001298 \times 35^2\right) = 311$$

Therefore, from Eq. (3.25),

$$U_t = \cfrac{1}{\cfrac{C}{T_p}\left[\cfrac{T_p - T_a}{N_g + f}\right]^{0.33} + \cfrac{1}{h_w}} + \cfrac{\sigma\left(T_p^2 + T_a^2\right)\left(T_p + T_a\right)}{\cfrac{1}{\varepsilon_p + 0.05N_g\left(1 - \varepsilon_p\right)} + \cfrac{2N_g + f - 1}{\varepsilon_g} - N_g}$$

$$= \cfrac{1}{\cfrac{311}{353}\left[\cfrac{(353 - 288)}{(2 + 0.723)}\right]^{0.33} + \cfrac{1}{11.294}} + \cfrac{5.67 \times 10^{-8}\left(353^2 + 288^2\right)(353 + 288)}{\cfrac{1}{0.1 + 0.05 \times 2(1 - 0.1)} + \cfrac{(2 \times 2) + 0.723 - 1}{0.88} - 2} = 2.306 \ W/m^2 \ K$$

The difference between this value and the one obtained in Example 3.2 is only 4.6%, but the latter was obtained with much less effort.

3.3.3 Temperature distribution between the tubes and collector efficiency factor

Under steady-state conditions, the rate of useful heat delivered by a solar collector is equal to the rate of energy absorbed by the heat transfer fluid minus the direct or indirect heat losses from the surface to the surroundings (see Fig. 3.29). As shown in Fig. 3.29, the absorbed solar radiation is equal to $G_t(\tau\alpha)$, which is similar to Eq. (3.6). The thermal energy lost from the collector to the surroundings by conduction, convection, and infrared radiation is represented by the product of the overall heat loss coefficient, U_L, times the difference between the plate temperature, T_p, and the ambient temperature, T_a. Therefore, in a steady state, the rate of useful energy collected from a collector of area A_c can be obtained from:

$$Q_u = A_c\left[G_t(\tau\alpha) - U_L\left(T_p - T_a\right)\right] = \dot{m}c_p[T_o - T_i] \tag{3.31}$$

As indicated above, Eq. (3.31) is an energy rate equation in Watts (W) when S is expressed in W/m^2 and U_L in W/m^2-K.

Equation (3.31) can also be used to give the amount of useful energy delivered in joules (not rate in watts), if the irradiance G_t (W/m^2) is replaced with irradiation I_t (J/m^2) and we multiply U_L, which is given in watts per square meter-degrees Centigrade ($W/m^2 \, °C$), by 3600 to convert to joules per square

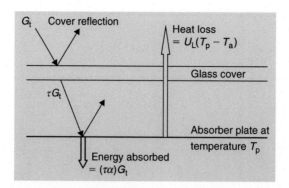

FIGURE 3.29

Radiation input and heat loss from a flat-plate collector.

meter-degrees Centigrade (J/m^2 °C) for estimations with step of 1 h. We could have indicated Eq. (3.31) with a rate symbol ($\dot{Q}_u$) and the hourly useful energy by (Q_u); however, it is always clear for the use of either G (in W/m^2) or I (in J/m^2) in the evaluation of S if we are dealing with a rate value or hourly energy value, so it is not necessary to use different symbols.

To model the collector shown in Fig. 3.29, a number of assumptions, which simplify the problem, need to be made. These assumptions are not against the basic physical principles and are as follows:

1. The collector is in a steady state.
2. The collector is of the header and riser type fixed on a sheet with parallel tubes.
3. The headers cover only a small area of the collector and can be neglected.
4. Headers provide uniform flow to the riser tubes.
5. Flow through the back insulation is one dimensional.
6. The sky is considered as a blackbody for the long-wavelength radiation at an equivalent sky temperature. Since the sky temperature does not affect the results much, this is considered equal to the ambient temperature.
7. Temperature gradients around tubes are neglected.
8. Properties of materials are independent of temperature.
9. No solar energy is absorbed by the cover.
10. Heat flow through the cover is one dimensional.
11. Temperature drop through the cover is negligible.
12. Covers are opaque to infrared radiation.
13. Same ambient temperature exists at the front and back of the collector.
14. Dust effects on the cover are negligible.
15. There is no shading of the absorber plate.

The collector efficiency factor can be calculated by considering the temperature distribution between two pipes of the collector absorber and assuming that the temperature gradient in the flow direction is negligible (Duffie and Beckman, 2006). This analysis can be performed by considering the sheet—tube configuration shown in Fig. 3.30a, where the distance between the tubes is W, the tube diameter is D, and the sheet thickness is δ. Since the sheet metal is usually made from copper or aluminum, which are good conductors of heat, the temperature gradient through the sheet is negligible; therefore, the region between the center line separating the tubes and the tube base can be considered as a classical fin problem.

FIGURE 3.30

Flat-plate sheet and tube configuration. (a) Schematic diagram. (b) Energy balance for the fin element. (c) Energy balance for the tube element.

The fin, shown in Fig. 3.30b, is of length $L = (W - D)/2$. An elemental region of width, Δx, and unit length in the flow direction are shown in Fig. 3.30c. The solar energy absorbed by this small element is $S\Delta x$ and the heat loss from the element is $U_L\Delta x(T_x - T_a)$, where T_x is the local plate temperature. Therefore, an energy balance on this element gives:

$$S\Delta x - U_L\Delta x(T - T_a) + \left(-k\delta\frac{dT}{dx}\right)\Big|_x - \left(-k\delta\frac{dT}{dx}\right)\Big|_{x+\Delta x} = 0 \tag{3.32}$$

where S is the absorbed solar energy. Dividing through with Δx and finding the limit as Δx approaches 0 gives:

$$\frac{d^2T}{dx^2} = \frac{U_L}{k\delta}\left(T - T_a - \frac{S}{U_L}\right) \tag{3.33}$$

The two boundary conditions necessary to solve this second-order differential equation are:

$$\left.\frac{dT}{dx}\right|_{x=0} = 0$$

and

$$T\big|_{x=L} = T_b$$

For convenience, the following two variables are defined:

$$m = \sqrt{\frac{U_L}{k\delta}} \tag{3.34}$$

$$\Psi = T - T_a - \frac{S}{U_L} \tag{3.35}$$

Therefore, Eq. (3.33) becomes:

$$\frac{d^2\Psi}{dx^2} - m^2\Psi = 0 \tag{3.36}$$

which has the boundary conditions:

$$\left.\frac{d\Psi}{dx}\right|_{x=0} = 0$$

and

$$\Psi\big|_{x=L} = T_b - T_a - \frac{S}{U_L}$$

Equation (3.36) is a second-order homogeneous linear differential equation whose general solution is:

$$\Psi = C_1' e^{mx} + C_2' e^{-mx} = C_1 \sinh(mx) + C_2 \cosh(mx) \tag{3.37}$$

The first boundary yields $C_1 = 0$, and the second boundary condition yields:

$$\Psi = T_b - T_a - \frac{S}{U_L} = C_2 \cosh(mL)$$

or

$$C_2 = \frac{T_b - T_a - S/U_L}{\cosh(mL)}$$

With C_1 and C_2 known, Eq. (3.37) becomes:

$$\frac{T - T_a - S/U_L}{T_b - T_a - S/U_L} = \frac{\cosh(mx)}{\cosh(mL)} \tag{3.38}$$

This equation gives the temperature distribution in the x direction at any given y.

The energy conducted to the region of the tube per unit length in the flow direction can be found by evaluating the Fourier's law at the fin base (Kalogirou, 2004):

$$q_{fin}' = -k\delta \left.\frac{dT}{dx}\right|_{x=L} = \frac{k\delta m}{U_L}[S - U_L(T_b - T_a)]\tanh(mL) \tag{3.39}$$

However, $k\delta m/U_L$ is just $1/m$. Equation (3.39) accounts for the energy collected on only one side of the tube; for both sides, the energy collection is:

$$q'_{\text{fin}} = (W - D)[S - U_L(T_b - T_a)]\frac{\tanh[m(W-D)/2]}{m(W-D)/2} \tag{3.40}$$

or with the help of fin efficiency,

$$q'_{\text{fin}} = (W - D)F[S - U_L(T_b - T_a)] \tag{3.41}$$

where factor F in Eq. (3.41) is the standard fin efficiency for straight fins with a rectangular profile, obtained from:

$$F = \frac{\tanh[m(W-D)/2]}{m(W-D)/2} \tag{3.42}$$

The useful gain of the collector also includes the energy collected above the tube region. This is given by:

$$q'_{\text{tube}} = D[S - U_L(T_b - T_a)] \tag{3.43}$$

Accordingly, the useful energy gain per unit length in the direction of the fluid flow is:

$$q'_u = q'_{\text{fin}} + q'_{\text{tube}} = [(W - D)F + D][S - U_L(T_b - T_a)] \tag{3.44}$$

This energy ultimately must be transferred to the fluid, which can be expressed in terms of two resistances as:

$$q'_u = \frac{T_b - T_f}{\dfrac{1}{h_{\text{fi}}\pi D_i} + \dfrac{1}{C_b}} \tag{3.45}$$

where $h_{\text{fi}} =$ heat transfer coefficient between the fluid and the tube wall (see Section 3.6.4 for details).

In Eq. (3.45), C_b is the bond conductance, which can be estimated from knowledge of the bond thermal conductivity, k_b, the average bond thickness, γ, and the bond width, b. The bond conductance on a per unit length basis is given by (Kalogirou, 2004):

$$C_b = \frac{k_b b}{\gamma} \tag{3.46}$$

The bond conductance can be very important in accurately describing the collector performance. Generally it is necessary to have good metal-to-metal contact so that the bond conductance is greater than 30 W/m K, and preferably the tube should be welded to the fin.

Solving Eq. (3.45) for T_b, substituting it into Eq. (3.44), and solving the resultant equation for the useful gain, we get:

$$q'_u = WF'[S - U_L(T_f - T_a)] \tag{3.47}$$

where F' is the collector efficiency factor, given by:

$$F' = \frac{\dfrac{1}{U_L}}{W\left\{\dfrac{1}{U_L[D + (W - D)F]} + \dfrac{1}{C_b} + \dfrac{1}{\pi D_i h_{\text{fi}}}\right\}} \tag{3.48}$$

A physical interpretation of F' is that it represents the ratio of the actual useful energy gain to the useful energy gain that would result if the collector-absorbing surface had been at the local fluid temperature. It should be noted that the denominator of Eq. (3.48) is the heat transfer resistance from the fluid to the ambient air. This resistance can be represented as $1/U_o$. Therefore, another interpretation of F' is:

$$F' = \frac{U_o}{U_L} \tag{3.49}$$

The collector efficiency factor is essentially a constant factor for any collector design and fluid flow rate. The ratio of U_L to C_b, the ratio of U_L to h_{fi}, and the fin efficiency, F, are the only variables appearing in Eq. (3.48) that may be functions of temperature. For most collector designs, F is the most important of these variables in determining F'. The factor F' is a function of U_L and h_{fi}, each of which has some temperature dependence, but it is not a strong function of temperature. Additionally, the collector efficiency factor decreases with increased tube center-to-center distances and increases with increase in both material thicknesses and thermal conductivity. Increasing the overall loss coefficient decreases F', while increasing the fluid–tube heat transfer coefficient increases F'.

It should be noted that if the tubes are centered in the plane of the plate and are integral to the plate structure as shown in Fig. 3.3c, the bond conductance term, $1/C_b$, is eliminated from Eq. (3.48).

EXAMPLE 3.4

For a collector having the following characteristics and ignoring the bond resistance, calculate the fin efficiency and the collector efficiency factor.

Overall loss coefficient = 6.9 W/m² °C.
Tube spacing = 120 mm.
Tube outside diameter = 15 mm.
Tube inside diameter = 13.5 mm.
Plate thickness = 0.4 mm.
Plate material = copper.
Heat transfer coefficient inside the tubes = 320 W/m² °C.

Solution

From Table A5.3, in Appendix 5, for copper, $k = 385$ W/m °C.
From Eq. (3.34),

$$m = \sqrt{\frac{U_L}{k\delta}} = \sqrt{\frac{6.9}{385 \times 0.0004}} = 6.69 \text{ m}^{-1}$$

From Eq. (3.42),

$$F = \frac{\tanh[m(W-D)/2]}{m(W-D)/2} = \frac{\tanh[6.69(0.12-0.015)/2]}{6.69[(0.12-0.015)/2]} = 0.961$$

Finally, from Eq. (3.48) and ignoring bond conductance,

$$F' = \frac{\dfrac{1}{U_L}}{W\left[\dfrac{1}{U_L[D+(W-D)F]} + \dfrac{1}{C_b} + \dfrac{1}{\pi D_i h_{fi}}\right]}$$

$$= \frac{1/6.9}{0.12\left[\dfrac{1}{6.9[0.015+(0.12-0.015)\times 0.961]} + \dfrac{1}{\pi \times 0.0135 \times 320}\right]} = 0.912$$

3.3.4 Heat removal factor, flow factor, and thermal efficiency

Consider an infinitesimal length δy of the tube as shown in Fig. 3.31. The useful energy delivered to the fluid is $q'_u \delta y$.

Under steady-state conditions, an energy balance for n tubes gives:

$$q'_u \delta y + \frac{\dot{m}}{n} c_p T_f - \frac{\dot{m}}{n} c_p \left(T_f + \frac{dT_f}{dy} \delta y \right) = 0 \qquad (3.50)$$

Dividing through by δy, finding the limit as δy approaches 0, and substituting Eq. (3.47) results in the following differential equation:

$$\dot{m} c_p \frac{dT_f}{dy} - nWF' \left[S - U_L (T_f - T_a) \right] = 0 \qquad (3.51)$$

Separating variables gives:

$$\frac{dT_f}{T_f - T_a - S/U_L} = \frac{nWF'U_L}{\dot{m} c_p} dy \qquad (3.52)$$

Assuming variables F', U_L, and c_p to be constants and performing the integrations gives:

$$\ln \left(\frac{T_{f,o} - T_a - S/U_L}{T_{f,i} - T_a - S/U_L} \right) = -\frac{nWLF'U_L}{\dot{m} c_p} \qquad (3.53)$$

The quantity nWL in Eq. (3.53) is the collector area A_c. Therefore,

$$\frac{T_{f,o} - T_a - S/U_L}{T_{f,i} - T_a - S/U_L} = \exp \left(-\frac{A_c F' U_L}{\dot{m} c_p} \right) \qquad (3.54)$$

It is usually desirable to express the collector total useful energy gain in terms of the fluid inlet temperature. To do this the collector heat removal factor needs to be used. Heat removal factor represents the ratio of the actual useful energy gain that would result if the collector-absorbing surface had been at the local fluid temperature. Expressed symbolically:

$$F_R = \frac{\text{Actual output}}{\text{Output for plate temperature} = \text{Fluid inlet temperature}} \qquad (3.55)$$

or

$$F_R = \frac{\dot{m} c_p (T_{f,o} - T_{f,i})}{A_c \left[S - U_L (T_{f,i} - T_a) \right]} \qquad (3.56)$$

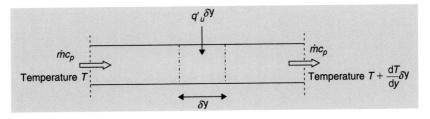

FIGURE 3.31

Energy flow through an element of riser tube.

Rearranging yields:

$$F_R = \frac{\dot{m}c_p}{A_c U_L}\left[1 - \frac{(S/U_L) - (T_{f,o} - T_a)}{(S/U_L) - (T_{f,i} - T_a)}\right] \tag{3.57}$$

Introducing Eq. (3.54) into Eq. (3.57) gives:

$$F_R = \frac{\dot{m}c_p}{A_c U_L}\left[1 - \exp\left(-\frac{U_L F' A_c}{\dot{m}c_p}\right)\right] \tag{3.58}$$

Another parameter usually used in the analysis of collectors is the flow factor. This is defined as the ratio of F_R to F', given by:

$$F'' = \frac{F_R}{F'} = \frac{\dot{m}c_p}{A_c U_L F'}\left(1 - \exp\left[-\frac{U_L F' A_c}{\dot{m}c_p}\right]\right) \tag{3.59}$$

As shown in Eq. (3.59), the collector flow factor is a function of only a single variable, the dimensionless collector capacitance rate, $\dot{m}c_p/A_c U_L F'$, shown in Fig. 3.32.

If we replace the nominator of Eq. (3.56) with Q_u and S with $G_t(\tau\alpha)$ from Eq. (3.6), then the following equation is obtained:

$$Q_u = A_c F_R[G_t(\tau\alpha) - U_L(T_i - T_a)] \tag{3.60}$$

This is the same as Eq. (3.31), with the difference that the inlet fluid temperature (T_i) replaces the average plate temperature (T_p) with the use of the F_R.

In Eq. (3.60), the temperature of the inlet fluid, T_i, depends on the characteristics of the complete solar heating system and the hot water demand or heat demand of the building. However, F_R is affected only by the solar collector characteristics, the fluid type, and the fluid flow rate through the collector.

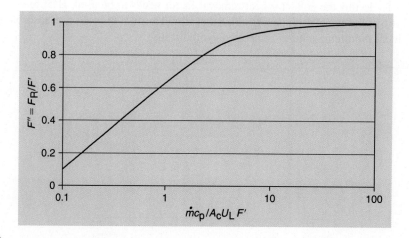

FIGURE 3.32

Collector flow factor as a function of the dimensionless capacitance rate.

From Eq. (3.60), the critical radiation level can also be defined. This is the radiation level where the absorbed solar radiation and loss term are equal. This is obtained by setting the term in the right-hand side of Eq. (3.60) equal to 0 (or $Q_u = 0$). Therefore, the critical radiation level, G_{tc}, is given by:

$$G_{tc} = \frac{F_R U_L (T_i - T_a)}{F_R(\tau\alpha)} \tag{3.61}$$

As in the collector performance tests, described in Chapter 4, the parameters obtained are the $F_R U_L$ and $F_R(\tau\alpha)$, it is preferable to keep F_R in Eq. (3.61). The collector can provide useful output only when the available radiation is higher than the critical one. The collector output can be written in terms of the critical radiation level as $Q_u = A_c F_R(\tau\alpha)(G_t - G_{tc})^+$, which implies that the collector produces useful output only when the absorbed solar radiation is bigger than the thermal losses and G_t is greater than G_{tc}.

Finally, the collector efficiency can be obtained by dividing Q_u, Eq. (3.60), by $(G_t A_c)$. Therefore,

$$\eta = F_R \left[(\tau\alpha) - \frac{U_L(T_i - T_a)}{G_t} \right] \tag{3.62}$$

For incident angles below about 35°, the product $\tau \times \alpha$ is essentially constant and Eqs. (3.60) and (3.62) are linear with respect to the parameter $(T_i - T_a)/G_t$, as long as U_L remains constant.

To evaluate the collector tube inside heat transfer coefficient, h_{fi}, the mean absorber temperature, T_p, is required. This can be found by solving Eqs. (3.60) and (3.31) simultaneously, which gives:

$$T_p = T_i + \frac{Q_u}{A_c F_R U_L}(1 - F_R) \tag{3.63}$$

EXAMPLE 3.5

For the collector outlined in Example 3.4, calculate the useful energy and the efficiency if collector area is 4 m², flow rate is 0.06 kg/s, $(\tau\alpha) = 0.8$, the global solar radiation for 1 h is 2.88 MJ/m², and the collector operates at a temperature difference of 5 °C.

Solution

The dimensionless collector capacitance rate is:

$$\frac{\dot{m}c_p}{A_c U_L F'} = \frac{0.06 \times 4180}{4 \times 6.9 \times 0.91} = 9.99$$

From Eq. (3.59),

$$F'' = 9.99\left(1 - e^{-1/9.99}\right) = 0.952$$

Therefore, the heat removal factor is ($F' = 0.912$ from Example 3.4):

$$F_R = F' \times F'' = 0.912 \times 0.952 = 0.868$$

From Eq. (3.60) modified to use I_t instead of G_t,

$$Q_u = A_c F_R [I_t(\tau\alpha) - U_L(T_i - T_a) \times 3.6] = 4 \times 0.868[2.88 \times 10^3 \times 0.8 - 6.9 \times 5 \times 3.6] = 7550 \text{ kJ}$$

$$= 7.55 \text{ MJ}$$

and the collector efficiency is:

$$\eta = Q_u / A_c I_t = 7.55/(4 \times 2.88) = 0.655, \quad \text{or} \quad 65.5\%$$

3.3.5 Serpentine collector

The analysis presented in Section 3.3.3 concerns the fin and tube assembly in a riser header config-uration. The same analysis applies also for a serpentine collector configuration, shown in the right side of Fig. 3.1a, if the pipes are fixed to separate fins as before. If, however, the absorber plate is a continuous plate then a reduced performance is obtained and the analysis is more complicated. The original analysis of this arrangement for a single bend (N) was made by Abdel-Khalik (1976) and subsequently Zhang and Lavan (1985) gave an analysis for up to four bends. In this analysis the heat removal factor, F_R, is given in terms of various dimensionless parameters as follows:

$$F_R = F_1 F_3 \left[1 + \frac{2F_4 F_5}{F_6 \mathrm{Exp}\left(-\dfrac{\sqrt{1 - F_2^2}}{F_3} \right) + F_4} - \frac{1}{F_2} - F_5 \right] \tag{3.64a}$$

where

$$F_1 = \frac{KNL}{U_L A_c} \frac{KR(1+\gamma)^2 - 1 - \gamma - KR}{[KR(1+\gamma) - 1]^2 - (KR)^2} \tag{3.64b}$$

$$F_2 = \frac{1}{KR(1+\gamma)^2 - 1 - \gamma - KR} \tag{3.64c}$$

$$F_3 = \frac{\dot{m}c_p}{F_1 U_L A_c} \tag{3.64d}$$

$$F_4 = \frac{1}{F_2} + F_5 - 1 \tag{3.64e}$$

$$F_5 = \sqrt{\frac{1 - F_2^2}{F_2^2}} \tag{3.64f}$$

$$F_6 = 1 - \frac{1}{F_2} + F_5 \tag{3.64g}$$

$$K = \frac{k\delta n}{(W - D)\sinh(n)} \tag{3.64h}$$

$$n = (W - D)m \tag{3.64i}$$

$$\gamma = -2\cosh(n) - \frac{DU_L}{K} \tag{3.64j}$$

$$R = \frac{1}{C_b} + \frac{1}{\pi D_i h_{fi}} \tag{3.64k}$$

It should be noted that m is given by Eq. (3.34) and C_b is given by Eq. (3.46). Recognizing that A_c is equal to NWL, Eq. (3.64b) can be written as:

$$F_1 = \frac{K}{U_L W} \frac{KR(1+\gamma)^2 - 1 - \gamma - KR}{[KR(1+\gamma) - 1]^2 - (KR)^2} \tag{3.65a}$$

Similarly, by applying Eq. (3.34) for m and applying Eq. (3.64i) to Eq. (3.64h):

$$K = \frac{\sqrt{k\delta U_L}}{\sinh(n)} \tag{3.65b}$$

Finally, a more simplified form of Eq. (3.64a) can be obtained by applying Eq. (3.64e) and performing some simple modifications:

$$F_R = F_1 F_3 F_4 \left[\frac{2F_5}{F_6 \mathrm{Exp}\left(-\dfrac{\sqrt{1 - F_2^2}}{F_3} \right) + F_4} - 1 \right] \tag{3.66}$$

3.3.6 Heat losses from unglazed collectors

When no glazing is used in a flat-plate collector there is no transmittance loss but the radiation and convection losses become very important. In this case the basic performance equation, by ignoring the ground-reflected radiation, is given by:

$$Q_u = A_c \left[\alpha G_t - h_c(T_i - T_a) - \varepsilon\sigma\left(T_i^4 - T_{sky}^4 \right) \right] \tag{3.67}$$

By adding and subtracting T_a^4 to the last term and performing some simple manipulations we get:

$$Q_u = A_c [\alpha G_t + \varepsilon G_L - h_c(T_i - T_a) - h_r(T_i - T_a)] \tag{3.68a}$$

or

$$Q_u = A_c [\alpha G_t + \varepsilon G_L - U_L(T_i - T_a)] \tag{3.68b}$$

where

$$U_L = h_c + h_r \tag{3.68c}$$

$$G_L = \sigma\left(T_{sky}^4 - T_a^4 \right) \tag{3.68d}$$

$$h_r = \varepsilon\sigma(T_i + T_a)\left(T_i^2 + T_a^2 \right) \tag{3.68e}$$

The term G_L given by Eq. (3.68d) is the longwave radiation exchange between the absorber and the sky. It should be noted that if the back and end losses from the collector are important, these should be added to the term U_L, although their magnitude is much lower than the convection and radiation losses. The efficiency can be obtained by diving Eq. (3.68b) by $A_c G_t$, which gives:

$$\eta = \alpha + \varepsilon\frac{G_L}{G_t} - U_L\frac{(T_i - T_a)}{G_t} \tag{3.69a}$$

or

$$\eta = \alpha - U_L\frac{(T_i - T_a)}{G_n} \tag{3.69b}$$

where

$$G_n = G_t + \frac{\varepsilon}{\alpha} G_L \qquad (3.69c)$$

The parameter G_n is called the *net incident radiation*. Typical values of ε/α are about 0.95.

By relating the sky emissivity in terms of ambient temperature, for clear sky conditions, Eq. (3.68d) becomes (Morrison, 2001):

$$G_L = \varepsilon_{sky}\sigma T_a^4 - \sigma T_a^4 = \sigma T_a^4 (\varepsilon_{sky} - 1) \qquad (3.70)$$

3.4 Thermal analysis of air collectors

A schematic diagram of a typical air-heating flat-plate solar collector is shown in Fig. 3.33. The air passage is a narrow duct with the surface of the absorber plate serving as the top cover. The thermal analysis presented so far applies equally well here, except for the fin efficiency and the bond resistance.

An energy balance on the absorber plate of area $(1 \times \delta x)$ gives:

$$S(\delta x) = U_t(\delta x)(T_p - T_a) + h_{c,p-a}(\delta x)(T_p - T) + h_{r,p-b}(\delta x)(T_p - T_b) \qquad (3.71)$$

where

$h_{c,p-a}$ = convection heat transfer coefficient from absorber plate to air (W/m² K);
$h_{r,p-b}$ = radiation heat transfer coefficient from absorber plate to back plate, which can be obtained from Eq. (2.73), (W/m² K).

An energy balance of the air stream volume $(s \times 1 \times \delta x)$ gives:

$$\left(\frac{\dot{m}}{W}\right) c_p \left(\frac{dT}{dx} \delta x\right) = h_{c,p-a}(\delta x)(T_p - T) + h_{c,b-a}(\delta x)(T_b - T) \qquad (3.72)$$

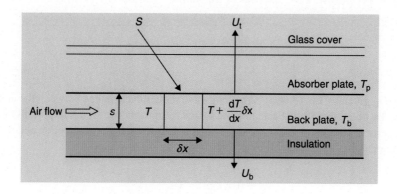

FIGURE 3.33

Schematic diagram of an air-heating collector.

where

$h_{c,b-a}$ = convection heat transfer coefficient from the back plate to air (W/m^2 K).

An energy balance on the back plate area $(1 \times \delta x)$ gives:

$$h_{r,p-b}(\delta x)(T_p - T_b) = h_{c,b-a}(\delta x)(T_b - T) + U_b(\delta x)(T_b - T_a) \tag{3.73}$$

As U_b is much smaller than U_t, $U_L \approx U_t$. Therefore, neglecting U_b and solving Eq. (3.73) for T_b gives:

$$T_b = \frac{h_{r,p-b}T_p + h_{c,b-a}T}{h_{r,p-b} + h_{c,b-a}} \tag{3.74}$$

Substituting Eq. (3.74) into Eq. (3.71) gives:

$$T_p(U_L + h) = S + U_L T_a + hT \tag{3.75}$$

where

$$h = h_{c,p-a} + \frac{1}{(1/h_{c,b-a}) + (1/h_{r,p-b})} \tag{3.76}$$

Substituting Eq. (3.74) into Eq. (3.72) gives:

$$hT_p = \left(\frac{\dot{m}}{W}\right)c_p\frac{dT}{dx} + hT \tag{3.77}$$

Finally, combining Eqs. (3.75) and (3.77) gives:

$$\left(\frac{\dot{m}}{W}\right)c_p\frac{dT}{dx} = F'[S - U_L(T - T_a)] \tag{3.78}$$

where F' = collector efficiency factor for air collectors, given by:

$$F' = \frac{1/U_L}{(1/U_L) + (1/h)} = \frac{h}{h + U_L} \tag{3.79}$$

The initial conditions of Eq. (3.78) are $T = T_i$ at $x = 0$. Therefore, the complete solution of Eq. (3.78) is:

$$T = \left(\frac{S}{U_L} + T_a\right) - \frac{1}{U_L}[S - U_L(T_i - T_a)]\exp\left[-\frac{U_L F'}{(\dot{m}/W)c_p}x\right] \tag{3.80}$$

This equation gives the temperature distribution of air in the duct. The temperature of the air at the outlet for the collector is obtained from Eq. (3.80), using $x = L$ and considering $A_c = WL$. Therefore,

$$T_o = T_i + \frac{1}{U_L}[S - U_L(T_i - T_a)]\left[1 - \exp\left(-\frac{A_c U_L F'}{\dot{m}c_p}\right)\right] \tag{3.81}$$

The energy gain by the air stream is then given by:

$$\frac{Q_u}{W} = \left(\frac{\dot{m}}{W}\right)c_p(T_o - T_i) = \frac{\dot{m}c_p}{A_c U_L}[S - U_L(T_i - T_a)]\left[1 - \exp\left(-\frac{A_c U_L F'}{\dot{m}c_p}\right)\right]$$ (3.82)

Using the equation for the heat removal factor given by Eq. (3.58), Eq. (3.82) gives:

$$Q_u = A_c F_R[S - U_L(T_i - T_a)]$$ (3.83)

Since $S = (\tau\alpha)G_t$, Eq. (3.83) is essentially the same as Eq. (3.60).

EXAMPLE 3.6

Estimate the outlet air temperature and efficiency of the collector shown in Fig. 3.33 for the following collector specifications.

Collector width, $W = 1.2$ m.
Collector length, $L = 4$ m.
Depth of air channel, $s = 15$ mm.
Total solar radiation, $G_t = 890$ W/m^2
Ambient temperature, $T_a = 15\,°C = 288$ K.
Effective $(\tau\alpha) = 0.90$.
Heat loss coefficient, $U_L = 6.5$ W/m^2 K.
Emissivity of absorber plate, $\varepsilon_p = 0.92$.
Emissivity of back plate, $\varepsilon_b = 0.92$.
Mass flow rate of air $= 0.06$ kg/s.
Inlet air temperature, $T_i = 50\,°C = 323$ K.

Solution

Here we need to start by assuming values for T_p and T_b. To save time the correct values are selected but in an actual situation the solution needs to be found by iteration. The values assumed are $T_p = 357$ K and $T_b = 316.2$ K. From these two temperatures, the mean air temperature can be determined from:

$$4(T_{m,air})^3 = (T_p + T_b)\left(T_p^2 + T_b^2\right)$$

from which

$$T_{m,air} = \sqrt[3]{\frac{(T_p + T_b)\left(T_p^2 + T_b^2\right)}{4}} = \sqrt[3]{\frac{(357 + 316.2)(357^2 + 316.2^2)}{4}} = 337 \text{ K}$$

The radiation heat transfer coefficient from the absorber to the back plate is given by:

$$h_{r,p-b} = \frac{\sigma(T_p + T_b)\left(T_p^2 + T_b^2\right)}{(1/\varepsilon_p) + (1/\varepsilon_b) - 1} = \frac{(5.67 \times 10^{-8})(357 + 316.2)(357^2 + 316.2^2)}{(1/0.92) + (1/0.92) - 1}$$

$$= 7.395 \text{ W/m}^2 \text{ K}$$

From $T_{m,\,air}$, the following properties of air can be obtained from Appendix 5:

$$\mu = 2.051 \times 10^{-5} \text{ kg/m s}$$

$$k = 0.029 \text{ W/m K}$$

$$c_p = 1008 \text{ J/kg K}$$

EXAMPLE 3.6—CONT'D

From fluid mechanics the hydraulic diameter of the air channel is given by:

$$D = 4\left(\frac{\text{Flow cross} - \text{sectional area}}{\text{Wetted perimeter}}\right) = 4\left(\frac{Ws}{2W}\right) = 2s = 2 \times 0.015 = 0.03$$

The Reynolds number is given by:

$$\text{Re} = \frac{\rho VD}{\mu} = \frac{\dot{m}D}{A\mu} = \frac{0.06 \times 0.03}{(1.2 \times 0.015) \times 2.051 \times 10^{-5}} = 4875.7$$

Therefore, the flow is turbulent, for which the following equation applies: $\text{Nu} = 0.0158(\text{Re})^{0.8}$. Since $\text{Nu} = (h_c D)/k$, the convection heat transfer coefficient is given by:

$$h_{c,p-a} = h_{c,b-a} = \left(\frac{k}{D}\right)0.0158(\text{Re})^{0.8} = \left(\frac{0.029}{0.03}\right)0.0158(4875.7)^{0.8} = 13.626 \text{ W}/\text{m}^2 \text{ K}$$

From Eq. (3.76),

$$h = h_{c,p-a} + \frac{1}{(1/h_{c,b-a}) + (1/h_{r,p-b})} = 13.626 + \frac{1}{(1/13.626) + (1/7.395)} = 18.4 \text{ W}/\text{m}^2 \text{ K}$$

From Eq. (3.79),

$$F' = \frac{h}{h + U_L} = \frac{18.4}{18.4 + 6.5} = 0.739$$

The absorbed solar radiation is:

$$S = G_t(\tau\alpha) = 890 \times 0.9 = 801 \text{ W}/\text{m}^2$$

From Eq. (3.81),

$$T_o = T_i + \frac{1}{U_L}[S - U_L(T_i - T_a)]\left[1 - \exp\left(-\frac{A_c U_L F'}{\dot{m}c_p}\right)\right]$$

$$= 323 + \left(\frac{1}{6.5}\right)[801 - 6.5(323 - 288)]\left[1 - \exp\left(-\frac{(1.2 \times 4) \times 6.5 \times 0.739}{0.06 \times 1007}\right)\right] = 351 \text{ K}$$

Therefore, the average air temperature is ½(351 + 323) = 337 K, which is the same as the value assumed before. If there is a difference in the two mean values, an iteration is required. This kind of problem requires just one iteration to find the correct solution by using the assumed values, which give the new mean temperature.

From Eq. (3.58),

$$F_R = \frac{\dot{m}c_p}{A_c U_L}\left\{1 - \exp\left[-\frac{U_L F' A_c}{\dot{m}c_p}\right]\right\}$$

$$= \frac{0.06 \times 1008}{(1.2 \times 4) \times 6.5}\left\{1 - \exp\left[-\frac{6.5 \times 0.739 \times (1.2 \times 4)}{0.06 \times 1008}\right]\right\} = 0.614$$

From Eq. (3.83),

$$Q_u = A_c F_R[S - U_L(T_i - T_a)] = (1.2 \times 4) \times 0.614[801 - 6.5(323 - 288)] = 1690 \text{ W}$$

Finally, the collector efficiency is:

$$\eta = \frac{Q_u}{A_c G_t} = \frac{1690}{(1.2 \times 4) \times 890} = 0.396$$

Finally, another check required is whether the two values of T_p and T_b are correct. In order to check that the assumptions are correct T_p needs to be estimated with Eq. (3.63), which in this case gives a value of 357.06 K (so the assumption also for T_p was correct). If there is a difference T_p is set to this value and T_b needs to be assumed to satisfy the correct $T_{m,air}$ value (337 K).

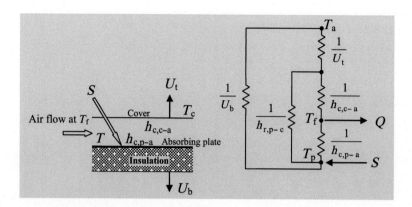

FIGURE 3.34

Solar air heater and its thermal resistance network.

Another case of air collector is to have airflow between the absorbing plate and the glass cover. This is shown graphically in Fig. 3.34 together with the thermal resistance network.

By following an energy balance of the cover, plate and fluid flowing through the collector the following set of equations can be obtained:

$$U_t(T_a - T_c) + h_{r,p-c}(T_p - T_c) + h_{c,c-a}(T_f - T_c) = 0 \tag{3.84a}$$

$$S + U_b(T_a - T_p) + h_{c,p-a}(T_f - T_p) + h_{r,p-c}(T_c - T_p) = 0 \tag{3.84b}$$

$$h_{c,c-a}(T_c - T_f) + h_{c,p-a}(T_p - T_f) = Q_u \tag{3.84c}$$

In these equations $h_{r,p-c}$ represent the radiation heat transfer coefficient from absorbing plate to cover and is given by Eq. (2.73) and

$h_{c,c-a}$ = convection heat transfer coefficient from cover to air (W/m^2 K);
$h_{c,p-a}$ = convection heat transfer coefficient from absorbing plate to air (W/m^2 K).

By performing some long algebraic manipulations to eliminate T_p and T_c the rate of useful energy can be obtained from:

$$Q_u = F'[S - U_L(T_f - T_a)] \tag{3.85}$$

where

$$F' = \frac{h_{r,p-c}h_{c,c-a} + U_t h_{c,p-a} + h_{c,p-a}h_{r,p-c} + h_{c,c-a}h_{c,p-a}}{(U_t + h_{r,p-c} + h_{c,c-a})(U_b + h_{c,p-a} + h_{r,p-c}) - h_{r,p-c}^2} \tag{3.86a}$$

$$U_L = \frac{(U_b + U_t)(h_{r,p-c}h_{c,c-a} + h_{c,p-a}h_{r,p-c} + h_{c,c-a}h_{c,p-a}) + U_b U_t(h_{c,c-a} + h_{c,p-a})}{h_{r,p-c}h_{c,c-a} + U_t h_{c,p-a} + h_{c,p-a}h_{r,p-c} + h_{c,c-a}h_{c,p-a}} \tag{3.86b}$$

3.5 Practical considerations for flat-plate collectors

For various reasons, the actual performance of an FPC may be different from the one obtained from the theoretical analysis presented in this section. The first reason is that the fluid flowing through the collector may not be uniform through all risers due to manufacturing errors. The section of the collector receiving a lower flow rate will have a lower F_R and therefore inferior performance. Leaks in air collectors are another reason for poorer performance. Additionally, for multi-panel collectors with serpentine absorbers, which are installed one next to the other, the edge losses are limited to the first and last collectors of the array, resulting in an improved U_L compared with that of a single collector.

Problems related to freeze protection of collectors are dealt with in Chapter 5. The effect of dust collected on the glass cover of the collector in an urban environment seems to have a negligible effect, and occasional rainfall is adequate to clean the surface. For those wishing to account for dust in temperate climates, it is suggested that radiation absorbed by the collector plate is reduced by 1% and for dry and for dusty climates by 2% (Duffie and Beckman, 2006). Degradation of the cover materials, however, can affect transmittance and seriously affect the long-term performance of the collector. This is more important in plastic collector covers. The same applies for the absorber plate coating. Additionally, the mechanical design of the collector may affect its performance, as for example the penetration of water or moisture into the collector, which would condense on the underside of the glass, thus significantly reducing its properties. A description of quality tests to verify the ability of the collector to withstand this and other effects is given in Chapter 4.

Concerning the manufacture of the collectors, it is important to have a collector casing that will withstand handling and installation and be able to enclose the collector elements and keep them clean from water and dust penetration for the life of the collector. In high latitudes, the collectors should be installed at an inclination to allow the snow to slide off their surface.

Installation of collectors is related to three elements: the transportation and handling of the collector, the installation of brackets, and manifolding. The first is related to the overall weight and size of the collector. For small (~ 2 m^2) collectors this can be done by hand; for bigger collectors, the help of machinery is required. The bracketing should be adequate to withstand wind loading, whereas manifolding can be the most time-consuming operation, although nowadays special bronze fittings are available that make the work easier. Attention is drawn here to the use of dissimilar materials, which can lead to electrolytic corrosion.

Attention to these factors can guarantee many years of trouble-free operation of the collectors, which is very important to both satisfy the customers and promote the use of solar energy.

3.6 Concentrating collectors

As we have seen in Section 3.2, concentrating collectors work by interposing an optical device between the source of radiation and the energy-absorbing surface. Therefore, for concentrating collectors, both optical and thermal analyses are required. In this book, only two types of concentrating collectors are analyzed: compound parabolic and parabolic trough collectors. Initially, the concentration ratio and its theoretical maximum value are defined.

The concentration ratio (C) is defined as the ratio of the aperture area to the receiver–absorber area; that is,

$$C = \frac{A_a}{A_r} \tag{3.87}$$

For FPCs with no reflectors, $C = 1$. For concentrators, C is always greater than 1. Initially the maximum possible concentration ratio is investigated. Consider a circular (three-dimensional) concentrator with aperture A_a and receiver area A_r located at a distance R from the center of the sun, as shown in Fig. 3.35. We saw in Chapter 2 that the sun cannot be considered a point source but a sphere of radius r; therefore, as seen from the earth, the sun has a half angle, θ_m, which is the acceptance half angle for maximum concentration. If both the sun and the receiver are considered to be blackbodies at temperatures T_s and T_r, the amount of radiation emitted by the sun is given by:

$$Q_s = (4\pi r^2)\sigma T_s^4 \tag{3.88}$$

A fraction of this radiation is intercepted by the collector, given by:

$$F_{s-r} = \frac{A_a}{4\pi R^2} \tag{3.89}$$

Therefore, the energy radiated from the sun and received by the concentrator is:

$$Q_{s-r} = A_a \frac{4\pi r^2}{4\pi R^2}\sigma T_s^4 = A_a \frac{r^2}{R^2}\sigma T_s^4 \tag{3.90}$$

A blackbody (perfect) receiver radiates energy equal to $A_r \sigma T_r^4$, and a fraction of this reaches the sun, given by:

$$Q_{r-s} = A_r F_{r-s}\sigma T_r^4 \tag{3.91}$$

Under this idealized condition, the maximum temperature of the receiver is equal to that of the sun. According to the second law of thermodynamics, this is true only when $Q_{r-s} = Q_{s-r}$. Therefore, from Eqs. (3.90) and (3.91),

$$\frac{A_a}{A_r} = \frac{R^2}{r^2}F_{r-s} \tag{3.92}$$

Since the maximum value of F_{r-s} is equal to 1, the maximum concentration ratio for three-dimensional concentrators is $[\sin(\theta_m) = r/R]$:

$$C_{max} = \frac{1}{\sin^2(\theta_m)} \tag{3.93}$$

A similar analysis for linear concentrators gives:

$$C_{max} = \frac{1}{\sin(\theta_m)} \tag{3.94}$$

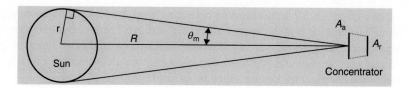

FIGURE 3.35

Schematic of the sun and a concentrator.

As was seen in Chapter 2, $2\theta_m$ is equal to $0.53°$ (or $32'$), so θ_m, the half acceptance angle, is equal to $0.27°$ (or $16'$). The half acceptance angle denotes coverage of one half of the angular zone within which radiation is accepted by the concentrator's receiver. Radiation is accepted over an angle of $2\theta_m$, because radiation incident within this angle reaches the receiver after passing through the aperture. This angle describes the angular field within which radiation can be collected by the receiver without having to track the concentrator.

Equations (3.93) and (3.94) define the upper limit of concentration that may be obtained for a given collector viewing angle. For a stationary CPC, the angle θ_m depends on the motion of the sun in the sky. For a CPC having its axis in an N–S direction and tilted from the horizontal such that the plane of the sun's motion is normal to the aperture, the acceptance angle is related to the range of hours over which sunshine collection is required; for example, for 6 h of useful sunshine collection, $2\theta_m = 90°$ (sun travels $15°$/h). In this case, $C_{max} = 1/\sin(45°) = 1.41$.

For a tracking collector, θ_m is limited by the size of the sun's disk, small-scale errors, irregularities of the reflector surface, and tracking errors. For a perfect collector and tracking system, C_{max} depends only on the sun's disk. Therefore,

For single-axis tracking, $C_{max} = 1/\sin(16') = 216$.

For full tracking, $C_{max} = 1/\sin^2(16') = 46{,}747$.

It can therefore be concluded that the maximum concentration ratio for two-axis tracking collectors is much higher. However, high accuracy of the tracking mechanism and careful construction of the collector are required with an increased concentration ratio, because θ_m is very small. In practice, due to various errors, much lower values than these maximum ones are employed.

EXAMPLE 3.7

From the diameter of the sun and the earth and the mean distance of sun from earth, shown in Fig. 2.1, estimate the amount of energy emitted from the sun, the amount of energy received by the earth, and the solar constant for a sun temperature of 5777 K. If the distance of Venus from the sun is 0.71 times the mean sun–earth distance, estimate the solar constant for Venus.

Solution

The amount of energy emitted from the sun, Q_s, is:

$$Q_s/A_s = \sigma T_s^4 = 5.67 \times 10^{-8} \times (5777)^4 = 63{,}152{,}788 \approx 63 \text{ MW/m}^2$$

or

$$Q_s = 63.15 \times 4\pi \left(1.39 \times 10^9/2\right)^2 = 3.83 \times 10^{20} \text{ MW}$$

From Eq. (3.90), the solar constant can be obtained as:

$$\frac{Q_{s-r}}{A_a} = \frac{r^2}{R^2}\sigma T_s^4 = \frac{\left(1.39 \times 10^9/2\right)^2}{\left(1.496 \times 10^{11}\right)^2} 63{,}152{,}788 = 1363 \text{ W/m}^2$$

The area of the earth exposed to sunshine is $\pi d^2/4$. Therefore, the amount of energy received from earth $= \pi(1.27 \times 10^7)^2 \times 1.363/4 = 1.73 \times 10^{14}$ kW. These results verify the values specified in the introduction to Chapter 2.

The mean distance of Venus from the sun is $1.496 \times 10^{11} \times 0.71 = 1.062 \times 10^{11}$ m. Therefore, the solar constant of Venus is:

$$\frac{Q_{s-r}}{A_a} = \frac{r^2}{R^2}\sigma T_s^4 = \frac{\left(1.39 \times 10^9/2\right)^2}{\left(1.062 \times 10^{11}\right)^2} 63{,}152{,}788 = 2705 \text{ W/m}^2$$

3.6.1 Optical analysis of a compound parabolic collector

The optical analysis of CPC collectors deals mainly with the way to construct the collector shape. A CPC of the Winston design (Winston and Hinterberger, 1975) is shown in Fig. 3.36. It is a linear two-dimensional concentrator consisting of two distinct parabolas, the axes of which are inclined at angles $\pm\theta_c$ with respect to the optical axis of the collector. The angle θ_c, called the collector *half acceptance angle*, is defined as the angle through which a source of light can be moved and still converge at the absorber.

The Winston-type collector is a non-imaging concentrator with a concentration ratio approaching the upper limit permitted by the second law of thermodynamics, as explained in previous section.

The receiver of the CPC does not have to be flat and parallel but, as shown in Fig. 3.5, can be bifacial, a wedge, or cylindrical. Figure 3.37 shows a collector with a cylindrical receiver; the lower portion of the reflector (AB and AC) is circular, while the upper portions (BD and CE) are parabolic. In this design, the requirement for the parabolic portion of the collector is that, at any point P, the normal to the collector must bisect the angle between the tangent line PG to the receiver and the incident ray at point P at angle θ_c with respect to the collector axis. Since the upper part of a CPC contributes little to the radiation reaching the absorber, it is usually truncated, forming a shorter version of the CPC, which is also cheaper. CPCs are usually covered with glass to avoid dust and other materials entering the collector and reducing the reflectivity of its walls. Truncation hardly affects the acceptance angle but results in considerable material saving and changes the height-to-aperture ratio, the concentration ratio, and the average number of reflections.

These collectors are more useful as linear or trough-type concentrators. The orientation of a CPC collector is related to its acceptance angle ($2\theta_c$, in Figs. 3.36 and 3.37). The two-dimensional CPC is an ideal concentrator, that is, it works perfectly for all rays within the acceptance angle, $2\theta_c$. Also, depending on the collector acceptance angle, the collector can be stationary or tracking. A CPC concentrator can be oriented with its long axis along either the north–south or east–west direction and its aperture tilted directly toward the equator at an angle equal to the local latitude. When oriented along the north–south direction, the collector must track the sun by turning its axis to face the sun continuously. Since the acceptance angle of the concentrator along its long axis is wide, seasonal tilt adjustment is not necessary. It can also be stationary, but radiation will be received only during the hours when the sun is within the collector acceptance angle.

When the concentrator is oriented with its long axis along the east–west direction, with a little seasonal adjustment in tilt angle, the collector is able to catch the sun's rays effectively through its wide acceptance angle along its long axis. The minimum acceptance angle in this case should be equal to the maximum incidence angle projected in a north–south vertical plane during the times when output is needed from the collector. For stationary CPC collectors mounted in this mode, the minimum acceptance angle is equal to 47°. This angle covers the declination of the sun from summer to winter solstices ($2 \times 23.5°$). In practice, bigger angles are used to enable the collector to collect diffuse radiation at the expense of a lower concentration ratio. Smaller (less than 3) concentration ratio CPCs are of greatest practical interest. These, according to Pereira (1985), are able to accept a large proportion of diffuse radiation incident on their apertures and concentrate it without the need to track the sun. Finally, the required frequency of collector adjustment is related to the collector concentration ratio. Thus, for $C \leq 3$ the collector needs only biannual adjustment, while for $C > 10$ the collector requires almost daily adjustment; these systems are also called *quasi-static*.

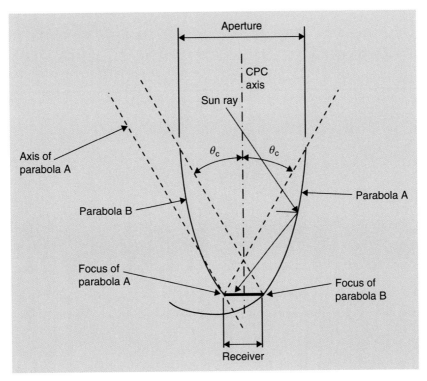

FIGURE 3.36

Construction of a flat receiver compound parabolic collector.

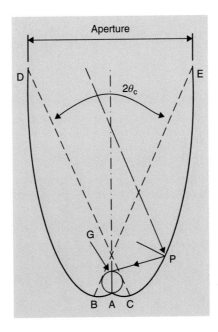

FIGURE 3.37

Schematic diagram of a compound parabolic collector.

Concentrators of the type shown in Fig. 3.5 have an area concentration ratio that is a function of the acceptance half angle, θ_c. For an ideal linear concentrator system, this is given by Eq. (3.94) by replacing θ_m with θ_c.

3.6.2 Thermal analysis of compound parabolic collectors

The instantaneous efficiency, η, of a CPC is defined as the useful energy gain divided by the incident radiation on the aperture plane; that is,

$$\eta = \frac{Q_u}{A_a G_t} \tag{3.95}$$

In Eq. (3.95), G_t is the total incident radiation on the aperture plane. The useful energy, Q_u, is given by an equation similar to Eq. (3.60), using the concept of absorbed radiation as:

$$Q_u = F_R[SA_a - A_r U_L(T_i - T_a)] \tag{3.96}$$

The absorbed radiation, S, is obtained from (Duffie and Beckman, 2006):

$$S = G_{B,CPC}\tau_{c,B}\tau_{CPC,B}\alpha_B + G_{D,CPC}\tau_{c,D}\tau_{CPC,D}\alpha_D + G_{G,CPC}\tau_{c,G}\tau_{CPC,G}\alpha_G \tag{3.97}$$

where

τ_c = transmittance of the CPC cover;
τ_{CPC} = transmissivity of the CPC to account for reflection loss.

The various radiation components in Eq. (3.97) come from radiation falling on the aperture within the acceptance angle of the CPC and are given as follows:

$$G_{B,CPC} = G_{Bn}\cos(\theta) \quad \text{if} \quad (\beta - \theta_c) \le \tan^{-1}[\tan(\Phi)\cos(z)] \le (\beta + \theta_c) \tag{3.98a}$$

$$G_{D,CPC} = \begin{cases} \dfrac{G_D}{C} & \text{if} \quad (\beta + \theta_c) < 90° \\[2mm] \dfrac{G_D}{2}\left(\dfrac{1}{C} + \cos(\beta)\right) & \text{if} \quad (\beta + \theta_c) > 90° \end{cases} \tag{3.98b}$$

$$G_{G,CPC} = \begin{cases} 0 & \text{if} \quad (\beta + \theta_c) < 90° \\[2mm] \dfrac{G_G}{2}\left(\dfrac{1}{C} - \cos(\beta)\right) & \text{if} \quad ((\beta + \theta_c) > 90° \end{cases} \tag{3.98c}$$

In Eqs. (3.98a)–(3.98c), β is the collector aperture inclination angle with respect to horizontal. In Eq. (3.98c), the ground-reflected radiation is effective only if the collector receiver "sees" the ground, that is, $(\beta + \theta_c) > 90°$.

It has been shown by Rabl et al. (1980) that the insolation, G_{CPC}, of a collector with a concentration C can be approximated very well from:

$$G_{CPC} = G_B + \frac{1}{C}G_D = (G_t - G_D) + \frac{1}{C}G_D = G_t - \left(1 - \frac{1}{C}\right)G_D \tag{3.99}$$

It is convenient to express the absorbed solar radiation, S, in terms of G_{CPC} in the following way:

$$S = G_{CPC}\tau_{cover}\tau_{CPC}\alpha_r = \left[G_t - \left(1 - \frac{1}{C}\right)G_D\right]\tau_{cover}\tau_{CPC}\alpha_r$$

$$\tag{3.100}$$

$$= G_t\tau_{cover}\tau_{CPC}\alpha_r\left[1 - \left(1 - \frac{1}{C}\right)\frac{G_D}{G_t}\right]$$

or

$$S = G_t\tau_{cover}\tau_{CPC}\alpha_r\gamma \tag{3.101}$$

where

τ_{cover} = transmissivity of the cover glazing;
τ_{CPC} = effective transmissivity of CPC;
α_r = absorptivity of receiver; and
γ = correction factor for diffuse radiation, given by:

$$\gamma = 1 - \left(1 - \frac{1}{C}\right)\frac{G_D}{G_t} \tag{3.102}$$

The factor γ, given by Eq. (3.102), accounts for the loss of diffuse radiation outside the acceptance angle of the CPC with a concentration C. The ratio G_D/G_t varies from about 0.11 on very clear sunny days to about 0.23 on hazy days.

It should be noted that only part of the diffuse radiation effectively enters the CPC, and this is a function of the acceptance angle. For isotropic diffuse radiation, the relationship between the effective incidence angle and the acceptance half angle is given by (Brandemuehl and Beckman, 1980):

$$\theta_e = 44.86 - 0.0716\theta_c + 0.00512\theta_c^2 - 0.00002798\theta_c^3 \tag{3.103}$$

The effective transmissivity, τ_{CPC}, of the CPC accounts for reflection loss inside the collector. The fraction of the radiation passing through the collector aperture and eventually reaching the absorber depends on the specular reflectivity, ρ, of the CPC walls and the average number of reflections, n, expressed approximately by:

$$\tau_{CPC} = \rho^n \tag{3.104}$$

This equation can also be used to estimate $\tau_{CPC,B}$, $\tau_{CPC,D}$, and $\tau_{CPC,G}$ in Eq. (3.97), which are usually treated as the same. Values of n for full and truncated CPCs can be obtained from Fig. 3.38. As noted before, the upper ends of CPCs contribute little to the radiation reaching the receiver, and usually CPCs are truncated for economic reasons. As can be seen from Fig. 3.38, the average number of reflections is a function of concentration ratio, C, and the acceptance half angle, θ_c. For a truncated concentrator, the line $(1 - 1/C)$ can be taken as the lower bound for the number of reflections for radiation within the acceptance angle. Other effects of truncation are shown in Figs. 3.39 and 3.40. Figures 3.38–3.40 can be used to design a CPC, as shown in the following example. For more accuracy, the equations representing the curves of Figs. 3.38–3.40 can be used as given in Appendix 6.

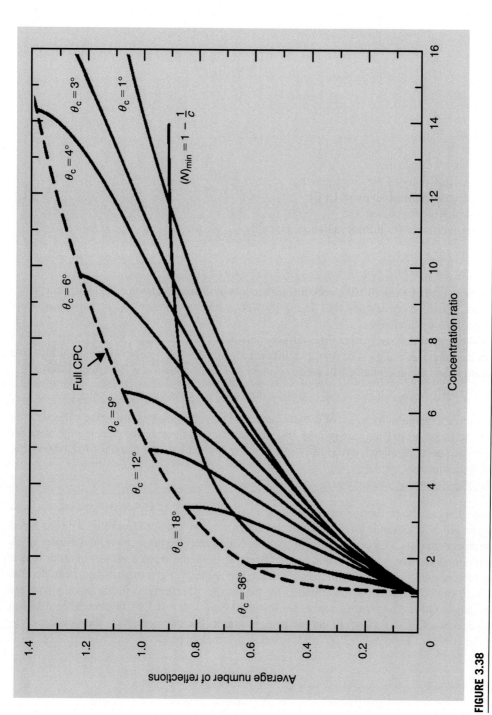

Reprinted from Rabl (1976), with permission from Elsevier.

FIGURE 3.38

Average number of reflections for full and truncated compound parabolic collectors.

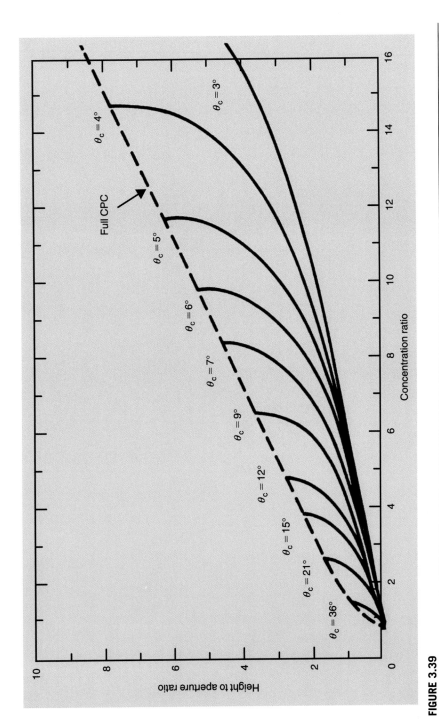

FIGURE 3.39

Ratio of height to aperture for full and truncated compound parabolic collectors.

Reprinted from Rabl (1976), with permission from Elsevier.

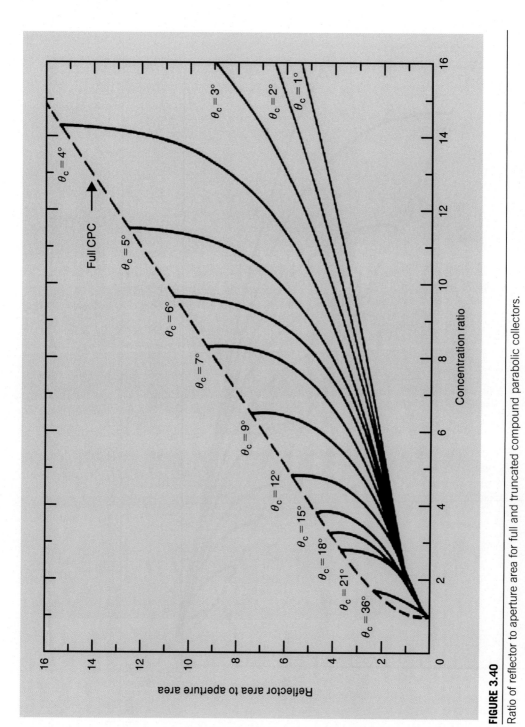

Reprinted from Rabl (1976), with permission from Elsevier.

FIGURE 3.40

Ratio of reflector to aperture area for full and truncated compound parabolic collectors.

EXAMPLE 3.8

Find the CPC characteristics for a collector with acceptance half angle $\theta_c = 12°$. Find also its characteristics if the collector is truncated so that its height-to-aperture ratio is 1.4.

Solution

For a full CPC, from Fig. 3.39 for $\theta_c = 12°$, the height-to-aperture ratio = 2.8 and the concentration ratio = 4.8. From Fig. 3.40, the area of the reflector is 5.6 times the aperture area; and from Fig. 3.38, the average number of reflections of radiation before reaching the absorber is 0.97.

For a truncated CPC, the height-to-aperture ratio = 1.4. Then, from Fig. 3.39, the concentration ratio drops to 4.2; and from Fig. 3.40, the reflector-to-aperture area drops to 3, which indicates how significant is the saving in reflector material. Finally, from Fig. 3.38, the average number of reflections is at least $1 - 1/4.2 = 0.76$.

EXAMPLE 3.9

A CPC has an aperture area of 4 m² and a concentration ratio of 1.7. Estimate the collector efficiency given the following.

Total radiation = 850 W/m².
Diffuse to total radiation ratio = 0.12.
Receiver absorptivity = 0.87.
Mirror reflectivity = 0.90.
Glass cover transmissivity = 0.90.
Collector heat loss coefficient = 2.5 W/m² K.
Circulating fluid = water.
Entering fluid temperature = 80 °C.
Fluid flow rate = 0.015 kg/s.
Ambient temperature = 15 °C.
Collector efficiency factor = 0.92.

Solution

The diffuse radiation correction factor, γ, is estimated from Eq. (3.102):

$$\gamma = 1 - \left(1 - \frac{1}{C}\right)\frac{G_D}{G_t} = 1 - \left(1 - \frac{1}{1.7}\right)0.12 = 0.95$$

From Fig. 3.38 for $C = 1.7$, the average number of reflections for a full CPC is $n = 0.6$. Therefore, from Eq. (3.104),

$$\tau_{CPC} = \rho^n = 0.90^{0.6} = 0.94$$

The absorbed radiation is given by Eq. (3.101):

$$S = G_t \tau_{cover} \tau_{CPC} \alpha_r \gamma = 850 \times 0.90 \times 0.94 \times 0.87 \times 0.95 = 594.3 \text{ W/m}^2$$

The heat removal factor is estimated from Eq. (3.58):

$$F_R = \frac{\dot{m}c_p}{A_c U_L}\left[1 - \exp\left(-\frac{U_L F' A_c}{\dot{m}c_p}\right)\right]\frac{0.015 \times 4180}{4 \times 2.5}\left[1 - \exp\left(-\frac{2.5 \times 0.92 \times 4}{0.015 \times 4180}\right)\right] = 0.86$$

The receiver area is obtained from Eq. (3.87):

$$A_r = A_a/C = 4/1.7 = 2.35 \text{ m}^2$$

The useful energy gain can be estimated from Eq. (3.96):

$$Q_u = F_R[SA_a - A_r U_L(T_i - T_a)] = 0.86[594.3 \times 4 - 2.35 \times 2.5(80 - 15)] = 1716 \text{ W}$$

The collector efficiency is given by Eq. (3.95):

$$\eta = \frac{Q_u}{A_a G_t} = \frac{1716}{4 \times 850} = 0.505 \text{ or } 50.5\%$$

3.6.3 Optical analysis of parabolic trough collectors

A cross-section of a parabolic trough collector is shown in Fig. 3.41, where various important factors are shown. The incident radiation on the reflector at the rim of the collector (where the mirror radius, r_r, is maximum) makes an angle, φ_r, with the center line of the collector, which is called the *rim angle*. The equation of the parabola in terms of the coordinate system is:

$$y^2 = 4fx \tag{3.105}$$

where f = parabola focal distance (m).

For specular reflectors of perfect alignment, the size of the receiver (diameter D) required to intercept all the solar image can be obtained from trigonometry and Fig. 3.41, given by:

$$D = 2r_r \sin(\theta_m) \tag{3.106}$$

where

θ_m = half acceptance angle (degrees).

For a parabolic reflector, the radius, r, shown in Fig. 3.41 is given by:

$$r = \frac{2f}{1 + \cos(\varphi)} \tag{3.107}$$

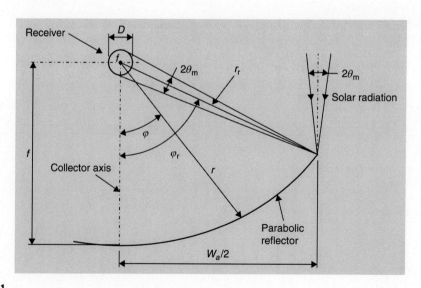

FIGURE 3.41

Cross section of a parabolic trough collector with circular receiver.

where

φ = angle between the collector axis and a reflected beam at the focus; see Fig. 3.41.

As φ varies from 0 to φ_r, r increases from f to r_r and the theoretical image size increases from $2f$ $\sin(\theta_m)$ to $2r_r \sin(\theta_m)/\cos(\varphi_r + \theta_m)$. Therefore, there is an image spreading on a plane normal to the axis of the parabola.

At the rim angle, φ_r, Eq. (3.107) becomes:

$$r_r = \frac{2f}{1 + \cos(\varphi_r)} \tag{3.108}$$

Another important parameter related to the rim angle is the aperture of the parabola, W_a. From Fig. 3.41 and simple trigonometry, it can be found that:

$$W_a = 2r_r \sin(\varphi_r) \tag{3.109}$$

Substituting Eq. (3.108) into Eq. (3.109) gives:

$$W_a = \frac{4f \sin(\varphi_r)}{1 + \cos(\varphi_r)} \tag{3.110}$$

which reduces to:

$$W_a = 4f \tan\left(\frac{\varphi_r}{2}\right) \tag{3.111}$$

The half acceptance angle, θ_m, used in Eq. (3.106) depends on the accuracy of the tracking mechanism and the irregularities of the reflector surface. The smaller these two effects, the closer θ_m is to the sun disk angle, resulting in a smaller image and higher concentration. Therefore, the image width depends on the magnitude of the two quantities. In Fig. 3.41, a perfect collector is assumed and the solar beam is shown striking the collector at an angle $2\theta_m$ and leaving at the same angle. In a practical collector, however, because of the presence of errors, the angle $2\theta_m$ should be increased to include the errors as well. Enlarged images can also result from the tracking mode used to transverse the collector. Problems can also arise due to errors in the positioning of the receiver relative to the reflector, which results in distortion, enlargement, and displacement of the image. All these are accounted for by the intercept factor, which is explained later in this section.

For a tubular receiver, the concentration ratio is given by:

$$C = \frac{W_a}{\pi D} \tag{3.112}$$

By replacing D and W_a with Eqs. (3.106) and (3.110), respectively, we get:

$$C = \frac{\sin(\varphi_r)}{\pi \sin(\theta_m)} \tag{3.113}$$

The maximum concentration ratio occurs when φ_r is 90° and $\sin(\varphi_r) = 1$. Therefore, by replacing $\sin(\varphi_r) = 1$ in Eq. (3.113), the following maximum value can be obtained:

$$C_{max} = \frac{1}{\pi \sin(\theta_m)} \tag{3.114}$$

The difference between this equation and Eq. (3.94) is that this one applies particularly to a PTC with a circular receiver, whereas Eq. (3.94) is the idealized case. So, by using the same sun half acceptance angle of 16′ for single-axis tracking, $C_{max} = 1/\pi\sin(16') = 67.5$.

In fact, the magnitude of the rim angle determines the material required for the construction of the parabolic surface. The curve length of the reflective surface is given by:

$$S = \frac{H_p}{2}\left\{\sec\left(\frac{\varphi_r}{2}\right)\tan\left(\frac{\varphi_r}{2}\right) + \ln\left[\sec\left(\frac{\varphi_r}{2}\right) + \tan\left(\frac{\varphi_r}{2}\right)\right]\right\} \tag{3.115}$$

where

H_p = lactus rectum of the parabola (m). This is the opening of the parabola at the focal point.

As shown in Fig. 3.42 for the same aperture, various rim angles are possible. It is also shown that, for different rim angles, the focus-to-aperture ratio, which defines the curvature of the parabola, changes. It can be demonstrated that, with a 90° rim angle, the mean focus-to-reflector distance and hence the reflected beam spread is minimized, so that the slope and tracking errors are less pronounced. The collector's surface area, however, decreases as the rim angle is decreased. There is thus a temptation to use smaller rim angles because the sacrifice in optical efficiency is small, but the saving in reflective material cost is great.

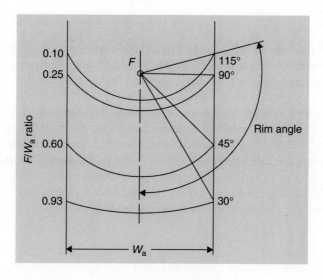

FIGURE 3.42

Parabola focal length and curvature.

EXAMPLE 3.10

For a PTC with a rim angle of 70°, aperture of 5.6 m, and receiver diameter of 50 mm, estimate the focal distance, the concentration ratio, the rim radius, and the length of the parabolic surface.

Solution

From Eq. (3.111),

$$W_a = 4f\tan\left(\frac{\varphi_r}{2}\right)$$

Therefore,

$$f = \frac{W_a}{4\tan(\varphi_r/2)} = \frac{5.6}{4\tan(35)} = 2 \text{ m}$$

From Eq. (3.112), the concentration ratio is:

$$C = W_a/\pi D = 5.6/0.05\pi = 35.7$$

The rim radius is given by Eq. (3.108):

$$r_r = \frac{2f}{1 + \cos(\varphi_r)} = \frac{2 \times 2}{1 + \cos(70)} = 2.98 \text{ m}$$

The parabola lactus rectum, H_p, is equal to W_a at $\varphi_r = 90°$ and $f = 2$ m. From Eq. (3.111),

$$H_p = W_a = 4f\tan\left(\frac{\varphi_r}{2}\right) = 4 \times 2\tan(45) = 8 \text{ m}$$

Finally, the length of the parabola can be obtained from Eq. (3.115) by recalling that $\sec(x) = 1/\cos(x)$:

$$S = \frac{H_p}{2}\left\{\sec\left(\frac{\varphi_r}{2}\right)\tan\left(\frac{\varphi_r}{2}\right) + \ln\left[\sec\left(\frac{\varphi_r}{2}\right) + \tan\left(\frac{\varphi_r}{2}\right)\right]\right\}$$

$$= \frac{8}{2}\{\sec(35)\tan(35) + \ln[\sec(35) + \tan(35)]\} = 6.03 \text{ m}$$

Optical efficiency

Optical efficiency is defined as the ratio of the energy absorbed by the receiver to the energy incident on the collector's aperture. The optical efficiency depends on the optical properties of the materials involved, the geometry of the collector, and the various imperfections arising from the construction of the collector. In equation form (Sodha et al., 1984):

$$\eta_o = \rho\tau\alpha\gamma\left[(1 - A_f\tan(\theta))\cos(\theta)\right] \tag{3.116}$$

where

ρ = reflectance of the mirror;
τ = transmittance of the glass cover;
α = absorptance of the receiver;
γ = intercept factor;
A_f = geometric factor; and
θ = angle of incidence.

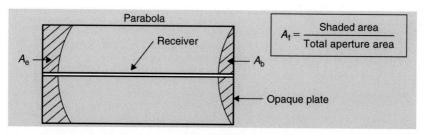

FIGURE 3.43

End effect and blocking in a parabolic trough collector.

The geometry of the collector dictates the geometric factor, A_f, which is a measure of the effective reduction of the aperture area due to abnormal incidence effects, including blockages, shadows, and loss of radiation reflected from the mirror beyond the end of the receiver. During abnormal operation of a PTC, some of the rays reflected from near the end of the concentrator opposite the sun cannot reach the receiver. This is called the *end effect*. The amount of aperture area lost is shown in Fig. 3.43 and given by:

$$A_e = f W_a \tan(\theta)\left[1 + \frac{W_a^2}{48f^2}\right] \qquad (3.117)$$

Usually, collectors of this type are terminated with opaque plates to preclude unwanted or dangerous concentration away from the receiver. These plates result in blockage or shading of a part of the reflector, which in effect reduces the aperture area. For a plate extending from rim to rim, the lost area is shown in Fig. 3.43 and given by:

$$A_b = \frac{2}{3} W_a h_p \tan(\theta) \qquad (3.118)$$

where

h_p = height of parabola (m).

It should be noted that the term $\tan(\theta)$ shown in Eqs. (3.117) and (3.118) is the same as the one shown in Eq. (3.116), and it should not be used twice. Therefore, to find the total loss in aperture area, A_l, the two areas, A_e and A_b, are added together without including the term $\tan(\theta)$ (Jeter, 1983):

$$A_l = \frac{2}{3} W_a h_p + f W_a\left[1 + \frac{W_a^2}{48f^2}\right] \qquad (3.119)$$

Finally, the geometric factor is the ratio of the lost area to the aperture area. Therefore,

$$A_f = \frac{A_l}{A_a} \qquad (3.120)$$

The most complex parameter involved in determining the optical efficiency of a PTC is the intercept factor. This is defined as the ratio of the energy intercepted by the receiver to the energy reflected by the focusing device, that is, the parabola. Its value depends on the size of the receiver, the surface angle errors of the parabolic mirror, and the solar beam spread.

The errors associated with the parabolic surface are of two types: random and non-random (Guven and Bannerot, 1985). *Random errors* are defined as those errors that are truly random in nature and, therefore, can be represented by normal probability distributions. Random errors are identified as apparent changes in the sun's width, scattering effects caused by random slope errors (i.e., distortion of the parabola due to wind loading), and scattering effects associated with the reflective surface. *Non-random errors* arise in manufacture assembly or the operation of the collector. These can be identified as reflector profile imperfections, misalignment errors, and receiver location errors. Random errors are modeled statistically, by determining the standard deviation of the total reflected energy distribution, at normal incidence (Guven and Bannerot, 1986), and are given by:

$$\sigma = \sqrt{\sigma_{sun}^2 + 4\sigma_{slope}^2 + \sigma_{mirror}^2} \tag{3.121}$$

Non-random errors are determined from a knowledge of the misalignment angle error β (i.e., the angle between the reflected ray from the center of the sun and the normal to the reflector's aperture plane) and the displacement of the receiver from the focus of the parabola (d_r). Since reflector profile errors and receiver mislocation along the Y axis essentially have the same effect, a single parameter is used to account for both. According to Guven and Bannerot (1986), random and non-random errors can be combined with the collector geometric parameters, concentration ratio (C), and receiver diameter (D) to yield error parameters universal to all collector geometries. These are called *universal error parameters*, and an asterisk is used to distinguish them from the already defined parameters. Using the universal error parameters, the formulation of the intercept factor, γ, is possible (Guven and Bannerot, 1985):

$$\gamma = \frac{1 + \cos(\varphi_r)}{2 \sin(\varphi_r)} \int_0^{\varphi_r} Erf\left\{ \frac{\sin(\varphi_r)[1 + \cos(\varphi)][1 - 2d^* \sin(\varphi)] - \pi\beta^*[1 + \cos(\varphi_r)]}{\sqrt{2}\pi\sigma^*[1 + \cos(\varphi_r)]} \right\}$$

$$- Erf\left\{ -\frac{\sin(\varphi_r)[1 + \cos(\varphi)][1 + 2d^* \sin(\varphi)] + \pi\beta^*[1 + \cos(\varphi_r)]}{\sqrt{2}\pi\sigma^*[1 + \cos(\varphi_r)]} \right\} \frac{d\varphi}{[1 + \cos(\varphi)]} \tag{3.122}$$

where

d^* = universal non-random error parameter due to receiver mislocation and reflector profile errors, $d^* = d_r/D$;
β^* = universal non-random error parameter due to angular errors, $\beta^* = \beta C$;
σ^* = universal random error parameter, $\sigma^* = \sigma C$;
C = collector concentration ratio, $= A_a/A_r$;
D = riser tube outside diameter (m);
d_r = displacement of receiver from focus (m); and
β = misalignment angle error (degrees).

Another type of analysis commonly carried out in concentrating collectors is ray tracing. This is the process of following the paths of a large number of rays of incident radiation through the optical system to determine the distribution and intensity of the rays on the surface of the receiver. Ray tracing determines the radiation concentration distribution on the receiver of the collector, called the *local concentration ratio* (LCR). As was seen in Fig. 3.41, the radiation incident on a differential element of reflector area is a cone having a half angle of 16′. The reflected radiation is a similar cone, having the

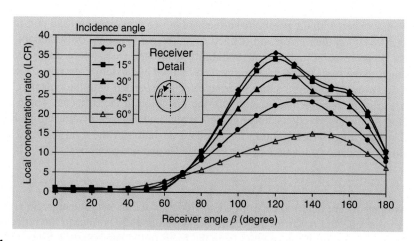

FIGURE 3.44

Local concentration ratio on the receiver of a parabolic trough collector.

same apex angle if the reflector is perfect. The intersection of this cone with the receiver surface determines the image size and shape for that element, and the total image is the sum of the images for all the elements of the reflector. In an actual collector, the various errors outlined previously, which enlarge the image size and lower the LCR, are considered. The distribution of the LCR for a parabolic trough collector is shown in Fig. 3.44. The shape of the curves depends on the random and non-random errors mentioned above and on the angle of incidence. It should be noted that the distribution for half of the receiver is shown in Fig. 3.44. Another more representative way to show this distribution for the whole receiver is in Fig. 3.45. As can be seen, the top part of the receiver essentially receives only direct sunshine from the sun and the maximum concentration for this particular collector, about 36 suns, occurs at 0 incidence angle and at an angle β of 120° (Fig. 3.44).

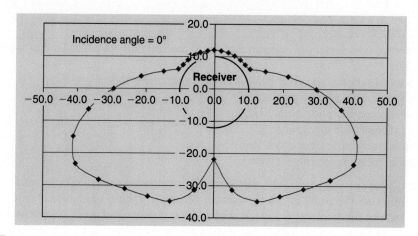

FIGURE 3.45

A more representative view of local concentration ratio for a collector with receiver diameter of 20 mm and rim angle of 90°.

EXAMPLE 3.11

For a PTC with a total aperture area of 50 m², aperture of 2.5 m, and rim angle of 90°, estimate the geometric factor and the actual area lost at an angle of incidence equal to 60°.

Solution

As $\varphi_r = 90°$, the parabola height $h_p = f$. Therefore, from Eq. (3.111),

$$h_p = f = \frac{W_a}{4\tan\left(\frac{\varphi_r}{2}\right)} = \frac{2.5}{4\tan(45)} = 0.625 \text{ m}$$

From Eq. (3.119):

$$A_l = \frac{2}{3}W_a h_p + fW_a\left[1 + \frac{W_a^2}{48f^2}\right] = \frac{2}{3}2.5 \times 0.625 + 0.625 \times 2.5\left[1 + \frac{2.5^2}{48(0.625)^2}\right] = 3.125 \text{ m}^2$$

The area lost at an incidence angle of 60° is:

$$\text{Area lost} = A_l \tan(60) = 3.125 \times \tan(60) = 5.41 \text{ m}^2$$

The geometric factor A_f is obtained from Eq. (3.120):

$$A_f = \frac{A_l}{A_a} = \frac{3.125}{50} = 0.0625$$

3.6.4 Thermal analysis of parabolic trough collectors

The generalized thermal analysis of a concentrating solar collector is similar to that of a flat-plate collector. It is necessary to derive appropriate expressions for the collector efficiency factor, F'; the loss coefficient, U_L; and the collector heat removal factor, F_R. For the loss coefficient, standard heat transfer relations for glazed tubes can be used. Thermal losses from the receiver must be estimated, usually in terms of the loss coefficient, U_L, which is based on the area of the receiver. The method for calculating thermal losses from concentrating collector receivers cannot be as easily summarized as for the flat-plate ones, because many designs and configurations are available. Two such designs are presented in this book: the PTC with a bare tube and the glazed tube receiver. In both cases, the calculations must include radiation, conduction, and convection losses.

For a bare tube receiver and assuming no temperature gradients along the receiver, the loss coefficient considering convection and radiation from the surface and conduction through the support structure is given by:

$$U_L = h_w + h_r + h_c \tag{3.123}$$

The linearized radiation coefficient can be estimated from:

$$h_r = 4\sigma\varepsilon T_r^3 \tag{3.124}$$

If a single value of h_r is not acceptable due to large temperature variations along the flow direction, the collector can be divided into small segments, each with a constant h_r.

For the wind loss coefficient, the Nusselt number can be used.
For $0.1 < \text{Re} < 1000$,

$$\text{Nu} = 0.4 + 0.54(\text{Re})^{0.52} \qquad (3.125a)$$

For $1000 < \text{Re} < 50{,}000$,

$$\text{Nu} = 0.3(\text{Re})^{0.6} \qquad (3.125b)$$

Estimation of the conduction losses requires knowledge of the construction of the collector, that is, the way the receiver is supported.

Usually, to reduce the heat losses, a concentric glass tube is employed around the receiver. The space between the receiver and the glass is usually evacuated, in which case the convection losses are negligible. In this case, U_L, based on the receiver area A_r, is given by:

$$U_L = \left[\frac{A_r}{(h_w + h_{r,c-a})A_g} + \frac{1}{h_{r,r-c}} \right]^{-1} \qquad (3.126)$$

where

$h_{r,c-a}$ = linearized radiation coefficient from cover to ambient estimated by Eq. (3.124) (W/m^2 K);
A_g = external area of glass cover (m^2); and
$h_{r,r-c}$ = linearized radiation coefficient from receiver to cover, given by Eq. (2.74):

$$h_{r,r-c} = \frac{\sigma\left(T_r^2 + T_g^2\right)\left(T_r + T_g\right)}{\dfrac{1}{\varepsilon_r} + \dfrac{A_r}{A_g}\left(\dfrac{1}{\varepsilon_g} - 1\right)} \qquad (3.127)$$

In the preceding equations, to estimate the glass cover properties, the temperature of the glass cover, T_g, is required. It should be noted that the subscripts c (cover) and g (glass) are used here interchangeably. This temperature is closer to the ambient temperature than the receiver temperature. Therefore, by ignoring the radiation absorbed by the cover, T_g may be obtained from an energy balance:

$$A_g\left(h_{r,c-a} + h_w\right)\left(T_g - T_a\right) = A_r h_{r,r-c}\left(T_r - T_g\right) \qquad (3.128)$$

Solving Eq. (3.128) for T_g gives:

$$T_g = \frac{A_r h_{r,r-c} T_r + A_g\left(h_{r,c-a} + h_w\right)T_a}{A_r h_{r,r-c} + A_g\left(h_{r,c-a} + h_w\right)} \qquad (3.129)$$

The procedure to find T_g is by iteration, that is, estimate U_L from Eq. (3.126) by considering a random T_g (close to T_a). Then, if T_g obtained from Eq. (3.129) differs from original value, iterate. Usually, no more than two iterations are required.

If radiation absorbed by the cover needs to be considered, the appropriate term must be added to the right-hand side of Eq. (3.126). The principles are the same as those developed earlier for the flat-plate collectors.

Next, the overall heat transfer coefficient, U_o, needs to be estimated. This should include the tube wall because the heat flux in a concentrating collector is high. Based on the outside tube diameter, this is given by:

$$U_o = \left[\frac{1}{U_L} + \frac{D_o}{h_{fi} D_i} + \frac{D_o \ln(D_o/D_i)}{2k} \right]^{-1} \tag{3.130}$$

where

D_o = receiver outside tube diameter (m);
D_i = receiver inside tube diameter (m); and
h_{fi} = convective heat transfer coefficient inside the receiver tube (W/m² K).

The convective heat transfer coefficient, h_{fi}, can be obtained from the standard pipe flow equation:

$$\text{Nu} = 0.023(\text{Re})^{0.8}(\text{Pr})^{0.4} \tag{3.131}$$

where

Re = Reynolds number = $\rho V D_i/\mu$;
Pr = Prandtl number = $c_p \mu/k_f$;
μ = fluid viscosity (kg/m s); and
k_f = thermal conductivity of fluid (W/m K).

It should be noted that Eq. (3.131) is for turbulent flow (Re > 2300). For laminar flow, Nu = 4.364 = constant. Properties of two nanofluids which can be used as heat transfer fluids are given in Appendix 5 (Tables A5.15 and A5.16).

A detailed thermal model of a PTC is presented by Kalogirou (2012). In this all modes of heat transfer were considered in detail and the set of equations obtained were solved simultaneously. For this purpose the program Engineering Equation Solver (EES) is used which includes routines to estimate the properties of various substances and can be called from TRNSYS (see Chapter 11, Section 11.5.1) which allows the development of a model that can use the capabilities of both programs.

The instantaneous efficiency of a concentrating collector may be calculated from an energy balance of its receiver. Equation (3.31) also may be adapted for use with concentrating collectors by using appropriate areas for the absorbed solar radiation (A_a) and heat losses (A_r). Therefore, the useful energy delivered from a concentrator is:

$$Q_u = G_B \eta_o A_a - A_r U_L (T_r - T_a) \tag{3.132}$$

Note that, because concentrating collectors can utilize only beam radiation, G_B is used in Eq. (3.132) instead of the total radiation, G_t, used in Eq. (3.31).

The useful energy gain per unit of collector length can be expressed in terms of the local receiver temperature, T_r, as:

$$q_u' = \frac{Q_u}{L} = \frac{A_a \eta_o G_B}{L} - \frac{A_r U_L}{L}(T_r - T_a) \tag{3.133}$$

In terms of the energy transfer to the fluid at the local fluid temperature, T_f (Kalogirou, 2004),

$$q'_u = \frac{\left(\frac{A_r}{L}\right)(T_r - T_f)}{\frac{D_o}{h_{fi}D_i} + \left(\frac{D_o}{2k}\ln\frac{D_o}{D_i}\right)} \tag{3.134}$$

If T_r is eliminated from Eqs. (3.133) and (3.134), we have:

$$q'_u = F'\frac{A_a}{L}\left[\eta_o G_B - \frac{U_L}{C}(T_f - T_a)\right] \tag{3.135}$$

where F' is the collector efficiency factor, given by:

$$F' = \frac{1/U_L}{\frac{1}{U_L} + \frac{D_o}{h_{fi}D_i} + \left(\frac{D_o}{2k}\ln\frac{D_o}{D_i}\right)} = \frac{U_o}{U_L} \tag{3.136}$$

As for the flat-plate collector, T_r in Eq. (3.132) can be replaced by T_i through the use of the heat removal factor, and Eq. (3.132) can be written as:

$$Q_u = F_R[G_B\eta_o A_a - A_r U_L(T_i - T_a)] \tag{3.137}$$

The collector efficiency can be obtained by dividing Q_u by $(G_B A_a)$. Therefore,

$$\eta = F_R\left[\eta_o - U_L\left(\frac{T_i - T_a}{G_B C}\right)\right] \tag{3.138}$$

where

C = concentration ratio, $C = A_a/A_r$ (Eq. 3.87).

For F_R, a relation similar to Eq. (3.58) is used by replacing A_c with A_r and using F', given by Eq. (3.136), which does not include the fin and bond conductance terms, as in FPCs. This equation explains why high temperatures can be obtained with concentrating collectors. This is because the heat losses term is inversely proportional to C, so the bigger the concentration ratio the smaller the losses.

Consideration of gas in annulus space
In case there is air or any other gas in the space between the receiver and the glass cover, Eq. (3.126) should also include the convection term for the gas in the annulus ($h_{c,r-c}$):

$$U_L = \left[\frac{A_r}{(h_w + h_{r,c-a})A_g} + \frac{1}{h_{c,r-c} + h_{r,r-c}}\right]^{-1} \tag{3.139}$$

Usually, the space in an enclosure (sealed) and the heat transfer in the annulus space is by free convection between concentric cylinders.

If the diameter of the internal cylinder (receiver) is D_r and the external cover cylinder (glass) is D_g and the characteristic length is the spacing between the two cylinders, $L_c=(D_g-D_r)/2$, the rate of heat transfer for a collector of length (L) through the annular space between the two cylinders by natural convection is:

$$Q = \frac{2\pi k_{eff} L}{\ln\left(D_g/D_r\right)}\left(T_r - T_g\right) \tag{3.140}$$

The recommended relation for effective thermal conductivity is (Raithby and Hollands, 1975):

$$\frac{k_{eff}}{k} = 0.386\left(\frac{Pr}{0.861 + Pr}\right)^{1/4}\left(F_{cyl}Ra_{L_c}\right)^{1/4} \tag{3.141}$$

where the geometric factor for concentric cylinders, F_{cyl}, is:

$$F_{cyl} = \frac{\left[\ln\left(D_g/D_r\right)\right]^4}{L_c^3\left(D_r^{-3/5} + D_g^{-3/5}\right)^5} \tag{3.142}$$

Note that the L_c indicator in the Rayleigh number indicates that this is the characteristic length to use in Eq. (3.12). The k_{eff} relation is applicable for $0.7 \le Pr \le 6000$ and $10^2 \le F_{cyl}Ra_L \le 10^7$. For $F_{cyl}Ra_L < 100$, natural convection currents are negligible, and thus $k_{eff} = k$. It should be noted that k_{eff} cannot be less than k, and thus $k_{eff} = k$ if $k_{eff}/k < 1$. All fluid properties are evaluated at the average temperature $(T_r + T_g)/2$.

It should be noted that Eq. (3.140) is also equal to:

$$Q = h_{c,r-c}A_r\left(T_r - T_g\right) \tag{3.143a}$$

By combining Eq. (3.140) with Eq. (3.143a):

$$h_{c,r-c} = \frac{2\pi k_{eff} L}{\ln\left(D_g/D_r\right)A_r} \tag{3.143b}$$

Or by using $A_r = \pi D_r L$ it can be simplified to:

$$h_{c,r-c} = \frac{2k_{eff}}{\ln\left(D_g/D_r\right)D_r} \tag{3.144}$$

The properties of various gases are given in Appendix 5 (Tables A5.11−A5.14).

Consideration of vacuum in annulus space

In the analysis presented so far, the convection losses in the annulus space are ignored. In fact convection heat transfer depends on the annulus pressure. At low pressures (<0.013 Pa), heat transfer is by molecular conduction, whereas at higher pressures is by free convection. When the annulus is under vacuum (pressure <0.013 Pa), the convection heat transfer between the receiver pipe and glass envelope occurs by free molecular convection and the heat transfer coefficient is given by (Ratzel et al., 1979):

$$h_{c,r-c} = \frac{k_{std}}{\frac{D_r}{2\ln\left(\frac{D_g}{D_r}\right)} + b\lambda\left(\frac{D_r}{D_g} + 1\right)} \quad (3.145)$$

This equation is applicable for: Ra $< (D_g/(D_g - D_r))^4$.
And

$$b = \frac{(2-a)(9\gamma - 5)}{2a(\gamma + 1)} \quad (3.146a)$$

$$\lambda = \frac{2.331 \times 10^{-20}(T_{r-g} + 273)}{(P_a\delta^2)} \quad (3.146b)$$

where

k_{std} = thermal conductivity of the annulus gas at standard temperature and pressure (W/m °C);
D_r = outside receiver pipe diameter (m);
D_g = inside glass envelope diameter (m);
b = interaction coefficient;
λ = mean free path between collisions of a molecule (cm);
a = accommodation coefficient;
γ = ratio of specific heats for the annulus gas (air);
T_{r-g} = average temperature $(T_r + T_g)/2$ (°C);
P_a = annulus gas pressure (mmHg); and
δ = molecular diameter of annulus gas (cm).

Equation (3.139) slightly overestimates the heat transfer for very small pressures (<0.013 Pa). The molecular diameters of air, δ, is equal to 3.55×10^{-8} cm (Marshal, 1976), the thermal conductivity of air is 0.02551 W/m °C, the interaction coefficient is 1.571, the mean free path between collisions of a molecule is 88.67 cm, and the ratio of specific heats for the annulus air is 1.39. These are for average fluid temperature of 300 °C and pressure equal to 0.013 Pa. Using these values, the convection heat transfer coefficient ($h_{c,r-c}$) is equal to 0.0001115 W/m² °C, which is the reason why usually it is ignored.

If for more accuracy this heat loss is considered Eq. (3.126) should include $h_{c,r-c}$ in the second term as well as Eq. (3.128) for the appropriate estimation of T_g.

If the receiver is filled or partially filled with ambient air or if the receiver annulus vacuum is lost, the convection heat transfer between the receiver pipe and glass envelope occurs by natural convection and for this purpose, the correlation for natural convection in an annular space (enclosure) between horizontal concentric cylinders can be used, as indicated in the previous section.

EXAMPLE 3.12

A 20 m long PTC with an aperture width of 3.5 m has a pipe receiver of 50 mm outside diameter and 40 mm inside diameter and a glass cover of 90 mm in diameter. If the space between the receiver and the glass cover is evacuated, estimate the overall collector heat loss coefficient, the useful energy gain, and the exit fluid temperature. The following data are given.

Absorbed solar radiation = 500 W/m^2.
Receiver temperature = 260 °C = 533 K.
Receiver emissivity, $\varepsilon_r = 0.92$.
Glass cover emissivity, $\varepsilon_g = 0.87$.
Circulating fluid, $c_p = 1350$ J/kg K.
Entering fluid temperature = 220 °C = 493 K.
Mass flow rate = 0.32 kg/s.
Heat transfer coefficient inside the pipe = 330 W/m^2 K.
Tube thermal conductivity, $k = 15$ W/m K.
Ambient temperature = 25 °C = 298 K.
Wind velocity = 5 m/s.

Solution

The receiver area $A_r = \pi D_o L = \pi \times 0.05 \times 20 = 3.14$ m^2. The glass cover area $A_g = \pi D_g L = \pi \times 0.09 \times 20 = 5.65$ m^2. The unshaded collector aperture area $A_a = (3.5 - 0.09) \times 20 = 68.2$ m^2.

Next, a glass cover temperature, T_g, is assumed to evaluate the convection and radiation heat transfer from the glass cover. This is assumed to be equal to 64 °C = 337 K. The actual glass cover temperature is obtained by iteration by neglecting the interactions with the reflector. The convective (wind) heat transfer coefficient $h_{c,c-a} = h_w$ of the glass cover can be calculated from Eq. (3.125). First, the Reynolds number needs to be estimated at the mean temperature ½(25 + 64) = 44.5 °C. Therefore, from Table A5.1 in Appendix 5, we get:

$$\rho = 1.11 \text{ kg/m}^3$$

$$\mu = 2.02 \times 10^{-5} \text{ kg/m s}$$

$$k = 0.0276 \text{ W/m K}$$

Now

$$\text{Re} = \rho V D_g / \mu = (1.11 \times 5 \times 0.09)/2.02 \times 10^{-5} = 24,728$$

Therefore, Eq. (3.125b) applies, which gives:

$$\text{Nu} = 0.3(\text{Re})^{0.6} = 129.73$$

and

$$h_{c,c-a} = h_w = (\text{Nu})k/D_g = 129.73 \times 0.0276/0.09 = 39.8 \text{ W/m}^2 \text{ K}$$

The radiation heat transfer coefficient, $h_{r,c-a}$, for the glass cover to the ambient is calculated from Eq. (2.75):

$$h_{r,c-a} = \varepsilon_g \sigma (T_g + T_a)(T_g^2 + T_a^2) = 0.87(5.67 \times 10^{-8})(337 + 298)(337^2 + 298^2)$$

$$= 6.34 \text{ W/m}^2 \text{ K}$$

The radiation heat transfer coefficient, $h_{r,r-c}$, between the receiver tube and the glass cover is estimated from Eq. (3.127):

$$h_{r,r-c} = \frac{\sigma\left(T_r^2 + T_g^2\right)(T_r + T_g)}{\dfrac{1}{\varepsilon_r} + \dfrac{A_r}{A_g}\left(\dfrac{1}{\varepsilon_g} - 1\right)} = \frac{(5.67 \times 10^{-8})(533^2 + 337^2)(533 + 337)}{\dfrac{1}{0.92} + \dfrac{0.05}{0.09}\left(\dfrac{1}{0.87} - 1\right)}$$

$$= 16.77 \text{ W/m}^2 \text{ K}$$

Continued

EXAMPLE 3.12–CONT'D

Since the space between the receiver and the glass cover is evacuated, there is no convection heat transfer. Therefore, based on the receiver area, the overall collector heat loss coefficient is given by Eq. (3.126):

$$U_L = \left[\frac{A_r}{(h_w + h_{r,c-a})A_g} + \frac{1}{h_{r,r-c}} \right]^{-1} = \left[\frac{0.05}{(39.8 + 6.34)0.09} + \frac{1}{16.77} \right]^{-1} = 13.95 \ \text{W/m}^2 \ \text{K}$$

Since U_L is based on the assumed T_g value, we need to check if the assumption made was correct. Using Eq. (3.129), we get:

$$T_g = \frac{A_r h_{r,r-c} T_r + A_g (h_{r,c-a} + h_w) T_a}{A_r h_{r,r-c} + A_g (h_{r,c-a} + h_w)} = \frac{3.14 \times 16.77 \times 260 + 5.65(6.34 + 39.8)25}{3.14 \times 16.77 + 5.65(6.34 + 39.8)}$$

$$= 64.49\,^{\circ}\text{C}$$

This is about the same as the value assumed earlier.

The collector efficiency factor can be calculated from Eq. (3.136):

$$F' = \frac{1/U_L}{\dfrac{1}{U_L} + \dfrac{D_o}{h_{fi} D_i} + \left(\dfrac{D_o}{2k} \ln \dfrac{D_o}{D_i} \right)} = \frac{1/13.95}{\dfrac{1}{13.95} + \dfrac{0.05}{330 \times 0.04} + \left(\dfrac{0.05}{2 \times 15} \ln \dfrac{0.05}{0.04} \right)} = 0.945$$

The heat removal factor can be calculated from Eq. (3.58) by using A_r instead of A_c:

$$F_R = \frac{\dot{m} c_p}{A_r U_L} \left[1 - \exp \left(-\frac{U_L F' A_r}{\dot{m} c_p} \right) \right] = \frac{0.32 \times 1350}{3.14 \times 13.95} \left[1 - \exp \left(-\frac{13.95 \times 0.95 \times 3.14}{0.32 \times 1350} \right) \right]$$

$$= 0.901$$

The useful energy is estimated from Eq. (3.137) using the concept of absorbed radiation:

$$Q_u = F_R [S A_a - A_r U_L (T_i - T_a)] = 0.901[500 \times 68.2 - 3.14 \times 13.95(220 - 25)] = 23{,}028 \ \text{W}$$

Finally, the fluid exit temperature can be estimated from:

$$Q_u = \dot{m} c_p (T_o - T_i) \quad \text{or} \quad T_o = T_i + \frac{Q_u}{\dot{m} c_p} = 220 + \frac{23{,}028}{0.32 \times 1350} = 273.3\,^{\circ}\text{C}$$

Another analysis usually performed for PTCs applies a piecewise two-dimensional model of the receiver by considering the circumferential variation of solar flux shown in Figs. 3.44 and 3.45. Such an analysis can be performed by dividing the receiver into longitudinal and isothermal nodal sections, as shown in Fig. 3.46, and applying the principle of energy balance to the glazing and receiver nodes (Karimi et al., 1986).

The generalized glazing and absorber nodes, showing the various modes of heat transfer considered, are shown in Fig. 3.47. It is assumed that the length of each section is very small so that the working fluid in that section stays in the inlet temperature. The temperature is adjusted in a stepwise fashion at the end of the longitudinal section. By applying the principle of energy balance to the glazing and absorber nodes, we get the following equations.

For the glazing node,

$$q_{G1} + q_{G2} + q_{G3} + q_{G4} + q_{G5} + q_{G6} + q_{G7} + q_{G8} = 0 \tag{3.147a}$$

Longitudinal division of the receiver into sections

Cross-section of the receiver showing the isothermal nodal sections

FIGURE 3.46

Piecewise two-dimensional model of the receiver assembly.

Reproduced from Karimi et al. (1986).

For the absorber node,

$$q_{A1} + q_{A2} + q_{A3} + q_{A4} + q_{A5} + q_{A6} + q_{A7} + q_{A8} = 0 \qquad (3.147b)$$

where

q_{G1} = solar radiation absorbed by glazing node i;

q_{G2} = net radiation exchange between glazing node i to the surroundings;

q_{G3} = natural and forced convection heat transfer from glazing node i to the surroundings;

q_{G4} = convection heat transfer to the glazing node from the absorber (across the gap);

q_{G5} = radiation emitted by the inside surface of the glazing node i;

q_{G6} = conduction along the circumference of glazing from node i to $i + 1$;

q_{G7} = conduction along the circumference of the glazing from node i to $i - 1$;

q_{G8} = fraction of the total radiation incident upon the inside glazing surface that is absorbed;

q_{A1} = solar radiation absorbed by absorber node i;

q_{A2} = thermal radiation emitted by outside surface of absorber node i;

q_{A3} = convection heat transfer from absorber node to glazing (across the gap);

q_{A4} = convection heat transfer to absorber node i from the working fluid;

q_{A5} = radiation exchange between the inside surface of absorber and absorber node i;

q_{A6} = conduction along the circumference of absorber from node i to $i + 1$;

q_{A7} = conduction along the circumference of the absorber from node i to $i - 1$; and

q_{A8} = fraction of the total radiation incident upon the inside absorber node that is absorbed.

For all these parameters, standard heat transfer relations can be used. The set of non-linear equations is solved sequentially to obtain the temperature distribution of the receiver, and the solution is obtained by an iterative procedure. In Eqs. (3.147a) and (3.147b), factors q_{G1} and q_{A1} are calculated by the optical model, whereas factor q_{A5} is assumed to be negligible.

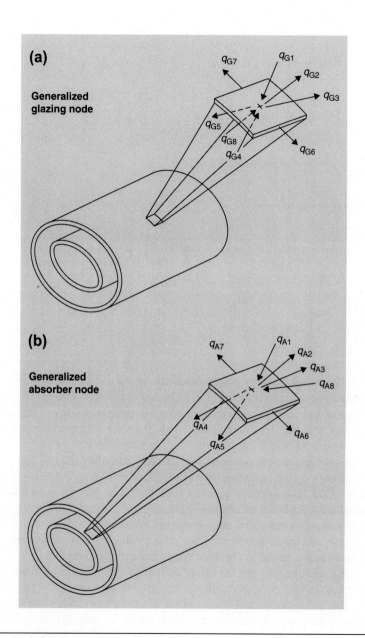

FIGURE 3.47

Generalized glazing and absorber nodes, showing the various modes of heat transfer.

Karimi et al. (1986).

This analysis can give the temperature distribution along the circumference and length of the receiver, so any points of high temperature, which might reach a temperature above the degradation temperature of the receiver selective coating, can be determined.

3.7 Second law analysis

The analysis presented here is based on Bejan's work (Bejan et al., 1981; Bejan, 1995). The analysis, however, is adapted to imaging collectors, because entropy generation minimization is more important to high-temperature systems. Consider that the collector has an aperture area (or total heliostat area), A_a, and receives solar radiation at the rate Q^* from the sun, as shown in Fig. 3.48. The net solar heat transfer, Q^*, is proportional to the collector area, A_a, and the proportionality factor, q^* (W/m^2), which varies with geographical position on the earth, the orientation of the collector, meteorological conditions, and the time of day. In the present analysis, q^* is assumed to be constant and the system is in a steady state; that is,

$$Q^* = q^* A_a \qquad (3.148)$$

For concentrating systems, q^* is the solar energy falling on the reflector. To obtain the energy falling on the collector receiver, the tracking mechanism accuracy, the optical errors of the mirror, including its reflectance, and the optical properties of the receiver glazing must be considered.

Therefore, the radiation falling on the receiver, q_o^*, is a function of the optical efficiency, which accounts for all these errors. For the concentrating collectors, Eq. (3.116) can be used. The radiation falling on the receiver is (Kalogirou, 2004):

$$q_o^* = \eta_o q^* = \frac{\eta_o Q^*}{A_a} \qquad (3.149)$$

The incident solar radiation is partly delivered to a power cycle (or user) as heat transfer Q at the receiver temperature, T_r. The remaining fraction, Q_o, represents the collector-ambient heat loss:

$$Q_o = Q^* - Q \qquad (3.150a)$$

FIGURE 3.48

Imaging concentrating collector model.

For imaging concentrating collectors, Q_o is proportional to the receiver-ambient temperature difference and to the receiver area as:

$$Q_o = U_r A_r (T_r - T_o) \tag{3.150b}$$

where U_r is the overall heat transfer coefficient based on A_r. It should be noted that U_r is a characteristic constant of the collector.

Combining Eqs. (3.150a) and (3.150b), it is apparent that the maximum receiver temperature occurs when $Q = 0$, that is, when the entire solar heat transfer Q^* is lost to the ambient. The maximum collector temperature is given in dimensionless form by:

$$\theta_{\max} = \frac{T_{r,\max}}{T_o} = 1 + \frac{Q^*}{U_r A_r T_o} \tag{3.151a}$$

Combining Eqs. (3.149) and (3.151a),

$$\theta_{\max} = 1 + \frac{q_o^* A_a}{\eta_o U_r A_r T_o} \tag{3.151b}$$

Considering that $C = A_a/A_r$, then:

$$\theta_{\max} = 1 + \frac{q_o^* C}{\eta_o U_r T_o} \tag{3.151c}$$

As can be seen from Eq. (3.151c), $\theta_{\max}$ is proportional to C, that is, the higher the concentration ratio of the collector, the higher are $\theta_{\max}$ and $T_{r,\max}$. The term $T_{r,\max}$ in Eq. (3.151a) is also known as the *stagnation temperature of the collector*, that is, the temperature that can be obtained at a no flow condition. In dimensionless form, the collector temperature, $\theta = T_r/T_o$, varies between 1 and $\theta_{\max}$, depending on the heat delivery rate, Q. The stagnation temperature, $\theta_{\max}$, is the parameter that describes the performance of the collector with regard to collector-ambient heat loss, since there is no flow through the collector and all the energy collected is used to raise the temperature of the working fluid to the stagnation temperature, which is fixed at a value corresponding to the energy collected equal to energy loss to ambient. Hence, the collector efficiency is given by:

$$\eta_c = \frac{Q}{Q^*} = 1 - \frac{\theta - 1}{\theta_{\max} - 1} \tag{3.152}$$

Therefore η_c is a linear function of collector temperature. At the stagnation point, the heat transfer, Q, carries zero *exergy*, or zero potential for producing useful work.

3.7.1 Minimum entropy generation rate

The minimization of the entropy generation rate is the same as the maximization of the power output. The process of solar energy collection is accompanied by the generation of entropy upstream of the collector, downstream of the collector, and inside the collector, as shown in Fig. 3.49.

The exergy inflow coming from the solar radiation falling on the collector surface is:

$$E_{in} = Q^* \left(1 - \frac{T_o}{T_*} \right) \tag{3.153}$$

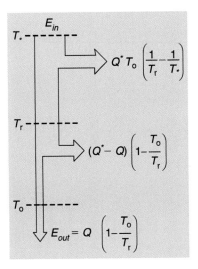

FIGURE 3.49

Exergy flow diagram.

where T^* is the apparent sun temperature as an exergy source. In this analysis, the value suggested by Petela (1964) is adopted, that is, T^* is approximately equal to $\frac{3}{4}T_s$, where T_s is the apparent blackbody temperature of the sun, which is about 5770 K. Therefore, the T^* considered here is 4330 K. It should be noted that, in this analysis, T^* is also considered constant; and because its value is much greater than T_o, E_{in} is very near Q^*. The output exergy from the collector is given by:

$$E_{out} = Q\left(1 - \frac{T_o}{T_r}\right) \tag{3.154}$$

whereas the difference between E_{in} and E_{out} represents the destroyed exergy. From Fig. 3.49, the entropy generation rate can be written as:

$$S_{gen} = \frac{Q_o}{T_o} + \frac{Q}{T_r} - \frac{Q^*}{T_*} \tag{3.155a}$$

This equation can be written with the help of Eq. (3.150a) as:

$$S_{gen} = \frac{1}{T_o}\left[Q^*\left(1 - \frac{T_o}{T_*}\right) - Q\left(1 - \frac{T_o}{T_r}\right)\right] \tag{3.155b}$$

By using Eqs. (3.153) and (3.154), Eq. (3.155b) can be written as:

$$S_{gen} = \frac{1}{T_o}(E_{in} - E_{out}) \tag{3.156}$$

or

$$E_{out} = E_{in} - T_o S_{gen} \tag{3.157}$$

Therefore, if we consider E_{in} constant, the maximization of the exergy output (E_{out}) is the same as the minimization of the total entropy generation, S_{gen}.

3.7.2 Optimum collector temperature

By substituting Eqs. (3.150a) and (3.150b) into Eq. (3.155b), the rate of entropy generation can be written as:

$$S_{gen} = \frac{U_r A_r (T_r - T_o)}{T_o} - \frac{Q^*}{T_*} + \frac{Q^* - U_r A_r (T_r - T_o)}{T_r} \tag{3.158}$$

By applying Eq. (3.151c) in Eq. (3.158) and performing various manipulations,

$$\frac{S_{gen}}{U_r A_r} = \theta - 2 - \frac{q_o^* C}{\eta_o U_r T_*} + \frac{\theta_{max}}{\theta} \tag{3.159}$$

The dimensionless term, $S_{gen}/U_r A_r$, accounts for the fact that the entropy generation rate scales with the finite size of the system, which is described by $A_r = A_a/C$.

By differentiating Eq. (3.159) with respect to θ and setting it to 0, the optimum collector temperature (θ_{opt}) for minimum entropy generation is obtained:

$$\theta_{opt} = \sqrt{\theta_{max}} = \left(1 + \frac{q_o^* C}{\eta_o U_r T_o}\right)^{1/2} \tag{3.160}$$

By substituting θ_{max} with $T_{r,max}/T_o$ and θ_{opt} with $T_{r,opt}/T_o$, Eq. (3.160) can be written as:

$$T_{r,opt} = \sqrt{T_{r,max} T_o} \tag{3.161}$$

This equation states that the optimal collector temperature is the geometric average of the maximum collector (stagnation) temperature and the ambient temperature. Typical stagnation temperatures and the resulting optimum operating temperatures for various types of concentrating collectors are shown in Table 3.4. The stagnation temperatures shown in Table 3.4 are estimated by considering mainly the collector radiation losses.

As can be seen from the data presented in Table 3.4 for high-performance collectors such as the central receiver, it is better to operate the system at high flow rates to lower the temperature around the value shown instead of operating at very high temperature to obtain higher thermodynamic efficiency from the collector system.

Table 3.4 Optimum collector temperatures for various types of concentrating collectors.

Collector type	Concentration ratio	Stagnation temperature (°C)	Optimal temperature (°C)
Parabolic trough	50	565	227
Parabolic dish	500	1285	408
Central receiver	1500	1750	503

Note: Ambient temperature considered = 25 °C.

FIGURE 3.50

Entropy generated and optimum temperatures against collector concentration ratio.

By applying Eq. (3.160) to Eq. (3.159), the corresponding minimum entropy generation rate is:

$$\frac{S_{gen,min}}{U_r A_r} = 2(\sqrt{\theta_{max}} - 1) - \frac{\theta_{max} - 1}{\theta_*} \tag{3.162}$$

where $\theta_* = T_*/T_o$. It should be noted that, for flat-plate and low-concentration ratio collectors, the last term of Eq. (3.162) is negligible, since θ_* is much bigger than $\theta_{max} - 1$; but it is not for higher concentration collectors such as the central receiver and the parabolic dish ones, which have stagnation temperatures of several hundreds of degrees.

By applying the stagnation temperatures shown in Table 3.4 to Eq. (3.162), the dimensionless entropy generated against the collector concentration ratios considered here, as shown in Fig. 3.50, is obtained.

3.7.3 Non-isothermal collector

So far, the analysis was carried out considering an isothermal collector. For a non-isothermal one, which is a more realistic model, particularly for long PTCs, and by applying the principle of energy conservation,

$$q^* = U_r(T - T_o) + \dot{m}c_p \frac{dT}{dx} \tag{3.163}$$

where x is from 0 to L (the collector length). The generated entropy can be obtained from:

$$S_{gen} = \dot{m}c_p \ln \frac{T_{out}}{T_{in}} - \frac{Q^*}{T_*} + \frac{Q_o}{T_o} \tag{3.164}$$

From an overall energy balance, the total heat loss is:

$$Q_o = Q^* - \dot{m}c_p(T_{out} - T_{in}) \tag{3.165}$$

Substituting Eq. (3.165) into Eq. (3.164) and performing the necessary manipulations, the following relation is obtained:

$$N_s = M\left(\ln\frac{\theta_{out}}{\theta_{in}} - \theta_{out} + \theta_{in}\right) - \frac{1}{\theta_*} + 1 \tag{3.166}$$

where $\theta_{out} = T_{out}/T_o$, $\theta_{in} = T_{in}/T_o$, N_s is the entropy generation number, and M is the mass flow number given by:

$$N_s = \frac{S_{gen}T_o}{Q^*} \tag{3.167}$$

and

$$M = \frac{\dot{m}c_pT_o}{Q^*} \tag{3.168}$$

If the inlet temperature is fixed, $\theta_{in} = 1$, then the entropy generation rate is a function of only M and θ_{out}. These parameters are interdependent because the collector outlet temperature depends on the mass flow rate.

Exercises

3.1 For a one-cover flat-plate collector (FPC) of dimensions 3×6 m, tilted at $40°$ from horizontal, determine the overall heat loss coefficient. The environmental temperature is $10\,°C$ and the wind blows at 4 m/s. The absorber plate is 0.5 mm thick with an emissivity of 0.92. The glass cover is 3.5 mm thick and located 35 mm from the absorbing plate, and the glass emissivity is 0.88. The insulation is fiberglass 45 mm thick at the back and 25 mm at the edges. The mean absorber temperature is $90\,°C$. Estimate the overall heat loss coefficient using both the detailed and the empirical methods and compare the results.

3.2 For a two-cover FPC of dimensions 3×6 m, tilted $45°$ from horizontal, determine the overall heat loss coefficient. The environmental temperature is $5\,°C$ and the wind blows at 5 m/s. The absorber plate is 0.6 mm thick with an emissivity of 0.15. The glass covers are 3.5 mm thick with a gap of 20 mm and the bottom one is located 50 mm from the absorbing plate. The glass emissivity is 0.88. The insulation is fiberglass 50 mm thick at the back and 30 mm at the edges. The mean absorber temperature is $90\,°C$. Estimate the overall heat loss coefficient using both the detailed and the empirical methods and compare the results.

3.3 An FPC 4 m^2 in area is tested during the night to measure the overall heat loss coefficient. Water at $60\,°C$ circulates through the collector at a flow rate of 0.06 L/s. The ambient temperature is $8\,°C$ and the exit temperature is $49\,°C$. Determine the overall heat loss coefficient.

3.4 For a two-cover 2×6 m FPC, tilted at $45°$ from horizontal, determine the overall heat loss coefficient. The environmental temperature is $-5\,°C$ and the wind blows at 8 m/s. The absorber plate is 0.1 cm thick with an emissivity of 0.93 maintained at 80 °C. The glass covers are 0.5 cm thick with a gap of 2.5 cm and the bottom one is located 6 cm from the absorbing plate. The glass emissivity is 0.88. The insulation is fiberglass 7 cm thick at the back and 3 cm at the edges.

3.5 A single-glazed 3 × 6 m FPC faces south, tilted 45° from horizontal. The collector is located at 35°N latitude, and on March 21 from 2:00 to 3:00 p.m., the solar radiation on the surface of the collector is 890 W/m^2 and the ambient temperature is 8 °C. Estimate the useful energy gain of the collector if:
Overall heat loss coefficient = 5.6 W/m^2 °C.
Water inlet temperature = 50 °C.
Mass flow rate through the collector = 0.25 kg/s.
Tube inside convection heat transfer coefficient = 235 W/m^2 K.
Single glazing with $n = 1.526$ and $KL = 0.037$.
Absorber plate is selective, with $\alpha_n = 0.92$, thickness = 0.5 mm.
Copper tubes are used for risers, with 13.5 mm inside diameter, 15 mm outside diameter, and distance between risers 12 cm.

3.6 A flat-plate solar collector with dimensions of 1 × 2 m has eight copper riser tubes of 13.5 mm inside diameter and 15 mm outside diameter, mounted on a copper absorbing plate 0.5 mm in thickness, which is at 85 °C. Inlet water temperature is 55 °C and the flow rate is 0.03 kg/s. Calculate the convection heat transfer coefficient inside the riser pipes, the outlet water temperature, and the absorbed solar radiation on the collector surface, assuming a fin efficiency of 95%.

3.7 The overall heat loss coefficient of an FPC is 6.5 W/m^2 K. The absorber plate is 0.4 mm thick and the riser tubes have 10 mm inside diameter and 12 mm outside diameter. If the distance between the centers of the riser tubes is 12 cm and the tube inside convection heat transfer coefficient is 250 W/m^2 K, estimate the collector efficiency factor when the material used is aluminum and copper.

3.8 A single-glazed air heating collector that has a flow channel behind the absorber that is 1.5 m wide, 3.5 m long, and 5 cm high. The mass flow rate of the air is 0.045 kg/s and the air inlet temperature is 45 °C. The solar radiation on the tilted collector surface is 920 W/m^2 and the collector effective $(\tau\alpha)$ is 0.87. When the ambient temperature is 12 °C, the overall heat loss coefficient is 4.5 W/m^2 K. If the emissivity of the surfaces of the airflow channel is 0.9, estimate the outlet air temperature and the efficiency of the collector.

3.9 A CPC has an acceptance half angle of 16° and its long axis is oriented along the east–west direction with a tilt of 45°. The collector is located at a latitude of 35°N where on March 10 at 1:00 to 2:00 p.m., the beam radiation on the horizontal is 1.3 MJ/m^2 and the diffuse is 0.4 MJ/m^2. A single glass cover is used on the concentrator, with $KL = 0.032$. Estimate the absorbed radiation for the hour indicated if specular reflectivity is 0.85 and the absorptivity values are 0.96 at normal incidence, 0.95 at incidence angle of 20°, 0.94 at 40°, and 0.89 at 60°. What is the useful energy output of the collector per unit aperture area if the overall heat loss coefficient is 7 W/m^2 K, the heat removal factor is 0.88, the ambient temperature is 10 °C, and the inlet fluid temperature is 55 °C?

3.10 A parabolic trough collector has a tubular steel receiver with a glass cover and the space between the receiver and the glass is evacuated. The receiver is 10 m long and has an outside diameter of 5 cm and an inside diameter of 4 cm. The glass cover diameter is 8 cm. If the receiver surface is selective with $\varepsilon = 0.11$ and is at 250 °C, determine the heat loss coefficient of the receiver when environmental temperature is 24 °C, the wind speed is 2 m/s, and the glass emissivity is 0.92.

3.11 For the previous problem, if the collector aperture is 4 m, the receiver tube is steel, the receiver inside convection coefficient is 280 W/m² K, and the absorbed solar radiation is 500 W/m² of aperture area, estimate the useful energy gain of the collector and the collector outlet temperature. The circulating fluid is oil with a specific heat of 1.3 kJ/kg K, circulating at a flow rate of 1 kg/s and entering the receiver at 210 °C.

Assignment: Design of a flat-plate solar collector

Student name: _____ Number of problem: _____

  Design a flat-plate solar collector with a single cover 2 m long operating in an environment with an ambient temperature of 20 °C using water as the heat transfer fluid. The absorbing plate and the pipes (risers) are from copper (k = 385 W/mK). The thickness of the absorbing plate is 0.5 mm and the thickness of the pipes is 1 mm. The collector cover is made from glass with $\eta = 1.526$, $K = 9.25 \text{ m}^{-1}$, and a thickness of 4 mm. The collector insulation is 40 mm thick and has a thermal conductivity of k = 0.05 W/mK. The distance between the riser tubes is 10 cm, and they are welded onto the absorbing plate, so $C_b = 195$ W/mK. The mass flow rate to the collector is 0.02 kg/s-m². Other characteristics are:

1. Collector inclination (β): _____ °
2. Collector area (A_c): _____ m²
3. Distance between the cover and the collector absorbing plate: _____ mm
4. Mean collector absorbing plate temperature (T_p): _____ °C
5. Absorbing plate emittance (ε_p): _____
6. Glass cover emittance (ε_g): _____
7. Wind velocity: _____ m/s
8. Riser pipes diameter (D) = _____ mm
9. Solar radiation (G_t): _____ W/m²
10. Absorbing plate absorptivity at normal incidence (α_n): _____

Estimate the thermal efficiency of the collector (η%)

Notes

1. You need to find the inlet temperature of the collector at which the result of Eqs. (3.31) and (3.60) are equal.
2. As a hint, you need to estimate in sequence: U_L, h_{fi}, number of riser tubes/flow rate in each pipe, F', F_R, $(\tau\alpha)_n$, Q_u (with T_i and T_p), and finally, estimate the efficiency, η (%).
3. For the solutions, you can use Excel to ease iterative solutions.

Problem options for various students

Student	Slope (β)	A_c (m^2)	Cover-plate distance (mm)	T_p (Deg. C)	ε_p (−)	ε_g (−)	Wind velocity (m/s)	D (mm)	G_t (W/m^2)	Alpha (α_n)
1	10	1.4	25	60	0.06	0.76	1.4	6	600	0.8
2	15	1.6	30	65	0.08	0.78	1.6	8	650	0.82
3	20	1.8	35	70	0.1	0.8	1.8	10	700	0.84
4	25	2	40	75	0.12	0.82	2	12	750	0.86
5	30	2.2	45	80	0.14	0.84	2.2	15	800	0.88
6	35	2.4	40	85	0.12	0.86	2.4	18	850	0.9
7	40	2.6	35	70	0.1	0.84	2.2	15	800	0.88
8	45	2.8	30	75	0.08	0.82	2	12	750	0.86
9	50	3	25	80	0.06	0.8	1.8	10	700	0.84
10	55	3.2	35	85	0.1	0.78	1.6	8	650	0.82
11	40	1.6	35	70	0.05	0.88	1.5	10	900	0.92
12	45	1.8	40	80	0.07	0.9	1.7	12	900	0.94
13	10	2	25	60	0.06	0.84	2	15	800	0.86
14	15	2.2	30	65	0.08	0.86	1.8	18	850	0.84
15	20	2.4	35	70	0.1	0.84	1.6	15	800	0.82

References

Abdel-Khalik, S.A., 1976. Heat removal factor for a flat-plate solar collector with serpentine tube. Sol. Energy 18 (1), 59−64.

ASHRAE, 2007. Handbook of HVAC Applications. ASHRAE, Atlanta. Note: The material quoted in this reference may also appear in subsequent ASHRAE handbooks, but here the year of the original version of the handbook used is shown.

Beckman, W.A., Klein, S.A., Duffie, J.A., 1977. Solar Heating Design. John Wiley & Sons, New York.

Bejan, A., 1995. Entropy Generation Minimization, second ed. CRC Press, Boca Raton, FL (Chapter 9).

Bejan, A., Kearney, D.W., Kreith, F., 1981. Second law analysis and synthesis of solar collector systems. J. Sol. Energy Eng. 103, 23−28.

Benz, N., Hasler, W., Hetfleish, J., Tratzky, S., Klein, B., 1998. Flat-plate solar collector with glass TI. In: Proceedings of EuroSun 98 Conference on CD ROM. Portoroz, Slovenia.

Berdahl, P., Martin, M., 1984. Emittance of clear skies. Sol. Energy 32, 663−664.

Boultinghouse, K.D., 1982. Development of a Solar-Flux Tracker for Parabolic-Trough Collectors. Sandia National Laboratory, Albuquerque, NM.

Brandemuehl, M.J., Beckman, W.A., 1980. Transmission of diffuse radiation through CPC and flat-plate collector glazings. Sol. Energy 24 (5), 511−513.

Briggs, F., 1980. Tracking−Refinement Modeling for Solar-Collector Control. Sandia National Laboratory, Albuquerque, NM.

De Laquil, P., Kearney, D., Geyer, M., Diver, R., 1993. Solar-thermal electric technology. In: Johanson, T.B., Kelly, H., Reddy, A.K.N., Williams, R.H. (Eds.), Renewable Energy: Sources for Fuels and Electricity. Earthscan. Island Press, Washington DC, pp. 213–296.

Dudley, V., 1995. SANDIA Report Test Results for Industrial Solar Technology Parabolic Trough Solar Collector, SAND94-1117. Sandia National Laboratory, Albuquerque, NM.

Duffie, J.A., Beckman, W.A., 2006. Solar Engineering of Thermal Processes. John Wiley & Sons, New York.

Feuermann, D., Gordon, J.M., 1991. Analysis of a two-stage linear Fresnel reflector solar concentrator. ASME J. Sol. Energy Eng. 113, 272–279.

Francia, G., 1961. A new collector of solar radiant energy. In: UN Conference New Sources of Energy Rome, vol. 4, p. 572.

Francia, G., 1968. Pilot plants of solar steam generation systems. Sol. Energy 12, 51–64.

Garg, H.P., Hrishikesan, D.S., 1998. Enhancement of solar energy on flat-plate collector by plane booster mirrors. Sol. Energy 40 (4), 295–307.

Geyer, M., Lupfert, E., Osuna, R., Esteban, A., Schiel, W., Schweitzer, A., Zarza, E., Nava, P., Langenkamp, J., Mandelberg, E., 2002. Eurotrough-parabolic trough collector developed for cost efficient solar power generation. In: Proceedings of 11th Solar PACES International Symposium on Concentrated Solar Power and Chemical Energy Technologies on CD ROM. Zurich, Switzerland.

Goetzberger, A., Dengler, J., Rommel, M., et al., 1992. New transparently insulated, bifacially irradiated solar flat-plate collector. Sol. Energy 49 (5), 403–411.

Grass, C., Benz, N., Hacker, Z., Timinger, A., 2000. Tube collector with integrated tracking parabolic concentrator. In: Proceedings of the EuroSun 2000 Conference on CD ROM. Copenhagen, Denmark.

Guven, H.M., Bannerot, R.B., 1985. Derivation of universal error parameters for comprehensive optical analysis of parabolic troughs. In: Proceedings of the ASME-ISES Solar Energy Conference. Knoxville, TN, pp. 168–174.

Guven, H.M., Bannerot, R.B., 1986. Determination of error tolerances for the optical design of parabolic troughs for developing countries. Sol. Energy 36 (6), 535–550.

Hollands, K.G.T., Unny, T.E., Raithby, G.D., Konicek, L., 1976. Free convection heat transfer across inclined air layers. J. Heat Trans. ASME 98, 189.

Jeter, M.S., 1983. Geometrical effects on the performance of trough collectors. Sol. Energy 30, 109–113.

Kalogirou, S.A., 1996. Design and construction of a one-axis sun-tracking mechanism. Sol. Energy 57 (6), 465–469.

Kalogirou, S.A., 2003. The potential of solar industrial process heat applications. Appl. Energy 76 (4), 337–361.

Kalogirou, S.A., 2004. Solar thermal collectors and applications. Prog. Energy Combust. Sci. 30 (3), 231–295.

Kalogirou, S.A., 2012. A detailed thermal model of a parabolic trough collector receiver. Energy 49, 278–281.

Kalogirou, S.A., Eleftheriou, P., Lloyd, S., Ward, J., 1994. Design and performance characteristics of a parabolic-trough solar-collector system. Appl. Energy 47 (4), 341–354.

Kalogirou, S.A., Eleftheriou, P., Lloyd, S., Ward, J., 1994. Low cost high accuracy parabolic troughs: construction and evaluation. In: Proceedings of the World Renewable Energy Congress III. Reading, UK, vol. 1, pp. 384–386.

Karimi, A., Guven, H.M., Thomas, A., 1986. Thermal analysis of direct steam generation in parabolic trough collectors. In: Proceedings of the ASME Solar Energy Conference, pp. 458–464.

Kearney, D.W., Price, H.W., 1992. Solar thermal plants—LUZ concept (current status of the SEGS plants). In: Proceedings of the Second Renewable Energy Congress. Reading, UK, vol. 2, pp. 582–588.

Kienzlen, V., Gordon, J.M., Kreider, J.F., 1988. The reverse flat-plate collector: a stationary non-evacuated, low technology, medium temperature solar collector. ASME J. Sol. Energy Eng. 110, 23–30.

Klein, S.A., 1975. Calculation of flat-plate collector loss coefficients. Sol. Energy 17 (1), 79–80.

Klein, S.A., 1979. Calculation of the monthly average transmittance—absorptance products. Sol. Energy 23 (6), 547—551.

Kreider, J.F., 1982. The Solar Heating Design Process. McGraw-Hill, New York.

Kreider, J.F., Kreith, F., 1977. Solar Heating and Cooling. McGraw-Hill, New York.

Kruger, D., Heller, A., Hennecke, K., Duer, K., 2000. Parabolic trough collectors for district heating systems at high latitudes—a case study. In: Proceedings of EuroSun 2000 on CD ROM. Copenhagen, Denmark.

Lupfert, E., Geyer, M., Schiel, W., Zarza, E., Gonzalez-Anguilar, R.O., Nava, P., 2000. Eurotrough—a new parabolic trough collector with advanced light weight structure. In: Proceedings of Solar Thermal 2000 International Conference, on CD ROM. Sydney, Australia.

Marshal, N., 1976. Gas Encyclopedia. Elsevier, New York.

Mills, D.R., 2001. Solar thermal electricity. In: Gordon, J. (Ed.), Solar Energy: The State of the Art. James and James, London, pp. 577—651.

Molineaux, B., Lachal, B., Gusian, O., 1994. Thermal analysis of five outdoor swimming pools heated by unglazed solar collectors. Sol. Energy 53 (1), 21—26.

Morrison, G.L., 2001. Solar collectors. In: Gordon, J. (Ed.), Solar Energy: The State of the Art. James and James, London, pp. 145—221.

Orel, Z.C., Gunde, M.K., Hutchins, M.G., 2002. Spectrally selective solar absorbers in different non-black colors. In: Proceedings of WREC VII, Cologne, on CD ROM.

Pereira, M., 1985. Design and performance of a novel non-evacuated 1.2x CPC type concentrator. In: Proceedings of Intersol Biennial Congress of ISES. Montreal, Canada, vol. 2, pp. 1199—1204.

Petela, R., 1964. Exergy of heat radiation. ASME J. Heat Trans. 68, 187.

Prapas, D.E., Norton, B., Probert, S.D., 1987. Optics of parabolic trough solar energy collectors possessing small concentration ratios. Sol. Energy 39, 541—550.

Rabl, A., 1976. Optical and thermal properties of compound parabolic concentrators. Sol. Energy 18 (6), 497—511.

Rabl, A., O'Gallagher, J., Winston, R., 1980. Design and test of non-evacuated solar collectors with compound parabolic concentrators. Sol. Energy 25 (4), 335—351.

Raithby, G.D., Hollands, K.G.T., 1975. A general method of obtaining approximate solutions to laminar and turbulent free convection problems. In: Irvine, F., Hartnett, J.P. (Eds.), Advances in Heat Transfer, Vol. II. Academic Press, New York, pp. 265—315.

Ratzel, A., Hickox, C., Gartling, D., 1979. Techniques for reducing thermal conduction and natural convection heat losses in annular receiver geometries. J. Heat Trans. 101 (1), 108—113.

Romero, M., Buck, R., Pacheco, J.E., 2002. An update on solar central receiver systems projects and technologies. J. Sol. Energy Eng. 124 (2), 98—108.

Shewen, E., Hollands, K.G.T., Raithby, G.D., 1996. Heat transfer by natural convection across a vertical cavity of large aspect ratio. J. Heat Transfer ASME 119, 993—995.

Sodha, M.S., Mathur, S.S., Malik, M.A.S., 1984. Wiley Eastern Limited, Singapore.

Spate, F., Hafner, B., Schwarzer, K., 1999. A system for solar process heat for decentralized applications in developing countries. In: Proceedings of ISES Solar World Congress on CD ROM. Jerusalem, Israel.

Swinbank, W.C., 1963. Long-wave radiation from clear skies. Q. J. R. Meteorol. Soc. 89, 339—348.

Tabor, H., 1966. Mirror boosters for solar collectors. Sol. Energy 10 (3), 111—118.

Tripanagnostopoulos, Y., Souliotis, M., Nousia, T., 2000. Solar collectors with colored absorbers. Sol. Energy 68 (4), 343—356.

Tripanagnostopoulos, Y., Yianoulis, P., Papaefthimiou, S., Zafeiratos, S., 2000. CPC solar collectors with flat bifacial absorbers. Sol. Energy 69 (3), 191—203.

Wackelgard, E., Niklasson, G.A., Granqvist, C.G., 2001. Selective solar-absorbing coatings. In: Gordon, J. (Ed.), Solar Energy: The State of the Art. James and James, London, pp. 109—144.

Wazwaz, J., Salmi, H., Hallak, R., 2002. Solar thermal performance of a nickel-pigmented aluminum oxide selective absorber. Renew. Energy 27 (2), 277–292.

Winston, R., 1974. Solar concentrators of novel design. Sol. Energy 16, 89–95.

Winston, R., Hinterberger, H., 1975. Principles of cylindrical concentrators for solar energy. Sol. Energy 17 (4), 255–258.

Winston, R., O'Gallagher, J., Muschaweck, J., Mahoney, A., Dudley, V., 1999. Comparison of predicted and measured performance of an integrated compound parabolic concentrator (ICPC). In: Proceedings of ISES Solar World Congress on CD ROM. Jerusalem, Israel.

Xie, W.T., Dai, Y.J., Wang, R.Z., Sumathy, K., 2011. Concentrated solar energy applications using Fresnel lenses: a review. Renew. Sustain. Energy Rev. 15 (6), 2588–2606.

Yianoulis, P., Kalogirou, S.A., Yianouli, M., 2012. Solar selective coatings. In: Sayigh, A. (Ed.), Comprehensive Renewable Energy, vol. 3, pp. 301–312.

Zhang, H.-F., Lavan, Z., 1985. Thermal performance of serpentine absorber plate. Sol. Energy 34 (2), 175–177.

Performance of Solar Collectors

The thermal performance of solar collectors can be determined by the detailed analysis of the optical and thermal characteristics of the collector materials and collector design, as outlined in Chapter 3, or by experimental performance testing under controlled conditions. It should be noted that the accuracy of the heat transfer analysis depends on uncertainties in the determination of the heat transfer coefficients, which is difficult to achieve, due to the non-uniform temperature boundary conditions that exist in solar collectors. Such analysis is usually carried out during the development of prototypes, which are then tested under defined environmental conditions. In general, experimental verification of the collector characteristics is necessary and should be done on all collector models manufactured. In some countries, the marketing of solar collectors is permitted only after test certificates are issued from accredited laboratories to protect the customers.

A number of standards describe the testing procedures for the thermal performance of solar collectors. The most well known are ISO 9806 (2017) and ANSI/ASHRAE Standard 93 (2010). These can be used to evaluate the performance of both flat-plate and concentrating solar collectors. The thermal performance of a solar collector is determined partly by obtaining values of instantaneous efficiency for different combinations of incident radiation, ambient temperature, and inlet fluid temperature. This requires experimental measurement of the rate of incident solar radiation falling onto the solar collector as well as the rate of energy addition to the transfer fluid as it passes through the collector, all under steady-state or quasi-steady-state conditions. In addition, tests must be performed to determine the transient thermal response characteristics of the collector. The variation of steady-state thermal efficiency with incident angles between the direct beam and the normal to collector aperture at various sun and collector positions is also required.

ISO 9806 (2017) and ANSI/ASHRAE Standard 93 (2010) give information on testing solar energy collectors using single-phase fluids with a fixed flow rate and no significant internal storage. The data can be used to predict the collector performance in any location and under any weather conditions where load (including load temperature), weather, and insolation are known.

Solar collectors can be tested by two basic methods: under steady-state conditions or using a dynamic test procedure. The former method is widely used and the test procedures are well documented in the aforementioned standards for glazed collectors. The same standard includes the procedure to test unglazed collectors. For steady-state testing, the environmental conditions and collector operation must be constant during the testing period. For clear dry locations, the required steady environmental conditions are easily satisfied and the testing period requires only a few days. In many locations of the world, however, steady conditions may be difficult to achieve and testing may be possible only in certain periods of the year, mainly during summertime, and even then, extended testing periods may be needed. For this reason, transient or dynamic test methods have been developed. Transient testing involves the monitoring of collector performance for a range or radiation and incident angle conditions. Subsequently, a time-dependent mathematical model is used to identify from the transient data the collector performance parameters. An advantage of the transient method is that it can be used to determine a wider range of collector performance parameters than the steady-state method. The dynamic test method was originally adopted by EN 12975-2 standard, which is now withdrawn and replaced by ISO 9806 (2017). The European standards are generally based on the ISO ones but are stricter. These are briefly introduced in Section 4.10.

To perform the required tests accurately and consistently, a test rig is required. Two such rigs can be used: closed and open loop collector test rigs, as shown in Figs. 4.1 and 4.2, respectively. For the tests, the following parameters need to be measured:

1. Global solar irradiance at the collector plane, G_t.
2. Diffuse solar irradiance at the collector aperture.
3. Air speed above the collector aperture.
4. Ambient air temperature, T_a.
5. Fluid temperature at the collector inlet, T_i.
6. Fluid temperature at the collector outlet, T_o.
7. Fluid flow rate, $\dot{m}$.

In addition, the gross collector aperture area, A_a, is required to be measured with certain accuracy. The collector efficiency, based on the gross collector aperture area, is given by:

$$\eta = \frac{\dot{m}c_p(T_o - T_i)}{A_a G_t} \qquad (4.1)$$

FIGURE 4.1

Closed loop test system.

FIGURE 4.2

Open loop test system.

In this chapter, the steady-state test method is thoroughly described. The dynamic method is presented later in the chapter.

4.1 Collector thermal efficiency

The collector performance test is performed under steady-state conditions, with steady radiant energy falling on the collector surface, a steady fluid flow rate, and constant wind speed and ambient temperature. When a constant inlet fluid temperature is supplied to the collector, it is possible to maintain a constant outlet fluid temperature from the collector. In this case, the useful energy gain from the collector is calculated from:

$$Q_u = \dot{m}c_p(T_o - T_i) \tag{4.2}$$

From Chapter 3, we have seen that the useful energy collected from a solar collector is given by:

$$Q_u = A_a F_R[G_t(\tau\alpha)_n - U_L(T_i - T_a)] \tag{4.3}$$

Moreover, the thermal efficiency is obtained by dividing Q_u by the energy input $(A_a G_t)$:

$$\eta = F_R(\tau\alpha)_n - F_R U_L \left(\frac{T_i - T_a}{G_t}\right) \tag{4.4}$$

During testing, the collector is mounted in such a way as to face the sun perpendicularly; as a result, the transmittance−absorptance product for the collector corresponds to that of beam radiation at normal incidence. Therefore, the term $(\tau\alpha)_n$ is used in Eqs. (4.3) and (4.4) to denote that the normal transmittance−absorptance product is used.

Similarly, for concentrating collectors, the following equations from Chapter 3 can be used for the useful energy collected and collector efficiency:

$$Q_u = F_R[G_B\eta_o A_a - A_r U_L(T_i - T_a)] \tag{4.5}$$

$$\eta = F_R\eta_o - \frac{F_R U_L(T_i - T_a)}{CG_B} \tag{4.6}$$

Notice that, in this case, G_t is replaced by G_B, since concentrating collectors can utilize only beam radiation (Kalogirou, 2004).

For a collector operating under steady irradiation and fluid flow rate, the factors F_R, $(\tau\alpha)_n$, and U_L are nearly constant. Therefore, Eqs. (4.4) and (4.6) plot as a straight line on a graph of efficiency versus the heat loss parameter $(T_i - T_a)/G_t$ for the case of flat-plate collectors (see Fig. 4.3a) and $(T_i - T_a)/G_B$ for the case of concentrating collectors (see Fig. 4.3b). The intercept (intersection of the line with the vertical efficiency axis) equals $F_R(\tau\alpha)_n$ for the FPCs and $F_R\eta_o$ for the concentrating ones. The slope of the line, i.e., the efficiency difference divided by the corresponding horizontal scale difference, equals $-F_R U_L$ and $-F_R U_L/C$, respectively. If experimental data on collector heat delivery at various temperatures and solar conditions are plotted with efficiency as the vertical axis and $\Delta T/G$ (G_t or G_B is

a) Flat-plate collector   **b) Concentrating collector**

c) Area used in efficiency calculations

FIGURE 4.3

Typical collector performance curves and area used in efficiency calculations.

used according to the type of collector) as the horizontal axis, the best straight line through the data points correlates the collector performance with solar and temperature conditions. The intersection of the line with the vertical axis is where the temperature of the fluid entering the collector equals the ambient temperature and collector efficiency is at its maximum. At the intersection of the line with the horizontal axis, collector efficiency is zero. This condition corresponds to such a low radiation level, or such a high temperature of the fluid into the collector, that heat losses equal solar absorption and the collector delivers no useful heat. This condition, normally called *stagnation*, usually occurs when no fluid flows in the collector. This maximum temperature (for an FPC) is given by:

$$T_{max} = \frac{G_t(\tau\alpha)_n}{U_L} + T_a \tag{4.7}$$

As can be seen from Fig. 4.3, the slope of the concentrating collectors is much smaller than the one for the flat plate. This is because the thermal losses are inversely proportional to the concentration ratio, C. This is the greatest advantage of the concentrating collectors, that is, the efficiency of concentrating collectors remains high at high inlet temperature; that is why this type of collector is suitable for high-temperature applications.

It should be noted that the efficiency curve depends on the magnitude of the irradiance level (G) and on the reference collector area (A) used. For FPCs, the area, i.e., gross aperture or absorber used, as shown in Fig. 4.3c, must be referenced when presenting the efficiency curve since the total irradiation used in the efficiency calculation is proportional to it. As already indicated, a given efficiency curve is also affected by the irradiation level (G). Therefore, different curves are produced at different solar radiation levels, and the standards specify the minimum value of G to be used during the performance testing, as shown in Section 4.1.3.

A comparison of the efficiency of various collectors at irradiance levels of 500 and 1000 W/m^2 is shown in Fig. 4.4 (Kalogirou, 2004). Five representative collector types are considered:

- Flat-plate collector (FPC).
- Advanced flat-plate collector (AFP). In this collector, the risers are ultrasonically welded to the absorbing plate, which is also electroplated with chromium selective coating.
- Stationary compound parabolic collector (CPC) oriented with its long axis in the east–west direction.
- Evacuated tube collector (ETC).
- Parabolic trough collector (PTC) with E–W tracking.

As seen in Fig. 4.4, the higher the irradiation level, the better is the efficiency, and the higher performance collectors, such as the CPC, ETC, and PTC, retain high efficiency, even at higher collector inlet temperatures. It should be noted that the radiation levels examined are considered as global radiation for all collector types except the PTC, for which the same radiation values are used but considered as beam radiation.

In reality, the heat loss coefficient, U_L, in Eqs. (4.3)–(4.6) is not constant but is a function of the collector inlet and ambient temperatures. Therefore,

$$F_R U_L = c_1 + c_2(T_i - T_a) \tag{4.8}$$

Comparison of the efficiency of various collectors at two irradiation levels: 500 and 1000 W/m².

Applying Eq. (4.8) in Eqs. (4.3) and (4.5), we have the following.
For flat-plate collectors:

$$Q_u = A_a \left[F_R(\tau\alpha)_n G_t - c_1(T_i - T_a) - c_2(T_i - T_a)^2 \right] \tag{4.9}$$

and for concentrating collectors:

$$Q_u = F_R G_B \eta_o A_a - A_r c_1(T_i - T_a) - A_r c_2(T_i - T_a)^2 \tag{4.10}$$

Therefore, for FPCs, the efficiency can be written as:

$$\eta = F_R(\tau\alpha)_n - c_1 \frac{(T_i - T_a)}{G_t} - c_2 \frac{(T_i - T_a)^2}{G_t} \tag{4.11}$$

and if we denote $c_o = F_R(\tau\alpha)_n$ and $x = (T_i - T_a)/G_t$, then:

$$\eta = c_o - c_1 x - c_2 G_t x^2 \tag{4.12}$$

And, for concentrating collectors, the efficiency can be written as:

$$\eta = F_R \eta_o - \frac{c_1(T_i - T_a)}{C G_B} - \frac{c_2(T_i - T_a)^2}{C G_B} \tag{4.13}$$

and if we denote $k_o = F_R \eta_o$, $k_1 = c_1/C$, $k_2 = c_2/C$, and $y = (T_i - T_a)/G_B$, then:

$$\eta = k_o - k_1 y - k_2 G_B y^2 \tag{4.14}$$

The difference in performance between flat-plate and concentrating collectors can also be seen from the performance equations. For example, the performance of a good FPC is given by:

$$\eta = 0.792 - 6.65\left(\frac{\Delta T}{G_t}\right) - 0.06\left(\frac{\Delta T^2}{G_t}\right) \tag{4.15}$$

whereas the performance equation of the Industrial Solar Technologies (IST) PTC is:

$$\eta = 0.762 - 0.2125\left(\frac{\Delta T}{G_B}\right) - 0.001672\left(\frac{\Delta T^2}{G_B}\right) \tag{4.16}$$

By comparing Eqs. (4.15) and (4.16), we can see that FPCs usually have a higher intercept efficiency because their optical characteristics are better (no reflection losses), whereas the heat loss coefficients of the concentrating collectors are much smaller because these factors are inversely proportional to the concentration ratio.

In the above Equations ΔT or $(T_i - T_a)$ is often termed *reduced temperature difference*. The standard ISO 9806 (2017) specifies the use of $(T_m - T_a)$, where $T_m = (T_i + T_o)/2$ is the mean temperature of the collector. In this case the collector efficiency equation is modified as shown in Section 4.6. It should be noted that the equation for the extracted power from a solar collector is modeled with an equation that includes eight terms to consider all possible design variations (e.g., unglazed collectors). For glazed collectors, this equation is reduced to Eq. (4.10).

Equations (4.11) and (4.13) include all important design and operational factors affecting steady-state performance, except collector flow rate and solar incidence angle. Flow rate inherently affects performance through the average absorber temperature. If the heat removal rate is reduced, the average absorber temperature increases and more heat is lost. If the flow is increased, collector absorber temperature and heat loss decrease. The effect of the solar incidence angle is accounted for by the incidence angle modifier, examined in Section 4.2.

4.1.1 Effect of flow rate

Experimental test data can be correlated to give values of $F_R(\tau\alpha)_n$ and $F_R U_L$ for a particular flow rate used during the test. If the flow rate of the collector is changed from the test value during normal use, it is possible to calculate the new F_R for the new flow rate using Eq. (3.58). A correction for the changed flow rate can be made if it is assumed that F' does not change with flow rate, because of changes in h_{fi}. The ratio r, by which the factors $F_R(\tau\alpha)_n$ and $F_R U_L$ are corrected, is given by (Duffie and Beckman, 1991):

$$r = \frac{F_R U_L|_{use}}{F_R U_L|_{test}} = \frac{F_R(\tau\alpha)_n|_{use}}{F_R(\tau\alpha)_n|_{test}} = \frac{\frac{\dot{m}c_p}{A_c F' U_L}\left[1 - \exp\left(-\frac{U_L F' A_c}{\dot{m}c_p}\right)\right]\Big|_{use}}{\frac{\dot{m}c_p}{A_c F' U_L}\left[1 - \exp\left(-\frac{U_L F' A_c}{\dot{m}c_p}\right)\right]\Big|_{test}} \tag{4.17a}$$

or

$$r = \frac{\frac{\dot{m}c_p}{A_c}\left[1 - \exp\left(-\frac{U_L F' A_c}{\dot{m}c_p}\right)\right]\Big|_{use}}{F_R U_L|_{test}} \tag{4.17b}$$

To use the above equations knowledge is required of the term $F'U_L$. For the test conditions this can be calculated from Eq. (3.58) as the term $F_R U_L$ is known from the performance testing. Therefore, by rearranging Eq. (3.58) we get:

$$F'U_L = -\frac{\dot{m}c_p}{A_c}\ln\left(1 - \frac{F_R U_L A_c}{\dot{m}c_p}\right) \tag{4.18}$$

For liquid collectors $F'U_L$ is approximately equal for both test and use conditions, so the value estimated by Eq. (4.18) can be used in both cases in Eq. (4.17a).

For air collectors or for liquid collectors where h_{fi} depends strongly on flow rate, F' needs to be estimated for the new value of h_{fi} using Eq. (3.48), (3.79) or (3.86a) according to the collector type. In this case h_{fi} is estimated from the Nusselt number and the type of flow determined from the Reynolds number. For example for turbulent flow, the Nusselt number given by Eq. (3.131) can be used for internal pipe flows. For other cases appropriate equations can be used from heat transfer textbooks.

4.1.2 Collectors in series

Performance data for a single panel cannot be applied directly to a series of connected panels if the flow rate through the series is the same as for the single panel test data. If, however, N panels of the same type are connected in series and the flow is N times that of the single panel flow used during the testing, then the single panel performance data can be applied. If two panels are considered connected in series and the flow rate is set to a single panel test flow, the performance will be less than if the two panels were connected in parallel with the same flow rate through each collector. The useful energy output from the two collectors connected in series is then given by (Morrison, 2001):

$$Q_u = A_c F_R[(\tau\alpha)G_t - U_L(T_i - T_a) + (\tau\alpha)G_t - U_L(T_{o1} - T_a)] \tag{4.19}$$

where T_{o1} = outlet temperature from first collector given by:

$$T_{o1} = \frac{A_c F_R[(\tau\alpha)G_t - U_L(T_i - T_a)]}{\dot{m}c_p} + T_i \tag{4.20}$$

Eliminating T_{o1} from Eqs. (4.19) and (4.20) gives:

$$Q_u = 2A_c F_{R1}\left(1 - \frac{K}{2}\right)[(\tau\alpha)_1 G_t - U_{L1}(T_i - T_a)] \tag{4.21}$$

where F_{R1}, U_{L1}, and $(\tau\alpha)_1$ are the factors for the single panel tested, and K is:

$$K = \frac{A_c F_{R1} U_{L1}}{\dot{m}c_p} \tag{4.22}$$

For N identical collectors connected in series with the flow rate set to the single panel flow rate,

$$F_R(\tau\alpha)|_{series} = F_{R1}(\tau\alpha)_1\left[\frac{1 - (1 - K)^N}{NK}\right] \tag{4.23}$$

$$F_R U_L|_{series} = F_{R1} U_{L1}\left[\frac{1 - (1 - K)^N}{NK}\right] \tag{4.24}$$

If the collectors are connected in series and the flow rate per unit aperture area in each series line of collectors is equal to the test flow rate per unit aperture area, then no penalty is associated with the flow rate other than an increased pressure drop from the circuit.

EXAMPLE 4.1

For five collectors in series, each 2 m^2 in area and $F_{R1}U_{L1} = 4$ W/m^2 °C at a flow rate of 0.01 kg/s, estimate the correction factor. Water is circulated through the collectors.

Solution

From Eq. (4.22),

$$K = \frac{A_c F_{R1} U_{L1}}{\dot{m} c_p} = \frac{2 \times 4}{0.01 \times 4180} = 0.19$$

The factor

$$\frac{1 - (1 - K)^N}{NK} = \frac{1 - (1 - 0.19)^5}{5 \times 0.19} = 0.686$$

This example indicates that connecting collectors in series without increasing the working fluid flow rate in proportion to the number of collectors results in significant loss of output.

4.1.3 Standard requirements

Here, the various requirements of the ISO standards for both glazed and unglazed collectors are presented. For a more comprehensive list of the requirements and details on the test procedures, the reader is advised to read the actual standard.

Glazed collectors

To perform the steady-state test satisfactorily, according to ISO 9806 (2017), certain environmental conditions are required:

1. Solar radiation greater than 700 W/m^2.
2. Wind speed must be maintained between 2 and 4 m/s (3 m/s $\pm$ 1 m/s). If the natural wind is less than 2 m/s, an artificial wind generator must be used.
3. Angle of incidence of direct radiation is within $\pm 2\%$ of the normal incident angle. The collector shall be tested at diffuse irradiance levels of always less than 30%.
4. Fluid flow rate should be set at 0.02 kg/s m^2 and the fluid flow must be stable within $\pm 1\%$ during each test but may vary up to $\pm 5\%$ between different tests. Other flow rates may be used, if specified by the manufacturer.

Data points that satisfy these requirements must be obtained for a minimum of four fluid inlet temperatures, which are evenly spaced over the operating range of the collector. The first must be within ± 3 K of the ambient temperature to accurately obtain the test intercept, and the last should be at the maximum collector operating temperature specified by the manufacturer. If water is the heat transfer fluid, 70 °C is usually adequate as a maximum temperature. At least four independent data points should be obtained for each fluid inlet temperature. If no continuous tracking is used, then an equal number of points should be taken before and after local solar noon for each inlet fluid temperature. Additionally, for each data point, a preconditioning period of at least 15 min is required,

Table 4.1 Tolerance of measured parameters for glazed collectors.

Parameter	Deviation from the mean
Total solar irradiance	± 50 W/m^2
Ambient air temperature	± 1.5 K
Wind speed	3 m/s ± 1 m/s
Fluid mass flow rate	$\pm 1\%$
Collector inlet fluid temperature	± 0.1 K
Collector outlet fluid temperature	± 0.4 K

using the stated inlet fluid temperature. The actual measurement period should be four times greater than the fluid transit time through the collector with a minimum test period of 15 min.

To establish that steady-state conditions exist, average values of each parameter should be taken over successive periods of 30 s and compared with the mean value over the test period. A steady-state condition is defined as the period during which the operating conditions are within the values given in Table 4.1.

Unglazed collectors

Unglazed collectors are more difficult to test, because their operation is influenced by not only the solar radiation and ambient temperature but also the wind speed. The last factor influences the collector performance to a great extent, since there is no glazing. Because it is very difficult to find periods of steady wind conditions (constant wind speed and direction), the ISO 9806 (2017) for unglazed collector testing recommends that an artificial wind generator is used to control the wind speed parallel to the collector aperture. The performance of unglazed collectors is also a function of the module size and may be influenced by the solar absorption properties of the surrounding ground (usually roof material). Data shall be measured for at least three fluid inlet temperatures spaced evenly over the operating temperature range, and at least four independent data points shall be obtained for each data set, i.e., 36 data points. In addition to the measured parameters listed at the beginning of this chapter, the longwave thermal irradiance at the collector plane needs to be measured. Alternatively, the dew point temperature could be measured, from which the longwave irradiance may be estimated.

Similar requirements for preconditioning apply here as in the case of glazed collectors. However, the length of the steady-state test period in this case should be more than four times the ratio of the thermal capacity of the collector to the thermal capacity flow rate $\dot{m}c_p$ of the fluid flowing through the collector. In this case, the collector is considered to operate under steady-state conditions if, over the testing period, the measured parameters deviate from their mean values by less than the limits given in Table 4.2.

Using a solar simulator

In countries with unsuitable weather conditions, the indoor testing of solar collectors with the use of a solar simulator is preferred. Solar simulators are generally of two types: those that use a point source of radiation mounted well away from the collector and those with large area multiple lamps mounted

Table 4.2 Tolerance of measured parameters for unglazed collectors.

Parameter	Deviation from the mean
Total solar irradiance	± 50 W/m^2
Longwave thermal irradiance	± 20 W/m^2
Ambient air temperature	± 1.5 K
Wind speed	± 3 m/s ± 1 m/s
Fluid mass flow rate	$\pm 2\%$
Collector inlet fluid temperature	± 1.5 K
Collector outlet fluid temperature	± 1.5 K

close to the collector. In both cases, special care should be taken to reproduce the spectral properties of the natural solar radiation. The simulator characteristics required are also specified in ISO 9806 (2017) and the main ones are:

1. The lamps shall be capable of producing a mean irradiance over the collector aperture area of at least 700 W/m^2, and measurements can start after reaching stable working conditions.
2. Radiation at any point on the collector aperture must not differ by more than $\pm 15\%$ from the mean radiation over the aperture.
3. The spectral distribution of the simulated solar radiation shall approximately equivalent to the spectrum at optical air mass 1.5. Measurement of the solar simulator's spectral qualities shall be made in the plane of the collector over the wavelength range of 0.3 and 3 μm, as indicated in ISO 9806 (2017).
4. As in multiple lamp simulators, the spectral characteristics of the lamp array change with time, and as the lamps are replaced, the characteristics of the simulator must be determined on a regular basis.

Solar simulators can be divided into three main types: continuous, flash, and pulse. The first type is described above. This type is mostly used for low-intensity testing, varying from less than one to several suns.

The second type is the flash simulator. This is mostly suitable for testing photovoltaic (PV) cells and panels. The measurement with this solar simulator is instantaneous and lasts approximately as long as the flash of a camera (several milliseconds). Some models allow also the current–voltage (I–V) measurement of the PV. With this simulator very high intensities of up to several thousand suns are possible. The main advantage of this simulator is that it avoids heat buildup in the test specimen. The disadvantage is that due to the rapid heating and cooling of the lamp, the intensity and light spectrum are not constant, which may create reliability problems in repeated measurements.

The third type of solar simulator is the pulse simulator. This uses a shutter between the lamp and the specimen under test to quickly block or unblock the light. Pulses are typically of the order of 100 ms. In this case the lamp remains ON during the whole duration of the test and thus this simulator offers a compromise between the continuous and flash types. The disadvantages of this simulator are the high power consumption and the relatively low light intensities of the continuous type. The advantages are the stable light intensity and spectrum output and low thermal load imposed on the test specimen.

4.2 Collector incidence angle modifier

4.2.1 Flat-plate collectors

The performance Eqs. (4.9) and (4.11) for FPCs assume that the sun is perpendicular to the plane of the collector, which rarely occurs. For the glass cover plates of an FPC, specular reflection of radiation occurs, thereby reducing the $(\tau\alpha)$ product. The incidence angle modifier, K_θ, is defined as the ratio of $(\tau\alpha)$ at some incident angle θ to $(\tau\alpha)$ at normal incidence $(\tau\alpha)_n$. According to ISO 9806 (2017) data are collected for angles of incidence between 0° and 90°, every 10°. A plot of incidence angle modified against incident angle is shown in Fig. 4.5.

If we plot the incidence angle modifier against $1/\cos(\theta) - 1$, it is observed that a straight line is obtained, as shown in Fig. 4.6, which can be described by the following expression:

$$K_\theta = \frac{(\tau\alpha)}{(\tau\alpha)_n} = 1 - b_o\left[\frac{1}{\cos(\theta)} - 1\right] \tag{4.25}$$

For a single glass cover, the factor b_o in Eq. (4.25), which is the slope of the line in Fig. 4.6, is about 0.1. A more general expression for the incidence angle modifier is a second-order equation given by:

$$K_\theta = 1 - b_o\left[\frac{1}{\cos(\theta)} - 1\right] - b_1\left[\frac{1}{\cos(\theta)} - 1\right]^2 \tag{4.26}$$

With the incidence angle modifier the collector efficiency, Eq. (4.11) can be modified as:

$$\eta = F_R(\tau\alpha)_n K_\theta - c_1\frac{(T_i - T_a)}{G_t} - c_2\frac{(T_i - T_a)^2}{G_t} \tag{4.27}$$

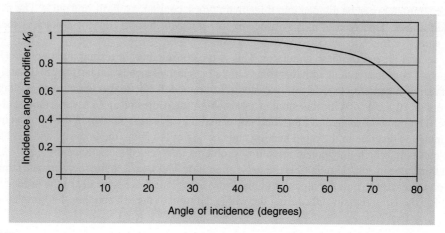

FIGURE 4.5

Incidence angle modified graph.

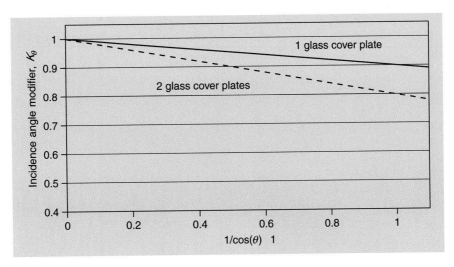

FIGURE 4.6

Plot of incidence angle modifier against $1/\cos(\theta) - 1$ for two types of flat-plate collectors.

The equation for the useful energy collected, Eq. (4.9), is also modified in a similar way.

The incidence angle modifier for collectors that do not have symmetric optical characteristics can be approximated by the product of two orthogonal incidence angle modifiers in the longitudinal (L) and transverse (t) planes of the collector $[K_\theta = K_{\theta,l}K_{\theta,t}]$, where in each plane the appropriate incidence angle is used. Collectors requiring this treatment are the evacuated tube, which have covers that are optically non-symmetrical.

4.2.2 Concentrating collectors

Similarly, for concentrating collectors, the performance Eqs. (4.10) and (4.13) described previously are reasonably well defined as long as the direct beam of solar irradiation is normal to the collector aperture. For off-normal incidence angles, the optical efficiency term (η_o) is often difficult to be described analytically, because it depends on the actual concentrator geometry, concentrator optics, receiver geometry, and receiver optics, which may differ significantly. As the incident angle of the beam radiation increases, these terms become more complex. Fortunately, the combined effect of these parameters at different incident angles can be accounted for with the incident angle modifier. This is simply a correlation factor to be applied to the efficiency curve and is a function of only the incident angle between the direct solar beam and the outward drawn normal to the aperture plane of the collector. It describes how the optical efficiency of the collector changes as the incident angle changes. With the incident angle modifier, Eq. (4.13) becomes:

$$\eta = F_R K_\theta \eta_o - \frac{c_1(T_i - T_a)}{CG_B} - \frac{c_2(T_i - T_a)^2}{CG_B} \tag{4.28}$$

If the inlet fluid temperature is maintained equal to ambient temperature, the incident angle modifier can be determined from:

$$K_\theta = \frac{\eta(T_i = T_a)}{F_R[\eta_o]_n} \qquad (4.29)$$

where $\eta(T_i = T_a)$ is the measured efficiency at the desired incident angle for an inlet fluid temperature equal to the ambient temperature. The denominator in Eq. (4.29) is the test intercept taken from the collector efficiency test with Eq. (4.13), with $[\eta_o]_n$ being the normal optical efficiency, that is, at a normal angle of incidence.

As an example, the results obtained from such a test are denoted by the small squares in Fig. 4.7. By using a curve-fitting method (second-order polynomial fit), the curve that best fits the points can be obtained (Kalogirou et al., 1994):

$$K_\theta = 1 - 0.00384(\theta) - 0.000143(\theta)^2 \qquad (4.30)$$

For the IST collector, the incidence angle modifier K_θ of the collector given by the manufacturer is:

$$K_\theta = \cos(\theta) + 0.0003178(\theta) - 0.00003985(\theta)^2 \qquad (4.31)$$

4.3 Concentrating collector acceptance angle

Another test required for the concentrating collectors is the determination of the collector acceptance angle, which characterizes the effect of errors in the tracking mechanism angular orientation.

FIGURE 4.7

Parabolic trough collector incidence angle modifier test results.

This can be found with the tracking mechanism disengaged and by measuring the efficiency at various out-of-focus angles as the sun is traveling over the collector plane. An example is shown in Fig. 4.8, where the angle of incidence measured from the normal to the tracking axis (i.e., out-of-focus angle) is plotted against the efficiency factor, that is, the ratio of the maximum efficiency at normal incidence to the efficiency at a particular out-of-focus angle.

A definition of the collector acceptance angle is the range of incidence angles (as measured from the normal to the tracking axis) in which the efficiency factor varies by no more than 2% from the value of normal incidence (ANSI/ASHRAE Standard 93, 2010). Therefore, from Fig. 4.8, the collector half acceptance angle, θ_m, is 0.5°. This angle determines the maximum error of the tracking mechanism.

4.4 Collector time constant

A last aspect of collector testing is the determination of the heat capacity of a collector in terms of a time constant. It is also necessary to determine the time response of the solar collector to be able to evaluate the transient behavior of the collector and select the correct time intervals for the quasi-steady-state or steady-state efficiency tests. Whenever transient conditions exist, Eqs. (4.9)–(4.14) do not govern the thermal performance of the collector, since part of the absorbed solar energy is used for heating up the collector and its components.

The time constant of a collector is the time required for the fluid leaving the collector to reach 63.2% of its ultimate steady value after a step change in incident radiation. The collector time constant is a measure of the time required for the following relationship to apply (ANSI/ASHRAE Standard 93, 2010):

$$\frac{T_{of} - T_{ot}}{T_{of} - T_i} = \frac{1}{e} = 0.368 \tag{4.32}$$

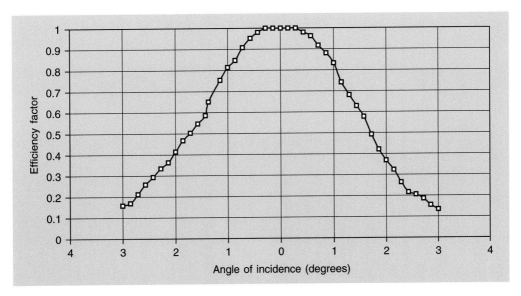

FIGURE 4.8

Parabolic trough collector acceptance angle test results.

where

T_{ot} = collector outlet water temperature after time t (°C);
T_{of} = collector outlet final water temperature (°C); and
T_i = collector inlet water temperature (°C).

The procedure for performing this test is as follows. The heat transfer fluid is circulated through the collector at the same flow rate as that used during collector thermal efficiency tests. The aperture of the collector is shielded from the solar radiation by means of a solar reflecting cover, or in the case of a concentrating collector, the collector is defocused and the temperature of the heat transfer fluid at the collector inlet is set approximately equal to the ambient air temperature. When a steady state has been reached, the cover is removed and measurements continue until steady-state conditions are achieved again. For the purpose of this test, a steady-state condition is assumed to exist when the outlet temperature of the fluid varies by less than 0.05 °C per minute (ISO 9806, 2017).

The difference between the temperature of the fluid at the collector outlet at time t and that of the surrounding air $(T_{ot} - T_a)$ (note that, for this test, $T_i = T_a$) is plotted against time, beginning with the initial steady-state condition $(T_{oi} - T_a)$ and continuing until the second steady state has been achieved at a higher temperature $(T_{of} - T_a)$, as shown in Fig. 4.9.

The time constant of the collector is defined as the time taken for the collector outlet temperature to rise by 63.2% of the total increase from $(T_{oi} - T_a)$ to $(T_{of} - T_a)$ following the step increase in solar irradiance at time 0.

The time constant specified in the standard ISO 9806 (2017), as described previously, occurs when the collector warms up. Another way to perform this test, specified in ANSI/ASHRAE Standard 93 (2010) and carried out in addition to the preceding procedure by some researchers, is to measure the time constant during cool-down. In this case, again, the collector is operated with the fluid inlet temperature maintained at the ambient temperature. The incident solar energy is then abruptly reduced to 0 by either shielding an FPC or defocusing a concentrating one. The temperatures of the transfer fluid are continuously monitored as a function of time until Eq. (4.33) is satisfied:

FIGURE 4.9

Time constant as specified in ISO 9806 (2017).

$$\frac{T_{ot} - T_i}{T_{oi} - T_i} = \frac{1}{e} = 0.368 \tag{4.33}$$

where T_{oi} = collector outlet initial water temperature (°C).

The graph of the difference between the various temperatures of the fluid in this case is as shown in Fig. 4.10.

The time constant of the collector, in this case, is the time taken for the collector outlet temperature to drop by 63.2% of the total increase from $(T_{oi} - T_a)$ to $(T_{of} - T_a)$ following the step decrease in solar irradiance at time 0 (ANSI/ASHRAE Standard 93, 2010).

4.5 Dynamic system test method

For locations that do not have steady environmental conditions for long periods of time, the transient or dynamic system test method can be used. This method involves monitoring the transient response of a collector over a number of days, which include both clear and cloudy conditions. The performance data obtained from the dynamic method allow a more detailed characterization of the collector performance in comparison with the steady-state method. The advantages of the dynamic test method are that the test period is much shorter and can be conducted at any time of the year under variable weather conditions. After testing, the data collected over the wide range of operating conditions are fitted to a transient mathematical model of the collector performance. The test data are measured every 5–10 min. For a glazed collector, the following model for the transient useful energy collection could be used (Morrison, 2001):

$$Q_u = \eta_o \left[K_{\theta,B} G_B + K_{\theta,D} G_D \right] - a_0 \left(\overline{T} - T_a \right) - a_1 \left(\overline{T} - T_a \right)^2 - c \frac{d\overline{T}}{dt} \tag{4.34}$$

where η_o, a_0, a_1, c, and the coefficients $K_{\theta,B}$ and $K_{\theta,D}$ are determined by the correlation of the test measured data.

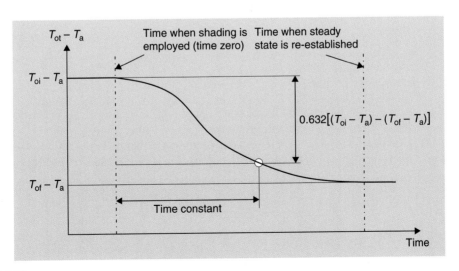

FIGURE 4.10

Time constant as specified in ANSI/ASHRAE Standard 93 (2010).

Equation (4.34) is similar to the second-order equations used for steady-state testing, presented earlier in this chapter, with the addition of a transient term and incident angle modifiers for both beam, $K_{\theta,B}$, and diffuse, $K_{\theta,D}$, radiation. The actual equation in ISO 9806 (2017) also includes eight terms. Most of these terms are eliminated for specific types of collectors. Standard FprEN 12975: 2021 shows a way to convert test results according to earlier standards. This applies to both steady-state and quasi-dynamic test methods (EN, 2021).

Models that are more complex can be used if the testing program can cover an extended range of operating conditions. In any case, the measured transient data are analyzed using a procedure that compares a set of model coefficients that minimize the deviation between the measured and predicted output. The method should be such that the various parameters should be determined as independently as possible. To be able to satisfy this requirement, sufficient data are needed; therefore, it is required to control the experimental conditions so that all variables independently influence the operation of the collector at various periods during testing. Additionally, a wide range of test conditions are required to determine the incident angle modifiers accurately. An added advantage of the method is that the equipment required is the same as the steady-state testing shown in Figs. 4.1 and 4.2, which means that a test center can have the same equipment and perform both steady-state and dynamic testing at different periods of the year, according to the prevailing weather conditions. The primary difference between the two methods is that, in the dynamic method, the data are recorded on a continuous basis over a day and averaged over 5–10 min.

Due to the wider range of collector parameters that can be determined with the dynamic method, it is likely that it may displace the steady-state testing method, even for locations that have clear and stable climatic conditions.

4.6 Efficiency parameters conversion

In the equations representing the performance of the solar collectors presented so far, the inlet fluid temperature is used. In various standards, however, the average temperature is used, T_m, defined as the arithmetic average of the inlet and outlet temperatures $[(T_i + T_o)/2]$ and the efficiency is plotted against $(T_m - T_a)/G_t$. In this case the collector instantaneous efficiency is given by:

$$\eta = F'_m(\tau\alpha) - F'_m U_L \left(\frac{T_m - T_a}{G_t} \right) \tag{4.35}$$

where F'_m is the mean temperature collector efficiency factor.

If the temperature rise through the collector is linear with distance, which is correct for most applications $F'_m = F'$ and the collector efficiency factor given by Eq. (3.48).

Therefore, in the case where T_m used for the collector performance, Eq. (4.35) plots as a straight line and the intercept $F'_m(\tau\alpha)_n$ and slope $F'_m U_L$ of the curve are related to $F_R(\tau\alpha)_n$ and $F_R U_L$, respectively, by eliminating η and T_o from Eqs. (4.2) and (4.4) and using the equation for T_m:

$$F_R(\tau\alpha)_n = F'_m(\tau\alpha)_n \left(1 + \frac{A_c F'_m U_L}{2\dot{m}c_p} \right)^{-1} \tag{4.36a}$$

$$F'_m(\tau\alpha)_n = F_R(\tau\alpha)_n \left(1 - \frac{A_c F_R U_L}{2\dot{m}c_p} \right)^{-1} \tag{4.36b}$$

And

$$F_R U_L = F'_m U_L \left(1 + \frac{A_c F'_m U_L}{2 \dot{m} c_p} \right)^{-1} \tag{4.36c}$$

$$F'_m U_L = F_R U_L \left(1 - \frac{A_c F_R U_L}{2 \dot{m} c_p} \right)^{-1} \tag{4.36d}$$

To apply these equations the fluid flow rate through the collector must be known.

Similarly in the case where T_o is used for the collector performance and the efficiency is plotted against $(T_o - T_a)/G_t$, sometimes applied in the case of air collectors. This is because in air-heating collectors, the difference between inlet and outlet temperatures can be as high as $30-40\,°C$ depending on the air mass flow, resulting in a small amount of heat being transmitted from the absorber to the air (Shams et al., 2016). This results in a non-linear increase in the air temperature along the absorber plate resulting in the arithmetic mean collector temperature not being representative of the heat loss. As a result, the physical mean temperature of the collector is often much closer to the outlet air temperature than the arithmetic mean temperature indicated above (Hastings and Morck, 2000). Therefore, in this case, the collector instantaneous efficiency is given by:

$$\eta = F'_o(\tau\alpha) - F'_o U_L \left(\frac{T_o - T_a}{G_t} \right) \tag{4.37}$$

Equation (4.37) also plots as a straight line and as above, the intercept $F'_o(\tau\alpha)_n$ and slope $F'_o U_L$ of the curve are related to $F_R(\tau\alpha)_n$ and $F_R U_L$, respectively, from:

$$F_R(\tau\alpha)_n = F'_o(\tau\alpha)_n \left(1 + \frac{A_c F'_o U_L}{\dot{m} c_p} \right)^{-1} \tag{4.38a}$$

$$F'_o(\tau\alpha)_n = F_R(\tau\alpha)_n \left(1 - \frac{A_c F_R U_L}{\dot{m} c_p} \right)^{-1} \tag{4.38b}$$

And

$$F_R U_L = F'_o U_L \left(1 + \frac{A_c F'_o U_L}{\dot{m} c_p} \right)^{-1} \tag{4.38c}$$

$$F'_o U_L = F_R U_L \left(1 - \frac{A_c F_R U_L}{\dot{m} c_p} \right)^{-1} \tag{4.38d}$$

4.7 Assessment of uncertainty in solar collector testing

With the procedure described earlier in this chapter for the performance evaluation of solar energy collectors, the collector performance equation is derived which can be used for the prediction of its output under any conditions. The basic aim of the collector testing is the determination of the collector thermal efficiency under specific conditions and as can be seen the behavior of the collector can be obtained with either a 2- or a 3-parameter single-node steady-state model given by Eqs. (4.4) and (4.11), respectively, for an FPC or from the quasi-dynamic state model or dynamic test method given by Eq. (4.34).

During the experimental phase, the inlet and outlet temperatures are measured, as well as the solar energy and the basic climatic quantities, whereas in analyzing the data a least-squares fit is performed on the measured data to determine the various parameters of the above equations. All these measurements and calculations have certain uncertainties which need to be considered according to Annex K of the previous EN 12975–2:2006 (this was withdrawn and replaced by ISO 9806, 2017). As mentioned in the standard the aim of the annex is to provide a general guidance for the assessment of uncertainty in the result of solar collector testing performed according to the standard. Uncertainty values are expressed in the same way as standard deviations.

It should be noted that the proposed methodology is one of the possible approaches for the assessment of uncertainty. The particular approach to be followed by a testing laboratory is often recommended by the accreditation body responsible for certifying the laboratory. The present approach is given here, however, as it is included in the above-mentioned standard.

Specifically, it is assumed that the behavior of the collector can be described by an M-parameter single-node steady-state or quasi-dynamic model given by:

$$\eta = c_1 p_1 + c_2 p_2 + \ldots + c_M p_M \tag{4.39}$$

where

$p_1, p_2, \ldots, p_M$ are quantities, the values of which are determined experimentally through testing; $c_1, c_2, \ldots, c_M$ are characteristic constants of the collector that are determined through testing.

In the case of the steady-state model given by Eq. (4.11), $M = 3$, $c_1 = \eta_0$, $c_2 = U_1$, $c_3 = U_2$, $p_1 = 1$, $p_2 = (T_i - T_a)/G_t$ and $p_3 = (T_i - T_a)^2/G_t$. During the experimental phase, the various quantities mentioned above are measured in J steady-state or quasi-dynamic state points, depending on the model used. From these primary measurements the values of parameters $\eta, p_1, p_2, \ldots, p_M$ are derived for each point of observation j, $j = 1, \ldots, J$. Generally, the experimental procedure of the testing leads to a formation of a group of J observations which comprise, for each one of the J testing points, the values of $\eta_j, p_{1,j}, p_{2,j}, \ldots, p_{M,j}$.

For the determination of uncertainties, it is required to calculate the respective combined standard uncertainties $u(\eta_j)$, $u(p_{1,j})$, $\ldots$, $u(p_{M,j})$ in each observations point. Additionally in practice, these uncertainties are almost never constant and the same for all points, but that each testing point has its own standard deviation which needs to be estimated.

According to the Standard and Mathioulakis et al. (1999), generally standard uncertainties in experimental data are of two types: Type A and Type B; the former are the uncertainties determined by statistical means while the latter are determined by other means.

The objective of a measurement is to determine the value of the measurand, that is, the value of the particular quantity to be measured. A measurement therefore begins with an appropriate specification of the measurand, the method of measurement, and the measurement procedure. It is now widely recognized that, when all of the known or suspected components of error have been evaluated and the appropriate corrections have been applied, there still remains an uncertainty about the correctness of the stated result, that is, a doubt about how well the result of the measurement represents the value of the quantity being measured (BIPM et al., 2008). The uncertainty of measurement is defined as the parameter associated with the result of a measurement characterizing the dispersion of the values that could reasonably be attributed to the measurand (Sabatelli et al., 2002).

Type A uncertainties $u_A(s)$ derive from the statistical analysis of repeated measurements at each point of the steady-state or quasi-dynamic state operation of the collector. When a measurement is repeated under repeatability conditions, one can observe a scatter of the measured values. Variations in repeated observations are assumed to arise because influence quantities that can affect the measurement result are not held completely constant. Therefore, for the steady-state test, Type A uncertainty is the standard deviation of the mean of N measurements taken for each quantity measured during testing given by:

$$u_A(s) = \left(\frac{\sum_{j=1}^{N} (x_j - \bar{x})^2}{N(N-1)} \right)^{0.5} \tag{4.40}$$

As is evident from Eq. (4.40), Type A uncertainty depends upon the specific conditions of the measurement and thus can be reduced by increasing the number of measurements (Sabatelli et al., 2002). In the case of the quasi-dynamic test, where no arithmetic mean of the repetitive measurements is used, uncertainty $u_A(s)$ can be equal to zero.

By nature, Type A uncertainties depend on the specific conditions of the test and they include the fluctuations of the measured quantities during the measurement which lie within the limits implied by the Standard and the fluctuations of the testing conditions such as the air speed and the global diffuse irradiance.

Type B uncertainties derive from a combination of uncertainties over the whole measurement, taking into account all available data such as sensor uncertainty, data logger uncertainty, and uncertainty resulting from the possible differences between the measured values perceived by the measuring device (Mathioulakis et al., 1999). Therefore, Type B uncertainties, being dependent upon the measuring instrument, cannot be reduced by augmenting the number of measurements. Relevant information should be obtained from calibration certificates or other technical specifications of the instruments used. The uncertainty $u(s)$ associated with a measurement s is the result of a combination of the Type B uncertainty $u_B(s)$, which is a characteristic feature of the calibration setup, and of the Type A uncertainty $u_A(s)$, which represents fluctuation during the sampling of data.

In some cases, it may be possible to estimate only bounds (upper and lower limits) for the value of a quantity X, in particular, to state that "the probability that the value of X lies within the interval a_- to a_+ for all practical purposes is equal to one and the probability that X lies outside this interval is essentially zero". If there is no specific knowledge about the possible values of i within the interval, one can only assume that it is equally probable for X to lie anywhere within it. Then x, the expectation or expected value of X, is the midpoint of the interval, $x = (a_- + a_+)/2$, with associated standard uncertainty (BIPM et al., 2008):

$$u_{B,x} = \frac{a}{\sqrt{3}} \tag{4.41}$$

If there is more than one independent source of uncertainty (Type A or Type B) u_k, the final uncertainty is calculated according to the general law of uncertainties combination:

$$u = \left[\sum_i u_i^2\right]^{0.5}$$

(4.42)

For example instruments that have two uncertainties, as, for example, the pyranometer that usually has a non-linearity error, of say ± 5 W/m², and temperature dependence, of say ± 5 W/m² within the actual operating range, by applying the law of propagation of errors, the respective standard uncertainty is given by:

$$u_B(x) = \left[\frac{(a_{g1})^2}{3} + \frac{(a_{g2})^2}{3}\right]^{0.5} = \left[\frac{(5)^2}{3} + \frac{(5)^2}{3}\right]^{0.5} = 4.1 \text{ W/m}^2$$

(4.43)

The term *combined standard uncertainty* means the standard uncertainty in a result when that result is obtained from the values of a number of other quantities (Mathioulakis et al., 1999). To evaluate the combined standard uncertainty of a measurand, it is necessary to consider all the possible error sources. This involves an evaluation not only of the sensor uncertainty but also of the whole data acquisition chain. In most cases a measured parameter Y is determined indirectly from N other directly measured quantities X_1, X_2, ..., X_N through a functional relationship $Y = f(X_1, X_2, ..., X_P)$. The standard uncertainty in the estimate y is given by the *law of error propagation*:

$$u(y) = \left(\sum_{i=1}^{P} \left(\frac{\partial f}{\partial x_i}\right)^2 (u(x_i))^2 + 2\sum_{i=1}^{P-1} \sum_{j=i+1}^{P} \frac{\partial f}{\partial x_i} \frac{\partial f}{\partial x_j} \text{cov}(x_i, x_j)\right)^{0.5}$$

(4.44)

In solar collector efficiency testing, an example of such indirect determination is the determination of instantaneous efficiency, η, which derives from the values of global solar irradiance in the collector level, fluid mass flow rate, temperature difference, collector area, and specific heat capacity. In this case the standard uncertainty $u(\eta_j)$ in each value η_j of instantaneous efficiency is calculated by the combination of standard uncertainties in the values of the primary measured quantities, taking into account their relation to the derived quantity η.

4.7.1 Fitting and uncertainties in efficiency testing results

For the analysis of the test data collected a least square fitting is performed, to determine the values of coefficients c_1, c_2, ..., c_M for which the model of Eq. (4.4) represents the series of J observations with the greatest accuracy. In reality the typical deviation is almost never constant and the same for all observations, but each data point $(\eta_j, p_{1,j}, p_{2,j}, ..., p_{M,j})$ has its own standard deviation σ_j, usually the *weighted least square* (WLS) method is used, which calculates on the base of the measured values and their uncertainties not only the model parameters but also their uncertainty. In the case of WLS, the maximum likelihood estimate of the model parameters is obtained by minimizing the χ^2 function:

$$\chi^2 = \sum_{j=1}^{J} \frac{\left(\eta_j - \left(c_1 p_{1,j} + c_2 p_{2,j} + ... + c_N p_{M,j}\right)\right)^2}{u_j^2}$$

(4.45)

where u_j^2 is the variance of the difference $\eta_j - \left(c_1 p_{1,j} + c_2 p_{2,j} + ... + c_N p_{M,j}\right)$ given by:

$$u_j^2 = \text{var}\left(\eta_j - \left(c_1 p_{1,j} + c_2 p_{2,j} + \ldots + c_N p_{M,j}\right)\right) = \left(u(\eta_j)\right)^2 + c_1^2 \left(u\left(p_{1,j}\right)\right)^2 + \ldots + c_M^2 \left(u\left(p_{M,j}\right)\right)^2$$

(4.46)

As can be understood from above, to find coefficients $c_1, c_2, \ldots, c_M$ and their standard uncertainties by minimizing χ^2 function is complicated, because of the non-linearity present in Eq. (4.45). A possible way to do it easily is to find these uncertainties numerically as follows (Press et al., 1996): Let K be a matrix whose $N \times M$ components $k_{i,j}$ are constructed from M basic functions evaluated at the N experimental values of $\Delta T / G_t \left[= T_i^*\right]$ and $(\Delta T)^2 / G_t \left[\, = G_t\left(T_i^*\right)^2\right]$, weighted by the uncertainty u_i:

$$k_{i,1} = 1 / u_i, \ k_{i,2} = T_i^* / u_i, \ k_{i,3} = G_t\left(T_i^*\right)^2 / u_i$$

$$K = \begin{vmatrix} k_{1,1} & . & . & k_{1,M} \\ . & . & . & . \\ . & . & . & . \\ k_{N,1} & . & . & k_{N,M} \end{vmatrix}$$

(4.47)

Let also **L** be a vector of length N whose components l_i are constructed from values of η_i to be fitted, weighted by the uncertainty u_i:

$$l_j = \frac{\eta_i}{u_u}, \quad \mathbf{L} = \begin{vmatrix} \eta_1 / u_1 \\ . \\ . \\ \eta_J / u_J \end{vmatrix}$$

(4.48)

The normal equation of the least square problem can be written as follows:

$$\left(K^T \cdot K\right) \cdot C = K^T \cdot \mathbf{L}$$

(4.49)

where **C** is a vector whose elements are the fitted coefficients.

For the calculation of variances u_i^2 the knowledge of coefficients $c_1, c_2, \ldots, c_M$ is needed, so a possible solution is to use the values of coefficients calculated by standard least squares fitting as the initial values. These initial values can be used in Eq. (4.46) for the calculation of $u_i^2, I = 1, \ldots, I$ and the formation of matrix K and of vector **L**.

The solution of Eq. (4.49) gives the new values of coefficients $c_1, c_2, \ldots, c_M$, which are not expected to differ noticeably from those calculated by standard least squares fitting and used as initial values for the calculation of u_i^2.

Furthermore, if $Z = INV(K^T \cdot K)$ is a matrix whose diagonal elements $z_{k,k}$ are the squared uncertainties (variances) and the off-diagonal elements $z_{k,l} = z_{l,k}$, for $k \neq l$ are the covariance between fitted coefficients:

$$u(c_m) = \sqrt{z_{m,m}}, \quad m = 1, 2, \ldots, M$$

(4.50)

where

$$\text{Cov}(c_k, c_l) = z_{k,l} = z_{l,k}, \quad \text{for } k = 1, \ldots, M; l = 1, \ldots, M \text{ and } k \neq l \qquad (4.51)$$

The knowledge of covariance between the fitted coefficients is necessary if one wishes to calculate, in a next stage, the uncertainty $u(\eta)$ in the predicted values of η using Eqs. (4.39) and (4.44).

It should be noted that Eq. (4.49) can be solved by a standard numerical method (e.g., Gauss–Jordan elimination). It is also possible to use matrix manipulation functions of spreadsheets.

For a more detailed review of the different aspects of determination of uncertainties in solar collector testing see also Mathioulakis et al. (1999), Müller-Schöll and Frei (2000), and Sabatelli et al. (2002).

4.8 Collector test results and preliminary collector selection

Collector testing is required to evaluate the performance of solar collectors and compare different collectors to select the most appropriate one for a specific application. As can be seen from Sections 4.1–4.5, the tests show how a collector absorbs solar energy and how it loses heat. They also show the effects of angle of incidence of solar radiation and the significant heat capacity effects, which are determined from the collector time constant.

Final selection of a collector should be made only after energy analyses of the complete system, including realistic weather conditions and loads, have been conducted for 1 year. In addition, a preliminary screening of collectors with various performance parameters should be conducted to identify those that best match the load. The best way to accomplish this is to identify the expected range of the parameter $\Delta T/G$ for the load and climate on a plot of efficiency η as a function of the heat loss parameter, as indicated in Fig. 4.11 (Kalogirou, 2004).

Collector efficiency curves may be used for preliminary collector selection. However, efficiency curves illustrate only the instantaneous performance of a collector. They do not include incidence

FIGURE 4.11

Collector efficiencies of various liquid collectors.

angle effects, which vary throughout the year; heat exchanger effects; and probabilities of occurrence of T_i, T_a, solar irradiation, system heat loss, or control strategies. Final selection requires the determination of the long-term energy output of a collector as well as performance cost-effectiveness studies. Estimating the annual performance of a particular collector and system requires the aid of appropriate analysis tools such as f-chart, WATSUN, or TRNSYS. These are presented in Chapter 11.

The collector performance equations can also be used to estimate the daily energy output from the collector. This is illustrated by means of the following example.

EXAMPLE 4.2

Consider an FPC with the following characteristics:

$$\eta = 0.76 - 5.6[(T_i - T_a)/G_t]$$

$$K_\theta = 1 - 0.12[1/\cos(\theta) - 1]$$

Find the energy collected during a day with the characteristics shown in Table 4.3.

Table 4.3 Data collected for Example 4.2.

Solar time	Ambient temperature, T_a (°C)	Solar radiation, G_t (W/m²)
6	25	100
7	26	150
8	28	250
9	30	400
10	32	600
11	34	800
12	35	950
13	34	800
14	32	600
15	30	400
16	28	250
17	26	150
18	25	100

The collector area is 2 m², located at 35 °N latitude, is south-facing and tilted at 45° from horizontal. The estimation is done on June 16 and collector inlet temperature is constant equal to 50 °C.

Solution

As the weather conditions are given for every hour, the estimation is performed on an hourly basis, during which it is considered that the weather conditions remain constant. The most difficult parameter to consider is T_i, the inlet temperature to the collector, which is dependent on the system and its location. In this example, this is considered as constant throughout the day and equal to 50 °C.

The efficiency η is equal to $Q_u/A_c I_t$. Therefore,

$$Q_u = A_c I_t [0.76 K_\theta - 5.6(T_i - T_a)/G_t]$$

Continued

EXAMPLE 4.2—CONT'D

The angle of incidence required for the estimation of the incidence angle modifier, K_θ, is obtained from Eq. (2.20). The declination on June 16 is 23.35°. It should be noted that, for the estimation of $\Delta T/G_t$, the radiation used is in W/m^2, whereas for the estimation of Q_u irradiation in kJ/m^2 is used, obtained by multiplying W/m^2 by 3.6. The results are shown in Table 4.4.

Table 4.4 Results of Example 4.2.

Solar time	T_a (°C)	I_t (kJ/m²)	$\Delta T/G_t$ (°C m²/W)	θ (degrees)	K_θ	Q_u (kJ)
6	25	360	0.250	93.9	0	0
7	26	540	0.160	80.5	0.393	0
8	28	900	0.088	67.5	0.806	215.6
9	30	1440	0.050	55.2	0.910	1185.4
10	32	2160	0.030	44.4	0.952	2399.8
11	34	2880	0.020	36.4	0.971	3605.5
12	35	3420	0.016	33.4	0.976	4460.8
13	34	2880	0.020	36.4	0.971	3605.5
14	32	2160	0.030	44.4	0.952	2399.8
15	30	1440	0.050	55.2	0.910	1185.4
16	28	900	0.088	67.5	0.806	215.6
17	26	540	0.160	80.5	0.393	0
18	25	360	0.250	93.9	0	0

Therefore, the total energy collected over the day = 19,273.4 kJ.
In this example, the use of a spreadsheet program greatly facilitates estimations.

4.9 Quality test methods

As we have seen in Chapter 3, the materials used for the construction of the collector should be able to withstand, in addition to the effects created because of the circulating fluid (corrosion, scale deposits, etc.), the adverse effects of the sun's ultraviolet radiation, and the collector should have an operation life of more than 20 years. Solar collectors are also required to withstand cyclic thermal operation many times a day and extreme operating conditions, such as freezing, overheating, thermal shocks, external impact due to hail or vandalism, and pressure fluctuations. Most of these factors occur simultaneously.

It is therefore required to perform tests on solar collectors to determine their quality. In particular, the ability of a collector to resist extreme operating conditions is examined as specified in International Standard ISO 9806 (2017). This standard applies to all types of solar collectors, including integral collector storage systems, except tracking concentrating collectors. Collectors are required to resist a number of influences, which can be clearly identified and quantified, such as high internal fluid pressures, high temperatures, and rain penetration, as shown in Table 4.5. The tests are required to be applied in the sequence specified in Table 4.5 so that possible degradation in one test will be exposed in

Table 4.5 Sequence of quality tests for solar collectors.

Sequence	Test
1	Air leakage rate test[a]
2	Rupture or collapse test[a]
3	Standard stagnation temperature
4	Exposure test[b]
5	External thermal shock test[b]
6	Internal thermal shock test
7	Rain penetration test
8	Freeze resistance test
9	Internal pressure test for liquid channels
10	Mechanical load test
11	Impact resistance test
12	Final inspection
13	Thermal performance test
14	Pressure drop measurement

[a]*Half-exposure test is required before the test. In case the heat transfer medium is in direct contact with polymeric materials, this test shall be performed at standard stagnation temperature.*
[b]*The external thermal shock test may be combined with the exposure test.*

a later test. The thermal performance tests described in this chapter are part of this procedure, as shown in Table 4.5.

For many quality tests, the collector is required to operate at the stagnation temperature. Provided that the collector was tested at a sufficiently high inlet water temperature, the performance equation can be used to determine stagnation temperature. By using Eq. (4.11) and denoting $F_R(\tau\alpha)_n$ as η_o,

$$T_{stag} = T_a + \frac{-c_1 + \sqrt{c_1^2 + 4\eta_o c_2 G_t}}{2c_2} \tag{4.52}$$

4.9.1 Internal pressure test

The absorber is pressure tested to assess the extent to which it can withstand the pressures it might meet in service. The fluid channels are pressure tested at ambient temperature, and the pressure must remain within ±5% of 1.5 times the maximum collector operating pressure specified by the manufacturer. The pressure must be maintained for at least 15 min.

For absorbers made of organic materials (plastics or elastomers), the test temperature is the maximum temperature the absorber will reach under stagnation conditions. This is because the properties of organic materials are temperature dependent. The test pressure should be 1.5 times the maximum collector operating pressure specified by the manufacturer and should be maintained for at least 1 h.

For air-heating collectors, the test conditions consist of two parts: a positive and a negative air leakage rate test. Therefore, the test must be performed with positive and negative pressure. The standard uncertainty must be better than $\pm 2\%$ for the volumetric airflow measurement and better than ± 10 Pa for the differential pressure measurement. The test must be performed at the maximum operating temperature specified by the manufacturer.

4.9.2 Rupture or collapse test

This test is intended for air-heating collectors only to determine their ability to withstand the pressure levels expected in the air duct systems with which they are incorporated. In this test, the pressure gauge is sealed into the collector inlet and connected to the inlet port through the flowmeter and pressure regulator to the air supply. Then the suitable positive or negative pressure is applied. Finally, the air supply is applied until 1.5 times the maximum (positive or negative) operating pressure is achieved as specified by the manufacturer is reached for 10 min.

4.9.3 Exposure test

The exposure test provides a simple reliability test sequence of operating conditions likely to occur during a longer period of natural aging. In addition, it allows the collector to settle, such that subsequent qualification tests are more likely to give repeatable results. An empty collector is mounted outdoors, and all of its fluid pipes are sealed to prevent cooling by natural circulation of air except one pipe, which is left open to permit free expansion of air in the absorber. One of the alternative sets of reference conditions given in Table 4.6 must be used, depending on the climate in which the collector will operate. For each class of reference conditions, the collector is exposed until at least 30 days (which need not be consecutive) have passed with the minimum irradiation shown in Table 4.6.

Table 4.6 Climate reference conditions for exposure test as well as for external and internal thermal shock tests.

Climate parameter	Class A+: Extremely sunny	Class A: Very sunny	Class B: Sunny
Minimum hemispherical solar irradiance on the collector plane (W/m^2)	1100	1000	900
Irradiation on the collector plane (MJ/m^2) for exposure test during minimum 30 days[a]	700	600	540
Minimum ambient temperature (°C)	40	20	15

[a]For half exposure sequences half values are used for 15 days.

4.9.4 External thermal shock test

Collectors from time to time may be exposed to sudden rainstorms on hot sunny days, causing a severe external thermal shock. This test is intended to assess the capability of a collector to withstand such thermal shocks without a failure. An empty collector is used here, as in previous tests prepared in the same way. An array of water jets is arranged to provide a uniform spray of water over the collector. The collector is maintained in steady-state operating conditions under a high level of solar irradiance for a period of 1 h before the water spray is turned on. It is then cooled by the water spray for 15 min before being inspected. Here again, one of the alternative sets of reference conditions given in Table 4.6 can be used, depending on the climate in which the collector will operate, and the heat transfer fluid must have a temperature of less than 25 °C.

4.9.5 Internal thermal shock test

Collectors from time to time may be exposed to a sudden intake of cold heat transfer fluid on hot sunny days, causing a severe internal thermal shock. This could happen, for example, after a period of shutdown, when the installation is brought back into operation while the collector is at its stagnation temperature. This test is intended to assess the capability of a collector to withstand such thermal shocks without failure. Here again, an empty collector is used, as in previous tests prepared in the same way; the same reference conditions given in Table 4.6 can be used, depending on the climate in which the collector will operate, and the heat transfer fluid must have a temperature of less than 25 °C.

4.9.6 Rain penetration

This test is intended to assess the extent to which collectors are substantially resistant to rain penetration. The collectors must not normally permit the entry of either free-falling rain or driving rain, either through the glazing seals or from ventilation holes or drain holes. For this test, the collector must be shaded and kept warm by circulating a hot fluid at 55 °C through the absorber. Collectors designed to be integrated into a roof structure must be mounted on a simulated roof and have their underside protected. Other collectors must be mounted in a conventional manner on an open frame. The collector must be sprayed on all sides using spray nozzles or showers for a test period of 4 h.

Water penetration into the collector can be determined only by visual inspection within 72 h after spraying. During this time, the collector remains shaded from direct sunlight but is not kept warm until the final inspection. Unnecessary transportation of the collector must be avoided.

4.9.7 Freeze resistance test

This test is intended to assess the extent to which water-heating collectors that are claimed to be freeze resistant can withstand freezing without damage. This test is not intended for use with collectors that are filled with antifreeze fluids. Two test procedures are specified: one for collectors claimed to be freeze resistant when filled with water and one for collectors claimed to resist freezing after being drained.

For collectors claimed to be able to withstand freezing, the collector is mounted in a cold chamber. The collector must be inclined at the shallowest angle to the horizontal recommended by the manufacturer. Unglazed collectors must be tested in a horizontal position, unless this is excluded by the manufacturer. Next, the collector is filled with water at the operating pressure. The cold chamber temperature is cycled, and at the end of each cycle, the collector is refilled with water at operating pressure.

For collectors claimed to resist freezing after being drained (i.e., they employ a drain down system to protect them from freezing), the collector is mounted in a cold chamber as before with the same provisions for the collector inclination. The test temperature is maintained at −20 °C (or as specified by the manufacturer) for at least 1 h per cycle. The collector is then thawed to at least 10 °C and kept warm for at least 1 h. The collector must be subjected to three freeze–thaw cycles.

4.9.8 Impact resistance test

Two test methods are suggested using either ice balls or steel balls. This is intended to assess the extent to which a collector can withstand the effects of heavy impacts, such as those caused by minor vandalism or likely to occur during installation. Heavy impacts may also be caused by hailstones.

The collector is mounted either vertically or horizontally on a stiff support that must have a negligible distortion or deflection at the time of impact. If the collector is mounted horizontally, then the steel balls are dropped vertically; if it is mounted vertically, then the impacts are directed horizontally by means of a pendulum.

The point of impact must be located inside a radius of 75 mm from the visible glass corner of the collector. For each series of shots (four shots of equal impact strength) a different corner must be chosen. Much more details are given in the standard (ISO 9806, 2017).

In addition to the preceding quality tests, the ISO developed a range of material and product quality test standards for solar collectors. The following specific material test methods standards have been developed:

- **ISO 9553:1997**. *Solar energy*—Methods of testing preformed rubber seals and sealing compounds used in collectors.
- **ISO 9808:1990**. *Solar water heaters*—Elastomeric materials for absorbers, connecting pipes, and fittings—Methods of assessment.
- **ISO/TR 10217:1989**. *Solar Energy—Water heating systems*—Guide to material selection with regard to internal corrosion.

4.10 European standards

In the framework of the European Committee for Standardization, CEN (Comité Européen de Normalisation), the operation of a technical committee dealing with solar thermal collectors and systems has been established. Specifically, CEN/TC 312, "Thermal solar systems and components", was created in 1994, following a request of the European Solar Thermal Industry Federation (ESTIF) to the CEN Central Secretariat. The scope of CEN/TC 312 is the preparation of European standards to cover terminology, general requirements, characteristics, and test methods of thermal solar systems and components.

The primary aim of the European standards is to facilitate the exchange of goods and services through the elimination of technical barriers to trade. The use of European standards by industry and social and economic partners is voluntary unless the European standard is related to European legislation (directives). Furthermore, conformity to such standards may be a presumption for solar projects to get a subsidy from national/regional renewable energy systems supporting programs (Kotsaki, 2001).

For the elaboration of European technical standards, corresponding national documents as well as international standards (ISO) have been taken into consideration. It should be noted that, compared with the existing standards, the European norms under consideration are performing a step forward, since they incorporate new features, such as quality and reliability requirements.

In April 2001, CEN started publishing a series of standards related to solar collectors and systems testing. With the publication of these European standards, all national standards related to the same topic and having conflicting provisions were (or have to be) withdrawn by the nations of the European Community. Some of these standards were revised in 2006 and are now under their second 5-year systematic review. A complete list of these standards is as follows:

- **EN 12975-1:2006 + A1:2010.** *Thermal solar systems and components—Solar collectors—Part 1: General requirements.* This European standard specifies requirements on durability (including mechanical strength), reliability, and safety for liquid-heating solar collectors. It also includes provisions for evaluation of conformity to these requirements. It should be noted that the A1: 2010 in the standard reference number denotes a minor amendment which was made in 2010 to change the scope of the standard so as to extend its application for concentrating collectors as well.
- **EN 12975-2:2006.** *Thermal solar systems and components—Solar collectors—Part 2: Test methods.* This European standard establishes test methods for validating the durability, reliability, and safety requirements for liquid-heating collectors as specified in EN 12975-1. This standard also includes three test methods for the thermal performance characterization for liquid-heating collectors. This standard was withdrawn and replaced by ISO 9806 (2017).
- **EN 12976-1:2021.** *Thermal solar systems and components—Factory-made systems—Part 1: General requirements.* This European standard specifies requirements on durability, reliability, and safety for factory-made solar systems. This standard also includes provisions for evaluation of conformity to these requirements.
- **EN 12976-2:2019.** *Thermal solar systems and components—Factory-made systems—Part 2: Test methods.* This European standard specifies test methods for validating the requirements for factory-made solar systems as specified in EN 12976-1. The standard also includes two test methods for the thermal performance characterization by means of whole-system testing.
- **EN 12977-1:2018.** *Thermal solar systems and components—Custom-built systems—Part 1: General requirements for solar water heaters and combisystems.* This European standard specifies requirements on durability, reliability, and safety of small and large custom-built solar heating and cooling systems with liquid heat transfer medium in the collector loop for residential buildings and similar applications. The standard also contains requirements on the design process of large custom-built systems.
- **EN 12977-2:2018.** *Thermal solar systems and components—Custom-built systems—Part 2: Test methods for solar water heaters and combisystems.* This European standard applies to small and large custom-built solar heating systems with liquid heat transfer medium for residential buildings and similar applications and specifies test methods for verification of the requirements specified in EN 12977-1. The standard also includes a method for thermal performance characterization and system performance prediction of small custom-built systems by means of component testing and system simulation.

- **EN 12977-3:2018**. *Thermal solar systems and components—Custom-built systems—Part 3: Performance test methods for solar water heating stores.* This European standard specifies test methods for the performance characterization of stores intended for use in small custom-built systems as specified in EN 12977-1.
- **EN 12977-4:2018**. *Thermal solar systems and components—Custom-built systems—Part 4: Performance test methods for solar combistores.* This European standard specifies test methods for the performance characterization of stores which are intended for use in small custom-built systems as specified in EN 12977-1. Stores tested according to this document are commonly used in solar combisystems.
- **EN 12977-5:2018**. *Thermal solar systems and components—Custom-built systems—Part 5: Performance test methods for control equipment.* This European standard specifies performance test methods for control equipment as well as requirements on accuracy, durability, and reliability of the control equipment.
- **EN ISO 9488:1999**. *Solar energy—Vocabulary (ISO 9488:2020).* This European-International standard defines basic terms relating to solar energy and has been elaborated in common with ISO.

The elaboration of these standards has been achieved through a wide European collaboration of all interested parties, such as manufacturers, researchers, testing institutes, and standardization bodies. Furthermore, these standards will promote a fair competition among producers of solar energy equipment on the market, since low-quality/low-price products will be easier to be identified by customers, based on uniform test reports comparable throughout Europe.

The increased public awareness of the environmental aspects is reinforced by these standards, which help ensure the quality level for the consumer and provide more confidence in the new solar heating technology and products available.

4.10.1 Solar Keymark

The Solar Keymark certification scheme was initiated by the ESTIF to avoid internal European trade barriers due to different requirements in national subsidy schemes and regulations.

Before the European standards and the Solar Keymark were established, solar thermal products had to be tested and certified according to different national standards and requirements. The Solar Keymark idea is that only one test and one certificate are necessary to fulfill all requirements in all EU member states.

The Solar Keymark certification scheme was introduced to harmonize national requirements for solar thermal products in Europe. The objective is that, once tested and certified, the product should have access to all national markets. This goal has now been achieved, except for some minor supplementary requirements in a few member states.

The CEN Solar Keymark certification scheme has been available for solar thermal products in Europe since 2003. The Solar Keymark states conformity with the European standards for solar thermal products. The CEN Keymark is the pan-European voluntary third-party certification mark, demonstrating to users and consumers that a product conforms to the relevant European standard (Nielsen, 2007).

The Solar Keymark is the keymark certification scheme applied specifically for solar thermal collectors and systems, stating conformity with the following European standards:

- EN12975. Thermal solar systems and components—Solar collectors.
- EN12976. Thermal solar systems and components—Factory-made systems.

Solar Keymark is the key to the European market because:

- Products with the Solar Keymark have access to all national subsidy schemes in EU member states.
- In some member states (e.g., Germany), it is now obligatory that solar collectors show the Keymark label.
- People expect the Solar Keymark; most collectors sold now are Keymark certified.

The main elements of the party Keymark certification are:

- Type testing according to European standards (test samples to be sampled by an independent inspector).
- Initial inspection of factory production control (quality management system at ISO 9001 level).
- Surveillance: annual inspection of factory production control.
- Biannual "surveillance test": detailed inspection of products.

4.11 Data acquisition systems

Today, most scientists and engineers use personal computers for data acquisition in laboratory research, test and measurement, and industrial automation. To perform the tests outlined in this chapter as well as whole-system tests, a computer data acquisition system (DAS) is required.

Many applications use plug-in boards to acquire data and transfer them directly to computer memory. Others use DAS hardware remote from the PC that is coupled via a parallel, serial, or USB port. Obtaining proper results from a PC-based DAS depends on each of the following system elements:

- The personal computer.
- Transducers.
- Signal conditioning.
- DAS hardware.
- Software.

The personal computer is integrated into every aspect of data recording, including sophisticated graphics, acquisition, control, and analysis. Modems connected to the Internet or an internal network allow easy access to remote personal computer-based data recording systems from virtually any place. This is very suitable when performing an actual solar system monitoring.

Almost every type of transducer and sensor is available with the necessary interface to make it computer compatible. The transducer itself begins to lose its identity when integrated into a system that incorporates such features as linearization, offset correction, and self-calibration. This has eliminated the concern regarding the details of signal conditioning and amplification of basic transducer outputs.

Many industrial areas commonly employ signal transmitters for control or computer data handling systems to convert the signal output of the primary sensor into a compatible common signal span. The system required for performing the various tests described in this chapter, however, needs to be setup by taking the standard requirements about accuracy of the instruments employed.

The vast selection of available DAS hardware makes the task of configuring a DAS difficult. Memory size, recording speed, and signal processing capability are major considerations in determining the correct recording system. Thermal, mechanical, electromagnetic interference, portability, and meteorological factors also influence the selection.

A digital DAS must contain an interface, which is a system involving one or several analog-to-digital converters and, in the case of multichannel inputs, a multiplexer. In modern systems, the interface also provides excitation for transducers, calibration, and conversion of units. Many DASs are designed to acquire data rapidly and store large records of data for later recording and analysis. Once the input signals have been digitized, the digital data are essentially immune to noise and can be transmitted over great distances.

One of the most frequently used temperature transducers is the thermocouple. These are commonly used to monitor temperature with PC-based DAS. Thermocouples are very rugged and inexpensive and can operate over a wide temperature range. A thermocouple is created whenever two dissimilar metals touch and the contact point produces a small open-circuit voltage as a function of temperature. This thermoelectric voltage is known as the *Seebeck voltage*, named after Thomas Seebeck, who discovered it in 1821. The voltage is non-linear with respect to temperature. However, for small changes in temperature, the voltage is approximately linear:

$$\Delta V \approx S \Delta T \qquad (4.53)$$

where

ΔV = change in voltage;
S = Seebeck coefficient; and
ΔT = change in temperature.

The Seebeck coefficient (S) varies with changes in temperature, causing the output voltages of thermocouples to be non-linear over their operating ranges. Several types of thermocouples are available; these thermocouples are designated by capital letters that indicate their composition. For example, a J-type thermocouple has one iron conductor and one constantan (a copper–nickel alloy) conductor.

Information from transducers is transferred to a computer recorder from the interface as a pulse train. Digital data are transferred in either serial or parallel mode. Serial transmission means that the data are sent as a series of pulses, 1 bit at a time. Although slower than parallel systems, serial interfaces require only two wires, which lowers their cabling cost. The speed of serial transmissions is rated according to the baud rate. In parallel transmission, the entire data word is transmitted at one time. To do this, each bit of a data word has to have its own transmission line; other lines are needed for clocking and control. Parallel mode is used for short distances or when high data transmission rates are required. Serial mode must be used for long-distance communications where wiring costs are prohibitive.

The most popular interface bus standards currently used for data transmission are IEEE488, RS232 serial interface, RS485, and the USB standard.

i. IEEE488 is considered a short-distance digital communications 8-bit parallel multimaster interface bus specification designed by Hewlett–Packard as HP-IB (Hewlett–Packard Interface Bus). Because of the way the IEEE488 bus system feeds data, its bus is limited to a cable length of 20 m and requires an interface connection at every meter for proper termination. The bus was created in the late 1960s to link automated test equipment but was also used during the 1970 and 1980s as a peripheral bus for early microcomputers. Newer standards have largely replaced IEEE488 for computer use, but it is still used in various test equipment (Institute of Electrical and Electronics Engineers, 1992).

ii. A serial communication standard feeds data serially down two wires, 1 bit at a time; RS232 is used in many data acquisition systems and is considered to be the most common serial communication. However, it is rather limited in the sense that it only supports communication to one device connected to the bus at a time and only transmits data as far as 300 m. For longer distances, it may feed a modem to send data over standard telephone lines (Electronic Industries Association. Engineering Department, 1981).

iii. Another serial communications standard found in many data acquisition systems is RS485. Although RS485 is not as widespread as RS232, it is more flexible in the sense that it provides communication to more than one device at a time and communicates data at distances of approximately 1220 m without using any repeaters or active devices (Telebyte Technology Inc., 1985).

iv. The universal serial bus (USB) is a relatively new industry standard for connecting PCs to exterior devices such as modems, monitors, printers, and data-acquisition devices (Hachman, 2019). The facility to provide power to the peripheral device provides various advantages over existing serial and parallel communications, including considerably higher bandwidth. There have been four generations of USB specifications: USB 1.x, USB 2.0, USB 3.x, and USB4, which can communicate up to 40 Gbits over 0.8 m. Extension cables self-powered USB hubs connected together can be used to extend the range of a USB device. Additionally, USB can be used over Ethernet to achieve distances of up to 100 m.

A local area network may also be available for transmitting information. With appropriate interfacing, transducer data are available to any computer connected to the local network.

4.11.1 Portable data loggers

Portable data loggers generally store electrical signals (analog or digital) to internal memory storage. The signal from connected sensors is typically stored to memory at timed intervals, which range from MHz to hourly sampling. Many portable data loggers can perform linearization, scaling, or other signal conditioning and permit logged readings to be either instantaneous or averaged values. Most modern portable data loggers have built-in clocks that record the time and date, together with transducer signal information. Portable data loggers range from single-channel input to 256 or more channels. Some general-purpose devices accept a multitude of analog or digital inputs or both; others are more specialized to a specific measurement (e.g., a portable pyranometer with built-in data logging capability) or for a specific application (e.g., temperature, relative humidity, wind speed, and solar radiation measurement with data logging for solar system testing applications).

4.11.2 IoT in data acquisition systems

In general, the Internet of Things, denoted IoT, is an environment where objects in everyday life (sensors, home appliances, vehicles, cameras, actuators, etc.) are connected to wired and wireless networks to share and exchange information. With progress and improvements in wired and wireless network technologies, mobile devices connected to the Internet, such as smartphones and tablets, are now in extensive use (Mohammed and Ahmed, 2017).

This technology supports the collection of detailed information concerning objects within a wide range of applications, including renewable energy systems, industrial and home automation, home energy management, healthcare, intelligent drive systems, medical aids, traffic maintenance, automotive industries, and many others.

Using the IoT technology to supervise solar thermal panels can greatly enhance solar plant performance, monitoring, and maintenance. With technological progress, renewable energy equipment costs have come down worldwide, thus encouraging large-scale PV and solar thermal installations. The enormous scale of solar installations requires advanced systems for the remote monitoring and control of plants with web-based interfaces, as a considerable number might be installed in non-accessible areas, which may make it impossible to supervise them from nearby areas.

Many standards have been developed by the IEEE Standards Association related to the environmental need for the IoT. The IEEE standardization activities mainly focus on physical and medium access control layers.

Numerous types of connections can be used by IoT sensors. The most popular are:

i. Radio-frequency identification (RFID) — technology using electromagnetic fields to read data stored in tags embedded in other items. It is used in several industrial and commercial applications, from tracking objects along a supply chain to keeping track of items checked out of a library (Hunt et al., 2007).

ii. Wi-Fi is a family of wireless network protocols based on IEEE 802.11 family of standards. They are commonly used for local area networking of devices and internet access (IEEE Computer Society, 2016). IEEE 802.11g is preferred, as it operates at 2.4 GHz and offers a data rate of about 54 Mbps, which is much higher than the ZigBee rate of 250 Kbps. Wi-Fi technology is well suited for controlling and remotely monitoring PV systems, especially in domestic applications (Adhya et al., 2016).

iii. Bluetooth is another wireless technology standard which is used for sending information between fixed and mobile devices over short distances at frequencies from 2.40 to 2.48 GHz for building personal area networks (PANs). It was initially considered as a wireless alternative to RS-232 wired connections (Woolley, 2019).

iv. ZigBee is an IEEE 802.15.4−based specification for a suite of high-level communication protocols used to set up personal area networks (PANs) using small, low-power digital radios, like data collection, home automation, medical devices, designed for small-scale projects that need wireless connections (Koubâa et al., 2005). Although Zigbee is considered to be a mature technology, it is not recommended for use in large-scale installations, as it cannot handle long-distance communications (Ben Belghith and Sbita, 2014).

Moreover, to facilitate connectivity over long distances (wide-area connections), other technologies should be considered, such as 3G, 4G, and 5G.

Exercises

4.1 For seven collectors in series, each $1.2 \, m^2$ in area, $F_{R1}U_{L1} = 7.5 \, W/m^2 \, °C$, and $F_{R1}(\tau\alpha)_1 = 0.79$ at a flow rate of $0.015 \, kg/s \, m^2$, estimate the useful energy collected if water is circulated through the collectors, the available solar radiation is $800 \, W/m^2$, and the ΔT ($= T_i - T_a$) is equal to $5 \, °C$.

4.2 Repeat Example 4.2 for September 15 considering that the weather conditions are the same.

4.3 Find the $F_R(\tau\alpha)_n$ and $F_R U_L$ for a collector $2.6 \, m^2$ in area with the following hour-long test results.

Q_u (MJ)	I_t (MJ/m²)	T_i (°C)	T_a (°C)
6.05	2.95	15.4	14.5
1.35	3.05	82.4	15.5

4.4 For a collector with $F_R(\tau\alpha)_n = 0.82$ and $F_R U_L = 6.05 \, W/m^2 \, °C$, find the instantaneous efficiency when $T_i = T_a$. If the instantaneous efficiency is equal to 0, $T_a = 25 \, °C$, and $T_i = 90 \, °C$, what is the value of solar radiation falling on the collector?

4.5 The data from an actual collector test are shown in the following table. If the collector area is $1.95 \, m^2$ and the test flow rate is $0.03 \, kg/s$, find the collector characteristics $F_R(\tau\alpha)_n$ and $F_R U_L$.

4.6 For a $5.6 \, m^2$ collector with $F' = 0.893$, $U_L = 3.85 \, W/m^2 \, °C$, $(\tau\alpha)_{av} = 0.79$, and flow rate $= 0.015 \, kg/m^2 \, s$, find F_R, Q_u, and efficiency when water enters at $35 \, °C$, the ambient temperature is $14.2 \, °C$, and I_t for the hour is $2.49 \, MJ/m^2$.

Number	G_t (W/m²)	T_a (°C)	T_i (°C)	T_o (°C)
1	851.2	24.2	89.1	93.0
2	850.5	24.2	89.8	93.5
3	849.1	24.1	89.5	93.3
4	855.9	23.9	78.2	83.1
5	830.6	24.8	77.9	82.9
6	849.5	24.5	77.5	82.5
7	853.3	23.9	43.8	52.1
8	860.0	24.3	44.2	52.4
9	858.6	24.5	44.0	51.9

4.7 The characteristics of a $2 \, m^2$ water-heating collector are $F_R(\tau\alpha)_n = 0.79$ and $F_R U_L = 5.05 \, W/m^2 \, °C$. If the test flow rate is $0.015 \, kg/m^2 \, s$, find the corrected collector characteristics when the flow rate through the collector is halved.

4.8 The characteristics of a water-heating collector are $F_R(\tau\alpha)_n = 0.77$, $F_R U_L = 6.05$ W/m^2 °C, and $b_o = 0.12$. The collector operates for a complete day, which has the characteristics shown in the following table. Find, for each hour, the useful energy collected per unit of aperture area and the collector efficiency. Also estimate the daily efficiency.

Time	I_t (kJ/m^2)	T_a (°C)	T_i (°C)	θ (°)
8–9	2090	18.5	35.1	60
9–10	2250	20.3	33.2	47
10–11	2520	22.6	30.5	35
11–12	3010	24.5	29.9	27
12–13	3120	26.5	33.4	25
13–14	2980	23.9	35.2	27
14–15	2490	22.1	40.1	35
15–16	2230	19.9	45.2	47
16–17	2050	18.1	47.1	60

4.9 For one cover collector system with $KL = 0.037$ and $\alpha_n = 0.92$, estimate incidence angle modifier constant (b_o) based on ($\tau\alpha$) at normal incidence and at $\theta = 60°$. The cover is made from glass with $n = 1.526$.

References

Adhya, S., Saha, D., Das, A., Jana, J., Saha, H., 2016. An IoT based smart solar photovoltaic remote monitoring and control unit. In: 2nd International Conference on Control, Instrumentation, Energy and Communication, CIEC 2016, pp. 432–436. https://doi.org/10.1109/CIEC.2016.7513793.

ANSI/ASHRAE Standard 93, 2010. Methods of Testing to Determine the Thermal Performance of Solar Collectors.

Ben Belghith, O., Sbita, L., 2014. Remote GSM module monitoring and Photovoltaic system control. In: 2014 1st International Conference on Green Energy. ICGE, pp. 188–192, 2014.

BIPM, IEC, IFCC, ILAC, ISO, IUPAC, IUPAP, OIML, 2008. Evaluation of Measurement Data — Guide to the Expression of Uncertainty in Measurement. http://www.bipm.org/utils/common/documents/jcgm/JCGM_100_2008_E.pdf.

Duffie, J.A., Beckman, W.A., 1991. Solar Engineering of Thermal Processes. John Wiley & Sons, New York.

Electronic Industries Association. Engineering Department, 1981. Interface between Data Terminal Equipment and Data Communication Equipment Employing Serial: Binary Data Interchange. Engineering Department, Electronic Industries Association.

FprEN 12975, 2021. Solar Collectors — General Requirements (EN, 2021).

Hachman, M., 2019. The New USB4 Spec Promises a Lot: Thunderbolt 3 Support, 40Gbps Bandwidth, and Less Confusion, PCWorld. Available at: https://www.pcworld.com/article/3347403/the-new-usb4-spec-promises-a-lot-thunderbolt-3-support-40gbps-bandwidth-and-less-confusion.html.

Hastings, S.R., Morck, O., 2000. Solar Air System: A Design Handbook. James and James Science, UK.

Hunt, V.D., Puglia, A., Puglia, M., 2007. RFID: A Guide to Radio Frequency Identification. John Wiley & Sons.

IEEE Computer Society, 2016. IEEE Standard for Information Technology—Telecommunications and Information Exchange between Systems Local and Metropolitan Area Networks—Specific Requirements - Part 11: Wireless LAN Medium Access Control (MAC) and Physical Layer (PHY) Specifications, IEEE Std 802.11-2016 (Revision of IEEE Std 802.11-2012), pp. 1—3534.

Institute of Electrical and Electronics Engineers, 1992. IEEE Standard Codes, Formats, Protocols, and Common Commands for Use with IEEE Std 488.1-1987. IEEE Standard Digital Interface for Programmable Instrumentation.

ISO 9806, 2017. Solar Energy — Solar Thermal Collectors, Test Methods (ISO, 2017).

Kalogirou, S.A., 2004. Solar thermal collectors and applications. Prog. Energy Combust. Sci. 30 (3), 231—295.

Kalogirou, S.A., Eleftheriou, P., Lloyd, S., Ward, J., 1994. Design and performance characteristics of a parabolic-trough solar-collector system. Appl. Energy 47 (4), 341—354.

Kotsaki, E., 2001. European solar standards. Refocus 2 (5), 40—41.

Koubâa, A., Alves, M., Tovar, E., 2005. IEEE 802.15.4 for Wireless Sensor Networks: A Technical Overview, Architecture (July). IPP-Hurray Group, pp. 1—21. Available at: http://www.hurray.isep.ipp.pt. (Accessed 26 January 2021).

Mathioulakis, E., Voropoulos, K., Belessiotis, V., 1999. Assessment of uncertainty in solar collector modeling and testing. Sol. Energy 66 (5), 337—347.

Mohammed, Z.K.A., Ahmed, E.S.A., 2017. Internet of Things Applications, Challenges and Related Future Technologies. World Scientific News. Available at: http://yadda.icm.edu.pl/yadda/element/bwmeta1.element. psjd-b638cb4d-d68f-4f4c-afa5-ad309a7c4838%0Ahttps://www.infona.pl/resource/bwmeta1.element.psjd-8c8e8b68-9180-4879-85d8-a7870d5644e9. (Accessed 25 January 2021).

Morrison, G.L., 2001. Solar collectors. In: Gordon, J. (Ed.), Solar Energy: The State of the Art. James and James, London, pp. 145—221.

Müller-Schöll, C., Frei, U., 2000. Uncertainty analyses in solar collector measurement. In: Proceedings of Eurosun 2000 on CD ROM, Copenhagen.

Nielsen, J.E., 2007. The key to the European market: the solar keymark. In: Proceedings of ISES Solar World Conference.

Press, W., Teukolsky, S.A., Vetterling, W.T., Flannery, B.P., 1996. Numerical Recipes, second ed. Cambridge University Press, Oxford.

Sabatelli, V., Marano, D., Braccio, G., Sharma, V.K., 2002. Efficiency test of solar collectors: uncertainty in the estimation of regression parameters and sensitivity analysis. Energy Convers. Manage. 43 (17), 2287—2295.

Shams, S.M.N., Mc Keever, M., Mc Cormack, S., Norton, B., 2016. Design and experiment of a new solar air heating collector. Energy 100, 374—383.

Telebyte Technology Inc., 1985. EIA Standard RS 485 Electrical Characteristics of Generators and Receivers for Use in Balanced Digital Multipoint Systems. Greenlawn, New York.

Woolley, M., 2019. Bluetooth Core Specification V5.0. Available at: https://www.esperienzandroid.it/wp-content/uploads/2020/06/1901_Feature_Overview_Brief_FINAL.pdf. (Accessed 26 January 2021).

Solar Water-Heating Systems

Perhaps the most popular application of solar systems is for domestic water heating. The popularity of these systems is based on the fact that relatively simple systems are involved and solar water-heating systems are generally viable. This category of solar systems belongs to the low-temperature heat applications.

The world's commercial low-temperature heat consumption is estimated to be about 10 EJ per year for hot water production, equivalent to 6 trillion m^2 of collector area (Turkenburg, 2000). In 2005, about 140 million m^2 of solar thermal collector area were in operation around the world, which is only 2.3% of the potential (Philibert, 2005). As indicated in Chapter 1, Section 1.3.6, the cumulative solar thermal capacity in operation at the end of 2021 was 522 GW_{th}, corresponding to 746 million square meters of collector area, which is 12.4% of the potential. This represented a net increase of 21 GW_{th}, or 31 million square meters of collector area, compared with the 2020 value. The annual solar thermal energy yield was 425 TWh, which equates to savings of 45.7 million tons of oil and 147.5 million tons of CO_2 (Weiss and Spörk-Dür, 2022).

A solar water heater is a combination of a solar collector array, an energy transfer system, and a storage tank. The main part of a solar water heater is the solar collector array, which absorbs the solar radiation and converts it to heat. This heat is then absorbed by a heat transfer fluid (water, non-freezing liquid, or air) that passes through the collector. This heat can then be stored or used directly. Because as it is understood, portions of the solar energy system are exposed to weather conditions, they must be protected from freezing and overheating caused by high insolation levels during periods of low energy demand.

Two types of solar water-heating systems are available:

- Direct or open loop systems, in which potable water is heated directly in the collector.
- Indirect or closed loop systems, in which potable water is heated indirectly by a heat transfer fluid that is heated in the collector and passes through a heat exchanger to transfer its heat to the domestic or service water.

Systems differ also with respect to the way the heat transfer fluid is transported:

- Natural (or passive) systems.
- Forced circulation (or active) systems.

Natural circulation occurs by natural convection (thermosiphoning), whereas forced circulation systems use pumps or fans to circulate the heat transfer fluid through the collector. Except for thermosiphon and integrated collector storage (ICS) systems, which need no control, solar domestic and service hot water systems are controlled using differential thermostats. Some systems also use a load-side heat exchanger between the potable water stream and the hot water tank.

Table 5.1 Solar water-heating systems.	
Passive systems	**Active systems**
Thermosiphon (direct and indirect) Integrated collector storage	Direct circulation (or open loop active) systems Indirect circulation (or closed loop active) systems, internal and external heat exchanger Air systems Heat pump systems Pool heating systems

Seven types of solar energy systems can be used to heat domestic and service hot water, as shown in Table 5.1. Thermosiphon and ICS systems are called *passive systems* because no pump is employed, whereas the others are called *active systems* because a pump or fan is employed to circulate the fluid. For freeze protection, recirculation and drain-down are used for direct solar water-heating systems and drain-back is used for indirect water-heating systems.

A wide range of collectors have been used for solar water-heating systems, such as flat plate, evacuated tube, and compound parabolic. In addition to these types of collectors, bigger systems can use more advanced types, such as the parabolic trough.

The amount of hot water produced by a solar water heater depends on the type and size of the system, the amount of sunshine available at the site, and the seasonal hot water demand pattern.

5.1 Passive systems

Two types of systems belong to this category: thermosiphon and the integrated collector storage systems. These are examined in the following sections.

5.1.1 Thermosiphon systems

Thermosiphon systems, shown schematically in Fig. 5.1, heat potable water or transfer fluid and use natural convection to transport it from the collector to storage. The thermosiphoning effect occurs because the density of water drops with the increase of the temperature. Therefore, by the action of solar radiation absorbed, the water in the collector is heated and thus expands, becoming less dense, and rises through the collector into the top of the storage tank. There it is replaced by the cooler water that has sunk to the bottom of the tank, from which it flows down the collector. Circulation continues as long as the sun is shining. Since the driving force is only a small density difference, larger than normal pipe sizes must be used to minimize pipe friction. Connecting lines must also be well insulated to prevent heat loss and sloped to prevent formation of air pockets, which would stop circulation.

The advantages of thermosiphon systems are that they do not rely on pumps and controllers, are more reliable, and have a longer life than forced circulation systems. Moreover, they do not require an electrical supply to operate and they naturally modulate the circulation flow rate in phase with the radiation levels.

FIGURE 5.1

Schematic diagram of a thermosiphon solar water heater.

The main disadvantage of thermosiphon systems is that because the storage tank should be above the collector they are comparatively tall units, which makes them not very attractive aesthetically. The two types of thermosiphon systems are pressurized and unpressurized. In pressurized thermosiphon units, the make-up water is from city mains or pressure units and the collectors and storage tanks must be able to withstand the working pressure. When the city water is used directly, pressure-reducing and relief valves must be installed to protect the system because the pressure can be greater than the working pressure of the collectors and storage tank. In gravity systems—usually installed where the city water supply is intermittent—a cold water storage tank is installed on top of the solar collector, supplying both the hot water cylinder and the cold water needs of the house. This makes the collector unit taller and less attractive.

Another disadvantage of the system is related to the quality of the water used. As the system is open, extremely hard or acidic water can cause scale deposits that clog or corrode the absorber fluid passages.

Typical collector configurations include flat plate, shown in Fig. 5.2a, and evacuated tube collectors, shown in Fig. 5.2b.

Thermosiphon systems can be built with freeze protection devices, ranging from dump valves or heaters in the bottom of the collector header for mild freeze areas to inherent freeze resistance by using a natural circulation antifreeze closed loop between the collector and the tank (Morrison, 2001).

Theoretical performance of thermosiphon solar water heaters

The performance of thermosiphon solar water heaters has been studied extensively, both experimentally and analytically, by numerous researchers. Among the first studies were those of Close (1962) and Gupta and Garg (1968), who developed one of the first models for the thermal performance of a natural circulation solar water heater with no load. They represented solar radiation and ambient temperature by Fourier series and were able to predict a day's performance in a manner that agreed substantially with experiments.

FIGURE 5.2

Thermosiphon system configurations. (a) Flat-plate collector configuration. (b) Evacuated tube collector configuration.

Ong (1974, 1976) performed two studies to evaluate the thermal performance of a solar water heater. He instrumented a relatively small system with five thermocouples on the bottom surface of the water tubes and six thermocouples on the bottom surface of the collector plate. A total of six thermocouples were inserted into the storage tank and a dye tracer mass flowmeter was employed. Ong's studies appear to be the first detailed ones on a thermosiphonic system.

Morrison and Braun (1985) studied the modeling and operation characteristics of thermosiphon solar water heaters with vertical or horizontal storage tanks. They found that the system performance is maximized when the daily collector volume flow is approximately equal to the daily load flow, and the system with a horizontal tank did not perform as well as that with a vertical one. This model has also been adopted by the TRNSYS simulation program (see Chapter 11, Section 11.5.1). According to this model, a thermosiphon system consisting of a flat-plate collector and a stratified tank is assumed to operate at a steady state. The system is divided into N segments normal to the flow direction, and the Bernoulli's equation for incompressible flow is applied to each segment. For steady-state conditions the sum of pressure drop at any segment is:

$$\Delta P_i = \rho_i g h_{fi} + \rho_i g H_i \tag{5.1a}$$

And the sum of the pressure changes around the loop is 0; that is,

$$\sum_{i=1}^{N} \rho_i h_{fi} = \sum_{i=1}^{N} \rho_i H_i \qquad (5.1b)$$

where

ρ_i = density of any node calculated as a function of local temperature (kg/m^3);
h_{fi} = friction head drop through an element (m); and
H_i = vertical height of the element (m).

For each time interval the thermosiphon flow rate must uniquely satisfy Eq. (5.1b).

The collector thermal performance can be modeled by dividing it into N_c equally sized nodes. The temperature at the midpoint of any collector mode k is given by:

$$T_k = T_a + \frac{I_t F_R(\tau\alpha)}{F_R U_L} + \left(T_i - T_a - \frac{I_t F_R(\tau\alpha)}{F_R U_L}\right) \exp\left(-\frac{F' U_L A_c}{\dot{m}_t c_p} \times \frac{(k - 1/2)}{N_c}\right) \qquad (5.2)$$

where

$\dot{m}_t$ = thermosiphonic flow rate (kg/s);
A_c = collector area (m^2).

The collector parameter, $F' U_L$, is calculated from the collector test data for $F_R U_L$ at test flow rate $\dot{m}_T$ by Eq. (4.18), which using the new symbol for test flow rate is:

$$F' U_L = \frac{-\dot{m}_T c_p}{A_c} \ln\left(1 - \frac{F_R U_L A_c}{\dot{m}_T c_p}\right) \qquad (5.3)$$

Finally, the useful energy from the collector is obtained from:

$$Q_u = r A_c [F_R(\tau\alpha) I_t - F_R U_L (T_i - T_a)] \qquad (5.4a)$$

Where the ratio r from Eq. (4.17a) becomes:

$$r = \frac{F_R(\dot{m}_t)}{F_R(\dot{m}_T)} = \frac{\dot{m}_t \left[1 - \exp\left(-\frac{F' U_L A_c}{\dot{m}_t c_p}\right)\right]}{\dot{m}_T \left[1 - \exp\left(-\frac{F' U_L A_c}{\dot{m}_T c_p}\right)\right]} \qquad (5.4b)$$

The temperature drop along the collector inlet and outlet pipes is usually very small (short distance, insulated pipes), and the pipes are considered to be single nodes, with negligible thermal capacitance. The first-law analysis gives the following expressions for the outlet temperature (T_{po}) of pipes:

$$T_{po} = T_a + (T_{pi} - T_a) \exp\left(-\frac{(UA)_p}{\dot{m}_t c_p}\right) \qquad (5.5)$$

The friction head loss in pipes is given by:

$$H_f = \frac{fLv^2}{2dg} + \frac{kv^2}{2g} \qquad (5.6)$$

where

d = pipe diameter (m);
v = fluid velocity (m/s);
L = length of pipe (m);
k = fitting loss coefficient; and
f = friction factor.

The friction factor, f, is equal to:

$$f = 64/\text{Re} \quad \text{for} \quad \text{Re} < 2000 \qquad (5.7a)$$

$$f = 0.032 \quad \text{for} \quad \text{Re} > 2000 \qquad (5.7b)$$

The fitting loss coefficient for various parts of the circuit can be estimated by using the data given in Table 5.2.

The friction factor for the developing flow in the connecting pipes and collector risers is given by:

$$f = 1 + \frac{0.038}{\left(\dfrac{L}{d\text{Re}}\right)^{0.964}} \qquad (5.7c)$$

Table 5.2 Fitting loss coefficient for various parts of the thermosiphon circuit.

Parameter	k Value
Entry from tank to connecting pipe to collector	0.5
Losses due to bends in connecting pipes	
Right-angle bend	Equivalent length of pipe increased by $30d$ for Re $\leq$ 2000 or $k = 1.0$ for Re $\geq$ 2000
45 degrees bend	Equivalent length of pipe increased by $20d$ for Re $\leq$ 2000 or $k = 0.6$ for Re > 2000
Cross-section change at junction of connecting pipes and header	
Sudden expansion	$k = 0.667(d_1/d_2)^4 - 2.667\,(d_1/d_2)^2 + 2.0$
Sudden contraction	$k = -0.3259(d_2/d_1)^4 - 0.1784\,(d_2/d_1)^2 + 0.5$
Entry of flow into tank	1.0

Note: For pipe diameters, d_1 = inlet diameter and d_2 = outlet diameter.

The collector header pressure drop, P_h, is equal to the average of pressure change along inlet and outlet headers for equal mass flow in each riser, N, given by:

$$S_1 = \sum_{i=1}^{N} \frac{N - i + 1}{N^2} \tag{5.8a}$$

$$S_2 = \sum_{i=1}^{N} \frac{(N - i + 1)^2}{N^2} \tag{5.8b}$$

$$A_1 = \frac{f L_h v_h^2}{2 d_h} \tag{5.9}$$

where, from Eq. (5.7a), $f = 64/\text{Re}$ (Re based on inlet header velocity and temperature) and

$$A_2 = A_1 \text{ if } f = 64/\text{Re} \tag{5.10}$$

Based on the outlet header velocity and temperature,

$$A_3 = \frac{\rho v_h^2}{2} \tag{5.11}$$

Finally,

$$P_h = \frac{-S_1 A_1 + 2(S_2 A_3) + S_1 A_2}{2} \tag{5.12}$$

To model the complete system, the interaction of the storage tank is required. This is modeled with the fully stratified storage tank model, which is presented in Section 5.3.3.

The procedure to model the complete system is as follows. Initially, the temperature distribution around the thermosiphon loop for the flow rate of the previous time step is evaluated. The inlet temperature to the collector is computed from the bulk mean temperature of the segments in the bottom of the tank with a volume equal to the collector volume flow (see Section 5.3.3). After allowance for heat loss from the inlet pipe, with Eq. (5.6), is made, the temperature of each of the N_c fixed nodes used to represent the collector temperature profile is evaluated from Eq. (5.2). Finally, the temperature of the new fluid segment returned to the tank is computed from the collector outlet temperature and the temperature drop across the return pipe to the tank. A new tank temperature profile is then evaluated (see Section 5.3.3).

The thermosiphon pressure head due to density differences around the loop is determined from the system temperature profile. The difference between the friction pressure drop around the circuit and the net thermosiphon pressure is evaluated for this flow rate. These values and those from the previous calculation, for the flow rate and net difference between the friction and static pressures, are then used to estimate the new flow; this process is repeated until Eq. (5.1b) is satisfied. This procedure is not suitable for hand calculations, but it is relatively easy to do with a computer.

A simple way to estimate the flow rate that will be created in a thermosiphon solar water heater is to assume that there is a constant temperature increase of water flowing through the collector and estimate the flow rate that will create this temperature difference for the estimated collector gain. Considering the basic collector performance Eq. (3.60) and using the concept of absorbed solar radiation:

$$Q_u = A_c F_R [S - U_L (T_i - T_a)] = \dot{m}_t c_p (T_o - T_i) \tag{5.13}$$

Solving for the flow rate we get:

$$\dot{m}_t = \frac{A_c F_R [S - U_L(T_i - T_a)]}{c_p(T_o - T_i)} \qquad (5.14)$$

By assuming that the collector efficiency factor F' is independent of flow rate, substituting Eq. (3.58) for F_R and rearranging gives:

$$\dot{m}_t = \frac{-U_L F' A_c}{c_p \ln\left[1 - \dfrac{U_L(T_o - T_i)}{S - U_L(T_i - T_a)}\right]} \qquad (5.15)$$

The obtained value of flow rate can then be used to estimate F' and if there is a difference a second iteration is made.

Close (1962) compared computed and experimental temperature differences in Australian-type collector systems and found that there exists a temperature difference of 10 °C when these are well designed and without serious flow restrictions.

More details on the thermosiphon head are given in Chapter 11, Section 11.1.4.

Reverse circulation in thermosiphon systems

At night or whenever the collector is cooler than the water in the tank, the direction of the thermosiphon flow reverses, thus cooling the stored water. It should be noted that thermosiphon collector loop circulation is driven by thermal stratification in the collector loop and the section of the tank below the collector flow return level. The major problem in thermosiphon system design is to minimize a heat loss due to reverse thermosiphon circulation at night, when the sky temperature is low. Norton and Probert (1983) recommend that, to avoid reverse flow, the tank-to-collector separation distance should be between 200 and 2000 mm. A practical way to prevent a reverse flow is to place the top of the collector about 300 mm below the bottom of the storage tank.

Nighttime heat loss from a collector is a function of an ambient air temperature and sky temperature. If the sky temperature is significantly below the ambient temperature, cooling of the collector will cause fluid to thermosiphon in the reverse direction through the collector, and the fluid may be cooled below the ambient temperature. When the reverse flow enters the return pipe to the bottom of the tank, it is mixed with the warmer water contained in the storage tank. The combination of cooling below the ambient temperature in the collector and heating in the return pipe causes reverse flow in all thermosiphon configurations, irrespective of the vertical separation between the top of the collector and the bottom of the tank (Morrison, 2001).

Vertical versus horizontal tank configurations

Because the operation of the thermosiphon system depends on the stratification of the water in the storage tank, vertical tanks are more effective. It is also preferable to have the auxiliary heater as high as possible in the storage tank, as shown in Fig. 5.1, to heat only the top of the tank with auxiliary energy when this is needed. This is important for three reasons:

1. It improves stratification.
2. Tank heat losses are increased linearly with the storage temperature.
3. As shown in Chapter 4, the collector operates at higher efficiency at a lower collector inlet temperature.

To reduce the overall height of the unit, however, horizontal tanks are frequently used. The performance of horizontal tank thermosiphon systems is influenced by the conduction between the high-temperature auxiliary zone at the top of the tank and the solar zone and by mixing of the flow injection points (Morrison and Braun, 1985). The performance of these systems can be improved by using separate solar and auxiliary tanks or by separating the auxiliary and preheat zones with an insulated baffle, as shown in Fig. 5.3. A disadvantage of the two tank systems or segmented tanks is that the solar input cannot heat the auxiliary zone until there is a demand.

Thermal stratification in shallow horizontal tanks also depends on the degree of mixing at the load, make-up water, and collector inlets to the tank. The load should be drawn from the highest possible point, whereas the make-up water flow should enter the tank through a distribution pipe or a diffuser so that it is introduced into the bottom of the tank without disturbing the temperature stratification or mixing the top auxiliary zone with the solar zone. The collector return flow to the tank also should enter through a flow distributor so that it can move to its thermal equilibrium position without mixing with intermediary fluid layers. Because the collector return is usually hot, many manufacturers make a small bend at the inlet pipe, facing upward.

Generally, the penalty associated with horizontal tanks is that the shallow tank depth degrades stratification because of conduction through the walls of the tank and water. Additionally, for in-tank auxiliary systems conduction between the auxiliary and solar zones influences the solar performance. For horizontal tanks with diameters greater than 500 mm, there is only a relatively small performance loss relative to a vertical tank, and the above effects increase significantly for smaller tank diameters (Morrison, 2001).

Freeze protection

For locations that have a mild climate, the open loop thermosiphon solar water heater is the most widely used system. With freeze protection, thermosiphon systems can also be used in locations that experience minor freeze conditions. This can be provided by water dump valves, electric heating in the collector header, or tapered riser tubes to control ice growth in the riser so that a rigid and expanding ice plug is avoided (Xinian et al., 1994). All these techniques have been used successfully by solar water heater manufacturers, and their suitability is proven in areas with mild freeze conditions. They are not suitable, though, in areas with hard freezing. In such cases, the only suitable design is the use of antifreeze collector loops with a heat exchanger between the collector and the tank and an antifreeze heat transfer fluid circulating in the collector and the heat exchanger. For horizontal tank configuration, the most widely adopted system is the mantle or annular heat exchanger concept, shown in Fig. 5.4.

Mantle heat exchanger tanks are easy to construct and provide a large heat transfer area. Mantle heat exchangers are also used in vertical tanks and forced circulation systems, as can be seen in Section 5.2.2. Manufacturers of horizontal tanks usually use as large a mantle as possible, covering almost the full

FIGURE 5.3

Configuration of a segmented tank with an insulating baffle.

FIGURE 5.4

Mantle heat exchanger concept.

circumference and full length of the storage tank. The usual heat transfer fluid employed in these systems is a water–ethylene glycol solution.

Tracking thermosiphons

The possibility of having either a movable thermosiphon solar water heater or a heater where only the inclination could be moved seasonally was investigated by the author and collaborators (Michaelides et al., 1999). The increased performance of the system was compared to the added cost to achieve the movement of the heaters, and it was found that even the simplest seasonal change of the collector inclination is not cost-effective compared to the traditional fixed system.

5.1.2 Integrated collector storage systems

Integrated collector storage (ICS) systems use the hot water storage as part of the collector, i.e., the surface of the storage tank is used as the collector absorber. As in other systems, to improve stratification, the hot water is drawn from the top of the tank and cold make-up water enters the bottom of the tank on the opposite side. Usually, the storage tank surface is selectively coated to avoid heat loss.

The main disadvantage of the ICS systems is the high thermal losses from the storage tank to the surroundings, since most of the surface area of the storage tank cannot be thermally insulated, because it is intentionally exposed to be able to absorb solar radiation. In particular, the thermal losses are greatest during the night and overcast days with low ambient temperatures. Due to these losses, the water temperature drops substantially during nighttime, especially during the winter. Various techniques have been used to keep this from happening. Tripanagnostopoulos et al. (2002) present a number of experimental units in which a reduction in thermal losses was achieved by considering single and double cylindrical horizontal tanks properly placed in truncated symmetric and asymmetric CPC reflector troughs. Alternatively, if a 24 h hot water supply is required, these systems can be used only for preheating and, in such a case, must be connected in series with a conventional water heater.

Details of an ICS unit developed by the author are presented here (Kalogirou, 1997). The system employs a non-imaging CPC cusp-type collector. A fully developed cusp concentrator for a cylindrical receiver is shown in Fig. 5.5. The particular curve illustrated has an acceptance half angle, θ_c, of 60 degrees or a full acceptance angle, $2\theta_c$, of 120 degrees. Each side of the cusp has two mathematically

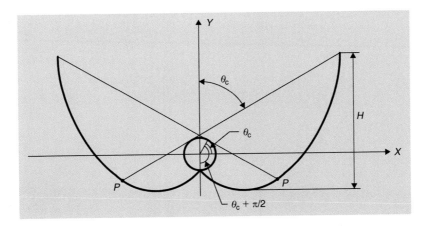

FIGURE 5.5

Fully developed cusp.

distinct segments, smoothly joined at a point P related to θ_c. The first segment, from the bottom of the receiver to point P, is the involute of the receiver's circular cross-section. The second segment is from point P to the top of the curve, where the curve becomes parallel to the y-axis (McIntire, 1979).

With reference to Fig. 5.6, for a cylindrical receiver, the radius, R, and the acceptance half angle, θ_c, the distance, ρ, along a tangent from the receiver to the curve, are related to the angle θ between the radius to the bottom of the receiver and the radius to the point of tangency, T, by the following expressions for the two sections of the curve (McIntire, 1979):

$$\rho(\theta) = R\theta, |\theta| \le \theta_c + \pi/2 \text{ (this is the involute part of the curve)}$$

$$\rho(\theta) = R\left\{ \frac{[\theta + \theta_c + \pi/2 - \cos(\theta - \theta_c)]}{1 + \sin(\theta - \theta_c)} \right\}, \quad \theta_c + \pi/2 \le \theta \le 3\pi/2 - \theta_c$$

(5.16)

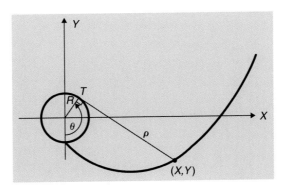

FIGURE 5.6

Mirror coordinates for ideal non-imaging cusp concentrator.

The two expressions for $\rho(\theta)$ are equivalent for the point P in Fig. 5.5, where $\theta = \theta_c + \pi/2$. The curve is generated by incrementing θ in radians, calculating ρ, then calculating the coordinates, X and Y, by:

$$X = R\sin(\theta) - \rho\cos(\theta)$$

$$Y = -R\cos(\theta) - \rho\sin(\theta) \qquad (5.17)$$

Figure 5.5 shows a full, untruncated curve, which is the mathematical solution for a reflector shape with the maximum possible concentration ratio. The reflector shape shown in Fig. 5.5 is not the most practical design for a cost-effective concentrator because reflective material is not effectively used in the upper portion of the concentrator. As in the case of the compound parabolic collector, a theoretical cusp curve should be truncated to a lower height and slightly smaller concentration ratio. Graphically, this is done by drawing a horizontal line across the cusp at a selected height and discarding the part of the curve above the line. Mathematically, the curve is defined to a maximum angle θ value less than $3\pi/2 - \theta_c$. The shape of the curve below the cutoff line is not changed by truncation, so the acceptance angle used for the construction of the curve, using Eq. (5.16), of a truncated cusp is equal to the acceptance angle of the fully developed cusp from which it was truncated.

A large acceptance angle of 75 degrees is used in this design so the collector can collect as much diffuse radiation as possible (Kalogirou, 1997). The fully developed cusp, together with the truncated one, is shown in Fig. 5.7. The receiver radius considered in the construction of the cusp is 0.24 m. The actual cylinder used, though, is only 0.20 m. This is done in order to create a gap at the underside of the receiver and the edge of the cusp in order to minimize the optical and conduction losses.

The final design is shown in Fig. 5.8. The collector aperture is 1.77 m², which, in combination with the absorber diameter used, gives a concentration ratio of 1.47 (Kalogirou, 1997). It should be noted that, as shown in Fig. 5.8, the system is inclined at the local latitude in order to work effectively.

Another possibility considered to reduce the night thermal losses for the design shown in Fig. 5.8, in view of the findings of Eames and Norton (1995) on the use of baffles to reduce convection currents flowing across the glazing that increase the thermal losses, is the insertion of a second cylinder of smaller diameter in the space between the main cylinder and the glass cover and using a small piece of insulation at the point of contact between the two cylinders and between the secondary cylinder and the

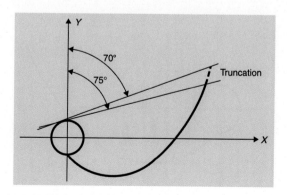

FIGURE 5.7

Truncation of non-imaging concentrator.

FIGURE 5.8

The complete solar ICS hot water system.

glass cover as shown in Fig. 5.8 with a dotted line. This modification offers a number of advantages: storage capacity increased by 30%; top cylinder provides some sort of insulation (for radiation heat loss) as the main cylinder does not see the sky directly; the top cylinder creates a restriction to the flow of the convection currents (just like the baffle does); and finally, the secondary cylinder is used as a preheating for the main one and thus the draw-off characteristics of the whole unit improved considerably, as the cold make-up water does not enter into the main cylinder directly. The extra cylinder increased the cost of the ICS system by 8%, whereas the performance of the system increased by about 7% (Kalogirou, 1998).

5.2 Active systems

In active systems, water or a heat transfer fluid is pumped through the collectors. These are usually more expensive and a little less efficient than passive systems, particularly if antifreeze measures are required. Additionally, active systems are more difficult to retrofit in houses, especially where there is no basement, because a space is required for the additional equipment, such as the hot water cylinder. Five types of systems belong in this category: direct circulation systems, indirect water-heating systems, air systems, heat pump systems, and pool heating systems. Before giving the details of these systems, the optimum flow rate is examined.

High flow rates have been used in pumped circulation solar water heaters to improve the heat removal factor, F_R, and thus maximize the collector efficiency. If the complete system performance is considered, however, rather than the collector as an isolated element of the system, it is found that the solar fraction can be increased if a low flow rate through the collector and a thermally stratified tank are used. Stratification can also be promoted by the use of flow diffusers in the tank and collector loop heat exchangers; for maximum effect, however, it is necessary to combine these features with a low flow rate.

The use of a low flow influences both the system initial capital cost and energy savings. The initial capital cost is affected because the system requires a low-power pump; piping to the collectors can be of smaller diameter (hence less expensive and easier to install), and the smaller tubes require lower thickness and lower-cost thermal insulation because the insulation R value depends on the ratio of outer to inner diameter of the insulation. Additionally, low-flow systems can use very small-diameter collector loop piping, and as a result, flexible annealed copper tubes can be used, which are much easier to install. In this case, the flexible pipe can be hand-bent to change the direction without the need for sharp bends, which lead to higher pressure drop.

According to Duff (1996), the flow in the collector loop should be in the range of $0.2-0.4$ L/min m^2 of collector aperture area. The effect of low flow rate is examined in Chapter 4, Section 4.1.1. In effect, the penalty for low flow rate is a reduction in collector efficiency due to higher collector temperature rise for a given inlet temperature. For example, for a reduction from 0.9 to 0.3 L/min m^2, the efficiency is reduced by about 6%; however, the reduction of the inlet temperature to the collectors because of the improved stratification in the tank more than compensates for the loss of collector efficiency. The pumps required for most of the active systems are low static head centrifugal (also called *circulators*), which for small domestic applications use $30-50$ W of electrical power to work.

5.2.1 Direct circulation systems

A schematic diagram of a direct circulation system is shown in Fig. 5.9. In this system, a pump is used to circulate potable water from storage to the collectors when there is enough available solar energy to increase its temperature and then return the heated water to the storage tank until it is needed. The

FIGURE 5.9

Direct circulation system.

pump is controlled from a differential thermostat that measures and compares the temperature at the collector outlet and the storage tank and operates the pump whenever a certain temperature difference exists. More details are given in Section 5.5.

Because a pump is used to circulate the water, the collectors can be mounted either above or below the storage tank. Direct circulation systems often use a single storage tank equipped with an auxiliary water heater, but two-tank storage systems can also be used. An important feature of this configuration is the spring-loaded check valve, which is used to prevent reverse thermosiphon circulation energy losses when the pump is not running.

Direct circulation systems can be used with water supplied from a cold-water storage tank or connected directly to city water mains. Pressure-reducing valves and pressure relief valves are required, however, when the city water pressure is greater than the working pressure of the collectors. Direct water-heating systems should not be used in areas where the water is extremely hard or acidic, because scale (calcium) deposits may clog or corrode the collectors.

Direct circulation systems can be used in areas where freezing is infrequent. For extreme weather conditions, a freeze protection is usually provided by recirculating warm water from the storage tank. This loses some heat but protects system. A special thermostat that operates the pump when temperature drops below a certain value is used in this case. Such recirculation freeze protection should be used only for locations where freezing occurs rarely (a few times a year), since stored heat is dumped in the process. A disadvantage of this system occurs in cases when there is power failure, in which case the pump will not work and the system could freeze. In such a case, a dump valve can be installed at the bottom of the collectors to provide additional protection.

For freeze protection, a variation of the direct circulation system, called the *drain-down system*, is used (shown schematically in Fig. 5.10). In this case, potable water is also pumped from storage to the collector array, where it is heated. When a freezing condition or a power failure occurs, the system drains automatically by isolating the collector array and exterior piping from the make-up water supply with the normally closed (NC) valve and draining it using the two normally open (NO) valves, shown in Fig. 5.10. It should be noted that the solar collectors and associated piping must be carefully sloped to drain the collector's exterior piping when circulation stops (see Section 5.4.2). The check valve shown on the top of the collectors in Fig. 5.10 is used to allow air to fill the collectors and piping during draining and to escape during a fill-up. The same comments about pressure and scale deposits apply here as for the direct circulation systems.

5.2.2 Indirect water-heating systems

A schematic diagram of indirect water-heating systems is shown in Fig. 5.11. In this system, the pump also operates with a differential thermostat (see Section 5.5) and a heat transfer fluid is circulated through the closed collector loop to a heat exchanger, where its heat is transferred to the potable water. The most commonly used heat transfer fluids are water–ethylene glycol solutions, although other heat transfer fluids such as silicone oils and refrigerants can be used. When fluids that are non-potable or toxic are used, double-wall heat exchangers should be employed; this can be two heat exchangers in series. The heat exchanger can be located inside the storage tank, around the storage tank (tank mantle), or external to the storage tank (see Section 5.3). It should be noted that the collector loop is closed; therefore, an expansion tank and a pressure relief valve are required. Additional over-temperature protection may be needed to prevent the collector heat-transfer fluid from decomposing or becoming corrosive.

FIGURE 5.10

Drain-down system.

FIGURE 5.11

Indirect water-heating system.

Systems of this type using water–ethylene glycol solutions are preferred in areas subject to extended freezing temperatures, because they offer good freeze protection. These systems are more expensive to construct and operate, since the solution should be checked every year and changed every few years, depending on the solution quality and system temperatures achieved.

Typical collector configurations include the internal heat exchanger shown in Fig. 5.11, an external heat exchanger shown in Fig. 5.12a, and a mantle heat exchanger shown in Fig. 5.12b. A general rule to follow is that the storage tank should be between 35 and 70 L/m^2 of collector aperture area, while the most widely used size is 50 L/m^2. More details on internal heat exchangers are given in Section 5.3.2.

For freeze protection, a variation of the indirect water-heating system, called the *drain-back system*, is used. Drain-back systems are generally indirect water-heating systems that circulate water through the closed collector loop to a heat exchanger, where its heat is transferred to potable water. Circulation continues as long as useable energy is available. When the circulation pump stops, the collector fluid drains by gravity to a drain-back tank. If the system is pressurized, the tank also serves as an expansion tank when the system is operating; in this case, it must be protected with temperature and pressure relief valves. In the case of an unpressurized system (Fig. 5.13), the tank is open and vented to the atmosphere. The second pipe directed from the collectors to the top of the drain-back tank is to allow air to fill the collectors during a drain-back.

Because the collector loop is isolated from the potable water, no valves are needed to actuate draining and scaling is not a problem; however, the collector array and exterior piping must be adequately sloped to drain completely. Freeze protection is inherent to the drain-back system because the collectors and the piping above the roof are empty whenever the pump is not running. A disadvantage of this system is that a pump with high static lift capability is required in order to fill the collector when the system starts up.

In drain-back systems, there is a possibility that the collectors will be drained during periods of insolation; it is therefore important to select collectors that can withstand prolonged periods of stagnation conditions. Such a case can happen when there is no load and the storage tank reaches a temperature that would not allow the differential thermostat to switch on the solar pump.

FIGURE 5.12

External and mantle heat exchangers. (a) External heat exchanger. (b) Mantle heat exchanger.

FIGURE 5.13

Drain-back system.

An alternative design to the one shown in Fig. 5.13, which is suitable for small systems, is to drain the water directly in the storage tank. In this case, the system is open (without a heat exchanger) and there is no need the have a separate drain-back tank; however, the system suffers from the disadvantages of the direct systems outlined in Section 5.2.1.

5.2.3 Air water-heating systems

Air systems are indirect water-heating systems because air, circulated through air collectors and via ductworks, is directed to an air-to-water heat exchanger. In the heat exchanger, heat is transferred to the potable water, which is also circulated through the heat exchanger and returned to the storage tank. Figure 5.14 shows a schematic diagram of a double storage tank system. This type of system is used most often because air systems are generally employed for preheating domestic hot water and hence the auxiliary heater is used in only one tank, as shown.

The advantages of this system are that air does not need to be protected from freezing or boiling, is non-corrosive, does not suffer from heat transfer fluid degradation, and is free. Additionally, the system is more cost-effective because no safety values or expansion vessels are required. The disadvantages are that air-handling equipment (ducts and fans) needs more space than piping and pumps, air leaks are difficult to detect, and parasitic power consumption (electricity used to drive the fans) is generally higher than that of liquid systems.

5.2.4 Heat pump systems

Heat pumps use mechanical energy to transfer thermal energy from a source at a lower temperature to a sink at a higher temperature. The bigger advantage of electrically driven heat pump heating systems, compared to electric resistance heating or expensive fuels, is that the heat pump's coefficient of

FIGURE 5.14

Air water-heating system.

performance (ratio of heating performance to electrical energy) is greater than unity for heating; so it yields 9–15 MJ of heat for each kilowatt hour of energy supplied to the compressor, which saves on purchase of energy.

The original system concept, proposed by Charters et al. (1980), was a system with direct evaporation of the working fluid of the heat pump in the solar collector. The condenser of the heat pump was actually a heat exchanger wrapped around the storage tank. In this way, the initial system cost and the parasitic energy requirements of the system are minimized. A possible disadvantage of this system is that the condenser heat transfer is limited by the free convection from the tank wall, which can be minimized by using a large heat transfer area in the tank. A more important disadvantage of this system is that the heat pump refrigeration circuit must be evacuated and charged on site, which requires special equipment and expertize. More details of these systems are given in Chapter 6, Section 6.3.5.

This disadvantage is removed by using compact solar heat pump systems. These incorporate an evaporator mounted outside the water storage tank with natural convection air circulation. This system needs to be installed outdoors, and if installed adjacent to the ventilation duct outlet of a building, it can also work as a waste heat recovery unit. The advantages of this system are that it has no parasitic energy requirement and, because the system is packaged, all its components are assembled in the factory and thus the system is precharged. The installation of this system is as simple as a conventional electric water heater because the unit requires no high-power electrical connection (Morrison, 2001).

5.2.5 Pool heating systems

Solar pool-heating systems require no separate storage tank, because the pool itself serves as storage. In most cases, the pool's filtration pump is used to circulate the water through solar panels or plastic pipes. For a daylong operation, no automatic controls are required, because the pool usually operates when the sun is shining. If such controls are employed, they are used to direct the flow of filtered water to the collectors only when solar heat is available. This can also be achieved by a simple manually operated valve. Normally, these kinds of solar systems are designed to drain down into the pool when the pump is switched off; thus the collectors are inherently freeze protected (ASHRAE, 2007).

The primary type of collector design used for heating swimming pools is the rigid black plastic panels made from polypropylene (see Chapter 3, Section 3.1.1). Additionally, plastic pipes or tube-on-sheet panels can be used. In all cases, however, a large area is required and the roof of a nearby building can be used for this purpose.

Recommendations for the design, installation, and commissioning of solar heating systems for swimming pools, using direct circulation of pool water to the solar collectors, are given in the technical report ISO/TR 12596 (1995). The report does not deal with the pool filtration system to which a solar heating system is often connected. The material presented in the report is applicable to all sizes of pools, both domestic and public, that are heated by solar energy, either alone or in conjunction with a conventional system. Additionally, the report includes details of the heating load calculations. The pool heating load is the total heat loss less any heat gains from incident radiation.

The total heat loss is the sum of losses due to evaporation, radiation, and convection. This calculation requires knowledge of the air temperature, wind speed, and relative humidity or partial vapor pressure. Other causes of heat losses, which have a much smaller effect, are turbulence caused by swimmers, conduction to the ground (usually neglected), and rainfall, which at substantial quantities can lower the pool temperature. The addition of make-up water should be considered if the temperature differs considerably from the pool operating temperature. Pools usually operate in a narrow temperature range of 24–32 °C. Since the pool has a large mass, its temperature does not change quickly.

The use of a pool cover reduces heat losses, particularly evaporative losses; however, when designing a solar pool-heating system, it is often not possible to know with certainty the times during which a cover will be in place. In addition, the cover may not have a perfect fit. Hence, a conservative approach should be taken when allowing for the effect of a cover (ISO/TR 12596, 1995).

Evaporation heat loss

The following analysis is for a still pool as per ISO/TR 12596 (1995). The evaporative heat loss from a still outdoor pool is a function of the wind speed and of the vapor pressure difference between the pool water and the atmosphere, given by:

$$q_e = (5.64 + 5.96v_{0.3})(P_w - P_a) \tag{5.18}$$

where

q_e = heat loss by evaporation (MJ/m^2 day);
P_w = saturation water vapor pressure at water temperature, t_w (kPa);
P_a = partial water vapor pressure in the air (kPa); and
$v_{0.3}$ = wind speed velocity at a height of 0.3 m above the pool (m/s).

If the wind velocity above the pool cannot be measured, it can be obtained from climatic data by the application of a reduction factor for the degree of wind shelter at the pool. Usually, the wind speed is measured at 10 m from the ground (v_{10}); therefore,

For normal suburban sites, $v = 0.30v_{10}$
For well-sheltered sites, $v = 0.15v_{10}$

For indoor pools, the low air velocity results in a lower evaporation rate than usually occurs in outdoor pools, and the evaporative heat loss is given by:

$$q_e = (5.64 + 5.96v_s)(P_w - P_{enc}) \tag{5.19}$$

where

P_{enc} = the partial water vapor pressure in the pool enclosure (kPa);
v_s = air speed at the pool water surface, typically 0.02−0.05 (m/s).

Partial water vapor pressure (P_a) can be calculated from the relative humidity,

$$P_a = \frac{P_s \times \text{RH}}{100} \tag{5.20}$$

where

P_s = saturation water vapor pressure at air temperature, t_a (kPa).

Saturation water vapor pressure can be obtained from:

$$P_s = 100\left(0.004516 + 0.0007178t_w - 2.649 \times 10^{-6}t_w^2 + 6.944 \times 10^{-7}t_w^3\right) \tag{5.21}$$

where

t_w = water temperature (°C).

The presence of swimmers in a pool significantly increases the evaporation rate. With five swimmers per 100 m², the evaporation rate has to increase by 25%−50%. With 20−25 swimmers per 100 m², the evaporation rate has to increase by 70%−100% more than the value for a still pool.

Radiation heat loss
Radiation heat loss is given by:

$$q_r = \frac{24 \times 3600}{10^6}\varepsilon_w\sigma\left(T_w^4 - T_s^4\right) = 0.0864\varepsilon_w h_r(T_w - T_s) \tag{5.22}$$

where

q_r = radiation heat loss (MJ/m² day);
ε_w = longwave emissivity of water = 0.95;
T_w = water temperature (K);
T_s = sky temperature (K); and
h_r = radiation heat transfer coefficient (W/m² K).

The radiation heat transfer coefficient is calculated from:

$$h_r = \sigma\left(T_w^2 + T_s^2\right)(T_w + T_s) \approx 2.268 \times 10^{-7}\left(\frac{T_w + T_s}{2}\right)^3 \tag{5.23}$$

For an indoor pool, $T_s = T_{enc}$, both in Kelvins, and T_{enc} is the temperature of the walls of the pool enclosure. For an outdoor pool,

$$T_s = T_a(\varepsilon_s)^{0.25} \tag{5.24}$$

where sky emissivity, ε_s, is a function of dew point temperature, t_{dp}, given by:

$$\varepsilon_s = 0.711 + 0.56\left(\frac{t_{dp}}{100}\right) + 0.73\left(\frac{t_{dp}}{100}\right)^2 \tag{5.25}$$

It should be noted that T_s might vary from $T_s \approx T_a$ for cloudy skies to $T_s \approx T_a - 20$ for clear skies.

Convection heat loss

Heat loss due to convection to ambient air is given by:

$$q_c = \frac{24 \times 3600}{10^6}(3.1 + 4.1v)(t_w - t_a) = 0.0864(3.1 + 4.1v)(t_w - t_a) \tag{5.26}$$

where

q_c = convection heat loss to ambient air (MJ/m^2 day);
v = wind velocity at 0.3 m above outdoor pools or over the pool surface for indoor pools (m/s);
t_w = water temperature (°C); and
t_a = air temperature (°C).

As can be seen from Eq. (5.26), the convective heat loss depends to a large extent on the wind velocity. During summer for outdoor pools, this may be negative, and in fact, the pool will gain heat by convection from the air.

Make-up water

If the make-up water temperature is different from the pool operating temperature, there will be a heat loss, given by:

$$q_{muw} = m_{evp}c_p(t_{muw} - t_w) \tag{5.27}$$

where

q_{muw} = make-up water heat loss (MJ/m^2 day);
m_{evp} = daily evaporation rate (kg/m^2 day);
t_{muw} = temperature of make-up water (°C); and
c_p = specific heat of water (J/kg °C).

The daily evaporation rate is given by:

$$m_{evp} = \frac{q_e}{h_{fg}} \tag{5.28}$$

where

h_{fg} = latent heat of vaporization of water (MJ/kg).

Solar radiation heat gain

Heat gain due to the absorption of solar radiation by the pool is given by:

$$q_s = \alpha H \tag{5.29}$$

where

q_s = rate of solar radiation absorption by the pool (MJ/m² day);

α = solar absorptance ($\alpha = 0.85$ for light-colored pools; $\alpha = 0.90$ for dark-colored pools); and

H = solar irradiation on a horizontal surface (MJ/m² day).

It should be noted that the solar absorptance, α, is dependent on the color, depth, and pool usage. For pools with continuous intensive use (public pools), an additional reduction of 0.05 should be made to the absorption factor (ISO/TR 12596, 1995).

EXAMPLE 5.1

A 500 m² light-colored swimming pool is located in a normal suburban site, where the measured wind speed at 10 m height is 3 m/s. The water temperature is 25 °C, the ambient air temperature is 17 °C, and relative humidity is 60%. There are no swimmers in the pool, the temperature of the make-up water is 22 °C, and the solar irradiation on a horizontal surface for the day is 20.2 MJ/m² day. How much energy must the solar system supply (Q_{ss}) to the pool to keep its temperature at 25 °C?

Solution

The energy balance of the pool is given by:

$$q_e + q_r + q_c + q_{muw} - q_s = q_{ss}$$

The velocity at 0.3 m above the pool surface is $0.3 \times 3 = 0.9$ m/s. The partial pressures for air and water are given by Eqs. (5.20) and (5.21). The saturation water vapor pressure at air temperature, t_a, is also given by Eq. (5.21); therefore,

$$P_s = 100\left(0.004516 + 0.0007178t_a - 2.649 \times 10^{-6}t_a^2 + 6.944 \times 10^{-7}t_a^3\right)$$

$$= 100\left(0.004516 + 0.0007178 \times 17 - 2.649 \times 10^{-6} \times 17^2 + 6.944 \times 10^{-7} \times 17^3\right)$$

$$= 1.936 \text{ kPa}$$

From Eq. (5.20),

$$P_a = \frac{P_s \times RH}{100} = \frac{1.936 \times 60}{100} = 1.162 \text{ kPa}$$

Saturation water vapor pressure can also be obtained from Eq. (5.21) by using t_w instead of t_a. Therefore,

$$P_w = 3.166 \text{ kPa}$$

From Eq. (5.18), evaporation heat losses are:

$$q_e = (5.64 + 5.96v_{0.3})(P_w - P_a) = (5.64 + 5.96 \times 0.9)(3.166 - 1.162)$$

$$= 22.052 \text{ MJ/m}^2 \text{ d}$$

From Eq. (5.25),

$$\varepsilon_s = 0.711 + 0.56\left(\frac{t_{dp}}{100}\right) + 0.73\left(\frac{t_{dp}}{100}\right)^2 = 0.711 + 0.56\left(\frac{17}{100}\right) + 0.73\left(\frac{17}{100}\right)^2 = 0.827$$

From Eq. (5.24),

$$T_s = T_a(\varepsilon_s)^{0.25} = 290(0.827)^{0.25} = 276.6 \text{ K}$$

From Eq. (5.22), radiation heat losses are:

$$q_r = \frac{24 \times 3600}{10^6} \varepsilon_w \sigma\left(T_w^4 - T_s^4\right) = 0.0864 \times 0.95 \times 5.67 \times 10^{-8}\left(298^4 - 276.6^4\right) = 9.46 \text{ MJ/m}^2 \text{ d}$$

Continued

EXAMPLE 5.1–CONT'D

From Eq. (5.26), convection heat losses are:

$$q_c = 0.0864(3.1 + 4.1v)(t_w - t_a) = 0.0864(3.1 + 4.1 \times 0.9)(25 - 17) = 4.693 \text{ MJ} / \text{m}^2 \text{ d}$$

From steam tables, h_{fg}, the latent heat of vaporization of water at 25 °C is equal to 2441.8 MJ/kg. Therefore, the daily evaporation rate is given by Eq. (5.28):

$$m_{evp} = \frac{q_e}{h_{fg}} = \frac{22.052 \times 10^3}{2441.8} = 9.031 \text{ kg/m}^2\text{-day}$$

From Eq. (5.27), the heat losses due to the make-up water are:

$$q_{muw} = m_{evp}c_p(t_{muw} - t_w) = 9.031 \times 4.18(22 - 25) \times 10^{-3}$$
$$= 0.1132 \text{ MJ/m}^2\text{-day (the negative sign is not used as all the above values are losses)}$$

From Eq. (5.29), solar radiation heat gain is:

$$q_s = \alpha H = 0.85 \times 20.2 = 17.17 \text{ MJ/m}^2\text{-day}$$

Therefore, the energy required by the solar system to keep the pool at 25 °C is:

$$q_{ss} = q_e + q_r + q_c + q_{muw} - q_s = 22.052 + 9.46 + 4.693 + 0.1132 - 17.17$$
$$= 19.148 \text{ MJ/m}^2\text{-day} \quad \text{or} \quad Q_{ss} = 9.57 \text{ GJ/day}$$

5.3 Heat storage systems

Thermal storage is one of the main parts of a solar heating, cooling, and power generating system. Because for approximately half the year any location is in darkness, heat storage is necessary if the solar system must operate continuously. For some applications, such as swimming-pool heating, daytime-air heating, and irrigation pumping, intermittent operation is acceptable, but most other uses of solar energy require operating at night and when the sun is hidden behind clouds.

Usually the design and selection of the thermal storage equipment is one of the most neglected elements of solar energy systems. It should be realized, however, that the energy storage system has an enormous influence on an overall system cost, performance, and reliability. Furthermore, the design of the storage system affects the other basic elements, such as the collector loop and the thermal distribution system.

A storage tank in a solar system has several functions, the most important of which are:

- Improvement of the utilization of collected solar energy by providing thermal capacitance to alleviate the solar availability and load mismatch and improve the system response to sudden peak loads or loss of solar input.
- Improvement of system efficiency by preventing the array heat transfer fluid from quickly reaching high temperatures, which lower the collector efficiency.

Generally, solar energy can be stored in liquids, solids, or phase-change materials (PCM). Water is the most frequently used storage medium for liquid systems, even though the collector loop may use water, oils, water–glycol mixtures, or any other heat transfer medium as the collector fluid. This is because water is inexpensive and non-toxic and it has a high storage capacity, based on both weight and volume. Additionally, as a liquid, it is easy to transport using conventional pumps and plumbing. For service water-heating applications and most building space heating, water is normally contained in some type of tank, which is usually circular. Air systems typically store heat in rocks or pebbles, but sometimes the structural mass of the building is used.

An important consideration is that the temperature of the fluid delivered to the load should be appropriate for the intended application. Additionally, the lower the temperature of the fluid supplied to the collectors, the higher is the efficiency of the collectors.

The location of the storage tank should also be given careful consideration. The best location is indoors, where thermal losses are minimal and weather deterioration will not be a factor. If the tank cannot be installed inside the building, then it should be located outside above the ground or on the roof. Such a storage tank should have a good insulation and good outside protection of the insulation. The storage tank should also be located as close as possible to the collector arrays to avoid long pipe runs.

5.3.1 Air system thermal storage

The most common storage media for air collectors are rocks. Other possible media include PCM, water, and the inherent building mass. Gravel is widely used as a storage medium because it is abundant and relatively inexpensive.

In cases where large interior temperature swings can be tolerated, the inherent structure of the building may be sufficient for a thermal storage. Loads requiring no storage are usually the most cost-effective applications of air collectors, and heated air from the collectors can be distributed directly to the space. Generally, storage may be eliminated in cases where the array output seldom exceeds the thermal demand (ASHRAE, 2004).

The main requirements for a gravel storage are good insulation, low air leakage, and low pressure drop. Many different designs can fulfill these requirements. The container is usually constructed from concrete, masonry, wood, or a combination of these materials. Airflow can be vertical or horizontal. A schematic diagram of a vertical flow bed is shown in Fig. 5.15. In this arrangement, the solar-heated air enters at the top and exits from the bottom. This tank can work as effectively as a horizontal flow bed. In these systems, it is important to heat the bed with the hot airflow in one direction and to retrieve the heat with airflow in the opposite direction. In this way, pebble beds perform as effective counterflow heat exchangers.

The size of the rocks for pebble beds range from 35 to 100 mm in diameter, depending on airflow, bed geometry, and desired pressure drop. The volume of the rocks needed depends on the fraction of collector output that must be stored. For residential systems, storage volume is typically in the range of $0.15–0.3\ m^3$ per square meter of collector area. For large systems, pebble beds can be quite large and their large mass and volume may lead to location problems.

Other storage options for air systems include PCMs and water. PCMs are functionally attractive because of their high volumetric heat storage capabilities, since they require only about one 10th the volume of a pebble bed (ASHRAE, 2004).

FIGURE 5.15

Vertical flow packed rock bed.

Water can also be used as a storage medium for air collectors through the use of a conventional air-to-water heat exchanger to transfer heat from the air to the water in the storage tank. This option has two advantages:

1. Water storage is compatible with hydronic heating systems.
2. It is relatively compact; the required storage water volume is roughly one third the pebble bed's volume.

5.3.2 Liquid system thermal storage

Two types of water storage for liquid systems are available: pressurized and unpressurized. Other differentiations include the use of an external or internal heat exchanger and single or multiple-tank configurations. Water may be stored in copper, galvanized metal, or concrete tanks. Whatever storage vessel is selected, however, this should be well insulated and large tanks should be provided with internal access for maintenance. Recommended U value is ≈ 0.16 W/m^2 K.

Pressurized systems are open to city mains water supply. Pressurized storage is preferred for small service water-heating systems, although in cases like Cyprus, where the water supply is intermittent, it is not suitable. Typical storage size is about 40–80 L per square meter of collector area. With pressurized storage, the heat exchanger is always located on the collector side of the tank. Either internal or external heat exchanger configurations can be used. Two principal types of internal heat exchangers exist: an immersed coil and a tube bundle, as shown in Fig. 5.16.

Sometimes, because of the required storage volume, more than one tank is used instead of one large one, if such a large-capacity tank is not available. Additional tanks offer, in addition to the extra storage volume, increased heat exchanger surface (when a heat exchanger is used in each tank) and reduced pressure drop in the collection loop. A multiple-tank configuration for pressurized storage is shown in Fig. 5.17. It should be noted that the heat exchangers are connected in a reverse-return mode to improve flow balance.

FIGURE 5.16

Pressurized storage with internal heat exchanger.

FIGURE 5.17

Multiple-tank storage arrangement with internal heat exchangers.

An external heat exchanger provides greater flexibility because the tank and the exchanger can be selected independently of other equipment (see Fig. 5.18). The disadvantage of this system is the parasitic energy consumption, in the form of electrical energy, that occurs because of the additional pump.

For small systems, an internal heat exchanger tank arrangement is usually used, which has the advantage of preventing the water side of the heat exchanger from freezing. However, the energy required to maintain the water above freezing is extracted from storage, and thus the overall system performance is decreased. With external heat exchanger, a bypass can be arranged to divert cold fluid around the heat exchanger until it has been heated to an acceptable level of about 25 °C (ASHRAE, 2004). When the heat transfer fluid is warmed to this level, it can enter the heat exchanger without causing freezing or extraction of heat from storage. If necessary, this arrangement can also be used with internal heat exchangers to improve performance.

For systems with sizes greater than about 30 m^3, unpressurized storage is usually more cost-effective than pressurized. This system, however, can also be employed in small domestic flat-plate

FIGURE 5.18

Pressurized storage system with external heat exchanger.

collector systems, and in this case, the make-up water is usually supplied from a cold water storage tank located on top of the hot water cylinder.

Unpressurized storage for water and space heating can be combined with the pressurized city water supply. This implies the use of a heat exchanger on the load side of the tank to isolate the high-pressure mains' potable water loop from the low-pressure collector loop. An unpressurized storage system with an external heat exchanger is shown in Fig. 5.19. In this configuration, heat is extracted from the top of the solar storage tank and the cooled water is returned to the bottom of the tank so as not to distract stratification. For the same reason, on the load side of the heat exchanger, the water to be heated flows from the bottom of the backup storage tank, where relatively cold water exists, and heated water returns to the top. Where a heat transfer fluid is circulated in the collector loop, the heat exchanger may have a double-wall construction to protect the potable water supply from contamination. A differential temperature controller (DTC) controls the two pumps on either side of the heat exchanger. When small pumps are used, both may be controlled by the same controller without overloading problems. The external heat exchanger shown in Fig. 5.19 provides good system flexibility and freedom in component selection. In some cases, system cost and parasitic power consumption may be reduced by an internal heat exchanger.

Stratification is the collection of hot water to the top of the storage tank and cold water to the bottom. This improves the performance of the tank because hotter water is available for use

FIGURE 5.19

Unpressurized storage system with external heat exchanger.

and colder water is supplied to the collectors, which enables the collector to operate at higher efficiency.

Another category of hot water stores is the so-called solar combistores. These are used mainly in Europe for combined domestic hot water preparation and space heating. More details on these devices are included in Chapter 6, Section 6.3.1.

5.3.3 Thermal analysis of storage systems

Here the water and air systems are examined separately.

Water systems

For fully mixed or unstratified energy storage, the capacity (Q_s) of a liquid storage unit at uniform temperature, operating over a finite temperature difference (ΔT_s), is given by:

$$Q_s = (Mc_p)_s \Delta T_s \tag{5.30}$$

where

M = mass of storage capacity (kg).

The temperature range over which such a unit operates is limited by the requirements of the process. The upper limit is also determined by the vapor pressure of the liquid.

An energy balance of the storage tank gives:

$$(Mc_p)_s \frac{dT_s}{dt} = Q_u - Q_l - Q_{tl} \tag{5.31}$$

where

Q_u = rate of collected solar energy delivered to the storage tank (W);
Q_l = rate of energy removed from storage tank to load (W); and
Q_{tl} = rate of energy loss from storage tank (W).

The rate of storage tank energy loss is given by:

$$Q_{tl} = (UA)_s (T_s - T_{env}) \tag{5.32}$$

where

$(UA)_s$ = storage tank loss coefficient and area product (W/°C);
T_{env} = temperature of the environment where the storage tank is located (°C).

To determine the long-term performance of the storage tank, Eq. (5.31) may be rewritten in finite difference form as:

$$(Mc_p)_s \frac{T_{s-n} - T_s}{\Delta t} = Q_u - Q_l - Q_{tl} \tag{5.33}$$

or

$$T_{s-n} = T_s + \frac{\Delta t}{(Mc_p)_s} [Q_u - Q_l - (UA)_s (T_s - T_{env})] \tag{5.34}$$

where

T_{s-n} = new storage tank temperature after time interval Δt (°C).

This equation assumes that the heat losses are constant in the period Δt. The most common time period for this estimation is an hour because the solar radiation data are also available on an hourly basis.

EXAMPLE 5.2

A fully mixed water storage tank contains 500 kg of water, has an *UA* product equal to 12 W/°C, and is located in a room that is at a constant temperature of 20 °C. The tank is examined in a 10 h period starting from 5 a.m. where the hourly values of Q_u are equal to 0, 0, 0, 10, 21, 30, 40, 55, 65, 55 MJ. The load is constant and equal to 12 MJ in the first 3 h, 15 MJ in the next 3 h, and 25 MJ the rest of time. Find the final storage tank temperature if the initial temperature is 45 °C.

Solution

The estimation time interval is 1 h. Using Eq. (5.34) and inserting the appropriate constants, we get:

$$T_{s-n} = T_s + \frac{1}{(500 \times 4.18 \times 10^{-3})}\left[Q_u - Q_l - 12 \times \frac{3600}{10^6}(T_s - 20)\right]$$

By inserting the initial storage tank temperature (45 °C), Q_u, and Q_l according to the problem, Table 5.3 can be obtained.

Table 5.3 Results for Example 5.2.

Hour	Q_u (MJ)	Q_l (MJ)	T_s (°C)	Q_u (MJ)
			45	
5	0	12	38.7	1.1
6	0	12	32.6	0.8
7	0	12	26.6	0.5
8	10	15	24.1	0.3
9	21	15	26.9	0.2
10	30	15	33.9	0.3
11	40	25	40.8	0.6
12	55	25	54.7	0.9
13	65	25	73.1	1.5
14	55	25	86.4	2.3

So for the first hour at 5 a.m.:

$$T_{s-n} = 45 + 0.4785[0 - 12 - 0.0432(45 - 20)] = 38.7\ °C$$

For the second hour at 6 a.m.:

$$T_{s-n} = 38.7 + 0.4785[0 - 12 - 0.0432(38.7 - 20)] = 32.6\ °C$$

The rest of calculations are shown in Table 5.3.

Therefore, the final storage tank temperature is 86.4 °C. For these calculations, the use of a spreadsheet program is recommended.

The collector performance equations in Chapter 4 can also be used with the more detailed determination of inlet fluid temperature to estimate the daily energy output from the collector. This is illustrated by the following example.

EXAMPLE 5.3

Repeat Example 4.2 by considering the system to have a fully mixed storage tank of 100 L and no load. The initial storage tank temperature at the beginning of the day is 40 °C and the environmental temperature at the area where the storage tank is located is equal to the ambient air temperature. The tank UA value is 12 W/°C. Calculate the useful energy collected over the day.

Solution

By using Eq. (5.34), the new storage tank temperature can be considered as the collector inlet. This is correct for the present example but is not very correct in practice because some degree of stratification is unavoidable in the storage tank.

$$T_{s-n} = T_s + \frac{1}{(100 \times 4.18)}\left[Q_u - 12 \times \frac{3600}{1000}(T_s - T_a)\right]$$

Table 5.4 Example 5.3 results.

Time	T_a (°C)	I_t (kJ/m²)	T_i (°C)	$\Delta t/G_t$ (°C m²/W)	θ (deg.)	$K\theta$	Q_u (kJ)
6	25	360	40.0	0.150	93.9	0	0
7	26	540	38.6	0.084	80.5	0.393	0.0
8	28	900	37.5	0.038	67.5	0.806	719.6
9	30	1440	38.4	0.021	55.2	0.910	1653.1
10	32	2160	41.7	0.016	44.4	0.952	2738.5
11	34	2880	47.5	0.017	36.4	0.971	3702.3
12	35	3420	55.1	0.021	33.4	0.976	4269.3
13	34	2880	63.1	0.036	36.4	0.971	3089.4
14	32	2160	67.3	0.059	44.4	0.952	1698.3
15	30	1440	67.5	0.094	55.2	0.910	475.8
16	28	900	64.6	0.146	67.5	0.806	0.0
17	26	540	60.6	0.231	80.5	0.393	0
18	25	360	56.9	0.319	93.9	0	0
Sum							18,346.3

The results in this case are shown in Table 5.4. It should be noted that for the new storage tank temperature the useful energy collected and T_i from the previous time step is used. For example at 9 a.m.:

$$T_{s-n} = 37.5 + 2.39 \times 10^{-3}[719.6 - 12 \times 3.6(37.5 - 30)] = 38.4 \,°C$$

where $Q_u = 719.6$ kJ (estimated at 8 a.m. using $K_\theta = 0.806$ and $I_t = 900$ kJ/m²) and $\Delta T/G_t = (37.5 - 28)/(900/3.6) = 0.038$.

Therefore, the total energy collected over the day = 18,346.3 kJ.

As can be seen from the results of this example, the collector performance is somewhat lower than those of Example 4.2 because a higher collector inlet temperature leads to lower collector efficiency. In this example, too, the use of a spreadsheet program greatly facilitates estimations.

The density of water (and other fluids) drops as its temperature increases. When hot water enters from the collectors and leaves for the load from the top of the tank and cool water flows (cold water returns to the collector and make-up water supply) occur at the bottom, the storage tank will stratify because of the density difference. Additionally, with cool water at the tank bottom, the temperature of water fed to the collector inlet is low, and thus the collector performance is enhanced. Moreover, water from the top of the tank, which is at the highest temperature, may meet the heating demand more effectively. The degree of stratification is measured by the temperature difference between the top and bottom of the storage tank and is crucial for the effective operation of a solar system.

There are basically two types of models developed to simulate stratification: the multimode and the plug flow. In the former, the tank modeled is divided into N nodes (or sections) and energy balances are written for each node. This results in a set of N differential equations, which are solved for the temperatures of the N nodes as a function of time. In the latter, segments of liquid of various temperatures are assumed to move through the storage tank in plug-flow and the models keep track of the size, temperature, and position of the segments. Neither of these methods is suitable for hand calculations; however, more details of the plug flow model are given here.

The procedure is presented by Morrison and Braun (1985) and is used in conjunction with TRNSYS thermosiphon model presented in Section 5.1.1. This model produces the maximum degree of stratification possible. The storage tank is initially represented by three fluid segments. Initially, the change of tank segment temperatures, due to heat loss to the surroundings and conduction between segments, is estimated. The energy input from the collector is determined by considering a constant temperature plug of fluid of volume $V_h (= \dot{m}\Delta t / \rho)$ entering the tank during the time step Δt. The plug of fluid entering the tank is placed between existing segments chosen to avoid developing a temperature inversion.

The load flow is considered in terms of another segment of fluid of volume, $V_L (= \dot{m}_L \Delta t / \rho)$, and temperature T_L, added either to the bottom of the tank or at its appropriate temperature level. Fluid segments are moved up the tank as a result of the addition of the new load flow segment. The net shift of the profile in the tank above the collector return level is equal to the load volume, V_L, and that below the collector return is equal to the difference between the collector and load volumes ($V_h - V_L$). After adjusting for the load flow, the auxiliary input is considered, and if sufficient energy is available, segments above the auxiliary input level are heated to the set temperature. According to the situation, the segment containing the auxiliary element is split so that only segments of the tank above the element are heated.

Segments and fractions of segments in the new tank profile that are outside the bounds of the tank are returned to the collector and load. The average temperature of the fluid delivered to the load is given by:

$$T_d = \sum_{i=1}^{i=j-1} \frac{(T_i V_i + a T_j V_j)}{V_L} \tag{5.35}$$

where j and a must satisfy:

$$V_L = \sum_{i=1}^{i=j-1} (V_i) + a V_j \tag{5.36}$$

and $0 \le a < 1$.

The average temperature of fluid returned to the collector is:

$$T_R = \sum_{i=1}^{i=n-1} \frac{(T_i V_i + b T_n V_n)}{V_h} \tag{5.37}$$

where n and b must satisfy:

$$V_R = \sum_{i=1}^{i=n-1} (V_i) + bV_n \tag{5.38}$$

and $0 \le b < 1$.

The main advantage of this tank model is that small fluid segments are introduced when stratification is developing, while zones of uniform temperature, such as those above the auxiliary heater, are represented by large fluid segments. Additionally, the size of fluid segments used to represent the tank temperature stratification varies with collector flow rate. If the collector flow rate is high, there will be little stratification in the preheat portion of the tank and the algebraic model will produce only a few tank segments. If the collector flow rate is low and the tank is stratified, then small tank segments will be generated. Generally, the number of segments generated in this model is not fixed but depends on many factors, such as the simulation time step, the size of the collector, load flow rates, heat losses, and auxiliary input. To avoid generating an excessive number of segments, adjacent segments are merged if they have a temperature difference of less than 0.5 °C.

Air systems

As we have seen before, in air systems, pebble beds are usually employed for energy storage. When solar radiation is available, hot air from the collectors enters the top of the storage unit and heats the rocks. As the air flows downward, heat transfer between the air and the rocks results in a stratified distribution of the pebbles, having a high temperature at the top and a low one at the bottom. This is the charging mode of the storage unit. When there is a heating demand, hot air is drawn from the top of the unit and cooler air is returned to the bottom of the unit, causing the bed to release its stored energy. This is the discharge mode of the pebble-bed storage unit. From this description, it can be realized that the two modes cannot occur at the same time. Unlike water storage, the temperature stratification in pebble-bed storage units can be easily maintained.

In the analysis of rock bed storage, it should be taken into account that both the rocks and air change temperature in the direction of airflow and there are temperature differentials between the rocks and air. Therefore, separate energy balance equations are required for the rocks and air. In this analysis, the following assumptions can be made:

1. Forced airflow is one-dimensional.
2. System properties are constant.
3. Conduction heat transfer along the bed is negligible.
4. Heat loss to the environment does not occur.

Therefore, the thermal behavior of the rocks and air can be described by the following two-coupled partial differential equations (Hsieh, 1986):

$$\rho_b c_b (1 - \varepsilon) \frac{\partial T_b}{\partial t} = h_v (T_a - T_b) \tag{5.39}$$

$$\rho_a c_a \varepsilon \frac{\partial T_a}{\partial t} = -\frac{\dot{m} c_a}{A} \frac{\partial T_a}{\partial x} - h_v (T_a - T_b) \tag{5.40}$$

where

A = cross-sectional area of storage tank (m^2);
T_b = temperature of the bed material (°C);
T_a = temperature of the air (°C);
ρ_b = density of bed material (kg/m^3);
ρ_a = density of air (kg/m^3);
c_b = specific heat of bed material (J/kg K);
c_a = specific heat of air (J/kg K);
t = time (s);
x = position along the bed in the flow direction (m);
$\dot{m}$ = mass flow rate of air (kg/s);
ε = void fraction of packing = void volume/total volume of bed (dimensionless); and
h_v = volumetric heat transfer coefficient (W/m^3 K).

An empirical equation for the determination of the volumetric heat transfer coefficient (h_v) is:

$$h_v = 650(G/d)^{0.7} \tag{5.41}$$

where

G = air mass velocity per square meter of bed frontal area (kg/s m^2).
d = rock diameter (m).

If the energy storage capacity of the air within the bed is neglected, Eq. (5.40) is reduced to:

$$\dot{m}c_a \frac{\partial T_a}{\partial x} = -Ah_v(T_a - T_b) \tag{5.42}$$

Equations (5.39) and (5.42) can also be written in terms of the number of transfer units (NTU) as:

$$\frac{\partial T_b}{\partial(\theta)} = \text{NTU}(T_a - T_b) \tag{5.43}$$

$$\frac{\partial T_a}{\partial(x/L)} = \text{NTU}(T_b - T_a) \tag{5.44}$$

where

L = bed length (m).

The dimensionless NTU is given by:

$$\text{NTU} = \frac{h_v AL}{\dot{m}c_a} \tag{5.45}$$

The parameter θ, which is also dimensionless in Eq. (5.43), is equal to:

$$\theta = \frac{t\dot{m}c_a}{\rho_b c_b(1 - \varepsilon)AL} \tag{5.46}$$

For the long-term study of solar air-storage systems, the two-coupled partial differential equations, Eqs. (5.43) and (5.44), can be solved by a finite difference approximation with the aid of a computer.

5.4 Module and array design
5.4.1 Module design

Most commercial and industrial systems require a large number of collectors to satisfy the heating demand. Connecting the collectors with just one set of manifolds makes it difficult to ensure drainability and low pressure drop. It would also be difficult to balance the flow so as to have the same flow rate through all collectors. A *module* is a group of collectors that can be grouped into parallel flow and combined series–parallel flow. Parallel flow is more frequently used because it is inherently balanced, has a low pressure drop, and can be drained easily. Figure 5.20 illustrates the two most popular collector header designs: external and internal manifolds.

Generally, flat-plate collectors are made to connect to the main pipes of the installation in one of the two methods shown in Fig. 5.20. The external manifold collector has a small-diameter connection because it is used to carry the flow for only one collector. Therefore, each collector is connected individually to the manifold piping, which is not part of the collector panel. The internal manifold collector incorporates several collectors with large headers, which can be placed side by side to form a continuous supply and return manifold, so the manifold piping is integral with each collector. The number of collectors that can be connected depends on the size of the header.

External manifold collectors are generally more suitable for small systems. Internal manifolding is preferred for large systems because it offers a number of advantages. These are cost savings because the system avoids the use of extra pipes (and fittings), which need to be insulated and properly supported, and the elimination of heat losses associated with external manifolding, which increases the thermal performance of the system.

It should be noted that the flow is parallel but the collectors are connected in series. When arrays must be greater than one panel high, a combination of series and parallel flow may be used, as shown in Fig. 5.21. This is a more suitable design in cases where collectors are installed on an inclined roof.

The choice of series or parallel arrangement depends on the temperature required from the system. Connecting collectors in parallel means that all collectors have as input the same temperature, whereas when a series connection is used, the outlet temperature from one collector (or row of collectors) is the input to the next collector (or row of collectors). The performance of such an arrangement can be obtained from the equations presented in Chapter 4, Section 4.1.2.

5.4.2 Array design

An array usually includes many individual groups of collectors, called *modules*, to provide the necessary flow characteristics. To maintain a balanced flow, an array or field of collectors should be

FIGURE 5.20

Collector manifolding arrangements for parallel flow modules. (a) External manifolding. (b) Internal manifolding.

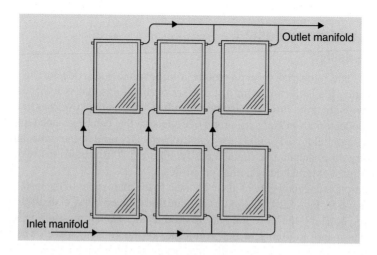

FIGURE 5.21

Collector manifolding arrangement for combined series-parallel flow modules.

built from identical modules. Basically, two types of systems can be used: direct return and reverse return. In direct return, shown in Fig. 5.22, balancing valves are needed to ensure uniform flow through the modules. The balancing valves must be connected at the module outlet to provide the flow resistance necessary to ensure filling of all modules on a pump start-up. Whenever possible, modules must be connected in a reverse-return mode, as shown in Fig. 5.23. The reverse return ensures that the

FIGURE 5.22

Direct-return array piping.

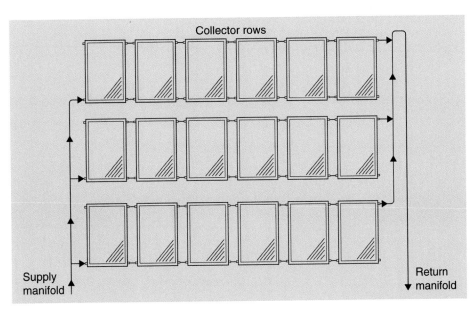

FIGURE 5.23

Reverse-return array piping.

array is self-balanced, as all collectors operate with the same pressure drop: i.e., the first collector in the supply manifold is the last in the return manifold, the second on the supply side is the second before the last in the return, and so on. With proper design, an array can drain, which is an essential requirement for drain-back and drain-down freeze protection. For this to be possible, piping to and from the collectors must be sloped properly. Typically, piping and collectors must slope to drain with an inclination of 20 mm per linear meter (ASHRAE, 2004).

External and internal manifold collectors have different mounting and plumbing considerations. A module with externally manifolded collectors can be mounted horizontally, as shown in Fig. 5.24a. In this case, the lower header must be pitched as shown. The slope of the upper header can be either horizontal or pitched toward the collectors, so it can drain through the collectors.

Arrays with internal manifolds are a little more difficult to design and install. For these collectors to drain, the entire bank must be tilted, as shown in Fig. 5.24b. Reverse return always implies an extra pipe run, which is more difficult to drain, so sometimes in this case it is more convenient to use direct return.

Solar collectors should be oriented and sloped properly to maximize their performance. A collector in the Northern Hemisphere should be located to face due south and a collector in the Southern Hemisphere should face due north. The collectors should face as south or as north, depending on the case, as possible, although a deviation of up to 10 degrees is acceptable. For this purpose, the use of a compass is highly recommended.

The optimum tilt angle for solar collectors depends on the longitude of the site. For maximum performance, the collector surface should be as perpendicular to the sun rays as possible. The optimum tilt can be calculated for each month of the year, but since a fixed inclination is used, an optimum slope throughout the year must be used. Some guidelines are given in Chapter 3, Section 3.1.1.

FIGURE 5.24

Mounting for drain-back collector modules. (a) External manifold. (b) Internal manifold.

Shading

When large collector arrays are mounted on flat roofs or level ground, multiple rows of collectors are usually installed. These multiple rows should be spaced so they do not shade each other at low sun angles. For this purpose, the method presented in Chapter 2, Section 2.2.3, could be used. Figure 5.25 shows the shading situation. It can be shown that the ratio of the row spacing to collector height, b/a, for a south facing array is given by:

$$\frac{b}{a} = \frac{\sin(\beta)}{\tan(\theta_s)} + \cos(\beta) \tag{5.47a}$$

A graphical solution of Eq. (5.47a) is shown in Fig. 5.26, from which the b/a ratio can be read directly if the shading angle, θ_s, and collector inclination, β, are known. It should be noted that this is the shading that occurs at the meridian (12.00 noon), where solar azimuth angle, z, is zero. At any other time the shading distance b_s, can be estimated using the solar azimuth angle, z, and solar altitude angle, α, from:

$$b_s = a \left[\frac{\sin(\beta)\cos(z)}{\tan(\alpha)} + \cos(\beta) \right] \tag{5.47b}$$

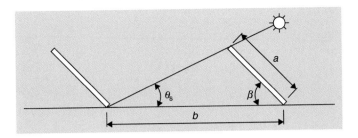

FIGURE 5.25

Row-to-row collector shading geometry.

FIGURE 5.26

Graphic solution of collector row shading.

Equations (5.47a) and (5.47b) neglect the thickness of the collector, which is small compared to dimensions a and b. If the piping, however, projects above the collector panels, it must be counted in the collector dimension a. The only unknown in Eq. (5.47a) is the shade angle, θ_s. To avoid shading completely, this can be found to be the minimum annual noon elevation, which occurs at noon on December 21. However, depending on the site latitude, this angle could produce very large row gaps (distance b), which might be not very practical. In this case, a compromise is usually made to allow some shading during winter months.

Thermal expansion

Another important parameter that needs to be considered is thermal expansion, which affects the modules of multi-collector array installations. Thermal expansion considerations deserve special attention in solar systems because of the temperature range within which the systems work. Thermal expansion (or contraction) of a module of collectors in parallel may be estimated by the following (ASHRAE, 2004):

$$\Delta = 0.0153n(t_{\max} - t_i) \tag{5.48}$$

where

Δ = expansion or contraction of the collector array (mm);
n = number of collectors in the array;
t_{max} = collector stagnation temperature (°C), see Chapter 4, Eq. (4.7); and
t_i = temperature of the collector when installed (°C).

Expansion considerations are very important, especially in the case of internal collector manifolds. These collectors should have a floating absorber plate, i.e., the absorber manifold should not be fastened to the collector casing, so it can move freely by a few millimeters within the case.

Galvanic corrosion

Galvanic corrosion is caused by the electrical contact between dissimilar metals in a fluid stream. It is therefore very important not to use different materials for the collector construction and piping manifolds. For example, if copper is used for the construction of the collector, the supply and return piping should also be made from copper. Where different metals must be used, dielectric unions between dissimilar metals must be used to prevent electrical contact. Of the possible metals used to construct collectors, aluminum is the most sensitive to galvanic corrosion, because of its position in the galvanic series. This series, shown in Table 5.5, indicates the relative activity of one metal against another. Metals closer to the anodic end of the series tend to corrode when placed in electrical contact with another metal that is closer to the cathodic end of the series in a solution that conducts electricity, such as water.

Array sizing

The size of a collector array depends on the cost, available roof or ground area, and percent of the thermal load required to be covered by the solar system. The first two parameters are straightforward and can easily be determined. The last, however, needs detailed calculations, which take into consideration the available radiation, performance characteristics of the chosen collectors, and other, less important parameters. For this purpose, methods and techniques that will be covered in other chapters of this book can be used, such as the f-chart method, utilizability method, and the use of computer simulation programs (see Chapter 11).

Table 5.5 Galvanic series of common metals and alloys.

Corroded end (anodic)
Magnesium
Zinc
Aluminum
Carbon steel
Brass
Tin
Copper
Bronze
Stainless steel
Protected end (cathodic)

It should be noted that, because the loads fluctuate on a seasonal basis, it is not cost-effective to have the solar system provide all the required energy, because if the array is sized to handle the months with the maximum load it will be oversized for the months with minimum load.

Heat exchangers

The function of a heat exchanger is to transfer heat from one fluid to another. In solar applications, usually one of the two fluids is the domestic water to be heated. In closed solar systems, it also isolates circuits operating at different pressures and separates fluids that should not be mixed. As was seen in the previous section, heat exchangers for solar applications may be placed either inside or outside the storage tank. The selection of a heat exchanger involves considerations of performance (with respect to heat exchange area), guaranteed fluid separation (double-wall construction), suitable heat exchanger material to avoid galvanic corrosion, physical size and configuration (which may be a serious problem in internal heat exchangers), pressure drop caused (influence energy consumption), and serviceability (providing access for cleaning and scale removal).

External heat exchangers should also be protected from freezing. The factors that should be considered when selecting an external heat exchanger for a system protected by a non-freezing fluid that is exposed to extreme cold are the possibility of freeze-up of the water side of the heat exchanger and the performance loss due to extraction of heat from storage to heat the low-temperature fluid.

The combination of a solar collector and a heat exchanger performs exactly like a collector alone with a reduced F_R. The useful energy gain from a solar collector is given by Eq. (4.3). The collector heat exchanger arrangement is shown in Fig. 5.27. Therefore, Eqs. (4.2) and (4.3), with the symbol convention shown in Fig. 5.27, can be written as:

$$Q_u = (\dot{m}c_p)_c (T_{co} - T_{ci})^+ \tag{5.49a}$$

$$Q_u = A_c F_R [G_t(\tau\alpha)_n - U_L(T_{ci} - T_a)]^+ \tag{5.49b}$$

The plus sign indicates that only positive values should be considered.

In addition to size and surface area, the configuration of the heat exchanger is important for achieving maximum performance. The heat exchanger performance is expressed in terms of its

FIGURE 5.27

Schematic diagram of a liquid system with an external heat exchanger between the solar collectors and storage tank.

effectiveness. By neglecting any piping losses, the collector energy gain transferred to the storage fluid across the heat exchanger is given by:

$$Q_{Hx} = Q_u = \varepsilon (\dot{m}c_p)_{min}(T_{co} - T_i)$$
(5.50)

where

$(\dot{m}c_p)_{min}$ = smaller of the fluid capacitance rates of the collector and tank sides of the heat exchanger (W/°C);
T_{co} = hot (collector loop) stream inlet temperature (°C); and
T_i = cold (storage) stream inlet temperature (°C).

The effectiveness, ε, is the ratio between the heat actually transferred and the maximum heat that could be transferred for a given flow and fluid inlet temperature conditions. The effectiveness is relatively insensitive to temperature, but it is a strong function of a heat exchanger design. A designer must decide what heat exchanger effectiveness is required for the specific application. The effectiveness for a counterflow heat exchanger is given by the following:
If $C \neq 1$,

$$\varepsilon = \frac{1 - e^{-NTU(1-C)}}{1 - C \times e^{-NTU(1-C)}}$$
(5.51)

If $C = 1$,

$$\varepsilon = \frac{NTU}{1 + NTU}$$
(5.52)

where NTU = number of transfer units given by:

$$NTU = \frac{UA}{(\dot{m}c_p)_{min}}$$
(5.53)

And the dimensionless capacitance rate, C, is given by:

$$C = \frac{(\dot{m}c_p)_{min}}{(\dot{m}c_p)_{max}}$$
(5.54)

For heat exchangers located in the collector loop, the minimum flow usually occurs on the collector side rather than the tank side.

Solving Eq. (5.49a) for T_{ci} and substituting into Eq. (5.49b) gives:

$$Q_u = \left[1 - \frac{A_c F_R U_L}{(\dot{m}c_p)_c}\right]^{-1} \{A_c F_R [G_t(\tau\alpha)_n - U_L(T_{co} - T_a)]\}$$
(5.55)

Solving Eq. (5.50) for T_{co} and substituting into Eq. (5.55) gives:

$$Q_u = A_c F_R'[G_t(\tau\alpha)_n - U_L(T_i - T_a)]$$
(5.56)

In Eq. (5.56), the modified collector heat removal factor takes into account the presence of the heat exchanger and is given by:

$$\frac{F_R'}{F_R}=\left\{1+\frac{A_cF_RU_L}{(\dot{m}c_p)_c}\left[\frac{(\dot{m}c_p)_c}{\varepsilon(\dot{m}c_p)_{min}}-1\right]\right\}^{-1} \tag{5.57}$$

In fact, the factor F_R'/F_R is the consequence, on the collector performance, that occurs because the heat exchanger causes the collector side of the system to operate at a higher temperature than a similar system without a heat exchanger. This can also be viewed as the increase of collector area required to have the same performance as a system without a heat exchanger.

EXAMPLE 5.4

A counterflow heat exchanger is located between a collector and a storage tank. The fluid in the collector side is a water–glycol mixture with $c_p = 3840$ J/kg °C and a flow rate of 1.35 kg/s, whereas the fluid in the tank side is water with a flow rate of 0.95 kg/s. If the UA of the heat exchanger is 5650 W/°C, the hot glycol enters the heat exchanger at 59 °C, and the water from the tank at 39 °C, estimate the heat exchange rate.

Solution

First, the capacitance rates for the collector and tank sides are required, given by:

$$C_c = (\dot{m}c_p)_c = 1.35 \times 3840 = 5184 \text{ W/°C}$$

$$C_s = (\dot{m}c_p)_s = 0.95 \times 4180 = 3971 \text{ W/°C}$$

From Eq. (5.54), the heat exchanger dimensionless capacitance rate is equal to:

$$C = \frac{(\dot{m}c_p)_{min}}{(\dot{m}c_p)_{max}} = \frac{3971}{5184} = 0.766$$

From Eq. (5.53),

$$NTU = \frac{UA}{(\dot{m}c_p)_{min}} = \frac{5650}{3971} = 1.423$$

From Eq. (5.51),

$$\varepsilon = \frac{1-e^{-NTU(1-C)}}{1-C \times e^{-NTU(1-C)}} = \frac{1-e^{-1.423(1-0.766)}}{1-0.766e^{-1.423(1-0.766)}} = 0.63$$

Finally, from Eq. (5.50),

$$Q_{Hx} = Q_u = \varepsilon(\dot{m}c_p)_{min}(T_{co}-T_i) = 0.63 \times 3971(59-39) = 50,035 \text{ W}$$

EXAMPLE 5.5

Redo the preceding example; if $F_RU_L = 5.71$ W/m² °C and collector area is 16 m², what is the ratio F_R'/F_R?

Solution

All data are available from the previous example. So, from Eq. (5.57),

$$\frac{F_R'}{F_R}=\left\{1+\frac{A_cF_RU_L}{(\dot{m}c_p)_c}\left[\frac{(\dot{m}c_p)_c}{\varepsilon(\dot{m}c_p)_{min}}-1\right]\right\}^{-1}=\left\{1+\frac{16 \times 5.71}{5184}\left[\frac{5184}{0.63(3971)}-1\right]\right\}^{-1}=0.98$$

This result indicates that 2% more collector area would be required for the system with a heat exchanger to deliver the same amount of solar energy as a similar system without a heat exchanger.

Pipe and duct losses

The collector performance equation can be modified to include the heat losses from the collector loop piping. The analysis is done by considering that there is a temperature drop ΔT_i from the outlet of the storage tank to the inlet of the collector (Beckman, 1978). Thus, the temperature inlet to the collector is $T_i - \Delta T_i$ and Eq. (3.60) becomes:

$$Q_u = A_c F_R [G_t(\tau\alpha) - U_L(T_i - \Delta T_i - T_a)] - Q_{pl} \tag{5.58}$$

where

Q_{pl} = rate of pipe losses [W].

The pipe losses can be obtained from the following integral for both the inlet and outlet portions:

$$Q_{pl} = U_p \int (T - T_a) dA \tag{5.59}$$

where

U_p = the loss coefficient from the pipe (W/m^2 K).

Equation (5.59) can be integrated but as pipes are usually well insulated the losses are very small and the integral can be approximated with good accuracy in terms of the collector inlet and outlet temperatures by:

$$Q_{pl} = U_p [A_{p,i}(T_i - T_a) + A_{p,o}(T_o - T_a)] \tag{5.60}$$

where

$A_{p,i}$ = inlet pipe area (m^2);
$A_{p,o}$ = outlet pipe area (m^2).

Estimating T_o from the right part of Eq. (3.31) and applying in Eq. (5.60), gives:

$$Q_{pl} = U_p(A_{p,i} + A_{p,o})(T_i - T_a) + \frac{U_p A_{p,o} Q_u}{(\dot{m} c_p)_c} \tag{5.61}$$

The temperature decrease, ΔT_i, due to the heat losses from storage tank to the inlet of the collector can be obtained with satisfactory accuracy from (Beckman, 1978):

$$\Delta T_i = \frac{U_p A_{p,i}(T_i - T_a)}{(\dot{m} c_p)_c} \tag{5.62}$$

Substituting Eqs. (5.61) and (5.62) into Eq. (5.58) and performing various manipulations, the rate of useful energy gain by considering the collector and its piping is given by:

$$Q_u = \frac{A_c F_R \left[G_t(\tau\alpha) - U_L \left(1 - \frac{U_p A_{p,i}}{(\dot{m} c_p)_c} + \frac{U_p(A_{p,i} + A_{p,o})}{A_c F_R U_L} \right)(T_i - T_a) \right]}{1 + \frac{U_p A_{p,o}}{(\dot{m} c_p)_c}} \tag{5.63}$$

Equation (5.63) can be written in the same way as Eq. (3.60) by using modified values of $(\tau\alpha)'$ and U_L' as:

$$Q_u = A_c F_R \left[G_t (\tau\alpha)' - U_L' (T_i - T_a) \right] \qquad (5.64a)$$

where

$$\frac{(\tau\alpha)'}{(\tau\alpha)} = \frac{1}{1 + \dfrac{U_p A_{p,o}}{(\dot{m}c_p)_c}} \qquad (5.64b)$$

and

$$\frac{U_L'}{U_L} = \frac{1 - \dfrac{U_p A_{p,i}}{(\dot{m}c_p)_c} + \dfrac{U_p (A_{p,i} + A_{p,o})}{A_c F_R U_L}}{1 + \dfrac{U_p A_{p,o}}{(\dot{m}c_p)_c}} \qquad (5.64c)$$

It should be noted that the same analysis can be applied to air collectors with respect to supply and return ducts and in this case the appropriate symbols are used for duct instead of pipe, i.e., U_d in place of U_p, $A_{d,i}$ in place of $A_{p,i}$ and $A_{d,o}$ in place of $A_{p,o}$.

Partially shaded collectors

In some cases, shading is either unavoidable or can be accepted in a few days during winter, especially for solar cooling applications, where the maximum requirement is during summer and shading effects are minimal due to the high sun altitude angle. Shading usually occurs at the bottom part of the collector in the second and subsequent rows of an array. The shaded part receives only diffuse radiation whereas the non-shaded part receives both direct and diffuse radiations. When shading occurs, the performance of the collectors can be estimated by considering an average value of radiation over the whole collector area or by considering the detailed analysis presented here. This situation is presented in Fig. 5.28 where as indicated the two parts of the collector receive radiation G_{t1} and G_{t2}. The temperature entering the first part of area A_1 is T_i and the temperature entering the second part of area A_2 is $T_{o,1}$, i.e., the hypothetical output of part 1. As the collector is "one unit" the values of F_R and U_L are the same for the two parts, whereas as the angle of incidence in the two parts are different (bottom

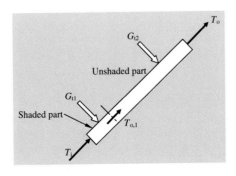

FIGURE 5.28

Partly shaded solar collector.

part receives only diffuse radiation from the sky dome it sees) the $(\tau\alpha)$ product for the two parts will be different. Applying Eq. (3.60) for the two parts we get:

$$\text{For the bottom part:} \quad Q_{u1} = A_1 F_R [G_{t1}(\tau\alpha)_1 - U_L(T_i - T_a)] \tag{5.65}$$

$$\text{For the top part:} \quad Q_{u2} = A_2 F_R [G_{t2}(\tau\alpha)_2 - U_L(T_{o,1} - T_a)] \tag{5.66}$$

The rate of useful energy from the bottom part is equal to:

$$Q_{u1} = \dot{m}c_p(T_{o,1} - T_i) \tag{5.67}$$

$$\text{From which:} \quad T_{o,1} = T_i + \frac{Q_{u1}}{\dot{m}c_p} \tag{5.68}$$

By applying Eq. (5.68) into Eq. (5.66) to eliminate $T_{o,1}$, and adding Eqs. (5.65) and (5.66), to get the total rate useful energy gain from the whole collector, we obtain:

$$Q_u = A_1 F_R(\tau\alpha)_1(1 - K_s)G_{t1} + A_2 F_R(\tau\alpha)_2 G_{t2} - F_R U_L[A_1(1 - K_s) + A_2](T_i - T_a) \tag{5.69}$$

where,

$$K_s = \frac{A_2 F_R U_L}{\dot{m}c_p} \tag{5.70}$$

In calculations and as the position of the sun in the sky changes continuously, the two areas A_1 and A_2 will be functions of time.

Over-temperature protection

Periods of high insolation and low load result in overheating of the solar energy system. Overheating can cause liquid expansion or excessive pressure, which may burst piping or storage tanks. Additionally, systems that use glycols are more problematic, since glycols break down and become corrosive at temperatures greater than 115 °C. Therefore, the system requires protection against this condition. The solar system can be protected from overheating by a number of methods, such as:

- Stopping circulation in the collection loop until the storage temperature decreases (in air systems);
- Discharging the overheated water from the system and replacing it with cold make-up water; and
- Using a heat exchanger coil for rejecting heat to the ambient air.

As will be seen in the next section, controllers are available that can sense over-temperature. The normal action taken by such a controller is to turn off the solar pump to stop heat collection. In a drain-back system, after the solar collectors are drained, they attain stagnation temperatures; therefore, the collectors used for these systems should be designed and tested to withstand over-temperature. In addition, drain-back panels should withstand the thermal shock of start-up when relatively cool water enters the solar collectors while they are at stagnation temperature.

In a closed loop antifreeze system that has a heat exchanger, if circulation stops, high stagnation temperatures occur. As indicated previously, these temperatures could break down the glycol heat transfer fluid. To prevent damage of equipment or injury due to excessive pressure, a pressure relief valve must be installed in the loop, as indicated in the various system diagrams presented earlier in this chapter, and a means of rejecting heat from the collector loop must be provided. The pressure relief valve should be set to relieve below the operating pressure of the component with the smallest operating pressure in the closed loop system.

It should be noted that, when the pressure relief valve is open, it discharges expensive antifreeze solution, which may damage roof membranes. Therefore, the discharge can be piped to containers to

FIGURE 5.29

Heat rejection by a solar heating system using a liquid-to-air heat exchanger.

save antifreeze, but the designer of such a system must pay special attention to safety issues because of the high pressures and temperatures involved.

Another point that should be considered is that, if a collector loop containing glycol stagnates, chemical decomposition raises the fusion point of the liquid and the fluid would not be able to protect the system from freezing.

The last option indicated previously is the use of a heat exchanger that dumps heat to the ambient air or other sink. In this system, fluid circulation continues, but this is diverted from storage through a liquid-to-air heat exchanger, as shown in Fig. 5.29. For this system, a sensor is used on the solar collector absorber plate that turns on the heat rejection equipment. When the sensor reaches the high-temperature set point, it turns on the pump and the fan. These continue to operate until the over-temperature controller senses that the temperature is within the safety limits and resets the system to its normal operating state.

5.5 Differential temperature controller

One of the most important components of an active solar energy system is the temperature controller because a faulty control is usually the cause of poor system performance. In general, control systems should be as simple as possible and should use reliable controllers, which are available nowadays. One of the critical parameters that need to be decided by the designer of the solar system is where to locate the collector, storage, over-temperature, and freezing-temperature sensors. The use of reliable, good-quality devices is required for many years of trouble-free operation. As was seen in the previous sections of this chapter, the control system should be capable of handling all possible system operating modes, including heat collection, heat rejection, power failure, freeze protection, and auxiliary heating.

The basis of solar energy system control is the differential temperature controller (DTC), shown as just DT in diagrams presented earlier in this chapter. This is simply a fixed temperature difference (ΔT) thermostat with hysteresis. The differential temperature controller is a comparing controller with at least two temperature sensors that control one or more devices. Typically, one of the sensors is located at the top side of the solar collector array and the second at the storage tank (Fig. 5.30). On unpressurized systems, other DTCs may control the extraction of heat from the storage tank. Most other controls used in solar energy systems are similar to those for building services systems.

The DTC monitors the temperature difference between the collectors and the storage tank. When the temperature of the solar collectors exceeds that of the tank by a predetermined amount (usually 4−11 °C), the DTC switches the circulating pump on. When the temperature of the solar collectors drops to 2−5 °C above the storage temperature, the DTC stops the pump. Instead of controlling the solar pump directly, the DTC can operate indirectly through a control relay to operate one or more pumps and possibly perform other control functions, such as the actuation of control valves.

The temperature differential set point of the differential temperature controller may be fixed or adjustable. If the controller set point is fixed, the controller selected should correspond to the requirements of the solar system. An adjustable differential set point makes the controller more flexible and allows it to be adjusted to the specific system or conditions of the solar system, i.e., different setting in summer and winter. The optimum differential on set point is difficult to calculate, because of the changing variables and conditions. Typically, the turn-on set point is 5−9 °C above the off set point. The optimum on set point is a balance between optimum energy collection and the avoidance of short starts and stops of the pump. The optimum turn-off temperature differential should be the minimum possible, which depends on whether there is a heat exchanger between the collectors and storage tank.

FIGURE 5.30

Basic collector control with a differential temperature controller.

Frequent starts and stops of the pump, also called *short cycling*, must be minimized because they can lead to premature pump failure. Short cycling depends on how quickly and how often the solar collector sensor temperature exceeds the on set point and drops below the off set point. This is influenced by the insolation intensity, the pump flow rate, the solar collector thermal mass, the response of the sensor, and the temperature of the fluid entering the collector. What happens in practice is that the water in the collector starts warming up as soon as the off-condition is reached and the flow stops. As the water heats up, it eventually reaches the on set point, at which point the pump is switched on and fluid circulates through the collector. Therefore, the hot fluid in the collector is pushed into the return manifold and replaced by relatively cool water from the supply manifold, which is warmed as it moves through the collector. A short cycle of the pump may mean that the hot water will never reach the storage tank, especially if the return pipe is long. The most common method of avoiding short cycling is the use of wide temperature difference between the on- and off set points. This, however, leads to the requirement of a lot of insolation to switch the pump on, which loses energy in the collector and may never reach the on set point in periods of low insolation. Therefore, the guidelines given in this section must be followed for deciding the correct setting.

If the system does not have a heat exchanger, a range of 1−4 °C is acceptable for the off set point. If the system incorporates a heat exchanger, a higher differential temperature set point is used to have an effective heat transfer, i.e., a higher-energy transfer between the two fluids. The minimum, or off, temperature differential is the point at which the cost for pumping the energy is equal to the cost of the energy being pumped, in which case the heat lost in the piping should also be considered. For systems with heat exchangers, the off set point is generally between 3 and 6 °C.

In closed-loop systems, a second temperature sensor may be used in the tank above the heat exchanger to switch the pump between low and high speed and hence provide some control of the return temperature to the tank heat exchanger. Furbo and Shah (1997) evaluated the use of a pump with a controller that varies the flow proportionally to the working fluid temperature and found that its effect on system performance is minor.

In the following analysis, the collector sensor is considered to be placed on the collector absorber plate. Using the concept of absorbed radiation, when the collector pump is off, the useful output from the collector is 0 and the absorber plate is at an equilibrium temperature given by:

$$\left[S - U_L \left(T_p - T_a\right)\right] = 0 \tag{5.71}$$

Therefore, the value of S when the plate temperature, T_p, is equal to $T_i + \Delta T_{ON}$ is:

$$S_{ON} = U_L(T_i + \Delta T_{ON} - T_a) \tag{5.72}$$

Using Eq. (3.60) with the absorbed solar radiation, when the pump is on, the useful gain from the collector is:

$$Q_u = A_c F_R \left[S_{ON} - U_L(T_i - T_a)\right] \tag{5.73}$$

If we substitute Eq. (5.72) into Eq. (5.73),

$$Q_u = A_c F_R U_L \Delta T_{ON} \tag{5.74}$$

However, the useful energy when the pump is on is also given by:

$$Q_u = \left(\dot{m} c_p\right)(T_o - T_i) \tag{5.75}$$

In fact, the temperature difference $(T_o - T_i)$, by ignoring heat losses from the pipes, is the difference seen by the DTC once the flow is turned on. Consequently, by combining Eqs. (5.74) and (5.75), the off set point must satisfy the following inequality because otherwise the system will become unstable:

$$\Delta T_{OFF} \leq \frac{A_c F_R U_L}{\dot{m} c_p} \Delta T_{ON} \tag{5.76a}$$

If the system has a collector heat exchanger then Eq. (5.50) is used instead of Eq. (5.75) and the inequality becomes:

$$\Delta T_{OFF} \leq \frac{A_c F_R U_L}{\varepsilon (\dot{m} c_p)_{min}} \Delta T_{ON} \tag{5.76b}$$

5.5.1 Placement of sensors

Proper placement of the collector temperature sensor is important for a good system operation. The sensor must have a good thermal contact with the collector plate or piping. Collector sensors may be located on the collector plate, on a pipe near the collector, or in the collector outlet pipe. The best of all is on the collector plate, but this is not the easiest, because dismantling and modification on one collector of the array is required, which would need to be done on site. The easiest and perhaps the best point for the location of the sensor is on the pipe leaving the collector. Usually a T piece is used and the sensor is placed in a deep well with a few drops of oil, which ensures good contact, as shown in Fig. 5.31a, or on the side of the T piece, as shown in Fig. 5.31b.

The storage tank sensor should be located near the bottom of the storage tank, at about one third of its height. If the system uses an internal heat exchanger, the sensor is located above the heat exchanger. Ideally, this sensor should identify if there is still water in the tank, which can be heated by solar energy. Therefore, the location indicated is considered a good compromise because a lower location would give a false reading even with the slightest demand, which will be replaced by make-up (cold) water, whereas a higher location would leave a lot of water at a low temperature, even if solar energy is available.

A freeze protection sensor, if used, should be located in such a position so as to detect the coldest liquid temperature. Two suitable locations are the back of the absorber plate and the entry pipe to the

FIGURE 5.31

Placement of collector sensor. (a) Deep well. (b) Side of T piece.

collector from the supply manifold. For the reasons indicated previously, the latter is preferred. The over-temperature sensor can be located either at the top part of the storage tank or on the collector exit pipe. For the latter, the sensor is located in a similar location and manner as the collector temperature sensor.

5.6 Hot water demand

The most important parameter that needs to be considered in the design of a water-heating system is the hot water demand over a certain period of time (hourly, daily, or monthly). The energy demand, D, required for the generation of sanitary hot water can be obtained if the volumetric consumption, V, is known for the required time period. Also required are the temperatures of the cold water supplied by public mains, T_m, and the water distribution, T_w. Then,

$$D = V\rho c_p(T_w - T_m) \tag{5.77}$$

If the two temperatures in Eq. (5.77) are known for a particular application, the only parameter on which the energy demand depends is the hot water volumetric consumption. This can be estimated according to the period of time investigated. For example, for the monthly water demand, the following equation can be used:

$$V = N_{days}N_{persons}V_{person} \tag{5.78}$$

where

N_{days} = number of days in a month;
$N_{persons}$ = number of persons served by the water-heating system; and
V_{person} = Volume of hot water required per person.

The volumetric consumption, V, varies considerably from person to person and from day to day. It has to do with the habits of the users, the weather conditions of a locality, and various socioeconomic conditions. It can be estimated by considering the hot water use for various operations. Typical operations and consumption for residential usage are given in Table 5.6. More details and other applications, such as water consumption in hotels, schools, and so forth, can be found in the ASHRAE *Handbook of Applications* (ASHRAE, 2007).

In addition to the quantities shown in Table 5.6, hot water is consumed in automatic dish washing and clothes washing, but these quantities of hot water are produced by the washer with electricity as part of the washing process.

Table 5.6 Typical residential usage of hot water per task.

Use	Flow (L)
Food preparation	10–20
Manual dish washing	12–18
Shower	10–20
Bath	50–70
Face and hand washing	5–15

Table 5.7 Hot water daily demand for a family of four persons in liters per person.			
Guideline	Low	Medium	High
Normal consumption	26	40	54
Maximum consumption	66	85	104

By using the data shown in Table 5.6 for a four-person family and normal daily tasks consisting of two food preparations, two manual dish washings, one shower for each person, and two face or hand washings per person per day, the low, medium, and high demand values in liters per person shown in Table 5.7 can be obtained. The maximum consumption case is where the shower for each person is replaced with a bath for each person per day.

EXAMPLE 5.6

Estimate the hot water energy demand for a family of four, with medium normal consumption, cold water mains supply of 18 °C, and water distribution temperature of 45 °C.

Solution

According to Table 5.7, the consumption per day per person is 40 L. Therefore, the daily demand, V, is 160 L/day or 0.16 m^3/day. From Eq. (5.65),

$$D = V \rho c_p (T_w - T_m) = 0.16 \times 1000 \times 4.18(45 - 18) = 18057.6 \ \text{kJ} / \text{day} = 18.06 \ \text{MJ} / \text{day}$$

In hourly simulations, the hourly distribution of hot water demand is required. Although the hot water demand is subject to a high degree of variation from day to day and consumer to consumer, it is impractical to use anything but a repetitive load profile. This is not quite correct during the summer period, when the consumption pattern is somewhat higher. However, during this period, the temperature requirement for hot water is not as high as during winter. Consequently, the total thermal energy requirement is reasonably constant throughout the year. The demand profile usually used in hour simulations is the Rand profile, illustrated in Fig. 5.32. This assumes a daily hot water consumption of 120 L at 50 °C for a family of four (30 L/person).

FIGURE 5.32

Hot-water daily consumption profile.

5.7 Solar water heater performance evaluation

Many test procedures have been proposed by various organizations to determine the thermal performance of solar water heaters. Testing of the complete system may serve a number of purposes. The main one is the prediction of the system's long-term thermal performance. System testing may also be used as a diagnostic tool to identify failure and causes of failure in system performance. Other purposes include the determination of the change in performance as a result of operation under different weather conditions or with a different load profile.

The International Organization for Standardization (ISO) publishes a series of standards, ranging from simple measurement and data correlation methods to complex parameter identification ones. ISO 9459 was developed by the Technical Committee, ISO/TC 180—Solar Energy, to help facilitate the international comparison of solar domestic water-heating systems. Because a generalized performance model, which is applicable to all systems, has not yet been developed, it has not been possible to obtain an international consensus for one test method and one standard set of test conditions. Therefore, each method can be applied on its own.

A total of five parts comprise ISO 9459 on solar domestic water-heater performance testing as described below with their current status:

ISO 9459-1:1993. *Solar heating, Domestic water-heating systems. Part 1. Performance rating procedure using indoor test methods.* → Active.

ISO 9459-2:1995. *Solar heating, Domestic water-heating systems. Part 2. Outdoor test methods for system performance characterization and yearly performance prediction of solar-only systems.* → Active (referenced in EN 12976).

ISO 9459-3:1997. *Solar heating, Domestic water-heating systems. Part 3. Performance test for solar plus supplementary systems.* → Withdrawn in 2005.

ISO 9459-4:2013. *Solar heating, Domestic water-heating systems. Part 4. System performance characterization by means of component tests and computer simulation.* → Active.

ISO 9459-5:2007. *Solar heating, Domestic water-heating systems. Part 5. System performance characterization by means of whole-system tests and computer simulation.* → Active (referenced in EN 12976).

Therefore, ISO 9459 has four active parts for the time being.

Rating test

- ISO 9459-1 describes test procedures for characterizing the performance of solar domestic water-heating systems operated without auxiliary boosting and for predicting annual performance in any given climatic and operating conditions. It is suitable for testing all types of systems, including forced circulation, thermosiphon, and Freon-charged collector systems. The results allow systems to be compared under identical solar, ambient, and load conditions.
- The test methods in this standard define procedures for the indoor testing of solar water heaters with a solar simulator. The characteristics of the solar simulator are defined in ISO 9845-1:1992 (see Chapter 4, Section 4.1.3). The entire test sequence usually takes 3—5 days and the result is the daily solar contribution for one set of conditions. An indoor test procedure in which the solar simulator is replaced by a controlled heat source, used to simulate the solar energy gain, is also described. This test has not been widely adopted.

Black box correlation procedures

- ISO 9459-2 is applicable to solar-only systems and solar preheat systems. The performance test for solar-only systems is a "black box" procedure, which produces a family of "input–output" characteristics for a system. The test results may be used directly with daily mean values of local solar irradiation, ambient air temperature, and cold-water temperature data to predict annual system performance.
- The results of tests performed in accordance with ISO 9459-2 permit performance predictions for a range of system loads and operating conditions, but only for an evening draw-off.
- ISO 9459-2 is one of the methods used in EN 12976 for performance testing.

Testing and computer simulation

- ISO 9459-4 standard suggests a procedure for characterizing annual system performance and uses measured component characteristics in the computer simulation program TRNSYS (described in Chapter 11, Section 11.5.1). The procedures for characterizing the performance of system components other than collectors are also presented in this part of ISO 9459. Procedures specified in ISO 9806 (see Chapter 4) are used to determine collector performance, whereas other tests are specified for characterizing the storage tank, heat exchangers (if used), and control system.
- ISO 9459-5 presents a procedure for the dynamic testing of complete systems to determine system parameters for use in a computer model. This model may be used with hourly values of local solar irradiation, ambient air temperature, and cold-water temperature data to predict an annual system performance.
- ISO 9459-5 specifies a method for outdoor laboratory testing of solar domestic hot water systems. The method may also be applied for in-situ tests and indoor tests by specifying appropriate draw-off profiles and irradiance profiles for indoor measurements. The system performance is characterized by means of whole-system tests using a "black box" approach, i.e., no measurements on the system components or inside the system are necessary. Detailed instructions are given on the measurement procedure, processing and analysis of the measurement data, and presentation of the test report.
- ISO 9459-5 is one of the methods used in EN 12976 for performance testing.
- The results of tests performed in accordance with ISO 9459-4 or ISO 9459-5 are directly comparable. These procedures permit performance predictions for a range of system loads and operating conditions. The disadvantage of these procedures is that a detailed computer simulation model of the system is required.

The procedures defined in ISO 9459-2, ISO 9459-4, and ISO 9459-5 for predicting annual performance allow the output of a system to be determined for a range of climatic conditions, whereas the results of tests performed in accordance with ISO 9459-1 provide a rating for a standard day.

One of the most used system test methods is standard ISO 9459-2. This is because it requires the least investment in equipment and operator skills. In this standard, the system is preconditioned at the start of each test day and charged to the required temperature, T_c, then it is left to operate with no loads applied and the only measurements required are those of solar radiation and ambient temperature. Energy monitoring is required at the end of the day, during the single draw-off, and this can be achieved with either a simple manual temperature and volume measurements or a data acquisition system. The daily energy gain is determined for a range of clear and cloudy days with irradiation between 8 and 25 MJ/m^2 day, with approximately the same $(T_a - T_c)$ value for each day. The correlation parameter $(T_a - T_c)$ is varied, however, by testing for a range of initial tank temperatures, T_c, for each day. The useful delivered energy at the end of the day, Q_u, is correlated to the test results by:

$$Q_u = \alpha_1 H + \alpha_2(T_a - T_c) + \alpha_3 \tag{5.79}$$

where

α_1, α_2, and α_3 = correlation coefficients.

The effects of thermal stratification and mixing in the storage tank are evaluated by a load calculation procedure using the temperature profiles measured during draw-off at the end of the day.

The long-term performance of the system is determined by a calculation procedure that accounts for the climatic conditions, energy carry-over from day to day, and the load volume. Additionally, a 1 h time step procedure is specified for the determination of the nighttime heat loss and the energy carry-over from day to day.

5.8 Simple system models

The equations presented in this chapter can be combined and used to model the whole system. The model includes all physical components of the system, such as the collector, storage tank, heat exchanger, loads, and heat losses from the system components, such as pipes and storage tank. Detailed models result in a set of coupled algebraic and differential equations, with time as the independent variable. The inputs to these equations are meteorological data and load variations (e.g., water draw-off profile). The time step for such a model is usually 1 h, and for annual calculations, a computer is required. More details on these models are given in Chapter 11. In this section, we deal with only simple models that can be solved by hand calculations or the help of a spreadsheet.

According to the system configuration, the equations presented in Section 4.1.1, for operating at a flow rate different from the one used in collector performance testing; Section 4.1.2 for collectors in series and previous sections for considering pipe loses in performance equations, for partially shaded collectors and the use of a heat exchanger in the collector circuit; and Section 5.5 for the differential temperature controller, need to be considered in a logical way. All these modifications were obtained from the basic collector performance model given by Eq. (3.60). Usually we start with the collector performance parameters and apply any corrections for different flow rate and/or connection of collectors in series, if needed, add the pipe losses, then the heat exchanger, if present. Generally, we start from the collector and move toward the storage tank.

A simple model considers a fully mixed or unstratified storage tank supplying hot water at a fixed flow rate and a make-up water constant temperature, T_{mu}. Therefore, by ignoring pipe losses and considering that the storage tank is at a uniform temperature, T_s, Eq. (5.31) for the storage tank can be combined with Eq. (4.3) for the collector and Eq. (5.32) for the storage tank losses, to give:

$$(Mc_p)_s \frac{dT_s}{dt} = A_c F_R[S - U_L(T_s - T_a)]^+ - \varepsilon_L (\dot{m}_L c_p)_{min}(T_s - T_{mu}) - (UA)_s(T_s - T_a) \tag{5.80}$$

The middle term of the right-hand side in this equation is the energy delivered to the load through a load heat exchanger, which has an effectiveness ε_L. If no load heat exchanger is used, the term

$\varepsilon_L (\dot{m}_L c_p)_{min}$ is replaced by $\dot{m}_L c_p$, where in both cases $\dot{m}_L$ is the load flow rate. This is, in fact, the same as Eq. (5.31) but with the various terms inserted in the equation.

To solve this equation, the collector parameters, storage tank size and loss coefficient, the effectiveness and mass flow rate of the heat exchanger, and the meteorological parameters are required. Once these are specified, the storage tank temperature can be estimated as a function of time. Additionally, the individual parameters, such as the useful energy gain from the collector and the losses from the storage tank, can be determined for a period of time by integrating the appropriate quantities. To solve Eq. (5.80), the simple Euler integration method can be used to express the temperature derivative dT_s/dt as $(T_{s-n}-T_s)/\Delta t$. This is similar to writing the equation in finite difference form, as indicated in Section 5.3.3. Therefore, Eq. (5.80) can be expressed as a change in storage tank temperature for the time period required as:

$$T_{s-n} = T_s + \frac{\Delta t}{(Mc_p)_s} \left[A_c F_R [S - U_L (T_s - T_a)]^+ - \varepsilon_L (\dot{m}_L c_p)_{min} (T_s - T_{mu}) - (UA)_s (T_s - T_a) \right] \quad (5.81)$$

The only caution required in using this integration scheme is to choose a small time step to ensure stability. Because meteorological data are available in hour increments, a time step of 1 h is also used in solving Eq. (5.81) if stability is kept. A good verification of the calculations is to check the energy balance of the tank by estimating the change of internal energy of the water, which must be equal to the summation of the useful energy supplied by the collector minus the summation of the energy to load and energy lost. In equation form,

$$Mc_p (T_{s,i} - T_{s,f}) = \sum Q_u - \sum Q_l - \sum Q_{tl} \quad (5.82)$$

where

$T_{s,i}$ = initial storage tank temperature (°C);
$T_{s,f}$ = final storage tank temperature (°C).

Problems for this kind of analysis are similar to Examples 5.2 and 5.3. In those examples, the load was considered to be known, whereas here it is calculated by the middle term of Eq. (5.81).

EXAMPLE 5.7

Estimate the energy balance in Example 5.2.

Solution

By summing up the various quantities in Table 5.3 of Example 5.2, we get:

$$\sum Q_u = 276 \text{ MJ}, \sum Q_l = 181 \text{ MJ} \text{ and } \sum Q_{tl} = 8.5 \text{ MJ}$$

Then, applying Eq. (5.82), we get:

$$500 \times 4.18(86.4 - 45) \times 10^{-3} = 276 - 181 - 8.5,$$

which gives:

$$86.53 \approx 86.5$$

which indicates that the calculations were correct.

5.9 Practical considerations

Installation of large collector arrays presents specific piping problems. This section examines issues related to the installation of pipes, supports, and insulation; pumps; valves; and instrumentation. Generally, the plumbing involved in solar energy systems is conventional, except in cases where a toxic or non-potable heat transfer fluid is circulated in the collector loop. A general guide is that the less complex the system is, the more trouble free its operation will be.

5.9.1 Pipes, supports, and insulation

The material of a solar energy system piping may be copper, galvanized steel, stainless steel, or plastic. All pipes are suitable for normal solar system operation except plastic piping, which is used only for low-temperature systems, such as swimming pool heating. Another problem related to plastic piping is its high coefficient of expansion, which is 3—10 times as high as that for copper pipes and causes deformation at high temperatures. Piping that carries potable water may be copper, galvanized steel, or stainless steel. Untreated steel pipes should not be used because they corrode rapidly.

System piping should be compatible with the collector piping material to avoid galvanic corrosion; for example, if the collector piping is copper the system piping should also be copper. If dissimilar metals must be joined, dielectric couplings must be used.

Pipes can be joined with a number of different methods, such as threaded, flared compression, hard soldered, and brazed. The method adopted also depends on the type of piping used; for example, a threaded connection is not suitable for copper piping but is the preferred method for steel pipes.

Pipes are usually installed on roofs; therefore, the piping layout should be designed in such a way as to allow expansion and contraction, have the minimum roof penetration, and keep the roof integrity and weather ability. A way to estimate the amount of expansion is indicated earlier in this chapter; the supports selected for the installation, however, have to allow for the free movement of the pipes to avoid deformation. An easy way to account for the expansion—contraction problem is to penetrate the roof at about the center of the solar array and allow for two equal lengths of loops on each side of the penetration point. If the pipes must be supported on the roof, this must be done in a way so as not to penetrate the weatherproof roof membrane. For this purpose, concrete pads can be constructed on which the pipe supports can be fitted.

Another important issue related to the installation of collector array piping is the pipe insulation. Insulation must be selected to have adequate R value to minimize heat losses. Other issues to be considered are insulation availability and workability, and because the insulation is exposed to the weather, it must have a high UV durability and low permeability by water. The last factors are usually obtained by installing a suitable protection of the insulation, such as aluminum waterproofing. Areas that require special attention in applying the waterproofing are joints between collectors and piping, pipe tees and elbows, and special places where valves and sensors protrude through the waterproofing. The types of insulation that can be used are glass fiber, rigid foam, and flexible foam.

5.9.2 Pumps

For solar energy systems, centrifugal pumps and circulators are used. Circulators are suitable for small domestic-size systems. Construction materials for solar system pumps depend on the particular application and fluid used in the circuit. Potable water and drain-down systems require pumps made from bronze, at least for the parts of the pump in contact with the water. Pumps should also be selected to be able to work at the operating temperature of the system.

5.9.3 Valves

Special attention must be paid to the proper selection and location of valves in solar energy systems. Careful selection and installation of a sufficient number of valves are required so that the system performs satisfactorily and is accessible for maintenance procedures. Using too many valves, however, should be avoided to reduce cost and pressure drop. The various types of valves required in these systems are isolation valves, balancing valves, relief valves, check valves, pressure-reducing valves, air vents, and drain valves. These are described briefly here:

- *Isolation valves.* Isolation or shutoff valves are usually gate of quarter-turn ball valves. These should be installed in such a way so as to permit certain components to be serviced without having to drain and refill the whole system. Special attention is required so as not to install isolation valves in a way that would isolate collectors from pressure relief valves.
- *Balancing valves.* Balancing or flow-regulating valves are used in multi-row installations to balance the flow in the various rows and ensure that all rows received the required quantity of flow. As already seen in this chapter, the use of these valves is imperative in direct return systems (see Section 5.4.2). The adjustment of these valves is done during commissioning of the system. For this purpose, flow rate or pressure may need to be measured for each row, so the system must have provisions for these measurements. After the balancing valves are adjusted, their setting must be locked to avoid accidental modification. The easiest way to do this is to remove the valve handle.
- *Relief valves.* Pressure safety or relief valves are designed to allow escape of water or heat transfer fluid from the system when the maximum working pressure of the system is reached. In this way, the system is protected from high pressure. This valve incorporates a spring, which keeps the valve closed. When the pressure of the circuit fluid exceeds the spring stiffness, the valve is opened (valve stem is lifted from its base) and allows a small quantity of the circulating fluid to escape so as to relieve the pressure. Two types of relief valves are available: the adjustable type and the preset type. The preset type comes in a number of relief pressure settings, whereas the adjustable type needs pressure testing to adjust the valve spring stiffness to the required relief pressure. The relief valve may be installed anywhere along the closed loop system. Attention should be paid to the fact that the discharge of such a valve will be very hot or even in a steam state, so the outlet should be piped to a drain or container. The latter is preferred because it gives an indication to the service personnel that the valve opened and they should look for possible causes or problems. The use of a tank is also preferred in systems with antifreeze, because the fluid is collected in the tank.
- *Check valves.* Check valves are designed to allow flow to pass in only one direction. In doing so, flow reversal is avoided. This valve comes in a number of variations, such as the swing valve and the spring-loaded valve. Swing valves require very little pressure difference to operate but are not suitable for vertical piping, whereas spring-loaded valves need more pressure difference to operate but can be installed anywhere in the circuit.

- *Pressure-reducing valves.* Pressure-reducing valves are used to reduce the pressure of make-up city water to protect the system from overpressure. These valves should be installed together with a check valve to avoid feeding the city circuit with water or antifreeze solution from the solar energy system.
- *Automatic air vents.* Automatic air vents are special valves used to allow air to escape from the system during fill-up. They are also used to eliminate air in a closed circuit system. This valve should be installed at the highest point of the collector circuit. Automatic air vent valves are of the float type, where water or the circulating fluid keeps the valve closed by forcing a bronze empty ball against the valve opening due to buoyancy. When air passes through the valve, the bronze empty ball is lowered because of its weight and allows the air to escape.
- *Drain valves.* Drain valves are used in drain-down systems. These are electro-mechanical devices, also called *solenoid valves*, that keep the valve closed as long as power is connected to the valve (normally open valves). When the valve is de-energized, a compression spring opens the valve and allows the drain of the system.

5.9.4 Instrumentation

Instrumentation used in solar energy systems varies from very simple temperature and pressure indicators, energy meters, and visual monitors to data collection and storage systems. It is generally preferable to have some kind of data collection to be able to monitor the actual energy collected from the solar energy system.

Visual monitors are used to provide instantaneous readings of various system parameters, such as temperatures and pressures at various locations in the system. Sometimes, these are equipped with a data storage. Energy meters monitor and report the time-integrated quantity of energy passing through a pair of pipes. This is done by measuring the flow rate and the temperature difference in the two pipes. Most energy meters must be read manually, but some provide an output to a recorder.

Automatic recording of data from a number of sensors in a system is the most versatile but also the most expensive system. This requires an electrical connection from the various sensors to a central recorder. Some recorders also allow processing of the data. More details on these systems are given in Chapter 4, Section 4.11. Nowadays, systems are available that collect and display results online on the Internet. These are very helpful in monitoring the state of the system, although they add to the total system cost. In countries where schemes such as the guaranteed solar results operate, where the solar energy system provider guarantees that the system will provide a certain amount of energy for a number of years, however, this is a must.

Exercises

5.1. Repeat Example 5.1 for an indoor swimming pool.

5.2. A 100 m² light-colored swimming pool is located in a well-sheltered site, where the measured wind speed at 10 m height is 4 m/s. The water temperature is 23 °C, the ambient air temperature is 15 °C, and relative humidity is 55%. There are no swimmers in the pool, the temperature of the make-up water is 20.2 °C, and the solar irradiation on a horizontal surface for the day is

19.3 MJ/m^2 day. If this pool is to be heated by solar energy, how many square meters of collectors would be required if their efficiency is 45%?

5.3. A water storage tank needs to be designed to hold enough energy to meet a load of 11 kW for 2 days. If the maximum storage temperature is 95 °C and the supply water must have at least a temperature of 60 °C, what size of tank is required?

5.4. A fully mixed water storage tank contains 1000 kg of water, has an UA product equal to 10 W/°C, and is located in a room that is at a constant 20 °C temperature. The tank is examined in a 10 h period starting from 7 a.m., where the hourly values of Q_u are equal to 0, 8, 20, 31, 41, 54, 64, 53, 39, 29 MJ. The load is constant and equal to 13 MJ in the first 3 h, 17 MJ in the next 3 h, 25 MJ in the next 2 h, and 20 MJ the rest of time. Find the final storage tank temperature if the initial temperature is 43 °C.

5.5. A storage tank needs to be designed to meet a load of 1.2 GJ. The temperature of the storage tank can vary by 30 °C. Determine the storage material volume if the material is water and concrete.

5.6. Repeat Example 5.3 by considering a storage tank of 150 kg and compare the results.

5.7. Repeat Example 5.3 for September 15, considering that the weather conditions are the same.

5.8. A solar water-heating system with a fully mixed tank has a capacity of 300 L and a UA value of 5.6 W/°C. The ambient temperature at the place where the tank is located is 21 °C. The solar system has a total area of 6 m^2, $F_R(\tau\alpha) = 0.82$, and $F_R U_L = 6.1$ W/m^2 °C. At the hour of estimation, the ambient temperature is 13.5 °C and the radiation on the collector plane is 16.9 MJ/m^2. If the temperature of the water in the tank is 41 °C, estimate the new tank temperature at the end of the hour.

5.9. A liquid-based solar heating system uses a heat exchanger to separate the collector loop from the storage loop. The collector overall heat loss coefficient is 6.3 W/m^2 °C, the heat removal factor is 0.91, and the collector area is 25 m^2. The heat capacity rate of the collector loop is 3150 W/°C and, for the storage loop, is 4950 W/°C. Estimate the thermal performance penalty that occurs because of the use of the heat exchanger if its effectiveness is 0.65 and 0.95.

5.10. A liquid-based solar heating system uses a heat exchanger to separate the collector loop from the storage loop. The flow rate of the water is 0.65 kg/s and that of the antifreeze is 0.85 kg/s. The heat capacity of the antifreeze solution is 3150 J/kg °C and the UA value of the heat exchanger is 5500 W/°C. The collector has an area of 60 m^2 and an $F_R U_L = 3.25$ W/m^2 °C. Estimate the factor F_R'/F_R.

5.11. Compare the performance of an air collector for the case of insulated and uninsulated ducts. The collector has an area of 30 m^2, $F_R U_L = 6.3$ W/m^2 °C and $F_R(\tau\alpha) = 0.7$. The product of mass flow rate-specific heat of air flowing through the collector is 450 W/°C and the areas of inlet and outlet ducts is 8 m^2. The loss coefficient U_d of the insulated duct is 0.95 W/m^2 °C whereas for the uninsulated duct it is 9 W/m^2 °C. In both cases the total radiation falling on the collector aperture is 650 W/m^2, inlet air temperature is 45 °C and the ambient temperature is 15 °C.

5.12. A collector is installed in an application that is partially shaded from beam radiation. The shaded part, which is 25% of the total collector area, receives 250 W/m^2 incident radiation whereas the rest of the collector is exposed to sunshine and receives 950 W/m². The flow rate of water flowing through the collector is 0.005 kg/m² s and the characteristics of the collector at this flow rate are $F_R U_L = 6.5$ W/m² °C and $F_R = 0.94$. The ambient temperature is 10 °C and the collector inlet temperature is 45 °C whereas the $(\tau\alpha)_{av}$ is equal to 0.75 for the shaded part and 0.91 for the unshaded part. Calculate the useful energy gain when the flow is from the low- to high-intensity zone and opposite.

5.13. A family of seven people live in a house. Two of them take a bath every day and the rest take showers. Estimate the daily hot water consumption of the family by considering two meal preparations, two manual dish washings, and two face- or hand-washings for each person.

5.14. Determine the solar collector area required to supply all the hot water needs of a residence of a family of six people in June, where the total insolation is 25,700 kJ/m² day, assuming a 45% collector efficiency. The demand hot water temperature is 60 °C, the cold water make-up temperature is 16 °C, and the consumption per person is 35 L/day. Estimate also the percentage of coverage (also called *solar fraction*) for heating the water in January, where the total insolation is 10,550 kJ/m² day.

5.15. A commercial building water-heating system uses a recirculation loop, which circulates hot water, to have hot water quickly available. If the temperature of hot water is 45 °C, the pipe UA is 32.5 W/°C, the tank UA is 15.2 W/°C, the make-up water temperature is 17 °C, and ambient temperature is 20 °C, estimate the weekly energy required to heat the water with continuous recirculation. The demand is 550 L/day for weekdays (Monday through Friday) and 150 L/day at weekends.

5.16. A solar collector system has a total area of 10 m², $F_R = 0.82$, and $U_L = 7.8$ W/m² °C. The collector is connected to a water storage tank of 500 L, which is initially at 40 °C. The storage tank loss coefficient-area product is 1.75 W/°C and the tank is located in a room at 22 °C. Assuming a load flow of 20 kg/h and a make-up water of 18 °C, calculate the performance of this system for the period shown in the following table and check the energy balance of the tank.

Hour	S (MJ/m²)	T_a (°C)
7–8	0	12.1
8–9	0.35	13.2
9–10	0.65	14.1
10–11	2.51	13.2
11–12	3.22	14.6
12–13	3.56	15.7
13–14	3.12	13.9
14–15	2.61	12.1
15–16	1.53	11.2
16–17	0.66	10.1
17–18	0	9.2

References

ASHRAE, 2004. Handbook of Systems and Equipment. ASHRAE, Atlanta, GA. Note: The material quoted in this reference may also appear in subsequent ASHRAE handbooks, but here the year of the original version of the handbook used is shown.

ASHRAE, 2007. Handbook of HVAC Applications. ASHRAE, Atlanta, GA. Note: The material quoted in this reference may also appear in subsequent ASHRAE handbooks, but here the year of the original version of the handbook used is shown.

Beckman, W.A., 1978. Duct and pipe losses in solar energy systems. Sol. Energy 21 (6), 531–532.

Charters, W.W.S., de Forest, L., Dixon, C.W.S., Taylor, L.E., 1980. Design and performance of some solar booster heat pumps. In: ANZ Solar Energy Society Annual Conference, Melbourne, Australia.

Close, D.J., 1962. The performance of solar water heaters with natural circulation. Sol. Energy 6 (1), 33–40.

Duff, W.S., 1996. Advanced Solar Domestic Hot Water Systems. International Energy Agency. Task 14. Final Report.

Eames, P.C., Norton, B., 1995. Thermal and optical consequences of the introduction of baffles into compound parabolic concentrating solar collector cavities. Sol. Energy 55 (2), 129–150.

Furbo, S., Shah, L.J., 1997. Smart Solar Tanks—Heat Storage of the Future. ISES Solar World Congress, Taejon, South Korea.

Gupta, G.L., Garg, H.P., 1968. System design in solar water heaters with natural circulation. Sol. Energy 12 (2), 163–182.

Hsieh, J.S., 1986. Solar Energy Engineering. Prentice-Hall, Englewood Cliffs, NJ.

ISO/TR 12596:1995 (E), 1995. Solar Heating—Swimming Pool Heating Systems—Dimensions, Design and Installation Guidelines.

Kalogirou, S.A., 1997. Design, construction, performance evaluation, and economic analysis of an integrated collector storage system. Renew. Energy 12 (2), 179–192.

Kalogirou, S.A., 1998. Performance Enhancement of an Integrated Collector Storage Hot Water System. Proceedings of the World Renewable Energy Congress V, Florence, Italy, pp. 652–655.

McIntire, W.R., 1979. Truncation of nonimaging cusp concentrators. Sol. Energy 23 (4), 351–355.

Michaelides, I.M., Kalogirou, S.A., Chrysis, I., Roditis, G., Hadjigianni, A., Kabezides, H.D., Petrakis, M., Lykoudis, A.D., Adamopoulos, P., 1999. Comparison of the performance and cost effectiveness of solar water heaters at different collector tracking modes, in Cyprus and Greece. Energy Convers. Manage. 40 (12), 1287–1303.

Morrison, G.L., 2001. Solar water heating. In: Gordon, J. (Ed.), Solar Energy: The State of the Art. James and James, London, pp. 223–289.

Morrison, G.L., Braun, J.E., 1985. System modeling and operation characteristics of thermosiphon solar water heaters. Sol. Energy 34 (4–5), 389–405.

Norton, B., Probert, S.D., 1983. Achieving thermal stratification in natural-circulation solar-energy water heaters. Appl. Energy 14 (3), 211–225.

Ong, K.S., 1974. A finite difference method to evaluate the thermal performance of a solar water heater. Sol. Energy 16 (3–4), 137–147.

Ong, K.S., 1976. An improved computer program for the thermal performance of a solar water heater. Sol. Energy 18 (3), 183–191.

Philibert, C., 2005. The Present and Future Use of Solar Thermal Energy as a Primary Source of Energy. International Energy Agency. Copyright by The Inter Academy Council, Paris, France.

Tripanagnostopoulos, Y., Souliotis, M., Nousia, T., 2002. CPC type integrated collector storage systems. Sol. Energy 72 (4), 327–350.

Turkenburg, W.C., 2000. Renewable Energy Technologies, World Energy Assessment, (Chapter 7). UNDP. Available from: www.undp.org/energy/activities/wea/drafts-frame.html.

Weiss, W., Spörk-Dür, M., 2022. Solar Heat Worldwide, Global Market Development and Trends in 2021, Solar Heating and Cooling Program. IEA, 2022.

Xinian, J., Zhen, T., Junshenf, L., 1994. Theoretical and experimental studies on sequential freezing solar water heaters. Sol. Energy 53 (2), 139–146.

Solar Space Heating and Cooling

6

The two principal categories of building solar heating and cooling systems are passive and active. The term *passive system* is applied to buildings that include, as integral parts of the building, elements that admit, absorb, store, and release solar energy and thus reduce the need for auxiliary energy for comfort heating. *Active systems* are the ones that employ solar collectors, storage tank, pumps, heat exchangers, and controls to heat and cool the building. The components and subsystems discussed in Chapter 5 may be combined to create a wide variety of building solar heating and cooling systems. Both types of systems are explained in this chapter. Initially, however, two methods of thermal load estimation are presented.

In 2021, 44 new large-scale solar heating systems ($>350\,\mathrm{kW_{th}}$; $500\,\mathrm{m^2}$) were built, with a capacity of $142\,\mathrm{MW_{th}}$. Twenty were installed in China, and 14 in Europe. The largest solar district heating system is in Praesto, Denmark, with a capacity of $5.6\,\mathrm{MW_{th}}$; a system in Nahwarme Friesach, Austria, and one in Muhlhausen, Germany, both have a capacity of $4\,\mathrm{MW_{th}}$. By the end of 2021, 530 large-scale solar thermal systems were in operation. The total installed capacity of these systems is $1970\,\mathrm{MW_{th}}$, equating to 2.8 million $\mathrm{m^2}$ of collectors (Weiss and Spörk-Dür, 2022).

6.1 Thermal load estimation

When estimating the building thermal load, adequate results can be obtained by calculating heat losses and gains based on a steady-state heat transfer analysis. For more accurate results and for energy analysis, however, transient analysis must be employed, since the heat gain into a conditioned space varies greatly with time, primarily because of the strong transient effects created by the hourly variation of a solar radiation. Many methods can be used to estimate the thermal load of buildings. The most well-known are the heat balance, weighting factors, thermal network, and radiant time series. In this book, only the heat balance method is briefly explained. Additionally, the degree-day method, which is a more simplified one used to determine the seasonal energy consumption, is described. Before proceeding, however, the three basic terms that are important in thermal load estimation are explained.

Heat gain

Heat gain is the rate at which energy is transferred to or generated within a space and consists of sensible and latent gain. Heat gains usually occur in the following forms:

1. Solar radiation passing through glazing and other openings;
2. Heat conduction with convection and radiation from the inner surfaces into the space;

3. Sensible heat convection and radiation from internal objects;
4. Ventilation and infiltration; and
5. Latent heat gains generated within the space.

Thermal load

The *thermal load* is the rate at which energy must be added or removed from a space to maintain the temperature and humidity at the design values.

The cooling load differs from the heat gain mainly because the radiant energy from the inside surfaces, as well as the direct solar radiation passing into a space through openings, is mostly absorbed in the space. This energy becomes part of the cooling load only when the room air receives the energy by convection and occurs when the various surfaces in the room attain higher temperatures than the room air. Hence, there is a time lag that depends on the storage characteristics of the structure and interior objects and is more significant when the heat capacity (product of mass and specific heat) is greater. Therefore, the peak cooling load can be considerably smaller than the maximum heat gain and occurs much later than the maximum heat gain period. The heating load behaves in a similar manner as the cooling load.

Heat extraction rate

The *heat extraction rate* is the rate at which energy is removed from the space by cooling and dehumidifying equipment. This rate is equal to the cooling load when the space conditions are constant and the equipment is operating. Since the operation of the control systems induces some fluctuation in the room temperature, the heat extraction rate fluctuates and this also causes fluctuations in the cooling load.

6.1.1 The heat balance method

The heat balance method is able to provide dynamic simulations of the building load. It is the foundation for all calculation methods that can be used to estimate the heating and cooling loads. Since all energy flows in each zone must be balanced, a set of energy balance equations for the zone air and the interior and exterior surfaces of each wall, roof, and floor must be solved simultaneously. The energy balance method combines various equations, such as equations for transient conduction heat transfer through walls and roofs, algorithms or data for weather conditions, and internal heat gains.

The method can be illustrated by considering a zone consisting of six surfaces, four walls, a roof, and a floor. The zone receives energy from solar radiation coming through windows, heat conducted through exterior walls and the roof, and internal heat gains due to lighting, equipment, and occupants. The heat balance on each of the six surfaces is generally represented by:

$$q_{i,\theta} = \left[h_{ci}\left(t_{a,\theta} - t_{i,\theta}\right) + \sum_{j=1, j \neq i}^{ns} g_{ij}\left(t_{j,\theta} - t_{i,\theta}\right) \right] A_i + q_{si,\theta} + q_{li,\theta} + q_{ei,\theta} \tag{6.1}$$

where

$q_{i,\theta}$ = rate of heat conducted into surface i at the inside surface at time θ (W);
i = surface number (1–6);

ns = number of surfaces in the room;

A_i = area of surface i (m²);

h_{ci} = convective heat transfer coefficient at interior of surface i (W/m² K);

g_{ij} = linearized radiation heat transfer factor between interior surface i and interior surface j (W/m² K);

$t_{a,\theta}$ = inside air temperature at time θ (°C);

$t_{i,\theta}$ = average temperature of interior surface i at time θ (°C);

$t_{j,\theta}$ = average temperature of interior surface j at time θ (°C);

$q_{si,\theta}$ = rate of solar heat coming through the windows and absorbed by surface i at time θ (W);

$q_{li,\theta}$ = rate of heat from the lighting absorbed by surface i at time θ (W); and

$q_{ei,\theta}$ = rate of heat from equipment and occupants absorbed by surface i at time θ (W).

The equations governing a conduction within the six surfaces cannot be solved independent of Eq. (6.1), since the energy exchanges occurring within the room affect the inside surface conditions, which in turn affect the internal conduction. Consequently, the aforementioned six formulations of Eq. (6.1) must be solved simultaneously with the equations governing conduction within the six surfaces to calculate the space thermal load. Among the possible ways to model this process are numerical finite element and time series methods. Most commonly, due to the greater computational speed and little loss of generality, conduction within the structural elements is formulated using conduction transfer functions (CTFs) in the general form:

$$q_{i,\theta} = \sum_{m=1}^{M} Y_{k,m} t_{o,\theta-m+1} - \sum_{m=1}^{M} Z_{k,m} t_{o,\theta-m+1} + \sum_{m=1}^{M} F_m q_{i,\theta-m} \tag{6.2}$$

where

i = inside surface subscript;

k = order of CTF;

m = time index variable;

M = the number of non-zero CTF values;

o = outside surface subscript;

t = temperature (°C);

θ = time;

Y = cross CTF values;

Z = interior CTF values; and

F_m = flux history coefficients.

Conduction transfer function coefficients generally are referred to as *response factors* and depend on the physical properties of the wall or roof materials and the scheme used for calculating them. These coefficients relate an output function at a given time to the value of one or more driving functions at a given time and at a set period immediately preceding (ASHRAE, 2005). The Y (cross-CTF) values refer to the current and previous flows of energy through the wall due to the outside conditions, the Z (interior CTF) values refer to the internal space conditions, and the F_m (flux history) coefficients refer to the current and previous heat flux to zone.

Equation (6.2), which utilizes the transfer function concept, is a simplification of the strict heat balance calculation procedure, which could be used in this case for calculating conduction heat transfer.

It must be noted that the interior surface temperature $t_{i,\theta}$ is present in both Eqs. (6.1) and (6.2), and therefore a simultaneous solution is required. In addition, the equation representing the energy balance on the zone air must also be solved simultaneously. This can be calculated from the cooling load equation:

$$q_\theta = \left[\sum_{i=1}^{m} h_{ci} \left(t_{i,\theta} - t_{\alpha,\theta} \right) \right] A_i + \rho c_p Q_{i,\theta} \left(t_{o,\theta} - t_{\alpha,\theta} \right) + \rho c_p Q_{v,\theta} \left(t_{v,\theta} - t_{\alpha,\theta} \right) + q_{s,\theta} + q_{1,\theta} + q_{e,\theta} \quad (6.3)$$

where

$t_{\alpha,\theta}$ = inside air temperature at time θ (°C);
$t_{o,\theta}$ = outdoor air temperature at time θ (°C);
$t_{v,\theta}$ = ventilation air temperature at time θ (°C);
ρ = air density (kg/m³);
c_p = specific heat of air (J/kg K);
$Q_{i,\theta}$ = volume flow rate of outdoor air infiltrating into the room at time θ (m³/s);
$Q_{v,\theta}$ = volume rate of flow of ventilation air at time θ (m³/s);
$Q_{s,\theta}$ = rate of solar heat coming through the windows and convected into the room air at time θ(W);
$q_{1,\theta}$ = rate of heat from the lights convected into the room air at time θ (W); and
$q_{e,\theta}$ = rate of heat from equipment and occupants convected into the room air at time θ (W).

6.1.2 The transfer function method

The ASHRAE Task Group on Energy Requirements developed the general procedure referred to as the *transfer function method* (TFM). This approach is a method that simplifies the calculations, can provide the loads originating from various parts of the building, and can be used to determine the heating and cooling loads.

The method is based on a series of conduction transfer functions (CTFs) and a series of room transfer functions (RTFs). The CTFs are used for calculating wall or roof heat conduction; the RTFs are used for load elements that have radiant components, such as lights and appliances. These functions are response time series, which relate a current variable to the past values of itself and other variables in periods of 1 h.

Wall and roof transfer functions

Conduction transfer functions are used by the TFM to describe the heat flux at the inside of a wall, roof, partition, ceiling, and floor. Combined convection and radiation coefficients on the inside (8.3 W/m² K) and outside surfaces (17.0 W/m² K) are utilized by the method. The approach uses sol–air temperatures to represent outdoor conditions and assumes constant indoor air temperature. Thus, the heat gain though a wall or roof is given by:

$$q_{e,\theta} = A \left[\sum_{n=0} b_n \left(t_{e,\theta-n\delta} \right) - t_{rc} \sum_{n=0} c_n - \sum_{n=1} d_n \left(q_{e,\theta-n\delta} / A \right) \right] \quad (6.4)$$

where

$q_{e,\theta}$ = heat gain through wall or roof, at calculation hour θ (W);
A = indoor surface area of wall or roof (m^2);
θ = time (s);
δ = time interval (s);
n = summation index (each summation has as many terms as there are non-negligible values of coefficients);
$t_{e,\theta-n\delta}$ = sol–air temperature at time $\theta-n\delta$ (°C);
t_{rc} = constant indoor room temperature (°C); and
b_n, c_n, d_n = conduction transfer function coefficients.

Conduction transfer function coefficients depend only on the physical properties of the wall or roof. These coefficients are given in tables (ASHRAE, 1997). The b and c coefficients must be adjusted for the actual heat transfer coefficient (U_{actual}) by multiplying them with the ratio $U_{actual}/U_{reference}$.

In Eq. (6.4), a value of the summation index n equal to 0 represents the current time interval, n equal to 1 is the previous hour, and so on.

The sol–air temperature is defined as:

$$t_e = t_0 + \alpha G_t / h_0 - \varepsilon \delta R / h_0 \tag{6.5}$$

where

t_e = sol–air temperature (°C);
t_0 = current hour dry-bulb temperature (°C);
α = absorptance of surface for solar radiation;
G_t = total incident solar load (W/m^2);
δR = difference between longwave radiation incident on the surface from the sky and surroundings and the radiation emitted by a blackbody at outdoor air temperature (W/m^2);
h_0 = heat transfer coefficient for convection over the building (W/m^2 K); and
$\varepsilon \delta R / h_0$ = longwave radiation factor = -3.9 °C for horizontal surfaces, 0 °C for vertical surfaces.

The term α/h_0 in Eq. (6.5) varies from about 0.026 m^2 K/W for a light-colored surface to a maximum of about 0.053 m^2 K/W. The heat transfer coefficient for convection over the building can be estimated from:

$$h_0 = 5.7 + 3.8 \, V \tag{6.6}$$

where h_0 is in W/m^2 K and V is the wind speed in m/s.

Partitions, ceilings, and floors

Whenever a conditioned space is adjacent to other spaces at different temperatures, the transfer of heat through the partition can be calculated from Eq. (6.4) by replacing the sol–air temperature with the temperature of the adjacent space.

When the air temperature of the adjacent space (t_b) is constant or the variations of this temperature are small compared to the difference of the adjacent space and indoor temperature difference, the rate of heat gains (q_p) through partitions, ceilings, and floors can be calculated from the formula:

$$q_p = UA(t_b - t_i) \tag{6.7}$$

where

A = area of element under analysis (m^2);
U = overall heat transfer coefficient (W/m^2 K); and
$(t_b - t_i)$ = adjacent space–indoor temperature difference (°C).

Glazing

The total rate of heat admission through glass is the sum of the transmitted solar radiation, the portion of the absorbed radiation that flows inward, and the heat conducted through the glass whenever there is an outdoor–indoor temperature difference. The rate of heat gain (q_s) resulting from the transmitted solar radiation and the portion of the absorbed radiation that flows inward is:

$$q_s = A(\text{SC})(\text{SHGC}) \tag{6.8}$$

where

A = area of element under analysis (m^2);
SC = shading coefficient; and
SHGC = solar heat-gain coefficient, varying according to orientation, latitude, hour, and month.

The rate of conduction heat gain (q) is:

$$q = UA(t_o - t_i) \tag{6.9}$$

where

A = area of element under analysis (m^2);
U = glass heat transfer coefficient (W/m^2 K); and
$(t_o - t_i)$ = outdoor–indoor temperature difference (°C).

People

The heat gain from people is in the form of sensible and latent heat. The latent heat gains are considered as instantaneous loads. The total sensible heat gain from people is not converted directly to cooling load. The radiant portion is first absorbed by the surroundings and convected to the space at a later time, depending on the characteristics of the room. The ASHRAE *Handbook of Fundamentals* (2005) gives tables for various circumstances and formulates the gains for the instantaneous sensible cooling load as:

$$q_s = N(\text{SHG}_p) \tag{6.10}$$

where

q_s = rate of sensible cooling load due to people (W);
N = number of people; and
SHG$_p$ = sensible heat gain per person (W/person).

The rate of latent cooling load is:

$$q_l = N(\text{LHG}_p) \tag{6.11}$$

where

q_l = latent cooling load due to people (W);
N = number of people; and
LHG_p = latent heat gain per person (W/person).

Lighting

Generally, lighting is often a major internal load component. Some of the energy emitted by the lights is in the form of radiation that is absorbed in the space and transferred later to the air by convection. The manner in which the lights are installed, the type of air distribution system, and the mass of the structure affect the rate of heat gain at any given moment. Generally, this gain can be calculated from:

$$q_{el} = W_1 F_{ul} F_{sa} \qquad (6.12)$$

where

q_{el} = rate of heat gain from lights (W);
W_1 = total installed light wattage (W);
F_{ul} = lighting use factor, ratio of wattage in use to total installed wattage; and
F_{sa} = special allowance factor (ballast factor in the case of fluorescent and metal halide fixtures).

Appliances

Considerable data are available for this category of cooling load, but careful evaluation of the operating schedule and the load factor for each piece of equipment is essential. Generally, the sensible heat gains from the appliances (q_a) can be calculated from:

$$q_a = W_a F_U F_R \qquad (6.13)$$

or

$$q_a = W_a F_L \qquad (6.14)$$

where

W_a = rate of energy input from appliances (W);
F_U, F_R, F_L = usage factors, radiation factors, and load factors.

Ventilation and infiltration air

Both sensible ($q_{s,v}$) and latent ($q_{l,v}$) rates of heat gain result from the incoming air, which may be estimated from:

$$q_{s,v} = m_a c_p (t_o - t_i) \qquad (6.15)$$

$$q_{l,v} = m_a (\omega_o - \omega_i) i_{fg} \qquad (6.16)$$

where

m_a = air mass flow rate (kg/s);
c_p = specific heat of air (J/kg K);

$(t_o - t_i)$ = temperature difference between incoming and room air (°C);
$(\omega_o - \omega_i)$ = humidity ratio difference between incoming and room air (kg/kg); and
i_{fg} = enthalpy of evaporation (J/kg K).

6.1.3 Heat extraction rate and room temperature

The cooling equipment, in an ideal case, must remove heat energy from the space's air at a rate equal to the cooling load. In this way, the space air temperature will remain constant. However, this is seldom true. Therefore, a transfer function has been devised to describe the process. The room air transfer function is:

$$\sum_{i=0}^{1} p_i \left(q_{x,\theta-i\delta} - q_{c,\theta-i\delta} \right) = \sum_{i=0}^{2} g_i \left(t_i - t_{r,\theta-i\delta} \right) \tag{6.17}$$

where

p_i, g_i = transfer function coefficients (ASHRAE, 1992);
q_x = heat extraction rate (W);
q_c = cooling load at various times (W);
t_i = room temperature used for cooling load calculations (°C); and
t_r = actual room temperature at various times (°C).

All g coefficients refer to a unit floor area. The coefficients g_0 and g_i depend also on the average heat conductance to the surroundings (UA) and the infiltration and ventilation rate to the space. The p coefficients are dimensionless.

The characteristic of the terminal unit usually is of the form:

$$q_{x,\theta} = W + S \times t_{r,\theta} \tag{6.18}$$

where W and S are parameters that characterize the equipment at time θ.

The equipment being modeled is actually the cooling coil and the associated control system (thermostat) that matches the coil load to the space load. The cooling coil can extract heat energy from the space air from some minimum to some maximum value.

Equations (6.17) and (6.18) may be combined and solved for $q_{x,\theta}$:

$$q_{x,\theta} = (W \times g_0 + S \times G_\theta)/(S + g_0) \tag{6.19}$$

where

$$G_\theta = t_i \sum_{i=0}^{2} g_i - \sum_{i=1}^{2} g_i \left(t_{y,\theta-i\delta} \right) + \sum_{i=0}^{1} p_i \left(q_{c,\theta-i\delta} \right) - \sum_{i=0}^{1} p_i \left(q_{x,\theta-i\delta} \right) \tag{6.20}$$

When the value of $q_{x,\theta}$ computed by Eq. (6.19) is greater than $q_{x,max}$, it is taken to be equal to $q_{x,max}$; when it is less than $q_{x,min}$, it is made equal to $q_{x,min}$. Finally, Eqs. (6.18) and (6.19) can be combined and solved for $t_{r,\theta}$:

$$t_{r,\theta} = \frac{(G_\theta - q_{x,\theta})}{g_0} \tag{6.21}$$

It should be noted that, although it is possible to perform thermal load estimation manually with both the heat balance and the transfer function methods, these are better suited for computerized calculation, due to the large number of operations that need to be performed.

6.1.4 Degree-day method

Frequently, in energy calculations, simpler methods are required. One such simple method, which can give comparatively accurate results, is the degree-day method. This method is used to predict the seasonal energy consumption. Each degree that the average outdoor air temperature falls below a balance temperature, T_b, of 18.3 °C (65 °F) represents a degree day. The number of degree days in a day is obtained approximately by the difference of T_b and the average outdoor air temperature, T_{av}, defined as $(T_{max} + T_{min})/2$. Therefore, if the average outdoor air temperature of a day is 15.3 °C, the number of heating degree-days $(DD)_h$ for the day is 3. The number of heating degree-days over a month is obtained by the sum of the daily values (only positive values are considered) from:

$$(DD)_h = \sum_m (T_b - T_{av})^+ \tag{6.22}$$

Similarly, cooling degree-days are obtained from:

$$(DD)_c = \sum_m (T_{av} - T_b)^+ \tag{6.23}$$

Degree days for both heating $(DD)_h$ and cooling $(DD)_c$ are published by the meteorological services of many countries. Appendix 7 lists the values of both heating and cooling degree-days for a number of countries. Using the degree-days' concept, the following equation can be used to determine the monthly or seasonal heating load or demand (D_h):

$$D_h = (UA)(DD)_h \tag{6.24}$$

where UA represents the heat loss characteristic of the building, given by:

$$(UA) = \frac{Q_h}{T_i - T_o} \tag{6.25}$$

where

Q_h = design rate or sensible heat loss (kW);
$T_i - T_o$ = design indoor−outdoor temperature difference (°C).

Substituting Eq. (6.25) into Eq. (6.24) and multiplying by $3600 \times 24 = 86,400$ to convert days into seconds, the following equation can be obtained for the monthly or seasonal heating load or demand in kJ:

$$D_h = \frac{86.4 \times 10^3 Q_h}{T_i - T_o}(DD)_h \tag{6.26}$$

For cooling, the balance temperature is usually 24.6 °C. Similar to the preceding, the monthly or seasonal cooling load or demand in kJ is given by:

$$D_c = \frac{86.4 \times 10^3 Q_c}{T_o - T_i}(DD)_c \tag{6.27}$$

EXAMPLE 6.1

A building has a peak heating load equal to 15.6 kW and a peak cooling load of 18.3 kW. Estimate the seasonal heating and cooling requirements if the heating degree-days are 1020 °C days, the cooling degree-days are 870 °C days, the winter indoor temperature is 21 °C, and the summer indoor temperature is 26 °C. The design outdoor temperature for winter is 7 °C and for summer is 36 °C.

Solution

Using Eq. (6.26), the heating requirement is:

$$D_h = \frac{86.4 \times 10^3 Q_h}{T_i - T_o}(DD)_h = \frac{86.4 \times 10^3 \times 15.6}{21 - 7}(1020) = 98.2 \times 10^6 \text{ kJ} = 98.2 \text{ GJ}$$

Similarly, for the cooling requirement, Eq. (6.27) is used:

$$D_c = \frac{86.4 \times 10^3 Q_c}{T_o - T_i}(DD)_c = \frac{86.4 \times 10^3 \times 18.3}{36 - 26}(870) = 137.6 \times 10^6 \text{ kJ} = 137.6 \text{ GJ}$$

6.1.5 Building heat transfer

The design of space-heating or cooling systems for a building requires the determination of the building thermal resistance. Heat is transferred in building components by all modes: conduction, convection, and radiation. In an electrical analogy, the rate of heat transfer through each building component can be obtained from:

$$Q = \frac{A \times \Delta T_{total}}{R_{total}} = UA \times \Delta T_{total} \tag{6.28}$$

where

ΔT_{total} = total temperature difference between inside and outside air (K);
R_{total} = total thermal resistance across the building element, $= \sum R_i (\text{m}^2 \text{ K}/\text{W})$; and
A = area of the building element perpendicular to the heat flow direction (m²).

It is obvious from Eq. (6.28) that the overall heat transfer coefficient, U, is equal to:

$$U = \frac{1}{R_{total}} \tag{6.29}$$

As in a collector heat transfer, described in Chapter 3, it is easier to apply an electrical analogy to evaluate the building thermal resistances. For conduction heat transfer through a wall element of thickness x (m) and thermal conductivity k (W/m K), the thermal resistance, based on a unit area, is:

$$R = \frac{x}{k} \tag{6.30}$$

The thermal resistance per unit area for convection and radiation heat transfer, with a combined convection and radiation heat transfer coefficient h (W/m² K), is:

$$R = \frac{1}{h} \tag{6.31}$$

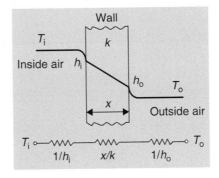

FIGURE 6.1

Heat transfer through a building element and equivalent electric circuit.

Figure 6.1 illustrates a single-element wall. The thermal resistance due to conduction through the wall is x/k, Eq. (6.30), and the thermal resistance at the inside and outside boundaries of the wall are $1/h_i$ and $1/h_o$, Eq. (6.31), respectively. Therefore, from the preceding discussion, the total thermal resistance based on the inside and outside temperature difference is the sum of the three resistances as:

$$R_{total} = R_i + R_w + R_o = \frac{1}{h_i} + \frac{x}{k} + \frac{1}{h_o} \tag{6.32}$$

or

$$U = \frac{1}{R_{total}} = \frac{1}{R_i + R_w + R_o} = \frac{1}{\frac{1}{h_i} + \frac{x}{k} + \frac{1}{h_o}} \tag{6.33}$$

The values of h_i, h_o, and k can be obtained from handbooks (e.g., ASHRAE, 2005). Values for typical materials are shown in Table A5.4 and for stagnant air and surface resistance in Table A5.5 in Appendix 5. For multilayer or composite walls like the one shown in Fig. 6.2, the following general equation can be used:

$$U = \frac{1}{\frac{1}{h_i} + \sum\limits_{i=1}^{m} \left(\frac{x_i}{k_i}\right) + \frac{1}{h_o}} \tag{6.34}$$

where

m = number of materials of the composite construction.

For the particular case shown in Fig. 6.2, where three layers of materials are used, the following equation applies:

$$U = \frac{1}{\frac{1}{h_i} + \frac{x_1}{k_1} + \frac{x_2}{k_2} + \frac{x_3}{k_3} + \frac{1}{h_o}} \tag{6.35}$$

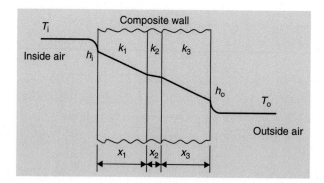

FIGURE 6.2

Multilayer wall heat transfer.

It should be noted that small thickness layers, like paints and glues, are often ignored from this calculation.

EXAMPLE 6.2

Find the overall heat transfer coefficient of the construction shown in Fig. 6.2 if components 1 and 3 are brick 10 cm in thickness and the center one is stagnant air 5 cm in thickness. Additionally, the wall is plastered with 25 mm plaster on each side.

Solution

Using the data shown in Tables A5.4 and A5.5 in Appendix 5, we get the following list of resistance values.
1. Outside surface resistance = 0.044.
2. Plaster 25 mm = 0.025/1.39 = 0.018.
3. Brick 10 cm = 0.10/0.25 = 0.4.
4. Stagnant air 50 mm = 0.18.
5. Brick 10 cm = 0.10/0.25 = 0.4.
6. Plaster 25 mm = 0.025/1.39 = 0.018.
7. Inside surface resistance = 0.12.
Total resistance = 1.18 m^2 K/W or $U = 1/1.18 = 0.847$ W/m^2 K.

In some countries minimum U values for the various building components are specified by law to prohibit building poorly insulated buildings, which require a lot of energy for their heating and cooling needs.

Another situation usually encountered in buildings is the pitched roof shown in Fig. 6.3.

Using the electrical analogy, the combined thermal resistance is obtained from:

$$R_{total} = R_{ceiling} + R_{roof}$$

or

$$\frac{1}{U_R A_c} = \frac{1}{U_c A_c} + \frac{1}{U_r A_r} \tag{6.36}$$

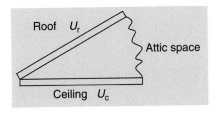

FIGURE 6.3

Pitched roof arrangement.

which gives:

$$U_R = \cfrac{1}{\cfrac{1}{U_c} + \cfrac{1}{U_r(A_r/A_c)}}$$ (6.37)

where

U_R = combined overall heat transfer coefficient for the pitched roof (W/m^2 K);
U_c = overall heat transfer coefficient for the ceiling per unit area of the ceiling (W/m^2 K);
U_r = overall heat transfer coefficient for the roof per unit area of the roof (W/m^2 K);
A_c = ceiling area (m^2); and
A_r = roof area (m^2).

6.2 Passive space-heating design

Passive solar heating systems require little, if any, non-renewable energy to function. Every building is passive in the sense that the sun tends to warm it during the day and it loses heat at night. Passive systems incorporate the solar energy collection, storage, and distribution into the architectural design of the building and make minimal or no use of mechanical equipment, such as fans, to deliver the collected energy. Passive solar heating, cooling, and lighting design must consider the building envelope and its orientation, the thermal storage mass, window configuration and design, the use of sun spaces, and natural ventilation.

As part of the design process, a preliminary analysis must be undertaken to investigate the possibilities for saving energy through solar energy and the selection of the appropriate passive technique. The first step to consider for each case investigated should include an analysis of the climatic data of the site and definition of the comfort requirements of the occupants and the way to meet them. The passive system can then be selected by examining both direct and indirect gains.

6.2.1 Building construction: thermal mass effects

Heat can be stored in the structural materials of the building to reduce the indoor temperature, reduce the cooling load peaks, and shift the time that maximum load occurs. The storage material is referred to as the *thermal mass*. In winter, during periods of high solar gain, energy is stored in the thermal mass, avoiding overheating. In the late afternoon and evening hours, when energy is needed, heat is released into the building, satisfying part of the heating load. In summer, the thermal mass acts in a similar way as in winter, reducing the cooling load peaks.

Heat gain in a solar house can be direct or indirect. Direct gain is the solar radiation passing through a window to heat the building interior, whereas an indirect gain is the heating of a building element by solar radiation and the use of this heat, which is transmitted inside the building, to reduce the heating load.

Indirect-gain solar houses use the south-facing wall surface of the structure to absorb solar radiation, which causes a rise in temperature that, in turn, conveys heat into the building in several ways. Glass has led to modern adaptations of the indirect-gain principle (Trombe et al., 1977).

By glazing a large south-facing, massive masonry wall, solar energy can be absorbed during the day and conduction of heat to the inner surface provides radiant heating at night. The mass of the wall and its relatively low thermal diffusivity delay the arrival of the heat at the indoor surface until it is needed. The glazing reduces the loss of heat from the wall back to the atmosphere and increases the collection efficiency of the system during the day.

Openings in the wall, near the floor, and near the ceiling allow convection to transfer heat to the room. The air in the space between the glass and the wall warms as soon as the sun heats the outer surface of the wall. The heated air rises and enters the building through the upper openings. Cool air flows through the lower openings, and convective heat gain can be established as long as the sun is shining (see Fig. 6.4, later in the chapter). This design is often called the *Trombe wall*, from the name of the engineer Felix Trombe, who applied the idea in France.

In most passive systems, control is accomplished by moving a shading device that regulates the amount of solar radiation admitted into the structure. Manually operated window shades or Venetian blinds are the most widely used because of their simple control.

The thermal storage capabilities inherent in building mass can have a significant effect on the temperature within the space as well as on the performance and operation of heating, ventilating, and air-conditioning (HVAC) systems.

Effective use of structural mass for thermal storage has been shown to reduce the building energy consumption, reduce and delay peak heating and cooling loads (Braun, 1990), and in some cases, improve comfort (Simmonds, 1991). Perhaps the best-known use of thermal mass to reduce energy consumption is in buildings that include passive solar techniques (Balcomb, 1983).

The effective use of thermal mass can be considered incidental and allowed for in the heating or cooling design, or it may be considered intentional and form an integral part of the system design.

Incidental thermal mass effects

The principal thermal mass effect on heating and cooling systems serving spaces in heavyweight buildings is that a greater amount of thermal energy must be removed or added to bring the room to a suitable condition than for a similar lightweight building. Therefore, the system must either start conditioning the spaces earlier or operate at a greater output. During the occupied period, a heavyweight building requires a lower output because a higher proportion of heat gains or losses are absorbed by the thermal mass.

Advantage can be taken of these effects if low-cost electrical energy is available during the night so as to operate the air-conditioning system during this period to pre-cool the building. This can reduce both the peak and total energy required during the following day but might not always be energy efficient.

Intentional thermal mass effects

To make the best use of a thermal mass, the building should be designed with this objective in mind. Intentional use of the thermal mass can be either passive or active. Passive solar heating is a common application that utilizes the thermal mass of the building to provide warmth when no solar energy is available. Passive cooling applies the same principles to limit the temperature rise during the day. The spaces can be naturally ventilated overnight to absorb surplus heat from the building mass. This technique works well in moderate climates with a wide diurnal temperature swing and low relative humidity, but it is limited by the lack of control over the cooling rate.

The effective use of building structural mass for thermal energy storage depends on (ASHRAE, 2007):

1. The physical characteristics of the structure.
2. The dynamic nature of the building loads.
3. The coupling between the mass and zone air.
4. The strategies for charging and discharging the stored thermal energy.

Some buildings, such as frame buildings with no interior mass, are inappropriate for thermal storage. Many other physical characteristics of a building or an individual zone, such as carpeting, ceiling plenums, interior partitions, and furnishings, affect thermal storage and the coupling of the building with zone air (Kalogirou et al., 2002).

The term *thermal mass* is commonly used to signify the ability of materials to store significant amounts of thermal energy and delay heat transfer through a building component. This delay leads to three important results:

- The slower response time tends to moderate indoor temperature fluctuations under outdoor temperature swings (Brandemuehl et al., 1990).
- In hot or cold climates, it reduces energy consumption in comparison to that for a similar low-mass building (Wilcox et al., 1985).
- It moves building energy demand to off-peak periods because energy storage is controlled through correct sizing of the mass and interaction with the HVAC system.

Thermal mass causes a time delay in the heat flow, which depends on the thermo-physical properties of the materials used. To store heat effectively, structural materials must have high density (ρ), thermal capacity (C), and conductivity (k), so that heat may penetrate through all the material during the specific time of heat charging and discharging. A low value of the $\rho C k$ product indicates a low heat-storage capacity, even though the material can be quite thick.

Thermal mass can be characterized by the thermal diffusivity (a) of the building material, which is defined as:

$$a = k / \rho c_p \qquad (6.38)$$

where c_p is the specific heat of the material (J/kg °C).

Heat transfer through a material with high thermal diffusivity is fast, the amount of heat stored in it is relatively small, and the material responds quickly to changes in temperature. The effect of thermal mass on building behavior varies primarily with the climate at the building site and the position of the wall insulation relative to the building mass.

Thermal diffussivity is the controlling transport property for a transient heat transfer. The time lag for some common building materials of 300 mm thickness is 10 h for common brick, 6 h for face brick, 8 h for heavyweight concrete, and 20 h for wood because of its moisture content (Lechner, 1991). Thermal storage materials can be used to store direct energy by solar radiation in the building envelope or in places where incident radiation enters through openings in the building envelope. Also, these materials can be used inside the building to store indirect radiation, i.e., infrared radiation and energy from a room air convection.

The ideal climate for taking advantage of thermal mass is one that has large daily temperature fluctuations. The mass can be cooled by natural ventilation at night and be allowed to "float" during the warmer day. When outdoor temperatures are at their peak, the inside of the building remains cool because the heat has not yet penetrated the mass. Often, the benefits are greater during spring and fall, when some climates closely approximate this ideal case. In climates where heating is used extensively, thermal mass can be used effectively to collect and store solar gains or to store heat provided by the mechanical system, allowing the heating system to operate during off-peak hours (Florides et al., 2002b).

The distribution of thermal mass depends on the orientation of the given surface. According to Lechner (1991), a surface with a north orientation has little need for time lag, since it exhibits only small heat gains. East orientation surfaces need either a very long time lag, greater than 14 h, so that heat transfer is delayed until the late evening hours or a very short one, which is preferable because of the lower cost. South orientations can operate with an 8 h time lag, delaying the heat from midday until the evening hours. For west orientations, an 8 h time lag is again sufficient since they receive radiation for only a few hours before sunset. Finally, the roof requires a very long time lag since it is exposed to solar radiation during most hours of the day. However, because it is very expensive to construct heavy roofs, the use of additional insulation is usually recommended instead.

The effectiveness of thermal mass also increases by increasing the allowable temperature swing in the conditioned space (without the intervention of HVAC systems), so the mass has the opportunity to charge during warm hours and discharge during cooler periods.

The performance of thermal mass is influenced by the use of insulation. Where heating of the building is the major concern, insulation is the predominant effective envelope factor. In climates where cooling is of primary importance, thermal mass can reduce energy consumption, provided the building is unused in the evening hours and the stored heat can be dissipated during the night. In this case, either natural or mechanical ventilation can be used during the night, to introduce cool outdoor air into the space and remove heat from the massive walls and roof.

To model the complex interactions of all envelope components, computer simulations are necessary. These programs account for material properties of the components, building geometry, orientation, solar gains, internal gains, and HVAC control strategy. The calculations are usually performed on an hourly basis, using a full year of weather data.

Numerous models have been devised in the past to simulate the effect of thermal walls (Duffin and Knowles, 1985; Nayak, 1987; Zrikem and Bilgen, 1987). Also, a number of modeling techniques have been used to estimate the heat flow through a thermal wall. A simple analytical model was suggested by Duffin and Knowles (1985), in which all parameters affecting the wall performance can be analyzed. Smolec and Thomas (1993) used a two-dimensional model to compute the heat transfer, whereas Jubran et al. (1993) based their model on the finite difference method to predict the transient response, temperature distribution, and velocity profile of a thermal wall. The transient response of the Trombe wall was also investigated by Hsieh and Tsai (1988).

Characteristics of a thermal storage wall

A thermal storage wall is essentially a high-capacitance solar collector directly coupled to the room. Absorbed solar radiation reaches the room either by conduction through the wall to the inside wall surface from which it is convected and radiated into the room or by the hot air flowing though the air gap. The wall loses energy to the environment by conduction, convection, and radiation through the glazing covers.

A thermal storage wall is shown diagrammatically in Fig. 6.4. Depending on the control strategy used, air in the gap can be exchanged with either the room air or the environment, or the flow through the gap can be stopped. The flow of air can be driven by a fan or be thermosiphonic, i.e., driven by higher air temperatures in the gap than in the room. Analytical studies of the thermosiphonic effect of air are confined to the case of laminar flow and neglect pressure losses in the inlet and outlet vents. Trombe et al. (1977) reported measurements of thermosiphon mass flow rates, which indicate that most of the pressure losses are due to expansion, contraction, and change of direction of flow, all associated with the inlet and outlet vents. For hot summer climates, a vent is provided at the upper part of the glazing (top-outdoor opening) to release the hot air produced in the gap between the glass and the thermal wall by drawing air from the inside of the room and by closing the top vent to the room. Alternatively the room can be completely isolated and use the bottom and top outdoor openings to cool the thermal wall without drawing air from the room.

FIGURE 6.4

Schematic of the thermal storage wall.

In the Trombe wall model used in TRNSYS (see Chapter 11, Section 11.5.1), the thermosiphon air flow rate is determined by applying Bernoulli's equation to the entire air flow system. For simplicity, it is assumed that the density and temperature of the air in the gap vary linearly with height. Solution of Bernoulli's equation for the mean air velocity in the gap yields (Klein et al., 2005):

$$\bar{v} = \sqrt{\frac{2gh}{C_1 \left(\dfrac{A_g}{A_v}\right)^2 + C_2} \cdot \frac{(T_m - T_s)}{|T_m|}} \tag{6.39}$$

where

A_g = total gap cross-sectional area (m^2);
A_v = total vent area (m^2);
C_1 = vent pressure loss coefficient;
C_2 = gap pressure loss coefficient;
g = acceleration due to gravity (m/s^2); and
T_m = mean air temperature in the gap (K).

The term T_s is either T_a or T_R, depending on whether air is exchanged with the environment (T_a) or the room (T_R). The term $C_1(A_g/A_v)^2 + C_2$ represents the pressure losses of the system. The ratio $(A_g/A_v)^2$ accounts for the difference between the air velocity in the vents and the air velocity in the gap.

The thermal resistance (R) to energy flow between the gap and the room when mass flow rate ($\dot{m}$) is finite is given by:

$$R = \frac{A \left\{ \left(\dfrac{\dot{m}c_{pa}}{2h_c A}\right) \left[\exp\left(-\dfrac{2h_c A}{\dot{m}c_{pa}}\right) - 1 \right] - 1 \right\}}{\dot{m}c_{pa} \left[\exp\left(-\dfrac{2h_c A}{\dot{m}c_{pa}}\right) - 1 \right]} \tag{6.40}$$

where

A = wall area (m^2);
c_{pa} = specific heat of air (J/kg °C); and
h_c = gap air heat transfer coefficient (W/m^2 K).

The value of h_c, the heat transfer coefficient between the gap air and the wall and glazing, depends on whether air flows through the gap (Klein et al., 2005). For a no-flow rate (Randal et al., 1979),

$$h_c = \frac{k_a}{L} \left[0.01711(\text{Gr Pr})^{0.29}\right] \tag{6.41}$$

where

k_a = air thermal conductivity (W/m °C);.
L = length (m);
Gr = Grashof number; and
Pr = Prandtl number.

For a flow condition and Reynolds number, Re > 2000 (Kays, 1966),

$$h_c = \frac{k_a}{L}\left(0.0158\ \mathrm{Re}^{0.8}\right) \tag{6.42}$$

For a flow condition and Re ≤ 2000 (Mercer et al., 1967),

$$h_c = \frac{k_a}{L}\left(4.9 + \frac{0.0606(x^*)^{-1.2}}{1 + 0.0856(x^*)^{-0.7}}\right) \tag{6.43a}$$

where

$$x^* = \frac{h}{\mathrm{Re}\ \mathrm{Pr}\dfrac{2A_g}{1+w}} \tag{6.43b}$$

According to Fig. 6.4, h is the distance between lower and upper openings (m) and w is the wall width (m).

Performance of thermal storage walls

A building with a thermal storage wall is shown in Fig. 6.5a, where L_m is the monthly energy loss from the building, Q_{aux} is the auxiliary energy required to cover the load, Q_D is the excess absorbed energy above what is required to cover the load that cannot be stored and must be dumped, and $\overline{T}_R$ is the mean room temperature, which is also equal to the low set point temperature setting of the room thermostat. The analysis of thermal storage walls is presented by Monsen et al. (1982) as part of the unutilizability method developed to design this type of systems, presented in Chapter 11, Section 11.4.2.

FIGURE 6.5

(a) Schematic of a thermal storage wall. (b) Equivalent electric circuit for the heat flow through the wall.

The monthly energy loss from the building, L_m, is defined as:

$$L_m = \int_{month} \left[(UA)(\overline{T}_R - \overline{T}_a) - \dot{g} \right]^+ dt = \int_{month} \left[(UA)(\overline{T}_b - \overline{T}_a) \right]^+ dt \tag{6.44}$$

where

(UA) = product of overall heat transfer coefficient and area of the building structure (W/°C);
$\dot{g}$ = rate of internal heat generation (W);
$\overline{T}_a$ = mean outdoor ambient temperature (°C); and
$\overline{T}_b$ = mean indoor balance temperature (°C), $\overline{T}_R - \dot{g}/(UA)$.

The variable of integration in Eq. (6.44) is time t, and the plus sign indicates that only positive values are considered. If (UA) and $\dot{g}$ are constant, L_m can be found from:

$$L_m = (UA)(DD)_b \tag{6.45}$$

where

$(DD)_b$ = monthly degree-days evaluated at $\overline{T}_b$.

The monthly energy loss from the building through the thermal storage wall, L_w, assuming that the glazing has zero transmissivity for solar radiation, can be found from:

$$L_w = U_w A_w (DD)_R \tag{6.46}$$

where

A_w = thermal storage wall area (m^2);
U_w = overall heat transfer coefficient of the thermal storage wall, including glazing (W/m^2 °C); and
$(DD)_R$ = monthly degree-days evaluated at $\overline{T}_R$.

From Fig. 6.5b, the overall heat transfer coefficient of the thermal storage wall, including glazing, is given from:

$$U_w = \cfrac{1}{\cfrac{1}{\overline{U}_o} + \cfrac{w}{k} + \cfrac{1}{h_i}} \tag{6.47}$$

where

w = wall thickness (m);
k = thermal conductivity of thermal storage wall (W/m °C);
h_i = inside wall surface film coefficient, = 8.33 W/m^2 °C [= 1/0.12], from Table A5.5, Appendix 5; and.
$\overline{U}_o$ = average overall heat transfer coefficient from the outer wall surface through the glazing to the ambient (W/m^2 °C).

Usually, night insulation is used to reduce the night heat losses. In this case, the average overall heat transfer coefficient $\overline{U}_o$ is estimated as the time average of the daytime and nighttime values from:

$$\overline{U}_o = (1 - F)U_o + F\left(\frac{U_o}{1 + R_{ins}U_o}\right) \tag{6.48}$$

where

U_o = overall coefficient with no night insulation (W/m^2 °C);
R_{ins} = thermal resistance of insulation (m^2 °C/W); and
F = fraction of time in which the night insulation is used.

A typical value of U_o for single glazing is 3.7 W/m^2 °C and for double glazing is 2.5 W/m^2 °C. The monthly energy balance of the thermal storage wall gives:

$$\overline{H}_t(\overline{\tau\alpha}) = U_k\left(\overline{T}_w - \overline{T}_R\right)\Delta t + \overline{U}_o\left(\overline{T}_w - \overline{T}_a\right)\Delta t \tag{6.49}$$

where

$\overline{H}_t$ = monthly average daily radiation per unit area incident on the wall (J/m^2);
$(\overline{\tau\alpha})$ = monthly average transmittance of glazing and absorptance of wall product;
$\overline{T}_w$ = monthly average outer wall surface temperature; see Fig. 6.5a (°C);
$\overline{T}_R$ = monthly average room temperature (°C);
$\overline{T}_a$ = monthly average ambient temperature (°C);
Δt = number of seconds in a day; and
U_k = overall heat transfer coefficient from outer wall surface to indoor space (W/m^2 °C).

The overall heat transfer coefficient from the outer wall surface to the indoor space can be obtained from:

$$U_k = \frac{1}{\dfrac{w}{k} + \dfrac{1}{h_i}} = \frac{h_i k}{wh_i + k} \tag{6.50}$$

Equation (6.49) can be solved for monthly average outer wall surface temperature:

$$\overline{T}_w = \frac{\overline{H}_t(\overline{\tau\alpha}) + \left(U_k\overline{T}_R + \overline{U}_o\overline{T}_a\right)\Delta t}{\left(U_k + \overline{U}_o\right)\Delta t} \tag{6.51}$$

Finally, the net monthly heat gain from the thermal storage wall to the building is obtained from:

$$Q_g = U_k A_w\left(\overline{T}_w - \overline{T}_R\right)N \times \Delta t \tag{6.52}$$

where

N = number of days in a month.

Methods for calculating the dump energy, Q_D, and auxiliary energy, Q_{aux}, are presented in Chapter 11, Section 11.4.2.

EXAMPLE 6.3

A building has a south-facing thermal storage wall with night insulation R_{ins} equal to 1.52 m² K/W, applied for 8 h. Estimate the monthly heat transfer through the wall into the indoor space with and without night insulation for the month of December. The following data are given.

1. $U_o = 3.7$ W/m² K.
2. $w = 0.42$ m.
3. $k = 2.0$ W/m K.
4. $h_i = 8.3$ W/m² K.
5. $\overline{H}_t = 9.8$ MJ/m² K.
6. $(\overline{\tau\alpha}) = 0.73$.
7. $\overline{T}_R = 20$ °C.
8. $\overline{T}_a = 1$ °C.
9. $A_w = 21.3$ m².

Solution

From Eq. (6.50), coefficient U_k can be calculated:

$$U_k = \frac{1}{\dfrac{w}{k} + \dfrac{1}{h_i}} = \frac{1}{\dfrac{0.42}{2.0} + \dfrac{1}{8.3}} = 3.026 \text{ W/m}^2 \text{ K}$$

The two cases are now examined separately.

Without Night Insulation.

From Eq. (6.51), we estimate the outer wall surface temperature $(\overline{U}_o = U_o)$:

$$\overline{T}_w = \frac{\overline{H}_t(\overline{\tau\alpha}) + (U_k\overline{T}_R + \overline{U}_o\overline{T}_a)\Delta t}{(U_k + \overline{U}_o)\Delta t} = \frac{9.8 \times 10^6 \times 0.73 + (3.026 \times 20 + 3.7 \times 1)86,400}{(3.026 + 3.7)86,400}$$

$$= 21.86 \text{ °C}$$

From Eq. (6.52),

$$Q_g = U_kA_w(\overline{T}_w - \overline{T}_R)N \times \Delta t = 3.026 \times 21.3(21.86 - 20) \times 31 \times 86,400 = 0.321 \text{ GJ}$$

With night insulation.

From Eq. (6.48) and by considering $F = 8/24 = 0.333$,

$$\overline{U}_o = (1 - F)U_o + F\left(\frac{U_o}{1 + R_{ins}U_o}\right) = (1 - 0.333) \times 3.7 + 0.333\left(\frac{3.7}{1 + 1.52 \times 3.7}\right)$$

$$= 2.65 \text{ W/m}^2 \text{ K}$$

From Eq. (6.51), we estimate the outer wall surface temperature:

$$\overline{T}_w = \frac{\overline{H}_t(\overline{\tau\alpha}) + (U_k\overline{T}_R + \overline{U}_o\overline{T}_a)\Delta t}{(U_k + \overline{U}_o)\Delta t} = \frac{9.8 \times 10^6 \times 0.73 + (3.026 \times 20 + 2.65 \times 1)86,400}{(3.026 + 2.65)86,400}$$

$$= 25.72 \text{ °C}$$

From Eq. (6.52).

$$Q_g = U_kA_w(\overline{T}_w - \overline{T}_R)N \times \Delta t = 3.026 \times 21.3(25.72 - 20) \times 31 \times 86,400 = 0.987 \text{ GJ}$$

So, by using night insulation, a considerable amount of loss is avoided and therefore more energy is transferred into the indoor space.

Use of phase change materials

Energy storage is enhanced with the use of phase change materials (PCMs), which change the phase of a material (usually from solid to liquid) and energy is stored in the form of latent heat, which is usually much larger than the sensible heat. Additionally, the change in phase is done at constant temperature according to the chemical composition of the material. These materials are usually microencapsulated into various building materials, like concrete, gypsum floor tiles and wallboard to increase thermal mass. Microencapsulation involves very small particles of PCM dispersed in the matrix of a building material. The usual material used for this purpose is paraffin wax, which contains *n*-octadecane with a melting point of 23 °C and heat of fusion of 184 kJ/kg, in gypsum.

The thermal improvements in a building due to the inclusion of PCMs depend on the climate, design and orientation of the construction, but also on the amount and type of PCM. Therefore, buildings employing PCM require a complete simulation of the thermal behavior of the designed space in the conditions of use carried out a priori.

Generally, PCM in buildings can be used for passive thermal energy storage in building envelopes and construction materials, for free-cooling and free-heating and for storing heat in solar air-heating systems.

Free-cooling is understood as a means to store ambient air coolness during the night, by solidifying the PCM, to supply cooling during the next day in summer. In fact the inside air of a building can be cooled down by exchanging heat with PCM. On the contrary, free-heating is understood as a means to store the solar radiation during daytime, by melting the PCM, to supply indoor heating during the night in winter. In principle, free-cooling or free-heating can keep the indoor air temperature in the comfortable region throughout the year if the thermo-physical properties of building envelope material are in the desired range. It has been found that with the help of PCM the indoor temperature fluctuations can be reduced significantly while maintaining desirable thermal comfort and decreasing the energy consumption.

In recent years the use of thermal energy storage in buildings with PCMs has become a topic with a lot of interest from the engineering community. Cabeza et al. (2011) presented a review of applications of PCMs in buildings. They also reviewed the requirements of the use of this technology, classification of materials, the phase change materials available and problems and possible solutions associated with the application of such materials in buildings.

Another important review is presented by Zhou et al. (2012), which covers investigations on thermal energy storage with PCMs in building applications, impregnation methods of PCM into construction materials, building applications and their thermal performance analyses, as well as numerical simulation of buildings with PCMs.

6.2.2 Building shape and orientation

The exposed surface area of a building is related to the rate at which the building gains or loses heat, while the volume is related to the ability of the building to store heat. Therefore, the ratio of volume to exposed surface area is widely used as an indicator of the rate at which the building heats up during the day and cools down at night. A high volume-to-surface ratio is preferable for a building that is desired to heat up slowly because it offers limited exposed surface for the control of both heat losses and gains (Dimoudi, 1997).

Building shape and orientation must be chosen in such a way so as to provide both heating and cooling. For heating, the designer must be careful to allow a solar access, i.e., allow the sun to reach the appropriate surfaces for the maximum possible hours, especially during the period from 9 a.m. to 3 p.m., which is the most useful energy period. For cooling, breeze and shading must be taken into consideration. The theoretical solar radiation impact for the various building surfaces can be obtained from appropriate tables, according to the time of the year and the surface orientation. From this analysis, the designer can select which surfaces should be exposed to or protected from the sun. Generally, south walls are the best solar collectors during wintertime but, together with the roof, they are the most problematic in summertime. With respect to shape, the best is the rectangular one with its long axis running in the east–west direction, because the south area receives three times more energy than the east or west. A square shape should be avoided, as well as the rectangular shape with its long axis running in the north–south direction.

One way to control the solar radiation reaching the building is to use trees that drop their leaves during winter in the sunlit area, such as south of the building. In this way, the sun reaches the surface in question during winter but the surface is in shade during summer.

6.2.3 Insulation

Insulation is a very important parameter to consider. In fact, before one considers any passive or active technique, the building must be well insulated to reduce thermal loads. The most important element of the building to insulate is the roof. This is very important for horizontal concrete roofs, which, during summertime, when the sun is higher in the sky, receive a considerable amount of radiation, which can increase the roof temperature considerably (Florides et al., 2000).

A good thermal insulation should ensure:

1. Healthy, comfortable and pleasant living without upsetting the thermal balance of the human body, which will satisfy the sense of comfort;
2. Economy in the consumption of energy by reducing the thermal losses from the building envelope;
3. Minimization of the initial cost for the installation of heating and cooling equipment;
4. Protection from noise as most thermal insulating materials are also sound insulators; and
5. Improved protection of the environment as a consequence of the reduced energy consumption of the building leading to lower GHG emissions.

The criteria that should be considered in the selection of the appropriate insulating material are:

1. Thermal characteristics:
   a. Thermal conductivity, k
   b. Dependence of thermal conductivity on temperature
   c. Dependence of thermal conductivity on humidity (the value of k increases substantially with moisture condensed in the mass of the thermal insulation)
   d. Specific heat
   e. Thermal expansion coefficient

2. Installation procedures:
   a. Prefabricated materials or construction in situ
   b. Protective measures required (protection from mechanical damages or environmental consequences)
3. Mechanical properties:
   a. Compression and bending stress
   b. Aging
   c. Density
   d. Elasticity, brittleness
4. Chemical behavior:
   a. Resistance in corrosion, micro-organisms, and insects
   b. Behavior in humidity (change of dimensions, permeance and moisture absorption)
   c. Fire resistance and maximum operating temperature
   d. Sensitivity to UV radiation exposure, various gases, seawater, etc.
5. Economic elements:
   a. Supply and installation cost
   b. Expenditure payback time
   c. Percentage of added value for the whole construction

A question often raised when thermal insulation is to be applied is whether the insulation should be installed on the inside, outside of a building element (wall or roof) or in the middle of a wall. There are certain advantages and certain disadvantages in each case. Generally, internal insulation is used in cases where the heating or cooling system is required to achieve the required indoor condition very quickly and we are not interested in the performance of the space when the heating or cooling system is switched off. Examples include schools, offices operating during daytime and holiday houses.

The advantages of internal insulation include:

1. Simple and fast construction;
2. Lower cost compared to external insulation;
3. Immediate response of the heating and cooling system; and
4. Insulating materials do not need protection from the external conditions (wind, humidity, solar radiation, etc.).

Disadvantages include:

1. Problem of indequate insulation and the creation of thermal bridges (items of the construction with inferior insulation allowing the passage of heat);
2. Quick loss of the indoor comfort conditions soon after an interruption of the heating or cooling system;
3. Possibility of surface condensation if vapor barrier is not installed;
4. Difficulty in hanging paintings, shelves, etc. on the walls;
5. Reduction of the useful internal building area; and
6. If applied on existing buildings, there is a disruption of the smooth operation of the building during installation.

External insulation is used in cases which do not require the immediate response of the building after switching on the heating or cooling system, but which demand good indoor comfort conditions long after the conditioning equipment has been switched off.

The advantages of external insulation include:

1. Conservation of good indoor conditions long after the heating or cooling equipment has been switched off due to the heat capacity of the building elements;
2. Bigger saving in energy due to the lowering of the operation time of the mechanical equipment;
3. Protection of the external surfaces from expansion and contraction because of the change in outdoor thermal conditions;
4. Minimization or non-existence of thermal bridges; and
5. In case of an installation on the existing buildings, there is no interaction with the building operation.

Disadvantages include:

1. Increased construction cost;
2. Care in construction is required (selection of proper materials and correct installation); and
3. Can be difficult to apply in buildings with many morphological features.

For the case of walls perhaps the more suitable installation for new buildings is to mount the insulation in between two layers of bricks either with or without an air gap. This solves most of the disadvantages of internal or external installation and combines the features of both constructions with respect to response time of the building to the mechanical heating and cooling, except the problem of thermal bridges. To avoid moisture migration in this construction an air gap is used between the external brick wall and the thermal insulation, which resists also the migration of moisture in the thermal insulation itself.

Another important measure often taken to increase the thermal insulation of buildings is to use a thermal barrier in double-glazed window aluminum frames. This is especially important in extreme environments and in cases where wide aluminum frames are used.

More details on this subject can be obtained from dedicated publications, such as the ASHRAE handbook series.

The way to account for insulation is by using Eq. (6.34), presented in Section 6.1.5, by incorporating the appropriate thickness and k value for the insulation used in a structure.

EXAMPLE 6.4

Consider the wall construction of Example 6.2 with the addition of 2.5 cm expanded polystyrene insulation. Estimate the new U value of the insulated wall construction.

Solution

The resistance value of the wall without insulation is 1.18 m^2 °C/W (from Example 6.2). From Table A5.4 in Appendix 5, the thermal conductivity of expanded polystyrene is 0.041 W/m °C. Therefore, its resistance is 0.025/0.041 = 0.610 m^2 °C/W. Therefore, the total resistance of the insulated wall construction is $1.18 + 0.610 = 1.79$ m^2 °C/W. Finally, the U value = 1/1.79 = 0.559 W/m^2 °C.

6.2.4 Windows: sunspaces

The direct gain system is the simplest way of achieving passive heating. In this system, sunlight enters through windows or skylights and is absorbed by the inside surfaces of the building. In a direct gain system, thermal collection, dissipation, storage, and transfer of energy take place within the habitable space. This is the most effective of all passive concepts from the point of view of energy collection and simplicity. Additionally, this system allows the use of sun for lighting (daylight). Control of the system is very simple with the use of curtains and venetian blinds. The most serious drawback of the system is the possible deterioration of certain materials from sunlight.

Generally, daylighting is the illumination of the building interiors with sunlight and sky light and is known to affect visual performance, lighting quality, health, human performance, and energy efficiency. In terms of energy efficiency, daylighting can provide substantial energy reductions, particularly in non-residential applications through the use of electric lighting controls. Daylight can displace the need for electric light at the perimeter zone with vertical windows and at the core zone with skylights (Kalogirou, 2007).

A modern way of controlling daylight entering the building is with electrochromic windows. These have the ability to change their transmittance according to an input voltage, so the system can easily be automated. Another way is by using thermochromic windows, in which the reflectance and transmittance properties change at a specific critical temperature. At this temperature, the material undergoes a semiconductor-to-metal transition. At low temperature, the window allows all the sun's energy to enter the room; above the critical temperature, it reflects the infrared portion of the sun's energy. Hence, thermochromic windows could be used to significantly reduce the thermal load of buildings.

The use of windows for space heating can be done with the normal building windows or with sunspaces. *Sunspaces* are usually special rooms made from glass attached to a normal building room, located in the south direction. The sunspace carries out the functions of thermal collection, storage, and transfer into the normal building spaces. There are three types of sunspaces, as shown in Fig. 6.6. The first uses just the south side of the building, the second uses the south side and part of the roof of the building, whereas the third is a semi-detached heating system for the main building and, in many cases, can be used as a greenhouse for growing plants. In the last type, because the sunspace is partially isolated from the main building, larger temperature swings can be accommodated in the sunspace than in a normal living room. The amount of energy received by the window surface is presented in the next section.

When designing a sunspace, the objective is to maximize the winter solar radiation received and minimize the summer one. When the sunspace is integrated into the house, good night insulation has to

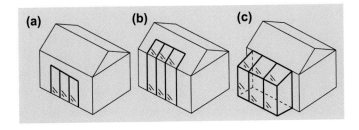

FIGURE 6.6

Various sunspace configurations.

be installed to protect the building spaces from excessive heating losses through the glass. If this is not possible, then double glazing should be used. The optimum orientation of a sunspace is due south, with variation of up to ±15° east or west being acceptable. Vertical glazing is preferred over sloped glazing, because it can be sealed more easily against leakages and reduces the tendency of sunspaces to overheat during summer. The performance of the vertical glazing, however, is about 15% lower than that of the optimally tilted glazing of equal area. A good compromise between vertical and sloped glazing is to use vertical glazing with some sloped portion (part of the roof), as shown in Fig. 6.6b.

In hot climates, where ventilation is a must, this can be done from the upper part of the sunspace. Therefore, the sunspace must be designed in such a way as to allow easy opening of some upper glass frames for ventilation.

6.2.5 Overhangs

Overhangs are devices that block direct solar radiation from entering a window during certain times of the day or the year. These are desirable for reducing the cooling loads and avoiding uncomfortable lighting in perimeter rooms due to excessive contrast. It would generally be advantageous to use long overhang projections in summer that could be retracted in winter, but in "real" buildings, the strategy is based not only on economic but also on aesthetic grounds.

To estimate the effect of the overhang length, the amount of shading needs to be calculated. For this purpose, the methodology presented in Chapter 2, Section 2.2.3, for the estimation of the profile angle, can be applied. By considering an overhang that extends far beyond the sides of the window, so as to be able to neglect the side effects, it is easier to estimate the fraction F $(0 \le F \le 1)$ of the window that will be shaded by the overhang. By considering an overhang with a perpendicular projection, P, and gap, G, above a window of height H, as shown in Fig. 6.7, the following relation developed by Sharp (1982) can be used:

$$F = \frac{P}{H} \frac{\sin(\alpha_c)}{\cos(\theta_c)} - \frac{G}{H} \tag{6.53}$$

where

α_c = solar altitude angle relative to the window aperture (°);
θ_c = incidence angle relative to the window aperture (°).

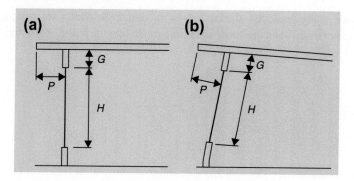

(a) **(b)**

FIGURE 6.7

Window with overhang: (a) vertical window, (b) general case of tilted window.

The fraction of the window that is sunlit can be obtained from:

$$FS = 1 + \frac{G}{H} - \frac{P}{H}\frac{\sin(\alpha_c)}{\cos(\theta_c)} \qquad (6.54)$$

The solar altitude and incidence angles relative to the window aperture are given by:

$$\sin(\alpha_c) = \sin(\beta)\cos(L)\cos(\delta)\cos(h) - \cos(\beta)\cos(Z_s)\sin(L)\cos(\delta)\cos(h)$$
$$- \cos(\beta)\sin(Z_s)\cos(\delta)\sin(h) + \sin(\beta)\sin(L)\sin(\delta) \qquad (6.55)$$
$$+ \cos(\beta)\cos(Z_s)\cos(L)\sin(\delta)$$

$$\cos(\theta_c) = \cos(\beta)\cos(L)\cos(\delta)\cos(h) + \sin(\beta)\cos(Z_s)\sin(L)\cos(\delta)\cos(h)$$
$$+ \sin(\beta)\sin(Z_s)\cos(\delta)\sin(h) + \cos(\beta)\sin(L)\sin(\delta) \qquad (6.56)$$
$$- \sin(\beta)\cos(Z_s)\cos(L)\sin(\delta)$$

where

β = surface tilt angle ($°$);
L = latitude ($°$);
δ = declination ($°$); and
Z_s = surface azimuth angle ($°$).

For the case of a vertical window shown in Fig. 6.7a, where the surface tilt angle is $90°$, the angle α_c is equal to the solar altitude angle α; therefore, Eqs. (6.55) and (6.56) become the same as Eqs. (2.12) and (2.19), respectively.

EXAMPLE 6.5

Estimate the shading fraction of a south-facing window 2 m in height, located in $40°$ latitude at 10 a.m. and 3 p.m. solar times on June 16. The overhang is wide enough to neglect the side effects and its length is 1 m, located 0.5 m above the top surface of the window. The window is tilted $15°$ from vertical and faces due south.

Solution

From Example 2.6, on June 16, $\delta = 23.35°$. The hour angle at 10 a.m. is $-30°$ and at 3 p.m. is $45°$. From the problem data, we have $P = 1$ m, $G = 0.5$ m, $H = 2$ m, $\beta = 75°$, and $Z_s = 0°$. Therefore, from Eqs. (6.55) and (6.56) we have the following:

At 10 a.m.:

$$\sin(\alpha_c) = \sin(75)\cos(40)\cos(23.35)\cos(-30) - \cos(75)\cos(0)\sin(40)\cos(23.35)\cos(-30)$$
$$- \cos(75)\sin(0)\cos(23.35)\sin(-30) + \sin(75)\sin(40)\sin(23.35)$$
$$+ \cos(75)\cos(0)\cos(40)\sin(23.35) = 0.7807$$

$$\cos(\theta_c) = \cos(75)\cos(40)\cos(23.35)\cos(-30) + \sin(75)\cos(0)\sin(40)\cos(23.35)\cos(-30)$$
$$+ \sin(75)\sin(0)\cos(23.35)\sin(-30) + \cos(75)\sin(40)\sin(23.35)$$
$$- \sin(75)\cos(0)\cos(40)\sin(23.35) = 0.4240$$

Continued

EXAMPLE 6.5–CONT'D

Therefore, from Eq. (6.53),

$$F = \frac{P}{H}\frac{\sin(\alpha_c)}{\cos(\theta_c)} - \frac{G}{H} = \frac{1 \times 0.7807}{2 \times 0.4240} - \frac{0.5}{2} = 0.671, \text{ or } 67.1\%$$

At 3 p.m.:

$$\sin(\alpha_c) = \sin(75)\cos(40)\cos(23.35)\cos(45) - \cos(75)\cos(0)\sin(40)\cos(23.35)\cos(45)$$
$$- \cos(75)\sin(0)\cos(23.35)\sin(45) + \sin(75)\sin(40)\sin(23.35)$$
$$+ \cos(75)\cos(0)\cos(40)\sin(23.35) = 0.6970$$

$$\cos(\theta_c) = \cos(75)\cos(40)\cos(23.35)\cos(45) + \sin(75)\cos(0)\sin(40)\cos(23.35)\cos(45)$$
$$+ \sin(75)\sin(0)\cos(23.35)\sin(45) + \cos(75)\sin(40)\sin(23.35)$$
$$- \sin(75)\cos(0)\cos(40)\sin(23.35) = 0.3045$$

Therefore, from Eq. (6.53),

$$F = \frac{P}{H}\frac{\sin(\alpha_c)}{\cos(\theta_c)} - \frac{G}{H} = \frac{1 \times 0.6970}{2 \times 0.3045} - \frac{0.5}{2} = 0.895, \text{ or } 89.5\%$$

The equation giving the area-average radiation received by the partially shaded window, by assuming that the diffuse and ground-reflected radiation is isotropic, is similar to Eq. (2.97):

$$I_w = I_B R_B F_w + I_D F_{w-s} + (I_B + I_D)\rho_G F_{w-g} \tag{6.57}$$

where the three terms represent the beam, diffuse, and ground-reflected radiation falling on the surface in question.

The factor F_w in the first term of Eq. (6.57) accounts for the shading of the beam radiation and can be estimated from Eq. (6.53) by finding the average of F for all sunshine hours. The third component of Eq. (6.57) accounts for the ground-reflected radiation and, by ignoring the reflections from the underside of the overhang, is equal to $[1 - \cos(90)]/2$, which is equal to 0.5. The second factor of Eq. (6.57) accounts for the diffuse radiation from the sky, and the view factor of the window, F_{w-s}, includes the effect of overhang. It should be noted that, for a window with no overhang, the value of F_{w-s} is equal to $[1 + \cos(90)]/2$, which is equal to 0.5 because half of the sky is hidden from the window surface. The values with an overhang are given in Table 6.1, where e is the relative extension of the overhang from the sides of the window, g is the relative gap between the top of the window and the overhang, w is the relative width of the window, and p is the relative projection of the overhang, obtained by dividing the actual dimensions with the window height (Utzinger and Klein, 1979).

A monthly average value of F_w can be calculated by summing the beam radiation with and without shading over a month:

$$\overline{F}_w = \frac{\int G_B R_B F_w dt}{\int G_B R_B dt} \tag{6.58}$$

Table 6.1 Window radiation view factor.

G	w	p = 0.1	p = 0.2	p = 0.3	p = 0.4	p = 0.5	p = 0.75	p = 1.0	p = 1.5	p = 2.0
Values for e = 0.00										
0.00	1	0.46	0.42	0.40	0.37	0.35	0.32	0.30	0.28	0.27
	4	0.46	0.41	0.38	0.35	0.32	0.27	0.23	0.19	0.16
	25	0.45	0.41	0.37	0.34	0.31	0.25	0.21	0.15	0.12
0.25	1	0.49	0.48	0.46	0.45	0.43	0.40	0.38	0.35	0.34
	4	0.49	0.48	0.45	0.43	0.40	0.35	0.31	0.26	0.23
	25	0.49	0.47	0.45	0.42	0.39	0.34	0.29	0.22	0.18
0.50	1	0.50	0.49	0.49	0.48	0.47	0.44	0.42	0.40	0.38
	4	0.50	0.49	0.48	0.46	0.45	0.41	0.37	0.31	0.28
	25	0.50	0.49	0.47	0.46	0.44	0.39	0.35	0.27	0.23
1.00	1	0.50	0.50	0.50	0.49	0.49	0.48	0.47	0.45	0.43
	4	0.50	0.50	0.49	0.49	0.48	0.46	0.43	0.39	0.35
	25	0.50	0.50	0.49	0.48	0.47	0.44	0.41	0.35	0.30
Values for e = 0.30										
0.00	1	0.46	0.41	0.38	0.33	0.33	0.28	0.25	0.22	0.20
	4	0.46	0.41	0.37	0.34	0.31	0.26	0.22	0.17	0.15
	25	0.45	0.41	0.37	0.34	0.31	0.25	0.21	0.15	0.12
0.25	1	0.49	0.48	0.46	0.43	0.41	0.37	0.34	0.30	0.28
	4	0.49	0.47	0.45	0.42	0.40	0.34	0.30	0.24	0.21
	25	0.49	0.47	0.45	0.42	0.39	0.33	0.29	0.22	0.18
0.50	1	0.50	0.49	0.48	0.47	0.45	0.42	0.39	0.35	0.33
	4	0.50	0.49	0.47	0.46	0.44	0.39	0.34	0.27	0.26
	25	0.50	0.49	0.47	0.46	0.44	0.39	0.34	0.27	0.22
1.00	1	0.50	0.50	0.49	0.49	0.48	0.47	0.45	0.42	0.40
	4	0.50	0.50	0.49	0.48	0.48	0.45	0.43	0.38	0.34
	25	0.50	0.50	0.49	0.48	0.47	0.44	0.41	0.35	0.30

Reprinted from Utzinger and Klein (1979), with permission from Elsevier.

Therefore, the mean monthly and area-average radiation on a shaded vertical window can be obtained by an equation similar to Eq. (2.107):

$$\overline{H}_w = \overline{H}\left[\left(1 - \frac{\overline{H}_D}{\overline{H}}\right)\overline{R}_B\overline{F}_w + \frac{\overline{H}_D}{\overline{H}}F_{w-s} + \frac{\rho_G}{2}\right] \tag{6.59}$$

Methods to estimate $\overline{H}_D/\overline{H}$ and $\overline{R}_B$ are described in Section 2.3.8 in Chapter 2. An easy way to estimate $\overline{F}_w$ is by using Eq. (6.53) and finding the average of F for all sunshine hours for the recommended average day for each month, shown in Table 2.1.

EXAMPLE 6.6

A window with height equal to 2 m and width equal to 8 m has an overhang with an extension equal to 0.5 m on both sides, gap 0.5 m, and the projection of 1.0 m. If $F_w = 0.3$, $R_B = 0.81$, $I_B = 3.05$ MJ/m^2, $I_D = 0.45$ MJ/m^2, and $\rho_G = 0.2$, estimate area–average radiation received by the window.

Solution

First, the relative dimensions are estimated. Therefore,

$$e = 0.5/2 = 0.25, \ w = 8/2 = 4, \ g = 0.5/2 = 0.25, \ p = 1.0/2 = 0.5$$

From what was said previously, $F_{w-g} = 0.5$, and from Table 6.1, $F_{w-s} = 0.40$.
From Eq. (6.57),

$$I_w = I_B R_B F_w + I_D F_{w-s} + (I_B + I_D)\rho_G F_{w-g}$$

$$= 3.05 \times 0.81 \times 0.3 + 0.45 \times 0.40 + (3.05 + 0.45) \times 0.2 \times 0.5 = 1.27 \ \text{MJ/m}^2$$

6.2.6 Natural ventilation

One way of achieving comfort is by direct evaporation of sweat with air movement through ventilation openings. In many parts of the earth, in some months of the year, local cooling of a building can be achieved with natural ventilation. This is achieved by allowing air that is colder than the building to enter the building. Doing so removes some of the heat stored in the building. The reduction of cooling load varies from 40% to 90%. The lower number applies in warm, humid areas and the larger in mild or dry areas. Natural ventilation also has some disadvantages: safety, noise, and dust. Safety is not a considerable issue but it creates extra cost to protect the ventilation openings from unauthorized entry; noise is a problem when the building is located near a road, whereas dust is always a problem but is more pronounced if a building is located near a road or in open fields.

The main objective of natural ventilation is to cool the building and not its occupants. Opening areas that are about 10% of a building floor area can give about 30 air changes per hour, which can remove considerable quantities of heat from the building. In some cases, such as in office buildings, natural ventilation is used during the night when the office is closed to remove excess heat; thus, the building requires less energy to cool during the next morning. This is usually combined with thermal mass effects, as explained in Section 6.2.1, to shift the maximum heat dissipation from the walls and roof during the night, where it can be removed by natural ventilation and thus lower the cooling load during the day.

When designing a natural ventilation system, it is important to consider the way air flows around the building. Generally, as shown in Fig. 6.8, wind creates a positive pressure on the windward side of the building, whereas on a side wall with respect to the windward side, the air creates a negative pressure, i.e., suction. A smaller amount of suction is also created on the leeward side of the building, due to eddies created from the wind. Therefore, an effective system is to allow openings in the windward and leeward sides, so as to create good cross-ventilation. It is usual to install insect screens to avoid having insects in the building. When these are used, they must be placed as far as possible from the window frame because they reduce the air flow by an amount equal to their blockage.

It should be noted that the presence of windows on two walls does not guarantee good cross-ventilation unless there is a significant pressure difference. Additionally, buildings with windows on only one external wall are difficult to ventilate, even if the wind strikes directly on these windows. In this case, ventilation is facilitated by having the two windows as far apart as possible and using devices such as wing walls, which can be added to the building exterior with fixed or movable structures

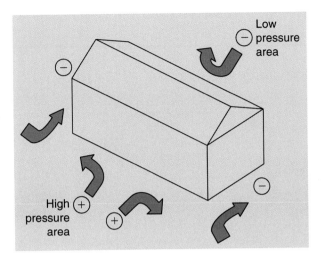

FIGURE 6.8

Pressure created because of the wind flow around a building.

(see Fig. 6.9). The objective of the wing wall is to cause one window to be in the positive pressure zone and the other in the negative.

If the inlet window is located in the center of the wall, the incoming air jet will maintain its shape for a length that is approximately equal to the window size and then disperse completely. If, however, the inlet window is located near a side wall, the air stream attaches to the wall. A similar effect is obtained when the window is very close to the floor or ceiling. In warm areas, the objective is to direct the air flow toward the hot room surfaces (walls and ceiling) to cool them. The relative location of the outlet window, high or low, does not appreciably affect the amount of air flow. For better effects, it is preferable to allow outside air to mix with the room air; therefore, the outlet should be in such a location as to make the air change direction before leaving the room.

FIGURE 6.9

Use of a wing wall to help natural ventilation of windows located on the same side of the wall.

To create higher inlet velocity, the inlet window should be smaller than the outlet one, whereas the air flow is maximized by having equal areas for the inlet and outlet windows.

6.3 Solar space heating and cooling

Systems for space heating are very similar to those for water heating, described in the previous chapter; and because the same considerations apply to both systems for combination with an auxiliary source, array design, over-temperature and freezing, controls, and so on, these will not be repeated. The most common heat transfer fluids are water, water and antifreeze mixtures, and air. The load is the building to be heated or cooled. Although it is technically possible to construct a solar heating or cooling system that can satisfy fully the design load, such a system would be non-viable, since it would be oversized most of the time. The size of the solar system may be determined by a life-cycle cost analysis described in Chapter 12.

Active solar space heating systems use collectors to heat a fluid, storage units to store solar energy until needed, and distribution equipment to provide the solar energy to the heated spaces in a controlled manner. Additionally, a complete system includes pumps or fans for transferring the energy to storage or to the load; these require a continuous availability of non-renewable energy, generally in the form of electricity.

The load can be space cooling, heating, or a combination of these two with hot water supply. When it is combined with conventional heating equipment, solar heating provides the same levels of comfort, temperature stability, and reliability as conventional systems.

During daytime, the solar energy system absorbs solar radiation with collectors and conveys it to storage using a suitable fluid. As the building requires heat, this is obtained from storage. Control of the solar energy system is exercised by differential temperature controllers, described in Chapter 5, Section 5.5. In locations where freezing conditions may occur, a low-temperature sensor is installed on the collector to control the solar pump when a preset temperature is reached. This process wastes some stored heat, but it prevents costly damage to the solar collectors. Alternatively, systems described in the previous chapter, such as the drain-down and drain-back, can be used, depending on whether the system is closed or open.

Solar cooling of buildings is an attractive idea because the cooling loads and availability of solar radiation are in phase. Additionally, the combination of solar cooling and heating greatly improves the use factors of collectors compared to heating alone. Solar air conditioning can be accomplished mainly by two types of systems: absorption cycles and adsorption (desiccant) cycles. Some of these cycles are also used in solar refrigeration systems. It should be noted that the same solar collectors are used for both space-heating and cooling systems when both are present.

A review of the various solar heating and cooling systems is presented by Hahne (1996) and a review of solar and low-energy cooling technologies is presented by Florides et al. (2002a).

6.3.1 Space heating and service hot water

Depending on the conditions that exist in a system at a particular time, the solar systems usually have five basic modes of operation:

1. When solar energy is available and heat is not required in the building, solar energy is added to storage.
2. When solar energy is available and heat is required in the building, solar energy is used to supply the building load demand.
3. When solar energy is not available, heat is required in the building, and the storage unit has stored energy, the stored energy is used to supply the building load demand.
4. When solar energy is not available, heat is required in the building, and the storage unit has been depleted, auxiliary energy is used to supply the building load demand.
5. When the storage unit is fully heated, there are no loads to meet, and the collector is absorbing heat, solar energy is discarded.

This last mode is achieved through the operation of pressure relief valves, or in the case of air collectors where the stagnant temperature is not detrimental to the collector materials, the flow of air is turned off; thus the collector temperature rises until the absorbed energy is dissipated by thermal losses.

In addition to the operation modes just outlined, the solar energy system is usually used to provide domestic hot water. These modes are usually controlled by thermostats. So, depending on the load of each service, heating, cooling, or hot water, the thermostat that is not satisfied gives a signal to operate a pump, provided that the collector temperature is higher than that of storage, as explained in Chapter 5, Section 5.5. Therefore, by using the thermostats, it is possible to combine modes and operate in more than one mode at a time. Some systems do not allow direct heating from a solar collector to heat the building but always transfer heat from collector to storage, whenever this is available, and from storage to load, whenever this is needed.

In Europe, solar heating systems for combined space and water heating are known as *combisystems*, and the storage tanks of these systems are called *combistores*. Many of these combistore tanks have one or more heat exchangers immersed directly in the storage fluid. The immersed heat exchangers are used for various functions, including charging via solar collectors or a boiler and discharging for domestic hot water and space heating uses.

For combisystems, the heat store is the key component, since it is used as short-term storage for solar energy and as buffer storage for the fuel or wood boiler. The storage medium used in solar combistores is usually the water of the space-heating loop and not the tap water used in conventional solar domestic hot water stores. The tap water is heated on demand by passing through a heat exchanger, which can be placed either inside or outside the tank containing the water of the heating loop. When the heat exchanger is in direct contact with the storage medium, the maximum tap water temperature at the start of the draw-off is similar to the temperature of the water inside the store. The tap water volume inside the heat exchanger can vary from a few liters for immersed heat exchangers to several hundred liters for tank-in-tank stores.

Three typical combistores are shown in Fig. 6.10. In the first, shown in Fig. 6.10a, an immersed heat exchanger is used, mounted on the whole inside surface of the mantle and top of the store. In the second store, shown in Fig. 6.10b, the preparation of hot water is based on a natural circulation (thermosiphoning) heat exchanger, which is mounted in the upper part of the store. The third one, shown in Fig. 6.10c, is the tank-in-tank case, where a conical hot water tank is placed inside the main tank, as shown, its bottom part reaching nearly the bottom of the store. Typical tap water volumes in heat exchanger are 15, 10, and 150−200 L for the three tanks, respectively (Druck and Hahne, 1998).

Legend: CW = Cold water, HW = Hot water, C = Collector, A = Auxiliary, SH = Space heating

FIGURE 6.10

(a) Immersed heat exchanger, (b) natural circulation heat exchanger, (c) tank-in-tank.

Adapted from Druck and Hahne (1998).

In the initial stages of design of a solar space-heating system, a number of factors need to be considered. Among the first ones is whether the system will be direct or indirect and whether a different fluid will be used in the solar system and the heat delivery system. This is decided primarily from the possibility of freezing conditions at the site of interest. Generally speaking, the designer must be aware that the presence of a heat exchanger in a system imposes a penalty of 5%–10% in the effective energy delivered to the system. This is usually translated as an extra percentage of collector area to allow the system to deliver the same quantity of energy as a system with no heat exchanger.

Another important parameter to consider is the time matching of the load and solar energy input. Over the annual seasonal cycle, energy requirements of a building are not constant. For the Northern Hemisphere, heating requirements start around October, the maximum heating load is during January or February, and the heating season ends around the end of April. Depending on the latitude, cooling requirements start in May, the maximum is about the end of July, and the cooling season ends around the end of September. The domestic hot water requirements are almost constant, with small variations due to variations in water supply temperature. Although it is possible to design a system that could cover the total thermal load requirements of a building, a very large collector area and storage would be required. Therefore, such a system would not be economically viable, because the system would be oversized for most of the year, i.e., it will collect energy that it could not use for most of the time.

As can be understood from above, the load is not constant and varies throughout the year and a space-heating system could be inoperative during many months of the year, which could create overheating problems in the solar collectors during summertime. To avoid this problem, a solar space-heating system needs to be combined with solar space cooling so as to utilize fully the solar system throughout the year. Solar heating systems are examined in this section, whereas solar cooling systems are examined in Section 6.4.

A space-heating system can use either air or liquid collectors, but the energy delivery system may use the same medium or a different one. Usually air systems use air for the collection, storage, and delivery system; but liquid systems may use water or water plus antifreeze solution for the collection circuit, water for storage, and water (e.g., a floor heating system) or air (e.g., a water-to-air heat exchanger and air handling unit) for the heat delivery.

Many variations of systems are used for both solar space heating and service hot water production. The basic configuration is similar to the solar water heating systems outlined in Sections 5.2.1–5.2.3. When used for both space and hot water production, these systems allow independent control of the solar collector-storage and storage-auxiliary-load loops, because solar-heated water can be added to storage at the same time that hot water is removed from storage to meet the building loads. Usually, a bypass is provided around the storage tank to avoid heating the storage tank, which can be of considerable size, with auxiliary energy. This is examined in more detail in Section 6.3.4.

6.3.2 Air systems

A schematic of a basic solar air-heating system with a pebble bed storage unit and auxiliary heating source is shown in Fig. 6.11. In this case, the various operation modes are achieved by the use of the dampers shown. Usually, in air systems, it is not practical to have simultaneous addition and removal of energy from the storage. If the energy supplied from the collector or storage is inadequate to meet the load, auxiliary energy can be used to top up the air temperature to cover the building load. As shown in Fig. 6.11, it is also possible to bypass the collector and storage unit when there is no sunshine and the storage tank is completely depleted, and use the auxiliary alone to provide the required heat. A more detailed schematic of an air space-heating system incorporating a subsystem for the preparation of domestic hot water is shown in Fig. 6.12. For the preparation of hot water, an air-to-water heat exchanger is used. Usually a preheat tank is used as shown. Details of controls are also shown in Fig. 6.12. Furthermore, the system can use air collectors and a hydronic space-heating system in an arrangement similar to the water-heating air system described in Section 5.2.3 and shown in Fig. 5.14.

FIGURE 6.11

Schematic of basic hot-air system.

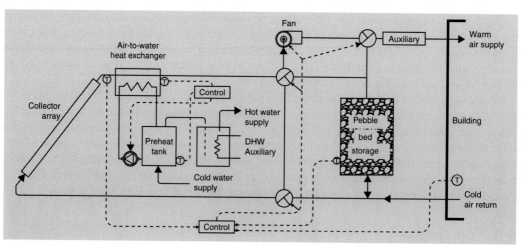

FIGURE 6.12

Detailed schematic of a solar air-heating system.

The advantages of using air as a heat transfer fluid are outlined in the section on water-heating air systems (Section 5.2.3). Other advantages include the high degree of stratification that occurs in the pebble bed, which leads to lower collector inlet temperatures. Additionally, the working fluid is air, and warm-air-heating systems are common in the building services industry. Control equipment that can be applied to these systems is also readily available from the building services industry. Amongst the disadvantages of water-heating air systems (see Section 5.2.3) is the difficulty of adding solar air conditioning to the system, higher storage costs, and noisier operation. Another disadvantage is that air collectors are operated at lower fluid capacitance rates and thus with lower values of F_R than the liquid-heating collectors.

Usually, air-heating collectors used in space-heating systems are operated at fixed air flow rates; therefore the outlet temperature varies through the day. It is also possible to operate the collectors at a fixed outlet temperature by varying the flow rate. When flow rates are low, however, they result in reduced F_R and therefore reduced collector performance.

6.3.3 Water systems

Many varieties of systems can be used for both solar space heating and domestic hot water production. The basic configurations are similar to the solar water heating systems outlined in Sections 5.2.1 and 5.2.2. When used for both space heating and hot water production and because solar-heated water can be added to storage at the same time that hot water is removed from storage to meet building loads, the system allows independent control of the solar collector-storage and storage-auxiliary-load loops. Usually, a bypass is provided around the storage tank to avoid heating the storage tank, which can be of considerable size, with auxiliary energy.

A schematic diagram of a solar heating and hot water system is shown in Fig. 6.13. Control of the solar heating system is based on two thermostats: the collector-storage temperature differential and the

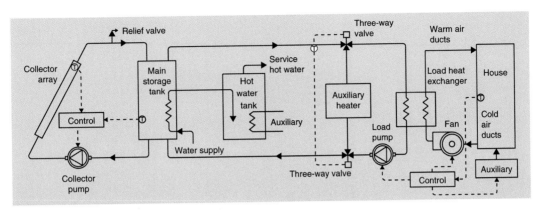

FIGURE 6.13

Schematic diagram of a solar space-heating and hot water system.

room temperature. The collector operates with a differential thermostat, as explained in Chapter 5, Section 5.5. When the room thermostat senses a low temperature, the load pump is activated, drawing the heated water from the main storage tank to meet the demand. If the energy stored in the tank cannot meet the load demand, the thermostat activates the auxiliary heater to supply the balance of the heating requirements. Usually, the controller also operates the three-way valves shown in Fig. 6.13 so that the flow is entirely through the auxiliary heater whenever the storage tank is depleted.

The solar heating system design shown in Fig. 6.13 is suitable for use in non-freezing climates. To use such a system in locations where freezing is possible, provisions for complete and dependable drainage of the collector must be made. This can be done with an automatic discharge valve, activated by the ambient air temperature, and an air vent. This is the usual arrangement of the drain-down system, described in Chapter 5, Section 5.2.1, in which the collector water is drained out of the system to waste. Alternatively, a drain-back system, described in Chapter 5, Section 5.2.2, can be used, in which the collector water is drained back to the storage whenever the solar pump stops. When this system drains, air enters the collector through a vent.

If the climate is characterized by frequent sub-freezing temperatures, positive freeze protection with the use of an antifreeze solution in a closed collector loop is necessary. A detailed schematic of such a liquid-based system is shown in Fig. 6.14. A collector heat exchanger is used between the collector and the storage tank, which allows the use of antifreeze solutions to the collector circuit. The usual solution is water plus glycol. Relief valves are also required to dump excess energy if over-temperature conditions exists. To "top off" the energy available from the solar energy system, auxiliary energy is required. It should be noted that the connections to the storage tank should be done in such a way as to enhance stratification, i.e., cold streams to be connected at the bottom and hot streams at the top. In this way, cooler water or fluid is supplied to the collectors to maintain the best possible efficiency. In this type of system, the auxiliary energy is never used directly in the solar storage tank. This is treated in more detail in the next section.

The use of a heat exchanger between the collector heat transfer fluid and the storage water imposes a temperature differential across the two sides, and thus the storage tank temperature is lowered. This is

FIGURE 6.14

Detailed schematic diagram of a solar space-heating and hot water system with antifreeze solution.

a disadvantage for system performance, as described in Chapter 5, Section 5.4.2; however, this system design is preferred in climates with frequent freezing conditions, to avoid the danger of malfunction in a self-draining system.

A load heat exchanger is also required, as shown in Fig. 6.14, to transfer energy from the storage tank to the heated spaces. It should be noted that the load heat exchanger must be adequately sized to avoid excessive temperature drops with a corresponding increase in the tank and collector temperatures.

The advantages of liquid heating systems are the high collector F_R, small storage volume requirement, and relatively easy combination with an absorption air conditioner for cooling (see Section 6.4.2).

The analysis of these systems is similar to the water heating systems presented in Chapter 5. When both space heating and hot water are considered, the rate of the energy removed from the storage tank to load (Q_l) in Eq. (5.31) is replaced by Q_{ls}, the space load supplied by solar energy through the load heat exchanger, and the term Q_{lw}, representing the domestic water heating load supplied through the domestic water heat exchanger, as shown in the following equation, which neglects stratification in the storage tank:

$$(Mc_p)_s \frac{dT_s}{dt} = Q_u - Q_{ls} - Q_{lw} - Q_{tl} \tag{6.60}$$

The terms Q_u and Q_{tl} can be obtained from Eq. (4.2) and Eq. (5.32), respectively.

The space-heating load, Q_{hl}, can be estimated from the following equation (positive values only):

$$Q_{hl} = (UA)_l (T_R - T_a)^+ \tag{6.61}$$

where

$(UA)_l$ = space loss coefficient and area product, given by Eq. (6.25).

The maximum rate of heat transferred across the load heat exchanger, $Q_{le(max)}$, is given by:

$$Q_{le(max)} = \varepsilon_l (\dot{m}c_p)_a (T_s - T_R) \tag{6.62}$$

where

ε_l = load heat exchanger effectiveness;
$(\dot{m}c_p)_a$ = air loop mass flow rate and specific heat product (W/K); and
T_s = storage tank temperature (°C).

It should be noted that, in Eq. (6.62), the air side capacitance rate of the water-to-air heat exchanger is considered to be the minimum because the c_p (~ 1.05 kJ/kg °C) of air is much lower than the c_p of water (~ 4.18 kJ/kg °C).

The space load, Q_{ls}, is then given by (positive values only):

$$Q_{ls} = [\min(Q_{le(max)}, Q_{hl})]^+ \tag{6.63}$$

The domestic water heating load, Q_w, can be estimated from:

$$Q_w = (\dot{m}c_p)_w (T_w - T_{mu}) \tag{6.64}$$

where

$(\dot{m}c_p)_w$ = domestic water mass flow rate and specific heat product (W/K);
T_w = required hot water temperature, usually 60 °C; and
T_{mu} = make-up water temperature from mains (°C).

The domestic water heating load supplied by solar energy through the domestic water heat exchanger, Q_{lw}, of effectiveness ε_w, can be estimated from:

$$Q_{lw} = \varepsilon_w (\dot{m}c_p)_w (T_s - T_{mu}) \tag{6.65}$$

Finally, the auxiliary energy required, Q_{aux}, to cover the domestic water heating and space loads is given by (positive values only):

$$Q_{aux} = (Q_{hl} + Q_{aux,w} + Q_{tl} - Q_{ls})^+ \tag{6.66}$$

where the auxiliary energy required to cover the domestic water heating load, $Q_{aux,w}$, is given by (positive values only):

$$Q_{aux,w} = (\dot{m}c_p)_w (T_w - T_s)^+ \tag{6.67}$$

EXAMPLE 6.7

A space is maintained at a room temperature $T_R = 21\ °C$ and has a $(UA)_l = 2500\ W/°C$. The ambient temperature is $1\ °C$ and the storage tank temperature is $80\ °C$. Estimate the space load, domestic water heating load, and auxiliary energy required if the following apply.

1. Heat exchanger effectiveness $= 0.7$.
2. Flow rate of air side of heat exchanger $= 1.1\ kg/s$.
3. Specific heat of air $= 1.05\ kJ/hg\ °C$.
4. Environmental temperature at the space where storage tank is located $= 15\ °C$.
5. UA of storage tank $= 2.5\ W/°C$.
6. Mass flow rate of domestic water $= 0.2\ kg/s$.
7. Required domestic water temperature $= 60\ °C$.
8. Make-up water temperature $= 12\ °C$.

Solution

The space-heating load, Q_{hl}, can be estimated from Eq. (6.61):

$$Q_{hl} = (UA)_l (T_R - T_a)^+ = 2500 \times (21 - 1) = 50\ kW$$

The maximum rate of heat transferred across the load heat exchanger, $Q_{le(max)}$, is given by Eq. (6.62):

$$Q_{le(max)} = \varepsilon_l (\dot{m}c_p)_a (T_s - T_R) = 0.7 \times (1.1 \times 1.05) \times (80 - 21) = 47.7\ kW$$

The space load, Q_{ls}, is then given by Eq. (6.63):

$$Q_{ls} = [\min(Q_{le(max)}, Q_{hl})]^+ = \min(47.7, 50)^+ = 47.7\ kW$$

The storage tank losses are estimated from Eq. (5.32):

$$Q_{tl} = (UA)_s (T_s - T_{env}) = (2.5) \times (80 - 15) = 162.5\ W = 0.16\ kW$$

The domestic water heating load, Q_w, can be estimated from Eq. (6.64):

$$Q_w = (\dot{m}c_p)_w (T_w - T_{mu}) = 0.2 \times 4.18 \times (60 - 12) = 40.1\ kW$$

The rate of auxiliary energy required to cover the domestic water heating load, $Q_{aux,w}$, is given by Eq. (6.67):

$$Q_{aux,w} = (\dot{m}c_p)_w (T_w - T_s)^+ = 0.2 \times 4.18 \times (60 - 80)$$

$$= -16.7\ kW\,(\text{not considered as it is negative})$$

The rate of auxiliary energy required, Q_{aux}, to cover the domestic water heating and space loads is given by Eq. (6.66):

$$Q_{aux} = (Q_{hl} + Q_{aux,w} + Q_{tl} - Q_{ls})^+ = 50 + 0 + 0.16 - 47.7 = 2.46\ kW$$

It should be noted that, in all cases where a heat exchanger is used, the penalty of using this heat exchanger can be estimated according to Eq. (5.57), as indicated by the following example.

EXAMPLE 6.8

A space-heating and hot water system has a collector with $F_R U_L = 5.71\ W/m^2\ °C$ and area of $16\ m^2$. What is the ratio F'_R / F_R if the flow rate of antifreeze is $0.012\ kg/s\ m^2$ and water is $0.018\ kg/s\ m^2$ in a collector–storage heat exchanger of effectiveness equal to 0.63, and the c_p of water is $4180\ J/kg\ °C$ and of antifreeze is $3350\ J/kg\ °C$?

EXAMPLE 6.8–CONT'D

Solution

First, the capacitance rates for the collector and tank sides are required, given by:

$$C_c = (\dot{m}c_p)_c = 0.012 \times 16 \times 3350 = 643.2 \ \text{W/}^\circ\text{C}$$

$$C_s = (\dot{m}c_p)_s = 0.018 \times 16 \times 4180 = 1203.8 \ \text{W/}^\circ\text{C}$$

Therefore, the minimum:

$$(\dot{m}c_p)_{min} = (\dot{m}c_p)_c = 643.2 \ \text{W/}^\circ\text{C}$$

From Eq. (5.57), we get:

$$\frac{F_R'}{F_R} = \left\{ 1 + \frac{A_c F_R U_L}{(\dot{m}c_p)_c} \left[\frac{(\dot{m}c_p)_c}{\varepsilon(\dot{m}c_p)_{min}} - 1 \right] \right\}^{-1} = \left[1 + \frac{16 \times 5.71}{643.2} \left(\frac{643.2}{0.63(643.2)} - 1 \right) \right]^{-1}$$

$$= 0.923$$

6.3.4 Location of auxiliary heater

One important consideration concerning the storage tank is the decision as to the best location for the auxiliary heater. This is especially important for solar space-heating systems because larger amounts of auxiliary energy are usually required and storage tank sizes are large. For maximum utilization of the energy supplied by an auxiliary source, the location of this energy input should be at the load, not at the storage tank. The supply of auxiliary energy at the storage tank will undoubtedly increase the temperature of fluid entering the collector, resulting in lower collector efficiency. When a water-based solar energy system is used in conjunction with a warm-air space-heating system, the most economical means of auxiliary energy supply is by the use of a fossil fuel-fired boiler. In case of bad weather, the boiler can take over the entire heating load.

When a water-based solar energy system is used in conjunction with a water space-heating system or to supply the heated water to an absorption air-conditioning unit, the auxiliary heater can be located in the storage-load loop, either in series or in parallel with the storage, as illustrated in Fig. 6.15. When auxiliary energy is employed to boost the temperature of solar energy heated water, as shown in Fig. 6.15a, maximum utilization of stored solar energy is achieved. This way of connecting the auxiliary supply, however, also has the tendency to boost the storage tank temperature because water returning from the load may be at a higher temperature than the storage tank. Increasing the storage temperature using auxiliary energy has the undesirable effect of lowering the collector effectiveness, in addition to diverting the storage capacity to the storage of auxiliary energy instead of solar energy. This, however, depends on the operating temperature of the heating system. Therefore, a low-temperature system is required. This can be achieved with a water-to-air heat exchanger, either centrally with an air-handling unit or in a distributed way with individual fan coil units in each room to be heated. This system has the advantage of connecting easily with space-cooling systems, as for example with an absorption system (see Section 6.4). By using this type of system, solar energy can be used more effectively, because a high-temperature system would imply that the hot water storage remains at high temperature, so solar collectors would work at lower efficiency.

Another possibility is to use an underfloor heating or an all-water system employing traditional heating radiators. In the latter case, provisions need to be made during the design stage to operate the system at low temperatures, which implies the use of bigger radiators. Such a system is also suitable for retrofit applications.

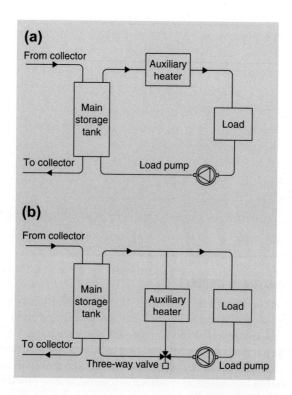

FIGURE 6.15

Auxiliary energy supply in water-based systems. (a) In series with load. (b) Parallel with load.

Figure 6.15b illustrates an arrangement that makes it possible to isolate the auxiliary heating circuit from the storage tank. Solar-heated storage water is used exclusively to meet load demands when its temperature is adequate. When the storage temperature drops below the required level, circulation through the storage tank is discontinued and hot water from the auxiliary heater is used exclusively to meet space-heating needs. This way of connecting the auxiliary supply avoids the undesirable increase in storage water temperature by auxiliary energy. However, it has the disadvantage that a stored solar energy at lower temperatures is not fully utilized, and this energy may be lost from the storage (through jacket losses). The same requirements for a low-temperature system apply here as well in order to be able to extract as much energy as possible from the storage tank.

6.3.5 Heat pump systems

Active solar energy systems can also be combined with heat pumps for domestic water heating or space heating. In residential heating, the solar energy system can be used in parallel with a heat pump, which supplies auxiliary energy when the sun is not available. Additionally, for domestic water systems requiring high water temperatures, a heat pump can be placed in series with the solar storage tank. Another type already described in Section 5.2.4 is the direct expansion solar-assisted heat pump.

A heat pump is a device that pumps heat from a low-temperature source to a higher-temperature sink. Heat pumps are usually vapor compression refrigeration machines, where the evaporator can take heat into the system at low temperatures and the condenser can reject heat from the system at high temperatures. In the heating mode, a heat pump delivers thermal energy from the condenser for space heating and can be combined with solar heating. In the cooling mode, the evaporator extracts heat from the air to be conditioned and rejects heat from the condenser to the atmosphere, with solar energy not contributing to the energy for cooling. The performance characteristics of an integral type of solar-assisted heat pump are given by Huang and Chyng (2001).

Heat pumps use mechanical energy to transfer thermal energy from a source at a lower temperature to a sink at a higher temperature. Electrically driven heat pump-heating systems have two advantages compared to electric resistance heating or expensive fuels. The first one, as was seen in Chapter 5, Section 5.2.4, is that the heat pump's ratio of heating performance to electrical energy (COP) is high enough to yield 9–15 MJ of heat for each kWh of energy supplied to the compressor, which saves on the purchase of energy. The second is the usefulness for air conditioning in the summer. Water-to-air heat pumps, which use solar-heated water from the storage tank as the evaporator energy source, are an alternative auxiliary heat source. Use of water involves freezing problems, which need to be taken into consideration.

Heat pumps have been used in combination with solar systems in residential and commercial applications. The additional complexity imposed by such a system and extra cost are offset by the high coefficient of performance and the lower operating temperature of the collector subsystem. A schematic of a common residential type heat pump system is shown in Fig. 6.16a. During favorable weather

FIGURE 6.16

Schematic diagram of a domestic water-to-water heat pump system (a) series arrangement. (b) parallel arrangement.

conditions, it is possible with this arrangement to have solar energy delivered directly to the forced air system while the heat pump is kept off.

The arrangement shown in Fig. 6.16a is a series configuration, where the heat pump evaporator is supplied with energy from the solar energy system, called a *water-to-air heat pump*. As can be seen, energy from the collector system is supplied directly to the building when the temperature of the water in the storage temperature is high. When the storage tank temperature cannot satisfy the load, the heat pump is operated; thus it benefits from the relatively high temperature of the solar energy system, which is higher than the ambient and this increases the heat pump's COP. A parallel arrangement is also possible, where the heat pump serves as an independent auxiliary energy source for the solar energy system, as shown in Fig. 6.16b. In this case, a *water-to-water heat pump* is used.

The series configuration is usually preferred because it allows all the solar collected power to be used, leaving the tank at a low temperature, which allows the solar energy system to work more effectively the next day. Additionally, the heat pump performance is higher with high evaporator temperatures. An added advantage of this system is that the solar energy system is conventional, using liquid collectors and a water storage tank. A dual-source heat pump can also be used, in which another form of renewable energy, such as a pellets boiler, can be used when the storage tank is completely depleted. In such a case, a control system selects the heat source to use for the best heat pump COP, i.e., it selects the higher of the two heat sources. Another possible design is to use an air solar heating system and an air-to-air heat pump.

Another type of solar-assisted heat pump system is the direct expansion unit, introduced briefly in Section 5.2.4. In a Direct Expansion SAHP system (DX-SAHP), solar energy is used directly to heat the working fluid of a heat pump, i.e., a refrigerant, and thus, the solar collector and the heat pump evaporator constitute one integrated unit. Two types of solar collectors can be used: unglazed and normal glazed with a single cover. In this system, the working fluid of the heat pump (refrigerant) flows through an integrated solar collector—evaporator. Due to solar radiation incident on the collector and the direct effect of the ambient surroundings, the refrigerant undergoes a phase change from liquid to vapor, so the evaporation process takes place at the location of the heat source, i.e., the solar collector, located at the roof of the building. The refrigerant is directly evaporated in the integrated solar collector—evaporator and then expanded.

A schematic diagram of the direct expansion of a solar-assisted heat pump is presented in Fig. 6.17. The thermal performance of such a system is improved if it operates under appropriate weather conditions (Kuang and Wang, 2006). A very important issue for DX-SAHP system operation is the selection of a suitable refrigerant, which seems to be one of the main problems hindering the widespread use of such systems (Gorozabel Chate et al., 2005). Another crucial and critical issue is the collector—evaporator panel sizing. The thermal capacity of solar collectors, which are also the evaporators of the heat pump, should be matched to the heat pumping capacity of the compressor. As the collector—evaporator temperature can fluctuate rapidly, flexible operation of the DX-SAHP system can be maintained through compressor capacity modulation using a variable speed compressor. The refrigerant, especially during phase change from liquid to vapor during the evaporation process, should operate under relatively stable conditions. However, in a DX-SAHP, the temperature level can change rapidly within a short time, i.e., during a few hours of a day or over a longer span. There are also large differences in the working fluid temperature between winter and summer—e.g., in summer, a refrigerant is evaporated at 30 °C or more, whereas a much lower temperature is used in winter.

FIGURE 6.17

Schematic diagram of a direct expansion solar assisted heat pump (DX-SAHP).

According to Omojaro and Breitkopf (2013), the COP of such systems can vary from 2 to 9, depending mainly on solar irradiation, ambient air temperature, and energy demand. Different methods and coefficients can be used to express the thermal performance of DX-SAHP systems. One coefficient used for whole systems is the seasonal performance factor (SPF), which can be expressed according to European Standard EN 14285. A new approach to determine this factor, based on a modified bin method and experimental results regarding DX-SAHP systems, has been proposed by Facao and Carvalho (2014).

6.4 Solar cooling

The quest for a safe and comfortable environment has always been one of the main preoccupations of the human race. In ancient times, people used the experience gained over many years to utilize in the best possible way the available resources to achieve adequate living conditions. Central heating was pioneered by the Romans, using double floors and passing the fumes of a fire through the floor cavity. Also in Roman times, windows were covered for the first time with materials such as mica or glass. Thus, light was admitted in the house without letting in wind and rain (Kreider and Rabl, 1994). The Iraqis, on the other hand, utilized the prevailing wind to take advantage of the cool night air and provide a cooler environment during the day (Winwood et al., 1997). Additionally, running water was employed to provide some evaporative cooling.

As late as the 1960s, though, house comfort conditions were only for the few. From then onward, central air-conditioning systems became common in many countries, due to the development of mechanical refrigeration and the rise in the standard of living. The oil crisis of the 1970s stimulated intensive research aimed at reducing energy costs. Also, global warming and ozone depletion and the escalating costs of fossil fuels over the last few years have forced governments and engineering bodies to re-examine the whole approach to building design and control. Energy conservation in the sense of fuel saving is also of great importance.

During recent years, research aimed at the development of technologies that can offer reductions in energy consumption, peak electrical demand, and energy costs without lowering the desired level of comfort conditions has intensified. Alternative cooling technologies that can be applied to residential and commercial buildings, under a wide range of weather conditions, are being developed. These include night cooling with ventilation, evaporative cooling, desiccant cooling, and slab cooling. The design of buildings employing low-energy cooling technologies, however, presents difficulties and requires advanced modeling and control techniques to ensure efficient operation.

Another method that can be used to reduce energy consumption is ground cooling. This is based on the heat loss dissipation from a building to the ground, which during the summer has a lower temperature than the ambient. This dissipation can be achieved either by direct contact of an important part of the building envelope with the ground or by blowing into the building air that has first been passed through an earth-to-air heat exchanger (Argiriou, 1997).

The role of designers and architects is very important, especially with respect to solar energy control, the utilization of thermal mass, and the natural ventilation of buildings, as was seen in Section 6.2.6. In effective solar energy control, summer heat gains must be reduced, while winter solar heat gains must be maximized. This can be achieved by proper orientation and shape of the building, the use of shading devices, and the selection of proper construction materials. Thermal mass, especially in hot climates with diurnal variation of ambient temperatures exceeding 10 °C, can be used to reduce the instantaneous high cooling loads, reduce energy consumption, and attenuate indoor temperature swings. Correct ventilation can enhance the roles of both solar energy control and thermal mass.

Reconsideration of the building structure; the readjustment of capital cost allocations, i.e., investing in energy conservation measures that may have a significant influence on thermal loads; and improvements in equipment and maintenance can minimize the energy expenditure and improve thermal comfort.

In intermediate seasons in hot, dry climates, processes such as evaporative cooling can offer energy conservation opportunities. However, in summertime, due to the high temperatures, low-energy cooling technologies alone cannot satisfy the total cooling demand of domestic dwellings. For this reason active cooling systems are required. Vapor compression cooling systems are usually used, powered by electricity, which is expensive and its production depends mainly on fossil fuel. In such climates, one source abundantly available is solar energy, which could be used to power an active solar cooling system based on the absorption cycle. The problem with solar absorption machines is that they are more expensive compared to vapor compression machines, and until recently, they were not readily available in the small-capacity range applicable to domestic cooling applications. Reducing the use of conventional vapor compression air-conditioning systems will also reduce their effect on both global warming and ozone layer depletion.

The integration of the building envelope with an absorption system should offer better control of the internal environment. Two basic types of absorption units are available: ammonia–water and lithium bromide (LiBr) water units. The latter are more suitable for solar applications since their operating (generator) temperature is lower and thus more readily obtainable with low-cost solar collectors (Florides et al., 2001).

The solar cooling of buildings is an attractive idea because the cooling loads and availability of solar radiation are in phase. Additionally, the combination of solar cooling and heating greatly improves the use factors of collectors compared to heating alone. Solar air conditioning can be accomplished by three types of systems: absorption cycles, adsorption (desiccant) cycles, and solar mechanical processes. Some of these cycles are also used in solar refrigeration systems and are described in the following sections.

Solar cooling can be considered for two related processes: to provide refrigeration for food and medicine preservation and to provide comfort cooling. Solar refrigeration systems usually operate at intermittent cycles and produce much lower temperatures (ice) than in air conditioning. When the same systems are used for space cooling they operate on continuous cycles. The cycles employed for solar refrigeration are absorption and adsorption. During the cooling portion of the cycles, the refrigerant is evaporated and re-absorbed. In these systems, the absorber and generator are separate vessels. The generator can be an integral part of the collector, with refrigerant absorbent solution in the tubes of the collector circulated by a combination of a thermosiphon and a vapor lift pump.

Many options enable the integration of solar energy into the process of "cold" production. Solar refrigeration can be accomplished by using either a thermal energy source supplied from a solar collector or electricity supplied from photovoltaics. This can be achieved by using either thermal adsorption or absorption units, or conventional vapor compression refrigeration equipment powered by photovoltaics. Solar refrigeration is employed mainly to cool vaccine stores in areas with no public electricity.

Photovoltaic refrigeration, although it uses standard refrigeration equipment, which is an advantage, has not achieved widespread use because of the low efficiency of the photovoltaic cells, although today with the reduced PV prices these are more viable (Li et al., 2018). Basic details of photovoltaics-operated vapor compression systems are given in Section 6.6. More details however are given on the solar adsorption and absorption units, with more emphasis on the latter.

Solar cooling is more attractive for the southern countries of the Northern Hemisphere and the northern countries of the Southern Hemisphere. Solar cooling systems are particularly applicable to large applications (e.g., commercial buildings) that have high cooling loads for large periods of the year. Such systems in combination with solar heating can make more efficient use of solar collectors, which would be idle during the cooling season. Generally, however, there is much less experience with solar cooling than solar heating systems.

Solar cooling systems can be classified into three categories: solar-sorption cooling, solar-mechanical systems, and solar-related systems examined briefly in the following sections (Florides et al., 2002a).

Solar-sorption cooling

Sorbents are materials that have an ability to attract and hold other gases or liquids. Desiccants are sorbents that have a particular affinity for water. The process of attracting and holding moisture is described as either absorption or adsorption, depending on whether the desiccant undergoes a chemical change as it takes on moisture. Absorption changes the desiccant the way, for example, table salt changes from a solid to a liquid as it absorbs moisture. Adsorption, on the other hand, does not change the desiccant except by the addition of the weight of water vapor, similar in some ways to a sponge soaking up the water (ASHRAE, 2005).

Compared to an ordinary cooling cycle, the basic idea of an absorption system is to avoid compression work. This is done by using a suitable working pair: a refrigerant and a solution that can absorb the refrigerant.

Absorption systems are similar to vapor-compression air-conditioning systems but differ in the pressurization stage. In general, an evaporating refrigerant is absorbed by an absorbent on the low-pressure side. Combinations include lithium bromide–water ($LiBr$–H_2O), where water vapor is the refrigerant, and ammonia–water (NH_3–H_2O) systems, where ammonia is the refrigerant (Keith et al., 1996).

Adsorption cooling is the other group of sorption air conditioners that utilizes an agent (the adsorbent) to adsorb the moisture from the air (or dry any other gas or liquid), then uses the evaporative cooling effect to produce cooling. Solar energy can be used to regenerate the drying agent. Solid adsorbents include silica gels, zeolites, synthetic zeolites, activated alumina, carbons, and synthetic polymers (ASHRAE, 2005). Liquid adsorbents can be triethylene glycol solutions of lithium chloride and LiBr solutions.

More details of these systems are given in separate sections further on.

Solar-mechanical systems

Solar-mechanical systems utilize a solar-powered prime mover to drive a conventional air-conditioning system. This can be done by converting solar energy into electricity by means of photovoltaic devices, then utilizing an electric motor to drive a vapor compressor. The photovoltaic panels, however, have a low field efficiency of about 10%–15%, depending on the type of cells used, which results in low overall efficiencies for the system. More details on these systems are given in Section 6.6.

The solar-powered prime mover can also be a Rankine engine. In a typical system, energy from the collector is stored, then transferred to a heat exchanger, and finally energy is used to drive the heat engine (see Chapter 10). The heat engine drives a vapor compressor, which produces a cooling effect at the evaporator. As shown in Fig. 6.18, the efficiency of the solar collector decreases as the operating temperature increases, whereas the efficiency of the heat engine of the system increases as the operating temperature increases. The two efficiencies meet at a point (A in Fig. 6.18), providing an optimum operating temperature for a steady-state operation. The combined system has overall efficiencies between 17% and 23%.

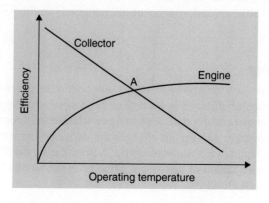

FIGURE 6.18

Collector and power cycle efficiencies as a function of operating temperature.

Due to the diurnal cycle, both the cooling load and the storage tank temperature vary through the day. Therefore, designing such a system presents appreciable difficulties. When a Rankine heat engine is coupled with a constant-speed air conditioner, the output of the engine seldom matches the input required by the air conditioner. Therefore, auxiliary energy must be supplied when the engine output is less than that required; otherwise, excess energy may be used to produce electricity for other purposes.

Solar-related air conditioning

Some components of systems installed for the purpose of heating a building can also be used to cool it but without the direct use of solar energy. Examples of these systems can be heat pumps, rock bed regenerators, and alternative cooling technologies or passive systems. Heat pumps were examined in Section 6.3.5. The other two methods are briefly introduced here:

- *Rock bed regenerator.* Rock bed (or pebble bed) storage units of solar air-heating systems can be night-cooled during summer to store "cold" for use the following day. This can be accomplished during the night when the temperatures and humidities are low by passing outside air through an optional evaporative cooler, through the pebble bed, and to the exhaust. During the day, the building can be cooled by passing the room air through the pebble bed. A number of applications using pebble beds for a solar energy storage are given by Hastings (1999). For such systems, airflow rates should be kept to a minimum so as to minimize fan power requirements without affecting the performance of the pebble bed. Therefore, an optimization process should be followed as part of the design.
- *Alternative cooling technologies or passive systems.* Passive cooling is based on the transfer of heat by natural means from a building to environmental sinks, such as clear skies, the atmosphere, the ground, and water. The transfer of heat can be by radiation, naturally occurring wind, airflow due to temperature differences, conduction to the ground, or conduction and convection to bodies of water. It is usually up to the designer to select the most appropriate type of technology for each application. The options depend on the climate type.

More details for the adsorption and absorption systems follow.

6.4.1 Adsorption units

Porous solids, called *adsorbents*, can physically and reversibly adsorb large volumes of vapor, called the *adsorbate*. Though this phenomenon, called *solar adsorption*, was recognized in nineteenth century, its practical application in the field of refrigeration is relatively recent. The concentration of adsorbate vapors in a solid adsorbent is a function of the temperature of the pair, i.e., the mixture of adsorbent and adsorbate and the vapor pressure of the latter. The dependence of adsorbate concentration on temperature, under constant pressure conditions, makes it possible to adsorb or desorb the adsorbate by varying the temperature of the mixture. This forms the basis of the application of this phenomenon in the solar-powered intermittent vapor sorption refrigeration cycle.

An adsorbent–refrigerant working pair for a solar refrigerator must have the following characteristics:

1. A refrigerant with a large latent heat of evaporation.
2. A working pair with high thermodynamic efficiency.

3. A low heat of desorption under the envisaged operating pressure and temperature conditions.
4. A low thermal capacity.

Water−ammonia has been the most widely used sorption refrigeration pair, and research has been undertaken to utilize the pair for solar-operated refrigerators. The efficiency of such systems is limited by the condensing temperature, which cannot be lowered without introduction of advanced and expensive technology. For example, cooling towers or desiccant beds have to be used to produce cold water to condensate ammonia at lower pressure. Among the other disadvantages inherent in using water and ammonia as the working pair are the heavy-gauge pipe and vessel walls required to withstand the high pressure, the corrosiveness of ammonia, and the problem of rectification, i.e., removing water vapor from ammonia during generation. A number of solid adsorption working pairs, such as zeolite−water, zeolite−methanol, and activated carbon−methanol, have been studied to find the one that performed best. The activated carbon−methanol working pair was found to perform the best (Norton, 1992).

Many cycles have been proposed for adsorption cooling and refrigeration (Dieng and Wang, 2001). The principle of operation of a typical system is indicated in Fig. 6.19. The process followed at the points from 1 to 9 of Fig. 6.19 is traced on the psychrometric chart depicted in Fig. 6.20. Ambient air is heated and dried by a dehumidifier from point 1 to 2, regeneratively cooled by exhaust air in 2−3, evaporatively cooled in 3−4, and introduced into the building. Exhaust air from the building is evaporatively cooled from points 5 to 6, heated to 7 by the energy removed from the supply air in the regenerator, heated by solar energy or another source to 8, then passed through the dehumidifier, where it regenerates the desiccant.

The selection of the adsorbing agent depends on the size of the moisture load and application.

Rotary solid desiccant systems are the most common for continuous removal of moisture from the air. The desiccant wheel rotates through two separate air streams. In the first stream, the process air is dehumidified by adsorption, which does not change the physical characteristics of the desiccant; in the second stream the reactivation or regeneration air, which is first heated, dries the desiccant. Figure 6.21 is a schematic of a possible solar-powered adsorption system.

FIGURE 6.19

Schematic of a solar adsorption system.

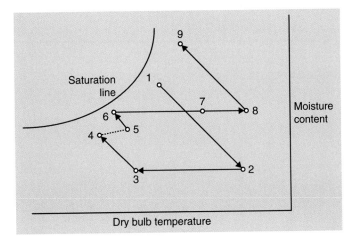

FIGURE 6.20

Psychrometric diagram of a solar adsorption process.

When the drying agent is a liquid, such as triethylene glycol, the agent is sprayed into an absorber, where it picks up moisture from the building air. Then, it is pumped through a sensible heat exchanger to a separation column, where it is sprayed into a stream of solar-heated air. The high-temperature air removes water from the glycol, which then returns to the heat exchanger and the absorber. Heat exchangers are provided to recover sensible heat, maximize the temperature in the separator, and minimize the temperature in the absorber. This type of cycle is marketed commercially and used in hospitals and large installations (Duffie and Beckman, 1991).

The energy performance of these systems depends on the system configuration, geometries of dehumidifiers, properties of adsorbent agent, and the like, but generally the COP of this technology is around 1.0. It should be noted, however, that in hot, dry climates the desiccant part of the system may not be required.

Because complete physical property data are available for only a few potential working pairs, the optimum performance remains unknown at the moment. In addition, the operating conditions of a solar-powered refrigerator, i.e., generator and condenser temperature, vary with its geographical location (Norton, 1992).

FIGURE 6.21

Solar adsorption cooling system.

The development of three solar-biomass adsorption air-conditioning and refrigeration systems is presented by Critoph (2002). All systems use active carbon–ammonia adsorption cycles and the principle of operation and performance prediction of the systems are given.

Thorpe (2002) presented an adsorption heat pump system that uses ammonia with a granular active adsorbate. A high COP is achieved and the cycle is suitable for the use of heat from high-temperature (150–200 °C) solar collectors for air conditioning.

6.4.2 Absorption units

Absorption is the process of attracting and holding moisture by substances called *desiccants*. Desiccants are sorbents, i.e., materials that have an ability to attract and hold other gases or liquids that have a particular affinity for water. During absorption, the desiccant undergoes a chemical change as it takes on moisture; an example mentioned before is table salt, which changes from a solid to a liquid as it absorbs moisture. The characteristic of the binding of desiccants to moisture makes the desiccants very useful in chemical separation processes (ASHRAE, 2005).

Absorption machines are thermally activated, and they do not require high input shaft power. Therefore, where power is unavailable or expensive, or where there is waste, geothermal, or solar heat available, absorption machines could provide reliable and quiet cooling. Absorption systems are similar to vapor compression air-conditioning systems but differ in the pressurization stage. In general, an absorbent, on the low-pressure side, absorbs an evaporating refrigerant. The most usual combinations of fluids include $LiBr–H_2O$, where water vapor is the refrigerant, and ammonia–water $(NH_3–H_2O)$ systems, where ammonia is the refrigerant.

Absorption refrigeration systems are based on extensive development and experience in the early years of the refrigeration industry, in particular for ice production. From the beginning, its development has been linked to periods of high energy prices. Recently, however, there has been a great resurgence of interest in this technology not only because of the rise in the energy prices but mainly due to the social and scientific awareness about the environmental degradation, which is related to the energy generation.

The pressurization is achieved by dissolving the refrigerant in the absorbent, in the absorber section (Fig. 6.22). Subsequently, the solution is pumped to a high pressure with an ordinary liquid pump. The addition of heat in the generator is used to separate the low-boiling refrigerant from the solution. In this

FIGURE 6.22

Basic principle of the absorption air-conditioning system.

way, the refrigerant vapor is compressed without the need for large amounts of mechanical energy that the vapor compression air-conditioning systems demand.

The remainder of the system consists of a condenser, expansion valve, and evaporator, which function in a similar way as in a vapor compression air-conditioning system.

Lithium bromide–water absorption systems

The LiBr–H_2O system operates at a generator temperature in the range of 70–95 °C, with water used as a coolant in the absorber and condenser, and has a COP higher than the NH_3–H_2O systems. The COP of this system is between 0.6 and 0.8. A disadvantage of the LiBr–H_2O systems is that their evaporator cannot operate at temperatures much below 5 °C, since the refrigerant is water vapor. Commercially available absorption chillers for air-conditioning applications usually operate with a solution of lithium bromide in water and use steam or hot water as the heat source. Two types of chillers are available on the market: the single effect and the double effect.

The single-effect absorption chiller is used mainly for building cooling loads, where chilled water is required at 6–7 °C. The COP will vary to a small extent with the heat source and the cooling water temperatures. Single effect chillers can operate with hot water temperature ranging from about 70 to 150 °C when water is pressurized (Florides et al., 2003).

The double-effect absorption chiller has two stages of generation to separate the refrigerant from the absorbent. Therefore, the temperature of the heat source needed to drive the high-stage generator is essentially higher than that needed for the single-effect machine and is in the range of 155–205 °C. Double-effect chillers have a higher COP of about 0.9–1.2 (Dorgan et al., 1995). Although double-effect chillers are more efficient than the single-effect machines, they are obviously more expensive to purchase. However, every individual application must be considered on its own merits, since the resulting savings in capital cost of the single-effect units can largely offset the extra capital cost of the double-effect chiller.

The Carrier Corporation pioneered lithium-bromide absorption chiller technology in the United States, with the early single-effect machines introduced around 1945. Due to the success of the product, soon other companies joined in production. The absorption business thrived until 1975. Then, the generally held belief that natural gas supplies were lessening led to U.S. government regulations prohibiting the use of gas in new constructions and, together with the low cost of electricity, led to the declination of the absorption refrigeration market (Keith, 1995). Today the major factor in the decision on the type of system to install for a particular application is the economic trade-off between different cooling technologies. Absorption chillers typically cost less to operate, but they cost more to purchase than vapor compression units. The payback period depends strongly on the relative cost of fuel and electricity, assuming that the operating cost for the needed heat is less than the operating cost for electricity.

The technology was exported to Japan from the United States early in the 1960s, and Japanese manufacturers set a research and development program to further improve the absorption systems. The program led to the introduction of the direct-fired double-effect machines with improved thermal performance.

Today gas-fired absorption chillers deliver 50% of the commercial space-cooling load worldwide but less than 5% in the United States, where electricity-driven vapor compression machines carry the majority of the load (Keith, 1995).

Many researchers have developed solar-assisted absorption refrigeration systems. Most of them have been produced as experimental units, and computer codes were written to simulate the systems. Some of these designs are presented here.

Hammad and Audi (1992) described the performance of a non-storage, continuous, solar-operated absorption refrigeration cycle. The maximum ideal coefficient of performance of the system was determined to be equal to 1.6, while the peak actual coefficient of performance was determined to be equal to 0.55.

Haim et al. (1992) performed a simulation and analysis of two open-cycle absorption systems. Both systems comprise a closed absorber and evaporator, as in conventional single-stage chillers. The open part of the cycle is the regenerator, used to re-concentrate the absorber solution by means of solar energy. The analysis was performed with a computer code developed for modular simulation of absorption systems under varying cycle configurations (open and closed cycle systems) and with different working fluids. Based on the specified design features, the code calculates the operating parameters in each system. Results indicate that there is a definite performance advantage of the direct regeneration system over the indirect one.

Hawlader et al. (1993) developed a LiBr absorption cooling system employing an 11×11 m collector–regenerator unit. They also developed a computer model, which they validated against real experimental values with good agreement. The experimental results showed a regeneration efficiency varying between 38% and 67% and the corresponding cooling capacities ranged from 31 to 72 kW.

Ghaddar et al. (1997) presented the modeling and simulation of a solar absorption system for Beirut. The results showed that each ton of refrigeration requires a minimum collector area of 23.3 m^2 with an optimum water-storage capacity ranging from 1000 to 1500 L for the system to operate solely on the solar energy for about 7 h/day. The monthly solar fraction of total energy use in cooling is determined as a function of the solar collector area and storage tank capacity. The economic analysis performed showed that the solar cooling system is marginally competitive only when it is combined with domestic water heating.

Erhard and Hahne (1997) simulated and tested a solar-powered absorption cooling machine. The main part of the device is an absorber–desorber unit, which is mounted inside a concentrating solar collector. Results obtained from field tests are discussed and compared with the results obtained from a simulation program developed for this purpose.

Hammad and Zurigat (1998) described the performance of a 1.5 ton solar cooling unit. The unit comprises a 14 m^2 flat-plate solar collector system and five shell and tube heat exchangers. The unit was tested in April and May in Jordan. The maximum value obtained for actual coefficient of performance was 0.85.

Zinian and Ning (1999) described a solar absorption air-conditioning system that uses an array of 2160 evacuated tubular collectors of total aperture area of 540 m^2 and a LiBr absorption chiller. Thermal efficiencies of the collector array are 40% for space cooling, 35% for space heating, and 50% for domestic water heating. It was found that the cooling efficiency of the entire system is around 20%.

Ameel et al. (1995) gave performance predictions of alternative low-cost absorbents for open-cycle absorption using a number of absorbents. The most promising of the absorbents considered was a mixture of two elements, lithium chloride and zinc chloride. The estimated capacities per unit absorber area were 50%–70% less than those of LiBr systems.

Calise (2012) presented a dynamic model of an innovative solar heating and cooling system (SHC) based on the coupling of parabolic trough collectors with a double-stage LiBr–H$_2$O absorption chiller, in which the auxiliary energy for both heating and cooling is supplied by a biomass-fired heater. The consumption of non-renewable energy resources is only due to the small amount of electrical energy consumed by some auxiliary devices. A case study is presented, in which the SHC provides space heating and cooling and domestic hot water for a small university hall, for the whole year. Both the SHC system and the building were dynamically simulated in TRNSYS. The analysis was also performed for a similar SHC in which the biomass heater was replaced by a gas-fired heater, in order to evaluate the influence of biomass on the overall system's economic and energetic performance.

A new family of ICPC designs developed by Winston et al. (1999) allows a simple manufacturing approach to be used and solves many of the operational problems of previous ICPC designs (see Section 3.1.3). A low concentration ratio that requires no tracking is used with an off-the-shelf, 20 ton, double-effect, LiBr, direct-fired absorption chiller, modified to work with hot water. The new ICPC design and double-effect chiller were able to produce cooling energy for the building using a collector field that was about half the size of that required for a more conventional collector and chiller.

A method to design, construct, and evaluate the performance of a single-stage LiBr–H$_2$O absorption machine is presented by Florides et al. (2003). In this work, the necessary heat and mass transfer relations and appropriate equations describing the properties of the working fluids are specified. Information on designing the heat exchangers of the LiBr–H$_2$O absorption unit is also presented. Single-pass, vertical tube heat exchangers have been used for the absorber and the evaporator. The solution heat exchanger was designed as a single-pass annulus heat exchanger. The condenser and the generator were designed using horizontal tube heat exchangers. Another valuable source of LiBr–H$_2$O system properties is the program EES (Engineering Equation Solver), which can also be used to solve the equations required to design such a system (Klein, 1992).

If power generation efficiency is considered, the thermodynamic efficiency of absorption cooling is very similar to that of the electrically driven compression refrigeration system. The benefits of the solar systems, however, are very obvious when environmental pollution is considered. This is accounted for by the total equivalent warming impact (TEWI) of the system. As proven by Florides et al. (2002c) in a study of domestic size systems, the TEWI of the absorption system was 1.2 times smaller than that of the conventional system.

Thermodynamic analysis

Compared to an ordinary cooling cycle, the basic idea of an absorption system is to avoid compression work by using a suitable working pair. The working pair consists of a refrigerant and a solution that can absorb the refrigerant. A more detailed schematic of the LiBr–H$_2$O absorption system is shown in Fig. 6.23 (Kizilkan et al., 2007), and a schematic presentation on a pressure–temperature diagram is illustrated in Fig. 6.24.

The main components of an absorption refrigeration system are the generator, absorber, condenser, and evaporator. In the model shown, Q_G is the heat input rate from the heat source to the generator, Q_C and Q_A are the heat rejection rates from condenser and absorber to the heat sinks, respectively, and Q_E is the heat input rate from the cooling load to the evaporator.

With reference to the numbering system shown in Fig. 6.23, at point 1, the solution is rich in refrigerant and a pump (1–2) forces the liquid through a heat exchanger to the generator. The temperature of the solution in the heat exchanger is increased (2–3).

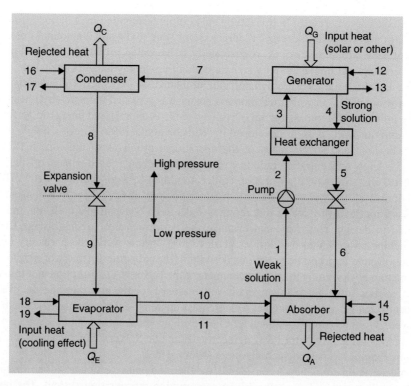

FIGURE 6.23

Schematic diagram of an absorption refrigeration system.

FIGURE 6.24

Pressure–temperature diagram of a single effect, LiBr–water absorption cycle.

In the generator, thermal energy is added and refrigerant boils off the solution. The refrigerant vapor (7) flows to the condenser, where heat is rejected as the refrigerant condenses. The condensed liquid (8) flows through a flow restrictor to the evaporator (9). In the evaporator, the heat from the load evaporates the refrigerant, which flows back to the absorber (10). A small portion of the refrigerant leaves the evaporator as a liquid spillover (11). At the generator exit (4), the stream consists of absorbent–refrigerant solution, which is cooled in the heat exchanger. From points 6 to 1, the solution absorbs refrigerant vapor from the evaporator and rejects heat through a heat exchanger. This procedure can also be presented in a Duhring chart (Fig. 6.25). This chart is a pressure–temperature graph, where diagonal lines represent constant LiBr mass fraction, with the pure water line at the left.

For the thermodynamic analysis of the absorption system, the principles of mass conservation and the first and second laws of thermodynamics are applied to each component of the system. Each component can be treated as a control volume with inlet and outlet streams, heat transfer, and work interactions. In the system, mass conservation includes the mass balance of each material of the solution. The governing equations of mass and type of material conservation for a steady-state, steady-flow system are (Herold et al., 1996):

$$\sum \dot{m}_i - \sum \dot{m}_o = 0 \tag{6.68}$$

$$\sum (\dot{m} \cdot x)_i - \sum (\dot{m} \cdot x)_o = 0 \tag{6.69}$$

where $\dot{m}$ is the mass flow rate and x is mass concentration of LiBr in the solution. The first law of thermodynamics yields the energy balance of each component of the absorption system as follows:

$$\sum (\dot{m} \cdot h)_i - \sum (\dot{m} \cdot h)_o + \left[\sum Q_i - \sum Q_o\right] + W = 0 \tag{6.70}$$

An overall energy balance of the system requires that the sum of the generator, evaporator, condenser, and absorber heat transfer must be 0. If the absorption system model assumes that the

FIGURE 6.25

Duhring chart of the water–lithium bromide absorption cycle.

system is in a steady state and that the pump work and environmental heat losses are neglected, the energy balance can be written as:

$$Q_C + Q_A = Q_G + Q_E \tag{6.71}$$

The energy, mass concentrations, and mass balance equations of the various components of an absorption system are given in Table 6.2 (Kizilkan et al., 2007). The equations of Table 6.2 can be used to estimate the energy, mass concentrations, and mass balance of a LiBr–H$_2$O system. In addition to these equations, the solution heat exchanger effectiveness is also required, obtained from (Herold et al., 1996):

$$\varepsilon_{SHx} = \frac{T_4 - T_5}{T_4 - T_2} \tag{6.72}$$

The absorption system shown in Fig. 6.23 provides chilled water for cooling applications. Furthermore, the system in Fig. 6.23 can also supply hot water for heating applications, by circulating the working fluids in the same fashion. The difference of operation between the two applications is the useful output energy and the operating temperature and pressure levels in the system. The useful output energy of the system for heating applications is the sum of the heat rejected from the absorber and the condenser while the input energy is supplied to the generator. The useful output energy of the system for the cooling applications is heat extracted from the environment by the evaporator while the input energy is supplied to the generator (Alefeld and Radermacher, 1994; Herold et al., 1996).

The cooling coefficient of performance of the absorption system is defined as the heat load in the evaporator per unit of heat load in the generator and can be written as (Herold et al., 1996; Tozer and James, 1997):

$$\text{COP}_{\text{cooling}} = \frac{Q_E}{Q_G} = \frac{\dot{m}_{10}h_{10} + \dot{m}_{11}h_{11} - \dot{m}_9 h_9}{\dot{m}_4 h_4 + \dot{m}_7 h_7 - \dot{m}_3 h_3} = \frac{\dot{m}_{18}(h_{18} - h_{19})}{\dot{m}_{12}(h_{12} - h_{13})} \tag{6.73}$$

Table 6.2 Energy and mass balance equations of absorption system components.

System components	Mass balance equations	Energy balance equations
Pump	$\dot{m}_1 = \dot{m}_2, x_1 = x_2$	$w = \dot{m}_2 h_2 - \dot{m}_1 h_1$
Solution heat exchanger	$\dot{m}_2 = \dot{m}_3, x_2 = x_3$	$\dot{m}_2 h_2 + \dot{m}_4 h_4 = \dot{m}_3 h_3 + \dot{m}_5 h_5$
	$\dot{m}_4 = \dot{m}_5, x_4 = x_5$	
Solution expansion valve	$\dot{m}_5 = \dot{m}_6, x_5 = x_6$	$h_5 = h_6$
Absorber	$\dot{m}_1 = \dot{m}_6 + \dot{m}_{10} + \dot{m}_{11}$	$Q_A = \dot{m}_6 h_6 + \dot{m}_{10} h_{10} + \dot{m}_{11} h_{11} - \dot{m}_1 h_1$
	$\dot{m}_1 x_1 = \dot{m}_6 x_6 + \dot{m}_{10} x_{10} + \dot{m}_{11} x_{11}$	
Generator	$\dot{m}_3 = \dot{m}_4 + \dot{m}_7$	$Q_G = \dot{m}_4 h_4 + \dot{m}_7 h_7 - \dot{m}_3 h_3$
	$\dot{m}_3 x_3 = \dot{m}_4 x_4 + \dot{m}_7 x_7$	
Condenser	$\dot{m}_7 = \dot{m}_8, x_7 = x_8$	$Q_C = \dot{m}_7 h_7 - \dot{m}_8 h_8$
Refrigerant expansion valve	$\dot{m}_8 = \dot{m}_9, x_8 = x_9$	$h_8 = h_9$
Evaporator	$\dot{m}_9 = \dot{m}_{10} + \dot{m}_{11}, x_9 = x_{10}$	$Q_E = \dot{m}_{10} h_{10} + \dot{m}_{11} h_{11} - \dot{m}_9 h_9$

where

h = specific enthalpy of working fluid at each corresponding state point (kJ/kg).

The heating COP of the absorption system is the ratio of the combined heating capacity, obtained from the absorber and condenser, to the heat added to the generator and can be written as (Herold et al., 1996; Tozer and James, 1997):

$$\text{COP}_{\text{heating}} = \frac{Q_C + Q_A}{Q_G} = \frac{(\dot{m}_7 h_7 - \dot{m}_8 h_8) + (\dot{m}_6 h_6 + \dot{m}_{10} h_{10} + \dot{m}_{11} h_{11} - \dot{m}_1 h_1)}{\dot{m}_4 h_4 + \dot{m}_7 h_7 - \dot{m}_3 h_3}$$

$$= \frac{\dot{m}_{16}(h_{17} - h_{16}) + \dot{m}_{14}(h_{15} - h_{14})}{\dot{m}_{12}(h_{12} - h_{13})}$$

(6.74)

Therefore, from Eq. (6.71), the COP for heating can be also written as:

$$\text{COP}_{\text{heating}} = \frac{Q_G + Q_E}{Q_G} = 1 + \frac{Q_E}{Q_G} = 1 + \text{COP}_{\text{cooling}}$$

(6.75)

Equation (6.75) shows that the heating COP is in all cases greater than the cooling COP.

The second-law analysis can be used to calculate the system performance based on exergy. Exergy analysis is the combination of the first and second laws of thermodynamics and is defined as the maximum amount of work potential of a material or an energy stream, in relation to the surrounding environment (Kizilkan et al., 2007). The exergy of a fluid stream can be defined as (Kotas, 1985; Ishida and Ji, 1999):

$$\varepsilon = (h - h_o) - T_o(s - s_o)$$

(6.76)

where

ε = specific exergy of the fluid at temperature T (kJ/kg).

The terms h and s are the enthalpy and entropy of the fluid, whereas h_o and s_o are the enthalpy and entropy of the fluid at environmental temperature T_o (in all cases absolute temperature is used in Kelvins).

The availability loss in each component is calculated by:

$$\Delta E = \sum \dot{m}_i E_i - \sum \dot{m}_o E_o - \left[\sum Q\left(1 - \frac{T_o}{T}\right)_i - \sum Q\left(1 - \frac{T_o}{T}\right)_o \right] + \sum W$$

(6.77)

where

ΔE = lost exergy or irreversibility that occurred in the process (kW).

The first two terms of the right-hand side of Eq. (6.77) are the exergy of the inlet and outlet streams of the control volume. The third and fourth terms are the exergy associated with the heat transferred from the source maintained at a temperature, T. The last term is the exergy of mechanical work added to the control volume. This term is negligible for absorption systems because the solution pump has very low power requirements.

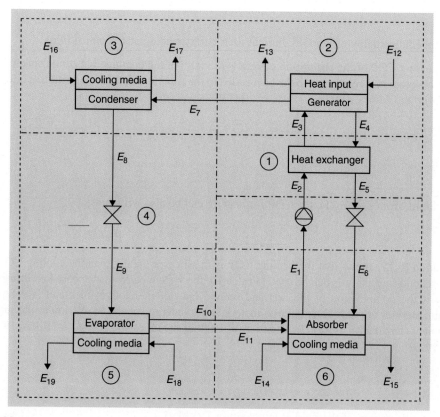

FIGURE 6.26

Availability flow balance of the absorption system.

The equivalent availability flow balance of the system is shown in Fig. 6.26 (Sencan et al., 2005). The total exergy loss of absorption system is the sum of the exergy loss in each component and can be written as (Talbi and Agnew, 2000):

$$\Delta E_T = \Delta E_1 + \Delta E_2 + \Delta E_3 + \Delta E_4 + \Delta E_5 + \Delta E_6 \qquad (6.78)$$

The second-law efficiency of the absorption system is measured by the exergetic efficiency, η_{ex}, which is defined as the ratio of the useful exergy gained from a system to that supplied to the system. Therefore, the exergetic efficiency of the absorption system for cooling is the ratio of the chilled water exergy at the evaporator to the exergy of the heat source at the generator and can be written as (Talbi and Agnew, 2000; Izquerdo et al., 2000):

$$\eta_{ex,cooling} = \frac{\dot{m}_{18}(E_{18} - E_{19})}{\dot{m}_{12}(E_{12} - E_{13})} \qquad (6.79)$$

The exergetic efficiency of absorption systems for heating is the ratio of the combined supply of hot water exergy at the absorber and condenser to the exergy of the heat source at the generator and can be written as (Lee and Sherif, 2001; Çengel and Boles, 1994):

$$\eta_{ex,heating} = \frac{\dot{m}_{16}(E_{17} - E_{16}) + \dot{m}_{14}(E_{15} - E_{14})}{\dot{m}_{12}(E_{12} - E_{13})} \qquad (6.80)$$

Design of single-effect LiBr–water absorption systems

To perform estimations of equipment sizing and performance evaluation of a single-effect water-lithium bromide absorption cooler, basic assumptions and input values must be considered. With reference to Figs. 6.23–6.25, usually the following assumptions are made:

1. The steady-state refrigerant is pure water.
2. There are no pressure changes except through the flow restrictors and the pump.
3. At points 1, 4, 8, and 11, there is only saturated liquid.
4. At point 10, there is only saturated vapor.
5. Flow restrictors are adiabatic.
6. The pump is isentropic.
7. There are no jacket heat losses.

A small 1 kW unit was designed and constructed by co-workers and the author (Florides et al., 2003). To design such a system, the design (or input) parameters must be specified. The parameters considered for the 1 kW unit are listed in Table 6.3.

The equations of Table 6.2 can be used to estimate the energy, mass concentrations, and mass balance of a LiBr–H$_2$O system. Some details are given in the following paragraphs so the reader will understand the procedure required to design such a system.

Table 6.3 Design parameters for the single-effect water–lithium bromide absorption cooler.

Parameter	Symbol	Value
Capacity	$\dot{Q}_E$	1.0 kW
Evaporator temperature	T_{10}	6 °C
Generator solution exit temperature	T_4	75 °C
Weak solution mass fraction	x_1	55% LiBr
Strong solution mass fraction	x_4	60% LiBr
Solution heat exchanger exit temperature	T_3	55 °C
Generator (desorber) vapor exit temperature	T_7	70 °C
Liquid carry-over from evaporator	$\dot{m}_{11}$	$0.025\dot{m}_{10}$

Since, in the evaporator, the refrigerant is a saturated water vapor and the temperature (T_{10}) is 6 °C, the saturation pressure at point 10 is 0.9346 kPa (from steam tables) and the enthalpy is 2511.8 kJ/kg. Since, at point 11, the refrigerant is a saturated liquid, its enthalpy is 23.45 kJ/kg. The enthalpy at point 9 is determined from the throttling process applied to the refrigerant flow restrictor, which yields $h_9 = h_8$. To determine h_8, the pressure at this point must be determined. Since, at point 4, the solution mass fraction is 60% LiBr and the temperature at the saturated state is assumed to be 75 °C, the LiBr–water charts (see ASHRAE, 2005) give a saturation pressure of 4.82 kPa and $h_4 = 183.2$ kJ/kg. Considering that the pressure at point 4 is the same as in 8, $h_8 = h_9 = 131.0$ kJ/kg (steam tables). Once the enthalpy values at all ports connected to the evaporator are known, mass and energy balances, shown in Table 6.2, can be applied to give the mass flow of the refrigerant and the evaporator heat transfer rate.

The heat transfer rate in the absorber can be determined from the enthalpy values at each of the connected state points. At point 1, the enthalpy is determined from the input mass fraction (55%) and the assumption that the state is a saturated liquid at the same pressure as the evaporator (0.9346 kPa). The enthalpy value at point 6 is determined from the throttling model, which gives $h_6 = h_5$.

The enthalpy at point 5 is not known but can be determined from the energy balance on the solution heat exchanger, assuming an adiabatic shell, as follows:

$$\dot{m}_2 h_2 + \dot{m}_4 h_4 = \dot{m}_3 h_3 + \dot{m}_5 h_5 \qquad (6.81)$$

The temperature at point 3 is an input value (55 °C) and since the mass fraction for points 1 to 3 is the same, the enthalpy at this point is determined as 124.7 kJ/kg. Actually, the state at point 3 may be a sub-cooled liquid. However, at the conditions of interest, the pressure has an insignificant effect on the enthalpy of the sub-cooled liquid and the saturated value at the same temperature and mass fraction can be an adequate approximation.

The enthalpy at state 2 can be determined from the equation for the pump shown in Table 6.2 or from an isentropic pump model. The minimum work input (w) can therefore be obtained from:

$$w = \dot{m}_1 v_1 (p_2 - p_1) \qquad (6.82)$$

In Eq. (6.82), it is assumed that the specific volume (v, m³/kg) of the liquid solution does not change appreciably from point 1 to point 2. The specific volume of the liquid solution can be obtained from a curve fit of the density (Lee et al., 1990) and noting that $v = 1/\rho$:

$$\rho = 1145.36 + 470.84x + 1374.79x^2 - (0.333393 + 0.571749x)(273 + T) \qquad (6.83)$$

This equation is valid for $0 < T < 200$ °C and $20 < x < 65\%$.

The temperature at point 5 can be determined from the enthalpy value. The enthalpy at point 7 can be determined, since the temperature at this point is an input value. In general, the state at point 7 is superheated water vapor and the enthalpy can be determined once the pressure and temperature are known.

A summary of the conditions at various parts of the unit is shown in Table 6.4; the point numbers are as shown in Fig. 6.23.

Ammonia–water absorption systems

Contrary to compression refrigeration machines, which need high-quality electric energy to run, ammonia–water absorption refrigeration machines use low-quality thermal energy. Moreover,

Table 6.4 LiBr–water absorption refrigeration system calculations based on a generator temperature of 75 °C and a solution heat exchanger exit temperature of 55 °C.

Point	h (kJ/kg)	ṁ (kg/s)	P (kPa)	T (°C)	% LiBr (x)	Remarks
1	83	0.00517	0.93	34.9	55	
2	83	0.00517	4.82	34.9	55	
3	124.7	0.00517	4.82	55	55	Sub-cooled liquid
4	183.2	0.00474	4.82	75	60	
5	137.8	0.00474	4.82	51.5	60	
6	137.8	0.00474	0.93	44.5	60	
7	2612.2	0.000431	4.82	70	0	Superheated steam
8	131.0	0.000431	4.82	31.5	0	Saturated liquid
9	131.0	0.000431	0.93	6	0	
10	2511.8	0.000421	0.93	6	0	Saturated vapor
11	23.45	0.000011	0.93	6	0	Saturated liquid

Description	Symbol	Power
Capacity (evaporator output power)	Q_S	1.0 kW
Absorber heat, rejected to the environment	Q_A	1.28 kW
Heat input to the generator	Q_G	1.35 kW
Condenser heat, rejected to the environment	Q_C	1.07 kW
Coefficient of performance	COP	0.74

because the temperature of the heat source does not usually need to be so high (80–170 °C), the waste heat from many processes can be used to power absorption refrigeration machines. In addition, an ammonia–water refrigeration system uses natural substances, which do not cause ozone depletion as working fluids. For all these reasons, this technology has been classified as environmentally friendly (Herold et al., 1996; Alefeld and Radermacher, 1994).

The NH_3–H_2O system is more complicated than the LiBr–H_2O system, since it needs a rectifying column to assure that no water vapor enters the evaporator, where it could freeze. The NH_3–H_2O system requires generator temperatures in the range of 125–170 °C with an air-cooled absorber and condenser and 80–120 °C when water cooling is used. These temperatures cannot be obtained with flat-plate collectors. The coefficient of performance, which is defined as the ratio of the cooling effect to the heat input, is between 0.6 and 0.7.

The single-stage ammonia–water absorption refrigeration system cycle consists of four main components—condenser, evaporator, absorber, and generator—as shown in Fig. 6.27. Other auxiliary components include expansion valves, pump, rectifier, and heat exchanger. Low-pressure, weak solution is pumped from the absorber to the generator through the solution heat exchanger operating at high pressure. The generator separates the binary solution of water and ammonia by causing the ammonia to vaporize and the rectifier purifies the ammonia vapor. High-pressure ammonia gas is passed through the expansion valve to the evaporator as low-pressure liquid ammonia. The high-pressure transport fluid (water) from the generator is returned to the absorber through the solution

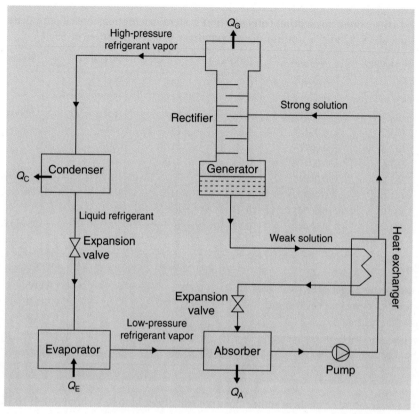

FIGURE 6.27

Schematic of the ammonia—water refrigeration system cycle.

heat exchanger and the expansion valve. The low-pressure liquid ammonia in the evaporator is used to cool the space to be refrigerated. During the cooling process, the liquid ammonia vaporizes and the transport fluid (water) absorbs the vapor to form a strong ammonia solution in the absorber (ASHRAE, 2005; Herold et al., 1996).

In some cases, a condensate pre-cooler is used to evaporate a significant amount of the liquid phase. This is, in fact, a heat exchanger located before the expansion valve, in which the low-pressure refrigerant vapor passes to remove some of the heat of the high-pressure and relatively high-temperature ($\sim$40 °C) ammonia. Therefore, some liquid evaporates and the vapor stream is heated, so there is additional cooling capacity available to further sub-cool the liquid stream, which increases the COP.

6.5 Solar cooling with absorption refrigeration

The greatest disadvantage of a solar heating system is that a large number of collectors need to be shaded or disconnected during summertime to reduce the danger of overheating. A way to avoid this problem and increase the viability of the solar system is to employ a combination of space heating and cooling and domestic hot water production system.

This is economically viable when the same collectors are used for both space heating and cooling. Flat-plate solar collectors are commonly used in solar space heating. Good-quality flat-plate collectors can attain temperatures suitable for LiBr–H_2O absorption systems. Another alternative is to use evacuated tube collectors, which can give higher temperatures; thus ammonia–water systems can be used, which need higher temperatures to operate.

A schematic diagram of a solar-operated absorption refrigeration system is shown in Fig. 6.28. The refrigeration cycle is the same as the ones described in Section 6.4.2. The difference between this system and the traditional fossil fuel-fired units is that the energy supplied to the generator is from the solar collector system shown on the left side of Fig. 6.28. Due to the intermittent nature of available solar energy, a hot water storage tank is needed; thus the collected energy is first stored in the tank and used as an energy source in the generator to heat the strong solution when needed. The storage tank of the solar heating system is used for this purpose. When the storage tank temperature is low, the auxiliary heater is used to top it off to the required generator temperature. Again, the same auxiliary heater of the space-heating system can be used, at a different set temperature. If the storage tank is completely depleted, the storage is bypassed, as in the space-heating system, to avoid boosting the storage temperature with auxiliary energy, and the auxiliary heater is used to meet the heating load of the generator. As in the case of space heating, the auxiliary heater can be arranged in parallel or in series with the storage tank. A collector heat exchanger can also be used to keep the collector liquid separate from the storage tank water (indirect system).

It should be noted that the operating temperature range of the hot water supplied to the generator of a LiBr–H_2O absorption refrigeration system is from 70 to 95 °C. The lower temperature limit is imposed from the fact that hot water must be at a temperature sufficiently high (at least 70 °C) to be effective for boiling the water off the solution in the generator. Also, the temperature of the concentrated LiBr solution returning to the absorber must be high enough to prevent crystallization of the LiBr. An unpressurized water storage tank system is usually employed in a solar energy system, and therefore an upper limit of about 95 °C is used to prevent water from boiling. For this type of system, the optimum generator temperature was found to be 93 °C (Florides et al., 2003).

Since in an absorption–refrigeration cycle heat must be rejected from the absorber and the condenser, a cooling water system must be employed in the cycle. Perhaps the most effective way of providing cooling water to the system is to use a cooling tower, as shown on the right side of Fig. 6.28. Because the absorber requires a lower temperature than the condenser, the cool water from the cooling tower is first directed to the absorber and then to the condenser. It should be noted that the use of a cooling tower in a small residential system is problematic with respect to both space and maintenance requirements; therefore, whenever possible, water drawn from a well or geothermal heat exchangers can be used.

A variation of the basic system shown in Fig. 6.28 is to eliminate the hot storage tank and the auxiliary heater and to supply the solar-heated fluid directly to the generator of the absorption unit. The advantage of this arrangement is that higher temperatures are obtained on sunny days, which increase the performance of the generator. The disadvantages are the lack of stored energy to produce cooling during evenings and on cloudy days and variations in the cooling load due to variations in the solar energy input. To minimize the intermittent effects of this arrangement, due to the absence of hot water storage, and make this system more effective, cold storage can be used. One way of doing this is to use

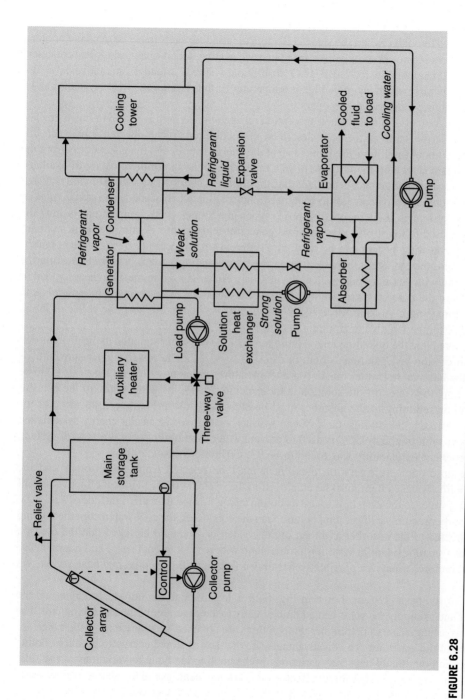

FIGURE 6.28

Schematic diagram of a solar-operated absorption refrigeration system.

the absorption machine to produce chilled water, which is then stored for cooling purposes (Hsieh, 1986). Such a solution would have the advantage of low-rate tank heat gains (actually is a loss in this case) because of the smaller temperature difference between the chilled water and its surroundings. An added disadvantage, however, is that the temperature range of a cool storage is small in comparison with that of a hot storage; therefore, larger storage volume of chilled water is needed to store the same amount of energy than in hot water storage. Because solar heating systems always employ a storage tank, the arrangement shown in Fig. 6.28 is preferred.

6.6 Solar photovoltaic powered heating and cooling

Solar photovoltaic power air-conditioning (PVAC) is one of the possible technologies to utilize solar energy for the heating and cooling of buildings. A typical solar PV powered heating and cooling system consists of a PV system for electricity generation, and an electric-driven air-conditioning system, usually of the vapor-compression type. Various studies show that the PVAC is able to generate enough electricity to balance the air-conditioning load in both summer and winter conditions (Li et al., 2018). When grid connection is considered, the PVAC can reduce about half of the electricity required from the power grid in sunny days. The energy performance of PVAC in summer is better than that in winter because the building cooling load is in phase with the PV production. The main difference between the PVAC and the solar thermal powered heating and cooling, is that the PVAC can be easily embedded in the electricity sector where the excess solar power can be fed to the power grid. The PVAC system also enhances the self-consumption of PV produced electricity which relieves the pressure on the utility electricity grids. The performance of PVAC is largely affected by the local solar radiation resource and the specific building heating and cooling demands, which highlights the importance of system modeling and simulation at the design stage.

The PV system of PVAC can be either grid-connected or stand-alone, as shown in Fig. 6.29. For grid-connected systems, shown schematically in Fig. 6.29a, electricity can be flexibly transmitted between PVACs and the power grid. When the PV generation is more than the energy demand, the excess PV electricity can be fed into the grid. When the PV generation is low to meet demand, the extra electricity can be provided from the grid.

For stand-alone or off-grid systems, energy storage can be usually integrated to store energy during the daytime and use the stored energy to power the PVAC when solar radiation is low. Stand-alone systems are suitable for applications such as in remote areas where the grid is not available. The main challenge when designing stand-alone PVAC systems is the energy balance between the generated and the consumed energy. The energy storage system is used when the PV generated power is not enough to meet the air-conditioning load power.

The energy storage system of PVAC can be categorized as battery storage and thermal storage is shown in Fig. 6.29b and c. These can be also integrated in grid-connected systems to leverage the price difference of the grid electricity. The battery storage system can store excess PV electricity or electricity from the grid and supply the electricity to the air-conditioning units when the PV power is not enough or the electricity price too high. Lead-acid and lithium-ion batteries are the most common types of batteries employed for this purpose. The integration of the battery can be either as dc

FIGURE 6.29

Energy flow diagrams for possible PVAC systems.

coupled or ac coupled. A higher system energy efficiency can be achieved if the air-conditioning unit is driven by dc motors as the dc/ac converter is not required. A thermal storage system is the other option for energy storage which is more favorable from the economic point of view. When there is surplus PV generation, the air-conditioning unit can provide more heating or cooling energy than the building actual demand and store it in the water, ice, or PCMs. In this way the stored energy can be supplied to the building when the PV does not produce enough energy to power the air-conditioning units of the building. However, thermal storage systems are relatively bulky requiring a larger installation space.

At the product level PVAC was initially introduced by Japanese company SANYO in 1993 (Ohnishi et al., 1995) but was not adopted by the industry due to the high price of PV systems at that time. However, after the substantial drop of the PV systems price the marketing of PVAC is now considered a feasible option, and many Chinese companies are now producing household PVAC systems.

Exercises

6.1 A building has a peak heating load equal to 18.3 kW and a peak cooling load of 23.8 kW. Estimate the seasonal heating and cooling requirements if the heating degree-days are 1240 °C-days, the cooling degree-days are 980 °C-days, the winter indoor temperature is 23 °C, and the summer indoor temperature is 25 °C. The design outdoor temperature for winter is 2 °C, and for summer, it is 39 °C.

6.2 Estimate the overall heat transfer coefficient of a wall that has the following layers:
Outside plaster, 2 cm.
Brick, 20 cm.
Air gap, 2 cm.
Polyurethane insulation, 3 cm.
Brick, 10 cm.
Inside plaster, 2 cm.

6.3 Estimate the overall heat transfer coefficient of the wall in Exercise 6.2 by replacing the 10 cm inside brick with the same thickness of medium-density concrete.

6.4 Estimate the overall heat transfer coefficient of the wall in Exercise 6.2 by replacing the air gap and polyurethane with 5 cm polyurethane.

6.5 Estimate the U value of a pitched roof that has a ceiling U value $= 1.56$ W/m^2 K, area of 65 m^2, and a roof U value $= 1.73$ W/m^2 K. The slope angle of the roof is 35°.

6.6 A building has a south-facing thermal storage wall with night insulation of $R_{ins} = 1.35$ m^2 K/W, applied for 6 h. Estimate the monthly heat transfer through the wall into the indoor space with and without night insulation for the month of January. The following data are given:
$U_o = 6.3$ W/m^2 K.
$w = 0.31$ m.
$k = 2.2$ W/m K.
$h_i = 8.3$ W/m^2 K.
$\overline{H}_t = 11.8$ MJ/m^2 K.
$(\tau\alpha) = 0.83$.
$\overline{T}_R = 21$ °C.
$\overline{T}_a = 3$ °C.
$A_w = 25.1$ m^2.

6.7 A house has a south-facing window of 1.8 m height, located in 35°N latitude. The overhang is wide enough so as to neglect the side effects and its length is 0.9 m located 0.6 m above the top surface of the window. Estimate the shading fraction for a vertical window facing due south and a window in the same direction and tilted 10° from vertical at 11 a.m. and 2 p.m. solar times on July 17.

6.8 A building with a south-facing window with height $= 2.5$ m and width $= 5$ m is located in an area where $\overline{K}_T = 0.574, h_{ss} = 80\,°, \overline{R}_B = 0.737, \overline{H} = 12.6$ MJ/m^2, and $\overline{F}_w = 0.705$. The ground reflectance is 0.3. Estimate the mean monthly area−average radiation received by the window when there is no shading and when an overhang with a gap of 0.625 m, extension 0.5 m on both sides of the window, and projection of 1.25 m is used.

6.9 A solar space and hot water heating system has a collector with $F_R U_L = 6.12$ W/m^2 °C and an area of 20 m^2. The flow rate in a collector−storage heat exchanger of antifreeze and water is 0.02 kg/s m^2, the heat exchanger has an effectiveness equal to 0.73. What is the ratio F'_R / F_R if the c_p of water is 4180 J/kg °C and that of antifreeze is 3350 J/kg °C?

6.10 A room is maintained at a temperature of $T_R = 22$ °C and has a $(UA)_l = 2850$ W/°C. The ambient temperature is 2 °C and the storage tank temperature is 75 °C. Estimate the space load, domestic water heating load, and rate of auxiliary energy required for the following conditions:

Heat exchanger effectiveness $= 0.75$.
Flow rate of air side of heat exchanger $= 0.95$ kg/s.
Specific heat of air $= 1.05$ kJ/hg °C.
Environmental temperature at the space where storage tank is located $= 18$ °C.
(UA) of storage tank $= 3.4$ W/°C.
Mass flow rate of domestic water $= 0.15$ kg/s.
Required domestic water temperature $= 55$ °C.
Make-up water temperature $= 14$ °C.

6.11 A liquid solar heating system has a 16 m^2 collector and is used to preheat the city water, which is at a temperature of 12 °C. If the tank is fully mixed and the capacitance of the collector side of the heat exchanger is 890 W/°C and the storage side is 1140 W/°C, estimate the final temperature in the storage tank at 3 p.m. and the energy balance of the system for the following parameters and conditions of the system:

$F_R(\tau\alpha) = 0.79$.
$F_R U_L = 6.35$ W/m^2 °C.
Heat exchanger effectiveness $= 0.71$.
Storage tank capacity $= 1100$ L.
Storage tank $UA = 4.5$ W/°C.
Initial tank water temperature $= 40$ °C.
Environmental temperature at the space where storage tank is located $= 18$ °C.
The meteorological conditions and load flow rate are given in the following table.

Hour	I_t (MJ/m^2)	T_a (°C)	Load flow rate (kg)
9−10	0.95	13	160
10−11	1.35	15	160
11−12	2.45	18	80
12−13	3.65	22	0
13−14	2.35	23	80
14−15	1.55	21	160

6.12 Using the data of the previous problem, estimate the effect of increasing the effectiveness of the heat exchanger to 0.92 and the city mains water temperature to 16 °C. Each modification should be considered separately and the result should be compared to those of the previous problem. In every case, the energy balance should be checked.

References

Alefeld, G., Radermacher, R., 1994. Heat Conversion Systems. CRC Press, Boca Raton, FL.

Ameel, T.A., Gee, K.G., Wood, B.D., 1995. Performance predictions of alternative, low cost absorbents for open-cycle absorption solar cooling. Sol. Energy 54 (2), 65−73.

Argiriou, A., 1997. Ground cooling. In: Santamouris, M., Asimakopoulos, D. (Eds.), Passive Cooling of Buildings. James and James, London, pp. 360−403.

ASHRAE, 1997. Handbook of Fundamentals. ASHRAE, Atlanta[*].

ASHRAE, 1992. Cooling and Heating Load Calculation Manual. ASHRAE, Atlanta[*].

ASHRAE, 2005. Handbook of Fundamentals. ASHRAE, Atlanta[*].

ASHRAE, 2007. Handbook of Applications. ASHRAE, Atlanta[*].

Balcomb, J.D., 1983. Heat Storage and Distribution inside Passive Solar Buildings. Los Alamos National Laboratory, Los Alamos, NM.

Brandemuehl, M.J., Lepore, J.L., Kreider, J.F., 1990. Modelling and testing the interaction of conditioned air with building thermal mass. ASHRAE Trans. 96 (2), 871−875.

Braun, J.E., 1990. Reducing energy costs and peak electrical demand through optimal control of building thermal storage. ASHRAE Trans. 96 (2), 876−888.

Cabeza, L.F., Castell, A., Barreneche, C., De Gracia, A., Fernández, A.I., 2011. Materials used as PCM in thermal energy storage in buildings: a review. Renew. Sustain. Energy Rev. 15 (3), 1675−1695.

Calise, F., 2012. High temperature solar heating and cooling systems for different Mediterranean climates: dynamic simulation and economic assessment. Appl. Therm. Eng. 32, 108−124.

Çengel, Y.A., Boles, M.A., 1994. Thermodynamics: An Engineering Approach. McGraw-Hill, New York.

Critoph, R.E., 2002. Development of three solar/biomass adsorption air conditioning refrigeration systems. In: Proceedings of the World Renewable Energy Congress VII on CD-ROM, Cologne, Germany.

Dieng, A.O., Wang, R.Z., 2001. Literature review on solar adsorption technologies for ice making and air conditioning purposes and recent development in solar technology. Renew. Sustain. Energy Rev. 5 (4), 313−342.

Dimoudi, A., 1997. Urban design. In: Santamouris, M., Asimakopoulos, D. (Eds.), Passive Cooling of Buildings. James and James, London, pp. 95−128.

Dorgan, C.B., Leight, S.P., Dorgan, C.E., 1995. Application Guide for Absorption Cooling/Refrigeration Using Recovered Heat. ASHRAE, Atlanta.

Druck, H., Hahne, E., 1998. Test and comparison of hot water stores for solar combisystems. In: Proceedings of EuroSun '98—The Second ISES-Europe Solar Congress on CD-ROM, Portoroz, Slovenia.

Duffie, J.A., Beckman, W.A., 1991. Solar Engineering of Thermal Processes, second ed. Wiley & Sons, New York.

Duffin, R.J., Knowles, G., 1985. A simple design method for the Trombe wall. Sol. Energy 34 (1), 69−72.

Erhard, A., Hahne, E., 1997. Test and simulation of a solar-powered absorption cooling machine. Sol. Energy 59 (4−6), 155−162.

Facao, J., Carvalho, M.J., 2014. New methodologies to analyze direct expansion solar assisted heat pumps for domestic hot water. Sol. Energy 100, 66−75.

[*]Note: The material quoted in these references may also appear in subsequent ASHRAE handbooks, but here the year of the original version of the handbook used is shown.

Florides, G., Kalogirou, S.A., Tassou, S., Wrobel, L., 2000. Modelling of the modern houses of Cyprus and energy consumption analysis. Energy Int. J. 25 (10), 915−937.

Florides, G., Kalogirou, S.A., Tassou, S., Wrobel, L., 2001. Modelling and simulation of an absorption solar cooling system for Cyprus. Sol. Energy 72 (1), 43−51.

Florides, G., Tassou, S., Kalogirou, S.A., Wrobel, L., 2002a. Review of solar and low energy cooling technologies for buildings. Renew. Sustain. Energy Rev. 6 (6), 557−572.

Florides, G., Tassou, S., Kalogirou, S.A., Wrobel, L., 2002b. Measures used to lower building energy consumption and their cost effectiveness. Appl. Energy 73 (3−4), 299−328.

Florides, G., Kalogirou, S.A., Tassou, S., Wrobel, L., 2002c. Modelling, simulation and warming impact assessment of a domestic-size absorption solar cooling system. Appl. Therm. Eng. 22 (12), 1313−1325.

Florides, G., Kalogirou, S.A., Tassou, S., Wrobel, L., 2003. Design and construction of a lithium bromide-water absorption machine. Energy Convers. Manage. 44 (15), 2483−2508.

Ghaddar, N.K., Shihab, M., Bdeir, F., 1997. Modelling and simulation of solar absorption system performance in Beirut. Renew. Energy 10 (4), 539−558.

Gorozabel Chata, F.B., Chaturvedi, S.K., Almogbel, A., 2005. Analysis of a direct expansion solar assisted heat pump using different refrigerants. Energy Conservat. Manage. 46, 2614−2624.

Hahne, E., 1996. Solar heating and cooling. In: Proceedings of EuroSun '96, vol. 1, pp. 3−19 (Freiburg, Germany).

Haim, I., Grossman, G., Shavit, A., 1992. Simulation and analysis of open cycle absorption systems for solar cooling. Sol. Energy 49 (6), 515−534.

Hammad, M.A., Audi, M.S., 1992. Performance of a solar LiBr-water absorption refrigeration system. Renew. Energy 2 (3), 275−282.

Hammad, M., Zurigat, Y., 1998. Performance of a second generation solar cooling unit. Sol. Energy 62 (2), 79−84.

Hastings, S.R., 1999. Solar Air Systems-Built Examples. James and James, London.

Hawlader, M.N.A., Noval, K.S., Wood, B.D., 1993. Unglazed collector/regenerator performance for a solar assisted open cycle absorption cooling system. Sol. Energy 50 (1), 59−73.

Herold, K.E., Radermacher, R., Klein, S.A., 1996. Absorption Chillers and Heat Pumps. CRC Press, Boca Raton, FL.

Huang, B.J., Chyng, J.P., 2001. Performance characteristics of integral type solar-assisted heat pump. Sol. Energy 71 (6), 403−414.

Hsieh, J.S., 1986. Solar Energy Engineering. Prentice-Hall, Englewood Cliffs, NJ.

Hsieh, S.S., Tsai, J.T., 1988. Transient response of the Trombe wall temperature distribution applicable to passive solar heating systems. Energy Convers. Manage. 28 (1), 21−25.

Ishida, M., Ji, J., 1999. Graphical exergy study on single state absorption heat transformer. Appl. Therm. Eng. 19 (11), 1191−1206.

Izquerdo, M., Vega, M., Lecuona, A., Rodriguez, P., 2000. Entropy generated and exergy destroyed in lithium bromide thermal compressors driven by the exhaust gases of an engine. Int. J. Energy Res. 24, 1123−1140.

Jubran, B.A., Humdan, M.A., Tashtoush, B., Mansour, A.R., 1993. An approximate analytical solution for the prediction of transient-response of the Trombe wall. Int. Commun. Heat Mass Transfer 20 (4), 567−577.

Kalogirou, S.A., 2007. Use of genetic algorithms for the optimum selection of the fenestration openings in buildings. In: Proceedings of the 2nd PALENC Conference and 28th AIVC Conference on Building Low Energy Cooling and Advanced Ventilation Technologies in the 21st Century, September 2007, Crete Island, Greece, pp. 483−486.

Kalogirou, S.A., Florides, G., Tassou, S., 2002. Energy analysis of buildings employing thermal mass in Cyprus. Renew. Energy 27 (3), 353−368.

Kays, W.M., 1966. Convective Heat and Mass Transfer. McGraw-Hill, New York.

Keith, E.H., 1995. Design challenges in absorption chillers. Mech. Eng. CIME 117 (10), 80—84.

Keith, E.H., Radermacher, R., Klein, S.A., 1996. Absorption Chillers and Heat Pumps. CRS Press, Boca Raton, FL, pp. 1—5.

Kizilkan, O., Sencan, A., Kalogirou, S.A., 2007. Thermoeconomic optimization of a LiBr absorption refrigeration system. Chem. Eng. Process 46 (12), 1376—1384.

Klein, S.A., et al., 2005. TRNSYS Version 16 Program Manual. Solar Energy Laboratory, University of Wisconsin, Madison et al.

Klein, S.A., 1992. Engineering Equation Solver. Details available from: www.fchart.com.

Kotas, T.J., 1985. The Exergy Method of Thermal Plant Analysis. Butterworth Scientific Ltd, Borough Green, Kent, Great Britain.

Kreider, J.F., Rabl, A., 1994. Heating and Cooling of Buildings—Design for Efficiency. McGraw-Hill, Singapore, pp. 1—21.

Kuang, Y.H., Wang, R.Z., 2006. Performance of multi-functional direct expansion solar assisted heat pump system. Sol. Energy 80, 795—803.

Lechner, N., 1991. Heating, Cooling and Lighting. Wiley & Sons, New York.

Lee, R.J., DiGuilio, R.M., Jeter, S.M., Teja, A.S., 1990. Properties of lithium bromide-water solutions at high temperatures and concentration. II. density and viscosity. ASHRAE Trans. 96, 709—728.

Lee, S.F., Sherif, S.A., 2001. Thermodynamic analysis of a lithium bromide/water absorption system for cooling and heating applications. Int. J. Energy Res. 25, 1019—1031.

Li, Y., Zhao, B.Y., Zhao, Z.G., Taylor, R.A., Wang, R.Z., 2018. Performance study of a grid-connected photovoltaic powered central air conditioner in the South China climate. Renew. Energy 126, 1113—1125.

Mercer, W.E., Pearce, W.M., Hitchcock, J.E., 1967. Laminar forced convection in the entrance region between parallel flat plates. J. Heat Trans. 89, 251—257.

Monsen, W.A., Klein, S.A., Beckman, W.A., 1982. The unutilizability design method for collector-storage walls. Sol. Energy 29 (5), 421—429.

Nayak, J.K., 1987. Transwall versus Trombe wall: relative performance studies. Energy Convers. Manage 27 (4), 389—393.

Norton, B., 1992. Solar Energy Thermal Technology. Springer-Verlag, London.

Ohnishi, M., Takeoka, A., Nakano, S., Kuwano, Y., 1995. Advanced photovoltaic technologies and residential applications. Renew. Energy 6 (3), 275—282.

Omojaro, P., Breitkopf, C., 2013. Direct expansion solar assisted heat pumps: a review of applications and recent research. Renew. Sustain. Energy Rev. 22, 33—45.

Randal, K.R., Mitchel, J.W., Wakil, M.M., 1979. Natural convection heat transfer characteristics of flat-plate enclosures. J. Heat Trans. 101, 120—125.

Sencan, A., Yakut, K.A., Kalogirou, S.A., 2005. Exergy analysis of LiBr/water absorption systems. Renew. Energy 30 (5), 645—657.

Sharp, K., 1982. Calculation of monthly average insolation on a shaded surface of any tilt and azimuth. Sol. Energy 28 (6), 531—538.

Simmonds, P., 1991. The utilization and optimization of building's thermal inertia in minimizing the overall energy use. ASHRAE Trans. 97 (2), 1031—1042.

Smolec, W., Thomas, A., 1993. Theoretical and experimental investigations of heat-transfer in a Trombe wall. Energy Convers. Manage 34 (5), 385—400.

Talbi, M.M., Agnew, B., 2000. Exergy analysis: an absorption refrigerator using lithium bromide and water as working fluids. Appl. Therm. Eng. 20 (7), 619—630.

Thorpe, R., 2002. Progress towards a highly regenerative adsorption cycle for solar thermal powered air conditioning. In: Proceedings of the World Renewable Energy Congress VII on CD-ROM, Cologne, Germany.

Tozer, R.M., James, R.W., 1997. Fundamental thermodynamics of ideal absorption cycles. Int. J. Refrig. 20 (2), 120–135.

Trombe, F., Robert, J.F., Cabanot, M., Sesolis, B., 1977. Concrete walls to collect and hold heat. Sol. Age 2 (8), 13–19.

Utzinger, M.D., Klein, S.A., 1979. A method of estimating monthly average solar radiation on shaded receivers. Sol. Energy 23 (5), 369–378.

Weiss, W., Spörk-Dür, M., 2022. Solar Heat Worldwide, Global Market Development and Trends in 2021, Solar Heating and Cooling Program. IEA, 2022.

Wilcox, B., Gumerlock, A., Barnaby, C., Mitchell, R., Huizerza, C., 1985. The effects of thermal mass exterior walls on heating and cooling loads in commercial buildings. In: Thermal Performance of the Exterior Envelopes of Buildings III. ASHRAE, pp. 1187–1224.

Winston, R., O'Gallagher, J., Duff, W., Henkel, T., Muschaweck, J., Christiansen, R., Bergquam, J., 1999. Demonstration of a new type of ICPC in a double-effect absorption cooling system. In: Proceedings of ISES Solar World Congress on CD-ROM, Jerusalem, Israel.

Winwood, R., Benstead, R., Edwards, R., 1997. Advanced fabric energy storage. Build. Serv. Eng. Res. Technol. 18 (1), 1–6.

Zhou, D., Zhao, C.Y., Tian, Y., 2012. Review on thermal energy storage with phase change materials (PCMs) in building applications. Appl. Energy 92, 593–605.

Zinian, H.E., Ning, Z., 1999. A solar absorption air-conditioning plant using heat-pipe evacuated tubular collectors. In: Proceedings of ISES Solar World Congress on CD-ROM, Jerusalem, Israel.

Zrikem, Z., Bilgen, E., 1987. Theoretical study of a composite Trombe-Michel wall solar collector system. Sol. Energy 39 (5), 409–419.

Industrial Process Heat, Chemistry Applications, and Solar Dryers

7.1 Industrial process heat: general design considerations

Beyond the low-temperature applications, there are several potential fields of application for solar thermal energy at a medium and medium–high temperature level (80–240°C). The most important of them is heat production for industrial processes, which represents a significant amount of heat. For example, industrial heat demand constitutes about 15% of the overall demand of final energy requirements in the southern European countries. In 2000 the demand in the EU for medium and medium–high temperatures was estimated to be about 300 TWh/a (Schweiger et al., 2000).

Many solar thermal systems for industrial process heat (IPH) have been reported worldwide. Currently, 61 systems based on parabolic trough collector (PTC) technology are installed and operating worldwide for IPH applications, with a total thermal power of 13.9 MW$_{th}$. The dominant sectors for the number of PTC plants are the dairy and beverage sectors, with 18 PTC plants installed, while in terms of installed power capacity, the dominant sector is food products with 4 MW$_{th}$ (http://ship-plants.info/solar-thermal-plants?collector_type=5). At least 74 solar process heat systems were added in 2020, which makes the total number of operating projects is at least equal to 891, with an overall installed capacity of 791 MW$_{th}$ (Weiss and Spörk-Dür, 2020).

From a number of studies on industrial heat demand, several industrial sectors have been identified as having favorable conditions for the application of solar energy. The most important industrial processes using heat at a medium temperature level are sterilizing, pasteurizing, drying, hydrolyzing, distillation and evaporation, washing and cleaning, and polymerization. Some of the most important processes and the range of the temperatures required for each are shown in Table 7.1 (Kalogirou, 2003).

Large-scale solar applications for process heat benefit from the effect of scale. Therefore, the investment costs should be comparatively low, even if the costs for the collector are higher. One way to ensure economical terms is to design systems with no heat storage, that is, the solar heat is fed directly into a suitable process, called a fuel saver. In this case, the maximum rate at which the solar energy system delivers energy must not be appreciably larger than the rate at which the process uses energy. This system, however, cannot be cost-effective in cases where heat is needed at the early or late hours of the day or at nighttime, when the industry operates on a double-shift basis.

The usual types of industries that use most of the energy are the food industry and non-metallic mineral product manufacturing. Particular types of food industries that can employ solar process heat are the milk (dairy) and cooked pork meats (sausage, salami, etc.) industries and breweries. Most of the process heat is used in the food and textile industries for such diverse applications as drying,

Table 7.1 Temperature ranges for various industrial processes.

Industry	Process	Temperature (°C)
Dairy	Pressurization	60–80
	Sterilization	100–120
	Drying	120–180
	Concentrates	60–80
	Boiler feed-water	60–90
Tinned food	Sterilization	110–120
	Pasteurization	60–80
	Cooking	60–90
	Bleaching	60–90
Textile	Bleaching, dyeing	60–90
	Drying, degreasing	100–130
	Dyeing	70–90
	Fixing	160–180
	Pressing	80–100
Paper	Cooking, drying	60–80
	Boiler feed-water	60–90
	Bleaching	130–150
Chemical	Soaps	200–260
	Synthetic rubber	150–200
	Processing heat	120–180
	Preheating water	60–90
Meat	Washing, sterilization	60–90
	Cooking	90–100
Beverages	Washing, sterilization	60–80
	Pasteurization	60–70
Flours and by-products	Sterilization	60–80
Timber by-products	Thermo-difussion beams	80–100
	Drying	60–100
	Preheating water	60–90
	Preparation pulp	120–170
Bricks and blocks	Curing	60–140
Plastics	Preparation	120–140
	Distillation	140–150
	Separation	200–220
	Extension	140–160
	Drying	180–200
	Blending	120–140

cooking, cleaning, and extraction. Favorable conditions exist in the food industry because food treatment and storage are processes with high energy consumption and high running time. Temperatures for these applications may vary from near ambient to those corresponding to low-pressure steam, and energy can be provided either from flat-plate or low-concentration-ratio concentrating collectors.

The principle of operation of collectors and other components of the solar systems outlined in the previous chapters also apply to industrial process heat applications. These applications, however, have some unique features; the main ones are the scale on which they are applied and the integration of the solar energy supply with an auxiliary energy source and the industrial process.

Generally, two primary problems need to be considered when designing an industrial process heat application. These concern the type of energy to be employed and the temperature at which the heat is to be delivered. For example, if hot water is needed for cleaning in food processing, the solar energy should be a liquid heater. If a process requires hot air for drying, an air heating system is probably the best solar energy system option. If steam is needed to operate a sterilizer, the solar energy system must be designed to produce steam, probably with concentrating collectors.

Another important factor required for the determination of the most suitable system for a particular application is the temperature at which the fluid will be fed to the collector array. Other requirements concern the fact that the energy may be needed at a particular temperature or over a range of temperatures, and possible sanitation requirements of the plant that must also be met, as, for example, in food processing applications.

The investments required in industrial solar application are generally large, and the best way to design the solar energy supply system can be done by modeling methods (see Chapter 11) that consider the transient and intermittent characteristics of the solar resource. In this way, designers can study various options in solar industrial applications at costs that are very small compared with the investments. For the preliminary design, the simple modeling methods presented in previous chapters apply here as well.

Another important consideration is that, in many industrial processes, large amounts of energy are required in small spaces. Therefore, there may be a problem for the location of collectors. If the need arises, collector arrays can be located on adjacent buildings or grounds. Locating the collectors in such areas, however, results in long runs of pipes or ducts, which cause heat losses that must be considered in the design of the system. Where feasible, when no land area is available, collectors can be mounted on the roof of a factory in rows. In this case, shading between adjacent collector rows should be avoided and considered. However, the collector area may be limited by the roof area, shape, and orientation. Additionally, roofs of existing buildings are not designed or oriented to accommodate arrays of collectors, and in many cases, structures to support collector arrays must be installed on existing roofs. It is usually much better and cost-effective if new buildings are readily designed to allow for collector mounting and access.

In a solar industrial process heat system, interfacing of the collectors with conventional energy supplies must be done in a way which is compatible with the process. The easiest way to accomplish this is by using heat storage, which can also allow the system to work in periods of low irradiation and nighttime.

The central system for heat supply in most factories uses hot water or steam at a pressure corresponding to the highest temperature needed in the different processes. Hot water or low-pressure steam at medium temperatures ($<150°C$) can be used for preheating water (or other fluids) used for processes

FIGURE 7.1

Possibilities of combining the solar energy system with the existing heat supply.

(washing, dyeing, etc.), for steam generation, or by direct coupling of the solar system to an individual process working at temperatures lower than that of the central steam supply. Various possibilities are shown in Fig. 7.1. In the case of water preheating, higher efficiencies are obtained due to the low input temperature to the solar system; thus low-technology collectors can work effectively and the required load supply temperature has no or little effect on the performance of the solar energy system.

Norton (1999) presented the history of solar industrial and agricultural process applications. The most common applications of industrial process heat and practical examples are described.

A system for solar process heat for decentralized applications in developing countries was presented by Spate et al. (1999). The system is suitable for community kitchens, bakeries, and post-harvest treatment. The system employs a fixed-focus parabolic collector, a high temperature flat-plate collector, and a pebble bed oil storage.

Benz et al. (1998) presented the planning of two solar thermal systems producing process heat for a brewery and a dairy in Germany. In both industrial processes, the solar yields were found to be comparable with the yields of solar systems for domestic solar water heating or space heating. Benz et al. (1999) also presented a study for the application of non-concentrating collectors for the food industry in Germany. In particular, the planning of four solar thermal systems producing process heat for a large and a small brewery, a malt factory, and a dairy was presented. In the breweries, the washing machines for the returnable bottles were chosen as a suitable process to be fed by solar energy; in the dairy, the spray dryers for milk and whey powder production were chosen; and in the malt factory, the wither and kiln processes. Up to 400 kWh/m^2/a was delivered from the solar collectors, depending on the type of collector.

Cotrado et al. (2014) presented a PTC system installed for energy production and reduced fuel consumption. The system covers an area of 2040 m^2, with thermal oil as the heat transfer fluid (HTF) and a maximum HTF temperature of 200°C. The HTF is fed to an indirect steam generator (SG) to produce and supply steam to the industry's steam network. The steam supports the milk pasteurization, washing processes, and washing machines when the milk drying process is not in operation. They estimated that the CO_2 reduction per year is around 290 tons.

In Texas, Johnson & Johnson manufacturing plant, a PTC plant is installed that produces steam at 174°C, covering an area of 1068 m^2. The HTF is pressurized water that flashes to an SG from where the steam is supplied to the industrial processes. The investigation of the system was carried out by

Brink et al. (1982), who presented the evaluation of the system on a clear day in September. The average daily efficiency was 30%−40%, and the average hourly efficiency varied from 38% to 42% on clear days.

NEP solar company has installed three PTC plants for IPH in different industries located in Switzerland. The first is a dairy industry in Bever in which a PTC system is installed on the factory's roof. It covers an area of 115 m^2 and produces steam at 190°C for milk processing. A bigger plant has been installed at a cheese manufacturing industry at Saignelegier. This system has an area of 627 m^2 and produces steam at 125°C. The system combines two oil-fired boilers (of 1 MW capacity) for preheating when solar radiation is low. The third solar plant is installed in Fribourg and covers 587 m^2. The HTF used is the water heated at 165°C for milk processing. Rittmann-Frank et al. (2018) carried out a technical evaluation of these three plants. At Bever, the annual efficiency of the system was 23% in 2014, and improvement tasks have been conducted to increase plant performance. At Saignelegier, the highest performance during the 3 years examined (2014, 2015, 2016) was 37% (year 2015), which corresponded to 263 MWh and met the expected production. The PTC production covers 12% of the load and has an annual CO_2 savings of 69 tons. At Fribourg, the annual performance was 40% (year 2016), which corresponded to more than 220 MWh and exceeded design expectations.

Another PTC system installed in Egypt covers 1958.4 m^2 and produces steam at a rate of 1.3 tons/h at 8 bar and 175°C. Abdel-Dayem et al. (2011) evaluated the system's performance by developing a mathematical model simulating the various processes. Based on the economic analysis (life cycle savings method), the authors concluded that the optimized area of the collector should be 538 m^2. This proposed system can cover 10% of the heat demand of the factory.

Many studies have also addressed the thermal energy storage (TES) system, which is essential for any solar thermal system, as it can collect energy and use it later (3−12 h). The main scope of a TES is to cover the thermal requirements of the industry during cloudy hours or insufficient solar radiation. It can also be used in the early morning for preheating when the sun is not shining. For steam production to be readily available by 7 a.m., most industries start preheating the SG and pipelines at least 2−3 h earlier when there is little or no solar radiation. So it is important to have a highly efficient TES that can supply heat during those hours and cover the thermal energy. A TES can also be used as a buffer to maintain a more stable system condition (temperature and pressure). It also minimizes the variabilities of the system behavior in case of significant fluctuation of solar radiation. By employing a TES in a solar system, the solar fraction is increased, leading to significant advantages due to lower fuel consumption, environmental impact, and less payback time (Powell and Edgar, 2012).

Kumaresan et al. (2012) showed an experimental setup of the PTC system, including a TES, a pump, and a tracking unit. The TES system consists of a vertical cylindrical storage tank (550 mm in diameter and 1100 mm in length) with a volume of 230 L. The TES is insulated with glass wool (0.15 m). The HTF used is Therminol 55 stored in a well-insulated tank. The authors examined the instantaneous efficiency, which varied from 50% to 62.5%, with the peak efficiency occurring at noon. As the TES is charged at 120°C and left undistributed for 45 h, they concluded that the average heat loss per hour is 1120 kJ. The efficiency of the energy collected to the energy stored from 8 a.m. to 2 p.m. varied from 41% to 60%, and the maximum overall system efficiency was close to 40%.

The high thermal conductivity and specific heat, the excellent mechanical properties, the similar thermal expansion rate of cement compared with steel pipes, and the low cost of the material makes concrete TES (CTES) very promising for future applications. The most severe challenge regarding TES is the high investment cost.

Many researchers are focused on achieving better TES performance at a lower price. At Plataforma Solar de Almeria in Spain, the German Aerospace Center tested the first TES using concrete, which is a cheaper storage technology (Liang et al., 2008). It should be noted that during the heating of concrete, residual water in the core of the concrete storage evaporates and can build up a critical vapor pressure if the permeability is too low. Therefore, the most important properties of concrete are storage capacity, thermal conductivity, and sufficient vapor permeability (so it will not be damaged). These were demonstrated and tested in a concrete storage module that consists of 132 tubes (9 m length 18 mm diameter), concrete, and pressure-resistant thermal insulation. The tests were performed for 4 months at a temperature range between 300 and 400°C. For 6 h of charging and discharging, the module was estimated to have a total capacity of 25.6 kWh/m^3 (Liang et al., 2008).

An IPH plant, which is the first industrial PTC system in Cyprus, is installed at the biggest soft drink factory (Ktistis et al., 2021). The system consists of 288 m^2 of PTC, an SG, and a concrete TES to keep the system dispatchable. Two operation strategies are developed, which are controlled automatically by the main processor of the system. The first strategy is enabled when there is a steam demand, and the second is when the energy can be stored directly in the CTES. Both strategies are tested, and it is shown that under Strategy 1, the PTC system can produce 940 litters of steam per day, and under Strategy 2, it can store 107.3 kWh$_{th}$. In the last 2 years, when the system has been in full operation, it has proven that it can supply the required amount of steam to the factory even when solar radiation is low, with support from the CTES.

7.1.1 Solar industrial air and water systems

The two types of applications employing solar air collectors are the open circuit and the recirculating applications. In the open circuit, heated ambient air is used in industrial applications where, because of contaminants, recirculation of air is not possible. Examples are paint spraying, drying, and supplying fresh air to hospitals. It should be noted that heating outside air is an ideal operation for the collector because it operates very close to ambient temperature, thus more efficiently.

In recirculating air systems, a mixture of recycled air from the dryer and ambient air is supplied to the solar collectors. Solar-heated air supplied to a drying chamber can be applied to a variety of materials, including lumber and food crops. In this case, adequate control of the rate of drying, which can be performed by controlling the temperature and humidity of the supply air, can improve product quality.

Similarly, the two types of applications employing solar water collectors are the once-through systems and the recirculating water heating applications. The latter are exactly similar to the domestic water heating systems presented in Chapter 5. Once-through systems are employed in cases where water is used for cleaning in food industries, and recycling the used water is not practical because of the contaminants picked up by the water in the cleaning process.

A solar energy system may deliver energy to the load either in series or parallel with the auxiliary heater. In a series arrangement, shown in Fig. 7.2, solar energy is used to preheat the load heat transfer fluid, which may be heated more, if necessary, by the auxiliary heater, to reach the required temperature. If the temperature of the fluid in the storage tank is higher than that required by the load, a three-way valve, also called a *tempering valve*, is used to mix it with cooler make-up or returning fluid. The parallel configuration is shown in Fig. 7.3. Since the energy cannot be delivered to the load at a

FIGURE 7.2

Simple industrial process heat system with a series configuration of auxiliary heater.

FIGURE 7.3

Simple industrial process heat system with a parallel configuration of auxiliary heater.

temperature lower than that of the load temperature, the solar system must be able to produce the required temperature before energy can be delivered.

Therefore, a series configuration is preferred over the parallel one because it provides a lower average collector operating temperature, which leads to higher system efficiency. The parallel feed, however, is common in steam-producing systems, as shown in Fig. 7.4, and is explained in the next section.

One of the most important design characteristics to consider when designing a solar industrial process heat system is the time matching of the solar energy source to the load. As was seen in the previous chapter, heating and cooling loads vary from day to day. In industrial process heat systems, however, the loads are pretty much constant and small variations are due to the seasonal variation of the make-up water temperature.

The thermal analysis of air and water solar industrial process heat systems is similar to the analysis presented in Chapter 5 for the solar water heating systems and will not be repeated here. The main difference is in the determination of the energy required by the load.

FIGURE 7.4

Simple industrial process heat steam system with a parallel configuration with an auxiliary steam boiler.

7.2 Solar steam generation systems

Parabolic trough collectors are frequently employed for solar steam generation because relatively high temperatures can be obtained without serious degradation in the collector efficiency. Low-temperature steam can be used in industrial applications, in sterilization, and for powering desalination evaporators.

Three methods have been employed to generate steam using parabolic trough collectors (Kalogirou et al., 1997):

1. The steam–flash concept, in which pressurized water is heated in the collector and flashed to steam in a separate vessel.
2. The direct or in situ concept, in which two-phase flow is allowed in the collector receiver so that steam is generated directly.
3. The unfired boiler concept, in which a heat transfer fluid is circulated through the collector and steam is generated via heat exchange in an unfired boiler.

All three steam generation systems have advantages and disadvantages. These are examined in the following section.

7.2.1 Steam generation methods

The steam–flash system is shown schematically in Fig. 7.5. In this system, water, pressurized to prevent boiling, is circulated through the collector and flashed across a throttling valve into a flash vessel. Treated feed-water input maintains the level in the flash vessel and the sub-cooled liquid is recirculated through the collector.

The in situ boiling concept, shown in Fig. 7.6, uses a similar system configuration with no flash valve. Sub-cooled water is heated to boiling and steam forms directly in the receiver tube. According to Hurtado and Kast (1984), capital costs associated with direct steam and flash–steam systems are approximately the same.

Although both systems use water, a superior heat transport fluid, the in situ boiling system is more advantageous. The flash system uses a sensible heat change in the working fluid, which makes the temperature differential across the collector relatively high. The rapid increase in water vapor pressure with temperature requires a corresponding increase in system operating pressure to prevent boiling.

FIGURE 7.5

The steam—flash steam generation concept.

FIGURE 7.6

The direct steam generation concept.

Increased operating temperatures reduce the thermal efficiency of the solar collector. Increased pressures within the system require a more robust design of collector components, such as receivers and piping. The differential pressure over the delivered steam pressure required to prevent boiling is supplied by the circulation pump and is irreversibly dissipated across the flash valve. When boiling occurs in the collectors, as in an in situ boiler, the system pressure drop and, consequently, electrical power consumption are greatly reduced. In addition, the latent heat transfer process minimizes the temperature rise across the solar collector. Disadvantages of in situ boiling are the possibility of a number of stability problems (Peterson and Keneth, 1982) and the fact that, even with a very good feed-water treatment system, scaling in the receiver is unavoidable. In multiple row collector arrays, the occurrence of flow instabilities could result in loss of flow in the affected row. This in turn could result in tube dry-out, with consequent damage to the receiver selective coating. No significant instabilities were reported, however, by Hurtado and Kast (1984) when experimentally testing a single-row 36 m system. Recently, once-through systems have been developed on a pilot scale for direct steam generation in which parabolic trough collectors inclined at 2—4 degrees are used (Zarza et al., 1999).

FIGURE 7.7

The unfired boiler steam generation concept.

A diagram of an unfired boiler system is shown in Fig. 7.7. In this system, a heat transfer fluid is circulated through the collector, which is non-freezing and non-corrosive and in which system pressures are low and control is straightforward. These factors largely overcome the disadvantages of water systems and are the main reasons for the predominant use of heat transfer oil systems in current industrial steam-generating solar systems.

The major disadvantage of the system results from the characteristics of the heat transfer fluid. These fluids are hard to contain, and most heat transfer fluids are flammable. Decomposition, when the fluids are exposed to air, can greatly reduce ignition point temperatures and leaks into certain types of insulation can cause combustion at temperatures that are considerably lower than measured self-ignition temperatures. Heat transfer fluids are also relatively expensive and present a potential pollution problem that makes them unsuitable for food industry applications (Murphy and Keneth, 1982). Heat transfer fluids have much poorer heat transfer characteristics than water. They are more viscous at ambient temperatures, are less dense, and have lower specific heats and thermal conductivities than water. These characteristics mean that higher flow rates, higher collector differential temperatures, and greater pumping power are required to obtain the equivalent quantity of energy transport when compared with a system using water. In addition, heat transfer coefficients are lower, so there is a larger temperature differential between the receiver tube and the collector fluid. Higher temperatures are also necessary to achieve cost-effective heat exchange. These effects result in reduced collector efficiency.

It should be noted that, for every application, a suitable system has to be selected by taking into consideration all these factors and constraints.

7.2.2 Flash vessel design

To separate steam at lower pressure, a flash vessel is used. This is a vertical vessel, as shown in Fig. 7.8, with the inlet of high-pressure high-temperature water located at about one-third of the way up its height. The standard design of flash vessels requires that the diameter of the vessel is chosen so that the steam flows toward the top outlet connection at no more than about 3 m/s. This should ensure that any water droplets could fall through the steam in a contra-flow, to the bottom of the vessel. Adequate height above the inlet is necessary to ensure separation. The separation is also facilitated by having the inlet projecting downward into the vessel. The water connection is sized to minimize the pressure drop from the vessel to the pump inlet to avoid cavitation.

FIGURE 7.8

Flash vessel schematic diagram.

7.3 Solar chemistry applications

Solar chemistry applications include a variety of fields; the main ones are the production of energy carriers (e.g., hydrogen), also called *reforming of fuels*; fuel cells; materials processing and detoxification; and recycling of waste materials. These are examined in this section.

7.3.1 Reforming of fuels

Solar energy is essentially unlimited and its utilization does not create ecological problems. However, solar radiation reaching the earth is intermittent and not distributed evenly. There is thus a need to collect and store solar energy and transport it from the sunny uninhabited regions, such as deserts, to industrialized populated regions, where great quantities of energy are needed. An effective way to achieve this process is by the thermochemical conversion of solar energy into chemical fuels. This method provides a thermochemically efficient path for storage and transportation. For this purpose, high-concentration-ratio collectors, similar to the ones used for power generation (see Chapter 10), are required. By concentrating solar radiation in receivers and reactors, one can supply energy to high-temperature processes to drive endothermic reactions.

Hydrogen is the main fuel (energy carrier) used in fuel cells (see next section). Today, however, no sources of hydrogen with a widespread delivery infrastructure are readily available. This issue can be solved by using fossil fuels to generate the hydrogen required. The transformation of a fossil fuel to hydrogen is generally called *fuel reforming*. Steam reforming is one example, in which steam is mixed with the fossil fuel at temperatures around 760°C. This high temperature can be obtained by burning conventional fuels or by high-concentration concentrating solar collectors. The chemical equation of this reforming reaction for natural gas composed primarily of methane (CH_4) is:

$$CH_4 + 2H_2O \rightarrow CO_2 + 4H_2 \tag{7.1}$$

Fuel reforming can be done in facilities of different sizes. It can be done in a central facility such as a chemical plant at a large scale. Such a plant produces pure hydrogen, which can be a high-pressure gas or liquid. Fuel reforming can also be performed on an intermediate scale in various facilities such as a gasoline station. In this case, refined gasoline or diesel fuel would be required, which can be delivered to the station with its current infrastructure. On-site equipment would then reform the fossil fuel into a mixture composed primarily of hydrogen and other molecular components, such as CO_2 and N_2. This hydrogen would most probably be delivered to customers as a high-pressure gas.

The fuel-reforming process can also be performed on a small scale, according to the requirements, immediately before its introduction into the fuel cell. For example, a fuel cell-powered vehicle can have a gasoline tank, which would use the existing infrastructure of gasoline delivery, and an onboard fuel processor, which would reform the gasoline into a hydrogen-rich stream that would be fed directly to the fuel cell.

In the future, it is anticipated that most of the hydrogen required to power fuel cells could be generated from renewable sources, such as wind or solar energy. For example, the electricity generated at a wind farm or with photovoltaics could be used to split water into hydrogen and oxygen through electrolysis. Electrolysis as a process could produce pure hydrogen and pure oxygen. The hydrogen thus produced could then be delivered by pipeline to the end users.

Chemistry applications include also the solar reforming of low-hydrocarbon fuels such as LPG and natural gas and upgrading them into a synthetic gas that can be used in gas turbines. Thus, weak gas resources diluted with carbon dioxide can be used directly as feed components for the conversion process. Therefore, natural gas fields currently not exploited due to high CO_2 content might be opened to the market. Furthermore, gasification products of unconventional fuels, such as biomass, oil shale, and waste asphaltenes, can also be fed into the solar upgrade process (Grasse, 1998).

A model for solar volumetric reactors for hydrocarbon-reforming operations at high temperature and pressure is presented by Yehesket et al. (2000). The system is based on two achievements; the development of a volumetric receiver tested at 5000–10,000 suns, gas outlet temperature of 1200°C, and pressure at 20 atm and a laboratory-scale chemical kinetics study of hydrocarbon reforming. Other related applications are a solar-driven ammonia-based thermochemical energy storage system (Lovegrove et al., 1999) and an ammonia synthesis reactor for a solar thermochemical energy storage system (Kreetz and Lovegrove, 1999).

Another interesting application is solar zinc and syngas production, both of which are very valuable commodities. Zinc finds application in zinc–air fuel cells (ZAFC) and batteries. Zinc can also react with water to form hydrogen, which can be further processed to generate heat and electricity. Syngas can be used to fuel highly efficient combined cycles or as the building block of a wide variety of synthetic fuels, including methanol, which is a very promising substitute for gasoline to fuel cars (Grasse, 1998).

7.3.2 Fuel cells

A *fuel cell* is an electrochemical device that converts the chemical energy of a fuel, such as hydrogen, natural gas, methanol, or gasoline, and an oxidant, such as air or oxygen, into electricity. Electrochemical devices generate electricity without combustion of the fuel and oxidizer, as opposed to what occurs with traditional methods of electricity generation. In principle, a fuel cell operates like a battery, but unlike a battery, it does not run down or require recharging. In fact, a fuel cell produces electricity and heat as long as fuel and an oxidizer are supplied. A fuel cell, like a battery, has a positively charged anode, a negatively charged cathode, and an ion-conducting material, called an *electrolyte*. The main fuel used in fuel cells is hydrogen. An introduction to hydrogen production and use is given in Chapter 1.

An electrochemical reaction is a reaction in which one species, the reducing agent, is oxidized (loses electrons) and another species, the oxidizing agent, is reduced (gains electrons).

The direct conversion of chemical energy to electrical energy is more efficient and generates much fewer pollutants than traditional methods that rely on combustion. Therefore, fuel cells can generate more electricity from the same amount of fuel. Furthermore, by avoiding the combustion process that occurs in traditional power-generating methods, the generation of pollutants during the combustion process is minimized. Some of the pollutants that are significantly lower for fuel cells are oxides of nitrogen and unburned hydrocarbons and carbon monoxide, which is a poisonous gas.

Basic characteristics

Fuel cell construction generally consists of a fuel electrode (anode) and an oxidant electrode (cathode) separated by an ion-conducting membrane. In the basic fuel cell, oxygen passes over one electrode and hydrogen over the other; in doing so, it generates electricity, water, and heat. Fuel cells chemically combine the molecules of a fuel and oxidizer without burning or having to dispense with the inefficiencies and pollution of traditional combustion.

Some other important characteristics of fuel cells are as follows:

- *Charge carrier.* The charge carrier is the ion that passes through the electrolyte. The charge carrier differs among different types of fuel cells. For most types of fuel cells, however, the charge carrier is a hydrogen ion, H^+, which has a single proton.
- *Contamination.* Fuel cells can be contaminated by different types of molecules. Such a contamination can lead to severe degradation in their performance. Because of the difference in electrolyte, catalyst, operating temperature, and other factors, different molecules can behave differently in various fuel cells. The major contamination agent for all types of fuel cells is sulfur-containing compounds, such as hydrogen sulfide (H_2S) and carbonyl sulfide (COS).
- *Fuels.* Hydrogen is currently the most popular fuel for fuel cells. Some gases, such as CO and CH_4, have different effects on fuel cells, depending on the type of fuel cell. For example, CO is a contaminant to fuel cells operating at relatively low temperatures, such as the proton exchange membrane fuel cell (PEMFC). However, CO can be used directly as a fuel for the high-temperature fuel cells, such as the solid oxide fuel cell (SOFC).
- *Performance factors.* The performance of a fuel cell depends on numerous factors, such as the electrolyte composition, the geometry of the fuel cell, the operating temperature, and gas pressure. The geometry of the fuel cell is affected mainly by the surface area of the anode and cathode.

A valuable source that covers highly technical information on different types of fuel cells is the *Fuel Cell Handbook* published by the U.S. Department of Energy. It is freely available on the Internet (U.S. Department of Energy, 2000) or from the fuel cell test and evaluation center of the U.S. Ministry of Defense (FCTec, 2008).

Fuel cell chemistry

Fuel cells generate electricity from a simple electrochemical reaction in which an oxidizer, typically oxygen from air, and a fuel, typically hydrogen, combine to form a product, which for the typical fuel cell is water. The basic principle of fuel cell operation is that it separates the oxidation and reduction into separate compartments, which are the anode and the cathode (separated by a membrane), thereby

FIGURE 7.9

Schematic diagram of a fuel cell.

forcing the electrons exchanged between the two half reactions to travel through the load. Oxygen (air) continuously passes over the cathode and hydrogen passes over the anode to generate electricity, while the by-products are heat and water. The fuel cell itself has no moving parts, so it is a quiet and reliable source of power. A schematic representation of a fuel cell with the reactant–product gases and the ion conduction flow directions through the cell is shown in Fig. 7.9. The basic physical structure or building block of a fuel cell consists of an electrolyte layer in contact with a porous anode and cathode on either side.

Figure 7.9 is a simplified diagram that demonstrates how the fuel cell works. In a typical fuel cell, gaseous fuels are fed continuously to the anode (negative electrode) compartment and an oxidant (i.e., oxygen from air) is fed continuously to the cathode (positive electrode) compartment. At electrodes, the electrochemical reactions take place and produce an electric current. The fuel cell is an energy conversion device that theoretically has the capability of producing electrical energy for as long as the fuel and oxidant are supplied to the electrodes. In reality, degradation, primarily corrosion, or malfunction of components limits the practical operating life of fuel cells (U.S. Department of Energy, 2000).

The electrolyte that separates the anode and cathode is an ion-conducting material. At the anode, hydrogen and its electrons are separated so that the hydrogen ions (protons) pass through the electrolyte while the electrons pass through an external electrical circuit as a direct current (DC) that can power useful devices, usually through an inverter, which converts the DC current into an AC one. The hydrogen ions combine with the oxygen at the cathode and are recombined with the electrons to form water. The reactions taking place in a fuel cell are as follows.

Anode half reaction (oxidation),

$$2H_2 \rightarrow 4H^+ + 4e^-$$ (7.2)

Cathode half reaction (reduction),

$$O_2 + 4H^+ + 4e^- \rightarrow 2H_2O$$ (7.3)

Overall cell reaction,

$$2H_2 + O_2 \rightarrow 2H_2O$$ (7.4)

To obtain the required power, individual fuel cells are combined into fuel cell stacks. The number of fuel cells in the stack determines the total voltage, and the surface area of each cell determines the total current (FCTec, 2008). Multiplying the voltage by the current yields the total electrical power generated as:

$$\text{Power(watts)} = \text{Voltage (volts)} \times \text{Current (amps)}$$ (7.5)

Porous electrodes, mentioned previously, are crucial for good electrode performance. Porous electrodes, used in fuel cells, achieve very high current densities, which are possible because the electrode has a high surface area relative to the geometric plate area, which significantly increases the number of reaction sites; and the optimized electrode structure has favorable mass transport properties. In an idealized porous gas fuel cell electrode, high current densities at reasonable polarization are obtained when the liquid (electrolyte) layer on the electrode surface is sufficiently thin, so that it does not significantly impede the transport of reactants to the electroactive sites and a stable three-phase (gas−electrolyte−electrode surface) interface is established (U.S. Department of Energy, 2000).

Types of fuel cells

Fuel cells are classified by their electrolyte material. Today, several types of fuel cells have been developed for applications ranging in size from a mobile phone (with under 1 W power) to a small power plant for an industrial facility or a small town (in the megawatt range). The main types of fuel cells that exist today are the following.

Alkaline fuel cell (AFC)

Alkaline fuel cells (AFCs) are one of the most developed technologies and have been used since the mid-1960s by NASA in the Apollo and space shuttle programs. The fuel cells on board these spacecraft provide electrical power for onboard systems, as well as drinking water. A schematic diagram of an AFC is shown in Fig. 7.10. AFCs are among the most efficient (nearly 70%) in generating electricity. The electrolyte in this fuel cell is an aqueous (water-based) solution of potassium hydroxide (KOH), which can be in concentrated (85 wt%) form for cells operated at high temperature ($\sim 250°C$) or less concentrated (35−50 wt%) for lower temperature ($<120°C$) operation. The electrolyte is retained in a matrix, usually made from asbestos. A wide range of electrocatalysts, such as Ni, Ag, metal oxides, and noble metals, can be used. One characteristic of AFCs is that they are very sensitive to CO_2 because this will react with the KOH to form K_2CO_3, thus altering the electrolyte. Therefore, the CO_2 reacts with the electrolyte, contaminating it rapidly and severely degrading the fuel cell performance. Even the smallest amount of CO_2 in the air must be considered with the alkaline cell. Therefore, AFCs must run on pure hydrogen and oxygen.

AFCs are the cheapest fuel cells to manufacture. This is because the catalyst required on the electrodes can be selected from a number of materials that are relatively inexpensive compared with the catalysts required for other types of fuel cells (U.S. Department of Energy, 2000).

FIGURE 7.10

Schematic diagram of an alkaline fuel cell (AFC).

The charge carrier for an AFC is the hydroxyl ion (OH−) transferred from the cathode to the anode, where it reacts with hydrogen to produce water and electrons. Water formed at the anode is transferred back to the cathode to regenerate hydroxyl ions. When operated, the AFC produces electricity and the by-product is heat.

Phosphoric acid fuel cell (PAFC)

Phosphoric acid fuel cells (PAFCs) were the first fuel cells to be commercialized. They were developed in the mid-1960s, have been field tested since the 1970s, and they have improved significantly in stability, performance, and cost. A schematic diagram of a PAFC is shown in Fig. 7.11. The efficiency of a PAFC in generating electricity is greater than 40%. Simple construction, low electrolyte volatility, and long-term stability are additional advantages. Phosphoric acid (H_3PO_4) concentrated to 100% is used for the electrolyte in this fuel cell, which operates at 150−220°C, since the ionic conductivity of phosphoric acid is low at low temperatures. The relative stability of concentrated phosphoric acid is high compared with other common acids. In addition, the use of concentrated acid minimizes the water

FIGURE 7.11

Schematic diagram of phosphoric acid fuel cell (PAFC).

vapor pressure, so water management in the cell is not difficult. The matrix universally used to retain the acid is silicon carbide and the electrocatalyst in both the anode and the cathode is platinum (Pt).

The charge carrier in this type of fuel cell is the hydrogen ion (H^+, proton). The hydrogen introduced at the anode is split into its protons and electrons. The protons are transferred through the electrolyte and combine with the oxygen, usually from air, at the cathode to form water. In addition, CO_2 does not affect the electrolyte or cell performance and PAFCs can therefore be easily operated with reformed fossil fuels.

Approximately 75 MW of PAFC generating capacity has been installed and is operating. Typical installations include hotels, hospitals, and electric utilities in Japan, Europe, and the United States (FCTec, 2008).

Molten carbonate fuel cell (MCFC)

Molten carbonate fuel cells (MCFCs) belong to the class of high-temperature fuel cells. The higher operating temperature allows them to use natural gas directly without the need for a fuel processor. A schematic diagram illustrating the working principle of an MCFC is shown in Fig. 7.12. MCFCs work quite differently from other fuel cells. The electrolyte in this fuel cell is composed of a molten mixture of carbonate salts. The fuel cell operates at 600–700°C, at which the alkali carbonates form a highly conductive molten salt, with carbonate ions providing ionic conduction. Two mixtures are currently used: the lithium carbonate and potassium carbonate or the lithium carbonate and sodium carbonate. These ions flow from the cathode to the anode, where they combine with hydrogen to yield water, carbon dioxide, and electrons. These electrons are routed through an external circuit back to the cathode, generating electricity and the by-product, heat. At the high operating temperatures in MCFCs, nickel (anode) and nickel oxide (cathode) are adequate to promote reaction, that is, noble metals are not required.

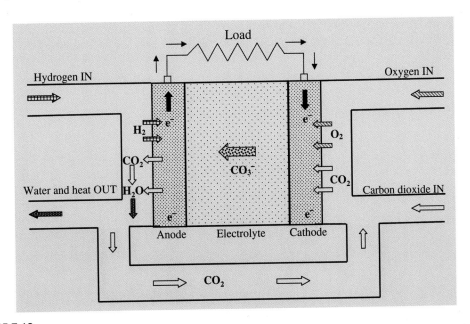

FIGURE 7.12

Schematic diagram of molten carbonate fuel cell (MCFC).

Compared with the lower temperature PAFCs and PEMFCs, the higher operating temperature of MCFCs has both advantages and disadvantages (FCTec, 2008). The advantages include:

1. At the higher operating temperature, fuel reforming of natural gas can occur internally, eliminating the need for an external fuel processor.
2. The ability to use standard materials for construction, such as stainless steel sheet, and allow the use of nickel-based catalysts on the electrodes.
3. The by-product heat from an MCFC can be used to generate high-pressure steam that can be used in many industrial and commercial applications.

The disadvantages are mainly due to the high temperatures and include:

1. High temperature requires significant time to reach operating conditions and responds slowly to changing power demands. These characteristics make MCFCs more suitable for constant power applications.
2. The carbonate electrolyte can cause electrode corrosion problems.
3. As CO_2 is consumed at the anode and transferred to the cathode, its introduction into the air stream and its control are problematic for achieving optimum performance.

Solid oxide fuel cell (SOFC)

Solid oxide fuel cells (SOFCs) can be operated over a wide temperature range, from 600 to 1000°C. They have been in development since the late 1950s and are the highest temperature fuel cells developed to allow a number of fuels to be used. A diagram of an SOFC is shown in Fig. 7.13. To operate at such high temperatures, the electrolyte is a thin, solid ceramic material (solid oxide) conductive to oxygen ions (O_2^-), which is the charge carrier. At the cathode, the oxygen molecules from the air are split into oxygen ions with the addition of four electrons. The oxygen ions are conducted through the electrolyte and combine with hydrogen at the anode, releasing four electrons. The electrons travel an external circuit providing electric power and producing heat as a by-product. The operating efficiency in generating electricity is among the highest of the fuel cells, at about 60% (FCTec, 2008).

FIGURE 7.13

Schematic diagram of solid oxide fuel cell (SOFC).

The solid electrolyte is impermeable to gas crossover from one electrode to another, in contrast to liquid electrolytes, where the electrolyte is contained in some porous supporting structure.

SOFCs operate at extremely high temperatures, so a significant amount of time is required to reach operating temperature. They also respond slowly to changes in electricity demand; thus, they are suitable for high-power applications, including industrial and large-scale central electricity generating stations.

The advantages of the high operating temperature of SOFCs are that it enables them to tolerate relatively impure fuels, such as those obtained from the gasification of coal or gases from industrial processes, and allows cogeneration applications, such as to create high-pressure steam that can be used in many applications. Furthermore, combining a high-temperature fuel cell with a turbine into a hybrid fuel cell further increases the overall efficiency of generating electricity, with a potential of an efficiency of more than 70% (FCTec, 2008). The disadvantage of SOFCs is that the high temperatures require more expensive construction materials.

Proton exchange membrane fuel cell (PEMFC)

Proton exchange membrane fuel cells (PEMFCs), also known as *polymer electrolyte membrane fuel cells*, are believed to be the best type of fuel cell for automobile applications that could eventually replace the gasoline and diesel internal combustion engines. First used in the 1960s for the NASA Gemini Program, PEMFCs are currently being developed and demonstrated for systems ranging from 1 W to 2 kW (FCTec, 2008). For a schematic diagram of this type of fuel cell please refer to the diagram for the PAFC, as it is very similar.

PEMFCs use a solid polymer membrane in the form of a thin plastic film as the electrolyte. This polymer is permeable to protons when it is saturated with water, but it does not conduct electrons. The fuel for the PEMFC is hydrogen and the charge carrier is the hydrogen ion (proton). At the anode, the hydrogen molecule is split into hydrogen ions (protons) and electrons. The hydrogen ions move across the electrolyte to the cathode while the electrons flow through an external circuit and produce electric power. Oxygen, usually in the form of air, is supplied to the cathode and combines with the electrons and the hydrogen ions to produce water (FCTec, 2008).

The electrolyte in these fuel cells is an ion exchange membrane (fluorinated sulfonic acid polymer or other similar polymer) that is an excellent proton conductor. The only liquid used in this fuel cell is water; thus, corrosion problems are minimal. Water management in the membrane is critical for efficient performance, as the fuel cell must operate under conditions where the by-product water does not evaporate faster than it is produced, because the membrane must be kept hydrated. Because of the limitation on the operating temperature imposed by the polymer, which is usually less than 120°C, and problems with water balance, an H_2-rich gas with minimal or no CO, which is a contaminant at low temperature, is used. Higher catalyst loading (Pt in most cases) is required for both the anode and the cathode (U.S. Department of Energy, 2000).

The advantage of PEMFCs is that they generate more power than other types of fuel cells, for a given volume or weight of fuel cell. This characteristic makes them compact and lightweight. Because the operating temperature is less than 100°C, rapid start up is achieved.

Since the electrolyte is a solid material, the sealing of the anode and cathode gases is simpler with a solid electrolyte, compared with a liquid; therefore, a lower cost is required to manufacture the cell. The solid electrolyte is also less sensitive to orientation and the corrosion problems are lower, compared with many of the other electrolytes, which leads to a longer cell and stack life.

One major disadvantage of the PEMFC is that the electrolyte must be saturated with water to operate optimally; therefore, careful control of the moisture of the anode and cathode streams is required. The high cost of platinum is another disadvantage.

Other types of fuels cells, not described in this book, include the direct methanol fuel cell (DMFC), regenerative fuel cell (RFC), zinc-air fuel cell (ZAFC), intermediate temperature solid oxide fuel cell (ITSOFC), and tubular solid oxide fuel cell (TSOFC). Interested readers can find information on these cells in other publications dedicated to the subject.

7.3.3 Materials processing

Solar energy material processing involves affecting the chemical conversion of materials by their direct exposure to concentrated solar energy. For this purpose, solar furnaces are used made of high-concentration, hence, high-temperature, collectors of the parabolic dish or heliostat type. Solar energy can also assist in the processing of energy-intensive high-temperature materials, as in the production of solar aluminum, the manufacture of which is one of the most energy-intensive processes. It also includes applications related to the production of high-added-value products, from fullerenes, which are large carbon molecules with major potential in commercial applications in semi- and superconductors, to commodity products such as cement (Norton, 2001). None of these processes, however, has achieved large-scale commercial adoption. Some pilot systems are described briefly here.

A solar thermochemical process developed by Steinfeld et al. (1996) combines the reduction of zinc oxide with reforming of natural gas, leading to the co-production of zinc, hydrogen, and carbon monoxide. At equilibrium, chemical composition in a blackbody solar reactor operated at a temperature of about 1000°C, atmospheric pressure and solar concentration of 2000, efficiencies between 0.4 and 0.65 have been obtained, depending on product heat recovery. A 5 kW solar chemical reactor was employed to demonstrate this technology in a high-flux solar furnace. Particles of zinc oxide were introduced continuously in a vortex flow, and natural gas contained within a solar cavity receiver was exposed to concentrated insolation from a heliostat field. The zinc oxide particles are exposed directly to the high radiative flux, avoiding the efficiency penalty and cost of heat exchangers.

A 2 kW concentrating solar furnace was used to study the thermal decomposition of titanium dioxide at temperatures of 2000–2500°C in an argon atmosphere (Palumbo et al., 1995). The decomposition rate was limited by the rate at which oxygen diffuses from the liquid–gas interface. It was shown that this rate is accurately predicted by a numerical model, which couples the equations of chemical equilibrium and steady-state mass transfer (Palumbo et al., 1995).

7.3.4 Solar detoxification

Another field of solar chemistry applications is solar photochemistry. Solar photochemical processes make use of the spectral characteristics of the incoming solar radiation to effect selective catalytic transformations, which find application in the detoxification of air and water and the processing of fine chemical commodities.

Solar detoxification achieves photocatalytic treatment of non-biodegradable persistent chlorinated water contaminants typically found in chemical production processes. For this purpose, parabolic trough collectors with glass absorbers are usually employed and the high intensity of solar radiation is used for the photocatalytic decomposition of organic contaminants. A development in photocatalytic detoxification and disinfection of water and air is presented by Goswami (1999). This process uses ultraviolet energy, available in sunlight, in conjunction with a photocatalyst (titanium dioxide), to decompose organic chemicals into non-toxic compounds (Mehos et al., 1992). Another application concerns the development of a prototype employing lower concentration compound parabolic collectors (Grasse, 1998).

The use of a compound parabolic concentrator technology for commercial solar detoxification applications is presented by Blanco et al. (1999). The objective is to develop a simple, efficient, and commercially competitive water treatment technology. A demonstration facility was erected at Plataforma Solar de Almeria in southern Spain.

7.4 Solar dryers

Generally, drying (or dewatering) is a simple process of excess water (moisture) removal from a natural or an industrial product to preserve it (foodstuff) or to reach a specified moisture content. Drying is an energy-intensive process, especially when used for food products, as these generally have a water content which is much higher (about 25%−80%) than is suitable for long preservation. Therefore the purpose of drying an agricultural product is to reduce its moisture content to a level that prevents its deterioration. It is essential to reduce the moisture content of foodstuff down to a certain level so as to slow down the action of enzymes, bacteria, yeasts and molds; this enables the food to be stored and preserved for long time without spoilage. Additionally, drying can be used for the total removal of moisture until the food has no moisture at all. When ready to use, the dehydrated food is re-watered and almost regains its initial conditions. Generally, crops are very sensitive to the drying conditions. Drying must be performed in a way that does not seriously affect their color, flavor, texture, or nutritional value. Thus the selection of drying conditions, especially temperature, is of major importance.

Solar drying is another very important application of solar energy. Solar dryers use air collectors to collect solar energy. The dryers are used primarily by the agricultural industry. In drying, two processes take place: one is a heat transfer to the product using energy from the heating source, and the other is a mass transfer of moisture from the interior of the product to its surface and from the surface to the surrounding air, in the form of water vapor.

Traditionally, farmers used the open-to-the-sun or natural drying technique, which achieves drying by using solar radiation, ambient temperature, relative humidity of ambient air, and natural wind.

This process has been used for millennia to preserve food. The technique involves the spread of the product to be dried in a thin layer on large outdoor threshing surfaces or concrete floors, where it is left until has been dried up to the desired moisture content. From time to time the material has to be turned over to accelerate drying by permitting trapped moisture to escape. Generally, the drying surface is made from concrete paved floors lined with polyethylene sheets; however sensitive food material is placed on perforated trays. Obviously the drying rate in the process is very slow, so the crops must remain outdoors for long periods of time, usually 10−30 days, depending on the product and the weather conditions of the site.

During drying, solar radiation is falling on the crop surface and simultaneously moisture is transferred from the material to the ambient air. A part of the solar radiation is lost to the atmosphere and to the surrounding soil. Heat and moisture transfer take place by natural convection and diffusion, respectively, and both depend on weather conditions and mostly on solar radiation intensity and wind velocity. According to Ramana (2009) more than 80% of food produced by small farmers in developing countries is dried by natural open-to-the-sun drying.

Capacity wise, and despite the very rudimentary nature of the process, natural drying remains the most common method of solar drying. This is because the energy requirements, which come from solar radiation and the air enthalpy, are readily available in the ambient environment and no capital investment in equipment is required. The process, however, has some serious limitations. Usually, the material must remain outdoors for a prolonged time-period. During this time, the farmer's goods are subject to weather changes and natural attacks. The most obvious limitations are that the crops suffer the undesirable effects of dust, dirt, atmospheric pollution, and insect and rodent attacks. Because of these limitations, the quality of the resulting product can be degraded, sometimes beyond edibility. Cases exist also of complete or partial deterioration of crops due to sudden storms, heavy rains, or hail that can harm even the plastic cover, if used. Very sensitive crops are spread on trays covered with transparent material and are dried by the sun's radiation and natural air circulation. All these disadvantages can be eliminated by using a solar dryer.

The technique of open-to-the-sun drying has changed very little since its early prehistoric uses. The sun's free energy for drying is offset by a number of disadvantages, which reduce not only the quantity but also the quality of the final product. The main ones are that the whole procedure depends on the experience of unskilled personnel; the lack of any scientific control of the final quality and moisture content which depends only on observations and experience; the very slow rate operation which, depending on the nature of product and the prevailing weather conditions, can take place from a few days up to 1 month; the fact that the product is exposed directly to all kinds of weather changes, as rain, hail, and strong winds that can rot or destroy totally the material; and the very large qualitative and quantitative losses due to the weather and natural attack conditions closely related to the open-air procedure (Belessiotis and Delyannis, 2011).

Irrespective of all these disadvantages, open-to-the-sun drying is an economic procedure that needs little initial capital. With good weather and continuous observation of the drying progress, especially for foodstuff that dries quickly, the final product can be of a very good quality.

Solar radiation, in the form of solar thermal energy, is an alternative source of energy for drying, especially to dry fruits, vegetables, agricultural grains, and other kinds of materials, such as wood and timber. It is estimated that in developing countries significant post-harvest losses of agricultural products exist, due to the lack of other preservation means, which can be saved by using solar dryers. For solar drying, many products need pre-treatment to facilitate drying or to keep their flavor and texture. Fruit's high sugar and acid content makes the direct sun drying safe. On the contrary, vegetables have low sugar and acid content which increases the risk of spoilage during open-air drying (Belessiotis and Delyannis, 2011).

The purpose of a dryer is to supply more heat to the product than that available naturally under ambient conditions, thus increasing sufficiently the vapor pressure of the crop moisture. Therefore, moisture migration from the crop is improved. The dryer also significantly decreases the relative humidity of the drying air, and by doing so, its moisture-carrying capability increases, thus ensuring a sufficiently low equilibrium moisture content.

There are two types of solar dryers: the ones that use solar energy as the only source of heat and the ones that use solar energy as a supplemental source. The airflow can be either natural convection or forced, generated by a fan. In the dryer, the product is heated by the flow of the heated air through the product, by directly exposing the product to solar radiation or a combination of both.

The transfer of heat to the moist product is by convection from the flowing air, which is at a temperature above that of the product, by direct radiation from the sun, and by conduction from heated surfaces in contact with the product.

Absorption of heat by the product supplies the energy necessary for vaporization of water from the product. From the surface of the product, the moisture is removed by evaporation. Moisture starts to vaporize from the surface of the product when the absorbed energy increases its temperature sufficiently and the vapor pressure of the crop moisture exceeds the vapor pressure of the surrounding air. Moisture replacement to the surface is by diffusion from the interior, and it depends on the nature of the product and its moisture content. If the diffusion rate is slow, it becomes the limiting factor in the drying process, but if it is fast enough, the controlling factor is the rate of evaporation from the surface, which occurs at the initiation of the drying process.

In direct radiation drying, part of the solar radiation penetrates the material, and it is absorbed within the product, thus generating heat both in the interior of the product and on its surface. Therefore, the solar absorptance of the product is an important factor in direct solar drying. Because of their color and texture, most agricultural materials have relatively high absorptance.

By considering product quality, the heat transfer and evaporation rates must be closely controlled to guarantee both optimum drying rates and product quality. The maximum drying rate is required so that drying is economically viable.

Solar energy dryers are generally considered as simple devices. They range from very primitive ones used in small, desert, or remote communities, up to more sophisticated industrial installations, although the latter are still very few and under development. Solar energy dryers are classified according to the heating mode employed, the way the solar heat is utilized, and their structural arrangement. With respect to the heating mode employed, the two main categories are active and passive dryers. In active systems, a fan is used to circulate air through the air collector to the product, whereas in passive or natural circulation solar energy dryers, solar-heated air is circulated through the crop by buoyancy forces as a result of wind pressure. Therefore, active systems require, in addition to solar energy, other non-renewable energy sources, usually electricity, for powering fans for forced air circulation or for auxiliary heating.

With respect to the mode of solar energy utilization and structural arrangements, the three major subclasses are distributed-, integral-, and mixed-mode-type dryers. These subclasses belong to both active and passive solar energy dryers. In a distributed-type solar energy dryer, the solar energy collector and the drying chamber are separate units. In an integral-type solar energy dryer, the same piece of equipment is used for both solar energy collection and drying, that is, the dryer is capable of collecting solar energy directly, and no solar collectors are required. In the mixed-mode type, the two systems are combined, that is, the dryer is able to absorb heat directly but the process is enhanced by the use of a solar collector. These types are explained in more detail in the following sections.

7.4.1 Active solar energy dryers

Active solar energy dryers are usually suitable for larger amounts of material and often they are hybrid units using auxiliary sources of energy as conventional fuels to operate during cloudy weather and/or nighttime. These are more complicated and more expensive than the passive systems as they require fans.

Distributed type

A typical distributed-type active solar dryer is shown in Fig. 7.14. It comprises four components: a drying chamber, a solar energy air heater, a fan, and ducting to transfer the hot air from the collector to the dryer. In this design, the crop is located on trays or shelves inside an opaque drying chamber, which does not allow the solar radiation to reach the product directly. Air, which is heated during its passage through an air solar collector by the action of a fan, is ducted to the drying chamber to dry the product. The advantage of not allowing the solar radiation to reach the product directly is outlined in the passive section.

Integral type

Large-scale, commercial, forced-convection, greenhouse-type dryers are like transparent-roof solar barns and are used for solar timber drying kilns (see Fig. 7.15). Small-scale forced dryers are often equipped with auxiliary heating.

A variation of the design shown in Fig. 7.15 is the active greenhouse-type dryer built inside a semi-cylindrical outer shell. The semi-cylindrical structure acts as a solar heater. It consists of an exterior transparent cover, which acts as the collector glazing, and an inner semi-cylindrical black absorber sheet. A fan circulates the hot air through the air duct formed by the two semi-circular sheets to the material and the moist air is finally exhausted from the top of the transparent cover.

Another variation of this type of dryer is the solar collector–roof/wall, in which the solar heat collector forms an integral part of the roof and/or wall of the drying chamber. A solar–roof dryer is shown in Fig. 7.16. A collector–wall system is like a Trombe wall, described in Chapter 6, where a black painted concrete block wall with outside glazing forms the solar collector and serves also as a thermal storage.

Mixed-mode type

The mixed-mode dryer is similar to the distributed type, with the difference that the walls and roof of the dryer are made from glass, to allow solar energy to warm the products directly, as shown in Fig. 7.17.

FIGURE 7.14

Schematic diagram of a distributed-type active solar dryer.

FIGURE 7.15

Schematic diagram of a forced-convection, transparent-roof solar barn.

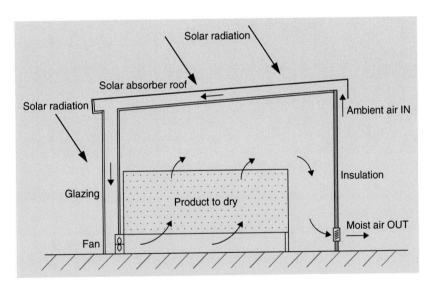

FIGURE 7.16

Schematic diagram of an active collector–roof solar energy storage dryer.

It should be noted that, because drying efficiency increases with temperature, in conventional dryers the maximum possible drying temperature that would not deteriorate the product quality is used. In solar dryers, however, the maximum drying temperature is determined by the solar collectors, because their efficiency decreases with higher operating temperatures and this may not yield an optimal dryer design.

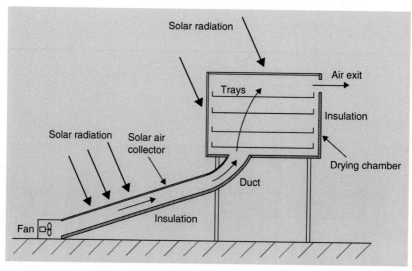

FIGURE 7.17

Schematic diagram of a mixed-mode-type active solar dryer.

Most air heaters use metal or wood absorbers, although black polythene absorbers have been used in a few designs in an attempt to minimize cost.

7.4.2 Passive solar energy dryers

Passive or direct drying of crops is still in common practice in many Mediterranean, tropical, and subtropical regions, especially in Africa and Asia or in small agricultural communities. Passive solar dryers are "hot box" units where the product in the hot box is exposed to the solar radiation through a transparent cover. Heating takes place by natural convection, through the dryer transparent cover or in a solar air heater.

The passive-type solar dryers are primitive inexpensive constructions, easy to install and operate, especially at sites where no electrical grid exists. Passive or natural circulation solar energy dryers operate by using entirely renewable sources of energy, such as solar and wind.

Distributed type

Distributed, natural circulation solar energy dryers are also called *indirect passive dryers*. A typical distributed natural circulation solar energy dryer comprises an air heating solar energy collector, appropriate insulated ducting, a drying chamber, and a chimney, as shown in Fig. 7.18. In this design, the crop is also located on trays or shelves inside an opaque drying chamber, which does not allow the solar radiation to reach the product directly. Air, which is heated during its passage through an air solar collector, is ducted to the drying chamber to dry the product. Because the crops do not receive direct sunshine, caramelization (formation of sugar crystals on the crop surface) and localized heat damage do not occur. Therefore, indirect dryers are usually used for some perishables and fruits, for which the vitamin content of the dried product is reduced by the direct exposure to sunlight. The color retention

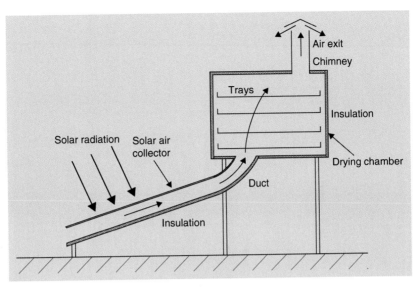

FIGURE 7.18

Schematic diagram of a distributed-type passive solar dryer.

in some highly pigmented commodities is also very adversely affected when they are exposed directly to the sun (Norton, 1992).

Higher operating temperatures are generally obtained in distributed natural circulation dryers than in direct dryers. They can generally produce higher quality products and are recommended for deep layer drying. Their disadvantages are that the fluctuation in the temperature of the air leaving the solar air collector makes constant operating conditions within the drying chamber difficult to maintain; they are relatively elaborate structures, requiring more capital investment in equipment; and they have higher running costs for maintenance than integral types. The efficiency of distributed-type dryers can be easily increased, because the components of the unit can be designed for optimal efficiency of their functions.

Integral type

Integral-type, natural circulation solar energy dryers are also called *direct passive solar energy dryers*. In this system, the crop is placed in a drying chamber, which is made with transparent walls; therefore, the necessary heat is obtained by the direct absorption of solar radiation at the product, from the internal surfaces of the chamber, and by convection from the heated air mass within the chamber. The heat removes the moisture from the product and, at the same time, lowers the relative humidity of the resident air mass, thus increasing its moisture-carrying capacity. The air in the chamber is also expanded because the density of the hot air is lower than the cold, thus generating natural circulation, which also helps in the removal of moisture, along with the warm air. Because heat is transferred to the crop by both convection and radiation, the rate of drying for direct dryers is greater than that for indirect dryers.

Integral-type, natural circulation solar energy dryers can be of a very simple construction, as shown in Fig. 7.19, which consists of a container insulated at its sides and covered with a single glazing or

FIGURE 7.19

Schematic diagram of an integral-type passive solar dryer.

roof. The interior walls are blackened; therefore, solar radiation transmitted though the cover is absorbed by the blackened interior surfaces as well as by the product, thus raising the internal temperature of the container. At the front, special openings provide ventilation, with warm air leaving via the upper opening under the action of buoyant forces. The product to be dried is placed on perforated trays inside the container. This type of dryer has the advantage of easy construction from cheap locally available materials and is used commonly to preserve fruits, vegetables, fish, and meat. The disadvantage is the poor air circulation obtained, which results in poor moist air removal and drying at high air temperatures (70—100°C), which is very high for most products, particularly perishables.

A variation of the dryer shown in Fig. 7.19 is the cabin dryer, which resembles an asymmetric solar still unit (see Fig. 8.1b, Chapter 8, Section 8.3) oriented north—south. The material to be dried is spread on perforated plates, through which the air circulates by natural convection and finally leaves the drying chamber from the upper north side. Bottom and side walls are opaque and well insulated. Cabin dryers are simple and inexpensive. They are suitable for drying agricultural products. They are normally constructed with a drying area of 1—2 m² and capacities of 10—20 kg.

Mixed-mode type

Mixed-mode, natural circulation solar energy dryers combine the features of the integral-type and the distributed-type natural circulation solar energy dryers. In this case, the combined action of solar radiation incident directly on the product to be dried and the air heated in a solar air collector provides the necessary heat required for the drying process. A mixed-mode, natural circulation solar energy dryer has the same structural characteristics as the distributed type, that is, a solar air heater, a separate drying chamber, and a chimney; in addition, the drying chamber walls are glazed so that the solar radiation can reach the product directly as in the integral-type dryers, as shown in Fig. 7.20.

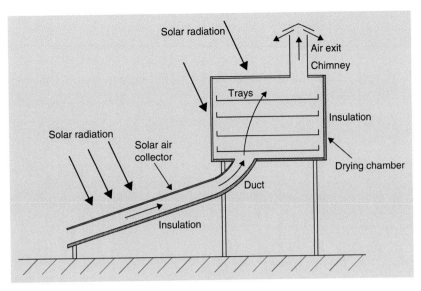

FIGURE 7.20

Schematic diagram of a natural circulation, mixed-mode solar energy dryer.

7.4.3 General remarks

The economics of solar dryers depend on the cost of the overall drying system and the gain from solar energy utilization, replacing conventional fuel. The main reasons for selecting solar energy for drying are the energy saving and the lack of availability of conventional energy sources to remote rural areas, or the high cost of transportation of fuel to those areas.

Another possibility considered recently is the use of mechanical dryers heated indirectly by solar energy. This is a rather new technique, not yet widely commercialized, which involves solar thermal energy collecting devices and off the shelf dryers, which are available in several types and sizes. The basic disadvantage of this indirect solar drying technique is the high initial capital cost for the dryer, solar collectors, auxiliary energy supply system, and the necessary supplementary equipment, such as ducts, pipes, pumps, blowers, control and measurement instruments. More skilled personnel are also required to operate the system. The advantages are many and include the high drying rate (15–30 h); the proper control of the drying process which ensures proper moisture content of the final product, so the dried product can be stored for longer times; the avoidance of losses as the product is not subject to any natural phenomena; the need for smaller surface areas for the same quantity of material as the trays can be accommodated in stacks inside the dryers; increased productivity, as dryers can be loaded again within few hours; and the flexibility of the dryer to accept similar seasonal crops, thus expanding operation of the system almost around the year (Belessiotis and Delyannis, 2011).

Energy storage is essential for places with either high or low radiation intensity and in cases where harvest products should be dried in continuous operation, immediately after harvest. Storage devices increase the initial capital cost as well as the operation costs, so to avoid unnecessary expenses storage must be applied in the following cases (Belessiotis and Delyannis, 2011):

- When solar intensity is high energy must be stored to avoid over-drying of the product at higher temperatures.
- When the agricultural product is very sensitive to temperature. By storing excess solar energy temperature can be easily regulated and controlled.
- When the drying operation has to be continued over night. This affects sensitive products that must be dried immediately after harvest.

Hybrid solar energy dryers were also developed in recent years, due to the significant increase in agricultural production. Hybrid solar energy dryers combine solar radiation energy with an auxiliary conventional source of energy. They can be operated solely by solar energy, solely by conventional energy sources, or by both. In most of the cases hybrid solar energy drying systems are medium to large capacity installations and operate by a solar fraction in the range of 50%–60%. Thus the drying operation, especially for large amounts of material, is nowadays satisfied with large active hybrid solar energy drying systems.

All dryers employing solar energy collection employ air flat-plate solar collectors. For passive solar dryers, simple air heaters made of transparent plastic films are sufficient. These are inexpensive, easily constructed, and operated. The design of suitable air solar collectors for drying systems is one of the most important tasks that affect the economy of the system. For low-temperature applications, used in the case of most foodstuffs, single glazed collectors of the type presented in Chapter 3, Section 3.1.1, are sufficient and appropriate.

Agricultural products are organic substances and due to the dependence on several parameters, the drying process is very complicated. Even for the same product, drying depends not only on the type of the crop but also on its physical condition, the site of culture, and the initial moisture content. The two most crucial drying condition are the drying temperature and the pre-treatment procedure. To establish suitable drying conditions, related to the dryer in use, experimentation is often necessary.

The most important condition for agricultural products is to keep the nutrient values, that is, vitamins sensitive to heat, and retain color and flavor is the drying temperature. The lower drying temperature starts from about 30°C, but at these low temperatures drying rate is very slow and there is a risk of spoilage or molding. Generally, the majority of food can be dried at a mean temperature of 60°C. Some products need lower drying temperatures at the beginning, and after being semi-dried the temperature can be raised up to a certain suitable point. This technique helps to keep the skin of the crop soft. In open-sun air drying, temperature depends on the intensity of solar radiation, is not easily controlled, and ranges from 40 to 80°C. In active solar energy dryers, in the event of high radiation intensities the temperature can be regulated by mixing hot air with the necessary amount of fresh air from the atmosphere.

Many crops, fruits, and vegetables are grown near the soil and are susceptible to the activity of various microorganisms. So immediately after harvest, if possible, they must be treated and dried, irrespective of the drying method employed. Pre-treatment helps to slow down the activity of microorganisms and soften the skin. Generally every crop has its own optimum conditions of pre-treatment when drying and these are simple methods based on experience. The main steps for agricultural products are (Jayaraman and Das Gupta, 1998):

1. Selection of the best quality of crops after harvest. They must be ripe, firm, and without scratches.
2. The products must be washed thoroughly to decrease microorganisms to a minimum, as they grow quickly when exposed to the atmosphere.

3. Depending on the product type, it must be shelled, peeled, and/or sliced.
4. Blanch the products to destroy enzymes and help retain color. Blanching consists of dipping the crops in boiling water or treating them by steam.
5. Sulfur the crops which help to prevent losses in color, flavor, and nutrients (vitamins) and acts as a disinfectant. Sulfuring is an old method of treating the crops by sodium sulfite solution or solutions of sodium bisulfite or metabisulfite. An old method usually applied is to burn sulfur and use the fumes for sulfuring.
6. Treatment with ascorbic acid solution to prevent browning of fruit or fruit slices.

Useful guidelines for solar drying are given by the El-Paso Solar Energy Association (EPSEA, 2012) which include the above pre-treatment procedure and other practical guidelines, such as the necessity to allow the product to cool down after drying and before storing, and the recommended material of the trays which is stainless steel rack, wood slats with cheesecloth cover, teflon, teflon-coated fiberglass, nylon, and food-grade plastics.

7.5 Greenhouses

Another application intended for the agricultural industry is the greenhouse. The basic function of a greenhouse is to provide environmental conditions that accelerate the process of photosynthesis. Photosynthesis is the driving force for plant growth, in which CO_2 is transformed into H_2O, using solar energy, to carbohydrates and oxygen. Photosynthesis is highly sensitive to environmental factors.

The requirements for the interior microclimate of a greenhouse vary according to the particular plant species and its stage of growth. This is characterized by the temperature, illumination, and the interior atmosphere, that is, water vapor, carbon dioxide, and pollutants (nitrogen oxides and sulfur).

The particular method required to create a specified environment and its economic viability depends on the prevailing ambient conditions and the value of the crop to be harvested in the particular greenhouse. It should be noted that a greenhouse designed for a particular climate can produce an environment suitable for a specific crop type, yet the same greenhouse in another location or at a different time of the year may be unsuitable for that same type of crop. Therefore, the plant varieties to be grown in a greenhouse should be chosen to suit the artificial environment that can be achieved economically within the greenhouse.

The main objective for the development of covered areas for growing food was the need for frost protection. Heat is usually obtained from solar radiation and auxiliary sources. As we saw in Chapter 2, by the expression *greenhouse effect*, we mean that the internal environment of a space is heated by the shortwave solar insolation transmitted through the cover and absorbed by its internal surfaces. These surfaces re-emit heat radiation, which is at longer wavelengths that cannot escape through the cover, and in this way, the heat is trapped into the space.

In places where summers are hot, greenhouses frequently need to be cooled. In areas where summers are not severe and the maximum ambient temperature remains less than 33°C, ventilation and shading techniques work well. In higher temperature environments, however, where ambient temperatures in summer generally exceed 40°C, evaporative cooling is usually applied, which is the most efficient means of greenhouse cooling. Evaporative cooling can lower the inside air temperature significantly below the ambient air, using fan-pads and fog or mist inside a greenhouse and roof cooling systems. Apart from these systems, two composite systems can be used for both heating and cooling greenhouses: the earth-to-air heat exchanger and the aquifer-coupled, cavity flow heat exchanger. A survey of these systems is given by Sethi and Sharma (2007).

7.5.1 Greenhouse materials

Traditionally, the first material used for greenhouse cover was glass. As an alternative cover material to glass, oiled paper was tried in the Netherlands during the late eighteenth century and was in common use in Japan well into the twentieth century (Norton, 1992). After the Second World War, plastic materials became more readily available. From the time clear plastic materials were first produced on a commercial scale, their potential for replacing glass in agricultural facilities has been recognized. Nowadays, PVC and polyethylene films are attached internally to the greenhouse framework, thus creating an insulating air gap between the outer cover and the protected artificial environment. Polyethylene is very popular for agricultural applications because it is available in wider sections than most other films and is low cost, although it has a short lifetime of about a year when exposed to typical weather conditions. Additionally, because polyethylene is the most common plastic film used, data for light transmission through this material are readily available.

Generally, plastic materials have inferior light transmission properties compared with glass. Additionally, since they degrade when exposed to heat and ultraviolet light, their useful life is much shorter, typically a few years compared with decades for glass. Condensation on the inner surface of the cover, which under some conditions could persist during the day, reduces the light transmission. This reduction is more pronounced with plastics than with glass because of the higher angle of conduct between the water bubbles and the plastic, leading to a higher proportion of reflected light. The advantages of plastic materials, however, are their low specific mass and high strength, requiring a lightweight structure and lower cost, resulting in lower initial investment.

Although polyethylene is the most widely used plastic film in agriculture, other materials are available, such as polymers containing fluorine compounds, whose radiation transmission properties and resistance to aging are superior to those of polyethylene films. These, however, are more expensive than polyethylene.

Exercises

7.1. An industrial process heating system uses air heated by both solar and auxiliary energy. It enters the duct supplying the air to the process at 37°C. The solar heat is supplied from a storage tank and transferred to the air via a water-to-air heat exchanger with an effectiveness equal to 0.95. The air temperature is topped by auxiliary energy to 60°C. The collector area is 70 m², $F_R(\tau\alpha) = 0.82$, $F_R U_L = 6.15$ W/m² °C. The fully mixed storage tank has 4.5 m³ capacity and its UA value is 195 W/°C; it is located in a room with temperature of 18°C. The capacitance of the collector side of the heat exchanger is 1150 W/°C and of the storage side is 910 W/°C. The load is required for 8 h a day, from 8 a.m. to 4 p.m., and the air has a constant flow rate of 0.25 kg/s. The heat capacity of air is 1012 J/kg °C and the flow rate of water through the load heat exchanger is 0.07 kg/s. For the period under investigation, the radiation and ambient temperatures shown in the following table apply. If the initial temperature of the storage tank on the day investigated was 42°C, estimate the energy supplied to the load and the amount of auxiliary energy required by the system to cover the load.

Time	Ambient temperature (°C)	I_t (MJ/m^2)
6–7	10	0
7–8	11	0
8–9	12	1.12
9–10	14	1.67
10–11	16	2.56
11–12	17	3.67
12–13	18	3.97
13–14	16	3.29
14–15	15	2.87
15–16	14	1.78
16–17	12	1.26
17–18	11	0

7.2. The system in Exercise 7.1 uses water at a temperature of 80°C instead of air. If the load is required for the same period of time, the flow rate of water is 0.123 kg/s, and the heat capacity of water is 4180 J/kg °C, estimate the energy supplied to the load and the amount of auxiliary energy required by the system to cover the load.

7.3. Repeat Exercise 7.1 for an effectiveness of the load heat exchanger of 0.66 and compare the results.

7.4. Repeat Exercise 7.2, but the load is required for only 30 min of each hour of operation.

Assignment

With the help of the System Advisor Model, the details of which are given in Appendix 10, design an IPH system using PTCs. The system can be selected by choosing Concentrating Solar Power/Parabolic Trough—Heat. A 1 MW system is to be designed for a location included in the program database or you own location, provided that suitable weather data are available. Try to use for the various input parameters as realistic values as possible. You can choose different types of financial models, and various types of incentives. For actual installation and operating costs of each combination, find the most cost-effective combination of equipment setup. For parameters not specified above you can use realistic figures (or the program default values) which need to be listed in the report produced.

References

Abdel-Dayem, A.M., 2011. Numerical-simulation and experimental-validation of the largest Egyptian solar process-heat system. J. Renew. Sustain. Energy 3, 043102.

Belessiotis, V., Delyannis, E., 2011. Solar drying. Sol. Energy 85 (8), 1665–1691.

Benz, N., Gut, M., Rub, W., 1998. Solar process heat in breweries and dairies. In: Proceedings of EuroSun 98 on CD ROM. Portoroz, Slovenia.

Benz, N., Gut, M., Beikircher, T., 1999. Solar process heat with non-concentrating collectors for food industry. In: Proceedings of ISES Solar World Congress on CD ROM, Jerusalem, Israel.

Blanco, J., Malato, S., Fernandez, P., Vidal, A., Morales, A., Trincado, P., et al., 1999. Compound parabolic concentrator technology development to commercial solar detoxification applications. In: Proceedings of ISES Solar World Congress on CD ROM, Jerusalem, Israel.

Brink, D.F., Youngblood, S.B., 1982. Operation and performance of the solar steam system at the Johnson & Johnson plant in Sherman, Texas. ASME Journal of Solar Energy Engineering 104 (3), 139–145.

Cotrado, M., Dalibard, A., Soll, R., Pietruschka, D., 2014. Design, control and first monitoring data of a large scale solar plant at the meat factory Berger, Austria. Energy Proc. 48, 1144–1151.

EPSEA, 2012. El-Paso Solar Energy Association. Available from: www.epsea.org/dry.html.

FCTec, 2008. Fuel Cell Test and Evaluation Center. U.S. Ministry of Defense. Available from: http://energy.nstl.gov.cn/MirrorResources/1922/.

Goswami, D.Y., 1999. Recent developments in photocatalytic detoxification and disinfection of water and air. In: Proceedings of ISES Solar World Congress on CD ROM, Jerusalem, Israel.

Grasse, W., 1998. Solar PACES Annual Report. DLR, Cologne, Germany.

Hurtado, P., Kast, M., 1984. Experimental Study of Direct In-Situ Generation of Steam in a Line Focus Solar Collector. SERI, Golden, CO.

Jayaraman, K.S., Das Gupta, D.K., 1998. Drying of fruits and vegetables. In: Mujumdar, A.E. (Ed.), Handbook of Industrial Drying, second ed., vol. 1. Marcel Dekker, Inc., pp. 643–690

Kalogirou, S.A., 2003. The potential of solar industrial process heat applications. Appl. Energy 76 (4), 337–361.

Kalogirou, S.A., Lloyd, S., Ward, J., 1997. Modelling, optimization and performance evaluation of a parabolic trough collector steam generation system. Sol. Energy 60 (1), 49–59.

Kreetz, H., Lovegrove, K., 1999. Theoretical analysis and experimental results of a 1 kWchem ammonia synthesis reactor for a solar thermochemical energy storage system. In: Proceedings of ISES Solar World Congress on CD ROM, Jerusalem, Israel.

Ktistis, P.K., Agathokleous, R.A., Kalogirou, S.A., 2021. Experimental performance of a parabolic trough collector system for an industrial process heat application. Energy 215, 119288.

Kumaresan, G., Sridhar, R., Velraj, R., 2012. Performance studies of a solar parabolic trough collector with a thermal energy storage system. Energy 47, 395–402.

Laing, D., Lehmann, D., Bahl, C., 2008. Concrete storage for solar thermal power plants and industrial process heat. In: Third International Renewable Energy Storage Conference (IRES 2008), Berlin.

Lovegrove, K., Luzzi, A., Kreetz, H., 1999. A solar driven ammonia based thermochemical energy storage system. In: Proceedings of ISES Solar World Congress on CD ROM, Jerusalem, Israel.

Mehos, M., Turchi, C., Pacheco, J., Boegel, A.J., Merrill, T., Stanley, R., 1992. Pilot-Scale Study of the Solar Detoxification of VOC-Contaminated Groundwater. NREL/TP-432-4981, Golden, CO.

Murphy, L.M., Keneth, E., 1982. Steam Generation in Line-Focus Solar Collectors: A Comparative Assessment of Thermal Performance, Operating Stability, and Cost Issues. SERI/TR-1311, Golden, CO.

Norton, B., 1992. Solar Energy Thermal Technology. Springer-Verlag, London.

Norton, B., 1999. Solar process heat: distillation, drying, agricultural and industrial uses. In: Proceedings of ISES Solar World Congress on CD ROM, Jerusalem, Israel.

Norton, B., 2001. Solar process heat. In: Gordon, J. (Ed.), Solar Energy: The State of the Art. James and James, London, pp. 477–496.

Palumbo, R., Rouanet, A., Pichelin, G., 1995. Solar thermal decomposition of TiO_2 at temperatures above 2200 K and its use in the production of Zn and ZnO. Energy 20 (9), 857–868.

Peterson, R.J., Keneth, E., 1982. Flow Instability during Direct Steam Generation in Line-Focus Solar Collector System. SERI/TR-1354, Golden CO.

Powell, K.M., Edgar, T.F., 2012. Modeling and control of a solar thermal power plant with thermal energy storage. Chem. Eng. Sci. 71, 138−145.

Ramana, M.V., 2009. A review of new technologies, modes and experimental investigations of solar dryers. Renewable Sustain. Energy Rev. 13, 835−844.

Rittmann-Frank, M.H., Mollenkamp, J., Caflisch, M., Haberle, A., 2018. Evaluation of solar process heat systems in Switzerland. In: AIP Conference Proceedings 2018; 2033 (November).

Schweiger, H., Mendes, J.F., Benz, N., Hennecke, K., Prieto, G., Gusi, M., Goncalves, H., 2000. The potential of solar heat in industrial processes. A state of the art review for Spain and Portugal. In: Proceedings of EuroSun 2000, Copenhagen, Denmark on CD ROM.

Sethi, V.P., Sharma, S.K., 2007. Survey of cooling technologies for worldwide agricultural greenhouse applications. Sol. Energy 81 (12), 1447−1459.

Spate, F., Hafner, B., Schwarzer, K., 1999. A system for solar process heat for decentralized applications in developing countries. In: Proceedings of ISES Solar World Congress on CD ROM, Jerusalem, Israel.

Steinfeld, A., Larson, C., Palumbo, R., Foley, M., 1996. Thermodynamic analysis of the co-production of zinc and synthesis gas using solar process heat. Energy 21 (3), 205−222.

U.S. Department of Energy, 2000. Fuel Cell Handbook, fifth ed. Available from: http://www.netl.doe.gov/technologies/coalpower/fuelcells/seca/pubs/fchandbook7.pdf.

Weiss, W., Spörk-Dür, M., 2020. Solar Heat Worldwide, Global Market Development and Trends in 2019, Solar Heating and Cooling Program. IEA, 2020.

Yehesket, J., Rubin, R., Berman, A., Karni, J., 2000. Chemical kinetics of high temperature hydrocarbons reforming using a solar reactor. In: Proceedings of EuroSun 2000 on CD ROM, Copenhagen, Denmark.

Zarza, E., Hennecke, K., Coebel, O., 1999. Project DISS (direct solar steam) update on project status and future planning. In: Proceedings of ISES Solar World Congress on CD ROM, Jerusalem, Israel.

Solar Desalination Systems

8.1 Introduction

The provision of freshwater is becoming an increasingly important issue in many areas of the world. In arid areas, potable water is very scarce and the establishment of a human habitat in these areas strongly depends on how such water can be made available. A brief historical introduction to solar desalination is given in Chapter 1, Section 1.5.2.

Water is essential to life. The importance of supplying potable water can hardly be overstressed. Water is one of the most abundant resources on earth, covering three-fourths of the planet's surface. About 97% of the earth's water is saltwater in the oceans and 3% (about 36 million km^3) is freshwater contained in the poles (in the form of ice), ground water, lakes, and rivers, which supply most human and animal needs. Nearly 70% from this tiny 3% of the world's freshwater is frozen in glaciers, permanent snow cover, ice, and permafrost. Thirty percent of all freshwater is underground, most of it in deep hard-to-reach aquifers. Lakes and rivers together contain just a little more than 0.25% of all freshwater; lakes contain most of it.

8.1.1 Water and energy

Water and energy are inseparable commodities that govern the lives of humanity and promote civilization. The history of humankind proves that water and civilization are inseparable entities. All great civilizations developed and flourished near large sources of water. Rivers, seas, oases, and oceans have attracted humankind to their coasts because water is the source of life. Maybe the most significant example of this influence is the Nile River in Egypt. The river provided water for irrigation and mud full of nutrients. Ancient Egyptian engineers were able to master the river water, and Egypt, as an agricultural nation, became the most important wheat-exporting country in the whole Mediterranean Basin (Delyannis, 2003). Due to the richness of the river, various disciplines of science, such as astronomy and mathematics, as well as law, justice, currency, and police protection, were created in the region at a time when no other human society held this knowledge or sophistication.

Energy is as important as water for the development of a good standard of life because it is the force that puts in operation all human activities. Water by itself is also a power-generating force. The first confirmed attempts to harness water power occurred more than 2000 years ago, at which time the energy gained was mainly used to grind grain (Major, 1990).

The Greeks were the first to express philosophical ideas about the nature of water and energy. Thales of Miletus (640–546 BC), one of the seven wise men of antiquity, wrote about water

Solar Energy Engineering. https://doi.org/10.1016/B978-0-323-99350-0.00012-0

(Delyannis, 1960) that it is fertile and molded (can take the shape of its container). The same philosopher said that seawater is the immense sea that surrounds the earth, which is the primary mother of all life. Later on, Embedokles (495–435 BC) developed the theory of the elements (Delyannis, 1960), describing that the world consists of four primary elements: fire, air, water, and earth. With today's knowledge, these elements may be translated to energy, atmosphere, water, and soil, which are the four basic constituents that affect the quality of our lives (Delyannis and Belessiotis, 2000).

Aristotle (384–322 BC), one of the greatest philosophers and scientists of antiquity, described in a surprisingly correct way the origin and properties of natural, brackish, and seawater. He also described accurately the water cycle in nature—a description that is still valid. In fact, the water cycle is a huge solar energy open distiller in a perpetual operational cycle.

Aristotle wrote that seawater becomes sweet when it turns into vapor, and the vapor does not form saltwater when it condenses again. In fact, Aristotle proved this experimentally.

8.1.2 Water demand and consumption

Humanity is dependent on rivers, lakes, and underground water reservoirs for freshwater requirements in domestic life, agriculture, and industry. However, rapid industrial growth and a worldwide population explosion have resulted in a large escalation of demand for freshwater, both for household needs and for crops to produce adequate quantities of food. Added to this is the problem of the pollution of rivers and lakes by industrial wastes and the large amounts of sewage discharge. On a global scale, human-made pollution of natural sources of water is becoming one of the greatest causes of freshwater shortage. Added to this is the problem of uneven distribution. For example, Canada has a 10th of the world's surface freshwater but less than 1% of the world's population.

Of the total water consumption, about 70% is used by agriculture, 20% is used by the industry, and only 10% of the water consumed worldwide is used for household needs. It should be noted that, before considering the application of any desalination method, water conservation measures should be considered. For example, drip irrigation, using perforated plastic pipes to deliver water to crops, uses 30%–70% less water than traditional methods and increases crop yield. This system was developed in the early 1960s, but today it is used in less than 1% of the irrigated land. In most countries, governments heavily subsidize irrigation water and farmers have no incentive to invest in drip systems or any other water-saving methods.

8.1.3 Desalination and energy

The only nearly inexhaustible sources of water are the oceans. Their main drawback, however, is their high salinity. Therefore, it would be attractive to tackle the water shortage problem by desalinizing this water. *Desalinize*, in general, means to remove salt from seawater or generally saline water.

According to the World Health Organization (WHO), the permissible limit of salinity in water is 500 parts per million (ppm) and for special cases up to 1000 ppm can be acceptable. Most of the water available on earth has salinity up to 10,000 ppm, and seawater normally has salinity in the range of 35,000–45,000 ppm in the form of total dissolved salts. Excess brackishness causes the problem of bad taste, stomach problems, and laxative effects. The purpose of a desalination system is to clean or purify brackish water or seawater and supply water with total dissolved solids (TDS) within the permissible limit of 500 ppm or less. This is accomplished by several desalination methods that are analyzed in this chapter.

Desalination processes require significant quantities of energy to achieve separation of salts from seawater. This is highly significant because it is a recurrent cost that few of the water-short areas of the world can afford. Many countries in the Middle East, because of oil income, have enough money to invest and run desalination equipment. However, people in many other areas of the world have neither the cash nor the oil resources to allow them to develop in a similar manner. The installed capacity of desalinated water systems in the year 2012 was about 75 million m^3/day, which is expected to increase drastically in the next decades. The dramatic increase of desalinated water supply will create a series of problems, the most significant of which are those related to energy consumption and environmental pollution caused by the use of fossil fuels. The production of 75 million m^3/day requires considerable amounts of energy. Given the current concern about the environmental problems related to the use of fossil fuels, if oil were much more widely available, it is questionable whether we could afford to burn it on the scale needed to provide everyone with freshwater. Given the current understanding of the greenhouse effect and the importance of CO_2 levels, this use of oil is debatable. Therefore, apart from satisfying the additional energy demand, environmental pollution would be a major concern. If desalination is accomplished by conventional technology, then it will require the burning of substantial quantities of fossil fuels. Given that conventional sources of energy are polluting, sources of energy that are not polluting must be developed. Fortunately, many parts of the world that are short of water have exploitable renewable sources of energy that could be used to drive desalination processes (Kalogirou, 2005).

Solar desalination is used by nature to produce rain, which is the main source of the freshwater supply. Solar radiation falling on the surface of the sea is absorbed as heat and causes evaporation of the water. The vapor rises above the surface and is moved by winds. When this vapor cools down to its dew point, condensation occurs and freshwater precipitates as rain. All available manmade distillation systems are small-scale duplications of this natural process.

Desalination of brackish water and seawater is one way to meet the water demand. Renewable energy systems (RESs) produce energy from sources that are freely available in nature. Their main characteristic is that they are friendly to the environment, that is, they do not produce harmful effluents. Production of freshwater using desalination technologies driven by RESs is thought to be a viable solution to the water scarcity at remote areas characterized by lack of potable water and conventional energy sources such as a heat and electricity grid. Worldwide, several renewable energy desalination pilot plants have been installed and the majority have been successfully operated for a number of years. Virtually all of them are custom designed for specific locations and utilize solar, wind, or geothermal energy to produce freshwater. Operational data and experience from these plants can be utilized to achieve higher reliability and cost minimization. Although renewable energy-powered desalination systems cannot compete with conventional systems in terms of the cost of water produced, they are applicable in certain areas and are likely to become more widely feasible solutions in the near future.

This chapter presents a description of the various methods used for seawater desalination. Only methods that are industrially mature are included. Other methods, such as freezing and humidification–dehumidification methods, are not included in this chapter, since they were developed at a laboratory scale and have not been used on a large scale for desalination. Special attention is given to the use of RESs in desalination. Among the various RESs, the ones that have been used, or can be used, for desalination are described. These include solar thermal collectors, solar ponds, photovoltaics, wind turbines, and geothermal energy.

8.2 Desalination processes

Desalination can be achieved using a number of techniques. Industrial desalination technologies either use phase change or involve semi-permeable membranes to separate the solvent or some solutes. Therefore, desalination techniques may be classified into the following categories: phase change or thermal processes and membrane or single-phase processes.

All processes require a chemical pre-treatment of raw seawater to avoid scaling, foaming, corrosion, biological growth, and fouling and also require a chemical post-treatment.

In Table 8.1, the most important technologies in use are listed. In the phase change or thermal processes, the distillation of seawater is achieved by utilizing a thermal energy source. The thermal energy may be obtained from a conventional fossil fuel source, nuclear energy, or a non-conventional solar energy source or geothermal energy. In the membrane processes, electricity is used for either driving high-pressure pumps or ionization of salts contained in the seawater.

Commercial desalination processes based on thermal energy are multi-stage flash (MSF) distillation, multiple-effect boiling (MEB), and vapor compression (VC), which could be thermal vapor compression (TVC) or mechanical vapor compression (MVC). MSF and MEB processes consist of a set of stages at successively decreasing temperature and pressure. The MSF process is based on the generation of vapor from seawater or brine due to a sudden pressure reduction when seawater enters an evacuated chamber. The process is repeated stage by stage at successively decreasing pressure. This process requires an external steam supply, normally at a temperature of around 100 °C. The maximum temperature is limited by the salt concentration to avoid scaling, and this maximum limits the performance of the process. In MEB, vapors are generated through the absorption of thermal energy by the seawater. The steam generated in one stage or effect can heat the salt solution in the next effect because the next effect is at lower temperature and pressure. The performance of the MEB and MSF processes is proportional to the number of stages or effects. MEB plants normally use an external steam supply at a temperature of about 70 °C. In TVC and MVC, after the initial vapor is generated from the saline solution, it is thermally or mechanically compressed to generate additional production. More details about these processes are given in Section 8.4.

Not only distillation processes but also freezing, membrane distillation and humidification–dehumidification processes involve phase change. The conversion of saline water to freshwater by

Table 8.1 Desalination processes.

Phase change processes	Membrane processes

1. Multi-stage flash (MSF)
2. Multiple-effect boiling (MEB)
3. Vapor compression (VC)
4. Freezing
5. Humidification–dehumidification
6. Solar stills
    Conventional stills
    Special stills
    Cascaded-type solar stills
    Wick-type stills
    Multiple-wick-type stills
	1. Reverse osmosis (RO)
	    RO without energy recovery
	    RO with energy recovery (ER-RO)
	2. Electrodialysis (ED)
	3. Membrane distillation (MD)

freezing has always existed in nature and has been known to humankind for thousands of years. In desalination of water by freezing, freshwater is removed and leaves behind a concentrated brine. It is a separation process related to the solid—liquid phase change phenomenon. When the temperature of saline water is reduced to its freezing point, which is a function of salinity, ice crystals of pure water are formed within the salt solution. These ice crystals can be mechanically separated from the concentrated solution, washed, and re-melted to obtain pure water. Therefore the basic energy input for this method is for the refrigeration system (Tleimat, 1980). The humidification—dehumidification method also uses a refrigeration system, but the principle of operation is different. The humidification—dehumidification process is based on the fact that air can be mixed with large quantities of water vapor. Additionally, the vapor-carrying capability of air increases with temperature (Parekh et al., 2003). In this process, seawater is added into an air stream to increase its humidity. Then this humid air is directed to a cool coil, on the surface of which water vapor contained in the air is condensed and collected as freshwater. These processes, however, exhibit technical problems that limit their industrial development. Because these technologies have not yet industrially matured, they are not described in this chapter.

The other category of industrial desalination processes does not involve phase change but membranes. These are reverse osmosis (RO) and electrodialysis (ED). The first one requires electricity or shaft power to drive the pump that increases the pressure of the saline solution to that required. The required pressure depends on the salt concentration of the resource of saline solution, and it is normally around 70 bar for seawater desalination.

ED also requires electricity for the ionization of water, which is cleaned by using suitable membranes located at the two oppositely charged electrodes. Both RO and ED are used for brackish water desalination, but only RO competes with distillation processes in seawater desalination. The dominant processes are MSF and RO, which account for 44% and 42% of worldwide capacity, respectively (Garcia-Rodriguez, 2003). The MSF process represents more than 93% of the thermal process production, whereas the RO process represents more than 88% of membrane processes production (El-Dessouky and Ettouney, 2000). The membrane processes are described in more detail in Section 8.4.

Another process that falls between the distillation and membrane systems is membrane distillation (MD). This is a relatively new and promising technology which can be used both for brackish and seawater desalination. Capacity wise, it is a technology of small to medium size desalination units, operating at atmospheric pressure and temperatures below 100 °C, and therefore is suitable for operation with low-temperature solar energy systems. MD is a non-isothermal membrane separation process, in which the water vapor molecules are transferred through a microporous hydrophobic membrane. Its operation is based on the partial pressure difference across the membrane that is created by the temperature difference between the two sides of the membrane, which is an attractive alternative solution for small to medium desalination systems, able to give a high-quality distillate, and to operate at low temperatures. Compared with other membrane separation processes, the MD operates at low pressures, lower than the pressures of the RO. Furthermore, its ability to operate with solar thermal energy or waste heat from other processes, makes MD an energy and cost efficient and environmentally friendly process (Belessiotis et al., 2016).

Solar energy can be used for seawater desalination by producing either the thermal energy required to drive the phase change processes or the electricity required to drive the membrane processes. Solar desalination systems are thus classified into two categories: direct and indirect collection systems. As their name implies, direct collection systems use solar energy to produce distillate directly in the solar collector, whereas in indirect collection systems, two subsystems are employed (one for solar energy collection and one for desalination). Conventional desalination systems are similar to solar energy

Table 8.2 RES desalination combinations.

RES technology	Feedwater salinity	Desalination technology
Solar thermal	Seawater	Multiple-effect boiling (MEB)
	Seawater	Multi-stage flash (MSF)
Photovoltaics	Seawater	Reverse osmosis (RO)
	Brackish water	Reverse osmosis (RO)
	Brackish water	Electrodialysis (ED)
Wind energy	Seawater	Reverse osmosis (RO)
	Brackish water	Reverse osmosis (RO)
	Seawater	Mechanical vapor compression (MVC)
Geothermal	Seawater	Multiple-effect boiling (MEB)

systems, since the same type of equipment is applied. The prime difference is that, in the former, either a conventional boiler is used to provide the required heat or public mains electricity is used to provide the required electric power; whereas in the latter, solar energy is applied. The most promising and applicable RES desalination combinations are shown in Table 8.2. These are obtained from a survey conducted under a European research project (THERMIE Program, 1998).

From 1990 to 2010, numerous desalination systems utilizing renewable energy were constructed. Almost all of these systems were built as research or demonstration projects and are consequently of a small capacity. It is not known how many of these plants still exist, but it is likely that only some remain in operation. The lessons learned, hopefully, have been passed on and are reflected in the plants currently being built and tested. A list of installed desalination plants operated with renewable energy sources is given by Tzen and Morris (2003).

8.2.1 Desalination systems exergy analysis

Although the first law of thermodynamics is an important tool in evaluating the overall performance of a desalination plant, such analysis does not take into account the quality of energy transferred. This is an issue of particular importance when both thermal and mechanical energy are employed, as they are in thermal desalination plants. First law analysis cannot show where the maximum loss of available energy takes place and would lead to the conclusion that the energy loss to the surroundings and the blowdown are the only significant losses. Second law (exergy) analysis is needed to place all energy interactions on the same basis and give relevant guidance for process improvement.

The use of exergy analysis in actual desalination processes from a thermodynamic point of view is of growing importance to identify the sites of greatest losses and improve the performance of the processes. In many engineering decisions, other facts, such as the impact on the environment and society, must be considered when analyzing the processes. In connection with the increased use of exergy analysis, second law analysis has come into more common usage in recent years. This involves a comparison of exergy input and exergy destruction along various desalination processes. In this section, initially the thermodynamics of saline water, mixtures, and separation processes is presented, followed by the analysis of multi-stage thermal processes. The former applies also to the analysis of RO, which is a non-thermal separation process.

   Saline water is a mixture of pure water and salt. A desalination plant performs a separation process in which the input saline water is separated into two output streams, those of brine and product water. The water produced from the process contains a low concentration of dissolved salts, whereas the brine contains the remaining high concentration of dissolved salts. Therefore, when analyzing desalination processes, the properties of salt and pure water must be taken into account. One of the most important properties in such analysis is salinity, usually expressed in parts per million (ppm), which is defined as: salinity = mass fraction (mf_s) × 10^6. Therefore, a salinity of 2000 ppm corresponds to a salinity of 0.2%, or a salt mass fraction of $mf_s = 0.002$. The mole fraction of salt, x_s, is obtained from (Cengel et al., 1999):

$$mf_s = \frac{m_s}{m_{sw}} = \frac{N_s M_s}{N_{sw} M_{sw}} = x_s \frac{M_s}{M_{sw}} \tag{8.1}$$

Similarly,

$$mf_w = x_w \frac{M_w}{M_{sw}} \tag{8.2}$$

where

   m = mass (kg);
   M = molar mass (kg/kmol);
   N = number of moles; and
   x = mole fraction.

   In Eqs. (8.1) and (8.2), the subscripts s, w, and sw stand for salt, water, and saline water, respectively. The apparent molar mass of the saline water is (Cerci, 2002):

$$M_{sw} = \frac{m_{sw}}{N_{sw}} = \frac{N_s M_s + N_w M_w}{N_{sw}} = x_s M_s + x_w M_w \tag{8.3}$$

   The molar mass of NaCl is 58.5 kg/kmol and the molar mass of water is 18.0 kg/kmol. Salinity is usually given in terms of mass fractions, but mole fractions are often required. Therefore, combining Eqs. (8.1)–(8.3) and considering that $x_s + x_w = 1$ gives the following relations for converting mass fractions to mole fractions:

$$x_s = \frac{M_w}{M_s \left(\dfrac{1}{mf_s} - 1 \right) + M_w} \tag{8.4}$$

and

$$x_w = \frac{M_s}{M_w \left(\dfrac{1}{mf_w} - 1 \right) + M_s} \tag{8.5}$$

   Solutions that have a concentration of less than 5% are considered to be dilute solutions, which closely approximate the behavior of an ideal solution, and thus the effect of dissimilar molecules on each other is negligible. Brackish underground water and even seawater are ideal solutions, since they have about a 4% salinity at most (Cerci, 2002).

EXAMPLE 8.1

Seawater of the Mediterranean Sea has a salinity of 35,000 ppm. Estimate the mole and mass fractions for salt and water.

Solution

From salinity, we get:

$$mf_s = \frac{\text{salinity (ppm)}}{10^6} = \frac{35,000}{10^6} = 0.035$$

From Eq. (8.4), we get:

$$x_s = \frac{M_w}{M_s\left(\frac{1}{mf_s} - 1\right) + M_w} = \frac{18}{58.5\left(\frac{1}{0.035} - 1\right) + 18} = 0.011$$

As $x_s + x_w = 1$, we have $x_w = 1 - x_s = 1 - 0.011 = 0.989$.

From Eq. (8.3),

$$M_{sw} = x_s M_s + x_w M_w = 0.011 \times 58.5 + 0.989 \times 18 = 18.45 \text{ kg/kmol}$$

Finally, from Eq. (8.2),

$$mf_w = x_w \frac{M_w}{M_{sw}} = 0.989 \frac{18}{18.45} = 0.965$$

Extensive properties of a mixture are the sum of the extensive properties of its individual components. Thus, the enthalpy (H) and entropy (S) of a mixture are obtained from:

$$H = \sum m_i h_i = m_s h_s + m_w h_w \tag{8.6}$$

and

$$S = \sum m_i s_i = m_s s_s + m_w s_w \tag{8.7}$$

Dividing by the total mass of the mixture gives the specific quantities (per unit mass of mixture) as:

$$h = \sum mf_i h_i = mf_s h_s + mf_w h_w \tag{8.8}$$

and

$$s = \sum mf_i s_i = mf_s s_s + mf_w s_w \tag{8.9}$$

The enthalpy of mixing of an ideal gas mixture is zero because no heat is released or absorbed during mixing. Therefore, the enthalpy of the mixture and the enthalpies of its individual components do not change during mixing. Thus, the enthalpy of an ideal mixture at a specified temperature and pressure is the sum of the enthalpies of its individual components at the same temperature and pressure (Klotz and Rosenberg, 1994). This also applies for the saline solution.

The brackish or seawater used for desalination is at a temperature of about 15 °C (288.15 K), pressure of 1 atm (101.325 kPa), and a salinity of 35,000 ppm. These conditions can be taken to be the conditions of the environment (dead state in thermodynamics).

The properties of pure water are readily available in tabulated water and steam properties tables. Those of salt are calculated by using the thermodynamic relations for solids, which require the set of the reference state of salt to determine the values of properties at specified states. For this purpose, the reference state of salt is taken at 0 °C, and the values of enthalpy and entropy of salt are assigned a value of 0 at that state. Then the enthalpy and entropy of salt at temperature T can be determined from:

$$h_s = h_{so} + c_{ps}(T - T_o) \tag{8.10}$$

and

$$s_s = s_{so} + c_{ps}\ln\left(\frac{T}{T_o}\right) \tag{8.11}$$

The specific heat of salt can be taken to be $c_{ps} = 0.8368$ kJ/kg K. The enthalpy and entropy of salt at $T_o = 288.15$ K can be determined to be $h_{so} = 12.552$ kJ/kg and $s_{so} = 0.04473$ kJ/kg K, respectively. It should be noted that, for incompressible substances, enthalpy and entropy are independent of pressure (Cerci, 2002).

EXAMPLE 8.2

Find the enthalpy and entropy of seawater at 40 °C.

Solution

From Eq. (8.10),

$$h_s = h_{so} + c_{ps}(T - T_o) = 12.552 + 0.8368(313.15 - 288.15) = 33.472 \text{ kJ} / \text{kg}$$

From Eq. (8.11),

$$s_s = s_{so} + c_{ps}\ln\left(\frac{T}{T_o}\right) = 0.04473 + 0.8368 \times \ln\left(\frac{313.15}{288.15}\right) = 0.11435 \text{ kJ} / \text{kg K}$$

Mixing is an irreversible process; hence, the entropy of a mixture at a specified temperature and pressure is greater than the sum of the entropies of the individual components at the same temperature and pressure before mixing. Therefore, since the entropy of a mixture is the sum of the entropies of its components, the entropies of the components of a mixture are greater than the entropies of the pure components at the same temperature and pressure. The entropy of a component per unit mole in an ideal solution at specified pressure P and temperature T is given by (Cengel and Boles, 1998):

$$s_i = s_{i,\text{pure}}(T, P) - R \ln (x_i) \tag{8.12}$$

where

$R = $ gas constant $= 8.3145$ kJ/kmol K.

It should be noted that $\ln(x_i)$ is a negative quantity, as $x_i < 1$, and therefore $-R \ln(x_i)$ is always positive. Equation (8.12) proves the statement made earlier that the entropy of a component in a mixture is always greater than the entropy of that component when it exists alone at a temperature and pressure equal to that of the mixture. Finally, the entropy of a saline solution is the sum of the entropies of salt and water in the saline solutions (Cerci, 2002):

$$s = x_s s_s + x_w s_w = x_s\left[s_{s,\text{pure}}(T, P) - R\ln(x_s)\right] + x_w\left[s_{w,\text{pure}}(T, P) - R\ln(x_w)\right]$$
$$= x_s s_{s,\text{pure}}(T, P) + x_w s_{w,\text{pure}}(T, P) - R[x_s\ln(x_s) + x_w\ln(x_w)] \tag{8.13}$$

The entropy of saline water per unit mass is determined by dividing the above quantity, which is per unit mole, by the molar mass of saline water. Therefore,

$$s = \mathrm{mf_s} s_{\mathrm{s,pure}}(T, P) + \mathrm{mf_w} s_{\mathrm{w,pure}}(T, P) - R[x_{\mathrm{s}} \ln(x_{\mathrm{s}}) + x_{\mathrm{w}} \ln(x_{\mathrm{w}})] \; [\mathrm{kJ / kg \; K}] \qquad (8.14)$$

The exergy of a flow stream is given as (Cengel and Boles, 1998):

$$e = h - h_{\mathrm{o}} - T_{\mathrm{o}}(s - s_{\mathrm{o}}) \qquad (8.15)$$

Finally, the rate of exergy flow associated with a fluid stream is given by:

$$E = \dot{m}e = \dot{m}[h - h_{\mathrm{o}} - T_{\mathrm{o}}(s - s_{\mathrm{o}})] \qquad (8.16)$$

Using the relations presented in this section, the specific exergy and exergy flow rates at various points of a reverse osmosis system can be evaluated. From the exergy flow rates, the exergy destroyed within any component of the system can be determined from exergy balance. It should be noted that the exergy of raw brackish or seawater is 0, since its state is taken to be the dead state. Also, exergies of brine streams are negative because they have salinities above the dead state level.

8.2.2 Exergy analysis of thermal desalination systems

From the first law of thermodynamics, the energy balance equation can be obtained as:

$$\sum_{\mathrm{in}} E_j + Q = \sum_{\mathrm{out}} E_j + W \qquad (8.17)$$

The mass, species, and energy balance equations for all the plant subsystems and a few associated state- and effect-related functions yield a set of independent equations. This set of simultaneous equations is solved by matrix algebra assuming equal temperature intervals for all effects and assuming that all effects have adiabatic walls. The boundary conditions are the specific seawater feed conditions (flow rate, salinity, temperature), the desired distillate production rate, and the specified maximum brine salinity and temperature. The matrix solutions obtained determine the distillation rates in the individual effects, the steam requirements, and hence the performance ratio (PR) (Hamed et al., 1996).

The steady-state exergy balance equation may be written as:

Total exergy transported into system = Total exergy transported out of system

+ Energy destroyed within system(or total irreversibility)

Therefore,

$$\sum E_{\mathrm{in}} = \sum E_{\mathrm{out}} + I_{\mathrm{T}} \qquad (8.18)$$

where

$$\sum E_{\mathrm{in}} = \sum E_{\mathrm{sw,in}} + \sum E_{\mathrm{steam}} + \sum E_{\mathrm{pumps}} \qquad (8.19)$$

and

$$\sum E_{\mathrm{out}} = \sum E_{\mathrm{cond}} + \sum E_{\mathrm{br}} \qquad (8.20)$$

The system overall irreversibility rate can be expressed as the summation of the subsystem irreversibility rate:

$$I_T = \sum_J I_i \qquad (8.21)$$

where J is the number of subsystems in the analysis and I_i is the irreversibility rate of subsystem i. The exergy (or second law) efficiency η_{II}, given by:

$$\eta_{II} = \frac{\sum E_{out}}{\sum E_{in}} \qquad (8.22)$$

is used as a criterion of performance, with E_{in} and E_{out} determined by Eqs. (8.19) and (8.20), respectively. The total loss of exergy is obtained from the individual exergy losses of the plant subsystems. The exergy efficiency defect, δ_i, of each subsystem is defined by (Hamed et al., 1996):

$$\delta_i = \frac{I_i}{\sum E_{in}} \qquad (8.23)$$

Combining Eqs. (8.22) and (8.23) gives:

$$\eta_{II} + \delta_1 + \delta_2 + \cdots + \delta_j = 1 \qquad (8.24)$$

The exergy of the working fluid at each point, calculated from its properties, is given by

$$E = M[(h - h_o) - T_o(s - s_o)] \qquad (8.25)$$

where the subscript o indicates the "dead state" or environment defined in the previous section.

8.3 Direct collection systems

Among the non-conventional methods to desalinate brackish water or seawater is solar distillation. This process requires a comparatively simple technology and can be operated by unskilled workers. Also, due to the low maintenance requirement, it can be used anywhere with a smaller number of problems.

A representative example of the direct collection system is the typical solar still, which uses the greenhouse effect to evaporate salty water. It consists of a basin in which a constant amount of seawater is enclosed in an inverted V-shaped glass envelope (see Fig. 8.1a). The sun's rays pass though the glass roof and are absorbed by the blackened bottom of the basin. As the water is heated, its vapor pressure is increased. The resultant water vapor is condensed on the underside of the roof and runs down into the troughs, which conduct the distilled water to the reservoir. The still acts as a heat trap because the roof is transparent to the incoming sunlight but opaque to the infrared radiation emitted by the hot water (greenhouse effect). The roof encloses the vapor, prevents losses, and keeps the wind from reaching and cooling the salty water. Another basic type of solar still is the single-slope one shown in Fig. 8.1b, also called an *asymmetric solar still* oriented in a north–south direction.

Figure 8.1a shows the various components of energy balance and thermal energy loss in a conventional double-slope symmetrical solar distillation unit (also known as a *roof-type* or *greenhouse-type solar still*). The still consists of an airtight basin, usually constructed out of concrete

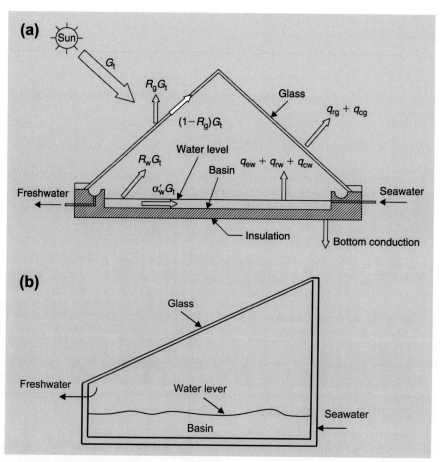

FIGURE 8.1

Schematic diagrams of the basic designs of solar stills. (a) Double-slope solar still showing also the energy flows. (b) Single-slope or asymmetric solar still.

galvanized iron sheet, or fiber-reinforced plastic, with a top cover of transparent material such as glass or plastic. The inner surface of the base, known as the *basin liner*, is blackened to efficiently absorb the solar radiation incident on it. There is also a provision to collect distillate output at the lower ends of the top cover. The brackish or saline water is fed inside the basin for purification using solar energy.

The stills require frequent flushing, which is usually done during the night. Flushing is performed to prevent salt precipitation. Design problems encountered with solar stills are brine depth, vapor tightness of the enclosure, distillate leakage, methods of thermal insulation, and cover slope, shape, and material (Eibling et al., 1971). A typical still efficiency, defined as the ratio of the energy utilized in vaporizing the water in the still to the solar energy incident on the glass cover, is 35% (maximum) and daily still production is about 3−4 L/m^2 (Daniels, 1974).

Talbert et al. (1970) gave an excellent historical review of solar distillation. Delyannis and Delyannis (1973) reviewed the major solar distillation plants around the world. This review also

FIGURE 8.2

Common designs of solar stills.

includes the work of Delyannis (1965), Delyannis and Piperoglou (1968), and Delyannis and Delyannis (1970). Malik et al. (1982) reviewed the work on passive solar distillation system until 1982, and this was updated up to 1992 by Tiwari (1992), who also included active solar distillation. Kalogirou (1997a) also reviewed various types of solar stills.

Several attempts have been made to use cheaper materials, such as plastics. These are less breakable, lighter in weight for transportation, and easier to set up and mount. Their main disadvantage is their shorter life. Many variations of the basic shapes shown in Fig. 8.1 have been developed to increase the production rates of solar stills (Eibling et al., 1971; Tleimat, 1978; Kreider and Kreith, 1981). Some of the most popular are shown in Fig. 8.2. Most of these designs also include provisions for rainfall collection.

8.3.1 Classification of solar distillation systems

On the basis of various modifications and modes of operation introduced in conventional solar stills, solar distillation systems are classified as passive or active. In active solar stills, an extra thermal energy by external equipment is fed into the basin of a passive solar still to achieve faster evaporation. The external equipment may be a collector—concentrator panel, waste thermal energy from an industrial plant, or a conventional boiler. If no such external equipment is used, then that type of solar still is known as a *passive solar still*. Types of solar stills available in literature are conventional solar stills, a single-slope solar still with passive condenser, a double-condensing chamber solar still, a vertical solar still (Kiatsiriroat, 1989), a conical solar still (Tleimat and Howe, 1967), an inverted absorber solar still (Suneja and Tiwari, 1999), and a multiple-effect solar still (Adhikari et al., 1995; Tanaka et al., 2000a;b).

Other researchers used different techniques to increase the production of stills. Rajvanshi (1981) used various dyes to enhance performance. These dyes darken the water and increase its solar radiation absorptivity. With the use of black naphthylamine at a concentration of 172.5 ppm, the still output could be increased by as much as 29%. The use of these dyes is safe because evaporation in the still occurs at 60 °C, whereas the boiling point of the dye is 180 °C.

Akinsete and Duru (1979) increased the production of a still by lining its bed with charcoal. The presence of charcoal leads to a marked reduction in start-up time. Capillary action by the charcoal partially immersed in a liquid and its reasonably black color and surface roughness reduce the system thermal inertia.

Lobo and Araujo (1978) developed a two-basin solar still. This still provides a 40%–55% increase in the freshwater produced as compared with a standard still, depending on the intensity of solar radiation. The idea is to use two stills, one on top of the other, the top one made completely from glass or plastic and separated into small partitions. Similar results were obtained by Al-Karaghouli and Alnaser (2004a;b), who compared the performance of single- and double-basin solar stills.

Frick and von Sommerfeld (1973), Sodha et al. (1981), and Tiwari (1984) developed a simple multiple-wick-type solar still, in which blackened wet jute cloth forms the liquid surface. Jute cloth pieces of increasing lengths were used, separated by thin black polyethylene sheets resting on foam insulation. Their upper edges were dipped in a saline water tank, where capillary suction provided a thin liquid sheet on the cloth, which was evaporated by solar energy. The results showed a 4% increase in still efficiency above conventional stills.

Evidently the distance of the gap between the evaporator tray and the condensing surface (glass cover) has a considerable influence on the performance of a solar still that increases with decreasing gap distance. This led to the development of a different category of solar stills, the cascaded-type solar still (Satcunanathan and Hanses, 1973). This consists mainly of shallow pools of water arranged in a cascade, as shown in Fig. 8.3, covered by a slopping transparent enclosure. The evaporator tray is usually made of a piece of corrugated aluminum sheet (similar to the one used for roofing) painted flat black.

Thermodynamic and economic analysis of solar stills is given by Goosen et al. (2000). Boeher (1989) reported on a high-efficiency water distillation of humid air with heat recovery, with a capacity range of 2–20 m³/day. Solar still designs in which the evaporation and condensing zones are separated are described in Hussain and Rahim (2001) and El-Bahi and Inan (1999). In addition, a device that uses a "capillary film distiller" was implemented by Bouchekima et al. (2001) and a solar still integrated in a greenhouse roof was reported by Chaibi (2000). Active solar stills in which the distillation temperature is increased by flat-plate collectors connected to the stills are described by Kumar and Tiwari (1998), Sodha and Adhikari (1990), and Voropoulos et al. (2001).

FIGURE 8.3

Schematic of a cascaded solar still.

8.3.2 Performance of solar stills

Solar stills are the most widely analyzed desalination systems. The performance of a conventional solar distillation system can be predicted by various methods, such as computer simulation, periodic and transient analysis, iteration methods, and numerical methods. In most of these methods, the basic internal heat and mass transfer relations, given by Dunkle (1961), are used.

Dunkle's (1961) procedure is summarized by Tiwari et al. (2003). According to this procedure, the hourly evaporation per square meter from a solar still is given by:

$$q_{ew} = 0.0163 h_{cw} (P_w - P_g) \left[\text{W/m}^2 \right] \tag{8.26}$$

where

P_w = partial vapor pressure at water temperature (N/m^2);
P_g = partial vapor pressure at glass temperature (N/m^2); and
h_{cw} = convective heat transfer coefficient from water surface to glass (W/m^2 °C).

The partial vapor pressures at the water and glass temperatures can be obtained from Eq. (5.21). The convective heat transfer coefficient can be obtained from:

$$\text{Nu} = \frac{h_{cw} d}{k} = C(\text{Gr} \times \text{Pr})^n \tag{8.27}$$

where

d = average spacing between water and glass surfaces (m);
k = thermal conductivity of humid air (W/m °C);
C = constant;
n = constant;
Gr = Grashof number (dimensionless); and
Pr = Prandl number (dimensionless).

The dimensionless quantities are given by:

$$\text{Gr} = \frac{g \beta \rho^2 (\Delta T) d^3}{\mu^2} = \frac{g \beta (\Delta T) d^3}{v^2} \tag{8.28}$$

$$\text{Pr} = \frac{c_p \mu}{k} \tag{8.29}$$

where

g = gravitational constant = 9.81 m/s^2;
β = coefficient of volumetric expansion of fluid (1/K);
ρ = density of fluid (kg/m^3);
ΔT = temperature difference between surface and fluid (K);
μ = dynamic viscosity of fluid (kg/m s);
v = kinetic viscosity of fluid (m^2/s); and
c_p = specific heat of fluid (J/kg K).

By using Eqs. (8.26) and (8.27), the hourly distillate output per square meter from a distiller unit ($\dot{m}_\text{w}$) is given by:

$$\dot{m}_\text{w} = 3600\frac{q_\text{ew}}{L_\text{v}} = 0.0163\left(P_\text{w} - P_\text{g}\right)\left(\frac{k}{d}\right)\left(\frac{3600}{L_\text{v}}\right)C(\text{Gr} \times \text{Pr})^n \tag{8.30}$$

or

$$\frac{\dot{m}_\text{w}}{R} = C(\text{Gr} \times \text{Pr})^n \tag{8.31}$$

where

$$R = 0.0163\left(P_\text{w} - P_\text{g}\right)\left(\frac{k}{d}\right)\left(\frac{3600}{L_\text{v}}\right) \tag{8.32}$$

L_v = latent heat of vaporization (kJ/kg).

It should be noted that, in the preceding equation, the product GrPr is known as the *Rayleigh number*, Ra. The constants C and n are calculated by regression analysis for known hourly distillate output (Dunkle, 1961), water and condensing cover temperatures, and design parameters for any shape and size of solar stills (Kumar and Tiwari, 1996).

According to Tiwari (2002), the instantaneous efficiency of a distiller unit is given as:

$$\eta_\text{i} = \frac{q_\text{ew}}{G_\text{t}} = \frac{h_\text{cw}\left(T_\text{w} - T_\text{g}\right)}{G_\text{t}} \tag{8.33}$$

Simplifying this equation, we can write:

$$\eta_\text{i} = F'\left[(\alpha\tau)'_\text{eff} + U_\text{L}\left(\frac{T_\text{w0} - T_\text{a}}{G_\text{t}}\right)\right]. \tag{8.34}$$

where

T_w0 = temperature of basin water at $t = 0$ (°C).

The preceding equation describes the characteristic curve of a solar still in terms of the solar still efficiency factor (F'), effective transmittance–absorptance product $(\tau\alpha)'_\text{eff}$, and overall heat loss coefficient (U_L) (Tiwari and Noor, 1996).

A detailed analysis of the equations of η_i justifies that the overall top loss coefficient (U_L) should be maximum for faster evaporation, which results in higher distillate output.

The meteorological parameters—wind velocity, solar radiation, sky temperature, ambient temperature—salt concentration, algae formation on water, and mineral layers on the basin liner affect significantly the performance of solar stills (Garg and Mann, 1976). For better performance of a conventional solar still, the following modifications were suggested by various researchers:

- Reducing the bottom loss coefficient.
- Reducing the water depth in a basin—multiwick solar still.
- Using a reflector.
- Using internal and external condensers.
- Using the back wall with cotton cloth.

- Using dyes.
- Using charcoal.
- Using an energy storage element.
- Using sponge cubes.
- Using a multiwick solar still.
- Condensing cover cooling.
- Using an inclined solar still.
- Increasing the evaporative area.

About a 10%−15% change in the overall daily yield of solar stills due to variations in climatic and operational parameters within the expected range has been observed.

EXAMPLE 8.3

A solar still has water and glass temperatures equal to 55 °C and 45 °C, respectively. The constants C and n are determined experimentally and found to be $C = 0.032$ and $n = 0.41$. If the convective heat transfer coefficient from water surface to glass is 2.48 W/m² K, estimate the hourly distillate output per square meter from the solar still.

Solution

From Eq. (5.21) and the temperatures of the water and glass, the partial pressures can be obtained as:

$$P_w = 100\left(0.004516 + 0.0007178 t_w - 2.649 \times 10^{-6} t_w^2 + 6.944 \times 10^{-7} t_w^3\right)$$

$$= 100\left(0.004516 + 0.0007178 \times 55 - 2.649 \times 10^{-6} \times 55^2 + 6.944 \times 10^{-7} \times 55^3\right)$$

$$= 15.15 \text{ kPa}$$

$$P_g = 100\left(0.004516 + 0.0007178 t_g - 2.649 \times 10^{-6} t_g^2 + 6.944 \times 10^{-7} t_g^3\right)$$

$$= 100\left(0.004516 + 0.0007178 \times 45 - 2.649 \times 10^{-6} \times 45^2 + 6.944 \times 10^{-7} \times 45^3\right)$$

$$= 9.47 \text{ kPa}$$

From Eq. (8.26),

$$q_{ew} = 0.0163 h_{cw}\left(P_w - P_g\right) = 0.0163 \times 2.48(15.15 - 9.47) \times 10^3 = 229.6 \text{ W/m}^2$$

From steam tables, the latent heat of vaporization at 55 °C (water temperature) is 2370.1 kJ/kg.
From Eq. (8.30),

$$\dot{m}_w = 3600 \frac{q_{ew}}{L_v} = 3600 \frac{229.6}{2370.1 \times 1000} = 0.349 \text{ kg/m}^2$$

8.3.3 General comments

Generally, the cost of water produced in solar distillation systems depends on the total capital investment to build the plant, the maintenance requirements, and the amount of water produced. No energy is required to operate the solar stills unless pumps are used to transfer the water from the sea. Therefore, the major share of the water cost in solar distillation is that of amortization of the capital cost. The production rate is proportional to the area of the solar still, which means that the cost per unit volume of water produced is nearly the same regardless of the size of the installation. This is in contrast with conditions for freshwater supplies as well as for most other desalination methods, where the capital cost of equipment per unit of capacity decreases as the capacity increases. This means that solar distillation may be more attractive than other methods for plants of small sizes. Howe and Tleimat (1974) reported that the solar distillation plants with capacity of less than 200 m^3/day are more economical than other plants.

Kudish and Gale (1986) presented the economic analysis of a solar distillation plant in Israel, assuming the maintenance cost of the system to be constant. An economic analysis for basin and multiple-wick solar stills was carried out by various scientists (Delyannis and Delyannis, 1985; Tiwari and Yadav, 1985; Mukherjee and Tiwari, 1986). Their economic analyses incorporated the effects of subsidy, rainfall collection, salvage value, and maintenance cost of the system.

Zein and Al-Dallal (1984) performed chemical analysis to find out its possible use as potable water and results were compared with tap water. They concluded that the condensed water can be mixed with well water to produce potable water and the quality of this water is comparable with that obtained from industrial distillation plants. The tests performed also showed that impurities such as nitrates, chlorides, iron, and dissolved solids in the water are completely removed by the solar still.

Although the yield of solar stills is very low, their use may prove to be economically viable if small water quantities are required and the cost of pipework and other equipment required to supply an arid area with naturally produced freshwater is high.

Solar stills can be used as desalinators for remote settlements where salty water is the only water available, power is scarce, and demand is less than 200 m^3/day (Howe and Tleimat, 1974). This is very feasible if setting of water pipelines for such areas is uneconomical and delivery by truck is unreliable or expensive. Since other desalination plants are uneconomical for low-capacity freshwater demand, solar stills are viewed as means for communities to attain self-reliance and ensure a regular supply of freshwater.

In conclusion, solar stills are the cheapest, with respect to their initial cost, of all available desalination systems in use today. They are direct collection systems, which are very easy to construct and operate. The disadvantage of solar stills is the very low yield, which implies that large areas of flat ground are required. It is questionable whether solar stills can be viable unless cheap desert-like land is available near the sea. However, obtaining freshwater from saline or brackish water with solar stills is useful for arid remote areas where no other economical means of obtaining water supply is available.

8.4 Indirect collection systems

The operating principle of indirect collection systems involves the implementation of two separate subsystems: a renewable energy collection system (solar collector, pressure vessel, wind turbine, etc.) and a plant for transforming the collected energy to freshwater. Some examples employing renewable energy to power desalination plants are presented in this section; a more extensive review is presented in Section 8.5. The plant subsystem is based on one of the following two operating principles:

- *Phase change processes*, for which multi-stage flash (MSF), multiple-effect boiling (MEB), or vapor compression (VC) is used.
- *Membrane processes*, for which reverse osmosis (RO), electrodialysis (ED) or membrane distillation (MD) is applied.

The operating principle of phase change processes entails reusing the latent heat of evaporation to preheat the feed while at the same time condensing steam to produce freshwater. The energy requirements of these systems are traditionally defined in terms of units of distillate produced per unit mass (kg or lb) of steam or per 2326 kJ (1000 Btu) heat input, which corresponds to the latent heat of vaporization at 73 °C. This dimensional ratio in kg/2326 kJ or lb/1000 Btu is known as the performance ratio (PR) (El-Sayed and Silver, 1980). The operating principle of membrane processes leads to the direct production of electricity from solar or wind energy, which is used to drive the plant. Energy consumption is usually expressed in kWh$_e$/m^3 (Kalogirou, 1997b).

8.4.1 The multi-stage flash process

The multi-stage flash (MSF) process is composed of a series of elements, called *stages*. In each stage, condensing steam is used to preheat the seawater feed. By fractionating the overall temperature differential between the warm source and seawater into a large number of stages, the system approaches ideal total latent heat recovery. Operation of this system requires pressure gradients in the plant. The principle of operation is shown in Fig. 8.4. Current commercial installations are designed with 10−30 stages (2 °C temperature drop per stage).

A practical cycle representing the MSF process is shown in Fig. 8.5. The system is divided into heat recovery and heat rejection sections. Seawater is fed through the heat rejection section, which rejects thermal energy from the plant and discharges the product and brine at the lowest possible temperature. The feed is then mixed with a large mass of water, which is recirculated around the plant. This water then passes through a series of heat exchangers to raise its temperature. The water next enters the solar collector array or a conventional brine heater to raise its temperature to nearly the saturation temperature at the maximum system pressure. The water then enters the first stage through an orifice and, in so doing, has its pressure reduced. Since the water was at the saturation temperature for a higher pressure, it becomes superheated and flashes into steam. The vapor produced passes through a wire mesh (demister) to remove any entrained brine droplets and then into the heat exchanger, where it is condensed and drips into a distillate tray. This process is repeated through the plant because both brine and distillate streams flash as they enter subsequent stages that are at successively lower pressures. In

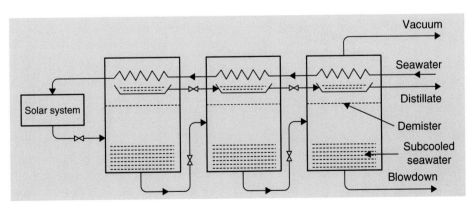

FIGURE 8.4

Principle of operation of the multi-stage flash (MSF) system.

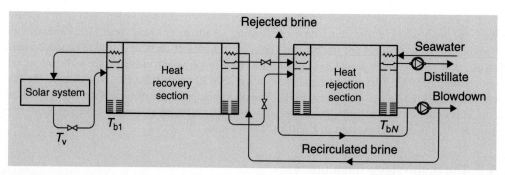

FIGURE 8.5

An multi-stage flash (MSF) process plant.

MSF, the number of stages is not tied rigidly to the PR required from the plant. In practice, the minimum must be slightly greater than the PR, whereas the maximum is imposed by the boiling point elevation. The minimum interstage temperature drop must exceed the boiling point elevation for flashing to occur at a finite rate. This is advantageous because, as the number of stages is increased, the terminal temperature difference over the heat exchangers increases and hence less heat transfer area is required, with obvious savings in plant capital cost (Morris and Hanbury, 1991).

MSF is the most widely used desalination process in terms of capacity. This is due to the simplicity of the process, performance characteristics, and scale control (Kalogirou, 1997b). A disadvantage of MSF is that precise pressure levels are required in the different stages; therefore, some transient time is required to establish the normal running operation of the plant. This feature makes the MSF relatively unsuitable for solar energy applications unless a storage tank is used for thermal buffering.

For the MSF system (El-Sayed and Silver, 1980),

$$\frac{M_f}{M_d} = \frac{L_v}{c_p \Delta F} + \frac{N-1}{2N} \tag{8.35}$$

where

M_d = mass rate of distillate (kg/h);
M_f = mass rate of feed (kg/h);
L_v = average latent heat of vaporization (kJ/kg);
c_p = mean specific heat under constant pressure for all liquid streams (kJ/kg K); and
N = total number of stages or effects.

The flashing temperature range, ΔF, according to the temperatures shown in Fig. 8.5, is given by:

$$\Delta F = T_h - T_{bN} = (T_{b1} - T_{bN})\frac{N}{N-1} \tag{8.36}$$

where

T_h = top brine temperature (K);
T_{bN} = temperature of brine in the last effect (K); and
T_{b1} = temperature of brine in first effect (K).

It should be noted that the rate of external feed per unit of product M_f/M_d is governed by the maximum brine concentration. Therefore,

$$\frac{M_f}{M_d} = \frac{y_{bN}}{y_{bN} - y_f} \tag{8.37}$$

where

y_{bN} = mass fraction of salts in brine in the last effect (dimensionless);
y_f = mass fraction of salts in feed (dimensionless).

The total thermal load per unit product is obtained by adding all loads, Q, and approximating $(N - 1)/N = 1$ and is given by (El-Sayed and Silver, 1980):

$$\frac{\sum Q}{M_d} = \frac{M_r}{M_d} c_p (T_h - T_o) = L_v \frac{T_h - T_o}{\Delta F} \tag{8.38}$$

where

M_r = mass rate of recirculated brine (kg/h);
T_o = environmental temperature (K).

EXAMPLE 8.4

Estimate M_f/M_d ratio for an MSF plant, which has 35 stages, brine temperature in first effect is 71 °C, and the temperature of the brine in the last effect is 35 °C. The mean latent heat is 2310 kJ/kg and the mean specific heat is 4.21 kJ/kg K.

Solution

From Eq. (8.36),

$$\Delta F = (T_{b1} - T_{bN}) \frac{N}{N - 1} = (71 - 35) \frac{35}{34} = 37.1 \,°C$$

From Eq. (8.35),

$$\frac{M_f}{M_d} = \frac{L_v}{c_p \Delta F} + \frac{N - 1}{2N} = \frac{2310}{4.21 \times 37.1} + \frac{(35 - 1)}{2 \times 35} = 15.3$$

The brine recycle flow rate can be obtained from the brine heater energy balance. The energy given by the heating steam (subscript s) is equal to:

$$M_s L_s = M_r c_p (T_v - T_{f1}) \tag{8.39}$$

By dividing Eq. (8.39) by the mass rate of distillate (M_d) and rearranging:

$$\frac{M_r}{M_d} = \frac{(M_s/M_d) L_s}{c_p (T_v - T_{f1})} \tag{8.40}$$

The term M_d/M_r is the system performance ratio (PR). The cooling water flow rate (subscript cw) is obtained from the energy balance of the whole system shown in Fig. 8.5 which gives:

$$M_{cw} = \frac{M_s L_s - M_f c_p (T_{bN} - T_{cw})}{c_p (T_{bN} - T_{cw})} \tag{8.41}$$

Dividing Eq. (8.41) by the mass rate of distillate (M_d):

$$\frac{M_{cw}}{M_d} = \frac{\left(\dfrac{M_s}{M_d}\right) L_s - \left(\dfrac{M_f}{M_d}\right) c_p (T_{bN} - T_{cw})}{c_p (T_{bN} - T_{cw})} \tag{8.42}$$

where as before M_d/M_r is the system PR and the ratio M_d/M_f is called the system *conversion ratio* (CR). Therefore, Eq. (8.42) can be written as:

$$\frac{M_{cw}}{M_d} = \frac{\left(\dfrac{L_s}{PR}\right) - \left(\dfrac{1}{CR}\right) c_p (T_{bN} - T_{cw})}{c_p (T_{bN} - T_{cw})} \tag{8.43}$$

It should be noted that the term 1/CR is given by Eq. (8.37).

Moustafa et al. (1985) report on the performance of a 10 m^3/day solar MSF desalination system tested in Kuwait. The system consisted of 220 m^2 parabolic trough collectors (PTCs), 7000 L of thermal storage, and a 12-stage MSF desalination system. The thermal storage system was used to level off the thermal energy supply and allowed the production of freshwater to continue during periods of low solar radiation and nighttime. The output of the system is reported to be more than 10 times the output of solar stills for the same solar collection area.

8.4.2 The multiple-effect boiling process

The multiple-effect boiling (MEB) process shown in Fig. 8.6 is also composed of a number of elements, which are called *effects*. The steam from one effect is used as heating fluid in another effect,

FIGURE 8.6

Principle of operation of a multiple-effect boiling (MEB) system.

which, while condensing, causes evaporation of a part of the salty solution. The produced steam goes through the following effect, where, while condensing, it makes some of the other solution evaporate, and so on. For this procedure to be possible, the heated effect must be kept at a pressure lower than that of the effect from which the heating steam originates. The solutions condensed by all effects are used to preheat the feed. In this process, vapor is produced by flashing and by boiling, but the majority of the distillate is produced by boiling. Unlike an MSF plant, the MEB process usually operates as a once-through system without a large mass of brine recirculating around the plant. This design reduces both pumping requirements and scaling tendencies (Kalogirou, 1997b).

As with the MSF plant, the incoming brine in the MEB process passes through a series of heaters, but after passing through the last of these, instead of entering the brine heater, the feed enters the top effect, where the heating steam raises its temperature to the saturation temperature for the effect pressure. Further amounts of steam, from either a solar collector system or a conventional boiler, are used to produce evaporation in this effect. The vapor then goes, in part, to heat the incoming feed and, in part, to provide the heat supply for the second effect, which is at a lower pressure and receives its feed from the brine of the first effect. This process is repeated all the way through (down) the plant. The distillate also passes down the plant. Both the brine and the distillate flash as they travel down the plant due to progressive reduction in pressure (Kalogirou, 1997b).

There are many possible variations of MEB plants, depending on the combinations of heat transfer configurations and flow sheet arrangements used. Early plants were of the submerged tube design and used only two or three effects. In modern systems, the problem of low evaporation rate has been resolved by using thin film designs with the feed liquid distributed on the heating surface in the form of a thin film instead of a deep pool of water. Such plants may have vertical or horizontal tubes. The vertical tube designs are of two types: climbing film, natural, and forced circulation type or long tube vertical (LTV), straight falling film type. In LTV plants, as shown in Fig. 8.7, the brine boils inside the tubes and the steam condenses outside. In the horizontal tube, falling film design, the steam condenses inside the tube with the brine evaporating on the outside. With multiple evaporation, the underlying

FIGURE 8.7

Long tube vertical MEB plant.

principle is to use the available energy of the leaving streams of a single-evaporation process to produce more distillate.

In the case of the MEB system, the ratio M_f/M_d is fixed by the maximum allowable brine concentration to a value on the order of two and is given by (El-Sayed and Silver, 1980):

$$\frac{M_f}{M_d} = \frac{\sum_1^N f_n}{M_d} \frac{L_v}{c_p N \Delta t_n} + \frac{N-1}{2N} \tag{8.44}$$

where

f_n = mass rate of distillate obtained by flashing per stage (kg/h);
Δt_n = temperature drop between two effects consisting of the heat transfer temperature difference and an augmented boiling point elevation (K).

The total thermal load per unit product obtained by adding all loads, Q, and dividing by M_d is given by (El-Sayed and Silver, 1980):

$$\frac{\sum Q}{M_d} = L_v + \frac{L_v}{N} + \frac{M_f}{M_d} c_p(\Delta t_t + \varepsilon) + \frac{1}{2} c_p(T_{b1} - T_{bN}) \tag{8.45}$$

where

ε = boiling point rise augmented by vapor frictional losses (K);
Δt_t = terminal temperature difference in feed heater condenser (K).

According to the analysis given by Al-Sahali and Ettouney (2007) who presented a modified analysis of Geankoplis (2003), the temperature drop across all effects is given by:

$$\Delta T_t = T_s - (N-1)\Delta T_\ell - T_{bN} \tag{8.46}$$

Where subscript t is for total and ΔT_ℓ represents temperature losses in each evaporation effect. Temperature drop in the first effect is given by:

$$\Delta T_1 = \frac{\Delta T_t}{U_1 \sum_{i=1}^N \frac{1}{U_i}} \tag{8.47}$$

And the temperature drop in effects 2-N is given by:

$$\Delta T_i = \Delta T_1 \frac{U_1}{U_i} \tag{8.48}$$

The brine temperature in the first effect is obtained from:

$$T_{b1} = T_s - \Delta T_1 \tag{8.49}$$

And the brine temperature in the effects 2-N is given by:

$$T_{bi} = T_{bi-1} - \Delta T_1 \frac{U_1}{U_i} - \Delta T_\ell \tag{8.50}$$

The distillate flow rate in the first effect is:

$$D_1 = \frac{M_d}{L_{v1}(1/L_{v1} + 1/L_{v2} + \cdots + 1/L_{vN-1} + 1/L_{vN})} \tag{8.51}$$

Distillate flow rate in the effects 2-N is:

$$D_i = D_1 \frac{L_{v1}}{L_{vi}} \tag{8.52}$$

The brine flow rate in various effects is given by:

$$B_i = D_i \frac{y_{cw}}{y_{bi} - y_{cw}} \tag{8.53}$$

where

y_{cw} = mass fraction of salts in cooling seawater (dimensionless).

And the feed rate in various effects is given by:

$$F_i = D_i + B_i \tag{8.54}$$

The heating steam flow rate is given by:

$$M_s = D_1 \frac{L_{v1}}{L_s} \tag{8.55}$$

The flow rate of the cooling seawater is given by:

$$M_{cw} = \frac{D_N L_{vN}}{c_p(T_f - T_{cw})} - M_f \tag{8.56}$$

With respect to the size of the equipment, the heat transfer area in the first effect is given by:

$$A_1 = \frac{D_1 L_{v1}}{U_1(T_s - T_{b1})} \tag{8.57}$$

The heat transfer area in the effects 2-N is given by:

$$A_i = \frac{D_i L_i}{U_i(T_{v_i-1} - T_{b_i})} \tag{8.58}$$

And the heat transfer area of the condenser (subscript c) is given by:

$$A_c = \frac{D_N L_{vN}}{U_c(\text{LMTD})_c} \tag{8.59a}$$

where

LMTD = log mean temperature difference given by:

$$\text{LMTD} = \frac{\Delta T_1 - \Delta T_2}{\ln\left(\dfrac{\Delta T_1}{\Delta T_2}\right)} \tag{8.59b}$$

in which ΔT_1 and ΔT_2 represent the temperature difference between the two fluids at the inlet and outlet of the condenser. It makes no difference which side of the condenser is designated as the inlet or the outlet.

Another type of MEB evaporator is the multiple-effect stack (MES). This is the most appropriate type for solar energy applications. It has a number of advantages, the most important of which is its stable operation between virtually 0% and 100% output, even when sudden changes are made, and its ability to follow a varying steam supply without upset. In Fig. 8.8, a four-effect MES evaporator is shown. Seawater is sprayed into the top of the evaporator and descends as a thin film over the horizontally arranged tube bundle in each effect. In the top (hottest) effect, steam from a steam boiler or a solar collector system condenses inside the tubes. Because of the low pressure created in the plant by the vent—ejector system, the thin seawater film boils simultaneously on the outside of the tubes, thus creating new vapor at a lower temperature than the condensing steam.

FIGURE 8.8

Schematic of the MES evaporator.

The seawater falling to the floor of the first effect is cooled by flashing through nozzles into the second effect, which is at a lower pressure. The vapor made in the first effect is ducted into the inside of the tubes in the second effect, where it condenses to form part of the product. Furthermore, the condensing warm vapor causes the external cooler seawater film to boil at the reduced pressure.

The evaporation—condensation process is repeated from effect to effect in the plant, creating an almost equal amount of product inside the tubes of each effect. The vapor made in the last effect is condensed on the outside of a tube bundle cooled by raw seawater. Most of the warmer seawater is then returned to the sea, but a small part is used as feedwater to the plant. After being treated with acid to destroy scale-forming compounds, the feedwater passes up the stack through a series of preheaters that use a little of the vapor from each effect to raise its temperature gradually, before it is sprayed into the top of the plant. The water produced from each effect is flashed in a cascade down the plant so that it can be withdrawn in a cool condition at the bottom of the stack. The concentrated brine is also withdrawn at the bottom of the stack. The MES process is completely stable in operation and automatically adjusts to changing steam conditions, even if they are suddenly applied, so it is suitable for load-following applications. It is a once-through process that minimizes the risk of scale formation without incurring a large chemical scale dosing cost. The typical product purity is less than 5 ppm TDS and does not deteriorate as the plant ages. Therefore, the MEB process with the MES-type evaporator appears to be the most suitable for use with solar energy.

Unlike the MSF plant, the PR for an MEB plant is more rigidly linked to and cannot exceed a limit set by the number of effects in the plant. For instance, a plant with 13 effects might typically have a PR of 10. However, an MSF plant with a PR of 10 could have 13—35 stages, depending on the design. MSF plants have a maximum PR of approximately 13. Normally, the figure is between 6 and 10. MEB plants commonly have PRs as high as 12—14 (Morris and Hanbury, 1991). The main difference between this process and the MSF is that the steam of each effect travels just to the following effect, where it is immediately used for preheating the feed. This process requires more complicated circuit equipment than the MSF; on the other hand, it has the advantage that it is suitable for solar energy utilization because the levels of operating temperature and pressure equilibrium are less critical.

A 14-effect MEB plant with a nominal output of 3 m^3/h coupled with 2672 m^2 PTCs was reported by Zarza et al. (1991a;b). The system is installed at the Plataforma Solar de Almeria in southern Spain. It also incorporates a 155 m^3 thermocline thermal storage tank. The circulated fluid through the solar collectors is a synthetic oil heat transfer fluid. The PR obtained by the system varies from 9.3 to 10.7, depending on the condition of the evaporator tube—bundle surfaces. The authors estimated that the efficiency of the system can be increased considerably by recovering the energy wasted when part of the cooling water in the final condenser is rejected. Energy recovery is performed with a double-effect absorption heat pump.

El-Nashar (1992) gives details of an MES system powered with 1862 m^2 evacuated tube collectors. The system is installed in Abu Dhabi, United Arab Emirates. A computer program was developed for the optimization of the operating parameters of the plant that affect its performance, that is, the collector area in service, the high-temperature collector set point, and the heating water flow rate. The maximum daily distillate production corresponding to the optimum operating conditions was found to be 120 m^3/day, which can be obtained for 8 months of the year.

Exergy analysis, based on actual measured data of the MES plant installed in the solar plant near Abu Dhabi, is presented by El-Nashar and Al-Baghdadi (1998). The exergy destruction was calculated for each source of irreversibility. The major exergy destruction was found to be caused by irreversibilities in the different pumps, with the vacuum pump representing the main source of destruction.

Major exergy losses are associated with the effluent streams of distillate, brine blowdown, and seawater. Exergy destruction due to heat transfer and pressure drop in the different effects, the preheaters, and the final condenser and in the flashing of the brine and distillate between the successive effects represents an important contribution to the total amount of exergy destruction in the evaporator.

8.4.3 The vapor compression process

In a vapor compression (VC) plant, heat recovery is based on raising the pressure of the steam from a stage by means of a compressor (see Fig. 8.9). The condensation temperature is thus increased and the steam can be used to provide energy to the same stage it came from or to other stages (Mustacchi and Cena, 1978). As in a conventional MEB system, the vapor produced in the first effect is used as the heat input to the second effect, which is at a lower pressure. The vapor produced in the last effect is then passed to the vapor compressor, where it is compressed and its saturation temperature is raised before it is returned to the first effect. The compressor represents the major energy input to the system, and since the latent heat is effectively recycled around the plant, the process has the potential for delivering high PR values (Morris and Hanbury, 1991).

Parametric cost estimates and process designs have been carried out and show that this type of plant is not particularly convenient, unless it is combined with an MEB system. Furthermore, it appears that the mechanical energy requirements have to be provided with a primary drive, such as a diesel engine, and cooling the radiator of such an engine provides more than enough heat for the thermal requirements of the process, making the solar collector system redundant (Eggers-Lura, 1979). Therefore, the VC system can be used in conjunction with an MEB system and operated at periods of low solar radiation or overnight.

FIGURE 8.9

Principle of operation of a vapor compression (VC) system.

Vapor compression systems are subdivided in two main categories: mechanical vapor compression (MVC) and thermal vapor compression (TVC) systems. MVC systems employ a mechanical compressor to compress the vapor, whereas thermal ones utilize a steam jet compressor. The main problems associated with the MVC process are (Morris and Hanbury, 1991):

1. Vapor-containing brine is carried over into the compressor and leads to corrosion of the compressor blades.
2. There are plant-size limitations because of limited compressor capacities.

Thermal vapor systems are designed for projects where steam is available. The required pressure is between 2 and 10 bar, and due to the relatively high cost of the steam, a large number of evaporative condenser heat recovery effects are normally justified.

The total thermal load per unit of distillate is simply the latent heat of vaporization and the heating of the feed all through the range $T_v - T_o$, given by (El-Sayed and Silver, 1980):

$$\frac{\sum Q}{M_d} = L_v + \frac{M_f}{M_d} c_p (T_v - T_o) \tag{8.60}$$

where

T_v = temperature of vapor entering the compressor as shown in Fig. 8.9 (K);
T_o = environmental temperature (K).

The analysis of the MVC is also given by Al-Sahali and Ettouney (2007). The work done by the vapor compressor of efficiency η and compressibility factor γ is given by:

$$W = \frac{\gamma}{\eta(\gamma - 1)} P_v V_v \left(\left(\frac{P_s}{P_v}\right)^{\left(\frac{\gamma-1}{\gamma}\right)} - 1 \right) \tag{8.61}$$

where

V_v = specific volume of vapor (m³/kg).

The evaporator mass and salt balance gives:

$$M_f = M_d + M_b \tag{8.62a}$$

$$M_b y_b = M_f y_f \quad \text{or} \quad \frac{M_b}{M_f} = \frac{y_f}{y_b} \tag{8.62b}$$

The energy balance of the evaporator actually equates the sensible heat for heating the feed and the latent heat for distillate evaporation with the latent heat of condensation of the compressed vapor and the sensible heat of superheating:

$$M_f c_{pf} (T_b - T_f) + M_d L_v = M_d L_d + M_d c_{pv} (T_s - T_d) \tag{8.63}$$

Finally, the evaporator heat transfer area is given in terms of the sensible and latent heat of the condensing vapor:

$$A_e = \frac{M_d L_d + M_d c_{pv}(T_s - T_d)}{U_e(T_d - T_b)} \tag{8.64}$$

A thermal performance and exergy analysis of a TVC system is presented by Hamed et al. (1996), who reached the following conclusions:

1. Operational data of a four-effect low-temperature TVC desalination plant revealed that PRs of 6.5–6.8 can be attained. Such ratios are almost twice those of a conventional four-effect boiling desalination plant.
2. The PRs of the TVC system increase with the number of effects and the entrainment ratio of the thermo-compressor and decrease with the top brine temperature.
3. Exergy analysis reveals that the TVC desalination plant is the most exergy efficient when compared with the MVC and multiple-effect boiling (MEB) ones.
4. The subsystem most responsible for exergy destruction in all three desalination systems investigated is the first effect, because of the high temperature of its heat input. In the TVC system, this amounts to 39%, with the second highest exergy defect being that of the thermo-compressor, equal to 17%.
5. Exergy losses can be significantly reduced by increasing the number of effects and the thermo-compressor entrainment ratio (vapor taken from evaporator and compressed by ejector) or by decreasing the top brine and first-effect heat input temperatures.

8.4.4 Reverse osmosis

The reverse osmosis system (RO) depends on the properties of semi-permeable membranes, which, when used to separate water from a salt solution, allow freshwater to pass into the brine compartment under the influence of osmotic pressure. If a pressure in excess of this value is applied to the salty solution, freshwater will pass from the brine into the water compartment. Theoretically, the only energy requirement is to pump the feedwater at a pressure above the osmotic pressure. In practice, higher pressures must be used, typically 50–80 atm, to have a sufficient amount of water pass through a unit area of membrane (Dresner and Johnson, 1980). With reference to Fig. 8.10, the feed is pressurized by a high-pressure pump and made to flow across the membrane surface. Part of this feed

FIGURE 8.10

Principle of operation of a reverse osmosis (RO) system.

passes through the membrane, where the majority of the dissolved solids are removed. The remainder, together with the remaining salts, is rejected at high pressure. In larger plants, it is economically viable to recover the rejected brine energy with a suitable brine turbine. Such systems are called *energy recovery reverse osmosis* (ER-RO) *systems*.

Solar energy can be used with RO systems as a prime mover source driving the pumps (Luft, 1982) or with the direct production of electricity through the use of photovoltaic panels (Grutcher, 1983). Wind energy can also be used as a prime mover source. Because the unit cost of the electricity produced from photovoltaic cells is high, photovoltaic-powered RO plants are equipped with energy recovery turbines. The output of RO systems is about $500-1500$ L/day/m^2 of membrane, depending on the amount of salts in the raw water and the condition of the membrane. The membranes are, in effect, very fine filters and very sensitive to both biological and non-biological fouling. To avoid fouling, careful pre-treatment of the feed is necessary before it is allowed to come in contact with the membrane surface.

One method used recently for the pre-treatment of seawater before directed to RO modules is nanofiltration (NF). NF was developed primarily as a membrane softening process, which offers an alternative to chemical softening. The main objectives of NF pre-treatment are (Adam et al., 2003):

1. Minimize particulate and microbial fouling of the RO membranes by removal of turbidity and bacteria.
2. Prevent scaling by removal of the hardness ions.
3. Lower the operating pressure of the RO process by reducing the feedwater TDS concentration.

The following model of the RO process is based on the two-parameter membrane model of Kimura and Sourirajan (1968) which is adapted according to membrane constructor practices as given by Vince et al. (2008). In the model the boron presence is not taken into account, the membrane is considered to be isothermal and the mass of salts in seawater is considered to be negligible in comparison with the mass of water, thus seawater density ρ is considered to be constant at 1000 kg/m^3.

If J_w is the permeate mass flux through the membrane (kg/m^2 s), J_s is the salts mass flux through the membrane (kg/m^2 s) and S_M is the membrane active area in m^2, then the membrane permeate mass flow rate (kg/s) is equal to:

$$\dot{m}_p = (J_w + J_s)S_M \tag{8.65}$$

The mass balance of water and salts across the membrane gives:

$$\dot{m}_f = \dot{m}_p + \dot{m}_c \tag{8.66}$$

$$\dot{m}_f C_f = \dot{m}_p C_p + \dot{m}_c C_c \tag{8.67}$$

where the subscripts f, p, and c are the feed, permeate, and concentrate water streams, respectively, all water mass flow rates are in kg/s, and the water stream salts mass concentration, C, is in kg of salts per kg of water.

The water recovery rate r is defined as the ratio of permeate to feed mass flow rate given by:

$$r = \frac{\dot{m}_p}{\dot{m}_f} \tag{8.68}$$

The membrane salts rejection rate R is given by:

$$R = 1 - \frac{C_p}{C_f} \qquad (8.69)$$

The permeate mass flux J_w (kg/m^2 s) through the membrane can be obtained by the Fick's law:

$$J_w = A(\Delta P - \Delta \pi) \qquad (8.70)$$

where

A = membrane permeability to pure water (kg/m^2 s Pa);
ΔP = trans-membrane pressure (Pa); and
$\Delta \pi$ = trans-membrane osmotic pressure (Pa).

Similarly, the salts mass flux J_s (kg/m^2 s) through the membrane is given by:

$$J_s = B(C_w - C_p) \qquad (8.71)$$

where

B = membrane permeability to salts (kg/m^2 s);
C_w = membrane wall mass salts concentration (kg of salts per kg of water).

The value in parenthesis in Eq. (8.71) is called the concentration polarization factor.
The assumption is made that the mass flux of solute through the membrane (kg/m^2 s) is equal to the mass flux of permeate multiplied by the permeate mass salts concentration:

$$J_s = J_w C_p \qquad (8.72)$$

The concentration polarization factor can be evaluated using the correlation developed by Taylor et al. (1994) and approximated for spiral-wound membranes by DOW (2006):

$$C_w - C_p = \left(\frac{C_f + C_c}{2} - C_p \right) e^{0.7r} \qquad (8.73)$$

where

C_c = the concentrate mass salts concentration (kg of salts per kg of water);
r = the water recovery rate of the membrane, given by Eq. (8.68).

The trans-membrane pressure ΔP (Pa) is calculated with:

$$\Delta P = P_f - P_p - \frac{\Delta p_{drop}}{2} \qquad (8.74)$$

where

P_f = the applied feed pressure (Pa);
P_p = the resulting permeate pressure (Pa); and
Δp_{drop} = the pressure drop along the membrane channel (Pa).

The pressure drop Δp_{drop} is approximated by the correlation defined by Schock and Miquel (1987) and adapted for spiral-wound membranes by DOW (2006):

$$\Delta p_{drop} = \lambda \left(\frac{\dot{m}_f + \dot{m}_c}{2\rho} \right)^{\alpha} \tag{8.75}$$

where the constants are: $\alpha = 1.7$ and $\lambda = 9.5 \times 10^8$.

Considering that seawater contains only NaCl salts, the transmembrane osmotic pressure (Pa) is approximated by the Van't Hoff relation:

$$\Delta \pi = \frac{2RT\rho}{M_{NaCl}} (C_w - C_p) \tag{8.76}$$

where

R = the universal gas constant (8.3145 J/mol K);
T = the water temperature (K); and
M_{NaCl} = the molar mass of NaCl (0.0585 kg/mol).

The membrane water permeability A is approximated as a function of feedwater temperature T, trans-membrane osmotic pressure Δp and fouling factor FF given by:

$$A = A_{ref}(\Delta \pi)\text{FF}\cdot\text{TCF} \tag{8.77}$$

where

$A_{ref}(\Delta \pi)$ = the reference permeability at $T_o = 298$ K without fouling (kg/m^2 s Pa);
TCF = the temperature correction factor at T; and
FF = the fouling factor which expresses the influence of membrane fouling on membrane permeability and varies between 100% for new membranes and 80% for 4-year-old membranes (DOW, 2006).

The temperature correction factor TCF, which expresses the influence of temperature on membrane permeability is obtained using Arrhenius-type correlation (Mehdizadeh et al., 1989):

$$\text{TCF} = \exp\left(\frac{e}{R} \left[\frac{1}{T_o} - \frac{1}{T} \right] \right) \tag{8.78}$$

where, T is the water temperature (K), T_o is the reference water temperature (298 K) and e is the membrane activation energy (J/mol), estimated for all RO membranes at 25,000 J/mol when $K \leq 298$ K and at 22,000 J/mol when $T > 298$ K (DOW, 2006).

Usually, the influence of the osmotic pressure $\Delta \pi$ on the reference pure water permeability $A_{ref}(\Delta \pi)$ is measured experimentally by membrane manufacturers at $T_o = 298$ K and FF = 1. As in most cases this relation is not given, $A_{ref}(\Delta \pi)$ is considered to be constant and equal to A_{ref}. The membrane permeability to salts B is also considered to be constant (Kimura and Sourirajan, 1968).

Usually a number of membrane elements n are installed in the same pressure vessel which is equivalent to having n membranes in series. Therefore, the pressure vessel permeate is a mixture of the n membrane module permeates and the pressure vessel recovery rate is defined by (Vince et al., 2008):

$$\bar{r} = r_1 + \sum_{k=2}^{n} \left[r_k \prod_{l=1}^{k-1} (1 - r_l) \right] \tag{8.79}$$

with r_k the water recovery rate of the membrane k for $k = 1 \ldots n$.

The pressure vessel is modeled as n successive membranes defined by Eqs. (8.65)–(8.78).

For $k = 2 \ldots n$, the concentrate flow rate, salinity, and pressure of membrane $k - 1$ are, respectively, the feedwater flow rate, salinity, and pressure of membrane k as given by:

$$J_w^k = \frac{J_w^{k-1}(1 - r_{k-1})}{r_{k-1}} \quad \text{for } k = 2 \ldots n \tag{8.80}$$

$$C_f^k = C_b^{k-1} \quad \text{for } k = 2 \ldots n \tag{8.81}$$

$$P_f^k = P_b^{k-1} \quad \text{for } k = 2 \ldots n \tag{8.82}$$

Tabor (1990) analyzed a system using an RO desalination unit driven by photovoltaic (PV) panels or from a solar thermal plant. He concluded that, due to the high cost of the solar equipment, the cost of freshwater is about the same as with an RO system operated from the main power supply.

Cerci (2002) performed an exergy analysis of a 7250 m^3/day RO desalination plant in California. The analysis of the system was conducted using actual plant operation data. The RO plant is described in detail, and the exergies across the major components of the plant are calculated and illustrated using exergy flow diagrams in an attempt to assess the exergy destruction distribution. He found that the primary locations of exergy destruction were the membrane modules in which the saline water is separated into the brine and the permeate and the throttling valves where the pressure of liquid is reduced, pressure drops through various process components, and the mixing chamber where the permeate and blend are mixed. The largest exergy destruction occurred in the membrane modules, and this amounted to 74.1% of the total exergy input. The smallest exergy destruction occurred in the mixing chamber. The mixing accounted for 0.67% of the total exergy input and presents a relatively small fraction. The second law efficiency of the plant was calculated to be 4.3%, which seems to be low. He shows that the second law efficiency can be increased to 4.9% by introducing a pressure exchanger with two throttling valves on the brine stream; this saved 19.8 kW of electricity by reducing the pumping power of the incoming saline water.

8.4.5 Electrodialysis

The electrodialysis (ED) system, shown schematically in Fig. 8.11, works by reducing salinity by transferring ions from the feedwater compartment, through membranes, under the influence of an electrical potential difference. The process utilizes a DC electric field to remove salt ions in the brackish water. Saline feedwater contains dissolved salts separated into positively charged sodium and negatively charged chlorine ions. These ions move toward an oppositively charged electrode immersed in the solution, that is, positive ions (cations) go to the negative electrode (cathode) and negative ions (anions) to the positive electrode (anode). If special membranes, alternatively cation permeable and anion permeable, separate the electrodes, the center gap between these membranes is depleted of salts (Shaffer and Mintz, 1980). In an actual process, a large number of alternating cation and anion membranes are stacked together, separated by plastic flow spacers that allow the passage of water. The streams of alternating flow spacers are a sequence of diluted and concentrated water, which flow

FIGURE 8.11

Principle of operation of electrodialysis (ED).

parallel to each other. To prevent scaling, inverters are used that reverse the polarity of the electric field approximately every 20 min.

Because the energy requirements of the system are proportional to the water's salinity, ED is more feasible when the salinity of the feedwater is no more than about 6000 ppm of dissolved solids. Similarly, due to the low conductivity, which increases the energy requirements of very pure water, the process is not suitable for water of less than about 400 ppm of dissolved solids.

Because the process operates with DC power, solar energy can be used with ED by directly producing the voltage difference required with photovoltaic panels.

8.4.6 Membrane distillation

Membrane distillation (MD) is a relatively simple process, which is a combination of the thermal distillation and the membrane technology. This is actually a thermal membrane separation process in which water vapor molecules are transferred from hot saline water, through a microporous hydrophobic membrane, because of the partial pressure difference created due to the temperature difference on the two sides of the membrane (Lawson and Lloyd, 1997). MD is a phase change process which requires significant quantities of energy to achieve separation. The coupling of MD with renewable energy sources, such as solar, wind and geothermal energy, or with low grade waste heat, makes MD an economic and energy efficient process for seawater desalination. The MD process is mainly used for the production of high purity water from the desalination of brackish or seawater.

One of the most crucial aspects of the MD is to have membranes with specific properties. An ideal MD membrane should exhibit high permeability, high liquid entry pressure of water, i.e., hydrophobicity, and low thermal conductivity. Moreover, it should have good thermal stability and excellent chemical resistance to seawater feed. The membrane itself does not substantially take part in the separation process but acts as a physical barrier between the two phases preventing the penetration of the liquid phase into the membrane pores and simultaneously provides the liquid-gas interface for mass and heat transfer.

The operation of the MD process is based on the principle of the liquid-vapor equilibrium. The hot aqueous solution flows to the hot membrane surface releasing vapor. The hydrophobic nature of membrane prevents the penetration of the liquid phase into the membrane pores, because of the surface tension forces, but allows the passage of the vapor molecules through it. As a result, a liquid-vapor interface is created at the membrane pores entrance, in which the liquid phase is separated from the gas phase. Since the process is non-isothermal, the temperature difference across the membrane creates a partial pressure gradient, which first causes the water vapor molecules to start to evaporate in the liquid-vapor interface at the entrance of the membrane pores and second the formed vapor to diffuse in the membrane. As the vapor pressure difference is maintained between the two membrane sides, the vapor molecules are transferred through the membrane pores from the high partial pressure area (the hot membrane surface) to the low partial pressure area (the cold membrane surface) and it is finally condensed either internally downstream of the membrane or externally outside the membrane module (Lawson and Lloyd, 1997; Curcio and Drioli, 2005).

The MD process constitutes an attractive alternative solution to the conventional desalination techniques, due to the production of high purity distillate and the possibility of operating at lower temperatures. Compared to other membrane technologies, the MD operates at relative low pressures leading to low equipment cost and increased process safety. The mechanical properties of the membranes used are more flexible than those of the other separation processes and their large available surface allows high concentration of vapor molecules:

Additional benefits of MD process are (Hsu et al., 2002):

 (i) reduced vapor space compared with conventional distillation,
 (ii) absence of limitations caused by the osmotic pressure effects,
(iii) feed water does not require extensive chemical pretreatment,
 (iv) negligible organic fouling,
  (v) compactness of process units, and their simple, fast assembly and installation.

Additionally, according to Schneider et al. (1988):

- All the component parts usually consist of plastic material and as a result there are no corrosion problems caused by seawater.
- The desalinated water is free of particles and bacteria.
- The process depends only on the pressure and consequently on the temperature developed on both membrane sides.
- The diffusion distance of the vapor molecules between the evaporation and condensation is extremely short, thus the membrane is very compact and thin (m^2/m^3).
- For systems with heat recovery the performance factor can reach up to 8.

Despite the important advantages presented above the process is not fully commercialized yet due to some disadvantages, the most important of which are:

- The low flux of the produced vapor compared with other membrane separation processes.
- The high sensitivity of the produced vapor flux with respect to the feed concentration and temperature conditions, which is due to the concentration and temperature polarization phenomena.

- The risk of membrane wetting from the liquid phase of the feed aqueous solution.
- The high membrane cost.
- The low thermal energy performance.
- The high energy consumption compared to the RO.

Solar-powered membrane distillation (SPMD)

The solar energy as an alternative energy source can be used as a thermal energy to preheat the aqueous feed solution through a solar collector. Hogan et al. (1991) were among the first that coupled the solar energy with an MD system. They described a pilot solar-powered MD installation, producing 50 kg of desalinated water per day using a solar collector of 3 m^2 surface area. Their system, which was tested in Sydney, Australia, and the solar-powered MD unit was found to be technically feasible and compatible with solar energy.

Within the frame of three European funded projects, SMADES, MEDESOL, and MEDIRAS, several compact systems have been installed with a nominal daily capacity equal to 100 L of distillated water in different countries, including Pozo Izquierdo and Tenerife (Grand Canary), Alexandria (Egypt), Irbid (Jordan) and Freiburg (Germany). Under the umbrella of SMADES project Banat et al. (2007a) designed a stand-alone solar driven MD system (compact SMADES), which was tested to generate a stable water supply in different remote areas (Jordan, Morocco and Egypt). The results from that compact system were used for designing and scaling-up a "large SMADES" system, which was assembled in Aqaba, Jordan (Banat et al., 2007b). The power supply in that pilot unit was given both by solar collectors and PV panels. The feed water was real seawater from the Red Sea. The unit can produce 120 L/day of desalinated water, i.e., 19 L/m^{2}day, with specific energy consumption in the range of 200−300 kWh/m^3.

A small-scale solar-driven MD system is also installed in Gran Canarias, Spain (Koschikowski et al., 2009). The system has four compact units of total distillate output of 1−1.5 m^3/day with temperatures at the evaporator inlet of 60−85 °C and gained output ratio (GOR) of 3−6. The heat required for heating is supplied by high performance flat-plate solar collectors, while PV panels supplied the energy for the pumps' operation. Another three compact systems have been installed in Tunisia and Tenerife.

The advantages of stand-alone MD plants with respect to RO are (Fath et al., 2008):

- Contrary to RO process specific pre-treatment of feed water is not required.
- The membranes that used in MD process are not sensitive to organic fouling as the RO ones.
- The MD modules operate at atmospheric pressure, contrary to RO modules that operate at high pressure.
- The feed water salinity has no influence on the process performance, while in the RO process the salinity determines the magnitude of the applied pressure and therefore the systems performance.

For solar-driven MD systems, the thermal recovery ratio (TRR) is defined as the ratio of the heat that is theoretically required to distill an amount of distillate to the total used energy in the system (Banat et al., 2007b):

$$TRR = \frac{0.69 V_{prod}}{IA_c} \tag{8.83}$$

The solar collector efficiency is:

$$\eta_{sc} = \frac{\dot{m}_f c_p (T_i - T_o)}{IA_c}$$

(8.84)

where V_{prod} is the produced water volume (m^3), I is the intensity of incident solar radiation (W/m^2), $\dot{m}_f$ is the mass flow rate of the hot feed water (kg/s), A_c is the collector surface area (m^2), and T_i, T_o are the temperatures at the inlet and the outlet of the collector (K), respectively.

Saffarini et al. (2012) define the Gain Output Ratio (GOR) as:

$$GOR = \frac{\dot{m}_v \Delta H_v}{q_{in}}$$

(8.85)

where $\dot{m}_v$ is the distillate mass flow rate (kg/s), ΔH_v is the latent heat of evaporation (J/kg), and q_{in} is the external heat flow rate in the system, which can be calculated as:

$$q_{in} = \dot{m}_f c_p (T_{f,in} - T_i)$$

(8.86)

where $T_{f,in}$, T_i are the inlet temperatures at the hot feed stream and at the solar collector (K), respectively.

Also, GOR can also be expressed as:

$$GOR = \frac{\dot{m}_v}{m_f} \frac{\Delta H_v}{c_p (T_{f,in} - T_i)}$$

(8.87)

The system performance ratio (PR) is then given as:

$$PR = \frac{\dot{m}_v}{q_{in}} = \frac{\dot{m}_v}{\dot{m}_f c_p (T_{f,in} - T_i)}$$

(8.88)

More details about this process can be found in Belessiotis et al. (2016).

8.5 Review of renewable energy desalination systems

Renewable energy systems (RESs) offer alternative solutions to decrease the dependence on fossil fuels. The total worldwide renewable energy desalination installations amount to capacities of less than 1% of that of conventional fossil-fueled desalination plants (Delyannis, 2003). This is mainly due to the high capital and maintenance costs required by renewable energy, making these desalination plants non-competitive with conventional fuel desalination plants.

Solar desalination plants coupled with conventional desalination systems have been installed in various locations of the world. The majority of these plants are of experimental or demonstration scale. A comprehensive review of renewable energy desalination systems is given by the author in Kalogirou (2005).

This section presents examples of desalination plants powered by RESs; they comprise systems not included in this book, such as wind energy and geothermal energy systems.

8.5.1 Solar thermal energy

A comprehensive review of the various types of collectors currently available is presented in Chapter 3. The solar system indicated in the various figures of Section 8.4 is composed of a solar collector array, a storage tank, and the necessary controls. A detailed diagram of such a system is shown in Fig. 8.12. In the storage tank, seawater flows through a heat exchanger to avoid scaling in the collectors. The solar collector circuit operates with a differential thermostat (not shown), as explained in Chapter 5, Section 5.5. The three-way valve, shown in Fig. 8.12, directs seawater either to the hot water storage tank heat exchanger or through the boiler when the storage tank is depleted.

In some desalination systems, such as the MEB, the solar system must be able to provide low-pressure steam. For this purpose, parabolic trough collectors are usually employed, and one of the three methods of solar steam generation outlined in Chapter 7 can be used.

Solar thermal energy is one of the most promising applications of renewable energy for seawater desalination. A solar distillation system may consist of two separate devices: the solar collector and the conventional distiller (indirect solar desalination). Indirect solar desalination systems usually consist of a commercial desalination plant connected to commercial or special solar thermal collectors. With respect to special solar thermal collectors, Rajvanshi (1980) designed such a solar collector to be connected to an MSF distillation plant. Hermann et al. (2000) reported on the design and testing of a corrosion-free solar collector for driving a multiple-effect humidification process. The pilot plant was installed at Pozo Izquierdo, Gran Canaria, Spain (Rommel et al., 2000).

8.5.2 Solar ponds

Details of solar ponds are given in Chapter 10. These are used mainly for electric power generation. Another use of the output from salt-gradient solar ponds, however, is to operate low-temperature distillation units to desalt seawater. This concept has applicability in desert areas near the oceans. Solar pond coupled desalination involves using hot brine from the pond as a thermal source to evaporate the water to be desalted at low pressure in a multiple-effect boiling (MEB) evaporator.

Matz and Feist (1967) proposed solar ponds as a solution to brine disposal at inland ED plants as well as a source of thermal energy to heat the feed of an ED plant, which can increase its performance.

FIGURE 8.12

Connection of a solar system with a desalination system.

8.5.3 Solar photovoltaic technology

Details of photovoltaic systems are given in Chapter 9. The photovoltaic (PV) technology can be connected directly to an RO system. The extent to which the PV energy is competitive with conventional energy depends on the plant capacity, the distance to the electricity grid, and the salt concentration of the feed. Kalogirou (2001) and Tzen et al. (1998) analyzed the cost of PV-RO desalination systems. Al-Suleimani and Nair (2000) presented a detailed cost analysis of a system installed in Oman. Thomson and Infield (2003) presented the simulation and implementation of a PV-driven RO for Eritrea with variable flow that is able to operate without batteries. The production capacity of the system was 3 m³/day with a PV array of 2.4 kWp. The model was validated with laboratory tests. The Canary Islands Technological Institute (ITC) developed a stand-alone system (DESSOL) with capacity of 1−5 m³/day of nominal output.

Another way of using PV is in combination with ED. The ED process is more suitable than RO for brackish water desalination in remote areas. Several pilot plants of ED systems connected to photovoltaic cells by means of batteries have been implemented. Gomkale (1988) analyzed solar desalination for Indian villages and concluded that solar cell-operated ED seems more advantageous for desalting brackish water than conventional solar stills. A PV-driven ED plant was installed at Spencer Valley, New Mexico. It was developed by the Bureau of Reclamation in the United States (Maurel, 1991). Experimental research in PV-ED was also performed at the Laboratory for Water Research at the University of Miami (Kvajic, 1981) as well as at the University of Bahrain (Al-Madani, 2003).

8.5.4 Wind power

A brief historical introduction to wind energy is given in Chapter 1, Section 1.6.1, whereas more details are given in Chapter 13. Since RO is the desalination process with the lowest energy requirements (see Section 8.6) and coastal areas present a high availability of wind power resources (Doucet, 2001), wind-powered desalination is one of the most promising alternatives for renewable energy desalination. A preliminary cost evaluation of wind-powered RO is presented by Garcia-Rodriguez et al. (2001). In particular, the influence of climatic conditions and plant capacity on product cost is analyzed for seawater RO driven by wind power. Additionally, the possible evolution of product cost due to possible future changes in wind power and RO technologies is evaluated. Finally, the influence on the competitiveness of wind-powered RO with conventional RO plants due to the evolution of financial parameters and cost of conventional energy is pointed out.

Another area of interest is the direct coupling of a wind energy system and an RO unit by means of shaft power. Research in this field has been carried out at the Canary Islands Technological Institute (ITC, 2001). On Coconut Island, off the northern coast of Oahu, Hawaii, a brackish water desalination wind-powered RO plant was installed. The system directly uses the shaft power production of a windmill with the high-pressure pump and RO. In particular, constant freshwater production of 13 L/min can be maintained for wind speed of 5 m/s (Lui et al., 2002).

Another possibility investigated is the use of wind power directly with an MVC. A detailed analysis of the influence of the main parameters of such systems was performed by Karameldin et al. (2003). On Borkum Island, in the North Sea, a pilot plant was erected with a freshwater production of about 0.3−2 m³/h (Bier et al., 1991). On Rügen Island, Germany, another pilot plant was installed with a 300 kW wind energy converter and 120−300 m³/day freshwater production (Plantikow, 1999).

Finally, another possibility investigated is the use of wind power with ED. Modeling and experimental test results of one such system with a capacity range of 72–192 m³/day, installed at the ITC, Gran Canaria, Spain, are presented by Veza et al. (2001).

8.5.5 Hybrid solar PV-wind power

The complementary features of wind and solar resources make the use of hybrid wind-solar systems to drive a desalination unit a promising alternative, since usually, when there is no sun, the wind is stronger, and vice versa. The Cadarache Center in France designed a pilot unit that was installed in 1980 at Borj-Cedria, Tunisia (Maurel, 1991). The system consists of a 0.1 m³/day compact solar distiller, a 0.25 m³/h RO plant, and an ED plant for 4000 ppm brackish water. The energy supply system consists of a photovoltaic field with a capacity of 4 kW peak and two wind turbines.

8.5.6 Geothermal energy

Measurements show that for shallow geothermal systems, the ground temperature below a certain depth remains relatively constant throughout the year. This is because the temperature fluctuations at the surface are diminished as the depth of the ground increases due to the high thermal inertia of the soil.

According to Popiel et al. (2001), from the point of view of the temperature distribution, three ground zones can be distinguished:

1. *Surface zone* reaching a depth of about 1 m, in which the ground temperature is very sensitive to short time changes of weather conditions.
2. *Shallow zone* extending from the depth of about 1–8 m (for dry, light soils) or 20 m (for moist, heavy, sandy soils), where the ground temperature is almost constant and close to the average annual air temperature; in this zone, the ground temperature distributions depend mainly on the seasonal cycle weather conditions.
3. *Deep zone* below the depth of the shallow zone, where the ground temperature is practically constant (and rises very slowly with depth according to the geothermal gradient).

For deep geothermal systems, there are different geothermal energy sources. They may be classified in terms of the measured temperature as low (<100 °C), medium ($100–150$ °C), and high temperature (>150 °C). The thermal gradient in the earth varies between 15 and 75 °C per kilometer depth; nevertheless, the heat flux is anomalous in different continental areas. Moreover, local centers of heat, between 6 and 10 km deep, were created by the disintegration of radioactive elements. Barbier (1997; 2002) presented a complete overview of geothermal energy technology. Baldacci et al. (1998) reported that the cost of electrical energy is generally competitive, 0.6–2.8 US cents/MJ (2–10 US cents/kWh), and that 0.3% of the world total electrical energy generated in 2000, that is, 177.5 billion MJ/a (49.3 billion kWh/a), is from geothermal resources. Geothermal energy can be used as a power input for desalination.

Energy from the earth, for shallow geothermal systems, is usually extracted with ground heat exchangers. These are made of a material that is extraordinarily durable but allows heat to pass through it efficiently. Ground heat exchanger manufacturers typically use high-density polyethylene, which is a tough plastic, with heat-fused joints. This material is usually warranted for as much as 50 years. The fluid in the loop is water or an environmentally safe antifreeze solution. Other types of heat exchangers utilize copper piping placed underground. The length of the loop depends on a number of factors, such as the type of loop configuration, the thermal load, the soil conditions, and local climate. A review of ground heat exchangers is given by Florides and Kalogirou (2004).

Low-temperature geothermal waters in the upper 100 m may be a reasonable energy source for desalination (Rodriguez et al., 1996). Ophir (1982) gave an economic analysis of geothermal desalination in which sources of 110–130 °C were considered. He concluded that the price of geothermal desalination is as low as the price of large multiple-effect dual-purpose plants.

Possibly the oldest paper found regarding desalination plants assisted by geothermal energy was written by Awerbuch et al. (1976). They reported that the Bureau of Reclamation of the U.S. Department of the Interior investigated a geothermal-powered desalination pilot plant near Holtville, California. Boegli et al. (1977), from the same department, reported experimental results of geothermal fluids desalination at the East Mesa test site. The processes analyzed included MSF distillation and high-temperature ED as well as different evaporation tubes and membranes.

Another possibility that can be investigated is the use of high-pressure geothermal power directly as shaft power on desalination. Moreover, there are commercial membranes that withstand temperatures up to 60 °C, which permits the direct use of geothermal brines for desalination (Houcine et al., 1999).

8.6 Process selection

During the design stage of a renewable energy-powered desalination system, the designer must select a process suitable for that particular application. The factors that should be considered for such a selection are the following (Kalogirou, 2005):

1. The suitability of the process for renewable energy application.
2. The effectiveness of the process with respect to energy consumption.
3. The amount of freshwater required in a particular application, in combination with the range of applicability of the various desalination processes.
4. The seawater treatment requirements.
5. The capital cost of the equipment.
6. The land area required or that could be made available for the installation of the equipment.

Before any process selection can start, a number of basic parameters should be investigated. The first is the evaluation of the overall water resources. This should be done in terms of both quality and quantity (for a brackish water resource). Should brackish water be available, then this may be more attractive, since the salinity is normally much lower (<10,000 ppm); hence, the desalination of the brackish water should be the more attractive option. In inland sites, brackish water may be the only option. On a coastal site, seawater is normally available. The identification and evaluation of the renewable energy resources in the area complete the basic steps to be performed toward the design of a renewable energy system (RES) to drive the desalination system. Renewable energy-driven desalination technologies mainly fall into two categories. The first includes distillation desalination technologies driven by heat produced by a RES; the second includes membrane and distillation desalination technologies driven by electricity or mechanical energy produced by a RES. Such systems should be characterized by robustness, simplicity of operation, low maintenance, compact size, easy transportation to the site, simple pretreatment and intake systems to ensure proper operation, and endurance of a plant at the difficult conditions often encountered in remote areas. Concerning their combination, the existing experience has shown no significant technical problems (Tzen and Morris, 2003).

Water production costs generally include the following items (Fiorenza et al., 2003):

- Fixed charges, which depend on the capital cost and a depreciation factor (determined from both plant life and financial parameters and consequently varying for each country).
- Variable charges, which depend on the consumption and cost of energy (related to the source employed and location selected), operational (personnel), maintenance cost (varying for each country), consumption and cost of chemicals used for pre- and post-treatment of water (especially in RO plants), and the rate at which the membranes are to be replaced in RO plants (both factors are site related).

Generally, the percentage of TDS in seawater has practically no effect in thermal processes but a remarkable effect in RO, where the energy demand increases linearly at a rate of more than 1 kWh/m^3 per 10,000 ppm (Fiorenza et al., 2003). If, however, the input pressures are left unchanged, the percentage of salts in the water produced could be intolerably high. Normally, this value for the RO process is expected to be around 300 ppm. The value, though lying well within the limit of 500 ppm (fixed by the WHO for drinking water), still results in at least one order of magnitude higher than the salinity of water produced from thermal processes. Also, for high salinity concentration, the use of RO technology is very problematic.

Renewable energy sources can provide thermal energy (solar collectors, geothermal energy), electricity (photovoltaics, wind energy, solar thermal power systems), or mechanical energy (wind energy). All these forms of energy can be used to power desalination plants.

Solar energy can generally be converted into useful energy either as heat, with solar collectors and solar ponds, or as electricity, with photovoltaic cells and solar thermal power systems. As was seen in previous sections, both methods have been used to power desalination systems. The direct collection systems can utilize solar energy only when it is available, and their collection is inefficient. Alternatively, in the indirect collection systems, solar energy is collected by more efficient solar collectors, in the form of hot water or steam. It should be noted, however, that solar energy is available for only almost half the day. This implies that the process operates for only half the time available, unless some storage device is used. The storage device, which is usually expensive, can be replaced by a backup boiler or electricity from the grid to operate the system during low insolation and nighttime. When such a system operates without thermal buffering, the desalination subsystem must be able to follow a variable energy supply, without upset. In all solar energy desalination systems, an optimum PR has to be calculated based on the solar energy collectors' cost, storage devices' cost (if used), and the cost of the desalination plant (Kalogirou, 2005). Probably the only form of stable energy supply is the solar pond, which, due to its size, does not charge or discharge easily, and thus is less sensitive to variations in the weather.

Wind energy is also a highly variable source of supply with respect to both wind speed and frequency. When wind energy is used for electricity generation, the variation of the wind source can be balanced by the addition of battery banks, which act in a way similar to a storage tank in solar thermal systems, that is, the batteries charge when wind is available and discharge to the load (desalination plant) when required. In the case of mechanical energy production from wind, the desalination plant can operate only when there is wind. In this case, the desalination plant is usually oversized with respect to water demand, and instead of storing the energy, the water produced when wind is available is stored.

In the technology selection, another parameter to be considered is the type of connection of the two technologies. An RO renewable desalination plant can be designed to operate coupled to the grid or off-grid (stand-alone autonomous system). When the system is grid connected, the desalination plant can operate continuously as a conventional plant, and the renewable energy source merely acts as a fuel substitute. Where no electricity grid is available, autonomous systems have to be developed that allow for the intermittent nature of the renewable energy source. Desalination systems have traditionally been designed to operate with a constant power input (Tzen et al., 1998). Unpredictable and insufficient power input forces the desalination plant to operate under non-optimal conditions and may cause operational problems (Tzen and Morris, 2003). Each desalination system has specific problems when it is connected to a variable power system. For instance, the RO system has to cope with the sensitivity of the membranes regarding fouling, scaling, and unpredictable phenomena due to start–stop cycles and partial load operation during periods of oscillating power supply. On the other hand, the VC system has considerable thermal inertia and requires sizable energy to get to the nominal working point. Thus, for autonomous systems, a small energy storage system, batteries or thermal stores, should be added to offer stable power to the desalination unit. Any candidate option resulting from the previous parameters should be further screened through constraints such as site characteristics (accessibility, land formation, etc.) and financial requirements (Tzen and Morris, 2003).

The energy required for various desalination processes, as obtained from a survey of manufacturers' data, is shown in Table 8.3. It can be seen from Table 8.3 that the process with the smallest energy requirement is RO with energy recovery. However, this is viable for only very large systems due to the high cost of the energy recovery turbine. The next lowest is the RO without energy recovery and the MEB. A comparison of the desalination equipment cost and the seawater treatment requirement, as obtained from a survey of manufacturers' data, is shown in Table 8.4. The cheapest of the considered systems is the solar still. This is a direct collection system, which is very easy to construct and operate.

Table 8.3 Energy consumption of desalination systems.

Process	Heat input (kJ/kg of product)	Mechanical power input (kWh/m³ of product)	Prime energy consumption (kJ/kg of Product)[a]
MSF	294	2.5–4 (3.7)[b]	338.4
MEB	123	2.2	149.4
VC	—	8–16 (16)	192
RO	—	5–13 (10)	120
ER-RO	—	4–6 (5)	60
ED	—	12	144
Solar still	2330	0.3	2333.6

[a]*Assumed conversion efficiency of electricity generation of 30%.*
[b]*Figure used for the prime energy consumption estimation shown in last column.*

Item	MSF	MEB	VC	RO	Solar still
Table 8.4 Comparison of desalination plants.					
Scale of application	Medium–large	Small–medium	Small	Small–large	Small
Seawater treatment	Scale inhibitor, antifoam chemical	Scale inhibitor	Scale inhibitor	Sterilizer Coagulant acid Deoxidizer	—
Equipment price (Euro/m^3)	950–1900	900–1700	1500–2500	900–2500, membrane replacement every 4–5 years	800–1000

Note: Low figures in equipment price refer to bigger size in the range indicated and vice versa.

The disadvantage of this process is the very low yield, which implies that large areas of flat ground are required. It is questionable whether such a process can be viable unless cheap desert-like land is available near the sea. The MEB is the cheapest of all the indirect collection systems and also requires the simplest seawater treatment. RO, although requiring a smaller amount of energy, is expensive and requires a complex seawater treatment.

Due to the development of RO technology, the energy consumption values of more than 20 kWh/m^3 during the year 1970 have been reduced today to about 5 kWh/m^3 (Fiorenza et al., 2003). This is due to improvements made in RO membranes. Research in this sector is ongoing worldwide, and we may see further reductions in both energy requirements and costs in the coming years. It should be noted that nearly 3 kg of CO_2 are generated for each cubic meter of water produced (at an energy consumption rate of 5 kWh/m^3 with the best technology currently used on a large scale), which could be avoided if the conventional fuel is replaced by a renewable one.

An alternative usually considered for solar-powered desalination is to use an RO system powered by photovoltaic cells. This is more suitable for intermittent operation than conventional distillation processes and has higher yields per unit of energy collected. According to Zarza et al. (1991b), who compared RO with photovoltaic-generated electricity with an MEB plant coupled to parabolic trough collectors, the following apply:

- The total cost of freshwater produced by an MEB plant coupled to parabolic trough collectors is less than that of the RO plant with photovoltaic cells, due to the high cost of the photovoltaic-generated electricity.
- The highly reliable MEB plant operation makes its installation possible in countries with high insolation levels but a lack of experienced personnel. Because any serious mistake during the operation of an RO plant can ruin its membranes, these plants must be operated by skilled personnel.

Also, since renewable energy is expensive to collect and store, an energy recovery turbine is normally fitted to recover energy from the rejected brine stream, which increases the RO plant cost considerably. Additionally, in polluted areas, distillation processes are preferred for desalination because water is boiled, which ensures that the distilled water does not contain any micro-organisms

(Kalogirou, 2005). In addition to the high salinity, specific water quality problems include manganese, fluoride, heavy metals, bacterial contamination, and pesticide—herbicide residues. In all these cases, thermal processes are preferred to membrane ones. Even the simple solar still can provide removal efficiencies on the order of 99% (Hanson et al., 2004).

If both RO and thermal processes are suitable for a given location, the renewable energy available and the electrical—mechanical—thermal energy required by the process limit the possible selection. Finally, the required plant capacity, the annual and daily distribution of the freshwater demand, the product cost, the technology maturity, and any problems related to the connection of the renewable energy and the desalination systems are factors that influence the selection.

If thermal energy is available, it can be used directly to drive a distillation process, such as MSF, MEB, or TVC. MEB plants are more flexible to operate at partial load, less sensible to scaling, cheaper, and more suitable for limited capacity than MSF plants. TVC has lower performance than MEB and MSF. In addition, the thermo-mechanical conversion permits the indirect use of thermal energy to drive RO, ED, or MVC processes.

If electricity or shaft power can be obtained from the available energy resources, RO, ED, or MVC can be selected. Fluctuations in the available energy would ruin the RO system. Therefore, an intermediate energy storage is required, but it would reduce the available energy and increase the costs. In remote areas, ED is most suitable for brackish water desalination, because it is more robust and its operation and maintenance are simpler than RO systems. In addition, the ED process can adapt to changes of available energy input. On the other hand, although MVC consumes more energy than RO, it presents fewer problems than RO due to the fluctuations of the energy resource. MVC systems are more suitable for remote areas, since they are more robust and need fewer skilled workers and fewer chemicals than RO systems (Garcia-Rodriguez, 2003). In addition, they need no membrane replacement and offer a better quality product than RO. Moreover, in case of polluted waters, the distillation ensures the absence of micro-organisms and other pollutants in the product.

It is believed that solar energy is best and most cheaply harnessed with thermal energy collection systems. Therefore, the two systems that could be used are the MSF and the MEB plants. As can be seen from previous sections, both systems have been used with solar energy collectors in various applications. According to Tables 8.3 and 8.4, the MEB process requires less specific energy, is cheaper, and requires only a very simple seawater treatment. In addition, the MEB process has advantages over other distillation processes. According to Porteous (1975), these are as follows:

1. Energy economy, because the brine is not heated above its boiling point, as is the case for the MSF process. The result is less irreversibility in the MEB process, since the vapor is used at the temperature at which it is generated.
2. The feed is at its lowest concentration at the highest plant temperature, so scale formation risks are minimized.
3. The feed flows through the plant in series, and the maximum concentration occurs only in the last effect; therefore, the worst boiling point elevation is confined to this effect.
4. The other processes have a high electrical demand because of the recirculation pump in the MSF or the vapor compressor in the VC systems.
5. MSF is prone to equilibrium problems, which are reflected by a reduction in PR. In MEB plants, the vapor generated in one effect is used in the next and PR is not subject to equilibrium problems.
6. Plant simplicity is promoted by the MEB process because fewer effects are required for a given PR.

Of the various types of MEB evaporators, the MES is the most appropriate for solar energy application. This has a number of advantages, the most important of which is stable operation between virtually 0% and 100% output, even when sudden changes are made, and the ability to follow a varying steam supply without upset (Kalogirou, 2005). For this purpose, collectors of proven technology such as the parabolic trough can be used to produce the input power to the MEB system in the form of low-pressure steam. The temperature required for the heating medium is between 70 and 100 °C, which can be produced with such collectors with an efficiency of about 65% (Kalogirou, 2005).

Exercises

8.1 Estimate the mole and mass fractions for the salt and water of seawater, which has a salinity of 42,000 ppm.

8.2 Estimate the mole and mass fractions for the salt and water of brackish water, which has a salinity of 1500 ppm.

8.3 Find the enthalpy and entropy of seawater at 35 °C.

8.4 A solar still has a water and glass temperature equal to 52.5 °C and 41.3 °C, respectively. The constants C and n are determined experimentally and are found to be $C = 0.054$ and $n = 0.38$. If the convective heat transfer coefficient from water surface to glass is 2.96 W/m^2 K, estimate the hourly distillate output per square meter from the solar still.

8.5 An MSF plant has 32 stages. Estimate the M_f/M_d ratio if the brine temperature in the first effect is 68 °C and the temperature of the brine in the last effect is 34 °C. The mean latent heat is 2300 kJ/kg and the mean specific heat is 4.20 kJ/kg K.

References

Adam, S., Cheng, R.C., Vuong, D.X., Wattier, K.L., 2003. Long Beach's dual-stage NF beats single-stage SWRO. Desalin. Water Reuse 13 (3), 18−21.

Adhikari, R.S., Kumar, A., Sodha, G.D., 1995. Simulation studies on a multi-stage stacked tray solar still. Sol. Energy 54 (5), 317−325.

Akinsete, V.A., Duru, C.U., 1979. A cheap method of improving the performance of roof type solar stills. Sol. Energy 23 (3), 271−272.

Al-Karaghouli, A.A., Alnaser, W.E., 2004a. Experimental comparative study of the performances of single and double basin solar-stills. Appl. Energy 77 (3), 317−325.

Al-Karaghouli, A.A., Alnaser, W.E., 2004b. Performances of single and double basin solar-stills. Appl. Energy 78 (3), 347−354.

Al-Madani, H.M.N., 2003. Water desalination by solar powered electrodialysis process. Renew. Energy 28 (12), 1915−1924.

Al-Sahali, M., Ettouney, H., 2007. Developments in thermal desalination processes: design, energy and costing aspects. Desalination 214, 227−240.

Al-Suleimani, Z., Nair, N.R., 2000. Desalination by solar powered reverse osmosis in a remote area of Sultanate of Oman. Appl. Energy 65 (1−4), 367−380.

Awerbuch, L., Lindemuth, T.E., May, S.C., Rogers, A.N., 1976. Geothermal energy recovery process. Desalination 19 (1−3), 325−336.

Baldacci, A., Burgassi, P.D., Dickson, M.H., Fanelli, M., 1998. Non-electric utilization of geothermal energy in Italy. In: Proceedings of World Renewable Energy Congress V. Part I, 20–25 September 1998, Florence, Italy. Pergamon, Oxford, UK, p. 2795.

Banat, F., Jwaied, N., Rommel, M., Koschikowski, J., Wieghaus, M., 2007a. Desalination by a "compact SMADES" autonomous solar powered membrane distillation unit. Desalination 117, 29–37.

Banat, F., Jwaied, N., Rommel, M., Koschikowski, J., Wieghaus, M., 2007b. Performance evaluation of a "large SMADES" autonomous desalination solar-driven membrane distillation plant in Aqaba, Jordan. Desalination 117, 17–28.

Barbier, E., 1997. Nature and technology of geothermal energy. Renewable Sustainable Energy Rev. 1 (1–2), 1–69.

Barbier, E., 2002. Geothermal energy technology and current status: an overview. Renewable Sustainable Energy Rev. 6 (1–2), 3–65.

Belessiotis, V., Kalogirou, S.A., Delyannis, E., 2016. Thermal Solar Desalination: Methods and Systems. Academic Press, London.

Bier, C., Coutelle, R., Gaiser, P., Kowalczyk, D., Plantikow, U., 1991. Sea-water desalination by wind-powered mechanical vapor compression plants. In: Seminar on New Technologies for the Use of Renewable Energies in Water Desalination. 26–28 September 1991, Athens, Commission of the European Communities, DG XVII for Energy, CRES (Centre for Renewable Energy Sources), Session II, 1991, pp. 49–64.

Boegli, W.J., Suemoto, S.H., Trompeter, K.M., 1977. Geothermal-desalting at the East Mesa test site. Desalination 22 (1–3), 77–90.

Boeher, A., 1989. Solar desalination with a high efficiency multi-effect process offers new facilities. Desalination 73, 197–203.

Bouchekima, B., Gros, B., Ouahes, R., Diboun, M., 2001. Brackish water desalination with heat recovery. Desalination 138 (1–3), 147–155.

Cengel, Y.A., Boles, M.A., 1998. Thermodynamics: An Engineering Approach, third ed. McGraw-Hill, New York.

Cengel, Y.A., Cerci, Y., Wood, B., 1999. Second law analysis of separation processes of mixtures. Proc. ASME Adv. Energy Syst. Div. 39, 537–543.

Cerci, Y., 2002. Exergy analysis of a reverse osmosis desalination plant. Desalination 142 (3), 257–266.

Chaibi, M.T., 2000. Analysis by simulation of a solar still integrated in a greenhouse roof. Desalination 128 (2), 123–138.

Curcio, E., Drioli, E., 2005. Membrane distillation and related operations — a review. Separ. Purif. Rev. 34, 35–86.

Daniels, F., 1974. Direct Use of the Sun's Energy, sixth ed. Yale University Press, New Haven, CT, and London (Chapter 10).

Delyannis, A., 1960. Introduction to Chemical Technology. History of Chemistry and Technology (In Greek). Athens (Chapter 1).

Delyannis, A.A., 1965. Solar stills provide an island's inhabitants with water. Sun Work 10 (1), 6.

Delyannis, E., 2003. Historic background of desalination and renewable energies. Sol. Energy 75 (5), 357–366.

Delyannis, E., Belessiotis, V., 2000. The history of renewable energies for water desalination. Desalination 128 (2), 147–159.

Delyannis, A.A., Delyannis, E.A., 1970. Solar desalting. J. Chem. Eng. 19, 136.

Delyannis, A., Delyannis, E., 1973. Solar distillation plant of high capacity. Proc. of Fourth Inter. Symp. on Fresh Water from Sea 4, 487.

Delyannis, E.E., Delyannis, A., 1985. Economics of solar stills. Desalination 52 (2), 167–176.

Delyannis, A., Piperoglou, E., 1968. The Patmos solar distillation plant: technical note. Sol. Energy 12 (1), 113–114.

Doucet, G., 2001. Energy for tomorrow's world. Renew. Energy 19–22.

DOW, 2006. Design a Reverse Osmosis System: Design Equations and Parameters. Technical Manual.

Dresner, L., Johnson, J., 1980. Hyperfiltration (reverse osmosis). In: Spiegler, K.S., Laird, A.D.K. (Eds.), Principles of Desalination, Part B, second ed. Academic Press, New York, pp. 401–560.

Dunkle, R.V., 1961. Solar water distillation; the roof type still and a multiple effect diffusion still, International Developments in Heat Transfer ASME. In: Proceedings of International Heat Transfer, Part V. University of Colorado, p. 895.

Eggers-Lura, A., 1979. Solar Energy in Developing Countries. Pergamon Press, Oxford, UK, pp. 35–40.

Eibling, J.A., Talbert, S.G., Lof, G.O.G., 1971. Solar stills for community use—digest of technology. Sol. Energy 13 (2), 263–276.

El-Bahi, A., Inan, D., 1999. A solar still with minimum inclination, coupling to an outside condenser. Desalination 123 (1), 79–83.

El-Dessouky, H., Ettouney, H., 2000. MSF development may reduce desalination costs. Water Wastewater Int. 20–21.

El-Nashar, A.M., 1992. Optimizing the operating parameters of a solar desalination plant. Sol. Energy 48 (4), 207–213.

El-Nashar, A.M., Al-Baghdadi, A.A., 1998. Exergy losses in multiple-effect stack seawater desalination plant. Desalination 116 (1), 11–24.

El-Sayed, Y.M., Silver, R.S., 1980. Fundamentals of distillation. In: Spiegler, K.S., Laird, A.D.K. (Eds.), Principles of Desalination, Part A, second ed. Academic Press, New York, pp. 55–109.

Fath, H.E.S., Elsherbiny, S.M., Alaa, A.H., Rommel, M., Wieghaus, M., Koschikowski, J., Vatansever, M., 2008. PV and thermally small-scale stand-alone solar desalination system with very low maintenance needs. Desalination 225, 58–69.

Fiorenza, G., Sharma, V.K., Braccio, G., 2003. Techno-economic evaluation of a solar powered water desalination plant. Energy Convers. Manage. 44 (4), 2217–2240.

Florides, G., Kalogirou, S., 2004. Ground heat exchangers—a review. In: Proceedings of Third International Conference on Heat Power Cycles. Larnaca, Cyprus, on CD ROM.

Frick, G., von Sommerfeld, J., 1973. Solar stills of inclined evaporating cloth. Sol. Energy 14 (4), 427–431.

Garcia-Rodriguez, L., 2003. Renewable energy applications in desalination: state of the art. Sol. Energy 75 (5), 381–393.

Garcia-Rodriguez, L., Romero, T., Gomez-Camacho, C., 2001. Economic analysis of wind-powered desalination. Desalination 137 (1–3), 259–265.

Garg, H.P., Mann, H.S., 1976. Effect of climatic, operational and design parameters on the year round performance of single sloped and double sloped solar still under Indian and arid zone condition. Sol. Energy 18 (2), 159–163.

Geankoplis, C.J., 2003. Transport Processes and Separation Process Principles, fourth ed. Prentice Hall, Englewood Cliffs, NJ.

Gomkale, S.D., 1988. Solar distillation as a means to provide Indian villages with drinking water. Desalination 69 (2), 171–176.

Goosen, M.F.A., Sablani, S.S., Shayya, W.H., Paton, C., Al-Hinai, H., 2000. Thermodynamic and economic considerations in solar desalination. Desalination 129 (1), 63–89.

Grutcher, J., 1983. Desalination a PV oasis. Photovoltaics Intern. (June/July) 24.

Hamed, O.A., Zamamiri, A.M., Aly, S., Lior, N., 1996. Thermal performance and exergy analysis of a thermal vapor compression desalination system. Energy Convers. Manage. 37 (4), 379–387.

Hanson, A., Zachritz, W., Stevens, K., Mimbella, L., Polka, R., Cisneros, L., 2004. Distillate water quality of a single-basin solar still: laboratory and field studies. Sol. Energy 76 (5), 635–645.

Hermann, M., Koschikowski, J., Rommel, M., 2000. Corrosion-free solar collectors for thermally driven seawater desalination. In: Proceedings of the EuroSun 2000 Conference on CD ROM, 19−22 June 2000, Copenhagen, Denmark.

Hogan, P.A., Sudjito, A., Fane, G., Morrison, G.L., 1991. Desalination by solar heated membrane distillation. In: Proc. 12th Int, Symposium on Desalination and Water Reuse, Malta, 15-18, vol. 1, pp. 81−90.

Houcine, I., Benjemaa, F., Chahbani, M.H., Maalej, M., 1999. Renewable energy sources for water desalting in Tunisia. Desalination 125 (1−3), 123−132.

Howe, E.D., Tleimat, B.W., 1974. Twenty years of work on solar distillation at the University of California. Sol. Energy 16 (2), 97−105.

Hsu, S.T., Cheng, K.T., Chiou, J.S., 2002. Seawater desalination by direct membrane distillation. Desalination 143, 279−287.

Hussain, N., Rahim, A., 2001. Utilization of new technique to improve the efficiency of horizontal solar desalination still. Desalination 138 (1−3), 121−128.

ITC (Canary Islands Technological Institute), 2001. Memoria de gestion (in Spanish).

Kalogirou, S.A., 1997a. Solar distillation systems: a review. In: Proceedings of First International Conference on Energy and the Environment, Vol. 2, Limassol, Cyprus, pp. 832−838.

Kalogirou, S.A., 1997b. Survey of solar desalination systems and system selection. Energy Int. J. 22 (1), 69−81.

Kalogirou, S.A., 2001. Effect of fuel cost on the price of desalination water: a case for renewables. Desalination 138 (1−3), 137−144.

Kalogirou, S.A., 2005. Seawater desalination using renewable energy sources. Prog. Energy Combust. Sci. 31 (3), 242−281.

Karameldin, A., Lotfy, A., Mekhemar, S., 2003. The Red Sea area wind-driven mechanical vapor compression desalination system. Desalination 153 (1−3), 47−53.

Kiatsiriroat, T., 1989. Review of research and development on vertical solar stills. ASEAN J. Sci. Technol. Dev. 6 (1), 15.

Kimura, S., Sourirajan, S., 1968. Concentration polarization effects in reverse osmosis using porous cellulose acetate membranes. Ind. Eng. Chem. Process Des. Dev. 7 (1), 41−48.

Klotz, I.M., Rosenberg, R.M., 1994. Chemical Thermodynamics: Basic Theory and Methods, fifth ed. Wiley, New York.

Koschikowski, J., Wieghaus, M., Rommel, M., 2009. Membrane distillation in cipollina, A. In: Micale, G., Rizzuti, L. (Eds.), Seawater Desalination. Springer − Verlag, Berlin & Heidelberg.

Kreider, J.F., Kreith, F., 1981. Solar Energy Handbook. McGraw-Hill, New York.

Kudish, A.I., Gale, J., 1986. Solar desalination in conjunction with controlled environment agriculture in arid zone. Energy Convers. Manage. 26 (2), 201−207.

Kumar, S., Tiwari, G.N., 1996. Estimation of convective mass transfer in solar distillation systems. Sol. Energy 57 (6), 459−464.

Kumar, S., Tiwari, G.N., 1998. Optimization of collector and basin areas for a higher yield active solar still. Desalination 116 (1), 1−9.

Kvajic, G., 1981. Solar power desalination, PV-ED system. Desalination 39, 175.

Lawson, K.W., Lloyd, D.R., 1997. Review: membrane distillation. J. Membr. Sci. 124, 1−25.

Lobo, P.C., Araujo, S.R., 1978. A simple multi-effect basin type solar still. In: SUN, Proceedings of the International Solar Energy Society, vol. 3. New Delhi, India, Pergamon, Oxford, UK, pp. 2026−2030.

Luft, W., 1982. Five solar energy desalination systems. Int. J. Sol. Energy 1 (21).

Lui, C.C.K., Park, J.W., Migita, R., Qin, G., 2002. Experiments of a prototype wind-driven reverse osmosis desalination system with feedback control. Desalination 150 (3), 277−287.

Major, J.K., 1990. Water wind and animal power. In: McNeil, J. (Ed.), An Encyclopedia of the History of Technology. Rutledge, R. Clay Ltd, Great Britain, Bungay, pp. 229−270.

Malik, M.A.S., Tiwari, G.N., Kumar, A., Sodha, M.S., 1982. Solar Distillation. Pergamon Press, Oxford, UK.

Matz, R., Feist, E.M., 1967. The application of solar energy to the solution of some problems of electrodialysis. Desalination 2 (1), 116–124.

Maurel, A., 1991. Desalination by reverse osmosis using renewable energies (solar-wind) Cadarache Centre Experiment. In: Seminar on New Technologies for the Use of Renewable Energies in Water Desalination. 26–28 September 1991, Athens. Commission of the European Communities, DG XVII for Energy, CRES (Centre for Renewable Energy Sources), Session II, pp. 17–26.

Mehdizadeh, H., Dickson, J.M., Eriksson, P.K., 1989. Temperature effects on the performance of thin-film composite, aromatic polyamide membranes. Ind. Eng. Chem. Res. 28, 814–824.

Morris, R.M., Hanbury, W.T., 1991. Renewable energy and desalination—a review. In: Proceedings of the New Technologies for the Use of Renewable Energy Sources in Water Desalination, Sec. I. Athens, Greece, pp. 30–50.

Moustafa, S.M.A., Jarrar, D.I., Mansy, H.I., 1985. Performance of a self-regulating solar multistage flush desalination system. Sol. Energy 35 (4), 333–340.

Mukherjee, K., Tiwari, G.N., 1986. Economic analysis of various designs of conventional solar stills. Energy Convers. Manage. 26 (2), 155–157.

Mustacchi, C., Cena, V., 1978. Solar Water Distillation, Technology for Solar Energy Utilization. United Nations, New York, pp. 119–124.

Ophir, A., 1982. Desalination plant using low-grade geothermal heat. Desalination 40 (1–2), 125–132.

Parekh, S., Farid, M.M., Selman, R.R., Al-Hallaj, S., 2003. Solar desalination with humidification–dehumidification technique—a comprehensive technical review. Desalination 160 (2), 167–186.

Plantikow, U., 1999. Wind-powered MVC seawater desalination-operational results. Desalination 122 (2–3), 291–299.

Popiel, C., Wojtkowiak, J., Biernacka, B., 2001. Measurements of temperature distribution in ground. Exp. Therm. Fluid Sci. 25, 301–309.

Porteous, A., 1975. Saline Water Distillation Processes. Longman, Essex, UK.

Rajvanshi, A.K., 1980. A scheme for large-scale desalination of seawater by solar energy. Sol. Energy 24 (6), 551–560.

Rajvanshi, A.K., 1981. Effects of various dyes on solar distillation. Sol. Energy 27 (1), 51–65.

Rodriguez, G., Rodriguez, M., Perez, J., Veza, J., 1996. A systematic approach to desalination powered by solar, wind and geothermal energy sources. In: Proceedings of the Mediterranean Conference on Renewable Energy Sources for Water Production. European Commission, EURORED Network, CRES, EDS, Santorini, Greece, 10–12 June 1996, pp. 20–25.

Rommel, M., Hermann, M., Koschikowski, J., 2000. The SODESA project: development of solar collectors with corrosion-free absorbers and first results of the desalination pilot plant. In: Mediterranean Conference on Policies and Strategies for Desalination and Renewable Energies. 21–23 June 2000, Santorini, Greece.

Saffarini, R.B., Summers, E.K., Arafat, H.A., Lienhard, V.J.H., 2012. Technical evaluation of standalone solar powered membrane distillation systems. Desalination 286, 332–341.

Satcunanathan, S., Hanses, H.P., 1973. An investigation of some of the parameters involved in solar distillation. Sol. Energy 14 (3), 353–363.

Schock, G., Miquel, A., 1987. Mass transfer and pressure loss in spiral wound modules. Desalination 64, 339–352.

Shaffer, L.H., Mintz, M.S., 1980. Electrodialysis. In: Spiegler, K.S., Laird, A.D.K. (Eds.), Principles of Desalination, Part A, second ed. Academic Press, New York, pp. 257–357.

Shneider, K., Holz, W., Wollbreck, R., Ripperger, S., 1988. Membranes and Modules for transmembrane distillation. J. Membr. Sci. 39 (1988), 25–42.

Sodha, M.S., Adhikari, R.S., 1990. Techno-economic model of solar still coupled with a solar flat-plate collector. Int. J. Energy Res. 14 (5), 533—552.

Sodha, M.S., Kumar, A., Tiwari, G.N., Tyagi, R.C., 1981. Simple multiple-wick solar still: analysis and performance. Sol. Energy 26 (2), 127—131.

Suneja, S., Tiwari, G.N., 1999. Optimization of number of effects for higher yield from an inverted absorber solar still using the Runge-Kutta method. Desalination 120 (3), 197—209.

Tabor, H., 1990. Solar energy technologies for the alleviation of fresh-water shortages in the Mediterranean basin, Euro-Med solar. In: Proceedings of the Mediterranean Business Seminar on Solar Energy Technologies, Nicosia, Cyprus, pp. 152—158.

Talbert, S.G., Eibling, J.A., Lof, G.O.G., 1970. Manual on Solar Distillation of Saline Water, R&D Progress Report No. 546. U.S. Department of the Interior, Battelle Memorial Institute, Columbus, OH.

Tanaka, H., Nosoko, T., Nagata, T., 2000a. A highly productive basin-type-multiple-effect coupled solar still. Desalination 130 (3), 279—293.

Tanaka, H., Nosoko, T., Nagata, T., 2000b. Parametric investigation of a basin-type-multiple-effect coupled solar still. Desalination 130 (3), 295—304.

Taylor, J.S., Hofman, J.A.M.H., Schippers, J.C., Duranceau, S.J., Kruithof, J.C., 1994. Simplified modeling of diffusion controlled membrane systems. Aqua J. Water Supply Res. Technol. 43 (5), 238—245.

THERMIE Program, 1998. Desalination Guide Using Renewable Energies. CRES, Athens, Greece.

Thomson, M., Infield, D., 2003. A photovoltaic-powered seawater reverse-osmosis system without batteries. Desalination 153 (1—3), 1—8.

Tiwari, G.N., 1984. Demonstration plant of multi-wick solar still. Energy Convers. Manage. 24 (4), 313—316.

Tiwari, G.N., 1992. Contemporary physics-solar energy and energy conservation, (Chapter 2). In: Recent Advances in Solar Distillation. Wiley Eastern Ltd, New Delhi, India.

Tiwari, G.N., 2002. Solar Energy. Narosa Publishing House, New Delhi, India.

Tiwari, G.N., Noor, M.A., 1996. Characterization of solar still. Int. J. Sol. Energy 18, 147.

Tiwari, G.N., Yadav, Y.P., 1985. Economic analysis of large-scale solar distillation plant. Energy Convers. Manage. 25 (14), 423—425.

Tiwari, G.N., Singh, H.N., Tripathi, R., 2003. Present status of solar distillation. Sol. Energy 75 (5), 367—373.

Tleimat, B.W., 1978. Solar Distillation: The State of the Art, Technology for Solar Energy Utilization. United Nations, New York, pp. 113—118.

Tleimat, M.W., 1980. Freezing methods. In: Spiegler, K.S., Laird, A.D.K. (Eds.), Principles of Desalination, Part B, second ed. Academic Press, New York, pp. 359—400.

Tleimat, B.W., Howe, E.D., 1967. Comparison of plastic and glass condensing covers for solar distillers. In: Proceedings of Solar Energy Society, Annual Conference. Arizona.

Tzen, E., Morris, R., 2003. Renewable energy sources for desalination. Sol. Energy 75 (5), 375—379.

Tzen, E., Perrakis, K., Baltas, P., 1998. Design of a stand-alone PV-desalination system for rural areas. Desalination 119 (1—3), 327—333.

Veza, J., Penate, B., Castellano, F., 2001. Electrodialysis desalination designed for wind energy (on-grid test). Desalination 141 (1), 53—61.

Vince, F., Marechal, F., Aoustin, E., Breant, P., 2008. Multi-objective optimisation of RO desalination plants. Desalination 222, 96—118.

Voropoulos, K., Mathioulakis, E., Belessiotis, V., 2001. Experimental investigation of a solar still coupled with solar collectors. Desalination 138 (1—3), 103—110.

Zarza, E., Ajona, J.I., Leon, J., Genthner, K., Gregorzewski, A., 1991a. Solar thermal desalination project at the Plataforma Solar de Almeria. In: Proceedings of the Biennial Congress of the International Solar Energy Society, 1, Part II, Denver, Colorado, pp. 2270—2275.

Zarza, E., Ajona, J.I., Leon, J., Genthner, K., Gregorzewski, A., Alefeld, G., Kahn, R., Haberle, A., Gunzbourg, J., Scharfe, J., Cord'homme, C., 1991b. Solar thermal desalination project at the Plataforma Solar de Almeria. In: Proceedings of the New Technologies for the Use of Renewable Energy Sources in Water Desalination, Athens, Greece, Sec. III, pp. 62−81.

Zein, M., Al-Dallal, S., 1984. Solar desalination correlation with meteorological parameters. In: Proceedings of the Second Arab International Conference, p. 288.

Photovoltaic Systems

Photovoltaic (PV) modules are solid-state devices that convert sunlight, the most abundant energy source on the planet, directly into electricity without an intervening heat engine or rotating equipment. PV equipment has no moving parts and, as a result, requires minimal maintenance and has a long life. It generates electricity without producing emissions of greenhouse or any other gases and its operation is virtually silent. Photovoltaic systems can be built in virtually any size, ranging from milliwatt to megawatt, and the systems are modular, i.e., more panels can be easily added to increase output. Photovoltaic systems are highly reliable and require little maintenance. They can also be set up as stand-alone systems.

A PV cell consists of two or more thin layers of semiconducting material, most commonly silicon. When the silicon is exposed to light, electrical charges are generated; and this can be conducted away by metal contacts as direct current. The electrical output from a single cell is small, so multiple cells are connected and encapsulated (usually glass covered) to form a module (also called a *panel*). The PV panel is the main building block of a PV system, and any number of panels can be connected together to give the desired electrical output. This modular structure is a considerable advantage of the PV system, where further panels can be added to an existing system as required.

In the early days of photovoltaics, some 60 years ago, the energy required to produce a PV panel was more than the energy the panel could produce during its lifetime. During the last decade, however, due to improvements in the efficiency of the panels and manufacturing methods, the payback times have fallen to 2–3 years for crystalline silicon PV systems, and to almost 1 year for some thin-film systems, under moderate levels of sunshine (Fthenakis and Kim, 2011).

Photovoltaic prices have fallen sharply since the mid-1970s. It is generally believed that, as photovoltaic prices fall, markets will expand rapidly. Worldwide photovoltaic sales in 2006 were about 2500 MWe annually and the increase from 2005 was 40% (Sayigh, 2008). The major problem limiting the widespread use of photovoltaics is the high cost of manufacturing the sheets of semiconductor materials needed for power systems. However, the cost of PV panels has also fallen markedly—both of crystalline silicon, as manufacturing output particularly in China has increased, and of thin-film, where large-scale manufacturing has commenced. In 2012 the market prices of PV modules had fallen to less than $0.80 per Watt, compared to around $2 per Watt in 2010 and $5 per Watt in 2000.

Falling module prices coupled with government incentives—notably feed-in tariffs in Europe and investment tax credits in the US—have caused the market for photovoltaics to grow rapidly in recent years. In 2000 the installed PV capacity worldwide was 1.4 GW (EPIA, 2010); at the end of 2011 it had reached 67 GW (Photon, 2012) whereas the installed capacity in 2021, as shown in Chapter 1, reached 942 GW.

Solar Energy Engineering. https://doi.org/10.1016/B978-0-323-99350-0.00018-1

A few years ago, despite achieving a mass-production scale and lower manufacturing costs, utility-scale PV is still more expensive than both fossil fuels and other renewable energies (Tidball et al., 2010). This holds for both its capital cost, where large PV projects typically cost $2.50–4 per Watt (Barbose et al., 2011), and its average cost of electricity of $0.17 per kWh. Residential PV systems are slightly more expensive, at roughly $5 in the US and $4 per Watt in Germany, without incentives (Barbose et al., 2011).

The gap is narrowing however. As of 2012, large PV plants in areas with high irradiance were estimated to generate electricity for as little as $0.10 per kWh. As a result, solar power has been found to be more economical than fossil fuels in an increasing number of real-world situations, such as for peak-load power or when compared to oil-burning power plants. It is also noteworthy that PV has become around a third cheaper than concentrating solar power, in terms of cost of per kWh of electricity, as PV module prices have fallen.

PV possesses other advantages that can justify a higher price: It produces no greenhouses gases, and relative to other renewables; it can be installed in a wider variety of places (for example on rooftops and parking shades); additional modules can be added incrementally; a utility-scale power plant can be constructed in months; and PV requires very little maintenance (Price and Margolis, 2010; Tidball et al., 2010). Also PV produces electricity during the afternoon when demand is highest, and it is often sited close to electricity users, which reduces transmission costs. These factors make electricity from PV more "valuable" than that from a traditional power station.

Costs can be reduced through several alternative paths. Systems based on thin films of materials, such as amorphous silicon alloys, cadmium telluride, or copper indium diselenide, are particularly promising because they are both well suited to mass production techniques and the amounts of active materials required are small.

Because of these merits, photovoltaics were traditionally applied in areas that are remote from utility grids, especially where the supply of power from conventional sources is impractical or costly, such as for telecommunication and weather stations. However, with falling costs and government incentives, PV has become an economic source of grid electricity in many places and the great majority of PV systems are now grid connected. These market conditions have also spurred large commercial PV systems on flat building rooftops and utility power plants. The largest plants are now in the order of hundreds of MW (RENI, 2012). For grid-connected distributed systems, the actual value of photovoltaic electricity can be high because this electricity is produced during periods of peak demand, thereby reducing the need for costly extra conventional capacity to cover the peak demand. Additionally, PV electricity is close to the sites where it is consumed, thereby reducing transmission and distribution losses and thus increasing system reliability.

Photovoltaic devices, or cells, are used to convert solar radiation directly into electricity. A review of possible materials that can be used for PV cells is given in Chapter 1, Section 1.5.1. Photovoltaic cells are made of various semiconductors, which are materials that are only moderately good conductors of electricity. The materials most commonly used are silicon (Si) and compounds of cadmium sulfide (CdS), cuprous sulfide (Cu_2S), and gallium arsenide (GaAs). These cells are packed into modules that produce a specific voltage and current when illuminated. Many review papers on cell and module technologies are published in literature following the comprehensive one published by Kazmerski (1997). PV modules can be connected in series or parallel to produce larger voltages or currents. PV systems rely on sunlight, have no moving parts, are modular to match power requirements on any scale, are reliable, and have a long life. The systems can be used independently or in conjunction with other electrical power sources. Applications powered by PV systems include communications (both on earth and in space), remote power, remote monitoring, lighting, water pumping, and battery charging. Some of these applications are analyzed in Section 9.4.

9.1 Semiconductors

To understand the photovoltaic effect, some basic theory about semiconductors and their use as photovoltaic energy conversion devices needs to be given as well as information on p–n junctions. These are explained in the following sections.

As is well known, an atom consists of the nucleus and electrons that orbit the nucleus. According to quantum mechanics, electrons of an isolated atom can have only specific discrete or quantized energy levels. In elements that have electrons in multiple orbitals, the innermost electrons have the minimum (maximum negative) energy and therefore require a large amount of energy to overcome the attraction of the nucleus and become free. When atoms are brought close together, the electronic energy of individual atoms is altered and the energy levels are grouped in energy bands. In some energy bands, electrons are allowed to exist, and in other bands electrons are forbidden. The electrons at the outermost shell are the only ones that interact with other atoms. This is the highest normally filled band, which corresponds to the ground state of the valence electrons in an atom and is called the *valence band*. The electrons in the valence band are loosely attached to the nucleus of the atom and, therefore, may attach more easily to a neighboring atom, giving that atom a negative charge and leaving the original atom as a positive charged ion. Some electrons in the valence band may possess a lot of energy, which enables them to jump into a higher band. These electrons are responsible for the conduction of electricity and heat, and this band is called the *conduction band*. The difference in the energy of an electron in the valence band and the innermost shell of the conduction band is called the *band gap*.

A schematic representation of the energy band diagrams of three types of materials is shown in Fig. 9.1. Materials whose valence gap is full and whose conduction band is empty have very high band gaps and are called *insulators* because no current can be carried by electrons in the filled band and the energy gap is so large that, under ordinary circumstances, a valence electron cannot accept energy, since the empty states in the conduction band are inaccessible to it. The band gap in these materials is greater than 3 eV.

Materials that have relatively empty valence bands and may have some electrons in the conduction band are called *conductors*. In this case, the valence and the conduction bands overlap. The valence electrons are able to accept energy from an external field and move to an unoccupied allowed state at slightly higher energy levels within the same band. Metals fall in this category, and the valence electrons in a metal can be easily emitted outside the atomic structure and become free to conduct electricity.

Materials with valence gaps partly filled have intermediate band gaps and are called *semiconductors*. The band gap in these materials is smaller than 3 eV. They have the same band structure as the insulators but their energy gap is much narrower. The two types of semiconductors are the pure ones, called *intrinsic semiconductors*, and those doped with small amounts of impurities, called *extrinsic semiconductors*. In intrinsic semiconductors, the valence electrons can easily be excited by thermal or optical means and jump the narrow energy gap into the conduction band, where the electrons have no atomic bonding and therefore are able to move freely through the crystal.

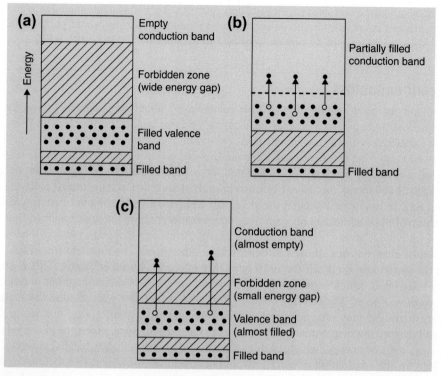

FIGURE 9.1

Schematic diagrams of energy bands for typical materials. (a) Insulator. (b) Conductor (metal). (c) Semiconductor.

9.1.1 p–n junction

Silicon (Si) belongs to group 4 of the periodic table of elements. In semiconductors, if the material that is doped has more electrons in the valence gap than the semiconductor, the doped material is called an *n-type semiconductor*. The n-type semiconductor is electronically neutral but has excess electrons, which are available for conduction. This is obtained when Si atoms are replaced with periodic table group 5 elements, such as arsenic (As) or antimony (Sb), and in so doing, form electrons that can move around the crystal. If these excess electrons are removed, the atoms will be left with positive charges.

In semiconductors, if the material that is doped has fewer electrons in the valence gap than the semiconductor, the doped material is called a *p-type semiconductor*. The p-type semiconductor is electronically neutral but it has positive holes (missing electrons) in its structure, which can accommodate excess electrons. This type of material is obtained when Si atoms are replaced with periodic table group 3 elements, such as gallium (Ga) or indium (In), and thereby form positive particles, called *holes*, that can move around the crystal through diffusion or drift. If additional electrons could fill the holes, the impurity atoms would fit more uniformly in the structure formed by the main semiconductor atoms, but the atoms would be negatively charged.

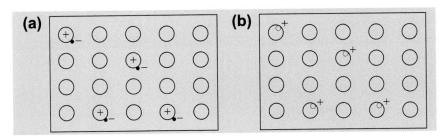

FIGURE 9.2

Schematic diagrams of n- and p-type semiconductors. (a) n-type, with excess electrons. (b) p-type, with excess positive holes.

Both types of semiconductors are shown schematically in Fig. 9.2. Both n- and p-type semiconductors allow the electrons and holes to move more easily in the semiconductors. For silicon, the energy needed to get an electron across a p–n junction is 1.11 eV. This is different for each semiconductor material.

What is described in the previous paragraph occurs when the p- and n-type semiconductors are joined together, i.e., form a junction, as shown in Fig. 9.3. As can be seen, when the two materials are joined, the excess electrons from the n-type jump to fill the holes in the p-type, and the holes from the p-type diffuse to the n-type side, leaving the n-side of the junction positively charged and the p-side negatively charged. The negative charges of the p-side restrict the movements of additional electrons from the n-side; however, the movement of additional electrons from the p-side is easier because of the positive charges at the junction on the n-side. Therefore the p–n junction behaves like a diode.

A schematic diagram of the energy bands of the n- and p-type semiconductors is shown in Fig. 9.4. In the n-type semiconductor, because the doped impurity donates additional electrons for the conduction of current, it is called the *donor* and its energy level is called the *donor level*. The n-type energy band diagram is shown in Fig. 9.4a, and as can be seen, the donor level is located within the forbidden band. In the p-type semiconductor, the doped impurity accepts additional electrons; therefore, it is called the *acceptor* and its energy level is called the *acceptor level*. Its energy band diagram is shown in Fig. 9.4b, and as can be seen, the acceptor level is located in the forbidden band.

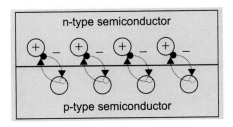

FIGURE 9.3

Schematic diagram of a p–n junction.

FIGURE 9.4

Energy band diagrams of n- and p-type semiconductors. (a) n-type semiconductor. (b) p-type semiconductor.

9.1.2 Photovoltaic effect

When a photon enters a photovoltaic material, it can be reflected, absorbed, or transmitted through. When this photon is absorbed by a valence electron of an atom, the energy of the electron is increased by the amount of energy of the photon. If, now, the energy of the photon is greater than the band gap of the semiconductor, the electron, which has excess energy, will jump into the conduction band, where it can move freely. Therefore, when the photon is absorbed, an electron is knocked loose from the atom. The electron can be removed by an electric field across the front and back of the photovoltaic material, and this is achieved with the help of a p–n junction. In the absence of a field, the electron recombines with the atom; whereas when there is a field, it flows through, thus creating a current. If the photon energy is smaller than that of the band gap, the electron will not have sufficient energy to jump into the conduction band, and the excess energy is converted into kinetic energy of the electrons, which leads to increased temperature. It should be noted that, irrespective of the intensity of the photon energy relative to the band gap energy, only one electron can be freed. This is the reason for the low efficiency of the photovoltaic cells.

The operation of a photovoltaic cell is shown in Fig. 9.5. These solar cells contain a junction of a p-type and an n-type semiconductor, i.e., a p–n junction. To some extent, electrons and holes diffuse across the boundary of this junction, setting up an electric field across it. The free electrons are generated in the n-layer by the action of the photons. When photons of sunlight strike the surface of a solar cell and are absorbed by the semiconductor, some of them create pairs of electrons and holes. If these pairs are sufficiently near the p–n junction, its electric field causes the charges to separate, electrons moving to the n-type side and holes to the p-type side. If the two sides of the solar cell are now connected through a load, an electric current will flow as long as sunlight strikes the cell.

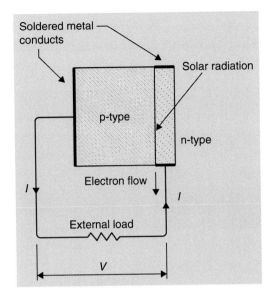

FIGURE 9.5

Photovoltaic effect.

The thickness of the n-type layer in a typical crystalline silicon cell is about 0.5 μm, whereas that of the p-type layer is about 0.25 mm. The speed of electromagnetic radiation is given by Eq. (2.31). The energy contained in a photon, E_p, is given by:

$$E_p = h\nu \tag{9.1}$$

where

h = Planck's constant, = 6.625×10^{-34} J s;
ν = frequency (s^{-1}).

Combining Eq. (2.31) with Eq. (9.1), we get:

$$E_p = \frac{hC}{\lambda} \tag{9.2}$$

Silicon has a band gab of 1.11 eV (1 eV = 1.6×10^{-19} J); therefore, by using Eq. (9.2), it can be found that photons with wavelength of 1.12 μm or less are useful in creating electron–hole pairs and thus electricity. By checking this wavelength on the distribution shown in Fig. 2.26, it can be seen that the majority of solar radiation can be used effectively in PVs. The number of photons, n_p, incident on a cell can be estimated from the intensity of light, I_p:

$$n_p = \frac{I_p}{E_p} \tag{9.3}$$

EXAMPLE 9.1

A beam of light with intensity of 3 mW and a wavelength of 743 nm is striking a solar cell. Estimate the number of photons incident on the cell.

Solution

Using Eq. (9.2) and speed of light equal to 300,000 [$=3 \times 10^8$ m/s],

$$E_p = \frac{hC}{\lambda} = \frac{6.625 \times 10^{-34} \times 3 \times 10^8}{743 \times 10^{-9}} = 2.675 \times 10^{-19} \text{J}$$

Using Eq. (9.3) for the intensity of 3×10^{-3} W or 3×10^{-3} J/s,

$$n_p = \frac{I_p}{E_p} = \frac{3 \times 10^{-3}}{2.675 \times 10^{-19}} = 1.12 \times 10^{16} \text{ photons/s}$$

A photovoltaic cell consists of the active photovoltaic material, metal grids, antireflection coatings, and supporting material. The complete cell is optimized to maximize both the amount of sunlight entering the cell and the power out of the cell. The photovoltaic material can be one of a number of compounds. The metal grids enhance the current collection from the front and back of the solar cell. The antireflection coating is applied to the top of the cell to maximize the light going into the cell. Typically, this coating is a single layer optimized for sunlight. As a result, photovoltaic cells range in color from black to blue. In some types of photovoltaic cells, the top of the cell is covered by a semi-transparent conductor that functions as both the current collector and the antireflection coating. A complete photovoltaic cell is a two-terminal device with positive and negative leads.

Silicon is an abundant chemical element covering 25% of the earth's crust. Silicon minerals are cheap, but silicon cells still must be individually fabricated by a long, complicated process that includes purifying the silicon, pulling a long crystal from a high-temperature melt, slicing the crystal into wafers, diffusing impurities into the wafers, and applying various coatings and electrical conducts. Labor now accounts for almost all the cost of a silicon cell. It is expected that fabrication techniques plus automation of the manufacturing process will radically lower further the price within the next few years.

9.1.3 PV cell characteristics

A photovoltaic PV generator is mainly an assembly of solar cells, connections, protective parts, and supports. As was seen already, solar cells are made of semiconductor materials, usually silicon, and are specially treated to form an electric field with positive on one side (backside) and negative on the other side (front side facing the sun). When solar energy (photons) hits the solar cell, electrons are knocked loose from the atoms in the semiconductor material, creating electron–hole pairs. If electrical conductors are attached to the positive and negative sides, forming an electrical circuit, the electrons are captured in the form of electric current, called *photocurrent*, I_{ph}. As can be understood from this description, during darkness the solar cell is not active and works as a diode, i.e., a p–n junction that does not produce any current or voltage. If, however, it is connected to an external, large voltage supply, it generates a current, called the *diode* or *dark current*, I_D. A solar cell is usually represented by an electrical equivalent one-diode model, shown in Fig. 9.6 (Lorenzo, 1994). This circuit can be used for an individual cell, a module consisting of a number of cells, or an array consisting of several modules.

FIGURE 9.6

Single solar cell model.

As shown in Fig. 9.6, the model contains a current source, I_{ph}, one diode, and a series resistance R_S, which represents the resistance inside each cell. The diode has also an internal shunt resistance, as shown in Fig. 9.6. The net current is the difference between the photocurrent, I_{ph}, and the normal diode current, I_D, given by:

$$I = I_{ph} - I_D = I_{ph} - I_0 \left\{ \exp\left[\frac{e(V + IR_S)}{kT_C} \right] - 1 \right\} - \frac{V + IR_S}{R_{SH}} \qquad (9.4a)$$

It should be noted that the shunt resistance is usually much bigger than a load resistance, whereas the series resistance is much smaller than a load resistance, so that less power is dissipated internally within the cell. Therefore, by ignoring these two resistances, the net current is the difference between the photocurrent, I_{ph}, and the normal diode current, I_D, given by:

$$I = I_{ph} - I_D = I_{ph} - I_0 \left[\exp\left(\frac{eV}{kT_C} \right) - 1 \right] \qquad (9.4b)$$

where

k = Boltzmann's gas constant, = 1.381×10^{-23} J/K;
T_C = absolute temperature of the cell (K);
e = electronic charge = 1.602×10^{-19} J/V;
V = voltage imposed across the cell (V); and
I_0 = dark saturation current, which depends strongly on temperature (A).

Figure 9.7 shows the $I-V$ characteristic curve of a solar cell for a certain irradiance (G_t) at a fixed cell temperature, T_C. The current from a PV cell depends on the external voltage applied and the amount of sunlight on the cell. When the cell is short-circuited, the current is at maximum (short-circuit current, I_{sc}), and the voltage across the cell is 0. When the PV cell circuit is open, with the leads not making a circuit, the voltage is at its maximum (open-circuit voltage, V_{oc}), and the current is 0. In either case, at open circuit or short circuit, the power (current times voltage) is 0. Between an open circuit and a short circuit, the power output is greater than 0. The typical current voltage curve shown in Fig. 9.7 presents the range of combinations of current and voltage. In this representation, a sign convention is used, which takes as positive the current generated by the cell when the sun is shining and a positive voltage is applied on the cell's terminals.

If the cell's terminals are connected to a variable resistance, R, the operating point is determined by the intersection of the $I-V$ characteristic of the solar cell with the load $I-V$ characteristics. As shown

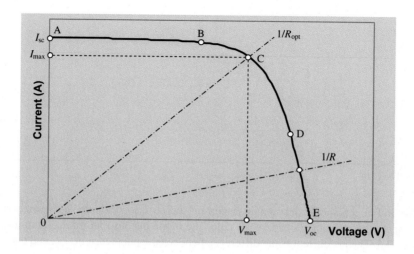

FIGURE 9.7

Representative current–voltage curve for photovoltaic cells.

in Fig. 9.7 for a resistive load, the load characteristic is a straight line with a slope $1/V = 1/R$. If the load resistance is small, the cell operates in the region AB of the curve, where the cell behaves as a constant current source, almost equal to the short-circuit current. On the other hand, if the load resistance is large, the cell operates on the region DE of the curve, where the cell behaves more as a constant voltage source, almost equal to the open-circuit voltage. The power can be calculated by the product of the current and voltage. If this exercise is performed and the results are plotted on a $P-V$ graph, then Fig. 9.8 can be obtained. The maximum power passes from a maximum power point (point

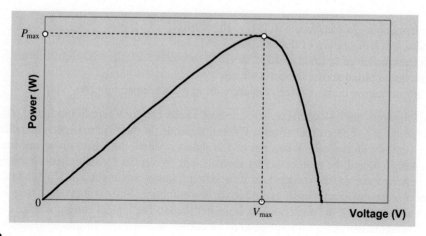

FIGURE 9.8

Representative power–voltage curve for photovoltaic cells.

C on Fig. 9.7), at which point the load resistance is optimum, R_{opt}, and the power dissipated in the resistive load is maximum and given by:

$$P_{max} = I_{max} V_{max} \tag{9.5}$$

Point *C* on Fig. 9.7 is also called the *maximum power point*, which is the operating point P_{max}, I_{max}, V_{max} at which the output power is maximized. Given P_{max}, an additional parameter, called the *fill factor*, FF, can be calculated such that:

$$P_{max} = I_{sc} V_{oc} \text{FF} \tag{9.6}$$

or

$$\text{FF} = \frac{P_{max}}{I_{sc} V_{oc}} = \frac{I_{max} V_{max}}{I_{sc} V_{oc}} \tag{9.7}$$

The fill factor is a measure of the real *I*−*V* characteristic. For good cells, its value is greater than 0.7. The fill factor decreases as the cell temperature increases.

Thus, by illuminating and loading a PV cell so that the voltage equals the PV cell's V_{max}, the output power is maximized. The cell can be loaded using resistive loads, electronic loads, or batteries. Typical parameters of a single-crystal solar cell are current density $I_{sc} = 32$ mA/cm^2, $V_{oc} = 0.58$ V, $V_{max} = 0.47$ V, FF = 0.72, and $P_{max} = 2273$ mW (ASHRAE, 2004).

Other fundamental parameters that can be obtained from Fig. 9.7 are the short-circuit current and the open-circuit voltage. The short-circuit current, I_{sc}, is the higher value of the current generated by the cell and is obtained under short-circuit conditions, i.e., $V = 0$, and is equal to I_{ph}. The open-circuit voltage corresponds to the voltage drop across the diode when it is traversed by the photocurrent, I_{ph}, which is equal to I_D, when the generated current is $I = 0$. This is the voltage of the cell during nighttime and can be obtained from Eq. (9.4b):

$$\exp\left(\frac{eV_{oc}}{kT_C}\right) - 1 = \frac{I_{sc}}{I_o} \tag{9.8}$$

which can be solved for V_{oc}:

$$V_{oc} = \frac{kT_C}{e} \ln\left(\frac{I_{sc}}{I_o} + 1\right) = V_t \ln\left(\frac{I_{sc}}{I_o} + 1\right) \tag{9.9}$$

where V_t = thermal voltage (V) given by:

$$V_t = \frac{kT_C}{e} \tag{9.10}$$

The output power, *P*, from a photovoltaic cell is given by:

$$P = IV \tag{9.11}$$

The output power depends also on the load resistance, *R*; and by considering that $V = IR$, it gives:

$$P = I^2 R \tag{9.12}$$

Substituting Eq. (9.4b) into Eq. (9.11) gives:

$$P = \left\{ I_{sc} - I_0 \left[\exp\left(\frac{eV}{kT_C}\right) - 1 \right] \right\} V \tag{9.13}$$

Equation (9.13) can be differentiated with respect to V. By setting the derivative equal to 0, the external voltage, V_{max}, that gives the maximum cell output power can be obtained:

$$\exp\left(\frac{eV_{max}}{kT_C}\right)\left(1 + \frac{eV_{max}}{kT_C}\right) = 1 + \frac{I_{sc}}{I_0} \tag{9.14}$$

This is an explicit equation of the voltage V_{max}, which maximizes the power in terms of the short-circuit current ($I_{sc} = I_{ph}$), the dark saturation current (I_0), and the absolute cell temperature, T_C. If the values of these three parameters are known, then V_{max} can be obtained from Eq. (9.14) by trial and error.

The load current, I_{max}, which maximizes the output power, can be found by substituting Eq. (9.14) into Eq. (9.4b):

$$I_{max} = I_{sc} - I_0 \left[\exp\left(\frac{eV_{max}}{kT_C}\right) - 1 \right] = I_{sc} - I_0 \left[\frac{1 + \dfrac{I_{sc}}{I_0}}{1 + \dfrac{eV_{max}}{kT_C}} - 1 \right] \tag{9.15}$$

which gives:

$$I_{max} = \frac{eV_{max}}{kT_C + eV_{max}}(I_{sc} + I_0) \tag{9.16}$$

By using Eq. (9.5),

$$P_{max} = \frac{eV_{max}^2}{kT_C + eV_{max}}(I_{sc} + I_0) \tag{9.17}$$

Efficiency is another measure of PV cells that is sometimes reported. *Efficiency* is defined as the maximum electrical power output divided by the incident light power. Efficiency is commonly reported for a PV cell temperature of 25 °C and incident light at an irradiance of 1000 W/m² with a spectrum close to that of sunlight at solar noon. An improvement in cell efficiency is directly connected to a cost reduction in photovoltaic systems. A series of R&D efforts have been made on each step of the photovoltaic process. Through this technological progress, the efficiency of a single crystalline silicon solar cell reaches 15%–20% and the polycrystalline silicon solar cells shows 13%–16% efficiency in the mass production lines (www.ases.org).

Another parameter of interest is the maximum efficiency, which is the ratio between the maximum power and the incident light power, given by:

$$\eta_{max} = \frac{P_{max}}{P_{in}} = \frac{I_{max} V_{max}}{A G_t} \tag{9.18}$$

where

A = cell area (m²).

EXAMPLE 9.2

If the dark saturation current of a solar cell is 1.7×10^{-8} A/m^2, the cell temperature is 27 °C, and the short-circuit current density is 250 A/m^2, calculate the open-circuit voltage, V_{oc}; voltage at maximum power, V_{max}; current density at maximum power, I_{max}; maximum power, P_{max}; and maximum efficiency, η_{max}. What cell area is required to get an output of 20 W when the available solar radiation is 820 W/m^2?

Solution

First the value of e/kT_C is evaluated, which is used in many relations:

$$\frac{e}{kT_C} = \frac{1.602 \times 10^{-19}}{1.381 \times 10^{-23} \times 300} = 38.67 \text{ V}^{-1}$$

Using Eq. (9.9),

$$V_{oc} = \frac{kT_C}{e} \ln\left(\frac{I_{sc}}{I_o} + 1\right) = \frac{1}{38.67} \ln\left(\frac{250}{1.7 \times 10^{-8}} + 1\right) = 0.605 \text{ V}$$

Voltage at maximum power can be found from Eq. (9.14) by trial and error:

$$\exp\left(\frac{eV_{max}}{kT_C}\right)\left(1 + \frac{eV_{max}}{kT_C}\right) = 1 + \frac{I_{sc}}{I_o}$$

or

$$\exp(38.67V_{max})(1 + 38.67V_{max}) = 1 + \frac{250}{1.7 \times 10^{-8}}$$

which gives $V_{max} = 0.526$ V.

Alternatively, the Excel function Goal Seek (see Appendix 9) can be used to find the value of V_{max} that makes the difference of the left- and right-hand sides of this equation zero.

The current density at maximum power point can be estimated from Eq. (9.16):

$$I_{max} = \frac{eV_{max}}{kT_C + eV_{max}}(I_{sc} + I_o)$$

$$= \frac{1.602 \times 10^{-19} \times 0.526}{1.381 \times 10^{-23} \times 300 + 1.602 \times 10^{-19} \times 0.526}(250 + 1.7 \times 10^{-8}) = 238.3 \text{ A/m}^2$$

Maximum power, P_{max}, is obtained from Eq. (9.5):

$$P_{max} = I_{max}V_{max} = 238.3 \times 0.526 = 125.3 \text{ W/m}^2$$

Maximum efficiency, η_{max}, is obtained from Eq. (9.18):

$$\eta_{max} = \frac{P_{max}}{P_{in}} = \frac{125.3}{820} = 15.3\%$$

Finally, the cell area required to get an output of 20 W is:

$$A = \frac{P_{req}}{P_{max}} = \frac{20}{125.3} = 0.16 \text{ m}^2$$

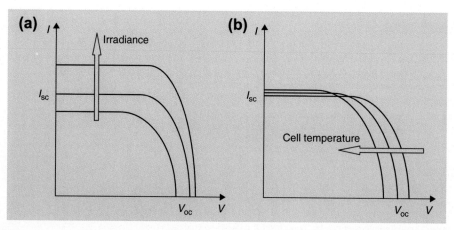

FIGURE 9.9

Influence of irradiation and cell temperature on PV cell characteristics. (a) Effect of increased irradiation. (b) Effect of increased cell temperature.

The $I-V$ characteristic of the solar cell, presented in Fig. 9.7, is only for a certain irradiance, G_t, and cell temperature, T_C. The influences of these two parameters on the cell characteristics are shown in Fig. 9.9. As shown in Fig. 9.9a, the open-circuit voltage increases logarithmically by increasing the solar radiation, whereas the short-circuit current increases linearly. The influence of the cell temperature on the cell characteristics is shown in Fig. 9.9b. The main effect of the increase in cell temperature is on open-circuit voltage, which decreases linearly with the cell temperature; thus the cell efficiency drops. As can be seen, the short-circuit current increases slightly with the increase of the cell temperature.

In practice solar cells can be connected in series or parallel. Figure 9.10 shows how the $I-V$ curve is modified in the case where two identical cells are connected in parallel and in series. As can be seen,

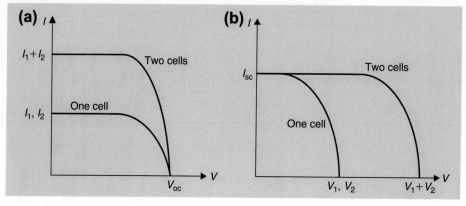

FIGURE 9.10

Parallel and series connection of two identical solar cells. (a) Parallel connection. (b) Series connection.

when two identical cells are connected in parallel, the voltage remains the same but the current is doubled; when the cells are connected in series, the current remains the same but the voltage is doubled.

9.2 Photovoltaic panels

PV modules are designed for outdoor use in such harsh conditions as marine, tropic, arctic, and desert environments. The choice of the photovoltaically active material can have important effects on system design and performance. Both the composition of the material and its atomic structure are influential. Photovoltaic materials include silicon, gallium arsenide, copper indium diselenide, cadmium telluride, indium phosphide, and many others. The atomic structure of a PV cell can be a single crystal, polycrystalline, or amorphous. The most commonly produced PV material is crystalline silicon, either single crystal or polycrystalline.

Cells are normally grouped into modules, which are encapsulated with various materials in order to protect the cells and the electrical connectors from the environment (Hansen et al., 2000). As shown in Fig. 9.11, PV cell modules consist of N_{PM} parallel branches and each branch has N_{SM} solar cells in series. In the following analysis, superscript M refers to the PV module and superscript C refers to the solar cell. Therefore, as shown in Fig. 9.11, the applied voltage at the module's terminals is denoted by V^M, whereas the total generated current is denoted by I^M.

A model of the PV module can be obtained by replacing each cell in Fig. 9.11 with the equivalent diagram from Fig. 9.6. The model, developed by Lorenzo (1994), has the advantage that it can be used by applying only standard manufacturer-supplied data for the modules and the cells. The PV module current I^M under arbitrary operating conditions can be described by:

$$I^M = I_{sc}^M \left[1 - \exp\left(\frac{V^M - V_{oc}^M + R_S^M I^M}{N_{SM} V_t^C} \right) \right] \tag{9.19}$$

FIGURE 9.11

Schematic diagram of a PV module consisting of N_{PM} parallel branches, each with N_{SM} cells in series.

Table 9.1 SRC and NOCT conditions.	
SRC conditions	**NOCT conditions**
Irradiation: $G_{t,o} = 1000$ W/m^2 Cell temperature: $T_o^C = 25$ °C	Irradiation: $G_{t,NOCT} = 800$ W/m^2 Ambient temperature: $T_{a,NOCT} = 20$ °C Wind speed: $W_{NOCT} = 1$ m/s

It should be noted that the PV module current, I^M, is an implicit function, which depends on:

1. The short-circuit current of the module, given by:

$$I_{sc}^M = N_{PM} I_{sc}^C$$

2. The open-circuit voltage of the module, given by:

$$V_{oc}^M = N_{SM} V_{oc}^C$$

3. The equivalent series resistance of the module, given by:

$$R_S^M = \frac{N_{SM}}{N_{PM}} R_S^C$$

4. The thermal voltage in the semiconductor of a single solar cell, given by:

$$V_t^C = \frac{kT^C}{e}$$

The current practice dictates that the performance of a PV module is determined by exposing it at known standard rating conditions (SRCs) of irradiance, $G_{t,o} = 1000$ W/m^2, and cell temperature, $T_o^C = 25$ °C. These conditions are different from the nominal operating cell temperature (NOCT), as indicated in Table 9.1.

9.2.1 PV arrays

The modules in a PV system are usually connected in arrays. An array with M_P parallel branches each with M_S modules in series is shown in Fig. 9.12. By using a superscript A to denote array characteristics, the applied voltage at the array's terminals is donated V^A, whereas the total current of the array is denoted I^A, given by:

$$I^A = \sum_{i=1}^{M_P} I_i \qquad (9.20)$$

If it is assumed that the modules are identical and the ambient irradiance is the same in all modules, then the array's current is given by:

$$I^A = M_p I^M \qquad (9.21)$$

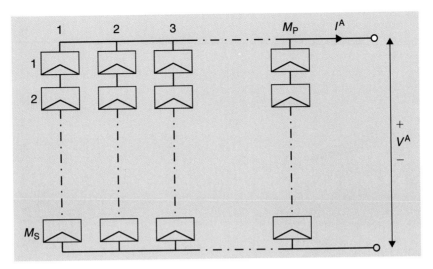

FIGURE 9.12

Cell array consisting of M_P parallel branches, each with M_S modules in series.

EXAMPLE 9.3

A PV system is required to produce 250 W at 24 V. Using the solar cells of Example 9.2, design the PV panel, working at the maximum power point, if each cell is 9 cm^2 in area.

Solution

From Example 9.2, $V_{max} = 0.526$ V. The current density at maximum power point is 238.3 A/m^2. Therefore, for the current cell,

$$I_{max} = 238.3 \times 9 \times 10^{-4} = 0.2145 \text{ A}$$

This yields a power per cell $= 0.526 \times 0.2145 = 0.1$ W.
Number of cells required $= 250/0.1 = 2500$.
Number of cells in series $=$ system voltage/voltage per cell $= 24/0.526 = 45.6 \approx 46$ (in fact with 46 cells, Voltage $= 24.2$ V).
Number of rows of 46 cells each, connected in parallel $= 2500/46 = 54.3 \approx 55$ (in fact this panel yields $55 \times 46 \times 0.1 = 253$ W).

PV cells are fragile and susceptible to corrosion by humidity or fingerprints and can have delicate wire leads. Also, the operating voltage of a single PV cell is about 0.5 V, making it unusable for many applications. A *module* is a collection of PV cells that provides a useable operating voltage and offers means that protects the cells. Depending on the manufacturer and the type of PV material, modules have different appearances and performance characteristics. Also, modules may be designed for specific conditions, such as hot and humid, desert, or frozen climates. Usually, the cells are series connected to other cells to produce an operating voltage around 30–60 V. These strings of cells are then encapsulated with a polymer, a front glass cover, and a back material. Also, a junction box is attached at the back of the module for convenient wiring to other modules or other electrical equipment.

9.2.2 Types of PV technology

Many types of PV cells are available today. The main types are crystalline silicon, which accounts for roughly 80% of the PV market; thin-films, which have expanded to around 20% market share; and triple-junction cells, which are used in the emerging field of concentrating PV. This section gives details of commercialized flat-plate PV technologies and an overview of the cells that are currently in the research and development stage. Concentrating PV is discussed in Section 9.7.

The choice between crystalline or thin-film PV modules for a given project depends heavily on climate and space. As we shall see below, crystalline modules are more efficient (i.e., give greater power output per unit area of module), while thin-film modules tend to have higher yield (i.e., give greater energy production for a given power rating)—especially in high temperatures (RENI, 2012). Assuming the same module price ($/W), crystalline modules are thus suited to space-constrained projects in moderate climates, and thin-film to hot climates and abundant space.

- *Monocrystalline silicon cells.* These cells are made from pure monocrystalline silicon. In these cells, the silicon has a single continuous crystal lattice structure with almost no defects or impurities. The main advantage of monocrystalline cells is their high efficiency, which is typically around 14%–15%. Premium modules are available in the market with efficiencies just over 20% (RENI, 2012). A disadvantage of these cells is the complicated manufacturing process that leads to relatively high costs, although greater production capacity of its raw material has decreased its price markedly in recent years, making it (and multicrystalline silicon) more price-competitive with thin-film modules. Compared to thin-film technologies, the power output of crystalline silicon decreases more rapidly with increasing cell temperature —its temperature coefficient is around minus 0.4%–0.5%/°C. Additionally, the efficiency of crystalline silicon decreases somewhat in low light conditions, whereas the efficiency of thin-film modules remains roughly constant (Marion, 2008). Being relatively expensive, monocrystalline modules are most often used where the most possible power is required from a confined space, such as residential and commercial rooftops.

- *Multicrystalline silicon cells.* Multicrystalline cells, also known as polycrystalline cells, are produced using numerous grains of monocrystalline silicon. In the manufacturing process, molten polycrystalline silicon is cast into ingots, which are subsequently cut into very thin wafers and assembled into complete cells. Multicrystalline cells are cheaper to produce than monocrystalline ones because of the simpler manufacturing process required. They are, however, slightly less efficient, with typical module efficiencies around 13%–15% (Price and Margolis, 2010) and high-end products up to 17% (RENI, 2010). They share monocrystalline cells' relatively high temperature coefficient. With a long track record, high efficiency and moderate cost, multicrystalline modules are widely used in a wide variety of applications including roof- and ground-mounted arrays.

- *Amorphous silicon.* The general characteristics of amorphous silicon solar cells are given in Chapter 1, Section 1.5.1. Unlike mono- and multicrystalline cells, the silicon atoms in amorphous silicon (a-Si) cells are arranged in a thin homogenous layer. Amorphous silicon absorbs light more effectively than crystalline silicon, which leads to thinner cells, giving rise to the name *thin film* photovoltaics. Advantages of these cells are their low manufacturing cost and high energy production per rated power capacity (kWh/kWp). The high yield stems from two attributes (Jardine et al., 2001; Kullmann, 2009):

1. Of all mass-market PV technologies, a-Si is the least impacted by heat, with a temperature coefficient of only around $-0.2\%/^{\circ}C$ (Marion, 2008).
2. Amorphous silicon is relatively effective at absorbing the blue wavelengths of light that are encountered in cloudy conditions. Thus over the course of a year a-Si modules will generally produce more electricity than crystalline silicon ones of the same peak power, especially in warm climates.

The efficiency of a-Si modules is only 6%–7% (Price and Margolis, 2010). Because of their low cost they are applied in a wide variety of PV systems; however, they face increasing competition from other thin-film technologies with higher efficiencies (see the following two types below). To increase their efficiency, a-Si is increasingly combined with layers of multicrystalline silicon or a variant, microcrystalline silicon. The properties of such "hybrid" modules lie between pure a-Si and crystalline silicon, with efficiencies around 9%–10%. Some large PV manufacturers have phased out pure a-Si product lines and replaced them with hybrid designs. Perhaps the greatest advantage of these cells is that amorphous silicon can be deposited on a wide range of substrates, both rigid and flexible. Nowadays, the panels made from amorphous silicon solar cells come in a variety of shapes, such as roof tiles, which can replace normal ceramic tiles in a solar roof.

- *Cadmium Telluride (CdTe).* The thin-film PV market was largely developed by a single manufacturer, First Solar. It accounted for 59% of the global thin-film market in 2008 (Schreiber, 2009), as a result of achieving low cost (it was the first company to manufacture PV modules at less than $1 per Watt) and large production capacity (Runyon, 2012). Although the price advantage of CdTe has fallen with the emergence of low-cost crystalline silicon production in China, it retains a sizable market presence with a manufacturing cost under $0.75/W (Runyon, 2012) and around 2 GW of annual production capacity. Like amorphous silicon, CdTe is relatively tolerant to heat (its temperature coefficient is around minus 0.25%–0.35%/$^{\circ}$C), yet it has a higher efficiency of 10%–11%. The record for laboratory CdTe solar cell efficiency is 22.1% by First Solar (https://www.energy.gov/eere/solar/cadmium-telluride). This combination of low cost, moderate efficiency, and large manufacturing volume has seen CdTe help create a new market for PV in recent years, that of utility-scale solar power plants. While CdTe modules are well suited to large ground-mounted arrays for commercial electricity production, they are also deployed on commercial rooftops.
- *Copper Indium Gallium Selenide (CIGS).* The most recent thin-film technology to be commercialized is the CIGS family. Like the other thin-films it can be manufactured at low cost and high volume in a continuous (as opposed to batch) manufacturing process. CIGS is also the main candidate for the commercial development of modules that do not use glass, making them flexible and very lightweight. Their electrical properties lie between crystalline silicon and CdTe, with conversion efficiency around 10%–13% and moderate temperature coefficient of minus 0.3%–0.4%/$^{\circ}$C. The moderate efficiency, low cost and light weight of CIGS make them ideally suited to rooftop installations, both residential and commercial. They are also a promising technology for building-integrated photovoltaic (BIPV) products. Numerous companies have been established to manufacture CIGS modules in recent years (Schreiber, 2009), although being new to the market they have not yet been deployed at the large scale of the other above-mentioned technologies (RENI, 2010).

- *Thermophotovoltaics.* These are photovoltaic devices that, instead of sunlight, use the infrared region of radiation, i.e., thermal radiation. A complete thermophotovoltaic (TPV) system includes a fuel, a burner, a radiator, a longwave photon recovery mechanism, a PV cell, and a waste heat recuperation system (Kazmerski, 1997). TPV devices convert radiation using exactly the same principles as photovoltaic devices, outlined in previous sections. The key differences between PV and TPV conversion are the temperatures of the radiators and the system geometries. In a solar cell, the radiation is received from the sun, which is at a temperature of about 6000 K and a distance of about 150×10^6 km. A TPV device, however, receives radiation, in either the broad or narrow band, from a surface at a much lower temperature of about 1300–1800 K and a distance of only a few centimeters. Although the blackbody power radiated by a surface varies at the fourth power of the absolute temperature, the inverse square law dependence of the power received by the detectors dominates. Therefore, although the power received by a non-concentrator solar cell is on the order of 0.1 W/cm^2, that received by a TPV converter is likely to be 5–30 W/cm^2, depending on the radiator temperature. Consequently, the power density output from a TPV converter is expected to be significantly greater than that from a non-concentrator PV converter. More details on TPVs can be found in the article by Coutts (1999).

Under development are polymer and organic solar cells. The attraction of these technologies is that they potentially offer fast production at low cost in comparison to crystalline silicon technologies, yet they typically have lower efficiencies, around 11% (https://www.energy.gov/eere/solar/organic-photovoltaics-research), and despite the demonstration of operational lifetimes and dark stabilities under inert conditions for thousands of hours, they suffer from stability and degradation problems. Organic materials are attractive, primarily due to the prospect of high-output manufacture using reel-to-reel or spray deposition. Other attractive features are the possibilities for ultrathin, flexible devices, which may be integrated into appliances or building materials, and tuning of color through the chemical structure (Nelson, 2002).

- *Perovskite solar cells.* Perovskite solar cells belong to the third-generation PV; the first generation is the crystalline silicon cells, and the second generation thin films. Perovskites are a family of materials with a specific crystal structure, named after the mineral with that structure. When used to create solar cells, they have shown potential for high performance and low production costs. Perovskite materials, such as methylammonium lead halides and all-inorganic cesium lead halide, are cheap to produce and simple to manufacture. Perovskite solar cell efficiencies of laboratory-scale devices increased from 3.8% in 2009 to 25.5% in 2020 in single-junction architectures and in silicon-based tandem cells to more than 29%. Perovskite solar cells are commercially attractive because they even have the potential of achieving even higher efficiencies and very low production costs.

The main advantage of these cells is that metal halide perovskites possess unique features such as low raw materials costs and low-cost fabrication—usually through various printing techniques (Razza et al., 2016). Moreover, their high absorption coefficient enables ultrathin films of around 500 nm to absorb the complete visible solar spectrum (Yin et al., 2014). These features enable the manufacture of low-cost, high-efficiency, thin, lightweight, and flexible solar modules. Their main problem they have is their short- and long-term stability. For example, the highest-performing perovskites solar cells are made from organic—inorganic hybrid perovskites with a power conversion efficiency of about 25%. The organic components used however, are inherently volatile and prone to destructive reactions with air, water, heat, and solar illumination. They thus require encapsulation, which is expensive and high energy demand process. One way to solve this problem is to use fully inorganic perovskites employing only chemically stable inorganic compounds. Other concerns include toxicity issues associated with the lead (Pb) content in perovskite solar cell strains. Therefore, perovskite recycling is necessary. The use of lead implies that it would be impossible to dispose of perovskite solar cells in landfills, as this would be a major health hazard.

9.3 Related equipment

Photovoltaic modules can be mounted on the ground or a building roof or can be included as part of the building structure, usually the façade. Wind and snow loading are major design considerations. The PV modules can last more than 25 years, in which case the support structures and building should be designed for at least as long as the same lifetime. Related equipment includes batteries, charge controllers, inverters, and peak-power trackers.

9.3.1 Batteries

Batteries are required in many PV systems to supply power at night or when the PV system cannot meet the demand. The selection of battery type and size depends mainly on the load and availability requirements. When batteries are used, they must be located in an area without extreme temperatures, and the space where the batteries are located must be adequately ventilated.

The main types of batteries available today include lead-acid, nickel cadmium, nickel hydride, and lithium. Deep-cycle lead-acid batteries are the most commonly used. These can be flooded or valve-regulated batteries and are commercially available in a variety of sizes. Flooded (or wet) batteries require greater maintenance but, with proper care, can last longer, whereas valve-regulated batteries require less maintenance.

The principal requirement of batteries for a PV system is that they must be able to accept repeated deep charging and discharging without damage. Although PV batteries have an appearance similar to car batteries, the latter are not designed for repeated deep discharges and should not be used. For more capacity, batteries can be arranged in parallel.

Batteries are used mainly in stand-alone PV systems to store the electrical energy produced during the hours when the PV system covers the load completely and there is excess or when there is sunshine but no load is required. During the night or during periods of low solar irradiation, the battery can supply the energy to the load. Additionally, batteries are required in such a system because of the fluctuating nature of the PV system output.

Batteries are classified by their nominal capacity (q_{max}), which is the number of ampere hours (Ah) that can be maximally extracted from the battery under predetermined discharge conditions. The efficiency of a battery is the ratio of the charge extracted (Ah) during discharge divided by the amount of charge (Ah) needed to restore the initial state of charge (SOC). Therefore, the efficiency depends on the SOC and the charging and discharging current. The SOC is the ratio between the present capacity of the battery and the nominal capacity; that is,

$$SOC = \frac{q}{q_{max}} \tag{9.22}$$

FIGURE 9.13

Schematic diagram of a battery.

As can be understood from the preceding definition and Eq. (9.22), SOC can take values between 0 and 1. If SOC $= 1$, then the battery is fully charged; and if SOC $= 0$, then the battery is totally discharged.

Other parameters related to batteries are the charge or discharge regime and the lifetime of the battery. The charge (or discharge) regime, expressed in hours, is the parameter that reflects the relationship between the nominal capacity of a battery and the current at which it is charged (or discharged)—for example, a discharge regime is 40 h for a battery with nominal capacity of 200 Ah that is discharged at 5 A. The lifetime of the battery is the number of charge–discharge cycles the battery can sustain before losing 20% of its nominal capacity.

In general, the battery can be viewed as a voltage source, E, in series with an internal resistance, R_o, as shown in Fig. 9.13. In this case, the terminal voltage, V, is given by:

$$V = E - IR_o \tag{9.23}$$

9.3.2 Inverters

An inverter is used to convert the direct current into alternating current electricity. The output of the inverter can be single or three phase. Inverters are rated by the total power capacity, which ranges from hundreds of watts to megawatts. Some inverters have good surge capacity for starting motors, and others have limited surge capacity. The designer should specify both the type and size of the load the inverter is intended to service.

The inverter is characterized by a power-dependent efficiency, η_{inv}. Besides changing the DC into AC, the main function of the inverter is to keep a constant voltage on the AC side and convert the input power, P_{in}, into the output power, P_{out}, with the highest possible efficiency, given by:

$$\eta_{inv} = \frac{P_{out}}{P_{in}} = \frac{V_{ac}I_{ac}\cos(\varphi)}{V_{dc}I_{dc}} \tag{9.24}$$

where

$\cos(\varphi) =$ power factor;
$I_{dc} =$ current required by the inverter from the DC side, i.e., controller (A); and
$V_{dc} =$ input voltage for the inverter from the DC side, i.e., controller (V).

Numerous types of inverters are available, but not all are suitable for use when feeding power back into the mains supply.

The efficiency of an inverter depends on the fraction of its rated power at which it operates. A PV system operates at high efficiency either when it has one inverter operating with a load large enough to maintain peak efficiency or is an interconnection of module-integrated inverters or master-slave configurations (Woyte et al., 2000). When one inverter is used, this is supplied with power from several series-connected PV modules connected in parallel on a DC bus. This configuration has a low-cost and provides high efficiency but requires a complex DC installation. In a module-integrated inverter, each PV module has its own individual inverter, called *micro-inverter* (see Section 9.3.4). These are more expensive than a central inverter; however, they prevent the use of expensive DC wiring (Woyte et al., 2000). A master-slave configuration requires multiple inverters connected together and generally can give greater PV output. At low solar irradiation, all the PVs of the array are connected to just a single inverter operating the inverter at its peak input power level, when solar irradiation increases the PV array is divided progressively into smaller sets of PVs, until every inverter operates independently at or near its peak rated capacity. Module-integrated inverters are located generally at the back of each module converting its DC output to AC power.

Inverter efficiency reaches its maximum, above 90%, for an input power level usually between 30% and 50% of its rated capacity. When a PV module is shaded, the PV output current decreases significantly, causing not only the particular module output power to drop, but the output power also drops which in turn affects inverter performance (Hashimoto et al., 2000).

The performance of an inverter depends on its point of work, threshold of operation, inverter output waveform, harmonic distortion and frequency, PV efficiency, maximum power point tracker (MPPT) and transformer (Norton et al., 2011). The main functions of an inverter are wave-shaping, regulation of output voltage, and operation near peak power point (Kjar et al., 2005). The three major types of inverter are sine wave, modified sine wave, and square wave. The major advantage of a sine wave inverter is that most appliances are designed for a sine-wave operation. A modified sine-wave inverter has a waveform more like a square wave, but with an extra step, can also operate with most appliances. Finally, a square-wave inverter can generally operate only simple devices with universal motors but its greatest advantage is that it is much cheaper than the sine-wave inverter. Additionally, using a power filter, the output square waveform can be converted to a sine waveform.

9.3.3 Charge controllers

Controllers regulate the power from PV modules to prevent the batteries from overcharging. The controller can be a shunt type or series type and also function as a low-battery voltage disconnect to prevent the battery from over-discharge. The controller is chosen for the correct capacity and desired features (ASHRAE, 2004).

Normally, controllers allow the battery voltage to determine the operating voltage of a PV system. However, the battery voltage may not be the optimum PV operating voltage. Some controllers can optimize the operating voltage of the PV modules independently of the battery voltage so that the PV operates at its maximum power point.

Any power system includes a controller and a control strategy, which describes the interactions between its components. In PV systems, the use of batteries as a storage medium implies the use of a charge controller. This is used to manage the flow of energy from PV system to batteries and load by

FIGURE 9.14

Operating principle of over-charge and over-discharge protection. (a) Over-charge. (b) Over-discharge.

using the battery voltage and its acceptable maximum and minimum values. Most controllers have two main modes of operation:

1. *Normal operating condition*, where the battery voltage varies between the acceptable maximum and minimum values.
2. *Over-charge or over-discharge condition*, which occurs when the battery voltage reaches a critical value.

The second mode of operation is obtained by using a switch with a hysteresis cycle, such as electromechanical or solid-state devices. The operation of this switch is shown in Fig. 9.14.

As shown in Fig. 9.14a, when the terminal voltage increases above a certain threshold, $V_{max,off}$, and when the current required by the load is less than the current supplied by the PV array, the batteries are protected from excessive charge by disconnecting the PV array. The PV array is connected again when the terminal voltage decreases below a certain value, $V_{max,on}$ (Hansen et al., 2000).

Similarly, as shown in Fig. 9.14b, when the current required by the load is bigger than the current delivered by the PV array, to protect the battery against excessive discharge the load is disconnected when the terminal voltage falls below a certain threshold, $V_{min,off}$. The load is connected to the system again when the terminal voltage is above a certain threshold, $V_{min,on}$ (Hansen et al., 2000).

9.3.4 Peak-power trackers

As was seen before, PV cells have a single operating point where the values of the current (I) and voltage (V) of the cell result in a maximum power output. These values correspond to a particular resistance, which is equal to V/I, as specified by Ohm's law. A PV cell has an exponential relationship between current and voltage, and there is only one optimum operating point, also called a *maximum power point* (MPP), on the power–voltage (or current) curve, as shown in Fig. 9.8. MPP changes according to the radiation intensity and the cell temperature, as shown in Fig. 9.9. Maximum power point trackers (MPPTs) utilize some type of control circuit or logic to search for this point and, thus, allow the converter circuit to extract the maximum power available from a cell. In fact, peak-power trackers optimize the operating voltage of a PV system to maximize the current. Typically, the PV system voltage is charged automatically. Simple peak-power trackers may have fixed operator-selected set points.

MPP trackers can either be applied at the array level or at the module level. In the former, a single tracker controls the current through all modules in the array. This function in fact is commonly integrated into the array's inverter, as already indicated in Section 9.3.2. The advantage of this approach is that a single large inverter/controller is used, which simplifies maintenance, minimizes cost, and allows high inverter efficiency. Its drawback is that the same current flows through all modules that are connected in series in the array, and because some modules may have different $I-V$ curves than others, not all modules will operate at their individual maximum power point. This is especially problematic when inconsistent manufacturing leads to variability between modules—"module mismatch"—and when modules in an array experience different amounts of shading or soiling. Modules may also age at different rates, causing further variation in $I-V$ curves.

To overcome this problem, MPP controllers can be applied to individual modules in an array, so that every module operates at its own maximum power point. These may either be DC–DC controllers, which still require a central inverter, or DC–AC "micro-inverters" which perform both MPP control and inversion as is seen also in Section 9.3.2. Manufacturers of both devices claim that they can increase the yield of a PV array by 5%–20%. In addition, they allow malfunctioning modules in a system to be more easily identified, as the output of each module is monitored. However multiple small controllers inherently cost more than a single centralized one. Therefore, a cost/benefit analysis needs to be carried out before choosing the right system to make sure that the extra energy collected offsets the extra cost of the MPP.

Micro-inverters have additional advantages: AC power is simpler to connect to a building's electrical system, they operate at safer voltages than central inverters (200–300 V, as opposed to 600 or 1000 V), and inverter failure means the loss of only a single module, not an entire array. However micro-inverters are not yet as reliable as large inverters, especially at high temperatures.

With these characteristics, MPP of individual modules is most suited to PV systems subject to shading or infrequent cleaning, and where the value of produced electricity is high. DC–DC and DC–AC module controllers are widely available in the market, and some modules now have DC–DC controllers built into them. They are used in both residential and commercial PV systems.

In PV systems designed to charge batteries for off-grid power systems, MPPT charge controllers are desirable to make the best use of all the energy generated by the panels. MPPT charge controllers are quickly becoming more affordable and more common. The benefits of MPPT regulators are greatest during cold weather, on cloudy or hazy days, or when the battery is deeply discharged. Peak-power trackers can be purchased separately or specified as an option with battery charge controllers or inverters. In all cases, however, the cost and complexity of adding a peak-power tracker should be balanced against the expected power gain and the impact on system reliability.

9.3.5 Remote fault detection, diagnosis, and cleaning of photovoltaic systems

These methods can be used to detect both types of faults, permanent and temporary. Permanent faults include discoloration, browning, breaking, and delamination, whereas temporary faults include surface soiling, snail trails, and hot spots caused by bird droppings. Drones are used to collect infrared and electroluminescence images and then used to localize and identify faults based on some advanced image processing algorithms. Recently, deep learning algorithms have been used for this purpose. These methods can identify and localize the faults; nevertheless, sophisticated equipment or instruments are required, such as thermal cameras and drones. Visual supervision by Operation and Maintenance (O&M) services is more expensive and time-consuming, and thus, it is not recommended for large-scale PV plants. In addition, even in small-scale PV plants, this method is not able to detect invisible faults such as hotspots. Several defects can be detected based on visual supervision, such as bubbles discoloration, browning, delamination, burning marks, cracked glass or cells, dirt points, loose, exposed, or damaged wiring, rusty or corroded interconnections, snail tails, and damaged or broken pieces. In these methods, the distance between the cameras from the PV array is determined by the drone positions relative to the PV panels. Image resolution should be considered to collect good-quality images for an accurate diagnosis based on images. More details can be found in the work of Mellit and Kalogirou (2021).

Generally, the most serious and usual problem is dust accumulation on solar PV modules, which reduces light transmission to the solar cells reducing photon absorption and thus contributing to performance reduction of PV systems and dirt on the panels keeps the production permanently reduced, until the module is cleaned. A decrease in power output is particularly noticeable in dusty, arid, and windy areas as the wind blows sand onto the panels. However, even in areas with more rainy weather, it is noticeable that dirt remains on the panels. This happens also when modules are installed at a large angle, which helps to collect less dust.

Power losses are not the only problem caused by soiling. Permanent soiling can result in hot spots forming over time. In urban areas bird droppings can also accumulate on modules. These droppings, apart from their acidity, also lead to hotspots forming on the module, a situation where the heat concentrates in one point, causing damage to the cell. Furthermore, the damaged cell will affect the yield of all the neighboring cells. Therefore, regular cleaning is a must to keep the solar PV installation in optimal operating condition.

The traditional method, and still one of the most widely used, is washing panels with handheld brushes. This technique is, however, only suitable for small installations. The growth in large-scale commercial installations, sometimes at hundreds of megawatts, is continuously increasing, making manual cleaning unsuitable. In such cases, labor costs would be excessively high and create safety issues for the labor force.

For ground-mounted PV installations, a possible solution is to mount long brushes onto vehicles, like slow-moving farmer tractors. A possible disadvantage is a need for enough space between rows so the vehicle can pass. It should be noted that the use of tractors requires a permanent supply of fuel, and the operation depends on the type of terrain and generally on ground conditions. Furthermore, the process is unsuitable for large-height clusters using PVs in multiple series. Generally, this is one of the mostly used methods to clean large-scale installations.

Another solution suitable for large-scale, ground-mounted PV installations is semiautomated long brushes of varying lengths that are fixed to the frame of the installation that move vertically along the array. These can be programmed to run at specific times of the day, mainly at night. These, however, are suitable for only a single array row and require labor to move the machines from one row to the next.

The most recent and possibly the most versatile solution for the cleaning of solar panels is cleaning with the use of robots. Numerous types of robots exist on the market today, varying in size and weight. They can clean almost any commercial installation, including PVs mounted on sloping roofs and floating solar plants.

Robotic systems are semiautomatic and are usually controlled by a single user. It is the most versatile solution, as it is relatively easy to transport from one site to another. Semi-automatic robots are the most cost-effective solution, adaptable to any type of commercial solar installation. Combining the automated inspection of the PV plant with drones equipped with normal infrared cameras can help to develop the cleaning schedule for large PV plants.

9.4 Applications

PV modules are designed for outdoor use under harsh conditions, such as marine, tropic, arctic, and desert environments. The PV array consists of a number of individual photovoltaic modules connected together to give a suitable current and voltage output. Common power modules have a rated power output of around 50−180 W each. As an example, a small system of 1.5 to 2 kWp may therefore comprise some 10−30 modules covering an area of around 15−25 m², depending on the technology used and the orientation of the array with respect to the sun.

Most power modules deliver direct current electricity at 12 V, whereas most common household appliances and industrial processes operate with alternating current at 240 or 415 V (120 V in the United States). Therefore, an inverter is used to convert the low-voltage DC to higher-voltage AC.

Other components in a typical PV system are the array mounting structure and various cables and switches needed to ensure that the PV generator can be isolated.

The basic principle of a PV system is shown in Fig. 9.15. As can be seen, the PV array produces electricity, which can be directed from the controller to either battery storage or a load. Whenever there is no sunshine, the battery can supply power to the load if it has a satisfactory capacity.

9.4.1 Direct-coupled PV system

In a direct-coupled PV system, the PV array is connected directly to the load. Therefore, the load can operate only whenever there is solar radiation, so such a system has very limited applications. The schematic diagram of such a system is shown in Fig. 9.16. A typical application of this type of system is for water pumping, i.e., the system operates as long as sunshine is available, and instead of storing electrical energy, water is usually stored.

9.4.2 Stand-alone applications

Stand-alone PV systems are used in areas that are not easily accessible or have no access to an electric grid. A stand-alone system is independent of the electricity grid, with the energy produced normally being stored in batteries. A typical stand-alone system would consist of a PV module or modules,

FIGURE 9.15

Basic principle of a PV solar energy system.

FIGURE 9.16

Schematic diagram of a direct-coupled PV system.

FIGURE 9.17

Schematic diagram of a stand-alone PV application.

batteries, and a charge controller. An inverter may also be included in the system to convert the direct current generated by the PV modules to the alternating current form required by normal appliances. A schematic diagram of a stand-alone system is shown in Fig. 9.17. As can be seen, the system can satisfy both DC and AC loads simultaneously.

9.4.3 Grid-connected system

Nowadays, it is usual practice to connect PV systems to the local electricity network. This means that, during the day, the electricity generated by the PV system can either be used immediately (which is normal for systems installed in offices, other commercial buildings, and industrial applications) or be sold to one of the electricity supply companies (which is more common for domestic systems, where the occupier may be out during the day). In the evening, when the solar system is unable to provide the electricity required, power can be bought back from the network. In effect, the grid is acting as an energy storage system, which means the PV system does not need to include battery storage. A schematic diagram of a grid-connected system is shown in Fig. 9.18.

9.4.4 Hybrid-connected system

In the hybrid-connected system, more than one type of electricity generator is employed. The second type of electricity generator can be renewable, such as a wind turbine, or conventional, such as a diesel engine generator or the utility grid. The diesel engine generator can also be a renewable source of electricity when the diesel engine is fed with biofuels. A schematic diagram of a hybrid-connected system is shown in Fig. 9.19. Again, in this system, both DC and AC loads can be satisfied simultaneously.

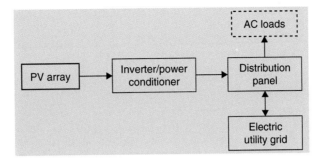

FIGURE 9.18

Schematic diagram of a grid-connected system.

FIGURE 9.19

Schematic diagram of a hybrid connected system.

9.4.5 Types of applications

These are some of the most common PV applications:

- *Remote-site electrification.* Photovoltaic systems can provide long-term power at sites far from utility grids. The loads include lighting, small appliances, water pumps (including small circulators of solar water heating systems), and communications equipment. In these applications, the load demand can vary from a few watts to tens of kilowatts. Usually, PV systems are preferred to fuel generators, since they do not depend on a fuel supply, which can be problematic, and they do avoid maintenance and environmental pollution problems.
- *Communications.* Photovoltaics can provide reliable power for communication systems, especially in remote locations, away from the utility grid. Examples include communication relay towers, travelers' information transmitters, cellular telephone transmitters, radio relay stations, emergency call units, and military communication facilities. Such systems range in size from a few watts for callbox systems to several kilowatts for relay stations. Obviously, these systems are stand-alone units in which PV-charged batteries provide a stable DC voltage that meets the varying current demand. Practice has shown that such PV power systems can operate reliably for a long time with little maintenance.

- *Remote monitoring.* Because of their simplicity, reliability, and capacity for unattended operation, photovoltaic modules are preferred in providing power at remote sites to sensors, data loggers, and associated meteorological monitoring transmitters, irrigation control, and monitoring highway traffic. Most of these applications require less than 150 W and can be powered by a single photovoltaic module. The batteries required are often located in the same weather-resistant enclosure as the data acquisition or monitoring equipment. Vandalism may be a problem in some cases; however, mounting the modules on a tall pole may solve the problem and avoid damage from other causes.

- *Water pumping.* Stand-alone photovoltaic systems can meet the need for small to intermediate-size water-pumping applications. These include irrigation, domestic use, village water supply, and livestock watering. Advantages of using water pumps powered by photovoltaic systems include low maintenance, ease of installation, and reliability. Most pumping systems do not use batteries but store the pumped water in holding tanks.

- *Charging vehicle batteries.* When not in use, vehicle batteries self-discharge over time. This is a problem for organizations that maintain a fleet of vehicles, such as the fire-fighting services. Photovoltaics battery chargers can help solve this problem by keeping the battery at a high state of charge by providing a trickle charging current. In this application, the modules can be installed on the roof of a building or car park (also providing shading) or on the vehicle itself. Another important application in this area is the use of PV modules to charge the batteries of electric vehicles.

- *Building-integrated photovoltaics.* BIPVs is a special application in which PVs are installed either on the façade or roof of a building and are an integral part of the building structure, replacing in each case the particular building component. To avoid an increase in the thermal load of the building, usually a gap is created between the PV and the building element (brick, slab, etc.), which is behind the PV, and in this gap, ambient air is circulated so as to remove the produced heat. During wintertime, this air is directed into the building to cover part of the building load; during summer, it is just rejected back to ambient at a higher temperature. A common example where these systems are installed is what is called *zero-energy houses*, where the building is an energy-producing unit that satisfies all its own energy needs. In another application related to buildings, PVs can be used as effective shading devices. As this is an important application it is examined in more detail in the following section.

Building-integrated photovoltaics

According to Sick and Erge (1996), approximately 25%−30% of energy consumed in buildings in industrialized countries is electricity. Photovoltaics can be integrated on virtually every structure. Grid-connected BIPV is the simplest low-voltage residential system which comprises a PV array and inverter. They feed electricity directly to an electricity grid and do not usually require batteries. The performance of a BIPV grid-connected system depends on PV efficiency, local climate, the orientation and inclination of the PV array, load characteristics and the inverter performance. A comprehensive review on BIPV systems is presented by Norton et al. (2011).

BIPV displaces conventional building materials, which leads to savings in the purchase and installation of conventional materials, thus the net cost of the BIPV is lower, which increases the cost effectiveness of the system. There would be some additional cost associated with the BIPV wiring, but this would be minimal in a new construction. BIPV walls, roofs, and sunshades provide fully integrated electricity generation while also serving as part of the weather protective building envelope (Archer and Hill, 2001). BIPV can serve as a shading device for a window, a semi-transparent glass facade, a building exterior cladding panel, a skylight, parapet unit or roofing system.

The sizing and design of a BIPV system is based on a building's electrical load profile, PV output and balance-of-system characteristics, but must also consider building design constraints and its location, the local climate, and possible future increase of the load. A realistic estimation of load profile to be satisfied is the first step in the design of a stand-alone BIPV system design. In grid-connected BIPV applications the economically optimal diurnal load that must be met by the PV may not correspond to the total load, particularly at night and during wintertime.

There are also regulatory requirements concerning buildings that must be met; most local building codes and product certification requirements will specify specific standards for BIPV mounting, fixing, and fire resistance. These will often vary with the location of the building to take account of possible differences in wind loading, earthquake risk, and the attendant risks associated with particular failure modes (Norton et al., 2011). For this purpose, a product certification is required, usually carried out in an independent testing laboratory after passing satisfactorily a prescribed set of testing procedures (e.g., cycling of humidity, freeze/thaw, temperature, rain).

An advantage of BIPV is that as some of the PV power could be used in the building, the demand on the power grid is reduced and the reliability of supplied power to the building is improved. Another potential significant advantage is that the heat collected by PV modules can also be used for space heating or hot water-heating (see Section 9.8).

From architectural, technical and financial perspectives, building-integrated photovoltaics:

- Reduce the initial investment costs by displacing facade/roof/shading elements.
- Are aesthetically appealing.
- Electricity is generated at the point of use, reducing the costs and losses associated with transmission and distribution.
- Are suitable for installation on roofs and facades in densely populated areas.
- Require no additional land area for the installation.
- Can satisfy all, or a considerable part, of the electricity consumption of the building.
- Can act as a shading device.
- Can act as a source of day lighting if semi-transparent PVs are used for fenestration.
- Can provide part of the hot water or space-heating loads of the building.

Roofs are an attractive location for BIPV (Norton et al., 2011):

- They offer unshaded solar access.
- Cost is partially offset by the displacement of roofing materials by BIPV modules.
- Flat roofs generally enable more optimal solar cell placement and orientation.
- In a pitched roof which is near optimally inclined, the need for and cost of a support frame is eliminated.

Fully integrated BIPV roofing systems must perform the function of a standard roof and provide water tightness, drainage, and insulation.

PV glass curtain-walls and PV metal curtain-walls are used for integration of PV modules with wall materials (Toyokawa and Uehara, 1997). BIPV can be integrated into the building facade as:

- Rainscreen overcladding.
- Structural glazing mullion/transom curtain-wall systems.
- Panel curtain-wall systems.
- Profile metal cladding.

The performance ratio (PR) expresses the performance of a PV system in comparison to a system with no losses of the same design and rating at the same location. It is equal to the system efficiency under realistic reporting conditions (RRCs) divided by the module efficiency under standard test conditions (STCs) (Simmons and Infield, 1996). It indicates how close a PV system approaches ideal performance during real operation (Blaesser, 1997). The PR is independent of location and is influenced by:

- Insolation (remember that the efficiency of PV array depends on irradiance).
- The efficiency of the various system components.
- Size of the inverter relative to the PV array.
- Utilization factor of the system (i.e., the extent to which the system output is used).

9.5 Design of PV systems

The electrical power output from a PV panel depends on the incident radiation, the cell temperature, the solar incidence angle, and the load resistance. In this section, a method to design a PV system is presented and all these parameters are analyzed. Initially, a method to estimate the electrical load of an application is presented, followed by the estimation of the absorbed solar radiation from a PV panel and a description of the method for sizing PV systems.

9.5.1 Electrical loads

As is already indicated, a PV system size may vary from a few watts to hundreds of kilowatts. In grid-connected systems, the installed power is not so important because the produced power, if not consumed, is fed into the grid. In stand-alone systems, however, the only source of electrical power is the PV system; therefore, it is very important at the initial stages of the system design to assess the electrical loads the system will cover. This is especially important in emergency warning systems. The main considerations that a PV system designer needs to address from the very beginning are:

1. According to the type of loads that the PV system will meet, which is more important, the total daily energy output or the average or peak power?
2. At what voltage will the power be delivered, and is it AC or DC?
3. Is a backup energy source needed?

Usually the first things the designer has to estimate are the load and the load profile that the PV system will meet. It is very important to be able to estimate precisely the loads and their profiles (time when each load occurs). Due to the initial expenditure needed, the system is sized at the minimum required to satisfy the specific demand. If, for example, three appliances exist, requiring 500, 1000, and 1500 W, respectively; each appliance is to operate for 1 h; and only one appliance is on at a time, then the PV system must have an installed peak power of 1500 W and 3000 Wh of energy requirement. If possible, when using a PV system, the loads should be intentionally spread over a period of time to keep the system small and thus cost-effective. Generally, the peak power is estimated by the value of the highest power occurring at any particular time, whereas the energy requirement is obtained by multiplying the wattage of each appliance by the operating hours and summing the energy requirements of all appliances connected to the PV system. The maximum power can easily be estimated with the use of a time-schedule diagram, as shown in the following example.

EXAMPLE 9.4

Estimate the daily load and the peak power required by a PV system that has three appliances connected to it with the following characteristics.

1. Appliance 1, 20 W operated for 3 h (10 a.m.−1 p.m.).
2. Appliance 2, 10 W operated for 8 h (9 a.m.−5 p.m.).
3. Appliance 3, 30 W operated for 2 h (2−4 p.m.).

Solution

The daily energy use is equal to:

$$(20W) \times (3h) + (10W) \times (8h) + (30W) \times (2h) = 200 \text{ Wh}$$

To find the peak power, a time schedule diagram is required (see Fig. 9.20).
As can be seen, the peak power is equal to 40 W.

FIGURE 9.20

Time schedule diagram.

EXAMPLE 9.5

A remote cottage has the loads listed in Table 9.2. Find the average load and peak power to be satisfied by a 12 V PV system with an inverter.

Table 9.2 Loads for cottage in Example 9.5.

Appliance type	Description	Power type	Period of operation
Lights	3, 25 W compact fluorescent bulbs, daily	DC	Nighttime 5 h each
Light	11 W compact fluorescent bulb, daily	AC	Nighttime 5 h
Water pump	50 W (6 A start current), daily	DC	Daytime 2 h
Oven	500 W, 3 times a week	AC	Daytime 1.5 h
Steam iron	800 W, once a week	AC	Daytime 1.5 h

Solution

In Table 9.3, the loads for this application are separated according to type of power. Because no information is given about the time schedule of the loads, these are assumed to occur simultaneously.

Table 9.3 Loads in Table 9.2, by type of power.

Appliance type	Power type	Power (W)	Run time (h)	Energy/day (Wh)	Energy/week (Wh)
Lights	DC	$3 \times 25 = 75$ W	5	375	2625
Lights	AC	11 W	5	55	385
Water pump	DC	50 W	2	100	700
Oven	AC	500 W	1.5	–	2250
Steam iron	AC	800 W	1.5	–	1200

From Table 9.3, the following can be determined:

$$\text{Average DC load} = 375 + 100 = 475 \text{ Wh/day}$$
$$\text{Average AC load} = (385 + 2250 + 1200)/7 = 547.9 \text{ Wh/day}$$
$$\text{Peak DC load} = 6 \times 12 + 75$$
$$= 147 \text{ W (the maximum occurs when the pump starts, } 6 \times 12 > 50 \text{ W)}$$
$$\text{Peak AC load} = 11 + 500 + 800 = 1311 \text{ W}$$

9.5.2 Absorbed solar radiation

The main factor affecting the power output from a PV system is the absorbed solar radiation, S, on the PV surface. As was seen in Chapter 3, S depends on the incident radiation, air mass, and incident angle. As in the case of thermal collectors, when radiation data on the plane of the PV are unknown, it is necessary to estimate the absorbed solar radiation using the horizontal data and information on incidence angle. As in thermal collectors, the absorbed solar radiation includes the beam, diffuse, and ground-reflected components. In the case of PVs, however, a spectral effect is also included. Therefore, by assuming that the diffuse and ground-reflected radiation is isotropic, S can be obtained from (Duffie and Beckman, 2006):

$$S = M \left\{ G_B R_B (\tau\alpha)_B + G_D (\tau\alpha)_D \left[\frac{1 + \cos(\beta)}{2} \right] + G\rho_G (\tau\alpha)_G \left[\frac{1 - \cos(\beta)}{2} \right] \right\} \quad (9.25)$$

where $M =$ air mass modifier.

The air mass modifier, M, accounts for the absorption of radiation by species in the atmosphere, which causes the spectral content of the available solar radiation to change, thus altering the spectral distribution of the incident radiation and the generated electricity. An empirical relation that accounts for the changes in the spectral distribution resulting from changes in the air mass, m, from the reference air mass of 1.5 (at sea level) is given by the following empirical relation developed by King et al. (2004):

$$M = \alpha_0 + \alpha_1 m + \alpha_2 m^2 + \alpha_3 m^3 + \alpha_4 m^4 \quad (9.26)$$

Constant α_i values in Eq. (9.26) depend on the PV material, although for small zenith angles, less than about 70°, the differences are small (De Soto et al., 2006). Table 9.4 gives the values of the α_i constants for various PV panels tested at the National Institute of Standards and Technology (NIST) (Fanney et al., 2002).

As was seen in Chapter 2, Section 2.3.6, the air mass, m, is the ratio of the mass of air that the beam radiation has to traverse at any given time and location to the mass of air that the beam radiation would traverse if the sun were directly overhead. This can be given from Eq. (2.81) or from the following relation developed by King et al. (1998):

$$m = \frac{1}{\cos(\Phi) + 0.5050(96.08 - \Phi)^{-1.634}} \quad (9.27)$$

As the incidence angle increases, the amount of radiation reflected from the PV cover increases. Significant effects of inclination occur at incidence angles greater than 65°. The effect of reflection and absorption as a function of incidence angle is expressed in terms of the incidence angle modifier, K_θ, defined as the ratio of the radiation absorbed by the cell at incidence angle θ divided by the radiation absorbed by the cell at normal incidence. Therefore, in equation form, the incidence angle modifier at angle θ is obtained by:

$$K_\theta = \frac{(\tau\alpha)_\theta}{(\tau\alpha)_n} \quad (9.28)$$

Table 9.4 Values of α_i constants for various PV panels tested at NIST.

Cell type	Silicon thin film	Monocrystalline	Polycrystalline	Three-junction amorphous
α_0	0.938110	0.935823	0.918093	1.10044085
α_1	0.062191	0.054289	0.086257	−0.06142323
α_2	−0.015021	−0.008677	−0.024459	−0.00442732
α_3	0.001217	0.000527	0.002816	0.000631504
α_4	−0.000034	−0.000011	−0.000126	-1.9184×10^{-5}

It should be noted that the incidence angle depends on the PV panel slope, location, and time of the day. As in thermal collectors, separate incidence angle modifiers are required for the beam, diffuse, and ground-reflected radiation. For the diffuse and ground-reflected radiation, the effective incidence angle given by Eq. (3.4) can be used. Although these equations were obtained for thermal collectors, they were found to give reasonable results for PV systems as well.

So, using the concept of incidence angle modifier and noting that:

$$K_{\theta,B} = \frac{(\tau\alpha)_B}{(\tau\alpha)_n}, K_{\theta,D} = \frac{(\tau\alpha)_D}{(\tau\alpha)_n}, K_{\theta,G} = \frac{(\tau\alpha)_G}{(\tau\alpha)_n}$$

Equation (9.25) can be written as:

$$S = (\tau\alpha)_n M \left\{ G_B R_B K_{\theta,B} + G_D K_{\theta,D} \left[\frac{1 + \cos(\beta)}{2} \right] + G\rho_G K_{\theta,G} \left[\frac{1 - \cos(\beta)}{2} \right] \right\}$$ (9.29)

It should be noted that, because the glazing is bonded to the cell surface, the incidence angle modifier of a PV panel differs slightly from that of a flat-plate collector and is obtained by combining the various equations presented in Chapter 2, Section 2.3.3:

$$(\tau\alpha)_\theta = e^{-[KL/\cos(\theta_r)]} \left\{ 1 - \frac{1}{2} \left[\frac{\sin^2(\theta_r - \theta)}{\sin^2(\theta_r + \theta)} + \frac{\tan^2(\theta_r - \theta)}{\tan^2(\theta_r + \theta)} \right] \right\}$$ (9.30)

where θ and θ_r are the incidence angle and refraction angle (same as angles θ_1 and θ_2 in Section 2.3.3). A typical value of the extinction coefficient, K, for PV systems is 4 m^{-1} (for water white glass), glazing thickness is 2 mm, and the refractive index for glass is 1.526.

A simpler way to obtain the incidence angle modifier is given by King et al. (1998), who suggested the following equation:

$$K_\theta = b_0 + b_1\theta + b_2\theta^2 + b_3\theta^3 + b_4\theta^4 + b_5\theta^5$$ (9.31)

Table 9.5 gives the values of the b_i constants for various PV panels tested at NIST (Fanney et al., 2002).

Table 9.5 Values of b_i constants for various PV panels tested at NIST.

Cell type	Silicon thin film	Monocrystalline	Polycrystalline	Three-junction amorphous
b_0	0.998980	1.000341	0.998515	1.001845
b_1	−0.006098	−0.005557	−0.012122	−0.005648
b_2	8.117×10^{-4}	6.553×10^{-4}	1.440×10^{-3}	7.250×10^{-4}
b_3	-3.376×10^{-5}	-2.733×10^{-5}	-5.576×10^{-5}	-2.916×10^{-5}
b_4	5.647×10^{-7}	4.641×10^{-7}	8.779×10^{-7}	4.696×10^{-7}
b_5	-3.371×10^{-9}	-2.806×10^{-9}	-4.919×10^{-9}	-2.739×10^{-9}

Therefore, Eq. (9.31) can be used directly for the specific type of cell to give the incidence angle modifier according to the incidence angle. Again, for the diffuse and ground-reflected radiation, the effective incidence angle given by Eq. (3.4) can be used.

EXAMPLE 9.6

A south-facing PV panel is installed at 30° in a location which is at 35°N latitude. If, on June 11 at noon, the beam radiation is 715 W/m^2 and the diffuse radiation is 295 W/m^2, both on a horizontal surface, estimate the absorbed solar radiation on the PV panel. The thickness of the glass cover on PV is 2 mm, the extinction coefficient K is 4 m^{-1}, and ground reflectance is 0.2.

Solution

From Table 2.1, on June 11, $\delta = 23.09°$. First, the effective incidence angles need to be calculated. For the beam radiation, the incidence angle is required, estimated from Eq. (2.20):

$$\cos(\theta) = \sin(L - \beta)\sin(\delta) + \cos(L - \beta)\cos(\delta)\cos(h)$$

$$= \sin(35 - 30)\sin(23.09) + \cos(35 - 30)\cos(23.09)\cos(0) = 0.951 \text{ or } \theta = 18.1°$$

For the diffuse and ground-reflected components, Eq. (3.4) can be used:

$$\theta_{e,D} = 59.68 - 0.1388\beta + 0.001497\beta^2 = 59.68 - 0.1388(30) + 0.001497(30)^2 = 56.9°$$

$$\theta_{e,G} = 90 - 0.5788\beta + 0.002693\beta^2 = 90 - 0.5788(30) + 0.002693(30)^2 = 75.1°$$

Next, we need to estimate the three incidence angle modifiers. At an incidence angle of 18.1°, the refraction angle from Eq. (2.44) is:

$$\sin(\theta_r) = \sin(\theta)/1.526 = \sin(18.1)/1.526 = 0.204 \text{ or } \theta_r = 11.75°$$

Using Eq. (9.30) with $K = 4$ m^{-1} and $L = 0.002$ m,

$$(\tau\alpha)_B = e^{-[KL/\cos(\theta_r)]}\left\{1 - \frac{1}{2}\left[\frac{\sin^2(\theta_r - \theta)}{\sin^2(\theta_r + \theta)} + \frac{\tan^2(\theta_r - \theta)}{\tan^2(\theta_r + \theta)}\right]\right\}$$

$$= e^{-[0.008/\cos(11.75)]}\left\{1 - \frac{1}{2}\left[\frac{\sin^2(11.75 - 18.1)}{\sin^2(11.75 + 18.1)} + \frac{\tan^2(11.75 - 18.1)}{\tan^2(11.75 + 18.1)}\right]\right\} = 0.9487$$

At normal incidence, as shown in Chapter 2, Section 2.3.3, Eq. (2.49), the term in the square bracket of Eq. (9.30) is replaced with $1 - [(n - 1)/(n + 1)]^2$. Therefore,

$$(\tau\alpha)_n = e^{-KL}\left[1 - \left(\frac{n - 1}{n + 1}\right)^2\right] = e^{-0.008}\left[1 - \left(\frac{1.526 - 1}{1.526 + 1}\right)^2\right] = 0.9490$$

And from Eq. (9.28),

$$K_{\theta,B} = \frac{(\tau\alpha)_B}{(\tau\alpha)_n} = \frac{0.9487}{0.9490} = 0.9997$$

For the diffuse radiation,

$$\sin(\theta_r) = \sin(\theta_{e,D})/1.526 = \sin(56.9)/1.526 = 0.5490 \text{ or } \theta_r = 33.3°$$

Using Eq. (9.30),

$$(\tau\alpha)_D = e^{-[KL/\cos(\theta_r)]}\left\{1 - \frac{1}{2}\left[\frac{\sin^2(\theta_r - \theta_{e,D})}{\sin^2(\theta_r + \theta_{e,D})} + \frac{\tan^2(\theta_r - \theta_{e,D})}{\tan^2(\theta_r + \theta_{e,D})}\right]\right\}$$

$$= e^{-[0.008/\cos(33.3)]}\left\{1 - \frac{1}{2}\left[\frac{\sin^2(33.3 - 56.9)}{\sin^2(33.3 + 56.9)} + \frac{\tan^2(33.3 - 56.9)}{\tan^2(33.3 + 56.9)}\right]\right\} = 0.9111$$

Continued

EXAMPLE 9.6—CONT'D

And from Eq. (9.28),

$$K_{\theta,D} = \frac{(\tau\alpha)_D}{(\tau\alpha)_n} = \frac{0.9111}{0.9490} = 0.9601$$

Using Eq. (9.31) for monocrystalline cells gives $K_{\theta,D} = 0.9622$; and for polycrystalline cells, $K_{\theta,D} = 0.9672$. Both values are close to the value just obtained, so even if the exact type of PV cell is not known, acceptable values can be obtained from Eq. (9.31) using either type of the cell.

For the ground-reflected radiation,

$$\sin(\theta_r) = \sin(\theta_{e,G}) / 1.526 = \sin(75.1) / 1.526 = 0.6333 \text{ or } \theta_r = 39.29°$$

Using Eq. (9.30),

$$(\tau\alpha)_G = e^{-[KL/\cos(\theta_r)]}\left\{1 - \frac{1}{2}\left[\frac{\sin^2(\theta_r - \theta_{e,G})}{\sin^2(\theta_r + \theta_{e,G})} + \frac{\tan^2(\theta_r - \theta_{e,G})}{\tan^2(\theta_r + \theta_{e,G})}\right]\right\}$$

$$= e^{-[0.008/\cos(39.29)]}\left\{1 - \frac{1}{2}\left[\frac{\sin^2(39.29 - 75.1)}{\sin^2(39.29 + 75.1)} + \frac{\tan^2(39.29 - 75.1)}{\tan^2(39.29 + 75.1)}\right]\right\} = 0.7325$$

And from Eq. (9.28),

$$K_{\theta,G} = \frac{(\tau\alpha)_G}{(\tau\alpha)_n} = \frac{0.7325}{0.9490} = 0.7719$$

Using Eq. (9.31) for monocrystalline cells gives $K_{\theta,G} = 0.7625$, and for polycrystalline cells, $K_{\theta,G} = 0.7665$. Both values, again, are close to the value obtained previously.

For the estimation of the air mass, the zenith angle is required, obtained from Eq. (2.12):

$$\cos(\Phi) = \sin(L)\sin(\delta) + \cos(L)\cos(\delta)\cos(h)$$

$$= \sin(35)\sin(23.09) + \cos(35)\cos(23.09)\cos(0) = 0.9785 \quad \text{or} \quad \Phi = 11.91°$$

The air mass is obtained from Eq. (9.27):

$$m = \frac{1}{\cos(\Phi) + 0.5050(96.08 - \Phi)^{-1.634}}$$

$$= \frac{1}{\cos(11.91) + 0.5050(96.08 - 11.91)^{-1.634}} = 1.022$$

It should be noted that the same result is obtained using Eq. (2.81):

$$m = \frac{1}{\cos(\Phi)} = 1.022$$

From Eq. (9.26),

$$M = \alpha_0 + \alpha_1 m + \alpha_2 m^2 + \alpha_3 m^3 + \alpha_4 m^4$$

$$= 0.935823 + 0.054289 \times (1.022) - 0.008677 \times (1.022)^2 + 0.000527 \times (1.022)^3$$

$$- 0.000011 \times (1.022)^4 = 0.9828$$

EXAMPLE 9.6–CONT'D

From Eq. (2.88),

$$R_B = \frac{\cos(\theta)}{\cos(\Phi)} = \frac{\cos(18.1)}{\cos(11.91)} = 0.971$$

Now, using Eq. (9.29),

$$S = (\tau\alpha)_n M \left\{ G_B R_B K_{\theta,B} + G_D K_{\theta,D} \left[\frac{1+\cos(\beta)}{2}\right] + G\rho_G K_{\theta,G}\left[\frac{1-\cos(\beta)}{2}\right] \right\}$$

$$= 0.9490 \times 0.9828 \left\{ 715 \times 0.971 \times 0.9997 + 295 \times 0.9601 \left[\frac{1+\cos(30)}{2}\right] + 1010 \times 0.2 \right.$$

$$\left. \times 0.7719\left[\frac{1-\cos(30)}{2}\right]\right\} = 903.5 \text{ W}/\text{m}^2$$

9.5.3 Cell temperature

As was seen in Section 9.1.3, the performance of the solar cell depends on the cell temperature. This temperature can be determined by an energy balance and considering that the absorbed solar energy that is not converted to electricity is converted to heat, which is dissipated to the environment. Generally, when operating solar cells at elevated temperatures, their efficiency is lowered. In cases where this heat dissipation is not possible, as in BIPVs and concentrating PV systems (see Section 9.7), the heat must be removed by some mechanical means, such as forced air circulation, or by a water heat exchanger in contact with the back side of the PV. In this case, the heat can be used to an advantage, as explained in Section 9.8; these systems are called *hybrid photovoltaic/thermal* (PV/T) systems. Because these systems offer a number of advantages, even normal roof-mounted PVs can be converted into hybrid PV/Ts.

The energy balance on a unit area of a PV module that is cooled by heat dissipation to ambient air is given by:

$$(\tau\alpha)G_t = \eta_e G_t + U_L(T_C - T_a) \tag{9.32}$$

For the $(\tau\alpha)$ product, a value of 0.9 can be used without serious error (Duffie and Beckman, 2006). The heat loss coefficient, U_L, includes losses by convection and radiation from the front and back of the PV to the ambient temperature, T_a.

By operating the module at the nominal operating cell temperature (NOCT) conditions (see Table 9.1) with no load, i.e., $\eta_e = 0$, Eq. (9.32) becomes:

$$(\tau\alpha)G_{t,NOCT} = U_L\left(T_{NOCT} - T_{a,NOCT}\right) \tag{9.33}$$

which can be used to determine the ratio:

$$\frac{(\tau\alpha)}{U_L} = \frac{T_{NOCT} - T_{a,NOCT}}{G_{t,NOCT}} \tag{9.34}$$

By substituting Eq. (9.34) into Eq. (9.32) and performing the necessary manipulations, the following relation can be obtained:

$$T_C = (T_{NOCT} - T_{a,NOCT}) \left[\frac{G_t}{G_{t,NOCT}}\right] \left[1 - \frac{\eta_e}{(\tau\alpha)}\right] + T_a \tag{9.35}$$

An empirical formula that can be used for the calculation of PV module temperature of poly-crystalline silicon solar cells was presented by Lasnier and Ang (1990). This is a function of the ambient temperature, T_a, and the incoming solar radiation, G_t, given by:

$$T_C = 30 + 0.0175(G_t - 300) + 1.14(T_a - 25) \tag{9.36}$$

When the temperature coefficient of the PV module is given, the following equation can be used to estimate the efficiency according to the cell temperature:

$$\eta_e = \eta_R[1 - \beta(T_C - T_{NOCT})] \tag{9.37}$$

where

β = temperature coefficient (per k^{-1});
η_R = reference efficiency.

EXAMPLE 9.7

If, for a PV module operating at NOCT conditions, the cell temperature is 42 °C, determine the cell temperature when this module operates at a location where $G_t = 683 \ W/m^2$, $V = 1 \ m/s$, and $T_a = 41°C$ and the module is operating at its maximum power point with an efficiency of 9.5%.

Solution

Using Eq. (9.35),

$$T_C = (T_{NOCT} - T_{a,NOCT}) \left[\frac{G_t}{G_{t,NOCT}}\right] \left[1 - \frac{\eta_e}{(\tau\alpha)}\right] + T_a$$

$$= (42 - 20) \left[\frac{683}{800}\right] \left(1 - \frac{0.095}{0.9}\right) + 41 = 57.8 \ °C$$

Using empirical Eq. (9.36),

$$T_C = 30 + 0.0175(683 - 300) + 1.14(41 - 25) = 54.9 \ °C$$

As can be seen, the empirical method is not as accurate but offers a good approximation.

It should be noted that, in Example 9.7, the module efficiency was given. If it was not given, then a trial-and-error solution needs to be applied. In this procedure, a value of module efficiency is assumed and T_C is estimated using Eq. (9.35). Provided that I_o and I_{sc} are known, the value of T_C is used to find V_{max} with Eq. (9.14). Subsequently, P_{max} and η_{max} are estimated with Eqs. (9.17) and (9.18), respectively. The initial guess value of η_e is then compared with η_{max}, and if there is a difference, iteration is used. Because the efficiency is strongly related to cell temperature, fast convergence is achieved.

9.5.4 Sizing of PV systems

Once the load and absorbed solar radiation are known, the design of the PV system can be carried out, including the estimation of the required PV panel's area and the selection of the other equipment, such as controllers and inverters. Detailed simulations of PV systems can be carried with the TRNSYS program (see Chapter 11, Section 11.5.1); however, usually a simple procedure needs to be followed to perform a preliminary sizing of the system. The simplicity of this preliminary design depends on the type of the application. For example, a situation in which a vaccine refrigerator is powered by the PV system, and a possible failure of the system to supply the required energy will destroy the vaccines is much different to a home system delivering electricity to a television and some lamps.

The energy delivered by a PV array, E_{PV}, is given by:

$$E_{PV} = A\eta_e \overline{G_t} \tag{9.38}$$

where

$\overline{G_t}$ = monthly average value of G_t, obtained from Eq. (2.97) by setting all parameters as monthly average values;
A = area of the PV array (m^2).

The energy of the array available to the load and battery, E_A, is obtained from Eq. (9.38) by accounting for the array losses, L_{PV}, and other power conditioning losses, L_C:

$$E_A = E_{PV}(1 - L_{PV})(1 - L_C) \tag{9.39}$$

Therefore, the array efficiency is defined as:

$$\eta_A = \frac{E_A}{A\overline{G_t}} \tag{9.40}$$

Grid-connected systems

The inverter size required for grid-connected systems is equal to the nominal array power. The energy available to the grid is simply what is produced by the array multiplied by the inverter efficiency:

$$E_{grid} = E_A \eta_{inv} \tag{9.41}$$

Usually, some distribution losses are present accounted by η_{dist} and, if not, all this energy can be absorbed by the grid, then the actual energy delivered, E_d, is obtained by accounting for the grid absorption rate, η_{abs}, from:

$$E_d = E_{grid}\eta_{abs}\eta_{dist} \tag{9.42}$$

Stand-alone systems

For stand-alone systems, the total equivalent DC demand, $D_{dc,eq}$, is obtained by summing the total DC demand, D_{dc}, and the total AC demand, D_{ac} (both expressed in kilowatt hours per day), converted to DC equivalent using:

$$D_{dc,eq} = D_{dc} + \frac{D_{ac}}{\eta_{inv}} \tag{9.43}$$

When the array supplies all energy to a DC load, the actual energy delivered, $E_{d,dc}$, is obtained by:

$$E_{d,dc} = E_A \eta_{dist} \tag{9.44}$$

When the battery directly supplies a DC load, the efficiency of the battery, η_{bat}, is accounted for, and the actual energy delivered, $E_{d,dc,bat}$, is obtained from:

$$E_{d,dc,bat} = E_A \eta_{bat} \eta_{dist} \tag{9.45}$$

When the battery is used to supply energy to an AC load, the inverter efficiency is also accounted for:

$$E_{d,ac,bat} = E_A \eta_{bat} \eta_{inv} \eta_{dist} \tag{9.46}$$

Finally, when the array supplies all energy to an AC load, the actual energy delivered, $E_{d,ac}$, is obtained by:

$$E_{d,ac} = E_A \eta_{inv} \eta_{dist} \tag{9.47}$$

This methodology is demonstrated by means of two examples. The first is a simple one and the second takes into account the various efficiencies.

EXAMPLE 9.8

A PV system is using 80 W, 12 V panels and 6 V, 155 Ah batteries in a good sunshine area. The battery efficiency is 73% and the depth of discharge is 70%. If, in wintertime, there are 5 h of daylight, estimate the number of PV panels and batteries required for a 24 V application with a load of 2600 Wh.

Solution

The number of PV panels required is obtained from:

Number of panels $= 2600(\text{Wh/day})/[5(\text{h/day}) \times 80(\text{W/panel})] = 6.5$, round off to 7 panels

Because the system voltage is 24 V and each panel produces 12 V, two panels need to be connected in series to produce the required voltage, so an even number is required; therefore, the number of PV panels is increased to eight.

If, for the location with good sunshine, we consider that 3 days of storage would be adequate, the storage required is:

$$2600(\text{Wh/day}) \times 3(\text{day})/(0.73 \times 0.7) = 15,264 \text{ Wh}$$

Number of batteries required $= 15,264(\text{Wh})/[155(\text{Ah}) \times 6(\text{V})]$

$$= 16.4, \text{rounded off to 17 batteries}$$

Again as the system voltage is 24 V and each battery is 6 V, we need to connect four batteries in series, so the number of batteries to use here is either 16 (very close to 16.4, with the possibility of not having enough power for the third day) or 20 (for more safety).

The second example uses the concept of efficiency of the various components of the PV system.

EXAMPLE 9.9

Using the data from Example 9.5, estimate the expected daily energy requirement. The efficiencies of the various components of the system are:

- Inverter = 90%.
- Battery = 75%.
- Distribution circuit = 95%.

Solution

From Example 9.5, the average DC load was 475 Wh and the average AC load was 547.9 Wh. These give a total load of 1022.9 Wh.

Expected daily loads are (from Example 9.5).

- Day DC = 100 Wh (from PV system).
- Night DC = 375 Wh (from battery).
- Night AC = 55 Wh (from battery).
- Day AC = 492.9 Wh, = (2250 + 1200)/7 (from PV system through the inverter).

The various energy requirements are obtained as follows.

- Day DC energy is obtained from Eq. (9.44): $E_{d,dc} = E_A \eta_{dist}$, so $E_A = 100/0.95 = 105.3$ Wh.
- Night DC energy is obtained from Eq. (9.45): $E_{d,dc,bat} = E_A \eta_{bat} \eta_{dist}$, so $E_A = 375/(0.75 \times 0.95) = 526.3$ Wh.
- Night AC energy is obtained from Eq. (9.46): $E_{d,ac,bat} = E_A \eta_{bat} \eta_{inv} \eta_{dist}$, so $E_A = 55/(0.75 \times 0.90 \times 0.95) = 85.8$ Wh.
- Day AC energy is obtained from Eq. (9.47): $E_{d,ac} = E_A \eta_{inv} \eta_{dist}$, so $E_A = 492.9/(0.90 \times 0.95) = 576.5$ Wh.
- Expected daily energy requirement = 105.3 + 526.3 + 85.8 + 576.5 = 1293.9 Wh.

Therefore the energy requirement is increased by 27% compared to 1022.9 Wh estimated before.

One way utilities historically have thought about generation reliability is loss-of-load probability (LLP). LLP is the probability that a generation will be insufficient to meet demand at some point over some specific time window, and this principle can also be used in sizing stand-alone PV systems. Therefore, the merit of a stand-alone PV system should be judged in terms of the reliability of the electricity supply to the load. Specifically, for stand-alone PV systems, LLP is defined as the ratio between the energy deficit and the energy demand both on the load and over a long period of time. Because of the random nature of the solar radiation, the LLP of even a trouble-free PV system is always greater than 0.

Any PV system consists mainly of two subsystems that need to be designed: the PV array (also called the *generator*) and the battery storage system (also called the *accumulator*). A useful definition of these parameters relates to the load. Therefore, on a daily basis, the PV array capacity, C_A, is defined as the ratio between the mean PV array energy production and the mean load energy demand. The storage capacity, C_S, is defined as the maximum energy that can be taken out from the accumulator divided by the mean load energy demand. According to Egido and Lorenzo (1992), the sizing pair C_A and C_S can be given by the following equations:

$$C_A = \frac{\eta_{PV} A H_t}{L} \tag{9.48}$$

$$C_S = \frac{C}{L} \tag{9.49}$$

where

A = PV array area (m^2);
η_{PV} = PV array efficiency;
H_t = mean daily irradiation on the PV array (Wh/m^2);

L = mean daily energy consumption (Wh); and
C = useful accumulator capacity (Wh).

The reliability of a PV system is defined as the percentage of load satisfied by the PV system, whereas the LLP is the percentage of the mean load (over large periods of time) not supplied by the PV system, i.e., it is the opposite of reliability.

As can be understood from Eqs. (9.48) and (9.49), it is possible to find many different combinations of C_A and C_S leading to the same LLP value. However, the larger the PV system size, the greater is the cost and the lower the LLP. Therefore, the task of sizing a PV system consists of finding the better trade-off between cost and reliability. Very often, the reliability is an a priori requirement from the user, and the problem is to find the pair of C_A and C_S values that lead to a given LLP value at the minimum cost.

Additionally, because C_A depends on the meteorological conditions of the location, this means that the same PV array for the same load can be "large" in one site and "small" in another site with lower solar radiation.

In cases where long-term averages of daily irradiation are available in terms of monthly means, Eq. (9.48) is modified as:

$$C_A' = \frac{\eta_{PV}A\overline{H}_t}{L} \tag{9.50}$$

where $\overline{H}_t$ = monthly average daily irradiation on the PV array (Wh/m^2).

In this case, C_A' is defined as the ratio of the average energy output of the generator in the month with worst solar radiation input divided by the average consumption of the load (assuming a constant consumption of load for every month).

Each point of the C_A–C_S plane represents a size of a PV system. This allows one to map the reliability, as is shown in Fig. 9.21. The curve is the loci of all the points corresponding to a same LLP

FIGURE 9.21

LLP curve for LLP = 0.01 and cost curve of a PV system.

value. Because of that, this type of curve is called an *iso-reliability curve*. In Fig. 9.21, an example LLP curve is represented for LLP equal to 0.01.

It should be noted that the definitions of C_A and C_S imply that this map is independent of the load and depends only on the meteorological behavior of the location. As can be seen from Fig. 9.21, the iso-reliability curve is very nearly a hyperbola with its asymptotes parallel to the x and y axes, respectively. For a given LLP value, the plot of the cost of the PV systems (dashed line in Fig. 9.21) corresponding to the iso-reliability curve is, approximately, a parabola having a minimum that defines the optimal solution to the sizing problem.

The LLP curve represents pairs of C_S and C_A values that lead to the same value of LLP. This means, for example, that for the pair $(C_S, C_A) = (2, 1.1)$, the proposed reliability is achieved by having a "big" generator and a "small" storage system. Similarly, for the same reliability, the pair $(C_S, C_A) = (9, 0.6)$ leads to a "small" generator and a "big" battery. As can be seen, the optimum size of the system is at $(C_S, C_A) = (7.5, 0.62)$, which gives the minimum PV system cost.

Many methods have been developed by researchers to establish relations between C_A, C_S, and LLP. The main ones are numerical methods that use detailed system simulations and analytical methods that use equations describing the behavior of the PV system. These methods are presented by Egido and Lorenzo (1992).

Fragaki and Markvart (2008) developed a new sizing approach applied to stand-alone PV systems design, based on system configurations without shedding load. The investigation is based on a detailed study of the minimum storage requirement and an analysis of the sizing curves. The analysis revealed the importance of using daily series of measured solar radiation data instead of monthly average values. Markvart et al. (2006) presented the system sizing curve as superposition of contributions from individual climatic cycles of low daily solar radiation for a location southeast of England.

Hontoria et al. (2005) used an artificial neural network (ANN) (see Chapter 11) to generate the sizing curve of stand-alone PV systems from C_S, LLP, and daily clearness index. Mellit et al. (2005) also used an ANN architecture for estimating the sizing coefficients of stand-alone PV systems based on the synthetic and measured solar radiation data.

Once the LLP curves are obtained, it is very simple to design both the capacity of the generator (C_A) and the accumulator capacity (C_S). Depending on the reliability needed for the PV system design, a specific value of the LLP is considered. For instance, Table 9.6 shows some usual values for typical PV systems.

Table 9.6 Recommended LLP values for various applications.

Application	LLP
Domestic appliances	10^{-1}
Rural home lighting	10^{-2}
Telecommunications	10^{-4}

9.6 Tilt and yield

Photovoltaic modules are usually characterized by their peak-power output, kWp, measured under Standard Test Condition (also known as Standard Rating Condition, SRC). Another useful measure is the cumulative energy that a module generates over a period of time, as in practice electricity is bought and sold in kWh. Achieving the highest energy output, or "yield", at the lowest cost is often a primary goal of the solar engineer.

The yield of a PV module is proportional to the amount of solar irradiation it receives, which can be increased by using various modes of tilting and tracking as described in Section 2.2.1. When multiple modules are used, however, tilting them can cause shading between the different rows. Therefore it is useful to examine the relationship between energy yield, tilting, and module spacing.

It should be noted that a module's conversion efficiency, and thus yield, is influenced by conditions specific to local climate such as temperature, light intensity and spectrum. Also, different types of PV vary in their response to these factors. Therefore it is difficult to accurately predict a module's yield at a given location—even if irradiation levels there are well known (Huld et al., 2010).

9.6.1 Fixed tilt

The simplest method for increasing the solar flux received by a PV module is to mount it on a fixed frame that is tilted relative to the horizontal. There are two angles to consider: the collector tilt angle, β (deviation from the horizontal plane) and the collector azimuth angle, Z_s (deviation from due south, in the northern hemisphere).

Tilting the module at an angle equal to the local geographic latitude, a practice known as latitude tilt, minimizes the average incidence angle throughout the year. In practice a smaller tilt angle is often used in order to reduce shading of adjacent modules, minimize wind load, and take greater advantage of summer months when there is more solar flux and the sun is higher in the sky.

Considering only the goal of maximizing annual electricity yield, the tilt angle (β) would be modified from latitude angle (L) to collect more of summer's higher irradiation. It has been empirically determined that the optimal tilt angle for annual yield can be approximated by (Chang, 2009):

$$\beta = 0.764L + 2.14°, \quad \text{for} \quad L \le 65° \tag{9.51a}$$

$$\beta = 0.224L + 33.65°, \quad \text{otherwise} \tag{9.51b}$$

Another option available to the engineer is to maximize energy output in the afternoon, when electricity demand is usually highest. In this case the azimuth angle of the module can be set somewhat to the west of due south. However, doing so would sacrifice morning electricity production and total yield for the day.

The yield benefit of tilting increases as one moves away from the equator and the sun is lower in the sky. The effect of global irradiation on an optimally tilted surface relative to a horizontal surface is quantified in Table 9.7 (Chang, 2009). The increase in irradiation reaches a maximum of 35% at latitude of 65° north, before declining again at higher latitudes and falling to zero at the poles.

The main cost of tilting a PV module from horizontal is the more elaborate mounting structure required, which must also be made stronger because of the higher wind loads occurring on an inclined module. Aesthetically, a raised structure on a roof is often considered to be less attractive than a flush

Table 9.7 Percentage of additional irradiation at optimal tilt relative to horizontal for various latitudes.

Latitude (°)	Additional global irradiation at optimal tilt relative to horizontal
0	0%
10	1%
20	3%
30	10%
40	17%
50	26%
60	33%
65	35%
70	33%
80	22%
90	0%

one. Tilting may also require wider spacing between module rows to prevent self-shading, as shown in Section 9.6.3.

For these reasons the most suitable locations for fixed tilting of PV modules tend to be large flat roofs of commercial buildings and ground-mounted systems, especially at higher latitudes, whereas residential houses usually take advantage of the existing slope of their roofs.

A further consideration when deciding whether to tilt a fixed PV module is the proportion of direct and diffuse irradiation at the location. On cloudy days, nearly all global irradiation is diffused. In such conditions a module's yield is in fact maximized by laying it flat, exposing it to the full dome of the sky (Kelly and Gibson, 2009; 2011).

9.6.2 Trackers

PV modules can also be mounted on single- or double-axis trackers, with the effect on solar flux received throughout the year described in Section 2.2.1. Single-axis tracking increases annual yield in the order of 25% relative to fixed PV modules and, more importantly, it greatly increases power output in the afternoon when a demand is high. At lower latitudes such trackers are typically aligned horizontally, which is mechanically simple and allows for long tracker rows with many PV modules attached to each tracker. As latitude increases there is greater yield benefit from tilting modules to the south, and greater use is made of inclined- or vertical-axis trackers. Trade-offs of inclining the axis include mechanical complexity and cost, higher wind loads, fewer PV modules per tracker, and greater distance between trackers to avoid shading.

Compared to single-axis trackers, double-axis ones achieve a modest additional increase in yield—typically in the range of 5%−10% (Kelly and Gibson, 2009; 2011)—yet they are significantly more mechanically complex. Therefore they are mainly used in concentrating PV systems whose focusing optics must be accurately aligned with the sun in order to function (see Section 9.7).

Single-axis trackers have traditionally been used with high-efficiency crystalline silicon modules, as trackers represent a fixed cost and there is incentive to maximize energy production from each one. The use of thin-film modules on horizontal trackers is uncommon but does occur, especially when suited by local climate. In deserts, for example, thin-film modules tend to yield more electricity than crystalline modules of the same rated power because of the lower reduction in their efficiency at high temperatures (Huld et al., 2010; Kullmann, 2009).

9.6.3 Shading

When PV modules are tilted there is the risk that modules may shade each other, especially when the sun is low in the sky. It is usually not practical to completely eliminate shading, however, as when the sun is just above the horizon a very large distance between modules would be needed. Allowing some shading is mitigated by the fact that there is little solar irradiation at dawn and dusk, so the yield sacrificed is small. Similarly there is less irradiation in winter, when shading is most severe.

Although these phenomena mitigate the effect of shading, module circuitry amplifies it because cells in a PV module are connected in series. Therefore, if 1 cell is shaded it operates as a diode in reverse bias, decreasing the voltage of its circuit and potentially making the module's maximum power point difficult to locate (Lisell and Mosey, 2010). Therefore shading can decrease a module's output by a much greater proportion than the area shaded—more than 30 times in certain scenarios (Deline, 2009). PV cells of different materials vary in their sensitivity to this effect, while (in crystalline silicon modules) electronics can be employed that bypass shaded or failed cells. Therefore the response of specific modules to partial shading is important to consider when designing the layout of a PV system, and good design of the modules is required to avoid having a large area of PVs working at low efficiency.

The optimal amount of allowed shading is also heavily influenced by variables specific to each solar project, both economic—the relative costs of land, electricity and PV modules—and environmental, such as the proportion of direct to diffuse irradiation and surroundings topography. Sufficient space must also be left between module rows for access for repairs and maintenance.

These numerous unknowns make it impossible to provide rules of thumb for spacing of PV-module rows. Nevertheless it is instructive to examine the geometric relation between spacing and shading, particularly for the common situations of fixed tilted and E—W horizontal-tracked modules. More details on the row shading are presented in Chapter 5, Section 5.4.2. For a fixed tilted PV module facing due south Eqs. (5.47a) and (5.47b) apply.

Consider a module at latitude 30°N and inclined at the same angle. The ratio of the northern shadow length to the module length (in the N—S direction) throughout the year and day is shown in Table 9.8. During the equinoxes, when all shadows on Earth move due eastward, its shadow falls 1.15 times the module height from its base. Throughout the day of the summer solstice, the shadow is shorter than the module length if it were laid flat. At the winter solstice the shadow cast is more than 1.5 times the module height throughout the day.

The above discussion applies to multiple rows of fixed, south-facing PV modules, where shadow length falling to the north dictates row spacing. We now consider rows of PV modules on single-axis trackers whose axes run north—south (a configuration seen in some large PV fields), and where shadow lengths to the west and east determine row spacing.

Time of day	Ratio of north-shadow length to module length		
	Summer solstice	Equinox	Winter solstice
6:00 a.m.	0.00	1.15	»1
7	0.59	1.15	»1
8	0.77	1.15	2.31
9	0.88	1.15	1.79
10	0.90	1.15	1.62
11	0.92	1.15	1.56
12:00 p.m.	0.92	1.15	1.54
13	0.92	1.15	1.56
14	0.90	1.15	1.62
15	0.88	1.15	1.79
16	0.77	1.15	2.31
17	0.59	1.15	»1
18	0.00	1.15	»1

Table 9.8 Ratio of north-shadow length to module height for different days of the year.

In this case the module tilt angle, β, varies throughout the day as the tracker follows the sun. Nominally the tilt angle would be set perpendicular to the sun's altitude, i.e., $\beta = \pi/2 - \alpha$ (or equal to the zenith angle). However when the sun is low on the horizon, the modules would be tilted almost vertical, creating severe shading between rows of any spacing. To overcome this, computer-controlled trackers can be employed that decrease the module's tilt in the early morning and late afternoon, a technique known as back-tracking. The increase in electricity yield from the avoided shading more than offsets the decrease in yield from greater incidence angle of sunlight on the modules during these times. Back-tracking allows module rows to be more closely spaced for the same yield, at the expense of a more sophisticated tracking control system.

9.6.4 Tilting versus spacing

Shading leads to a fundamental trade-off in the design of PV systems between tilting (fixed or tracking) and spacing (Denholm and Margolis, 2007). Tilting maximizes energy yield per unit area of module. However because tilting generally requires greater module spacing—both to reduce shading and to allow access for maintenance—it leads to a net decrease in energy yield per unit area of land, while also increasing mechanical complexity and capital cost of the system.

This effect is quantified in a simulation of tilting systems at Kansas City, which represents an energy density equal to the USA average value (Denholm and Margolis, 2007). The results are shown in Table 9.9, in which 3.5 to 5 m of spacing is assumed for the ground-mounted systems. It is seen that, in terms of yield per land area, the greater yield per module from tilting and tracking is outweighed by the greater spacing.

It follows that a key relationship that determines the attractiveness of tilting is that between the price of PV modules and the price of land (or roof space). Expensive modules and cheap land make it

Table 9.9 Results of a simulation for tilting PV modules in Kansas City, USA. Denholm and Margolis (2007).

System type	PV array power density (W/m² of land)	Incident solar radiation (kWh/m² of modules/day)	System energy density (kWh/m² of land/year)
Flat (rooftop)	135	4.31	150
10° tilt south (rooftop)	118	4.64	139
25° tilt south (ground based)	65	4.86	83
1-axis tracking, no tilt	48	5.70	73
2-axis tracking	20	6.60	35

attractive to tilt those modules; expensive land and cheap modules encourages maximizing yield per land area through less tilt and closer spacing. The dramatic fall in PV module prices in recent years has thus decreased the relative attractiveness of tilting and tracking.

9.7 Concentrating PV

A way to increase the effectiveness of PVs is to concentrate sunlight on small, highly efficient photovoltaic cells using inexpensive reflective material, lenses, or mirrors. These are known as *concentrating photovoltaics* (CPVs). Today, the technology takes up a very small portion of the solar industry; however, it is expected that the CPV industry will grow as technology improves and cost comes down and further field tests and demonstrations are conducted.

The solar spectrum has photons ranging up to 4 eV. A single-material PV cell can convert only about 15% of the available energy to useful electrical power. To improve this performance, multiple cells with different band gaps, which are more complex and therefore more expensive, can be used. These are called *multi-junction PVs*. Particularly, a triple-junction PV produced recently achieved a remarkable 40% efficiency (Noun, 2007). This PV consists of three layers of PV material placed one atop the other. Each of the three materials captures a separate portion of the solar spectrum (see Fig. 2.26) and the objective is to capture as much of the solar spectrum as possible. These are much more expensive than other silicon solar cells, but their efficiency offsets their high cost, and in concentrating systems, a small area of these cells is required.

The advantages of CPV systems are the following:

1. They replace expensive PV material with lower-cost mirrors or reflective materials.
2. Solar cells are more efficient at high-irradiation levels.
3. Due to tracking, production of energy starts earlier in the morning and extends later in the day.
4. They have high efficiencies of around 30%−40% at the module level (RENI, 2012) and 25% at the system level (i.e., including losses from inverters and tracking).
5. Due to 2-axis tracking and high efficiency modules, they produce a large amount of energy from a given surface area.

Disadvantages of CPV systems include:

1. At high concentration, cells heat up and lose efficiency, so they must be cooled.
2. Concentrating systems use only direct solar radiation, "wasting" diffuse radiation.
3. The system must track the sun; higher concentration requires more accurate tracking.
4. Concentrating systems are more complex than flat-plate ones and less reliable, because they have moving parts.
5. Two-axis trackers require relatively wide spacing to avoid shading, which reduces CPV's power output per area of land.

Compared to flat-plate PV, concentrating PV has a greater capital cost of $4–$6 per Watt and a higher conversion efficiency of direct irradiation. With this profile, CPV systems are suited to utility-scale power generation in locations with significant insolation and clear skies, such as deserts far from coasts. In such settings CPV is able to produce electricity at among the lowest cost per kWh of solar technologies.

Usually CPV uses lenses to concentrate sunlight onto small-size photovoltaic cells. Because a CPV module needs much less cell material than a traditional PV module, it is cost-effective to use higher-quality cells to increase efficiency. For CPVs, all concentrating systems presented in Chapter 3 can be used. The most popular system of CPV, however, is the Fresnel lens system. As in all concentrating systems, a tracking mechanism is required to follow the sun trajectory. Usually, a number of PVs are installed in a single box and atop each a Fresnel lens is installed. A CPV system can include a number of boxes, all put in a single tracking frame. For this type of system, two-axis tracking is required. A schematic diagram of a CPV Fresnel system is shown in Fig. 9.22a and a photograph of an actual system is shown in Fig. 9.22b. It should be noted that, in CPVs, the distribution of solar radiation on cells has to be as uniform as possible to avoid hot spots.

Because the temperature developed in CPV systems is high, some means of removing the heat energy must be provided to avoid reduction in the PV efficiency and prolong the life of the PVs. In some systems, this extra heat is used to provide thermal energy input to other processes, as in the hybrid PV/Ts analyzed in the next section.

FIGURE 9.22

Schematic diagram and a photograph of a CPV fresnel system. (a) Schematic diagram. (b) Photograph of an actual system.

9.8 Hybrid PV/T systems

A system that can provide both electrical and thermal energy simultaneously would be a very interesting application. Such a system could cover part of the electrical and thermal energy needs for a number of applications in industry and buildings (hospitals, schools, hotels, and houses).

The market for PV/T collectors and systems is growing fast, with significant global growth of about 13% in 2021. The total installed PV/T collector area at the end of 2021 was 1,442,596 m^2, with capacity divided into 751 MW$_{th}$ (thermal) and 254 MW$_{peak}$ (electric) (Weiss and Spörk-Dür, 2022).

Photovoltaic panels convert solar radiation to electricity with peak efficiencies in the range of 5%–20%, depending on the type of the PV cell. The efficiency of the solar cells drops with increasing operating temperatures. The temperature of PV modules increases by the absorbed solar radiation that is not converted into electricity, causing a decrease in their efficiency. For monocrystalline (c-Si) and polycrystalline (pc-Si) silicon solar cells, the efficiency decreases by about 0.45% for every degree rise in temperature. For a-Si cells, the effect is less, with a decrease of about 0.2% per degree rise in temperature, depending on the module design (see Section 9.2.2 for more details).

This undesirable effect can be partially avoided by a proper heat extraction with a fluid circulation. Natural circulation of air is the easiest way to remove heat from the PV modules and avoid the resulting efficiency drop. Hybrid PV/T collector systems may be applied, however, to achieve maximum energy output by simultaneous electricity and heat generation. In this way, the energy efficiency of the systems is increased considerably and the cost of the total energy output is expected to be lower than that of plain photovoltaic modules. The produced heat can be used to heat the building, for the production of hot water for the needs of the occupants or for low-temperature industrial applications. Stabilizing the temperature of the PV modules at a lower level is highly desirable and offers two additional advantages: an increase of the effective life of the PV modules and the stabilization of the current–voltage characteristic curve of the solar cells. Also, the solar cells act as good heat collectors and are fairly good selective absorbers (Kalogirou, 2001).

In hybrid PV/T solar systems the reduction of the PV module temperature can be combined with a useful fluid heating. Therefore, hybrid PV/T systems can simultaneously provide electrical and thermal energy, achieving a higher energy conversion rate of the absorbed solar radiation. These systems consist of PV modules coupled to heat extraction devices, in which air or water of lower temperature than that of PV modules is heated at the same time the PV module temperature is reduced. In PV/T system applications, the production of electricity is the main priority; therefore, it is necessary to operate the PV modules at low temperature to keep the PV cell electrical efficiency at a sufficient level. Natural or forced air circulation is a simple, low-cost method to remove heat from the PV modules, but it is less effective if the ambient air temperature is over 20 °C. To overcome this effect, the heat can be extracted by circulating water through a heat exchanger mounted at the rear surface of the PV module. PV/T systems provide a higher energy output than standard PV modules and could be cost-effective if the additional cost of the thermal unit is low. The water-type PV/T systems can be practical devices for water heating (mainly domestic hot water). Details of water PV/T systems are shown in Fig. 9.23. For air systems, a similar design is used but, instead of the heat exchanger shown in Fig. 9.23, the heat is removed by flowing air, as shown in Fig. 9.24.

Two basic types of PV/T systems can be considered, depending on the heat extraction fluid used, the water-type and the air-type PV/T systems, as shown in Figs. 9.23 and 9.24, respectively. The air-

FIGURE 9.23

Details of a water PV/T collector.

FIGURE 9.24

Details of an air PV/T collector.

type PV/T systems are of lower cost than the water-type PV/T ones and are suitable for building applications in medium- and high-latitude countries. In low-latitude countries, the ambient air temperature during the day is over 20 °C for almost half of the year, limiting the application of air-type PV/T systems to a shorter period in terms of effective electricity production. The water-type PV/T systems can be used effectively in all seasons, mainly in low-latitude countries, since water from public mains is usually under 20 °C.

Usually, the water-type PV/T models consist of silicon PV modules and the heat extraction unit is a metallic sheet with pipes for the water circulation, to avoid the direct contact of water with the PV rear surface. The heat exchanger is in thermal contact with the PV module rear surface and thermally insulated on the rear side of the heat exchanger element and the panel edges, as shown in Fig. 9.23. The heat exchanger in these systems is similar to the fin and tube arrangement used in flat-plate solar collectors, so the technology of this type of system is well known to the solar industry.

In the systems shown in Figs. 9.23 and 9.24, glazing is used and the final panels look like a conventional flat-plate collector. The systems, however, can also be unglazed, which is more suitable to very low-temperature applications. In the case of unglazed systems, satisfactory electrical output is obtained, depending on the operating conditions. The thermal efficiency, however, is reduced for higher operating temperatures, due to the increased thermal losses from the PV module front surface to the ambient. The addition of a glazing (like the glazing of the typical solar thermal collectors) increases significantly the thermal efficiency for a wider range of operating temperatures, but the additional optical losses from the glazing (from the additional absorption and reflection of the solar radiation) reduce the electrical output of the PV/T system.

The thermal analysis of the water and air PV/T systems are presented in the following two sections.

PV/T collectors with liquid heat recovery

The PV/T collector can be considered as a kind of solar thermal collector, which has as an absorption plate, the PV cells and a fluid heat extraction unit, in which the heat removal fluid circulates. In a water PV/T collector the heat extraction unit is usually a heat conductive plate with pipes for the circulation of the water, which is in thermal contact with PV rear side, while in air PV/T collectors it is usually an air duct placed at the rear side of the photovoltaic panel. In addition, a glazing can be used to reduce PV/T collector thermal losses, or the collector can be unglazed, to avoid the reduction in electrical output due to reflection optical losses and absorption heat losses from the glazing. The PV/T collector has also thermal insulation at the non-illuminated collector parts, similar to the way this is applied to a typical solar thermal collector. The flat-plate PV/T collector with water heat extraction can be analyzed in a similar way as a flat-plate thermal liquid collector using the basic collector model modified by Florschuetz (1979) and Tonui and Tripanagnostopoulos (2007).

Total thermal losses of PV/T collector U_L include top losses U_t, back losses U_b, and edge losses U_e, given by Eq. (3.9). These thermal losses are calculated using the same equations as the ones presented in Chapter 3, Section 3.3.2. Concerning the PV/T collector a modified heat losses coefficient $\overline{U}_L$ is used to give the reduced thermal losses due to the energy converted into electricity, given by:

$$\overline{U}_L = U_L - (\tau\alpha)\eta_{ref}\beta_{ref}G_t \qquad (9.52)$$

The electrical efficiency of the PV module η_{el} depends on the temperature T_{pv} and is given by (Florschuetz, 1979):

$$\eta_{el} = \eta_{ref}\left[1 - \beta_{ref}\left(T_{pv} - T_{ref}\right)\right] \qquad (9.53)$$

where

β_{ref} = temperature factor of PV efficiency (°C);
η_{ref} = electrical efficiency for the reference temperature T_{ref}.

The thermal efficiency of the collector η_{th} is obtained by dividing the useful energy collected Q_u by the available solar energy $(A_{pv}G_t)$. The useful energy is given by Eq. (3.31) or by using the collector inlet fluid temperature, T_i instead of collector plate temperature, T_p from Eq. (3.60) modified by Tonui and Tripanagnostopoulos (2007) for PV/T collectors:

$$\eta_{th} = \overline{F}_R \left[(\tau\alpha)(1 - n_{el}) - \overline{U}_L \left(\frac{T_i - T_a}{G_t} \right) \right] = \frac{\dot{m}c_p(T_o - T_i)}{A_{pv}G_t} \tag{9.54}$$

where the modified heat removal factor $\overline{F}_R$ is described by the modified collector efficiency factor $\overline{F}'$ and the two parameters differ from those of the flat-plate thermal collectors because the modified overall heat loss coefficient $\overline{U}_L$ is used instead of U_L. The relationship between $\overline{F}'$ and $\overline{F}_R$ is given by Florschuetz (1979):

$$\overline{F}_R = \frac{\dot{m}c_p}{A_{pv}\overline{U}_L} \left[1 - \exp\left(-\frac{A_{pv}\overline{U}_L\overline{F}'}{\dot{m}c_p} \right) \right] \tag{9.55}$$

PV/T collectors with air-heat recovery

In most air solar collectors, the air circulates through a channel formed between the PV solar radiation absorber and collector thermal insulation, and in some other systems through channels on both absorber (PV) sides, in one- or two-pass system (see Fig. 9.24). The usual heat extraction mode is the direct air heating from the absorber rear surface by natural or forced convection. The thermal efficiency depends on channel depth, air flow mode, and air flow rate. Small channel depth and high flow rate increase heat extraction, but increase also the pressure drop, which reduces the system net energy output in case of forced air flow, because of the increased power for the fan. In applications with natural air circulation, the small channel depth reduces air flow and therefore the heat extraction. In these systems a relatively large depth of air channel of about 0.1 m is necessary.

In the analysis of air PV/T collector performance the energy balance and thermal losses equations used in water PV/T collectors can be also applied. In a detailed analysis the air duct dimensions and other air circulation channel geometrical and air flow characteristics should be considered, similar to the analysis of air thermal collectors presented in Chapter 3, Section 3.4. The modified overall heat loss coefficient $\overline{U}_L$ and heat removal factor $\overline{F}_R$ for the air PV/T collectors can be also obtained from the formulas of Florschuetz (1979). For the air PV/T collector, the collector modified efficiency factor $\overline{F}'$ is calculated from Eq. (3.79) by replacing U_L by $\overline{U}_L$.

The steady-state thermal efficiency of the air PV/T collector can be calculated from Eq. (9.54).

9.8.1 Hybrid PV/T applications

The hybrid PV/T systems are considered an alternative to plain PV modules in several applications. They can be used effectively for converting the absorbed solar radiation into electricity and heat, therefore increasing their total energy output. In these systems, PV modules are coupled to heat extraction devices, in which water or air is heated and at the same time the PV module temperature is reduced to keep electrical efficiency at a sufficient level. Water-cooled PV/T systems are practical systems for water heating. These new solar energy systems are of practical interest for many applications, as they can effectively contribute to cover both the electrical and thermal loads.

It should be noted that the cost of the thermal unit remains the same irrespective of the type of PV material used, but the ratio of the additional cost of the thermal unit per PV module cost is almost double when amorphous silicon modules are used rather than the crystalline silicon ones. In addition, amorphous silicon PV modules present lower electrical efficiency, although the total energy output (electrical plus thermal) is almost equal to that of crystalline silicon PV modules.

The additional thermal output provided from the PV/T systems makes them cost-effective compared to separate PV and thermal units of the same total aperture surface area. In PV/T system applications, the production of electricity is the main priority; therefore, it is more effective to operate the PV modules at low temperature to keep a PV cell electrical efficiency at a sufficient level.

The daily and monthly performance of a hybrid PV/T system is investigated through modeling and simulation using the TRNSYS program (see Chapter 11, Section 11.5.1). Such a system provides more electrical energy than a standard photovoltaic system because it operates at a lower temperature; in addition, thermal energy is obtained, which can be used for water heating. As shown in Fig. 9.25, the system consists of a series of PV panels, a battery bank, and an inverter, whereas the thermal system consists of a hot water storage cylinder, a pump, and a differential thermostat (Kalogirou, 2001). In each case, the TRNSYS-type number used is indicated.

A copper heat exchanger is installed at the back of the photovoltaic panel, and the whole system is enclosed in a casing in which insulation is installed at the back and sides and a single low-iron glass is

FIGURE 9.25

Hybrid PV/T system schematic.

installed at the front to reduce the thermal losses (see Fig. 9.23). Water is used as a heat transfer medium. The system also employs eight batteries connected in a 4 × 2 mode, i.e., four batteries in parallel and two in series.

The findings from this investigation are very promising. In addition to the increased electrical energy produced by the system, almost 50% of the hot water needs of a four-person family are satisfied with such a system, and because of the heat removal, the PV cells' annual efficiency increased considerably.

In another case the application of water PV/T systems in industry has been studied. Water-type PV/T systems were also considered for this application. The PV/T systems can be used in several industrial applications, but the most suitable are applications that need heat in low (60–80°C) and mainly very low (<50°C) temperatures, since in these cases, both the electrical and the thermal efficiency of the PV/T system can be kept at an acceptable level. It should be noted that the fraction of heat demand at low temperatures is high, especially in the food, brewery and beverage industries and in the paper and textile industries, where its shares could be up to 80% of the overall thermal energy needs. For example, water-cooled PV/T systems could heat water for washing or cleaning processes. The PV/T collectors could be installed on the ground or on either flat or saw-tooth roofs, or on the façade of a factory (Kalogirou and Tripanagnostopoulos, 2007).

Finally, the performance and financial improvement of the PV/T systems was compared to the standard PV systems for building applications and proved very beneficial (Kalogirou and Tripanagnostopoulos, 2006). Additionally, it was proven that PV/T systems could be beneficial to the greater diffusion of PV units. This is especially important for countries with good penetration of solar water heaters, where it is a habit to produce hot water with solar energy. In these cases, it would be difficult to convince potential customers to install a PV system, whereas a hybrid system producing both electricity and hot water has better chances of success.

9.8.2 Water and air-heating BIPV/T

Building-integrated PV, called BIPV, is examined in a previous section. When this is combined with heat extraction it is called BIPV/T. This can be used with water or air fluid circulation. When a BIPV/T heat removal uses water (usually an aqueous propylene glycol solution) as the working fluid, the cost is much higher due to the required plumbing, more complex facade and the need to build hydronic systems integration and greater weight. Careful system output optimization of water-heating BIPV/T systems is therefore required to justify the initial capital cost investment. Although BIPV hydronic heat removal will improve PV efficiency, if the water circulation stops, for example from a pump failure, the very high consequential panel temperatures will adversely affect PV durability (Affolter et al., 2000). Furthermore the thermal efficiency of a BIPV/T collector is lower than a solar thermal flat-plate collector due to the lack of an aperture cover to inhibit convective heat loss. The solution would be to add a glazed cover which however, as seen before, will reduce the PV performance due to increase of PV optical losses and raise of PV temperature. For a large façade area structural and cost constraints may also be a problem.

Flat-plate air-heating BIPV/T collectors also yield an optimal combination of both electrical and thermal conversion efficiencies. A BIPV/T facade could act as an unglazed thermosiphon photovoltaic-thermal air-heating collector to provide natural ventilation in summer, preheated air in winter and electrical output throughout the year. A duct arranged behind the PV module allows air flow induced by buoyancy from the back of the PV panel and this air movement is governed by a combination of natural convection and wind-induced flow (Batagiannis and Gibbons, 2001). The temperature attained by the BIPV/T system depends on the incident solar energy, surface area, ambient air temperature, flow conditions, radiant surfaces, and the flow and temperature distribution (Mosfegh and Sandberg, 1998; Brinkworth et al., 1997; Tonui and Tripanagnostopoulos, 2008). It should be noted that for air flow in a PV rear-duct section, wind blowing from one direction may assist the air movement, and achieve greater cooling of the modules, whereas wind blowing from another direction may act against the required air flow direction, and reduce the cooling potential (Batagiannis and Gibbons, 2001). Additionally, for a naturally ventilated BIPV/T cladding element, buoyancy forces are balanced by the pressure drops due to the friction at the entrance and exit (Brinkworth et al., 2000). For zero wind velocity, the flow through the ventilated stack is driven only by buoyancy forces, whereas in all the other cases the flow inside the duct is due to a mixture of free and forced convection. Buoyancy induced air flow in a duct between the BIPV and the wall, even with a low mean air velocities can reduce BIPV operating temperature by about 15–20 °C giving also a 15% increase in electrical conversion efficiency (Brinkworth et al., 1997).

Exercises

9.1 Find the wavelength of radiation whose photons have energy equal to the band gap of cuprous sulfide (Cu_2S) cell (1.80 eV), compounds of cadmium sulfide (CdS) cell (2.42 eV), and gallium arsenide (GaAs) cell (1.40 eV).

9.2 A beam of blue light with wavelength of 0.46 μm and intensity of 1 mW strikes a solar cell. Estimate the number of photons incident on the cell.

9.3 The dark saturation current of a solar cell is 1.75×10^{-8} A when the cell is at 35°C and the short-circuit current when in sunlight is 4 A. Estimate the open-circuit voltage, the maximum power output of the cell, and the number and arrangement of cells required to make a panel to supply 90 W at 12 V.

9.4 A PV system gives 9 A when the solar radiation is 750 W/m². How many amperes will it give at 850 W/m²?

9.5 A 6 m² PV system gives 24 V and 18 A when exposed to solar radiation of 750 W/m². Estimate the cells' efficiency.

9.6 A PV system is required to produce 96 W at 12 V. Using solar cells that have I_{max} equal to 250 A/m² and V_{max} equal to 0.4 V, design the PV panel, working at the maximum power point, if each cell is 80 cm² in area.

9.7 Estimate the daily load and the peak power required by a PV system that has the following equipment connected:

   Four lamps, 15 W each, operated from 6 p.m. to 11 p.m.
   Television, 80 W, operated from 6 p.m. to 11 p.m.
   Computer, 150 W, operated from 4 p.m. to 7 p.m.
   Radio, 25 W, operated from 11 a.m. to 6 p.m.
   Water pump, 50 W, operated from 7 a.m. to 10 a.m.

9.8 A remote cottage has the following loads. Estimate the daily load and peak power to be satisfied by a 24 V PV system.

Appliance	Type	Power (W)	Daytime run (h)	Nighttime run (h)
5 lamps	DC	11 W each	0	5
Television	AC	75 W	2	4
Computer	AC	160 W	4	3
Radio	DC	25 W	3	1
Water pump	AC	60 W (6 A start current)	1	1
Stove	AC	1200 W	2	1

9.9 Using the loads of Exercise 9.8, estimate the expected daily energy requirement if the efficiency of the inverter is 91%, of the battery is 77%, and of the distribution circuit is 96%.

9.10 If the array and the power conditioning system losses are 10%, find the total energy delivered for a grid-connected system, assuming the efficiency of the inverter equals 90%, of the distribution circuit equals 95%, and a grid absorption rate of 90%. The energy delivered by the PV array is 500 Wh.

9.11 A south-facing PV panel is installed at 35° in a location that is at 40°N latitude. If, on May 15 noon, the beam radiation is 685 W/m^2 and the diffuse radiation is 195 W/m^2, both on a horizontal surface, estimate the absorbed solar radiation on the PV panel. The thickness of glass cover on PV is 2 mm, the extinction coefficient K is 4 m^{-1}, and ground reflectance is 0.2.

9.12 If, for a PV module operating under NOCT conditions, the cell temperature is 44 °C, determine the cell temperature when this module operates at a location where $G_t = 725$ W/m^2, $V = 1$ m/s, $T_a = 35$ °C, and the module is operating at its maximum power point. The dark saturation current of a solar module is 1.7×10^{-8} A/m^2 and the short-circuit current is 250 A/m^2.

9.13 Using the simple design method, design a PV system using 60 W, 12 V panels and 145 Ah, 6 V batteries. The PV system is required to offer 3 days of storage, the battery efficiency is 75%, and the depth of discharge is 70%. The location where the system is located has 6 h of daylight during wintertime and the application is 24 V with a load of 1500 Wh.

Assignment 1

With the help of the System Advisor Model (SAM), details of which are given in Appendix 10, design a grid-connected PV system for a small house in one of the locations given in the program weather database or your own location, provided that suitable weather data are available. Use the electrical energy cost rate ($/kWh) of your location and annual electrical energy consumption of 6000 kWh. The system can be simulated using the Detailed PV Model/Distributed/Residential Owner. The electric load system should be scaled to a total of 6000 kWh annual load, PV panels, and an inverter included in the program database. As part of this simulation, determine the number of modules in series and parallel and choose an inverter that will match the array voltage and power.

Assignment 2

With the help of the SAM, details of which are given in Appendix 10, design a PV system of 50 MW capacity for a location included in the program database or your own location, provided that suitable weather data are available. Try to use for the various input parameters the most realistic values possible. You can choose different types of financial models, various types of PV modules, various types of inverters, tracking modes, and incentives. From the installation and operating costs of each combination, find the most cost-effective equipment setup combination. For parameters not specified above, you can use realistic figures (or the program default values), which need to be listed in the report produced.

Assignment 3

Using TRNSYS V.18 demo, downloaded as explained in Section 11.5.1, use the example PV/IV curve and data from a real PV panel to produce the IV curve of the panel, compare it with the manufacturer curve, and estimate the maximum power.

Assignment 4

Using TRNSYS V.18 demo, downloaded as explained in Section 11.5.1, use example photovoltaic/PV-Inverter data to simulate and examine the performance of a grid-connected PV system with the following characteristics:

N-series = 8
N-Parallel = 2
Area = 1.2 m^2

Note: These can be changed by choosing parameters (calculator sign), and "Edit all equations" at the bottom of the window.

You will also need to change the "Type 65d" plotter left axis maximum from 5,000 to 20,000 to properly see the plotted values.

References

Affolter, P., Ruoss, D., Tuggweiler, P., Haller, A., 2000. New Generation of Hybrid Solar PV/T Collectors. Final Report DIS56360/16868, June.

Archer, M.D., Hill, R. (Eds.), 2001. Clean Electricity from Photovoltaics, vol. 1. Imperial College Press, London, UK.

ASHRAE, 2004. Handbook of Systems and Applications. ASHRAE, Atlanta. Note: The material quoted in this reference may also appear in subsequent ASHRAE handbooks, but here the year of the original version of the handbook used is shown.

Barbose, G., Darghouth, N., Wiser, R., Seel, J., 2011. Tracking the Sun IV: An Historical Summary of the Installed Cost of Photovoltaics in the United States from 1998 to 2010. Lawrence Berkeley National Laboratory, p. 61.

Batagiannis, P., Gibbons, C., 2001. Thermal assessment of silicon-based composite materials used in photovoltaics. In: Conference Proceedings of Renewable Energy in Maritime Island Climates, Belfast, UK, pp. 151–157.

Blaesser, G., 1997. PV system measurements and monitoring the European experience. Sol. Energy Mater. Sol. Cells 47, 167–176.

Brinkworth, B.J., Cross, B.M., Marshall, R.H., Hongxing, Y., 1997. Thermal regulation of photovoltaic cladding. Sol. Energy 61 (3), 169–178.

Brinkworth, B.J., Marshall, R.H., Ibarahim, Z., 2000. A validated model of naturally ventilated PV cladding. Sol. Energy 69 (1), 67–81.

Chang, T.P., 2009. The Sun's apparent position and the optimal tilt angle of a solar collector in the northern hemisphere. Sol. Energy 83 (8), 1274–1284.

Coutts, T.J., 1999. A review of progress in thermophotovoltaic generation of electricity. Renew. Sustain. Energy Rev. 3 (2–3), 77–184.

De Soto, W., Klein, S.A., Beckman, W.A., 2006. Improvement and validation of a model for photovoltaic array performance. Sol. Energy 80 (1), 78–88.

Deline, P., 2009. Partially shaded operation of a grid-tied PV system. In: Proceedings of Photovoltaic Specialists Conference (PVSC) IEEE, 7–12 June 2009, Philadelphia, Pennsylvania, USA.

Denholm, P., Margolis, R., 2007. The Regional Per-Capita Solar Electric Footprint for the United States. National Renewable Energy Laboratory. Technical Report NREL/TP-670-42463.

Duffie, J.A., Beckman, W.A., 2006. Solar Engineering of Thermal Processes, third ed. Wiley & Sons, New York.

Egido, M., Lorenzo, E., 1992. The sizing of stand-alone PV systems: a review and a proposed new method. Sol. Energy Mater. Sol. Cells 26 (1–2), 51–69.

EPIA, 2010. Global Market Outlook for Photovoltaics until 2014. European Photovoltaic Industry Association, p. 28. May 2010 update.

Fanney, A.H., Dougherty, B.P., Davis, M.W., 2002. Evaluating building integrated photovoltaic performance models. In: Proceedings of the 29th IEEE Photovoltaic Specialists Conference (PVSC), New Orleans, LA.

Florschuetz, L.W., 1979. Extension of the Hottel -Whillier model to the analysis of combined photovoltaic/thermal flat plate collectors. Sol. Energy 22 (2), 361–366.

Fragaki, A., Markvart, T., 2008. Stand-alone PV system design: results using a new sizing approach. Renew. Energy 33 (1), 162–167.

Fthenakis, V.M., Kim, H.C., 2011. Photovoltaics: life-cycle analyses. Sol. Energy 85 (8), 1609–1628.

Hansen, A.D., Sorensen, P., Hansen, L.H., Binder, H., 2000. Models for a Stand-Alone PV System. Riso National Laboratory, Roskilde, Denmark. Riso-R-1219(EN)/SEC-R-12.

Hashimoto, O., Shimizu, T., Kimura, G., 2000. A novel high performance utility interactive photovoltaic inverter system. In: 35th IAS Annual Meeting and World Conference on Industrial Applications of Electrical Energy, October, Rome, Italy, pp. 2255–2260.

Hontoria, L., Aguilera, J., Zufiria, P., 2005. A new approach for sizing stand-alone photovoltaic systems based in neural networks. Sol. Energy 78 (2), 313–319.

Huld, T., Gottschalg, R., Beyer, H.G., Topic, M., 2010. Mapping the performance of PV modules, effects of module type and data averaging. Sol. Energy 84 (2), 324–328.

Jardine, C.N., Lane, K., Conibeer, G.J., 2001. PV-Compare: direct comparison of eleven PV technologies at two locations in Northern and Southern Europe. In: Proceedings of the 17th European Conference on Photovoltaic Solar Energy Conversion, Munich, 2001.

Kalogirou, S.A., 2001. Use of TRNSYS for modelling and simulation of a hybrid PV-thermal solar system for Cyprus. Renew. Energy 23 (2), 247–260.

Kalogirou, S.A., Tripanagnostopoulos, Y., 2006. Hybrid PV/T solar systems for domestic hot water and electricity production. Energy Convers. Manage. 47 (18–19), 3368–3382.

Kalogirou, S.A., Tripanagnostopoulos, Y., 2007. Industrial application of PV/T solar energy systems. Appl. Therm. Eng. 27 (8–9), 1259–1270.

Kazmerski, L., 1997. Photovoltaics: a review of cell and module technologies. Renewable Sustainable Energy Rev. 1 (1–2), 71–170.

Kelly, N., Gibson, T., 2009. Improved photovoltaic energy output for cloudy conditions. Sol. Energy 83 (11), 2092–2102.

Kelly, N., Gibson, T., 2011. Increasing the solar photovoltaic energy capture on sunny and cloudy days. Sol. Energy 85 (1), 111–125.

King, D.L., Kratochvil, J.E., Boyson, W.E., Bower, W.I., 1998. Field experience with a new performance characterization procedure for photovoltaic arrays. In: Proceedings of the Second World Conference and Exhibition on Photovoltaic Solar Energy Conversion on CD ROM, Vienna, Austria.

King, D.L., Boyson, W.E., Kratochvil, J.E., 2004. Photovoltaic Array Performance Model. Sandia National Laboratories. Report SAND 2004–3535.

Kjar, S.B., Pederson, J.K., Blaabjerg, F., 2005. A review of simple-phase grid-connected inventors for photovoltaic modules. IEE Trans. Ind. Appl. 41, 1292–1306.

Kullmann, S., 2009. Specific Energy Yield of Low-Power Amorphous Silicon and Crystalline Silicon Photovoltaic Modules in a Simulated Off-Grid, Battery-Based System. MSc Thesis. Humboldt State University, p. 123.

Lasnier, F., Ang, T.G., 1990. Photovoltaic Engineering Handbook. Adam Higler, Princeton, NJ, p. 258.

Lisell, L., Mosey, G., 2010. Feasibility Study of Economics and Performance of Solar Photovoltaics at the Former St. Marks Refinery. St. Marks, Florida. National Renewable Energy Laboratory. Technical Report, NREL/TP-6A2–48853.

Lorenzo, E., 1994. Solar Electricity Engineering of Photovoltaic Systems. Artes Graficas Gala, S.L., Madrid, Spain.

Marion, B., 2008. Comparison of Predictive Models for PV Module Performance. National Renewable Energy Laboratory, p. 9. May 2008, NREL/CP-520–42511.

Markvart, T., Fragaki, A., Ross, J.N., 2006. PV system sizing using observed time series of solar radiation. Sol. Energy 80 (1), 46–50.

Mellit, A., Kalogirou, S., 2021. Artificial intelligence and internet of things to improve efficacy of diagnosis and remote sensing of solar photovoltaic systems: challenges, recommendations and future directions. Renew. Sustain. Energy Rev. 143, 110889.

Mellit, A., Benghanem, M., Hadj Arab, A., Guessoum, A., 2005. An adaptive artificial neural network for sizing of stand-alone PV system: application for isolated sites in Algeria. Renew. Energy 3 (10), 1501–1524.

Mosfegh, B., Sandberg, M., 1998. Flow and heat transfer in air gap behind photovoltaic panels. Renew. Sustain. Energy Rev. 2, 287–301.

Nelson, J., 2002. Organic photovoltaic films. Curr. Opin. Solid State Mater. Sci. 6 (1), 87–95.

Norton, B., Eames, P.C., Mallick, T.K., Huang, M.J., McCormack, S.J., Mondol, J.D., Yohanis, Y.G., 2011. Enhancing the performance of building integrated photovoltaics. Sol. Energy 85 (8), 1629–1664.

Noun, B., 2007. Solar Power. . Renewable Energy 2007/2008. WREN, pp. 62–63.

Photon, January 2012. 27.7 GW of Grid-Connected PV Added in 2011. Photon.

Price, S., Margolis, R., January 2010. 2008 Solar Technologies Market Report. National Renewable Energy Laboratory, p. 131. Primary Authors.

Razza, S., Castro-Hermosa, S., Di Carlo, A., Brown, T.M., 2016. Research Update: large-area deposition, coating, printing, and processing techniques for the upscaling of perovskite solar cell technology. Apl. Mater. 4 (91508), 091508.

RENI, April 2010. PV Power Plants 2010 — Industry Guide. RENI Renewables Insight, p. 45.

RENI, April 2012. PV Power Plants 2012 — Industry Guide. RENI Renewables Insight, p. 108.

Runyon, J., 2012. Staying Alive: Could Thin-Film Manufacturers Come Out Ahead in the PV Wars? Part 2. Available from: www.RenewableEnergyWorld.com.

Sayigh, A.A.M., 2008. Renewable Energy 2007/2008. World Renewable Energy Network, pp. 9–15.

Schreiber, D., 2009. PV Market Overview: Where Does the Thin Film Market Stand and Could Go? EuPD Research, Munich, p. 23.

Sick, F., Erge, T., 1996. Photovoltaics in Buildings: A Design Handbook for Architects and Engineers. James & James Limited, London, UK.

Simmons, A.D., Infield, D.G., 1996. Grid-connected amorphous silicon photovoltaic array. Prog. Photovoltaics Res. Appl. 4, 381–388.

Tidball, R., Bluestein, J., Rodriguez, N., Knoke, S., November, 2010. Cost and Performance Assumptions for Modeling Electricity Generation Technologies. National Renewable Energy Laboratory, p. 211. NREL/SR-6A20-48595.

Tonui, J.K., Tripanagnostopoulos, Y., 2007. Performance improvement of PV/T solar collectors with natural air flow operation. Sol. Energy 81 (4), 498–511.

Tonui, J.K., Tripanagnostopoulos, Y., 2008. Performance improvement of PV/T solar collectors with natural air flow operation. Sol. Energy 82 (1), 1–12.

Toyokawa, S., Uehara, S., 1997. Overall evaluation for R&D of PV modules integrated with construction materials. In: Conference Record of the IEEE Photovoltaic Specialists Conference, October, Anaheim, CA, USA, pp. 1333–1336.

Weiss, W., Spörk-Dür, M., 2022. Solar Heat Worldwide, Global Market Development and Trends in 2021, Solar Heating and Cooling Program. IEA, 2022.

Woyte, A., van den Keybus, J., Belmans, R., Nijs, J., 2000. Grid connected photovoltaics in the urban environment – an experimental approach to system optimisation. In: IEEE Proceedings for Power Electronics and Variable Speed Drives, September, pp. 548–553.

Yin, W.-J., Shi, T., Yan, Y., 2014. Unique properties of halide perovskites as possible origins of the superior solar cell performance. Adv. Mater. 26 (27), 4653–4658.

Solar Thermal Power Systems

10.1 Introduction

As was seen in Chapter 1, Section 1.5, solar thermal power systems were among the very first applications of solar energy. During the eighteenth century, solar furnaces capable of melting iron, copper, and other metals were constructed of polished iron, glass lenses, and mirrors. The furnaces were in use throughout Europe and the Middle East. The most notable examples were the solar furnace built by the well-known French chemist Lavoisier in 1774, various concentrators built by the French naturalist Bouffon (1747–48), and a steam-powered printing press exhibited at the Paris Exposition by Mouchot in 1872. This last application utilized a concentrating collector to supply steam to a heat engine.

Many of the early applications of solar thermal-mechanical systems were for small-scale uses, such as water pumping, with output ranging up to 100 kW. During the last 40 years, several large-scale experimental power systems have been constructed and operated, which led to the commercialization of some types of systems, and plants of 30–80 MW electric generating capacity have been in operation for many years.

Although the thermal processes for conversion of solar to mechanical and electrical energy operate at higher temperatures than those treated in earlier chapters, these are fundamentally similar to other solar thermal processes.

As was discussed in Chapter 9, the direct conversion of solar to electrical energy can be done with photovoltaics, which are solid-state devices. Electricity can also be produced with geothermal energy and wind power. However, with concentrating solar power systems, there are no complicated silicon-manufacturing processes, as in the case of PVs; no deep holes to drill, as in the case of geothermal systems; and no turbine housings that need to be kept greased at high elevations from the ground, as in wind-power systems. This chapter deals with the generation of mechanical and subsequently electrical energy from solar energy by heat engines powered with concentrating solar collectors. The use of solar ponds and solar updraft tower for power production is also examined. The cost of thermal power systems is much lower than that for photovoltaics, but most of them are suitable only for large-scale systems. Concentrating solar power plants use mirrors to generate high-temperature heat that drives steam turbines traditionally powered from conventional fossil fuels.

The basic schematic of conversion of solar to mechanical energy is shown in Fig. 10.1. In these systems, solar thermal energy, usually collected by concentrating solar collectors, is used to operate a heat engine. Some of these systems also incorporate heat storage, which allows them to operate during cloudy weather and nighttime. The main challenge in designing these systems is to select the correct

Solar Energy Engineering. https://doi.org/10.1016/B978-0-323-99350-0.00002-8

FIGURE 10.1

Schematic diagram of a solar thermal energy conversion system.

operating temperature. This is because the efficiency of the heat engine rises as its operating temperature rises, whereas the efficiency of the solar collector reduces as its operating temperature rises. Concentrating solar collectors are used exclusively for such applications because the maximum operating temperature for flat-plate collectors is low relative to the desirable input temperature for heat engines, and therefore system efficiencies would be very low.

Six system architectures have been used for such applications. The first four are high-temperature systems: the parabolic trough collector system, the linear Fresnel reflector, the power tower system, and the dish system. The last two are the solar pond and the solar updraft tower, which are low-temperature systems. These, except the linear Fresnel reflector system, which has not yet reached industrial maturity, are analyzed in this chapter, together with models of heat engines derived from basic thermodynamic principles.

In concentrating solar power (CSP) systems, sunlight is concentrated using mirrors to create heat, then the heat is used to create steam, which is used to drive turbines and generators, just like in a conventional power station. Such plants have been operating successfully in California since the mid-1980s and currently provide power for about 100,000 homes. About 10 years ago, a CSP plant, Nevada Solar I, started operating in Nevada, and another two, called PS10 (10 MW) and PS20 (20 MW), started operating in Spain. More CSP plants are under construction in several other countries of the world. Apparently, the Spanish government has realized the huge potential of the CSP industry and is subsidizing the electricity produced with a feed-in tariff scheme. PS10 and PS20 together provide electricity for 200,000 homes.

CSP projects contributing more than 1.3 GW were under construction in 2020 in the United Arab Emirates, China, Chile, and India. At the end of 2020, plants under construction worldwide comprised just over 1 GW of parabolic trough systems, just under 0.3 GW of tower systems, and 14 MW of a Fresnel system. Apart from the two hybrid CSP-natural gas plants, all the other plants included thermal energy storage (TES).

Because of the large area required for the CSP plants, these are usually located on non-fertile ground, such as deserts. According to the Trans-Mediterranean Renewable Energy Corporation (TREN), each square kilometer of the desert receives solar energy equivalent to 1.5 million barrels of oil. It has also been estimated that, if an area of desert measuring 65,000 km^2, which is less than 1% of the Sahara Desert, were covered with CSP plants, it could produce electricity equal to the year 2000 world electricity consumption (Geyer and Quaschning, 2000). One fifth of this area could produce the current electricity consumption of the European Union. Similar studies in the United States predict that the solar resource in southwestern states could produce about 7000 GW with CSP, which is about seven times the current total U.S. electric capacity (Wolff et al., 2008).

The main technologies used in CSP plants are the parabolic trough collectors, power towers, and dish/Stirling-engine systems. Mainly due to the plants operating in California for more than 20 years, parabolic troughs are the most proven technology, and today they produce electricity at about US$0.10/kWh. The success and durability of these plants demonstrates the robustness and reliability of the parabolic trough technology. An interesting feature of parabolic troughs and power tower systems is their capacity to store heat, which enables them to continue producing electricity during the night or cloudy days. For this purpose, concrete, molten salts, ceramics, or phase-change media can be used, and this method is currently much cheaper than storing electricity in batteries. Fossil and renewable fuels such as oil, gas, coal, and biomass can be used for backup energy in these plants. The flexibility of heat storage combined with backup fuel operation enables the plants to provide both base-load power and peak power, which can be used to cover the air-conditioning load usually occurring in midday during summer, when the plants produce higher output.

Table 10.1 gives an overview of some of the performance characteristics of the concentrating solar power concepts (Muller-Steinhagen and Trieb, 2004). Parabolic troughs, linear Fresnel reflectors, and power towers can be coupled to steam cycles of 10–200 MW electric capacity, with thermal cycle efficiencies of 30%–40%. The same efficiency range applies for Stirling engines coupled to dish systems. The conversion efficiency of the power block remains essentially the same as in fuel-fired power plants. Overall solar-electric efficiencies, defined as the net power generation over incident beam radiation, are lower than the conversion efficiencies of conventional steam or combined cycles, because they include the conversion of solar radiative energy to heat within the collector and the conversion of the heat to electricity in the power block.

Due to the higher levels of concentration, dish systems usually achieve higher efficiencies than the parabolic trough system and are better suited for stand-alone, small power-producing systems; however, for higher outputs, many dish systems could be used.

Table 10.1 Performance characteristics of various CSP technologies.

Technology	Capacity range (MW)	Concentration	Peak solar efficiency (%)	Solar-electric efficiency (%)	Land use (m²/MWh a)
Parabolic trough	10–200	70–80	21	10–15	6–8
Fresnel reflector	10–200	25–100	20	9–11	4–6
Power tower	10–150	300–1000	20	8–10	8–12
Dish-Stirling	0.01–0.4	1000–3000	29	16–18	8–12

10.2 Parabolic trough collector systems

Details of this type of collector are given in Chapter 3, Section 3.2.1. As was seen in Chapter 3, parabolic trough collectors are the most mature solar technology to generate heat at temperatures up to 400 °C for solar thermal electricity generation or process heat applications. The biggest and oldest application of this type of system is the nine southern California power plants known as solar electric generating systems (SEGS), which have a total installed capacity of 354 MWe (Kearney and Price, 1992). Details on these plants are given in Table 10.2 (LUZ, 1990). As can be seen, SEGS I is 13.8 MWe, SEGS II–VII are 30 MWe each, and SEGS VIII and IX are 80 MWe each. These have been designed, installed, and operated in the Mojave Desert of southern California, the first one since 1985 and the last one since 1991. These plants are based on large parabolic trough concentrators providing steam to Rankine power plants. They generate peaking power, which is sold to the Southern California Edison utility. These plants were built in response to the 1970s oil crises, when the U.S. government gave tax and investment incentives on alternative energy, totaling to nearly 40% of their costs. Due to the research and development, economies of scale, and accumulated experience, there was a fall in cost of the parabolic trough generated electricity from US$0.30/kWh in 1985, when the first plant was built, to US$0.14/kWh in 1989, when the seventh plant of the cluster was built—a fall of more than 50% in 4 years. By 2008, California's parabolic trough plants have generated well over 15,000 GWh of utility-scale electricity with 12,000 GWh from solar energy alone, which is more than half of all solar electricity ever generated (Taggart, 2008a). This represented about US$2 billion worth of electricity sold over the last 20 years. The nine plants continue to perform as well or even better than when first installed. These plants have accumulated more than 290 plant-years of operating experience.

Parabolic solar collectors focus sunlight onto a receiver pipe through which a synthetic oil circulates. The current synthetic oil is an aromatic hydrocarbon, biphenyl-diphenyl oxide, trademark Monsanto Therminol VP-1. The synthetic oil is then piped through a heat exchanger to produce steam that drives a conventional electricity-generating turbine. As with other renewable technologies, no

Table 10.2 Characteristics of SEGS plants.

SEGS plant	Year operation began	Net output (MWe)	Solar outlet temp. (°C)	Luz collector used	Solar field area (m²)	Solar turbine efficiency (%)	Fossil turbine efficiency (%)	Annual output (MWh)
I	1985	13.8	307	LS-1	82,960	31.5	—	30,100
II	1986	30	316	LS-2	190,338	29.4	37.3	80,500
III	1987	30	349	LS-2	230,300	30.6	37.4	92,780
IV	1987	30	349	LS-2	230,300	30.6	37.4	92,780
V	1988	30	349	LS-2	250,500	30.6	37.4	91,820
VI	1989	30	390	LS-2	188,000	37.5	39.5	90,850
VII	1989	30	390	LS-2 + LS-3	194,280	37.5	39.5	92,646
VIII	1990	80	390	LS-3	464,340	37.6	37.6	252,750
IX	1991	80	390	LS-3	483,960	37.6	37.6	256,125

pollutants are emitted in the process of generating electricity. A natural gas system hybridizes the plants and contributes 25% of their output. The plants can supply peaking power, using solely solar energy, solely natural gas, or a combination of the two, regardless of time or weather. Fossil fuel can be used to superheat solar-generated steam (SEGS I) in a separate fossil-fired boiler to generate steam when insufficient solar energy is available (SEGS II–VII), or in an oil heater in parallel with the solar field when insufficient solar energy is available (SEGS VIII–IX). The most critical time for power generation and delivery, and the time in which the selling price of the power per kWh is highest, is between noon and 6 p.m. in the summer months of June to September. The operating strategy is designed to maximize solar energy use. The turbine-generator efficiency is best at a full load; therefore, the natural gas supplement is also used to allow full load operation, which maximizes plant output. A photograph of a typical system is shown in Fig. 10.2.

The basic component of the solar field is the solar collector assembly. Such an assembly is an independently tracking parabolic trough collector made of the metal support structure on which the parabolic reflectors (mirrors) are installed, together with the receiver tubes and supports. The tracking system includes the drive, sensors, and controller. Table 10.3 shows the design characteristics of the Luz collectors used in the California plants and Eurotrough, which is the product of a European research project. By combining the data shown in Table 10.3 with those of the nine plants shown in Table 10.2 and it can be seen that the general trend was to build larger collectors with higher concentration ratios so as to maintain high collector efficiency at higher fluid outlet temperatures.

The major components of the systems are the collectors, the fluid transfer pumps, the power generation system, the natural gas auxiliary subsystem, and the controls. The reflectors are made of black-silvered, low-iron float glass panels, which are shaped over parabolic forms. Metallic and lacquer protective coatings are applied to the back of the silvered surface. The glass is mounted on truss

FIGURE 10.2

Photograph of a SEGS plant.

http://www.sandia.gov/.

Table 10.3 Luz and eurotrough solar collector characteristics.

Collector	LS-1	LS-2	LS-2	LS-3	Eurotrough
Year	1984	1985	1988	1989	2004
Area (m^2)	128	235	235	545	545/817.5
Aperture (m)	2.5	5	5	5.7	5.77
Length (m)	50	48	48	99	99.5/148.5
Receiver diameter (m)	0.042	0.07	0.07	0.07	0.07
Concentration ratio	61	71	71	82	82
Optical efficiency	0.734	0.737	0.764	0.8	0.78
Receiver absorptance	0.94	0.94	0.99	0.96	0.95
Receiver emittance at (°C)	0.3 (300)	0.2 (300)	0.1 (350)	0.1 (350)	0.14 (400)
Mirror reflectance	0.94	0.94	0.94	0.94	0.94
Operating temperature (°C)	307	349	390	390	390

structures and the position of large arrays of modules is adjusted by hydraulic drive motors. The receivers are 70 mm in diameter steel tubes (except for LS-1) with cement selective surfaces surrounded by a vacuum glass jacket in order to minimize heat loss.

Maintenance of high reflectance is critical to plant operation. With a total of 2315×10^3 m^2 of mirror area, mechanized equipment has been developed for cleaning the reflectors, which is done regularly, at intervals of about 2 weeks.

Tracking of the collectors is controlled by sun sensors that utilize an optical system to focus solar radiation on two light-sensitive diodes. Any imbalance between the two sensors causes the controller to give a signal to correct the positioning of the collectors; the resolution of the sensor is 0.5°. There is a sensor and controller on each collector row, which rotate about horizontal north–south axes, an arrangement that results in slightly less energy incident on them over the year but favors summertime operation when peak power is needed.

Parabolic trough technology proved to be tough, dependable, and proven. They are sophisticated optical instruments, and today, the second-generation parabolic troughs have more precise mirror curvature and alignment, which enables them to have higher efficiency than the first plants erected in California. Other improvements include the use of a small mirror on the backside of the receiver to capture and reflect any scattered sun rays back onto the receiver, direct steam generation into the receiver tube to simplify the energy conversion and reduce heat loss, and the use of more advanced materials for the reflectors and selective coatings of the receiver. Particularly, research and development, which aims to reduce the cost in half in the coming years, includes:

- Higher-reflectivity mirrors.
- More sophisticated sun-tracking systems.
- Better receiver selective coatings, with higher absorptance and lower emittance.
- Better mirror-cleaning techniques.
- Better heat-transfer techniques by adopting direct steam generation.
- Optimized hybrid integrated solar combined-cycle system (ISCCS) designs to allow maximum solar input.

- Development of trough system designs that provide the best combination of low initial cost and low maintenance.
- Development of thermal storage options that allow nighttime dispatch of solar-only trough plants.

In the previous section, the possibility of using heat storage is mentioned. Parabolic trough collector systems produce heat at about 400 °C. This heat can be stored in an insulated container and used during nighttime. Currently molten salt is used for this purpose, and this system was used in the California plants, the newly built plants, and the plants that are under construction or planned for the near future. Since research in this field is ongoing, this could change in the coming years.

10.2.1 Description of the PTC power plants

Parabolic trough solar collector technology is currently the most proven solar thermal collector technology. This is primarily due to the nine plants operating in California's Mojave Desert since the mid-1980s. In these plants, large fields of parabolic trough collectors supply the thermal energy used to produce steam supplied to a Rankine steam turbine-generator cycle to produce electricity. Each collector has a linear parabolic reflector, which focuses the sun's direct (or beam) radiation on a linear receiver located at the focus of the parabola. Figure 10.3 shows a process flow diagram, representative of the majority of plants operating today in California. The collector field consists of many large

FIGURE 10.3

Schematic diagram of a solar Rankine parabolic trough system.

single-axis-tracking PTC collectors, installed in parallel rows aligned on a north–south horizontal axis and tracking the sun from east to west during the day to ensure that the sun is continuously focused on the linear receiver. A heat transfer fluid is circulated through the receiver, where it is heated by solar energy and returns to a series of heat exchangers in the power block to generate high-pressure superheated steam and back to the solar field. This steam is used in a conventional reheat steam turbine-generator to produce electricity. As shown in Fig. 10.3, the steam from the turbine is piped to a standard condenser and returns to the heat exchangers with pumps so as to be transformed again into steam. The type of condenser depends on whether a large source of water is available near the power station. Because all plants in California are installed in a desert, cooling is provided with a mechanical draft wet cooling towers.

Parabolic trough plants are designed to use primarily solar energy to operate; if it is sufficient, this solar energy alone can operate the plants at the full rated power. During summer months, the plants operate at full rated electric output for 10–12 h/day. Because the technology can be easily hybridized with fossil fuels, the plants can be designed to provide firm peaking to intermediate load power. All plants in California, however, are hybridized to use natural gas as backup to produce electricity that supplements the solar output during periods of low solar radiation and nighttime. As shown in Fig. 10.3, the natural gas-fired heater is situated in parallel to the solar field, or the optional gas-fired boiler steam reheater is located in parallel with the solar heat exchangers, to allow operation with either or both of the energy resources.

A synthetic heat transfer fluid is heated in the collectors and piped to the solar steam generator and superheater, where it generates the steam to supply the turbine. Reliable high-temperature circulating pumps are critical to the success of the plants, and substantial engineering effort has gone into assuring that pumps will stand the high fluid temperatures and temperature cycling. The normal temperature of the fluid returned to the collector field is 304 °C and that leaving the field is 390 °C.

As shown in Fig. 10.3, the power generation system consists of a conventional Rankine cycle reheat steam turbine with feed-water heaters, deaerators, and other standard equipment. The condenser cooling water is cooled in forced draft cooling towers.

The evaporator generates saturated steam and demands feed-water flow from the feed-water pump. In a steam generator, the heat transfer oil is used to produce slightly superheated steam at 5–10 MPa (50–100 bar) pressure, which then feeds a steam turbine connected to a generator to produce electricity.

Usually, the SEGS plants incorporate a turbine that has both high- and low-pressure stages, with reheat of the steam occurring between the stages. The turbine operational limitations are, at minimum, 16.2 bar steam pressure and 22.2 °C superheat. When these conditions are not met, steam is diverted around the turbine to the condenser via a bypass circuit and fractional splitter. During start-up or shutdown, the fraction sent to the turbine varies linearly between 0 and 1. A throttling valve in the bypass loop provides the equivalent pressure drop as the turbine would provide under the same conditions.

The feed-water heaters are heat exchangers that condense steam extracted from the turbine to heat feed-water, thereby increasing the Rankine cycle efficiency.

The deaerator is a type of feed-water heater, where steam is mixed with sub-cooled condensate to produce saturated water at the outlet. This helps purge oxygen from the feed-water, controlling corrosion.

Steam exiting the turbine is condensed so that it can be pumped through the steam generation system.

A new design concept, which integrates a parabolic trough plant with a gas turbine combined-cycle plant, called the *integrated solar combined-cycle system* (ISCCS), is shown schematically in Fig. 10.4. Such a system offers a possibility to reduce cost and improve the overall solar-to-electricity efficiency. As shown, the ISCCS uses solar heat to supplement the waste heat from a gas turbine to augment power in the steam Rankine bottoming cycle. In this system, solar energy is used to generate additional steam, and the gas turbine waste heat is used for preheating and steam superheating.

One of the most serious problems when working in a desert environment is cleaning dust from the parabolic mirrors. As a general rule, the reflectivity of glass mirrors can be returned to design levels with good washing. After considerable experience gained over the years, operating and maintenance procedures nowadays includes deluge washing and direct and pulsating high-pressure sprays, which use demineralized water for good effectiveness. Such operations are carried out during nighttime. Another measure that's applied is periodic monitoring of mirror reflectivity, which can be a valuable

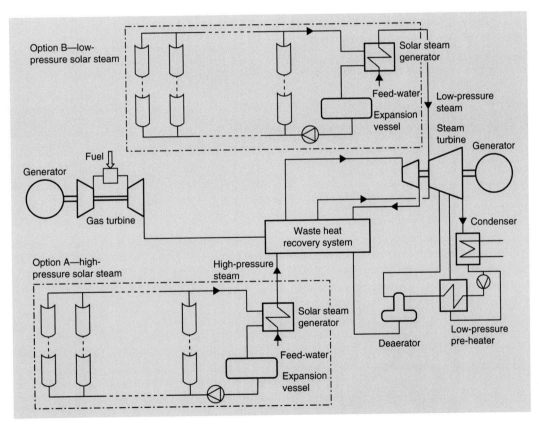

FIGURE 10.4

Schematic diagram of the integrated solar combined-cycle plant.

quality control tool to optimize mirror-washing frequency and the labor costs associated with this operation.

The benefits of direct steam generation were outlined previously. This method was demonstrated in Plataforma Solar de Almeria, Spain, in a 500 m long test loop providing superheated steam at 400 °C and 10 MPa (100 bar). To keep a two-phase steam-water flow in a large number of long, parallel, and horizontal absorber tubes is a major technical challenge. The system must be able to maintain a constant turbine inlet conditions and avoid flow instabilities, even during spatially and temporally changing insolation. Control strategies have been developed based on extensive experimentation and modeling of two-phase flow phenomena.

10.2.2 Outlook for the technology

The experience of the California plants has shown some benefits and some negative impacts that should be considered in designing new plants. The benefits include:

* The provision of the lowest-cost solar-generated electricity for many years of operation;
* Daytime peaking power coverage and with hybridization the ability to provide firm power, even during cloudy periods and night;
* Enhanced environmental protection, because no emissions occur during solar operation; and
* Positive impacts on local economy because systems are labor intensive during both construction and operation.

The negative impacts include:

* Heat transfer fluids could spill and leak, which can create problems in the soil;
* Water availability can be a significant issue in the arid regions that are best suited for trough plants. The majority of this water is required for the cooling towers;
* Parabolic trough plants require a considerable amount of land that cannot be used concurrently for other purposes; and
* Emissions occur when plants are operated with conventional fuels during hybrid operation.

Generally, in economic terms, by increasing the parabolic trough plant size, the cost of solar electricity is reduced. Cost reductions typically occur because of increased manufacturing volume, which lowers the cost per square meter, the relative initial cost to build a bigger plant, and operating and maintenance cost on a per-kilowatt-hour basis. Additionally, hybridization offers a number of potential benefits to solar plants, including reduced risk to investors and improved solar-to-electric conversion efficiency. Since fossil fuels are currently cheap, hybridization also provides a good opportunity to reduce the average cost of electricity from the plant.

The last issue to consider is thermal storage. The availability of low-cost thermal storage is important for the long-term cost reduction of solar trough technology and could significantly increase the potential market opportunities. For example, a plant located in California without fossil-fuel backup and thermal storage would produce electricity at an annual load factor of only 25%. The addition of thermal storage could increase this factor to about 50% because the plant would be able to work at non-solar times of the day and by allowing the solar field to be oversized. It should be noted, however, that attempting to increase the factor much above 50% would result in significant dumping of solar energy during summer months.

In 2007, the Nevada Solar One started operating in the state of Nevada. The Nevada Solar One plant uses the parabolic trough technology, and provides energy to Las Vegas. Nevada Solar One is a PTC system located near Boulder City, Nevada. It is a collaboration of Nevada Power Company and Sierra Pacific Resources and the construction completed in June 2007. It occupies 400 acres and the total capacity is 64 MW, harnessing solar energy to power more than 14,000 homes every year. It represents a major renewable-energy success story and has the potential to compete directly with conventional fossil fuel-powered technologies. Nevada Solar One is an environmentally friendly, renewable utility-scale power solution that creates power with near-zero carbon emissions. It is estimated that the CO_2 emissions avoided are the equivalent to taking approximately 20,000 cars off the road annually.

It is estimated that an energy project utilizing concentrating solar power technology deployed over an area of approximately 100 × 100 miles in the Southwest U.S. could produce enough power for the entire U.S. annually.

A significant parabolic trough plant constructed in 2013, is the 280 MW Solana Generating Station installed at Gila Bend, near Phoenix, Arizona. The Solana Generating Station (*solana* means a sunny place in Spanish) required 7.7 km^2 PTCs to be installed and produce enough electricity to satisfy about 70,000 homes. This is the largest parabolic trough plant in the United States and is expected to be 20%–25% more efficient than the California plants, mainly due to a reduction in heat loss from the receiver. At about the same time, Andasol 1 and 2, the first parabolic trough power plants in Europe, were put in operation near Guadix in Andalusia, Spain (2009). Their name is a combination of *Anda*lusia and *Sol* (Sun in Spanish). The plants are located at high altitude (1100 m) in a semi-arid climate, so the site receives exceptionally high annual direct insolation of about 2200 kWh/m^2 per year. Each plant has a gross electricity output of 50 MWe, producing around 180 GWh per year. Each collector has a surface of 51 ha (hectares) and needs about 200 ha of land. The Andasol power plants are used to meet the Spanish power demand caused primarily by air-conditioning units. This peaks in the early afternoon at the time when both solar radiation and the plants' power output are also at their peak.

The Andasol plants have a thermal storage system that absorbs part of the heat produced in the solar collector field during the day. This heat is stored in a molten salt mixture (60% sodium nitrate and 40% potassium nitrate). The power system produces electricity using this heat during the evening, or cloudy weather. The thermal reservoir holds 1010 MWh of heat, enough to run the steam turbine for about 7.5 h at full-load. The thermal reservoirs consist of two tanks, 14 m in height and 36 m in diameter, containing molten salt. Andasol 1, 2, and 3 are able to supply environmentally friendly solar electricity for up to 600,000 people. In the United Arab Emirates, Mohammed bin Rashid Al Maktoum Solar Park is under construction which includes a 600 MW parabolic trough facility. Other plants are planned for many countries of the world.

As indicated previously, the best location of CSP plants is in isolated desert regions. Very few people live in these places, however, so the transmission of electricity to cities could be problematic and costly. Therefore, a long-term development of the technology depends on the willingness of governments or utility companies to erect high-capacity power lines to transfer the CSP electricity produced in remote regions to cities. A possibility to alleviate this problem is to use solar energy to produce hydrogen as an energy carrier, which can be transported more easily. Hydrogen can then be used in a fuel cell to produce electricity (see Chapter 7). Another attractive possibility to increase the penetration of CSP in the energy supply system is to combine it with other systems requiring thermal or electrical energy to operate. A good example is desalination, using either electricity with a reverse osmosis system or thermal energy with multiple-effect or multi-stage evaporators (see Chapter 8). Such a synergy could create large-scale environmentally friendly electricity and water solutions for many sunny places of the world.

10.3 Power tower systems

As was explained in Chapter 3, Section 3.2.4, power towers or central receiver systems use thousands of individual sun-tracking mirrors, called *heliostats*, to reflect solar energy onto a receiver located atop a tall tower. The receiver collects the sun's heat in a heat transfer fluid (molten salt) that flows through the receiver. This is then passed optionally to storage and finally to a power conversion system, which converts the thermal energy into electricity and supplies it to the grid. Therefore, a central receiver system is composed of five main components: heliostats, including their tracking system; receiver; heat transport and exchange; thermal storage; and controls. In many solar power studies, it has been observed that the collector represents the largest cost in the system; therefore, an efficient engine is justified to obtain maximum useful conversion of the collected energy. The power tower plants are quite large, generally 10 MWe or more, while the optimum sizes lie between 50 and 400 MW. It is estimated that power towers could generate electricity at around US$0.04/kWh by 2020 (Taggart, 2008b).

The salt's heat energy is used to make steam to generate electricity in a conventional steam generator, located at the foot of the tower. The molten salt storage system retains heat efficiently, so it can be stored for hours or even days before being used to generate electricity.

The heliostats reflect solar radiation to the receiver at the desired flux density at minimal cost. A variety of receiver shapes has been considered, including cylindrical receivers and cavity receivers. The optimum shape of the receiver is a function of radiation intercepted and absorbed, thermal losses, cost, and design of the heliostat field. For a large heliostat field, a cylindrical receiver is best suited to be used with Rankine cycle engines. Another possibility is to use Brayton cycle turbines, which require higher temperatures (of about 1000 °C) for their operation; in this case, cavity receivers with larger tower height to heliostat field area ratios are more suitable.

For gas turbine operation, the air to be heated must pass through a pressurized receiver with a solar window. Combined-cycle power plants using this method could require 30% less collector area than the equivalent steam cycles. A first prototype of this system was built within a European research project, and three receiver units were coupled to a 250 kW gas turbine and tested.

Brayton cycle engines provide high engine efficiencies but are limited by the fact that a cavity receiver is required, which reduces the numbers of heliostats that can be used. Rankine cycle engines, driven from steam generated in the receiver and operated at 500–550 °C, have two important advantages over the Brayton cycle. The first is that the heat transfer coefficients in the steam generator are high, allowing the use of high-energy densities and smaller receivers. The second is that they employ cylindrical receivers, which permit larger heliostat fields to be used.

The U.S. Department of Energy and a consortium of U.S. utilities and industry built the first large-scale, demonstration solar power tower, called the Solar One, in the desert near Barstow, California. The plant operated successfully from 1982 to 1988, and the main outcome of the project was to prove that power towers could work efficiently to produce utility-scale power from sunlight. The system had the capacity to produce 10 MW of power. This plant used water-steam as the heat transfer fluid in the receiver, which presented several problems in terms of storage and continuous turbine operation.

These problems were addressed by Solar Two, which was an upgrade of Solar One. Solar Two operated from 1996 to 1999. Solar Two demonstrated how solar energy can be stored efficiently and economically as heat in tanks of molten salt, so that power can be produced even when the sun is not shining. The Solar Two plant used nitrate salt (molten salt, 60% $NaNO_3$ + 40% KNO_3) as both the heat transfer fluid in the receiver and the heat storage media. In this plant, the molten nitrate salt at 290 °C was pumped from a cold storage tank through the receiver, where it was heated to approximately 565 °C and then traveled to a storage tank, which had a capacity of 3 h of storage. A schematic of the system is shown in Fig. 10.5.

When power is needed from the plant, the hot salt is pumped to a generator that produces steam. The steam activates a turbine-generator system that creates electricity. From the steam generator, the salt is returned to the cold storage tank, where it is stored and can be eventually reheated in the receiver.

By using thermal storage, power tower plants can potentially operate for 65% of the year with no need for a backup fuel source. Without energy storage, solar technologies such as the parabolic trough plants are limited to annual capacity factors near 25%.

A commercial 10 MW solar tower, called PS10 and a 20 MW system called PS20 are currently operating near Seville, Spain. The plants are located 20 km west of Seville, which receives at least 9 h of sunshine 320 days per year, with 15 h per day in mid-summer. Each of the 624 heliostats of PS10 has a surface area of 120 m^2 that concentrates the sun's rays to the top of a 115 m high tower where a solar receiver and a steam turbine are located. The solar receiver produces saturated steam at 275 °C. The energy conversion efficiency is approximately 17%. PS20 consists of a solar field of 1255 heliostats designed by Abengoa Solar. Each heliostat with a surface area of 120 m^2 (same as PS10) reflects the

FIGURE 10.5

Schematic of the solar two plant.

solar radiation onto the receiver, located on the top of a 165 m high tower, producing steam which is converted into electricity by a turbine generator.

Another plant in Spain is Gemasolar. This power plant, which is also near Seville, is the first commercial solar thermal power plant to supply uninterrupted power for a full 24 h. Owned by Torresol Energy, Gemasolar, originally named Solar Tres, is the first commercial plant in the world to use molten salt thermal storage in a central tower configuration with a heliostat field (see Fig. 10.6). Gemasolar is seen as a breakthrough in the CSP sector, and its commercial operation is expected to lead the way for other central tower plants with molten salt receiver technology, an efficient system that improves the dispatchability of electric power from renewable sources.

The Gemasolar Concentrated Solar Power system has a capacity of 19.9 MW and consists of an array of 2650 heliostats that aim solar radiation at the top of a 140 m height central tower. The radiation heats molten salt that circulates inside the receiver to temperatures of more than 500 °C. The hot molten salt is then stored in tanks that are specially designed to maintain the high temperatures. These high temperatures in turn generate hotter, with respect to other CSP systems, pressurized steam, which significantly increases the plant's efficiency. The heat storage system enables the power plant to run steam turbines and generate electricity for up to 15 h without any incoming solar radiation. Therefore, the salt storage capacity enables the plant to supply energy to the grid based on demand, regardless of whether there is solar radiation.

Gemasolar's ability to generate 24 h of electricity marks an important step toward demonstrating the reliability of solar technology, which is one of the industry's biggest challenges. In addition to providing continuous power on cloudy days and at night, Gemasolar's storage capacity makes it possible to manage the supply of electricity sent to the network, and to respond to sudden increases in demand. In this way, the reliability of solar energy becomes comparable to that of conventional fossil-fueled power plants, which is important for the future adoption of renewable energy systems.

FIGURE 10.6

Gemasolar power system located near Seville, Spain.

Image by Torresol Energy.

The power generated by Gemasolar is supplied to a nearby power substation, where it is fed into the grid. Gemasolar can supply electricity to about 27,500 households in the South of Spain, as well as reduce carbon dioxide emissions by more than 30,000 tons per year. The plant is expected to produce a net total of over 110 GWh per year, by operating for a total of 6450 h a year at full capacity.

In United Arab Emirates, Mohammed bin Rashid Al Maktoum Solar Park is under construction which includes also a 100 MW tower facility.

Another important system is the construction of an experimental 1.5 MWe solar tower power plant in Germany. It has been operational since December 2008 and started the production of electricity in spring 2009. The plant, shown in Fig. 10.7, was built by Kraftanlagen München and is operated by the local utility Stadtwerke Jülich. The location of Jülich was chosen because of its fluctuating direct solar radiation, which allows, and requires, investigation of the system operation strategy under transient conditions—especially in combination with thermal storage. The objective of the solar power tower project in Jülich is to demonstrate the entire system.

The concentrator system consists of 2150 sun-tracking heliostats of about 8 m^2 reflective surface each and reflects the sunlight onto a 22 m^2 receiver aperture that has the shape of an inclined segment of a cylinder. The open volumetric receiver technology applied is used to heat up ambient air to high temperatures for steam generation in a boiler of a conventional steam turbine cycle. The porous absorber located at the top of a 60 m tower collects the highly concentrated solar radiation inside a honeycomb structure allowing the heat to be transferred to the air very effectively.

The advantages of this technology are simplicity and scalability, the ability to include a thermal storage, the low thermal capacity (quick start-up), and a high efficiency potential due to high achievable temperatures. A thermal storage unit is integrated into the air cycle, through which the operation of the power plant can be held for a certain time at constant power, depending on the storage dimensions. Generally, this thermal storage can be designed with an unlimited capacity, securing a continuous power plant operation. It is used as a buffer, storing energy in times of high irradiation, and enables operation of the plant after sunset or during periods of cloudy weather.

FIGURE 10.7

Solar tower Jülich.

10.3.1 System characteristics

Central receiver (or power tower) systems use a field of distributed mirrors, called *heliostats*, that individually track the sun and focus the sunlight on the top of a tower. By concentrating the sunlight 300–1500 times, they can achieve temperatures from 800 to over 1000 °C. The solar energy is absorbed by a working fluid, which is subsequently used to generate steam to power a conventional turbine. The average solar flux impinging on the receiver has values between 200 and 1000 kW/m^2. This high flux allows working at relatively high temperatures and integrating solar thermal energy in more efficient cycles.

Central receiver systems can easily be hybridized in a wide variety of options and have the potential to operate more than half the hours of each year at nominal power using a thermal energy storage. The central receiver plant is characterized by the heat transfer fluid, thermal storage medium, and power conversion cycle used. The heat transfer fluid may either be water-steam, liquid sodium, or molten nitrate salt (sodium nitrate–potassium nitrate), whereas the thermal storage medium may be oil mixed with crushed rock, molten nitrate salt, or liquid sodium.

In the initial central receiver plants, a receiver made from bundles of steel tubes on top of the tower was used to absorb the concentrated solar heat coming from the heliostat field. The Solar Two plant in California used molten salt as the heat transfer fluid and the thermal storage medium for nighttime operation. In Europe, air is preferred as the heat transfer medium, but tube receivers are not appropriate for this purpose because of a poor heat transfer and local overheating of the tubes. Therefore, within the PHOEBUS project in the 1990s, a volumetric receiver was developed using a wire mesh directly exposed to the incident radiation and cooled by air flowing through that mesh. In this project, the receiver achieved 800 °C and was used to operate a 1 MW steam cycle. A ceramic thermal heat storage was used for nighttime operation. A larger plant of 2.5 MW (thermal), based on this concept, was tested at the Plataforma Solar in the Almeria research center. In this plant, the solar energy is collected by 350 heliostats, each 40 m^2 in area. For even higher temperatures, the wire mesh screens are replaced by porous SiC or Al$_2$O$_3$ structures.

A European industry group, the PHOEBUS consortium, is leading the way with air-based systems. Air heat transfer receivers allow operation at significantly higher outlet temperatures, require higher operating pressures, but have relatively high heat losses compared to water-steam receivers. For these reasons, the PHOEBUS consortium developed a novel technology solar air (TSA) receiver, which is a volumetric air receiver that distributes the heat-exchanging surface over a three-dimensional volume and operates at ambient pressures. The greatest advantages of this system are its relative simplicity and safety. These make it ideal for applications in developing countries.

A photograph of the Solar Two system is shown in Fig. 10.8. The heliostat system consists of 1818 individually oriented reflectors made of back-silvered glass, each consisting of 12 concave panels with a total area of 39.13 m^2 (see Fig. 10.9), for a total mirror area of 71,100 m^2. The receiver of the plant is a single-pass superheated boiler, which is cylindrical in shape, 13.7 m in height, and 7 m in diameter. It is an assembly of 24 elements, each 0.9 m wide and 13.7 m long. Six of the elements on the south side, which receives the least radiation, are used as feed-water preheaters, and the rest are used as boilers. The top of the tower is 90 m above the ground. The receiver was designed to produce 50,900 kg/h of steam at 565 °C with the absorber operating at 620 °C. A detail of the receiver of Solar Two is also shown in Fig. 10.9.

The general requirements to install a solar tower plant are a site with high direct normal insolation, and the site needs to be level and have available water for the cooling towers.

FIGURE 10.8

Photograph of the solar two central receiver plant.

http://www.sandia.gov/.

FIGURE 10.9

Heliostat detail of the solar two plant.

http://www.sandia.gov/.

10.4 Dish systems

As was seen in Chapter 3, Section 3.2.3, dish systems use dish-shaped parabolic mirrors as reflectors to concentrate and focus the sun's rays onto a receiver, which is mounted above the dish at the dish focal point. The receiver absorbs the energy and converts it into thermal energy. This can be used directly as heat or can support chemical processes, but its most common application is in power generation. The thermal energy can be either transported to a central generator for conversion or converted directly into electricity at a local generator coupled to the receiver.

A dish-engine system is a stand-alone unit composed primarily of a collector, a receiver, and an engine, as shown in Fig. 10.10. It works by collecting and concentrating the sun's energy with a dish-shaped surface onto a receiver that absorbs the energy and transfers it to the engine. The heat is then converted in the engine to mechanical power, in a manner similar to conventional engines, by compressing the working fluid when it is cold, heating the compressed working fluid, and expanding it through a turbine or with a piston to produce mechanical power. An electric generator converts the mechanical power into electrical power.

Dish-engine systems use a dual-axis tracking system to follow the sun and so are the most efficient collector systems because they are always pointing at the sun. Concentration ratios usually range from 600 to 2000, and they can achieve temperatures in excess of 1500 °C. While Rankine cycle engines, Brayton cycle engines, and sodium-heat engines have all been considered for systems using dish-mounted engines, greatest attention has been paid to Stirling-engine systems (Schwarzbözl et al., 2000; Chavez et al., 1993).

The ideal concentrator shape is parabolic, created either by a single reflective surface (as shown in Fig. 3.23b) or multiple reflectors or facets (as shown in Fig. 10.10). Each dish produces 5–25 kW of electricity and can be used independently or linked together to increase generating capacity. A 650 kW plant composed of 25-kW dish-engine systems requires about a hectare of land.

The focus of current developments in the United States and Europe is on 10 kWe systems for remote applications. Three dish-Stirling systems are demonstrated at Plataforma Solar de Almeria in Spain. Within the European project EURODISH, a cost-effective 10 kW dish-Stirling engine for decentralized electric power generation was developed by a European consortium with partners from industry and academia.

FIGURE 10.10

Photograph of a dish concentrator with stirling engine.

http://www.sandia.gov/.

10.4.1 Dish collector system characteristics

Systems that employ small generators at the focal point of each dish provide energy in the form of electricity rather than heated fluid. The power conversion unit includes the thermal receiver and the heat engine. The thermal receiver absorbs the concentrated beam of solar energy, converts it to heat, and transfers the heat to the heat engine. A thermal receiver can be a bank of tubes with a cooling fluid circulating through it. The heat transfer medium usually employed as the working fluid for an engine is hydrogen or helium. Alternate thermal receivers are heat pipes wherein the boiling and condensing of an intermediate fluid is used to transfer the heat to the engine.

The heat engine system uses the heat from the thermal receiver to produce electricity. The engine-generators include basically the following components:

- A receiver to absorb the concentrated sunlight to heat the working fluid of the engine, which then converts the thermal energy into mechanical work.
- A generator attached to the engine to convert the work into electricity.
- A waste-heat exhaust system to vent excess heat to the atmosphere.
- A control system to match the engine's operation to the available solar energy.

The distributed parabolic dish system lacks thermal storage capabilities but can be hybridized to run on fossil fuel during periods without sunshine. The Stirling engine is the most common type of heat engine used in dish-engine systems. Other possible power conversion unit technologies that are evaluated for future applications are microturbines and concentrating photovoltaics (Pitz-Paal, 2002).

Solar dish systems are the most efficient solar energy systems. They provide economical power for utility line support and distributed and remote applications and are capable of fully autonomous operation. Their size typically ranges from 5 to 15 m in diameter or 5–25 kW per dish. Because of their size, they are particularly well suited for decentralized power supply and remote, stand-alone power systems, such as water pumping or village power applications, or grouped to form megawatt-scale power plants. Like all concentrating systems, they can additionally be powered by fossil fuel or biomass, providing constant capacity at any time.

10.5 Thermal analysis of solar power plants

Thermal solar power plants are similar to the conventional ones with the exception that a field of concentrating solar collectors replaces the conventional steam boiler. In hybrid plants, a conventional boiler is also present, operating on conventional fuel, usually natural gas, whenever there is a need. Therefore, the thermal analysis of solar power plants is similar to that of any other plant and the same thermodynamic relations are applied. The analysis is greatly facilitated by drafting the cycle on a T-s diagram. In these cases, the inefficiencies of pump and steam turbine should be considered. In this section, the equations of the basic Rankine power cycle are given and two of the more practical cycles, the reheat and the regenerative Rankine cycles, are analyzed through two examples. To solve the problems of these cycles, steam tables are required. Alternatively, the curve fits shown in Appendix 5 can be used. The problems that follow were solved by using steam tables.

The basic Rankine cycle is shown in Fig. 10.11a and its T-s diagram in Fig. 10.11b.

As can be seen in Fig. 10.11, the actual pumping process is 1–2′ and the actual turbine expansion process is 3–4′. The various parameters are as follows:

FIGURE 10.11

Basic Rankine power plant cycle. (a) Basic Rankine cycle schematic. (b) *T-s* diagram.

Turbine efficiency,

$$\eta_{\text{turbine}} = \frac{h_3 - h_{4'}}{h_3 - h_4} \tag{10.1}$$

Pump efficiency,

$$\eta_{\text{pump}} = \frac{h_2 - h_1}{h_{2'} - h_1} \tag{10.2}$$

Net work output,

$$W = (h_3 - h_{4'}) - (h_{2'} - h_1) \tag{10.3}$$

Heat input,

$$Q = h_3 - h_{2'} \tag{10.4}$$

Pump work,

$$_1W_{2'} = h_{2'} - h_1 = \frac{v(P_2 - P_1)}{\eta_{\text{pump}}} \tag{10.5}$$

Cycle efficiency,

$$\eta = \frac{W}{Q} = \frac{(h_3 - h_{4'}) - (h_{2'} - h_1)}{(h_3 - h_{2'})} \tag{10.6}$$

where

h = specific enthalpy (kJ/kg);
v = specific volume (m^3/kg); and
P = pressure (bar) = 10^5 N/m^2.

Generally, the efficiency of a Rankine cycle can be improved by increasing the pressure in the boiler. To avoid the increase of moisture in the steam coming out from the turbine, steam is expanded to an intermediate pressure and reheated in the boiler. In a reheat cycle, the expansion takes place in two turbines. The steam expands in the high-pressure turbine to some intermediate pressure, then passes back to the boiler, where it is reheated at constant pressure to a temperature that is usually equal to the original superheat temperature. This reheated steam is directed to the low-pressure turbine, where is expanded until the condenser pressure is reached. This process is shown in Fig. 10.12.

The reheat cycle efficiency is given by:

$$\eta = \frac{(h_3 - h_{4'}) + (h_5 - h_{6'}) - (h_{2'} - h_1)}{(h_3 - h_{2'}) + (h_5 - h_{4'})} \tag{10.7}$$

EXAMPLE 10.1

The steam in a reheat Rankine cycle leaves the boiler and enters the turbine at 60 bar and 390 °C. It leaves the condenser as a saturated liquid. The steam is expanded in the high-pressure turbine to a pressure of 13 bar and reheated in the boiler at 390 °C. It then enters the low-pressure turbine, where it expends to a pressure of 0.16 bar. Estimate the efficiency of the cycle if the pump and turbine efficiency is 0.8.

Solution

At point 3, P_3 = 60 bar and T_3 = 390 °C. From superheated steam tables, h_3 = 3151 kJ/kg and s_3 = 6.500 kJ/kg K.

At point 4, $s_4 = s_3$ = 6.500 kJ/kg K. From the problem definition, P_4 = 13 bar. From steam tables, h_4 = 2787 kJ/kg. To find $h_{4'}$, we need to use Eq. (10.1) for turbine efficiency:

$$\eta_{turbine} = \frac{h_3 - h_{4'}}{h_3 - h_4}$$

or $h_{4'} = h_3 - \eta_{turbine}(h_3 - h_4) = 3151 - 0.8(3151 - 2787) = 2860$ kJ/kg.

At point 5, P_5 = 13 bar and T_5 = 390 °C. From superheated steam tables, h_5 = 3238 kJ/kg and s_5 = 7.212 kJ/kg K.

At point 6, $s_6 = s_5$ = 7.212 kJ/kg K. From the problem definition, P_6 = 0.16 bar. From steam tables, s_{6f} = 0.772 kJ/kg K and s_{6g} = 7.985 kJ/kg K. Therefore, at this point, we have a wet vapor and its dryness fraction is:

$$x = \frac{s - s_f}{s_{fg}} = \frac{7.212 - 0.772}{7.985 - 0.772} = 0.893$$

At a pressure of 0.16 bar, h_f = 232 kJ/kg and h_{fg} = 2369 kJ/kg; therefore, $h_6 = h_f + x h_{fg} = 232 + 0.893 \times 2369 = 2348$ kJ/kg. To find $h_{6'}$, we need to use Eq. (10.1) for turbine efficiency:

$$h_{6'} = h_5 - \eta_{turbine}(h_5 - h_6) = 3238 - 0.8(3238 - 2348) = 2526 \text{ kJ / kg}$$

At point 1, the pressure is also 0.16 bar. Therefore, from steam tables at saturated liquid state, we have v_1 = 0.001015 m^3/kg and h_1 = 232 kJ/kg.

From Eq. (10.5),

$$h_{2'} - h_1 = \frac{v(P_2 - P_1)}{\eta_{pump}} = \frac{0.001015(60 - 0.16) \times 10^2}{0.8} = 7.592 \text{ kJ / kg}$$

Therefore, $h_{2'}$ = 232 + 7.592 = 239.6 kJ/kg.
Finally, the cycle efficiency is given by Eq. (10.7):

$$\eta = \frac{(h_3 - h_{4'}) + (h_5 - h_{6'}) - (h_{2'} - h_1)}{(h_3 - h_{2'}) + (h_5 - h_{4'})} = \frac{(3151 - 2860) + (3238 - 2526) - (239.6 - 232)}{(3151 - 239.6) + (3238 - 2860)}$$

$$= 30.3\%$$

FIGURE 10.12

Reheat Rankine power plant cycle. (a) Reheat Rankine cycle schematic. (b) T-s diagram.

The efficiency of the simple Rankine cycle is much less than the Carnot efficiency, because some of the heat supplied is transferred while the temperature of the working fluid varies from T_3 to T_1. If some means could be found to transfer this heat reversibly from the working fluid in another part of the cycle, then all the heat supplied from an external source would be transferred at the upper temperature and efficiencies close to the Carnot cycle efficiency could be achieved. The cycle where this technique is used is called a *regenerative cycle*.

In a regenerative cycle, expended steam is extracted at various points in the turbine and mixed with the condensed water to preheat it in the feed-water heaters. This process, with just one bleed point, is shown in Fig. 10.13, in which the total steam flow rate is expanded to an intermediate point 6, where a fraction, f, is bled off and taken to a feed-water heater; the remaining $(1 - f)$ is expanded to the condenser pressure and leaves the turbine at point 7. After condensation to state 1, the $(1 - f)$ kg of water is compressed in the first feed pump to the bleeding pressure, P_6. It is then mixed in the feed-water heater with f kg of bled steam in state 6 and the total flow rate of the mixture leaves the heater in state 3 and is pumped to the boiler, 4.

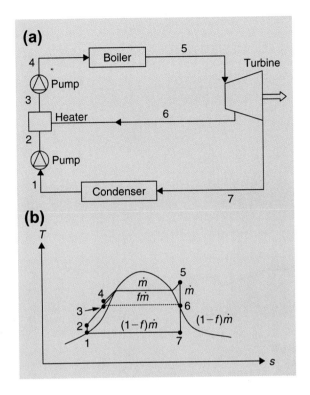

FIGURE 10.13

Rankine power plant cycle with regeneration. (a) Regenerative Rankine cycle schematic. (b) T-s diagram.

Although one feed-water heater is shown in Fig. 10.13, in practice, a number of them can be used; the exact number depends on the steam conditions. Because this is associated with additional cost, however, the number of heaters and the proper choice of bleed pressures is a matter of lengthy optimization calculations. It should be noted that if x number of heaters are used, $x + 1$ number of feed pumps are required.

The regenerative cycle efficiency is given by:

$$\eta = \frac{(h_5 - h_6) + (1 - f)(h_6 - h_7) - (1 - f)(h_2 - h_1) - (h_4 - h_3)}{(h_5 - h_4)} \tag{10.8}$$

where $f =$ fraction of steam in the turbine bled at state 6 to mix with the feed-water.

In this cycle, the enthalpy at state 3 can be found by an energy balance as:

$$\dot{m}h_3 = f\dot{m}h_6 + (1 - f)\dot{m}h_2 \tag{10.9}$$

from which:

$$h_3 = f(h_6 - h_2) + h_2 \tag{10.10}$$

EXAMPLE 10.2

In a regenerative cycle, steam leaves the boiler to enter a turbine at a pressure of 60 bar and a temperature of 500 °C. In the turbine, it expands to 5 bar, then a part of this steam is extracted to preheat the feed-water in a heater that produces saturated liquid, also at 5 bar. The rest of the steam is further expanded in the turbine to a pressure of 0.2 bar. Assuming a pump and turbine efficiency of 100%, determine the fraction of steam used in the feed-water heater and the cycle efficiency.

Solution

At point 5, $P_5 = 60$ bar and $T_5 = 500$ °C. From superheated steam tables, $s_5 = 6.879$ kJ/kg K and $h_5 = 3421$ kJ/kg.

At point 6, $s_6 = s_5 = 6.879$ kJ/kg K and $P_6 = 5$ bar. Again from superheated steam tables by interpolation, $h_6 = 2775$ kJ/kg.

At point 7, $P_7 = 0.2$ bar and s_7 is also equal to $s_5 = 6.879$ kJ/kg K. At this pressure, $s_f = 0.832$ kJ/kg K and $s_g = 7.907$ kJ/kg K. Therefore, the dryness fraction is:

$$x = \frac{s - s_f}{s_{fg}} = \frac{6.879 - 0.832}{7.907 - 0.832} = 0.855$$

At the same pressure, $h_f = 251$ kJ/kg and $h_{fg} = 2358$ kJ/kg. Therefore, $h_7 = h_f + x h_{fg} = 251 + 0.855 \times 2358 = 2267$ kJ/kg.

At point 1, the pressure is 0.2 bar, and because we have saturated liquid, $h_1 = h_f = 251$ kJ/kg and $v_1 = 0.001017$ m³/kg.

At point 2, $P_2 = 5$ bar and as $h_2 - h_1 = v_1(P_2 - P_1)$, $h_2 = 251 + 0.001017(5 - 0.2) \times 10^2 = 251.5$ kJ/kg.

At point 3, $P_3 = 5$ bar. From the problem definition, the water at this point is a saturated liquid. So, $v_3 = 0.001093$ m³/kg and $h_3 = 640$ kJ/kg. Using Eq. (10.9), $h_3 = f h_6 + (1 - f)h_2$, or:

$$f = \frac{h_3 - h_2}{h_6 - h_2} = \frac{640 - 251.5}{2775 - 251.5} = 0.154$$

At point 4, $P_4 = 60$ bar. Therefore, $h_4 - h_3 = v_3(P_4 - P_3)$ or $h_4 = h_3 + v_3(P_4 - P_3) = 640 + 0.001093(60 - 5) \times 10^2 = 646$ kJ/kg.

Finally, the cycle efficiency is obtained from Eq. (10.8):

$$\eta = \frac{(h_5 - h_6) + (1 - f)(h_6 - h_7) - (1 - f)(h_2 - h_1) - (h_4 - h_3)}{(h_5 - h_4)}$$

$$= \frac{(3421 - 2775) + (1 - 0.154)(2775 - 2267) - (1 - 0.154)(251.5 - 251) - (646 - 640)}{(3421 - 646)}$$

$$= 38.5\%$$

10.6 Solar updraft towers

The solar updraft tower is a renewable-energy power plant for generating electricity from solar power. The principle of operation of a solar chimney or updraft power plant is that because of its lower density hot air rises and creates a draft. The system consists of a chimney, a solar energy collector and wind turbines. In the collector, which is a large area of covered land, air is heated by solar radiation under a transparent (glass) or translucent (plastic) roof. This heat is forced upward through the chimney thereby creating a wind force as the joint between the roof and the tower base is airtight. Suction from the tower then draws in more hot air from the collector, and cold air comes in from the outer perimeter.

By placing wind turbines at the base of the tower the wind force can be used to produce electricity. One advantage of the system is that the collector itself functions as a greenhouse and could be used for

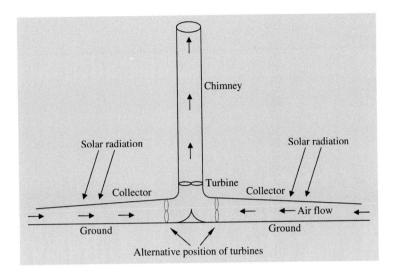

FIGURE 10.14

Schematic diagram of a solar updraft tower.

growing various crops. The solar updraft power technology has relatively low conversion efficiency and relatively high investment costs per MWh of electricity produced, but the operating costs are very low. The technology is particularly suitable in remote areas where low-value land can be used for the heat collection.

A schematic diagram of a solar updraft tower is shown in Fig. 10.14. Solar updraft towers make use of differences in temperatures of air near the ground and at the top of the tower or chimney. Through the chimney effect, i.e., forcing the air through a relatively small opening, the wind force becomes very strong. By placing a vertical wind turbine at the base of the tower or a number of horizontal turbines in a ring around the base of the tower, this updraft force can be used to produce electricity. As the only source of energy is solar energy, the solar updraft tower functions as a solar thermal power plant. Over the day the ground gets heated and over the night it gives its warmth to the ascending collector air. Continuous 24 h operation can be best achieved by placing water-filled tubes or bags under the roof. The water heats up during daytime and releases its heat at night. The tubes are filled with water only once (no further refilling is required) to increase the effect. Black colored bodies have the ability to absorb short wavelength radiation during the day to heat the water and emit long wavelength heat at night to warm the air.

Hot air for the solar tower is produced by the greenhouse effect in a simple air collector consisting of a nearly horizontal glass or plastic glazing installed several meters above the ground. The height of the glazing increases near the tower base, which has a smooth entrance so that the air is diverted to vertical movement with minimum friction loss. Therefore, the ground under the transparent roof is heated and transfers its heat to the air flowing radially inside the tower.

The requirement of a relatively large area implies that the technology is particularly suitable in those countries which have large areas of desert-type land.

A good review on the developments and advancement of solar chimney power plants is presented by Bernandes (2010).

The energy efficiency of the solar updraft tower is low. Therefore, a relatively large area is required for collecting the heated air in combination with a tall chimney. It is estimated that for a 200 MW capacity plant, the solar collector should have an area of 38 km^2 and the chimney height is 1000 m (Schlaich et al., 2005). The conversion efficiency of such a plant would be about 0.5% (or 1 kWh/m^2). The capital costs for such a plant would be relatively high as it requires a specialized expertise mainly for the tower construction, but the running costs would be very low and concern mainly maintenance of the wind turbines. However, when plastic is chosen as the cover for the solar collector, then this will require replacement every few years. Producing electricity with such a 200 MW solar tower would cost €0.07 per kWh whereas it would cost €0.21 per kWh for a smaller 5 MW plant due to economies of scale (Schlaich et al., 2005).

10.6.1 Initial steps and first demonstration

A Spanish artillery colonel Isidoro Cabanyes in 1903 provided one of the earliest descriptions of a solar chimney power station. He proposed a solar engine project in which an air heater was attached to a house with a chimney. In the house a kind of wind propeller was placed to produce electricity (Bernandes, 2010).

In 1926 Bernard Dubos proposed to the French Academy of Sciences the construction of a Solar Aero-Electric Power Plant in North Africa with its solar chimney on the slope of the high mountain. The author claimed that an air speed of 50 m/s can be reached in the chimney, and that this enormous amount of energy can be extracted by wind turbines (Bernandes, 2010).

Following these original concepts, the first prototype of a solar updraft tower was erected in 1982 in Manzanares, 150 km south of Madrid, Spain (see Fig. 10.15). The chimney height was 195 m and its diameter was 5.08 m. The collector area was 46,000 m^2 (about 11 acres, or 244 m diameter). This prototype is considered as a small-scale experimental model and although it was not intended for power generation, the peak power output was about 50 kW. During the testing phase (1982–89) different glazing materials were tested. Additionally, parts of the collector area were used as an actual greenhouse, growing plants under the glass. The prototype was built with financial support from the Federal Republic of Germany (Haaf et al., 1983). Preliminary test results including energy balances,

FIGURE 10.15

Photograph of the manzanares solar updraft tower plant in Spain.

collector efficiency values, pressure losses due to friction and losses in the turbine section were presented by Haaf (1984). Unfortunately the tower's guy-wires were not protected against corrosion and failed due to rust and storm winds. The tower blew over and was decommissioned in 1989.

The aim of this research project was to verify, through field measurements, the performance estimated from calculations based on theory, and to examine the influence of individual components on the plant's output and efficiency under realistic engineering and meteorological conditions. The main dimensions and technical data for the experimental plant are shown in Table 10.4.

Completion of the construction phase in 1982 was followed by an experimental phase, the purpose of which was to demonstrate the operating principle of a solar tower. The goals of this phase of the project were to (Schlaich et al., 2005):

1. Obtain data on the efficiency of the technology developed;
2. Demonstrate fully automatic, power plant-like operation with a high degree of reliability; and
3. Record and analyze operational behavior and physical relationships on the basis of long-term measurements.

The tower tube stood on a supporting ring 10 m above ground level. A pre-stressed membrane of plastic-coated fabric, shaped to provide good flow characteristics, formed the transition between the roof and the tower. The tower was guyed at four levels, and in three directions, to foundations secured with rock anchors. The turbine was supported independently of the tower on a steel framework 9 m above the ground level. The four blades of the turbine were adjustable according to the face velocity of the air in order to achieve an optimal pressure drop across the turbine blades. Maximum vertical wind velocity during turbine operation was 12 m/s. The collector roof of the solar tower not only has to have a transparent or translucent covering, it must also be durable and of low cost. A variety of plastic sheets and glass was considered in order to determine which was the best. Glass can resist heavy storms for

Table 10.4 Main dimensions and technical specifications of the manzanares prototype.

Item	Value
Nominal output	50 kW
Tower height	194.6 m
Tower radius	5.08 m
Mean collector radius	122.0 m
Mean roof height	1.85 m
Number of turbine blades	4
Operation modes	Stand-alone or grid connected
Typical collector air temperature increase	20 K
Collector covered with plastic membrane	40,000 m^2
Collector covered with glass	6000 m^2

many years without harm and prove to be self-cleaning due to the occasional rains. The plastic membranes were clamped to a frame and stressed down to the ground at the center by use of a plate with drain holes. The initial investment cost of plastic membranes is lower than that of glass; however, in the Manzanares pilot plant the membranes got brittle with time due to the UV radiation and thus tended to tear. Materials with temperature and UV stability and design improvements achieved in recent years may overcome this disadvantage.

In developing countries, the technology was tested in Botswana in 2005 where a small 22 m tall chimney was built with a collector area of 160 m^2. The chimney was made of polyester material and the roof of the collector area was made of glass.

The solar updraft tower plant has notable advantages in comparison with other power production technologies (Schlaich, 1995):

- The collector uses both direct and diffuse radiation.
- The ground provides a natural heat storage, which allows 24 h operation. This can be enhanced with additional water tubes or bags placed under the collector roof to absorb part of the radiated energy during the day and release it into the collector at night.
- The low number of rotating parts ensures good reliability.
- No cooling water is necessary for its operation in arid areas.
- Simple materials and known technologies are used for its construction.
- Developed countries are able to implement such technology without costly technological efforts and without high foreign currency expenditure using local resources and work force.

Additionally, these plants can have a much longer life time than conventional power plants, they have minimum operation and maintenance requirements, they are environmentally friendly as there are no CO_2 emissions during operation and they provide sustainable and affordable source of electricity.

Solar updraft towers also have features that make them less suitable for some sites (Schlaich et al., 2005):

- They require large areas of flat land, which should be available at low cost with no competing usage, like for example for intensive agriculture.
- They are not safe in earthquake prone areas, as in this case tower costs would increase drastically.
- Zones with frequent sand storms should also be avoided, as either collector performance losses or collector operation and maintenance costs would be substantial.

10.6.2 Thermal analysis of solar updraft tower plants

The fundamental analysis of solar updraft tower plants that influence the power output of a solar tower is presented by Schlaich et al. (2005). Generally, the power output P of the solar tower can be calculated as the solar input $\dot{Q}_{solar}$ multiplied by the efficiencies of collector (η_c), tower (η_{tr}) and turbine/s (η_t), given by:

$$P = \dot{Q}_{solar}\eta_c\eta_{tr}\eta_t = \dot{Q}_{solar}\eta_p \tag{10.11}$$

where η_p is the total efficiency of the plant comprising the collector, tower, and turbine. The solar energy input $\dot{Q}_{solar}$ into the system can be written as the product of global horizontal radiation G and collector area A_c:

$$\dot{Q}_{solar} = GA_c \tag{10.12}$$

Due to the chimney effect, the tower essentially converts the heat flow produced by the collector into kinetic and potential energies. Therefore the driving force is due to the density difference of the air caused by the temperature rise in the collector. The pressure difference Δp_{tot} produced between the tower base (or collector outlet) and the ambient air is given by:

$$\Delta p_{tot} = g \int_0^{H_t} (\rho_a - \rho_t)dh \tag{10.13}$$

Therefore, Δp_{tot} increases with tower height. Neglecting friction losses, the pressure difference Δp_{tot} can be subdivided into a static and a dynamic component:

$$\Delta p_{tot} = \Delta p_s + \Delta p_d \tag{10.14}$$

The static pressure difference drops at the turbine, whereas the dynamic component represents the kinetic energy of the airflow. With the total pressure difference and the volume flow of the air at $\Delta p_s = 0$ the total power (P_{tot}) that the air flow carries is given by:

$$P_{tot} = \Delta P_{tot} V_{t,max} A_c \tag{10.15}$$

With the help of Eq. (10.15) the efficiency of the tower can be obtained, given by:

$$\eta_t = \frac{P_{tot}}{\dot{Q}} = \frac{\Delta P_{tot} V_{t,max} A_c}{\dot{Q}} \tag{10.16}$$

It should be noted that the actual separation of the pressure difference into static and dynamic components depends on the energy taken up by the turbine. Without a turbine, the maximum flow speed of $V_{t,max}$ is obtained and the whole pressure difference is used to accelerate the air and thus all energy is converted into kinetic energy:

$$P_{tot} = \frac{1}{2}\dot{m}V_{t,max}^2 \tag{10.17}$$

Using the Boussinesq approximation, the speed reached by free convection currents can be obtained from:

$$V_{t,max} = \sqrt{2gH_t\frac{\Delta T}{T_o}} \tag{10.18}$$

where ΔT is the temperature rise between ambient and collector outlet [=tower inlet].

Finally, by replacing Eq. (10.18) into Eq. (10.17) and then into Eq. (10.16), and considering that $\dot{Q} = \dot{m}c_p\Delta T$ the tower efficiency is given by:

$$\eta_t = \frac{gH_t}{c_p T_o} \qquad (10.19)$$

where, g is the acceleration due to gravity (m^2/s), H_t is the tower height (m), c_p is the air heat capacity (J/kg K) and T_o is the ambient temperature (K). For instance, with a chimney height of 1000 m and standard conditions for temperature and pressure, the tower efficiency achieves a maximum value of 3%. Considering a collector efficiency (η_c) of 60% and a turbine efficiency (η_{tr}) of 80%, the total efficiency of the plant (η_p) reaches 1.4%, as $\eta_p = \eta_c \, \eta_{tr} \, \eta_t = 0.6 \times 0.8 \times 0.03 = 0.014$.

The simplified analysis presented above indicates that the tower efficiency actually depends only on its height, which is the basic characteristic of the solar updraft tower. Additionally, for heights of 1000 m, the deviation from the exact solution, caused by the Boussinesq approximation (Eq. 10.18), is negligible (Schlaich et al., 2005). By using Eqs. (10.11), (10.12) and (10.19) it can be found that solar tower power output is proportional to the collector area and tower height. Additionally, as the electrical output of the solar tower is proportional to the volume included within the tower height and collector area, the same output may result from a large tower with a small collector area and vice versa (Schlaich et al., 2005). Although this simplified analysis is not strictly valid when friction losses in the collector are considered, it is a good rule of thumb as long as the solar collector diameter is not too large (Schlaich et al., 2005).

The tower (or chimney) is the plant's actual thermal engine. It is a pressure tube with low friction loss. The updraft velocity of the air is approximately proportional to the air temperature rise (ΔT) in the collector and to the tower height as indicated by Eq. (10.18). In a large multi-megawatt solar updraft tower, the collector could rise the air temperature by about 30–35 K, which could produce a relatively small updraft velocity in the tower of about 15 m/s at nominal electric output.

Concerning the technology, towers of 1000 m height are a challenge, but they can be built today as complicated skyscrapers are now built even in earthquake countries like Japan. The solar updraft tower is much simpler, as a large diameter hollow cylinder which has much fewer demands in comparison with inhabited buildings, is required. The respective structural approaches are well known and have been used for many years in cooling towers. Recent detailed static and structural research showed that it is appropriate to stiffen the tower at several heights with cables arranged like spoke wheels within the tower and this allows thinner walls to be used (Schlaich et al., 2005).

10.7 Solar ponds

Salt gradient lakes, which exhibit an increase in temperature with depth, occur naturally. A salt gradient solar pond is a body of saline water in which the salt concentration increases with depth, from a very low value at the surface to near saturation at the depth of usually 1–2 m (Tabor, 1981). The density gradient inhibits free convection, and the result is that solar radiation is trapped in the lower region. Solar ponds are wide-surfaced collectors in which the basic concept is to heat a large pond or lake of water in such a way as to suppress the heat losses that would occur if less dense heated water is allowed to rise to the surface of the pond and lose energy to the environment by convection and radiation (Sencan et al., 2007). As shown in Fig. 10.16, this objective can be accomplished if a stagnant, highly transparent insulating zone is created in the upper part of the pond to contain the hot fluid in the lower part of the pond. In a non-convecting solar pond, part of the incident insolation is absorbed and converted to heat, which is stored in the lower regions of the pond. Solar ponds are both solar energy

FIGURE 10.16

Schematic vertical section through a salt-gradient solar pond.

collectors and heat stores. A salt-gradient non-convecting solar pond consists of three zones (Norton, 1992; Hassairi et al., 2001):

1. *Upper convecting zone* (UCZ). This is a zone, typically 0.3 m thick, of almost constant low salinity, which is at close to ambient temperature. The UCZ is the result of evaporation, wind-induced mixing, and surface flushing. Usually this layer is kept as thin as possible by the use of wave-suppressing surface meshes or by placing wind-breaks near the pond.
2. *Non-convecting zone* (NCZ). In this zone, both salinity and temperature increase with depth. The vertical salt gradient in the NCZ inhibits convection and thus gives the thermal insulation effect. The temperature gradient is formed due to the absorption of solar insolation at the pond base.
3. *Lower convecting zone* (LCZ). This is a zone of almost constant, relatively high salinity (typically 20% by weight) at a high temperature. Heat is stored in the LCZ, which should be sized to supply energy continuously throughout the year. As the depth increases, the thermal capacity increases and annual variations of temperature decrease. Large depths, however, increase the required initial capital expenditure and exhibit longer start-up times.

In solar ponds, it is necessary to suppress natural convection. Many techniques have been considered for this purpose; the most common method used is a salt stratification. Salinity increases with depth in the NCZ until the LCZ is reached (see Fig. 10.16). In the LCZ, solar radiation heats the high-salinity water, but because of its high relative density, hot, salty water cannot rise into the lower salinity layers, thus the heat is trapped and stored for use.

Chemically stable salts, as well as any natural brine, can be used in salt gradient solar ponds. A selected salt must be safe to handle, non-toxic, relatively cheap, and readily available, and its solubility should be temperature-dependent and should not reduce significantly the insolation transmission characteristics of water. Sodium and magnesium chlorides, though satisfying most criteria, have solubilities that are modestly temperature-dependent (Norton, 1992). Due to its low cost, sodium chloride remains the most popularly used salt.

Inorganic dirt brought by the wind can enter the pond, but generally the dirt causes no problem as it settles at the bottom. Various species of freshwater and saltwater algae grow under the conditions of temperature and salt concentration that exist in a stratified solar pond. Algae growth is undesirable because it reduces solar transmissivity. Most of these algae species are introduced by rainwater and airborne dust. An effective way to prevent algae formation is to add copper sulfate at a concentration of about 1.5 mg/L.

The thermal efficiency of a solar pond depends on the stability of its salt gradient. The pond cannot function without the proper maintenance of the stratification. The salt gradient is maintained by:

1. Control of the overall salinity difference among the three convecting layers;
2. Reducing internal convection currents in the NCZ; and
3. Limiting the growth of the UCZ.

Additionally, the efficiency of a solar pond is limited by some intrinsic physical properties. The first thing that reduces the efficiency is the reflection losses at the surface of the pond. After penetrating the surface, in the first few centimeters of water, the insolation is rapidly attenuated by about 50%, since half the solar spectrum is in the infrared region, for which water is almost opaque. This is the reason the shallow ponds yield negligible temperature rise. Practical efficiency values for ponds of 1 m depth are of the order of 15%–25% (Tabor, 1981). These figures are lower than for flat-plate collectors, but the lower cost, the built-in storage capability, and collection over large areas make solar ponds attractive under suitable environmental conditions. Generally, because the economics of solar ponds improve with size, large ponds are preferred.

10.7.1 Practical design considerations

In evaluating a particular site for a solar pond application, several factors need to be considered. The main ones are:

1. Since solar ponds are horizontal solar collectors, sites should be at low to moderate northern and southern latitudes, i.e., latitudes between ±40 °.
2. Each potential site has to be evaluated for its geological soil characteristics because the underlying earth structure should be free of stresses, strains, and fissures, which could cause differential thermal expansions, resulting in earth movement if the structure is not homogeneous.
3. Since the thermal conductivity of soil increases greatly with moisture content, the water table of the prospective site should be at least a few meters below the bottom of the pond to minimize heat losses.
4. A source of cheap salt- or seawater should be available locally.
5. The site should be fairly flat to avoid moving large quantities of earth.
6. A cheap source of water must be available to make up for evaporation losses.

Generally, two types of leakages occur in solar ponds: leakage of the saline water from the bottom of the pond and leakage of heat into the ground. The loss of hot saline water is the most serious, since it results in the loss of heat and salt. Additionally, the solar pond must not pollute the aquifers, and any continuous drain of hot water lowers the pond's storage capacity and effectiveness. Therefore, the selection of a liner for the pond is very important. Although it is possible to build a soil liner by compacting clay, in most cases, the permeability is unacceptable because the resultant loss of hot fluid to the soil increases thermal losses, requires replenishment of salt and water, and may present an environmental problem. All ponds constructed up to today have a plastic or elastomer liner, which is a reinforced polymer material 0.75–1.25 mm in thickness. The lining represents a considerable but not critical cost item that should be considered in cost analysis.

Evaporation is caused by insolation and wind action. The evaporation rate depends on the temperature of the UCZ and the humidity above the pond's surface. The higher the temperature of the water in the UCZ and the lower the humidity of ambient air, the greater is the evaporation rate. Excessive evaporation results in a growth of the UCZ downward into the NCZ (Onwubiko, 1984). Evaporation can be counter-balanced by surface water washing, called *surface flushing*, which could compensate for evaporated water as well as reduce the temperature of the pond's surface, especially during periods of high insolation. In fact, surface flushing is an essential process in maintaining the pond's salt gradient. Its effect on the growth of UCZ is reduced if the velocity of the surface washing water is small. Surface temperature fluctuations will result in heat being transferred upward through the UCZ by convection, especially at night, and downward by conduction. The thickness of the UCZ varies with the intensity of the incident radiation (Norton, 1992).

Another method to reduce the evaporation rate is by reducing the wind velocity over the water's surface by using windbreaks. The sheer forces of wind on a large area of water generate waves and surface drift. The kinetic energy transferred to the water is consumed partly by viscous losses and partly by mixing of the top surface water with the somewhat denser water just below the surface. When light to moderate winds exist, evaporation can be the dominant mechanism in a surface layer mixing. Under strong winds, however, evaporation becomes of secondary importance because wind-induced mixing can contribute significantly to the deepening of the UCZ (Elata and Levien, 1966). Another effect of wind is that it induces horizontal currents near the top surface of the pond, thus increasing convection in the UCZ. Wind mixing has been reduced by floating devices such as plastic pipes and plastic grids.

The pond is filled in layered sections, one after the other, each layer having a different salt concentration, as indicated above. Usually, these layers are built from the bottom upward, with the densest bottom layer filled first and subsequent lighter layers floated on the denser layer. Shortly after the stepwise filling process, the pond gradient smoothes itself, due to the diffusion and kinetic energy of liquid flow injected into the pond during the filling process (Tabor, 1981).

Salt slowly diffuses upward at an annual average rate of about $20 \, kg/m^2$ as a result of its concentration gradient (Norton, 1992). The diffusion rate depends on the ambient environmental conditions, type of salt, and temperature gradient. A combination of surface washing with freshwater and the injection of adequate density brines at the bottom of the pond are usually sufficient to maintain an almost stationary gradient.

A solar pond is usually constructed by flattening the site and building a retaining wall around the perimeter of the pond, not by digging out the earth; thus only a small fraction of the earth is moved, which reduces costs drastically. To avoid the use of wall supports, the earth walls thus built are tapered with a slope of one in 3, which gives an inclination of about $20°$ (Tabor, 1981). Preferred sites for solar ponds are near the sea, where saline water is locally available; otherwise, a large quantity of salt needs to be purchased. Sufficient quantities of low-salinity or fresh water are also required for the UCZ and for surface washing.

Thermal efficiency is defined as the ratio of total heat removed from the solar pond to the total amount of solar radiation which has fallen on the surface of the solar pond during a defined period. For this definition to be meaningful, the considered period should be long enough such that the stored or lost heat from the pond can be ignored compared with the received solar energy during that period. Numerous theoretical and experimental investigations have been carried out to determine the thermal efficiency of solar ponds. A transient computational analysis on the performance of solar ponds under varying conditions carried out by Wang and Akbarzadeh (1982) showed that a solar pond could provide heat at an efficiency of 15% and at an average temperature of about 87 °C or at 20% efficiency and an average temperature not higher than 65 °C.

10.7.2 Methods of heat extraction

Basically, two methods are used to extract heat accumulated to the bottom of the solar pond. The first uses a heat exchanger in the LCZ, which is in the form of a series of parallel pipes; the second is to use an external heat exchanger, which is supplied with hot saline water from the LCZ and returns the fluid to the other end of the pond at the same layer. For this purpose, horizontal nozzles that keep the velocity of efflux low are usually used. The same nozzles can also be used for filling the pond.

Heat has been successfully extracted from the lower convective zone of solar ponds for industrial process heating, space heating, and power generation (Andrews and Akbarzadeh, 2002; Rabl and Nielsen, 1975; Tabor and Doron, 1986).

The conventional method of heat extraction from a solar pond is to draw the heat from the LCZ only. This can be done using a heat exchanger located in the LCZ. A heat transfer fluid circulates in a closed cycle through the internal heat exchanger and transfers its thermal energy through an external heat exchanger. Figure 10.17 shows such a system for a heating application. The heat exchanger is usually a series of polyethylene tubes passing through the pond's LCZ connected to a manifold of larger diameter pipes outside the pond. This method was applied to the solar pond at Pyramid Hill, Australia. Heat extraction can also be performed by pumping the hot brine from the top of the LCZ through an external heat exchanger and then returning the brine at a reduced temperature to the bottom of the LCZ (see Fig. 10.18). The velocity of the brine being pumped needs to be regulated to prevent erosion of the gradient layer. In the salt-gradient solar pond at El Paso, USA, the heat is extracted from the pond using an external heat exchanger (Lu et al., 2004).

FIGURE 10.17

Heat extraction from a solar pond with an internal heat exchanger in the LCZ.

FIGURE 10.18

Heat extraction from a solar pond with an external heat exchanger using hot brine from the LCZ.

An alternative way of extracting heat from solar ponds was investigated theoretically with the aim of improving the overall energy efficiency by Andrews and Akbarzadeh (2005). In this method, heat is extracted from the NCZ as well as, or instead of, the LCZ. The theoretical analysis showed that the extraction of heat from the gradient layer can decrease heat losses to the surface and hence results in an increase in the overall energy efficiency.

A good survey of heat extraction methods from salinity-gradient solar ponds is given by Leblanc et al. (2011).

10.7.3 Transmission estimation

As was seen previously, when solar radiation falls on the surface of a solar pond, part of it is reflected at the water surface and part is absorbed at the bottom. Since water is a spectrally selective absorber, only shorter wavelengths reach the bottom of the pond. Because the absorption phenomena differ widely with wavelength, the absorption of solar energy in solutions with inorganic salts used in solar ponds can be represented by the sum of four exponential terms. The transmittance of water at depth x, $\tau(x)$, can be related to x by (Nielsen, 1976):

$$\tau(x) = \sum_{i=1}^{4} \alpha_i e^{-b_i x} \tag{10.20}$$

where the coefficients α_i and b_i are given in Table 10.5.

It should be noted that Eq. (10.20) does not include the transmission in the infrared part of the spectrum ($\lambda > 1.2$ μm, row 5 in Table 10.5), since this part is of no interest in solar pond analysis. Additionally, detailed analysis of heat transfer in a solar pond is very complex; it includes effects of volumetric absorption and variation of conductivity and density with salinity. The interested reader is referred to the articles by Tsilingiris (1994) and Angeli et al. (2006).

Table 10.5 Coefficients for Eq. (10.11).

i	Wavelength (μm)	α_i	b_i (m^{-1})
1	0.2–0.6	0.237	0.032
2	0.6–0.75	0.193	0.45
3	0.75–0.90	0.167	3.0
4	0.90–1.20	0.179	35.0
5	Over 1.20	0.224	225.0

EXAMPLE 10.3

Find the transmittance of a solar pond for a depth of 0.6 m.

Solution

From Eq. (10.20),

$$\tau(x) = \sum_{i=1}^{4} \alpha_i e^{-b_i x} = 0.237 e^{-0.032 \times 0.6} + 0.193 e^{-0.45 \times 0.6} + 0.167 e^{-3 \times 0.6} + 0.179 e^{-35 \times 0.6}$$

$$= 0.407$$

For more accuracy the fifth term in Table 10.5 can be considered which gives the same answer as the last term is equal to 5.25×10^{-60}. This is why this last term is usually ignored even at shallow depths, except perhaps at the very few millimeters.

10.7.4 Experimental solar ponds

Two important experimental solar pond applications are the El Paso solar pond in Texas, USA, and the Pyramid Hill solar pond, in Victoria, Australia.

The El Paso solar pond is a 3000 m^2 research, development and demonstration project operated by the University of Texas at El Paso. This project, initiated in 1983, has been operated intermittently since 1985 through the end of 2003, and became one of the longest operational solar pond projects in the world.

Through 16 years of operation and research, the El Paso solar pond project has provided valuable experiences and demonstrated various applications including desalination, waste brine management, industrial process heat production, and electricity generation (Leblanc et al., 2011).

The El Paso solar pond has a depth of about 3.25 m. The UCZ, NCZ, and LCZ are approximately 0.7, 1.2, and 1.35 m, respectively. The pond uses an aqueous solution of predominately sodium chloride (NaCl). The typical operating temperature of the pond ranged from 70 °C in winter to 90 °C in early fall. The highest temperature observed was 93 °C, and the maximum temperature difference between the LCZ and UCZ was above 70 °C.

The El Paso solar pond is a research, development and demonstration project operated by the University of Texas at El Paso and funded by the US Bureau of Reclamation and the State of Texas. This project is located on the property of Bruce Foods, Inc., a food canning company, and it was the first in the world to deliver industrial process heat to a commercial manufacturer, the first solar pond electric power generation facility in the US, and the nation's first experimental solar-pond-powered desalination facility. The El Paso solar pond ceased its operation and was decommissioned at the end of 2003. In order to make salinity-gradient solar pond technology more reliable, productive and economic, a series of techniques have been developed. These techniques include: automated instrumentation system for solar pond monitoring; stability analysis strategy and high-temperature (60–90 °C) gradient maintenance methods; scanning injection technique for improved salinity gradient construction and maintenance; new liner technology; and improved heat extraction system (Lu et al., 2004). Additionally, different lining systems had been used, including flexible membrane liners and compacted clay/plastic buried liners. During the 16 operational years, the El Paso solar pond had experienced liner failures (Robbins et al., 1995; Lu and Swift, 1996), and three different liners have been used. Details of the liners used and the causes of failure after some years of operation are outlined by Leblanc et al. (2011).

In February 2000, Royal Melbourne Institute of Technology (RMIT) University in partnership with two Australian companies began the "Pyramid Hill Solar Pond Project" (Andrews and Akbarzadeh, 2002) to demonstrate and commercialize a solar pond system as an innovative, cost-effective method of capturing and storing solar energy for a range of applications, including heating, electricity generation and combined heat and power. The project was made possible through a grant under the Renewable Energy Commercialization program from the Australian Greenhouse Office. The 3000 m^2 salt-gradient solar pond was constructed at Pyramid Salt's facility in northern Victoria, Australia, hence its name. The initial stage of the project has focused on industrial process heating, providing heat for use in high-grade salt production. The heat was used also for aquaculture, specially producing brine shrimps for a stock feed. Supply of heat for commercial salt production began in June 2001. The solar pond facility was also used to demonstrate inland desalting in December 2006.

The pond was designed with a depth of 2.3 m. The LCZ was designed to be 0.8 thick, the NCZ 1.2 m thick, and the UCZ 0.3 m thick. Pyramid Salt is an established commercial producer of salt from saline groundwater that is pumped to the surface as part of a salinity mitigation scheme. The solar pond has been integrated into the scheme where the surface of the pond is flushed with saline groundwater (about 3% salinity) and the overflow is used in the salt production process. The pond was positioned about 200 m from Pyramid Hill's salt production plant to minimize heat losses.

The pond was lined with a 1 mm thick Nylex Millennium polypropylene liner. This particular lining was selected for its ability to withstand saturated brine at temperatures up to 100 °C and its resistance to ultraviolet (UV) radiation.

10.7.5 Applications

Solar ponds can be used to provide energy for many different types of applications. The smaller ponds have been used mainly for space heating and cooling and domestic hot water production, whereas the larger ponds are proposed for industrial process heat, electric power generation, and desalination.

Solar ponds are very attractive for space heating and cooling and domestic hot water production because of their intrinsic storage capabilities. To increase the economic viability, large solar ponds can be used for district heating and cooling, and such a system can offer also seasonal storage. However, no such project has been undertaken so far. Cooling is achieved with the use of absorption chillers, which require heat energy to operate (see Chapter 6, Section 6.4.2). For this purpose, temperatures of about

90 °C are required, which can easily be obtained from a solar pond with little fluctuation during the summer period.

Although many feasibility studies have been made for the generation of electric power from solar ponds, the only operational systems are in Israel (Tabor, 1981). These include a 1500 m^2 pond used to operate a 6 kW Rankine cycle turbine-generator and a 7000 m^2 pond producing 150 kW peak power. Both of these ponds operate at about 90 °C. A schematic of the power plant design, working with an organic fluid, is shown in Fig. 10.19.

For power production in the multi-megawatt range, a solar pond of several square kilometers surface area is needed. However, this is not feasible economically, since excavation and preparation account for more than 40% of the total capital cost of the power-generating station (Tabor, 1981). So, it would appear logical to employ a natural lake and convert a shallow portion of it to a solar pond.

Another use of the output from a salt gradient solar pond is to operate a low-temperature distillation unit to desalt seawater, such as MSF (see Chapter 8, Section 8.4.1). Such systems operate at a top temperature of 70 °C, which can easily be obtained with a solar pond. This concept has applicability in desert areas near oceans. Solar pond coupled desalination also involves the use of the hot brine from the pond as a thermal source to evaporate the water to be desalted at low pressure in a multiple-effect boiling (MEB) evaporator. The low pressure is produced by vacuum pumps powered by the electricity produced by the organic Rankine cycle (ORC) engine.

Matz and Feist (1967) propose solar ponds as a solution to brine disposal at inland Electrodialysis (ED) plants as well as a source of thermal energy to heat the feed of an ED plant, which can increase its performance.

Desalination of brackish water appears to be an extremely useful application of solar pond technology, as discussed above. In addition to providing clean renewable energy to power the desalination processes, salinity-gradient solar ponds can utilize the waste brine. The El Paso solar pond project showed that the cost of water produced by a zero discharge desalination system—which combines a solar pond with membrane filtration, thermal desalination and brine concentrator—ranges from US$

FIGURE 10.19

Schematic of a solar pond power generation system.

$1.06/m^3$ for a 3800 m^3/day desalination plant to US$ 0.92/m^3 for a 75,000 m^3/day desalination plant (Swift et al., 2002).

For industrial process heat at medium temperatures (50–90 °C), the levelized energy cost (LEC), which is an economic measure that is useful for comparing and ranking technology options because it is a cost that accounts for the purchase, financing, tax and operation of a power system over its lifetime, ranges from US$ 6.60 per GJ for a 1-hectare (ha) pond to US$ 1.30/GJ for a 100 ha pond at sites having similar climate conditions to El Paso, Texas (Leblanc et al., 2011). The unit cost of supplying industrial process heat from a salinity-gradient solar pond is less expensive than either natural gas or coal for even modest pond sizes. For generating electricity, maximum revenue is obtained by maximizing operating hours of the engine or in a base load mode of operation. This is due to the relative high cost of the ORC generation equipment. The production of base load electricity using solar ponds is more expensive than current electrical base load generation technologies. However, it can become more cost competitive when the pond size is larger, greater than about 100 ha, and if the impact of environmental costs associated with burning fossil fuels are considered.

Exercises

10.1 In a simple Rankine steam cycle, the boiler gives steam at 60 bar and 550 °C. The condenser operates at 0.1 bar. If both pump and turbine efficiencies are 90%, estimate the cycle efficiency.

10.2 In a reheat Rankine cycle, a CSP system gives steam at 50 bar and 400 °C. After expansion in the high-pressure turbine to 6 bar, the steam is reheated again to 400 °C and expands again in the low-pressure turbine to a pressure of 0.1 bar in a dry, saturated steam condition. If the steam leaves the condenser as a saturated liquid and the pump and high-pressure turbine efficiency is 85%, determine the efficiency of a low-pressure turbine, the work output of the two turbines, the work input to the pump (both per unit mass of working fluid), the heat added by the CSP system, and the cycle efficiency. If the solar radiation is 900 W/m^2 and the CSP system operates at 40% efficiency, how many square meters of collectors are required if the steam flow rate is 1 kg/s?

10.3 Repeat Example 10.2 for a turbine and pump efficiency of 90%. Also estimate the boiler heat required.

10.4 Find the transmittance of a solar pond for a maximum depth of 2 m.

Assignment

With the help of the System Advisor Model, details of which are given in Appendix 10, design a CSP system of 50 MW capacity for a location included in the program database or your own location, provided that suitable weather data are available. For the various input parameters, try to use values that are as realistic as possible. You can choose different types of concentrators (parabolic troughs, power tower, linear Fresnel system), various types of financial models (power purchase agreement, merchant plant, LCOE calculator), various sizes and duration of storage, tracking modes, and incentives. For actual installation and operating costs of each combination, find the most cost-effective combination of equipment setup. For parameters not specified above, you can use realistic figures (or the program default values), which need to be listed in the report produced. This can also be repeated for a 100 MW system.

References

Andrews, J., Akbarzadeh, A., 2002. Solar pond project: stage 1 — solar ponds for industrial process heating. In: End of Project Report for Project Funded under Renewable Energy Commercialisation Program, Australian Greenhouse Office. RMIT, Melbourne.

Andrews, J., Akbarzadeh, A., 2005. Enhancing the thermal efficiency of solar ponds by extracting heat from the gradient layer. Sol. Energy 78 (6), 704–716.

Angeli, C., Leonardi, E., Maciocco, L., 2006. A computational study of salt diffusion and heat extraction in solar pond plants. Sol. Energy 80 (11), 1498–1508.

Bernandes, M.A., 2010. Solar chimney power plants — developments and advancements. (Chapter 9). In: Rugescu, R.D. (Ed.), Solar Energy. Intech, pp. 171–186.

Chavez, J.M., Kolb, G.J., Meineck, W., 1993. In: Becker, M., Klimas, P.C. (Eds.), Second Generation Central Receiver Technologies—A Status Report. Verlag C.F. Müller GmbH, Karlsruhe, Germany.

Elata, C., Levien, O., 1966. Hydraulics of the solar ponds. In: Eleventh International Congress, Leningrad Report 2.3. International Association for Hydraulic Research, pp. 1–14.

Geyer, M., Quaschning, V., July–August 2000. Renewable Energy World 184–191.

Haaf, W., 1984. Solar Chimneys — Part II: Preliminary Test Results from the Manzanares Pilot Plant. Taylor & Francis, pp. 141–161.

Haaf, W., Friedrich, K., Mayr, G., Schlaich, J., 1983. Solar chimneys, part I: principle and construction of the pilot plant in Manzanares. Int. J. Sol. Energy 2, 3–20.

Hassairi, M., Safi, M.J., Chibani, S., 2001. Natural brine solar pond: an experimental study. Sol. Energy 70 (1), 45–50.

Kearney, D.W., Price, H.W., 1992. Solar thermal plants—LUZ concept (current status of the SEGS plants). In: Proceedings of the Second Renewable Energy Congress, Reading, UK, vol. 2, pp. 582–588.

Leblanc, J., Akbarzadeh, A., Andrews, J., Lu, H., Golding, P., 2011. Heat extraction methods from salinity-gradient solar ponds and introduction of a novel system of heat extraction for improved efficiency. Sol. Energy 85 (12), 3103–3142.

Lu, H., Swift, A.H.P., 1996. Reconstruction and operation of the El Paso solar pond with a geosynthetic clay liner system. In: Davidson, J., Chavez, J. (Eds.), Solar Engineering — Proceedings of ASME International Solar Energy Conference, San Antonio, TX.

Lu, H., Swift, A.H.P., Hein, H.D., Walton, J.C., May, 2004. Advancements in salinity gradient solar pond technology based on sixteen years of operational experience. ASME J. Sol. Energy Eng. 126, 759–767.

LUZ, 1990. Solar Electric Generating System IX Technical Description. LUZ International Limited.

Matz, R., Feist, E.M., 1967. The application of solar energy to the solution of some problems of electrodialysis. Desalination 2 (1), 116–124.

Muller-Steinhagen, H., Trieb, F., 2004. Concentrating solar power: a review of the technology. Ingenia 18, 43–50.

Nielsen, C.E., 1976. Experience with a prototype solar pond for space heating. Sharing the Sun 5, 169–182. ISES, Winnipeg, Canada.

Norton, B., 1992. Solar Energy Thermal Technology. Springer-Verlag, London.

Onwubiko, C., 1984. Effect of evaporation on the characteristic performance of the salt-gradient solar pond. In: Solar Engineering, Proceedings of ASME Solar Energy Division, Sixth Annual Conference, Las Vegas, NV, pp. 6–11.

Pitz-Paal, R., 2002. Concentrating solar technologies—the key to renewable electricity and process heat for a wide range of applications. In: Proceedings of the World Renewable Energy Congress VII on CD-ROM, Cologne, Germany.

Rabl, A., Nielsen, C.E., 1975. Solar ponds for space heating. Sol. Energy 17, 1–12.

Robbins, M.C., Lu, H., Swift, A.H.P., 1995. Investigation of the suitability of a geosynthetic clay liner system for the El Paso solar pond. In: Proceedings of the 1995 Annual Conference of American Solar Energy Society, Minneapolis, MN, pp. 63–68.

Schlaich, J., Bergermann, R., Schiel, W., Weinrebe, G., 2005. Design of commercial solar updraft tower systems—utilization of solar induced convective flows for power generation. J. Sol. Energy Eng. 127 (1), 117–124.

Schlaich, J., 1995. The Solar Chimney: Electricity from the Sun. Edition Axel Menges, Stuttgart.

Schwarzbözl, P., Pitz-Paal, R., Meinecke, W., Buck, R., 2000. Cost-optimized solar gas turbine cycles using volumetric air receiver technology. In: Proceedings of the Renewable Energy for the New Millennium, Sydney, Australia, pp. 171–177.

Sencan, A., Kizilkan, O., Bezir, N., Kalogirou, S.A., 2007. Different methods for modeling an absorption heat transformer powered from a solar pond. Energy Convers. Manage. 48 (3), 724–735.

Swift, A.H.P., Lu, H., Becerra, H., 2002. Zero Discharge Waste Brine Management for Desalination Plants, Final Report. Agreement No. 99-FC-81–0181. US Department of Interior, Denver, CO.

Tabor, H., 1981. Solar ponds. Sol. Energy 27 (3), 181–194.

Tabor, H., Doron, B., 1986. Solar ponds — lessons learned from the 150 KWe power plant at Ein Boqek. In: Proceedings of ASME Solar Energy Division, Anaheim, CA.

Taggart, S., 2008a. Parabolic troughs: CSP's quiet achiever. Renew. Energy Focus 46–50.

Taggart, S., 2008b. Hot stuff: CSP and the power tower. Renew. Energy Focus 51–54.

Tsilingiris, P.T., 1994. Steady-state modeling limitations in solar ponds design. Sol. Energy 53 (1), 73–79.

Wang, Y.F., Akbarzadeh, A., 1982. A study on the transient behaviour of solar ponds. Energy Int. J. 7 (12), 1005–1017.

Wolff, G., Gallego, B., Tisdale, R., Hopwood, D., January–February 2008. CSP concentrates the mind. Renew. Energy Focus 42–47.

Designing and Modeling Solar Energy Systems

The proper sizing of the components of a solar energy system is a complex problem, which includes both predictable (collector and other components' performance characteristics) and unpredictable (weather data) components. In this chapter, various design methods are presented as well as an overview of the simulation techniques and programs suitable for solar heating and cooling systems. A brief review of artificial intelligence (AI) methods and their applications in solar energy systems is also presented.

The design methods presented include the f-chart, utilizability Φ, the $\overline{\Phi}$ f-chart and the unutilizability method. The f-chart is based on the correlation of the results of a large number of simulations in terms of easily calculated dimensionless variables. The utilizability method is used in cases where the collector operating temperature is known or can be estimated and for which critical radiation levels can be established. The utilizability method is based on the analysis of hourly weather data to obtain the fraction of the total month's radiation that is above a critical level. The $\overline{\Phi}$ f-chart method is a combination of the utilizability and f-chart methods, applied in systems where the energy supplied to a load is above a minimum useful temperature and the temperature of this energy supply has no effect on the performance of the system as long as it is greater than the minimum temperature.

For more detailed results, modeling and simulation are used. In recent years, because of the increase of computational speed of personal computers, annual simulations have begun replacing design methods. Design methods, however, are much faster; therefore, they are still useful in early design studies. The software programs described briefly in this book include TRNSYS, WATSUN, Polysun, and AI techniques applied in solar-energy systems modeling and prediction.

11.1 *F*-chart method and program

The f-chart method is used for estimating the annual thermal performance of active building heating systems using a working fluid, which is either liquid or air, and where the minimum temperature of energy delivery is near 20 °C. The system configurations that can be evaluated by the f-chart method are common in residential applications. With the f-chart method, the fraction of the total heating load that can be supplied by the solar energy system can be estimated. Let the purchased energy for a fuel-only system or the energy required to cover the load be L, the purchased auxiliary energy for a solar system be L_{AUX}, and the solar energy delivered be Q_S. For a solar energy system, $L = L_{AUX} + Q_S$. For a month, i, the fractional reduction of purchased energy when a solar energy system is used, called the *solar fraction, f*, is given by the ratio:

$$f = \frac{L_i - L_{AUX,i}}{L_i} = \frac{Q_{S,i}}{L_i} \tag{11.1}$$

Solar Energy Engineering. https://doi.org/10.1016/B978-0-323-99350-0.00007-7

601

The *f*-chart method was developed by Klein et al. (1976, 1977) and Beckman et al. (1977). In the method, the primary design variable is the collector area, while the secondary variables are storage capacity, collector type, load and collector heat-exchanger size, and fluid flow rate. The method is a correlation of the results of many hundreds of thermal performance simulations of solar heating systems performed with TRNSYS, in which the simulation conditions were varied over specific ranges of parameters of practical system designs shown in Table 11.1 (Klein et al., 1976, 1977). The resulting correlations give *f*, i.e., the fraction of the monthly load supplied by solar energy, as a function of two dimensionless parameters. The first is related to the ratio of collector losses to heating load, and the second to the ratio of absorbed solar radiation to heating load. The heating load includes both space heating and hot water loads. The *f*-chart method has been developed for three standard system configurations: liquid and air systems for space and hot water heating and systems for service hot water only.

Based on the fundamental equation presented in Chapter 6, Section 6.3.3, Klein et al. (1976) analyzed numerically the long-term thermal performance of solar heating systems of the basic configuration shown in Fig. 6.14. When Eq. (6.60) is integrated over a time period, Δt, such that the internal energy change in the storage tank is small compared to the other terms (usually 1 month), we get:

$$(Mc_p)_s \int_{\Delta t} \frac{dT_s}{dt} = \int_{\Delta t} Q_u - \int_{\Delta t} Q_{ls} - \int_{\Delta t} Q_{lw} - \int_{\Delta t} Q_{tl} \tag{11.2}$$

The sum of the last three terms of Eq. (11.2) represents the total heating load (including space load and hot water load) supplied by solar energy during the integration period. If this sum is denoted by Q_S, using the definition of the solar fraction, *f*, from Eq. (11.1), we get:

$$f = \frac{Q_S}{L} = \frac{1}{L} \int_{\Delta t} Q_u^+ dt \tag{11.3}$$

where

L = total heating load during the integration period (MJ).

Using Eq. (5.56) for Q_u and replacing G_t by H_t, the total (beam and diffuse) insolation over a day, Eq. (11.3) can be written as:

$$f = \frac{A_c F_R'}{L} \int_{\Delta t} [H_t(\tau\alpha) - U_L(T_s - T_a)]dt \tag{11.4}$$

Table 11.1 Range of design variables used in developing *f*-charts for liquid and air systems.

Parameter	Range
$(\tau\alpha)_n$	0.6–0.9
$F_R A_c$	5–120 m^2
U_L	2.1–8.3 W/m^2 °C
β (collector slope)	30–90°
$(UA)_h$	83.3–666.6 W/°C

Reprinted from Klein et al. (1976, 1977), with permission from Elsevier.

The last term of Eq. (11.4) can be multiplied and divided by the term $(T_{ref} - T_a)$, where T_{ref} is a reference temperature chosen to be $100\,°C$, so the following equation can be obtained:

$$f = \frac{A_c F_R'}{L} \int_{\Delta t} \left[H_t(\tau\alpha) - U_L (T_{ref} - T_a) \frac{(T_s - T_a)}{(T_{ref} - T_a)} \right] dt \tag{11.5}$$

The storage tank temperature, T_s, is a complicated function of H_t, L, and T_a; therefore, the integration of Eq. (11.5) cannot be explicitly evaluated. This equation, however, suggests that an empirical correlation can be found, on a monthly basis, between the f factor and the two dimensionless groups mentioned above as follows:

$$X = \frac{A_c F_R' U_L}{L} \int_{\Delta t} (T_{ref} - \overline{T}_a) dt = \frac{A_c F_R' U_L}{L} (T_{ref} - \overline{T}_a) \Delta t \tag{11.6}$$

$$Y = \frac{A_c F_R'}{L} \int_{\Delta t} H_t(\tau\alpha) dt = \frac{A_c F_R'}{L} (\overline{\tau\alpha}) \overline{H}_t N \tag{11.7}$$

where

L = monthly heating load or demand (MJ);
N = number of days in a month;
$\overline{T}_a$ = monthly average ambient temperature (°C);
$\overline{H}_t$ = monthly average daily total radiation on the tilted collector surface (MJ/m^2); and
$(\overline{\tau\alpha})$ = monthly average value of $(\tau\alpha)$, = monthly average value of absorbed over incident solar radiation = $\overline{S}/\overline{H}_t$.

For the purpose of calculating the values of the dimensionless parameters X and Y, Eqs. (11.6) and (11.7) are usually rearranged to read:

$$X = F_R U_L \frac{F_R'}{F_R} (T_{ref} - \overline{T}_a) \Delta t \frac{A_c}{L} \tag{11.8}$$

$$Y = F_R(\tau\alpha)_n \frac{F_R'}{F_R} \left[\frac{(\overline{\tau\alpha})}{(\tau\alpha)_n} \right] \overline{H}_t N \frac{A_c}{L} \tag{11.9}$$

The reason for the rearrangement is that the factors $F_R U_L$ and $F_R(\tau\alpha)_n$ are readily available from standard collector tests (see Chapter 4, Section 4.1). The ratio F_R'/F_R corrects the collector performance because a heat exchanger is used between the collector and the storage tank, which causes the collector side of the system to operate at higher temperature than a similar system without a heat exchanger and is given by Eq. (5.57) in Chapter 5. For a given collector orientation, the value of the factor $(\overline{\tau\alpha})/(\tau\alpha)_n$ varies slightly from month to month. For collectors tilted and facing the equator with a slope equal to latitude plus 15°, Klein (1976) found that the factor is equal to 0.96 for a one-cover collector and 0.94 for a two-cover collector for the whole heating season (winter months). Using the preceding definition of $(\overline{\tau\alpha})$, we get:

$$\frac{(\overline{\tau\alpha})}{(\tau\alpha)_n} = \frac{\overline{S}}{\overline{H}_t(\tau\alpha)_n} \tag{11.10}$$

If the isotropic model is used for $\overline{S}$ and substituted in Eq. (11.10), then:

$$\frac{(\overline{\tau\alpha})}{(\tau\alpha)_n} = \frac{\overline{H}_B\overline{R}_B}{\overline{H}_t}\frac{(\overline{\tau\alpha})_B}{(\tau\alpha)_n} + \frac{\overline{H}_D}{\overline{H}_t}\frac{(\overline{\tau\alpha})_D}{(\tau\alpha)_n}\left(\frac{1+\cos(\beta)}{2}\right) + \frac{\overline{H}\rho_G}{\overline{H}_t}\frac{(\overline{\tau\alpha})_G}{(\tau\alpha)_n}\left(\frac{1-\cos(\beta)}{2}\right) \tag{11.11}$$

In Eq. (11.11), the $(\overline{\tau\alpha})/(\tau\alpha)_n$ ratios can be obtained from Fig. 3.27 for the beam component at the effective angle of incidence, $\overline{\theta}_B$, which can be obtained from Figure A3.8 in Appendix 3, and for the diffuse and ground-reflected components at the effective incidence angles at β from Eqs. (3.4a) and (3.4b).

The dimensionless parameters, X and Y, have some physical significance. The parameter X represents the ratio of the reference collector total energy loss to total heating load or demand (L) during the period Δt, whereas the parameter Y represents the ratio of the total absorbed solar energy to the total heating load or demand (L) during the same period.

As was indicated, f-chart is used to estimate the monthly solar fraction, f_i, and the energy contribution for the month is the product of f_i and monthly load (heating and hot water), L_i. To find the fraction of the annual load supplied by the solar energy system, F, the sum of the monthly energy contributions is divided by the annual load, given by:

$$F = \frac{\sum f_i L_i}{\sum L_i} \tag{11.12}$$

The method can be used to simulate standard solar water and air system configurations and solar energy systems used only for hot water production. These are examined separately in the following sections.

EXAMPLE 11.1

A standard solar-heating system is installed in an area where the average daily total radiation on the tilted collector surface is 12.5 MJ/m^2, average ambient temperature is 10.1 °C, and it uses a 35 m^2 aperture area collector (net collector aperture area), which has $F_R(\tau\alpha)_n = 0.78$ and $F_R U_L = 5.56$ W/m^2 °C, both determined from the standard collector tests. If the space heating and hot water load is 35.2 GJ, the flow rate in the collector is the same as the flow rate used in testing the collector, $F_R'/F_R = 0.98$, and $(\overline{\tau\alpha})/(\tau\alpha)_n = 0.96$ for all months, estimate the parameters X and Y.

Solution

Using Eqs. (11.8) and (11.9) and noting that Δt is the number of seconds in a month, equal to 31 days $\times$ 24 h $\times$ 3600 s/h, we get:

$$X = F_R U_L \frac{F_R'}{F_R}(T_{ref} - \overline{T}_a)\Delta t\frac{A_c}{L}$$

$$= 5.56 \times 0.98(100 - 10.1) \times 31 \times 24 \times 3600 \times \frac{35}{35.2 \times 10^9} = 1.30$$

$$Y = F_R(\tau\alpha)_n \frac{F_R'}{F_R}\left[\frac{(\overline{\tau\alpha})}{(\tau\alpha)_n}\right]\overline{H}_t N\frac{A_c}{L}$$

$$= 0.78 \times 0.98 \times 0.96 \times 12.5 \times 10^6 \times 31 \times \frac{35}{35.2 \times 10^9} = 0.28$$

It should be noted that $F_R' U_L$ and $F_R'(\tau\alpha)_n$ can be given instead of $F_R U_L$, $F_R(\tau\alpha)_n$ and F_R'/F_R, given in this problem.

11.1.1 Performance and design of liquid-based solar heating systems

Knowledge of the system thermal performance is required in order to be able to design and optimize a solar heating system. The *f*-chart for liquid-based systems is developed for a standard solar liquid-based solar energy system, shown in Fig. 11.1. This is the same as the system shown in Fig. 6.14, drawn without the controls, for clarity. The typical liquid-based system shown in Fig. 11.1 uses an antifreeze solution in the collector loop and water as the storage medium. A water-to-water load-heat exchanger is used to transfer heat from the storage tank to the domestic hot water (DHW) system. Although in Fig. 6.14 a one-tank DHW system is shown, a two-tank system could be employed, in which the first tank is used for preheating.

The fraction *f* of the monthly total load supplied by a standard liquid-based solar energy system is given as a function of the two dimensionless parameters, *X* and *Y*, and can be obtained from the *f*-chart in Fig. 11.2 or the following equation (Klein et al., 1976):

$$f = 1.029Y - 0.065X - 0.245Y^2 + 0.0018X^2 + 0.0215Y^3 \tag{11.13}$$

Application of Eq. (11.13) or Fig. 11.2 allows the simple estimation of the solar fraction on a monthly basis as a function of the system design and local weather conditions. This is for a standard storage capacity per square meter of collector area of 75 l/m² and a standard heat exchanger of *Z* = 2 for the range of *X* and *Y* parameters indicated in Fig. 11.2. The annual value can be obtained by summing up the monthly values using Eq. (11.12). As will be shown in the next chapter, to determine the economic optimum collector area, the annual load fraction corresponding to different collector areas is required. Therefore, the present method can easily be used for these estimations.

It should be noted that the *f*-chart was developed using fixed nominal values of storage capacity per unit of collector area, collector liquid flow rate per unit of collector area, and load heat-exchanger size relative to a space-heating load. Therefore, it is important to apply various corrections for the particular system configuration used.

FIGURE 11.1

Schematic diagram of a standard liquid-based solar heating system.

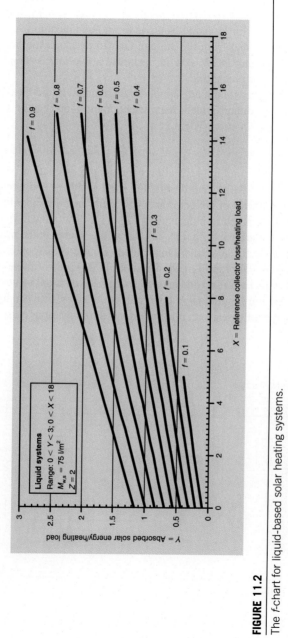

FIGURE 11.2

The *f*-chart for liquid-based solar heating systems.

EXAMPLE 11.2

If the solar heating system given in Example 11.1 is liquid-based, estimate the annual solar fraction if the collector is located in an area having the monthly average weather conditions and monthly heating and hot water loads shown in Table 11.2.

Table 11.2 Average monthly weather conditions and heating and hot water loads for Example 11.2.

Month	$\overline{H}_t$ (MJ/m²)	$\overline{T}_a$ (°C)	L (GJ)
January	12.5	10.1	35.2
February	15.6	13.5	31.1
March	17.8	15.8	20.7
April	20.2	19.0	13.2
May	21.5	21.5	5.6
June	22.5	29.8	4.1
July	23.1	32.1	2.9
August	22.4	30.5	3.5
September	21.1	22.5	5.1
October	18.2	19.2	12.7
November	15.2	16.2	23.6
December	13.1	11.1	33.1

Solution

The values of the dimensionless parameters X and Y found from Example 11.1 are equal to 1.30 and 0.28, respectively. From the weather and load conditions shown in Table 11.2, these correspond to the month of January. From Fig. 11.2 or Eq. (11.13), $f = 0.188$. The total load in January is 35.2 GJ. Therefore, the solar contribution in January is $fL = 0.188 \times 35.2 = 6.62$ GJ. The same calculations are repeated from month to month, as shown in Table 11.3.

Table 11.3 Monthly calculations for Example 11.2.

Month	$\overline{H}_t$ (MJ/m²)	$\overline{T}_a$ (°C)	L (GJ)	X	Y	f	fL
January	12.5	10.1	35.2	1.30	0.28	0.188	6.62
February	15.6	13.5	31.1	1.28	0.36	0.259	8.05
March	17.8	15.8	20.7	2.08	0.68	0.466	9.65
April	20.2	19.0	13.2	3.03	1.18	0.728	9.61
May	21.5	21.5	5.6	7.16	3.06	1	5.60
June	22.5	29.8	4.1	8.46	4.23	1	4.10
July	23.1	32.1	2.9	11.96	6.34	1	2.90
August	22.4	30.5	3.5	10.14	5.10	1	3.50
September	21.1	22.5	5.1	7.51	3.19	1	5.10
October	18.2	19.2	12.7	3.25	1.14	0.694	8.81
November	15.2	16.2	23.6	1.76	0.50	0.347	8.19
December	13.1	11.1	33.1	1.37	0.32	0.219	7.25
		Total load = 190.8				Total contribution = 79.38	

Continued

EXAMPLE 11.2—CONT'D

It should be noted that the values of f marked in bold are outside the range of the f-chart correlation and a fraction of 100% is used, as during these months, the solar energy system covers the load fully. From Eq. (11.12), the annual fraction of load covered by the solar energy system is:

$$F = \frac{\sum f_i L_i}{\sum L_i} = \frac{79.38}{190.8} = 0.416 \quad \text{or} \quad 41.6\%$$

Storage capacity correction

It can be proven that the annual performance of liquid-based solar energy systems is insensitive to the storage capacity, as long as this is more than 50 L of water per square meter of collector area. For the f-chart of Fig. 11.2, a standard storage capacity of 75 L of stored water per square meter of collector area was considered, as shown in the figure. Other storage capacities can be used by modifying the factor X by a storage size correction factor X_c/X, given by (Beckman et al., 1977):

$$\frac{X_c}{X} = \left(\frac{M_{w,a}}{M_{w,s}}\right)^{-0.25} \tag{11.14}$$

where

$M_{w,a}$ = actual storage capacity per square meter of collector area (L/m^2);
$M_{w,s}$ = standard storage capacity per square meter of collector area (= 75 L/m^2).

Equation (11.14) is applied in the range of $0.5 \le (M_{w,a}/M_{w,s}) \le 4.0$ or $37.5 \le M_{w,a} \le 300$ L/m^2. The storage correction factor can also be determined from Fig. 11.3 directly, obtained by plotting Eq. (11.14) for this range.

EXAMPLE 11.3

Estimate the solar fraction for the month of March of Example 11.2 if the storage tank capacity is 130 L/m^2.

Solution

First the storage correction factor needs to be estimated. By using Eq. (11.14),

$$\frac{X_c}{X} = \left(\frac{M_{w,a}}{M_{w,s}}\right)^{-0.25} = \left(\frac{130}{75}\right)^{-0.25} = 0.87$$

For March, the corrected value of X is then $X_c = 0.87 \times 2.08 = 1.81$. The value of Y remains as estimated before, i.e., $Y = 0.68$. From the f-chart, $f = 0.481$ compared to 0.466 before the correction, an increase of about 2%.

FIGURE 11.3

Storage correction factor for liquid-based systems.

Collector flow rate correction

The *f*-chart of Fig. 11.2 has been generated using a collector antifreeze solution flow rate equal to 0.015 L/s m². A lower flow rate can reduce the energy collection rate significantly, especially if the low flow rate leads to fluid boiling and relief of pressure through the relief valve. Although the product of the mass flow rate and the specific heat of the fluid flowing through the collector strongly affects the performance of the solar energy system, the value used is seldom lower than the value used for the *f*-chart development. Additionally, since an increase in the collector flow rate beyond the nominal value has a small effect on the system performance, Fig. 11.2 is applicable for all practical collector flow rates.

Load heat exchanger size correction

The size of the load heat exchanger strongly affects the performance of the solar energy system. This is because the rate of heat transfer across the load heat exchanger directly influences the temperature of the storage tank, which consequently affects the collector inlet temperature. As the heat exchanger used to heat the building air is reduced in size, the storage tank temperature must increase to supply the same amount of heat energy, resulting in higher collector inlet temperatures and therefore reduced collector performance. To account for the load heat exchanger size, a new dimensionless parameter is specified, Z, given by (Beckman et al., 1977):

$$Z = \frac{\varepsilon_{\mathrm{L}} (\dot{m}c_{\mathrm{p}})_{\min}}{(UA)_{\mathrm{L}}} \tag{11.15}$$

where

ε_{L} = effectiveness of the load heat exchanger;

$(\dot{m}c_{\mathrm{p}})_{\min}$ = minimum mass flow rate–specific heat product of heat exchanger (W/K); and

$(UA)_{\mathrm{L}}$ = building loss coefficient and area product used in degree-day space-heating load model (W/K).

In Eq. (11.15), the minimum capacitance rate is that of the air side of the heat exchanger. System performance is asymptotically dependent on the value of Z; and for $Z > 10$, the performance is essentially the same as for an infinitely large value of Z. Actually, the reduction in performance due to a small-size load heat exchanger is significant for values of Z lower than 1. Practical values of Z are between 1 and 3, whereas a value of $Z = 2$ was used for the development of the f-chart of Fig. 11.2, as indicated in the figure. The performance of systems having other values of Z can be estimated by multiplying the dimensionless parameter Y by the following correction factor:

$$\frac{Y_{\mathrm{c}}}{Y} = 0.39 + 0.65 \exp \left(-\frac{0.139}{Z} \right) \tag{11.16}$$

Equation (11.16) is applied in the range of $0.5 \leq Z \leq 50$. The load heat-exchanger size correction factor can also be determined from Fig. 11.4 directly, obtained by plotting Eq. (11.16) for this range.

FIGURE 11.4

Load heat exchanger size correction factor.

EXAMPLE 11.4

If the liquid flow rate in Example 11.2 is 0.525 L/s, the air flow rate is 470 L/s, the load heat-exchanger effectiveness is 0.65, and the building overall loss coefficient–area product, $(UA)_L$, is 422 W/K, find the effect on the solar fraction for the month of November.

Solution

The minimum value of capacitance needs to be found first. Therefore, if we assume that the operating temperature is 350 K (77 °C) and the properties of air and water at that temperature are as from Tables A5.1 and A5.2 in Appendix 5, respectively,

$$C_{air} = 470 \times 0.998 \times 1009/1000 = 473.3 \ \text{W/K}$$

$$C_{water} = 0.525 \times 974 \times 4190/1000 = 2142.6 \ \text{W/K}$$

Therefore, the minimum capacitance is for the air side of the load heat exchanger. From Eq. (11.15),

$$Z = \frac{\varepsilon_L \left(\dot{m} c_p\right)_{min}}{(UA)_L} = \frac{0.65 \times (473.3)}{422} = 0.729$$

The correction factor is given by Eq. (11.16):

$$\frac{Y_c}{Y} = 0.39 + 0.65 \exp\left(-\frac{0.139}{Z}\right) = 0.39 + 0.65 \exp\left(-\frac{0.139}{0.729}\right) = 0.93$$

From Example 11.2, the value of the dimensionless parameter Y is 0.50. Therefore, $Y_c = 0.50 \times 0.93 = 0.47$. The value of the dimensionless parameter X for this month is 1.76, which from Eq. (11.13) gives a solar fraction $f = 0.323$, a drop of about 2% from the previous value.

Although in the examples in this section only one parameter was different from the standard one, if both the storage size and the size of the heat exchanger are different than the standard ones, both X_c and Y_c need to be calculated for the determination of the solar fraction. Additionally, most of the required parameters in this section are given as input data. In the following example, most of the required parameters are estimated from the information given in earlier chapters.

EXAMPLE 11.5

A liquid-based solar space and domestic water-heating system is located in Nicosia, Cyprus (35°N latitude). Estimate the monthly and annual solar fraction of the system, which has a total collector area of 20 m² and the following information is given.

1. The collectors face south, installed with 45° inclination. The performance parameters of the collectors are $F_R(\tau\alpha)_n = 0.82$ and $F_R U_L = 5.65$ W/m² °C, both determined from the standard collector tests.
2. The flow rate of both the water and antifreeze solution through the collector heat exchanger is 0.02 L/s m² and the factor $F'_R/F_R = 0.98$.
3. The storage tank capacity is equal to 120 L/m².
4. The $(\overline{\tau\alpha})/(\tau\alpha)_n = 0.96$ for October through March and 0.93 for April through September.
5. The building UA value is equal to 450 W/K. The water-to-air load heat exchanger has an effectiveness of 0.75 and air flow rate is 520 L/s.
6. The ground reflectivity is 0.2.
7. The climatic data and the heating degree days for Nicosia, Cyprus, are taken from Appendix 7 and reproduced in Table 11.4 with the hot-water load.

Continued

EXAMPLE 11.5–CONT'D

Table 11.4 Climate data and heating degree days for Example 11.5.

Month	$\overline{H}$ (MJ/m^2)	$\overline{T}_a$ (°C)	Clearness index $\overline{K}_T$	Heating °C degree days	Hot water load, D_w(GJ)
January	8.96	12.1	0.49	175	3.5
February	12.38	11.9	0.53	171	3.1
March	17.39	13.8	0.58	131	2.8
April	21.53	17.5	0.59	42	2.5
May	26.06	21.5	0.65	3	2.1
June	29.20	29.8	0.70	0	1.9
July	28.55	29.2	0.70	0	1.8
August	25.49	29.4	0.68	0	1.9
September	21.17	26.8	0.66	0	2.0
October	15.34	22.7	0.60	1	2.7
November	10.33	17.7	0.53	36	3.0
December	7.92	13.7	0.47	128	3.3

Solution

The loads need to be estimated first. For the month of January, from Eq. (6.24):

$$D_h = (UA)(DD)_h = 450(W\,/\,K) \times 24(h\,/\,day) \times 3600(J\,/\,Wh) \times 175(°C\ days) = 6.80\ \ GJ$$

The monthly heating load (including hot-water load) = 6.80 + 3.5 = 10.30 GJ. The results for all the months are shown in Table 11.5.

Next, we need to estimate the monthly average daily total radiation on the tilted collector surface from the daily total horizontal solar radiation, $\overline{H}$. For this estimation, the average day of each month is used, shown in Table 2.1, together with the declination for each day. For each of those days, the sunset hour angle, h_{ss}, is required, given by Eq. (2.15), and the sunset hour angle on the tilted surface, h'_{ss}, given by Eq. (2.109). The calculations for the month of January are as follows.

From Eq. (2.15),

$$h_{ss} = \cos^{-1}[-\tan(L)\tan(\delta)] = \cos^{-1}[-\tan(35)\tan(-20.92)] = 74.5°$$

EXAMPLE 11.5—CONT'D

Table 11.5 Heating load for all the months in Example 11.5.

Month	Heating °C degree days	D_h (GJ)	D_w (GJ)	L (GJ)
January	175	6.80	3.5	10.30
February	171	6.65	3.1	9.75
March	131	5.09	2.8	7.89
April	42	1.63	2.5	4.13
May	3	0.12	2.1	2.22
June	0	0	1.9	1.90
July	0	0	1.8	1.80
August	0	0	1.9	1.90
September	0	0	2.0	2.00
October	1	0.04	2.7	2.74
November	36	1.40	3.0	4.40
December	128	4.98	3.3	8.28
				Total = 57.31

From Eq. (2.109),

$$h'_{ss} = \min\{h_{ss}, \cos^{-1}[-\tan(L-\beta)\tan(\delta)]\}$$

$$= \min\{74.5°, \cos^{-1}[-\tan(35-45)\tan(-20.92)]\} = \min\{74.5°, 93.9°\} = 74.5°$$

From Eq. (2.105b),

$$\frac{\overline{H_D}}{\overline{H}} = 0.775 + 0.00653(h_{ss} - 90) - [0.505 + 0.00455(h_{ss} - 90)]\cos(115\overline{K}_T - 103)$$

$$= 0.775 + 0.00653(74.5 - 90) - [0.505 + 0.00455(74.5 - 90)]\cos(115 \times 0.49 - 103)$$

$$= 0.38$$

From Eq. (2.108),

$$\overline{R}_B = \frac{\cos(L-\beta)\cos(\delta)\sin(h'_{ss}) + (\pi/180)h'_{ss}\sin(L-\beta)\sin(\delta)}{\cos(L)\cos(\delta)\sin(h_{ss}) + (\pi/180)h_{ss}\sin(L)\sin(\delta)}$$

$$= \frac{\cos(35-45)\cos(-20.92)\sin(74.5) + (\pi/180)74.5\sin(35-45)\sin(-20.92)}{\cos(35)\cos(-20.92)\sin(74.5) + (\pi/180)74.5\sin(35)\sin(-20.92)} = 2.05$$

Continued

EXAMPLE 11.5−CONT'D

From Eq. (2.107),

$$\bar{R} = \frac{\bar{H}_t}{\bar{H}} = \left(1 - \frac{\bar{H}_D}{\bar{H}}\right)\bar{R}_B + \frac{\bar{H}_D}{\bar{H}}\left(\frac{1 + \cos(\beta)}{2}\right) + \rho_G\left(\frac{1 - \cos(\beta)}{2}\right)$$

$$= (1 - 0.38)2.05 + 0.38\left[\frac{1 + \cos(45)}{2}\right] + 0.2\left[\frac{1 - \cos(45)}{2}\right] = 1.62$$

Table 11.6 Monthly average calculations for Example 11.5.

Month	N	δ (°)	h_{ss} (°)	h'_{ss} (°)	$\bar{H}_D/\bar{H}$	$\bar{R}_B$	$\bar{R}$	$\bar{H}_t$ (MJ/m²)
January	17	−20.92	74.5	74.5	0.38	2.05	1.62	14.52
February	47	−12.95	80.7	80.7	0.37	1.65	1.38	17.08
March	75	−2.42	88.3	88.3	0.36	1.27	1.15	20.00
April	105	9.41	96.7	88.3	0.38	0.97	0.96	20.67
May	135	18.79	103.8	86.6	0.36	0.78	0.84	21.89
June	162	23.09	107.4	85.7	0.35	0.70	0.78	22.78
July	198	21.18	105.7	86.1	0.34	0.74	0.81	23.13
August	228	13.45	99.6	87.6	0.34	0.88	0.90	22.94
September	258	2.22	91.6	89.6	0.33	1.14	1.07	22.65
October	288	−9.6	83.2	83.2	0.34	1.52	1.32	20.25
November	318	−18.91	76.1	76.1	0.36	1.94	1.58	16.32
December	344	−23.05	72.7	72.7	0.38	2.19	1.71	13.54

And finally, $\bar{H}_t = \bar{R}\bar{H} = 1.62 \times 8.96 = 14.52$ MJ/m². The calculations for all months are shown in Table 11.6.
We can now move along in the *f*-chart estimation. The dimensionless parameters *X* and *Y* are estimated from Eqs. (11.8) and (11.9):

$$X = F_R U_L \frac{F'_R}{F_R}(T_{ref} - \bar{T}_a)\Delta t \frac{A_c}{L}$$

$$= 5.65 \times 0.98(100 - 12.1) \times 31 \times 24 \times 3600 \times \frac{20}{10.30 \times 10^9} = 2.53$$

$$Y = F_R(\tau\alpha)_n \frac{F'_R}{F_R}\left[\frac{(\overline{\tau\alpha})}{(\tau\alpha)_n}\right]\bar{H}_t N \frac{A_c}{L}$$

$$= 0.82 \times 0.98 \times 0.96 \times 14.52 \times 10^6 \times 31 \times \frac{20}{10.30 \times 10^9} = 0.67$$

The storage tank correction is obtained from Eq. (11.14):

$$\frac{X_c}{X} = \left(\frac{M_{w,a}}{M_{w,s}}\right)^{-0.25} = \left(\frac{120}{75}\right)^{-0.25} = 0.89$$

Then, the minimum capacitance value needs to be found (at an assumed temperature of 77 °C):

$$C_{air} = 520 \times 0.998 \times 1009/1000 = 523.6 \text{ W/K}$$

$$C_{water} = (0.02 \times 20) \times 974 \times 4190/1000 = 1632 \text{ W/K}$$

EXAMPLE 11.5—CONT'D

Therefore, the minimum capacitance is for the air side of the load heat exchanger. From Eq. (11.15),

$$Z = \frac{\varepsilon_L (\dot{m}c_p)_{min}}{(UA)_L} = \frac{0.75 \times (523.6)}{450} = 0.87$$

The correction factor is given by Eq. (11.16):

$$\frac{Y_c}{Y} = 0.39 + 0.65\exp\left(-\frac{0.139}{Z}\right) = 0.39 + 0.65\exp\left(-\frac{0.139}{0.87}\right) = 0.94$$

Therefore,

$$X_c = 2.53 \times 0.89 = 2.25$$

and

$$Y_c = 0.67 \times 0.94 = 0.63$$

When these values are used in Eq. (11.13), they give $f = 0.419$. The complete calculations for all months of the year are shown in Table 11.7.

Table 11.7 Complete monthly calculations for the *f*-chart for Example 11.5.

Month	X	Y	X_c	Y_c	f	fL
January	2.53	0.67	2.25	0.63	0.419	4.32
February	2.42	0.76	2.15	0.71	0.483	4.71
March	3.24	1.21	2.88	1.14	0.714	5.63
April	5.73	2.24	5.10	2.11	1	4.13
May	10.49	4.57	9.34	4.30	1	2.22
June	11.21	5.38	9.98	5.06	1	1.90
July	11.67	5.95	10.39	5.59	1	1.80
August	11.02	5.59	9.81	5.25	1	1.90
September	10.51	5.08	9.35	4.78	1	2.00
October	8.37	3.53	7.45	3.32	1	2.74
November	5.37	1.72	4.78	1.62	0.846	3.72
December	3.09	0.78	2.75	0.73	0.464	3.84
						Total = 38.91

From Eq. (11.12), the annual fraction of load covered by the solar energy system is:

$$F = \frac{\sum f_i L_i}{\sum L_i} = \frac{38.91}{57.31} = 0.679 \quad \text{or} \quad 67.9\%$$

11.1.2 Performance and design of air-based solar heating systems

Klein et al. (1977) developed for air-based systems a design procedure similar to that for liquid-based systems. The *f*-chart for air-based systems is developed for the standard solar air-based solar energy system, shown in Fig. 11.5. This is the same as the system shown in Fig. 6.12, drawn without the

FIGURE 11.5

Schematic diagram of the standard air-based solar heating system.

controls, for clarity. As can be seen, the standard configuration of air-based solar heating system uses a pebble-bed storage unit. The energy required for the DHW is provided through the air-to-water heat exchanger, as shown. During summertime, when heating is not required, it is preferable not to store heat in the pebble bed, so a bypass is usually used, as shown in Fig. 11.5 (not shown in Fig. 6.12), which allows the use of the collectors for water heating only.

The fraction f of the monthly total load supplied by a standard solar air-based solar energy system, shown in Fig. 11.5, is also given as a function of the two parameters, X and Y, and can be obtained from the f-chart given in Fig. 11.6 or from the following equation (Klein et al., 1977):

$$f = 1.040Y - 0.065X - 0.159Y^2 + 0.00187X^2 - 0.0095Y^3 \qquad (11.17)$$

EXAMPLE 11.6

A solar air-heating system of the standard configuration is installed in the same area as the one in Example 11.2 and the building has the same load. The air collectors have the same area as in Example 11.2 and are double glazed, with $F_R U_L = 2.92 \ \text{W/m}^2 \ ^\circ\text{C}$, $F_R(\tau\alpha)_n = 0.52$, and $(\overline{\tau\alpha})(\tau\alpha)_n = 0.93$. Estimate the annual solar fraction.

Solution

A general condition of air systems is that no correction factor is required for the collector heat exchanger and ducts are well insulated; therefore, heat losses are assumed to be small, so $F'_R/F_R = 1$. For the month of January and from Eqs. (11.8) and (11.9).

$$X = F_R U_L \frac{F'_R}{F_R} \left(T_{ref} - \overline{T}_a\right) \Delta t \frac{A_c}{L} = 2.92(100 - 10.1) \times 31 \times 24 \times 3600 \times \frac{35}{35.2 \times 10^9} = 0.70$$

$$Y = F_R(\tau\alpha)_n \frac{F'_R}{F_R} \left[\frac{(\overline{\tau\alpha})}{(\tau\alpha)_n}\right] \overline{H}_t N \frac{A_c}{L} = 0.52 \times 0.93 \times 12.5 \times 10^6 \times 31 \times \frac{35}{35.2 \times 10^9} = 0.19$$

From Eq. (11.17) or Fig. 11.6, $f = 0.147$. The solar contribution is $fL = 0.147 \times 35.2 = 5.17$ GJ. The same calculations are repeated for the other months and tabulated in Table 11.8.

EXAMPLE 11.6–CONT'D

Table 11.8 Solar contribution and *f*-values for all months for Example 11.6.

Month	$\overline{H}_t$ (MJ/m²)	$\overline{T}_a$ (°C)	L (GJ)	X	Y	f	fL
January	12.5	10.1	35.2	0.70	0.19	0.147	5.17
February	15.6	13.5	31.1	0.69	0.24	0.197	6.13
March	17.8	15.8	20.7	1.11	0.45	0.367	7.60
April	20.2	19.0	13.2	1.63	0.78	0.618	8.16
May	21.5	21.5	5.6	3.84	2.01	**1**	5.60
June	22.5	29.8	4.1	4.54	2.79	**1**	4.10
July	23.1	32.1	2.9	6.41	4.18	**1**	2.90
August	22.4	30.5	3.5	5.44	3.36	**1**	3.50
September	21.1	22.5	5.1	4.03	2.10	**1**	5.10
October	18.2	19.2	12.7	1.74	0.75	0.587	7.45
November	15.2	16.2	23.6	0.94	0.33	0.267	6.30
December	13.1	11.1	33.1	0.74	0.21	0.164	5.43
			Total load = 190.8		**Total contribution = 67.44**		

It should be noted here that, again, the values of *f* marked in bold are outside the range of the *f*-chart correlation and a fraction of 100% is used because during these months, the solar energy system covers the load fully. From Eq. (11.12), the annual fraction of load covered by the solar energy system is:

$$F = \frac{\sum f_i L_i}{\sum L_i} = \frac{67.44}{190.8} = 0.353 \quad \text{or} \quad 35.3\%$$

Therefore, compared to the results from Example 11.2, it can be concluded that, due to the lower collector optical characteristics, *F* is lower.

This is for a standard storage capacity of 0.25 m³/m² and a standard collector flow rate per square meter of collector area of 10 l/s.m² for the range of X and Y parameters indicated in Fig. 11.6.

Air systems require two correction factors: one for the pebble-bed storage size and one for the air flow rate, which affects stratification in the pebble bed. There are no load heat exchangers in air systems, and care must be taken to use the collector performance parameters $F_R U_L$ and $F_R(\tau\alpha)_n$, determined at the same air flow rate as used in the installation; otherwise, the correction outlined in Chapter 4, Section 4.1.1, needs to be used.

Pebble-bed storage size correction
For the development of the *f*-chart of Fig. 11.6, a standard storage capacity of 0.25 cubic meters of pebbles per square meter of collector area was considered, as indicated in the figure, which corresponds to 350 kJ/m² °C for typical void fractions and rock properties. Although the performance of air-based systems is not as sensitive to the storage capacity as in liquid-based systems, other storage capacities can be used by modifying the factor X by a storage size correction factor, X_c/X, as given by (Klein et al., 1977):

$$\frac{X_c}{X} = \left(\frac{M_{b,a}}{M_{b,s}}\right)^{-0.30} \tag{11.18}$$

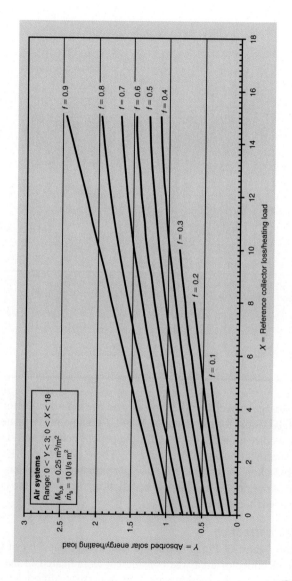

FIGURE 11.6

The *f*-chart for air-based solar heating systems.

where

$M_{b,a}$ = actual pebble storage capacity per square meter of collector area (m³/m²);
$M_{b,s}$ = standard storage capacity per square meter of collector area = 0.25 m³/m².

Equation (11.18) is applied in the range of $0.5 \leq (M_{b,a}/M_{b,s}) \leq 4.0$ or $0.125 \leq M_{b,a} \leq 1.0$ m³/m². The storage correction factor can also be determined from Fig. 11.7 directly, obtained by plotting Eq. (11.18) for this range.

Air flow rate correction

Air-based heating systems must also be corrected for the flow rate. An increased air flow rate tends to increase system performance by improving F_R, but it tends to decrease performance by reducing the pebble bed thermal stratification. The standard collector flow rate is 10 L/s of air per square meter of collector area, as indicated in Fig. 11.6. The performance of systems having other collector flow rates can be estimated by using appropriate values of F_R and Y, then modifying the value of X by a collector air flow rate correction factor, X_c/X, to account for the degree of stratification in the pebble bed (Klein et al., 1977):

$$\frac{X_c}{X} = \left(\frac{\dot{m}_a}{\dot{m}_s}\right)^{0.28} \tag{11.19}$$

where

$\dot{m}_a$ = actual collector flow rate per square meter of collector area (L/s m²);
$\dot{m}_s$ = standard collector flow rate per square meter of collector area = 10 L/s m².

Equation (11.19) is applied in the range of $0.5 \leq (\dot{m}_a / \dot{m}_s) \leq 2.0$ or $5 \leq \dot{m}_s \leq 20$ L/s m². The air flow rate correction factor can also be determined from Fig. 11.8 directly, obtained by plotting Eq. (11.19) for this range.

FIGURE 11.7

Storage size correction factor for air-based systems.

FIGURE 11.8

Correction factor for air flow rate to account for stratification in pebble bed.

EXAMPLE 11.7

If the air system of Example 11.6 uses a flow rate equal to 17 L/s m^2, estimate the solar fraction for the month of January. At the new flow rate, the performance parameters of the collector are $F_R U_L = 3.03$ W/m^2 °C, $F_R(\tau\alpha) = 0.54$.

Solution

From Eq. (11.19),

$$\frac{X_c}{X} = \left(\frac{\dot{m}_a}{\dot{m}_s}\right)^{0.28} = \left(\frac{17}{10}\right)^{0.28} = 1.16$$

The increased air flow rate also affects the F_R and the performance parameters, as shown in the problem definition. Therefore, the value of X to use is the value from Example 11.6 corrected for the new air flow rate through the collector and the pebble bed. Hence,

$$X_c = X\left(\frac{F_R U_L|_{new}}{F_R U_L|_{test}}\right)\frac{X_c}{X} = 0.70\left(\frac{3.03}{2.92}\right)1.16 = 0.84$$

The dimensionless parameter Y is affected only by the F_R. So,

$$Y_c = Y\left(\frac{F_R(\tau\alpha)|_{new}}{F_R(\tau\alpha)|_{test}}\right) = 0.19\left(\frac{0.54}{0.52}\right) = 0.20$$

Finally, from the f-chart of Fig. 11.6 or Eq. (11.17), $f = 0.148$ or 14.8%. Compared to the previous result of 14.7%, there is no significant reduction for the increased flow rate, but there will be an increase in fan power.

If, in a solar energy system, both air flow rate and storage size are different from the standard ones, two corrections must be done on dimensionless parameter *X*. In this case, the final *X* value to use is the uncorrected value multiplied by the two correction factors, as shown in the following example.

EXAMPLE 11.8

If the air system of Example 11.6 uses a pebble storage tank equal to 0.35 m^3/m^2 and the flow rate is equal to 17 L/s m^2, estimate the solar fraction for the month of January. At the new flow rate, the performance parameters of the collector are as shown in Example 11.7.

Solution

The two correction factors need to be estimated first. The correction factors for *X* and *Y* for the increased flow rate are as shown in Example 11.7. For the increased pebble bed storage, from Eq. (11.18),

$$\frac{X_c}{X} = \left(\frac{M_{b,a}}{M_{b,s}}\right)^{-0.30} = \left(\frac{0.35}{0.25}\right)^{-0.30} = 0.90$$

The correction for the air flow rate is given in Example 11.7 and is equal to 1.16. By considering also the correction for the flow rate on F_R and the original value of *X*,

$$X_c = X\left(\frac{F_R U_L|_{new}}{F_R U_L|_{test}}\right)\left.\frac{X_c}{X}\right|_{flow}\left.\frac{X_c}{X}\right|_{storage} = 0.70\left(\frac{3.03}{2.92}\right)1.16 \times 0.90 = 0.76$$

The dimensionless parameter *Y* is affected only by the F_R. So, the value of Example 11.7 is used here (= 0.20). Therefore, from the *f*-chart of Fig. 11.6 or Eq. (11.17), *f* = 0.153 or 15.3%.

11.1.3 Performance and design of solar service water systems

The *f*-chart of Fig. 11.2 or Eq. (11.13) can also be used to estimate the performance of solar service water-heating systems with a configuration like that shown in Fig. 11.9. Although a liquid-based system is shown in Fig. 11.9, air or water collectors can be used in the system with the appropriate heat exchanger to transfer heat to the preheat storage tank. Hot water from the preheat storage tank is then fed to a water heater where its temperature can be increased, if required. A tempering valve may also be used to maintain the supply temperature below a maximum temperature, but this mixing can also be done at the point of use by the user.

FIGURE 11.9

Schematic diagram of the standard of water-heating system configuration.

The performance of solar water heating systems is affected by the public mains water temperature, T_m, and the minimum acceptable hot water temperature, T_w; both affect the average system operating temperature and thus the collector energy losses. Therefore, the parameter X, which accounts for the collector energy losses, needs to be corrected. The additional correction factor for the parameter X is given by (Beckman et al., 1977):

$$\frac{X_c}{X} = \frac{11.6 + 1.18T_w + 3.86T_m - 2.32\overline{T}_a}{100 - \overline{T}_a} \tag{11.20}$$

where

T_m = mains water temperature (°C);
T_w = minimum acceptable hot water temperature (°C); and
$\overline{T}_a$ = monthly average ambient temperature (°C).

The correction factor X_c/X is based on the assumption that the solar preheat tank is well insulated. Tank losses from the auxiliary tank are not included in the f-chart correlations. Therefore, for systems supplying hot water only, the load should also include the losses from the auxiliary tank. Tank losses can be estimated from the heat loss coefficient and tank area (UA) based on the assumption that the entire tank is at the minimum acceptable hot water temperature, T_w.

The solar water heater performance is based on storage capacity of 75 L/m² of collector aperture area and a typical hot water load profile, with little effect by other distributions on the system performance. For different storage capacities, the correction given by Eq. (11.14) applies.

EXAMPLE 11.9

A solar water heating system is installed in an area where, for the 31-day month under investigation, the average daily total radiation on the tilted collector surface is 19.3 MJ/m², average ambient temperature is 18.1 °C, and it uses a 5 m² aperture area collector that has $F_R(\tau\alpha)_n = 0.79$ and $F_R U_L = 6.56$ W/m² °C, both determined from the standard collector tests. The water heating load is 200 L/day, the public mains water temperature, T_m, is 12.5 °C, and the minimum acceptable hot water temperature, T_w, is 60 °C. The storage capacity of the preheat tank is 75 L/m² and auxiliary tank has a capacity of 150 L, a loss coefficient of 0.59 W/m² °C, diameter 0.4 m, and height of 1.1 m; it is located indoors, where the environment temperature is 20 °C. The flow rate in the collector is the same as the flow rate used in testing the collector, the $F'_R/F_R = 0.98$, and the $(\overline{\tau\alpha})/(\tau\alpha)_n = 0.94$. Estimate the solar fraction.

Solution
The monthly water-heating load is the energy required to heat the water from T_m to T_w plus the auxiliary tank losses. For the month investigated, the water heating load is:

$$200 \times 31 \times 4190(60 - 12.5) = 1.234 \text{ GJ}$$

The auxiliary tank loss rate is given by $UA(T_w - T_a)$. The area of the tank is:

$$\pi d^2/2 + \pi dl = \pi(0.4)^2/2 + \pi \times 0.4 \times 1.1 = 1.63 \text{ m}^2$$

Thus, auxiliary tank loss = $0.59 \times 1.63(60 - 20) = 38.5$ W. The energy required to cover this loss in a month is:

$$38.5 \times 31 \times 24 \times 3600 = 0.103 \text{ GJ}.$$

Therefore,

$$\text{Total heating load} = 1.234 + 0.103 = 1.337 \text{ GJ}$$

EXAMPLE 11.9–CONT'D

Using Eqs. (11.8) and (11.9), we get:

$$X = F_R U_L \frac{F_R'}{F_R} \left(T_{\text{ref}} - \overline{T}_a\right) \Delta t \frac{A_c}{L}$$

$$= 6.56 \times 0.98(100 - 18.1) \times 31 \times 24 \times 3600 \times \frac{5}{1.337 \times 10^9} = 5.27$$

$$Y = F_R (\tau\alpha)_n \frac{F_R'}{F_R} \left[\frac{(\tau\alpha)}{(\tau\alpha)_n}\right] \overline{H}_t N \frac{A_c}{L}$$

$$= 0.79 \times 0.98 \times 0.94 \times 19.3 \times 10^6 \times 31 \times \frac{5}{1.337 \times 10^9} = 1.63$$

From Eq. (11.20), the correction for *X* is:

$$\frac{X_c}{X} = \frac{11.6 + 1.18 T_w + 3.86 T_m - 2.32\overline{T}_a}{100 - \overline{T}_a}$$

$$= \frac{11.6 + 1.18 \times 60 + 3.86 \times 12.5 - 2.32 \times 18.1}{100 - 18.1} = 1.08$$

Therefore, the corrected value of *X* is:

$$X_c = 5.27 \times 1.08 = 5.69$$

From Fig. 11.2 or Eq. (11.13), for X_c and *Y*, we get $f = 0.808$ or 80.8%.

11.1.4 Thermosiphon solar water-heating systems

It is a fact that globally, the majority of solar water heaters installed are of the thermosiphon type. Therefore, it is important to develop a simple method to predict their performance similar to the forced circulation or active systems. This method may be a modification of the original *f*-chart method presented in the previous sections to account for the natural circulation occurring in thermosiphon systems, which exhibit also a strong stratification of the hot water in the storage tank.

In fact, the original *f*-chart cannot be used as it is to predict the thermal performance of thermosiphon solar water heaters with good accuracy, for two reasons (Fanney and Klein, 1983; Malkin et al., 1987):

1. The *f*-chart design tool was developed for pumped systems and the assumption that the fluid flow rate through the collector loop is known and it is fixed. The varying flow rates encountered in thermosiphon solar water-heating systems, driven by the strength of solar radiation, lead to different F_R and $F_R U_L$ values from those encountered in active systems. These parameters are determined experimentally with the procedures outlined in Chapter 4 at fixed flow rate.

2. Additionally, a major assumption made for the development of the *f*-chart design method is that the hot-water storage tank is fully mixed. Therefore, the thermal performance of thermosiphon solar-water heaters would be greatly underestimated due to the enhanced thermal stratification produced because of the lower mass flow rate from the collector to storage tank. If this is ignored, *f*-chart would give wrong results leading to system oversizing and would predict reduced cost effectiveness of the thermosiphon solar water-heating system.

The modification to the *f*-chart design method presented here is suggested by Malkin et al. (1987). In fact they used a correction factor to take into account on the system performance the effect of the

Table 11.9 Range of design variables used in developing the modified *f*-chart method for thermosiphon units.

Parameter	Range
Load draw	150–500 L
Hot-water storage tank size	100–500 L
Collector slope	30–90°
$F_R U_L$	3.6–8.6 W/m^2 °C
$F_R(\tau\alpha)$	0.7–0.8

Malkin (1985).

enhanced thermal stratification within the hot-water storage tank. This was obtained by using, similarly as in the original *f*-chart method, a number of TRNSYS simulations for various thermosiphon solar water-heating systems operating in three different locations in the USA (Albuquerque, Madison and Seattle). The characteristics of these three systems are shown in Table 11.9. In addition to these characteristics, in the simulations the overall loss coefficient for the hot-water storage tank was assumed to be constant and equal to 1.46 W/K.

According to Malkin et al. (1987) it is possible to modify the *f*-chart method to enable prediction of the improved performance of systems exhibiting stratified storage tanks. This can be done by considering the varying flow rate through the thermosiphon system to be approximated by an "equivalent average" fixed flow rate for each month in a corresponding active system of the same size. In this case, the active system operating at this fixed flow rate may yield similar results for monthly fractional energy savings by solar as the thermosiphon system. Therefore, in this way the long-term performance of a thermosiphon system may be easily predicted, using the modified form of the *f*-chart method but the same chart shown in Fig. 11.2 for liquid systems.

Once the equivalent monthly average fluid flow rate has been calculated from the density differences in the thermosiphon fluid flow loop then the modified values of $F_R(\tau\alpha)_n$ and $F_R U_L$ need to be calculated using the corrective factor (r) given for different test and use flow rates in Chapter 4, Section 4.1.1 by Eq. (4.17) or for the thermosiphon flow in Chapter 5, Section 5.1.1 by Eq. (5.4b) against the normal (or test) flow rate.

It should be noted that thermally stratified storage tanks return the fluid to the collector at a temperature below that of the average temperature of the storage tank. This results in improved collector efficiency as the temperature entering the collector is lower, closer to ambient and thus the thermal losses from the collector are lower. Thermosiphon systems usually circulate domestic water through the collectors, so they do not include a heat exchanger, therefore by combining Eqs. (11.8) and (11.20) the X parameter, called X_{mix}, is given by:

$$X_{mix} = \frac{A_c F_R U_L \left(11.6 + 1.18 T_w + 3.86 T_m - 2.32 \overline{T}_a\right) \Delta t}{L} \tag{11.21}$$

Similarly the Y parameter without the presence of the heat exchanger term can be given by modifying Eq. (11.9) as:

$$Y_{mix} = \frac{A_c F_R (\overline{\tau\alpha}) \overline{H}_t N}{L} \tag{11.22}$$

It should be pointed again that the X_{mix} and Y_{mix} parameters shown in Eqs. (11.21) and (11.22) assume that the hot-water storage tank is fully mixed. Copsey (1984) was the first to develop a modification of the *f*-chart to account for stratified storage tanks. He has shown that the solar fraction of a stratified tank system (*f*) can be obtained by analyzing an otherwise identical fully mixed tank system with a reduced collector loss coefficient (U_L). As shown by Eq. (3.58) the collector heat-removal factor F_R is a function of the collector heat-loss coefficient and the flow rate through the collector, therefore a modification of the *f*-chart method that is based on the collector loss coefficient will also require modification of the heat removal factor (F_R).

The predicted value of the solar fraction of a thermosiphon solar water heater exhibiting stratified storage (f_{str}) would be between the solar fraction that could be met with a fully mixed hot-storage tank (f_{mix}) using a loss coefficient equal to that reported at test conditions and the solar fraction that could be supplied from a solar water-heating system with a fully mixed hot-water storage tank with U_L equal to zero. If the collector had no thermal losses then the *X* parameter calculated using Eq. (11.21) would be zero and the maximum value of *Y*, denoted as Y_{max}, for which $F_R = 1$, can be estimated from:

$$Y_{max} = \frac{A_c(\overline{\tau\alpha})\overline{H}_t N}{L} \tag{11.23}$$

As reported from Malkin et al. (1987) the relationship between the mixed and stratified coordinates used to predict the solar fraction of the heating load could be related using an *X*-parameter corrective factor, denoted as ($\Delta X/\Delta X_{max}$), for stratified storage. This factor is a function of the monthly average collector to load flow ratio ($\overline{M}_c/M_L$) and mixed tank solar fraction (f_{mix}) and is given by (Malkin, 1985):

$$\frac{\Delta X}{\Delta X_{max}} = \frac{X_{mix} - X_{str}}{X_{mix}} = \frac{Y_{str} - Y_{mix}}{Y_{max} - Y_{mix}} = \frac{1.040\left(\overline{M}_c/M_L\right)}{\left[0.726\left(\overline{M}_c/M_L\right) + 1.564f_{mix} - 2.760f_{mix}^2\right]^2 + 1} \tag{11.24}$$

where the coefficients used are as reported by Malkin (1985) which minimize the root mean square error between the predicted thermal performance of a solar water heater with a stratified hot-water storage tank simulated using TRNSYS and that predicted using the *f*-chart procedure modified for stratified hot-water storage tanks. Equation (11.24) is valid when the value of ($\overline{M}_c/M_L$) is greater than 0.3, which is usual.

The daily average flow through the collector could be estimated from a correlation equivalent to the number of hours (N_p) that a solar collector would have a useful energy output. According to Mitchell et al. (1981), for a zero degree differential temperature controller:

$$N_p = -\overline{H}_t\frac{d\overline{\phi}}{dI_c} \tag{11.25}$$

Where

$\overline{\phi}$ = monthly average daily utilizability;

I_c = critical radiation level (W/m²); and

$\overline{H}_t$ = solar radiation incident of the collector (kJ/m² day).

It should be noted that by dividing $\overline{H}_t$ by 3.6 the units of N_p is hours.

Utilizability is defined as the fraction of the incident solar radiation that can be converted into useful heat by a collector having $F_R(\tau\alpha) = 1$ and operating at a fixed collector inlet to ambient temperature difference. It should be noted that although the collector has no optical losses and has a heat removal factor of 1, the utilizability is always less than 1 since the collector has thermal losses (see Section 11.2.3). Evans et al. (1982) developed an empirical relationship to calculate the monthly average daily utilizability as a function of the critical radiation level, the monthly average clearness index $(\overline{K}_T)$, given by Eq. (2.82a), the collector tilt angle (β) and the latitude (L):

$$\overline{\phi} = 0.97 + A\overline{I}_c + B\overline{I}_c^2 \tag{11.26a}$$

where:

$$A = -4.86 \times 10^{-3} + 7.56 \times 10^{-3}\overline{K}'_T - 3.81 \times 10^{-3}\left(\overline{K}'_T\right)^2 \tag{11.26b}$$

$$B = 5.43 \times 10^{-6} - 1.23 \times 10^{-5}\overline{K}'_T + 7.62 \times 10^{-6}\left(\overline{K}'_T\right)^2 \tag{11.26c}$$

For

$$\overline{K}'_T = \overline{K}_T \cos[0.8(\beta_m - \beta)] \tag{11.27}$$

where β_m is the monthly optimal collector tilt shown in Table 11.10.

Differentiating Eq. (11.26a) and substituting into Eq. (11.25), yields an expression for the monthly average daily collector operating time as:

$$\overline{N}_p = -\overline{H}_t\left(A + 2B\overline{I}_c\right) \tag{11.28}$$

The monthly average critical radiation level is defined as the level above which useful energy may be collected and is given by:

$$\overline{I}_c = \frac{F_R U_L\left(\overline{T}_i - \overline{T}_a\right)}{F_R(\overline{\tau\alpha})} \tag{11.29}$$

Table 11.10 Recommended monthly optimal tilt angle.

Month	β_m (deg.)
January	$L + 29$
February	$L + 18$
March	$L + 3$
April	$L - 10$
May	$L - 22$
June	$L - 25$
July	$L - 24$
August	$L - 10$
September	$L - 2$
October	$L + 10$
November	$L + 23$
December	$L + 30$

$L = $ *local latitude (deg.).*
Evans et al. (1982).

As can be seen from Eq. (11.29) in order to find the monthly average critical radiation level, a value for the monthly average collector inlet temperature $(\overline{T}_i)$ must be known. This however cannot be determined analytically and it is a function of the thermal stratification in the storage tank. It may be approximated by the Phillip's stratification coefficient, which is described subsequently. By assuming that the storage tank remains stratified, the monthly average collector inlet temperature may initially be estimated to be the mains water temperature.

To apply the method an initial estimated value for a flow rate is used and the solar fraction of an active system stratified tank is estimated using the *f*-chart method by applying the Copsey's modification for stratified storage. Subsequently, the collector parameters $F_R U_L$ and $F_R(\tau\alpha)$ are corrected for the estimated flow using Eq. (4.17a). An iterative method is required to determine the "equivalent average" flow rate of a thermosiphon solar water-heating system using the initial estimated value of flow rate (Malkin et al., 1987) as explained subsequently. The average temperature in the storage tank is calculated from a correlation developed between the solar fraction of a thermosiphon system and a non-dimensional form of the monthly average tank temperature obtained from the simulations with TRNSYS indicated above. The least squares regression equation obtained is (Malkin et al., 1987):

$$\frac{\overline{T}_{tank} - T_m}{T_{set} - T_m} = 0.117 f_{str} + 0.356 f_{str}^2 + 0.424 f_{str}^3 \tag{11.30}$$

The average temperature distribution in the system can be predicted from the fraction calculated using the initial estimate of the flow rate allowing an approximate calculation to be made of the thermosiphon head for the system geometry shown in Fig. 11.10 (Malkin et al., 1987). This value is then compared to the frictional losses in the collector loop. Iterations need to be made until the agreement between the thermosiphon head and frictional losses are within 1%.

The temperature at the bottom of the storage tank is between the mains temperature (T_m) and the average tank temperature $(\overline{T}_{tank})$. The proximity of the two temperatures depends on the degree of

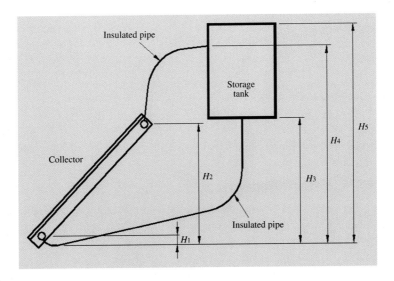

FIGURE 11.10

Thermosiphon system configuration and geometry.

stratification present. This may be measured approximately using the stratification coefficient, K_{str}, defined by Phillips and Dave (1982):

$$K_{str} = \frac{A_c(I_t F_R(\tau\alpha) - F_R U_L(T_i - T_a))}{A_c(I_t F_R(\tau\alpha) - F_R U_L(\overline{T}_{tank} - T_a))} \tag{11.31}$$

The stratification coefficient is also a function of two dimensionless variables; the mixing number, M, and the collector effectiveness, E, given by:

$$M = \frac{A_{cyl}k}{\dot{m}c_p H} \tag{11.32}$$

$$E = \frac{F_R U_L}{\dot{m}c_p} \tag{11.33}$$

where k is the thermal conductivity of water and H is the storage tank height. Physically, the mixing number is the ratio of conduction to convection heat transfer in the storage tank and in the limit as conduction becomes negligible M approaches zero. Phillips and Dave (1982) showed that:

$$K_{str} = \frac{\ln\left(1/1 - E\right)}{E\left(1 + M\ln\left(1/1 - E\right)\right)} \tag{11.34}$$

The temperature of the return fluid from the tank to the collector can be found by solving Eq. (11.31) for T_i. Thus:

$$T_i = K_{str} T_{tank} + (1 - K_{str})\left(\frac{F_R(\tau\alpha)}{F_R U_L} I_t + T_a\right) \tag{11.35}$$

Therefore, by using the pump-operating time, estimated from Eq. (11.28), the monthly average temperature of return fluid is approximated as:

$$\overline{T}_i = K_{str}\overline{T}_{tank} + (1 - K_{str})\left(\frac{F_R(\overline{\tau\alpha})}{(F_R U_L)(\overline{N}_p)}\overline{H}_t + \overline{T}_a\right) \tag{11.36}$$

It should be noted that if a value is obtained from Eq. (11.36) which is lower than T_m, then T_m is used instead as it is considered impossible to have a storage tank temperature which is lower than the mains temperature.

The collector outlet temperature is found by equating Eq. (3.60), by using I_t instead of G_t, and an energy balance across the collector:

$$\dot{m}c_p(T_o - T_i) = A_c F_R[I_t(\tau\alpha) - U_L(T_i - T_a)] \tag{11.37}$$

By integrating Eq. (11.37) on a monthly basis, the monthly average collector fluid outlet temperature can be obtained:

$$\overline{T}_o = \overline{T}_i + \frac{A_c}{\dot{m}c_p\overline{N}_p}\left[\overline{H}_t F_R(\overline{\tau\alpha}) - F_R U_L \overline{N}_p(\overline{T}_i - \overline{T}_a)\right] = \overline{T}_i + \frac{A_c}{\dot{m}c_p\overline{N}_p}\left[\overline{\phi}\,\overline{H}_t F_R(\overline{\tau\alpha})\right] \tag{11.38}$$

This procedure is undertaken on a monthly basis allowing the "equivalent average" flow rate to be determined and the solar fraction calculated from this value.

Once the monthly average collector fluid inlet and outlet temperatures are known, an estimate of the thermosiphon head may be found based on the relative positions of the tank and the flat-plate collectors as shown in Fig. 11.10. Close (1962) has shown that the thermosiphon head generated by the differences in density of fluid in the system may be approximated by making the following assumptions:

1. There are no thermal losses in the connecting pipes.
2. Water from the collector rises to the top of the tank.
3. The temperature distribution in the tank is linear.

Therefore, according to the dimensions indicated in Fig. 11.10, the thermosiphon head generated is given by (Close, 1962):

$$h_T = \frac{1}{2}(S_i - S_o)\left[2(H_3 - H_1) - (H_2 - H_1) - \frac{(H_3 - H_5)^2}{(H_4 - H_5)}\right] \tag{11.39}$$

where S_i and S_o are the specific gravities of the fluid at the collector inlet and outlet respectively. Here only direct circulation thermosiphon systems are considered in which water is the collection fluid. The specific gravity according to the temperature (in °C) of water is given by:

$$S = 1.0026 - 3.906 \times 10^{-5}T - 4.05 \times 10^{-6}T^2 \tag{11.40}$$

The equivalent average flow rate is that which balances the thermosiphon buoyancy force with the frictional resistance in the flow circuit on a monthly average basis. As indicated in Chapter 5 (Section 5.1.1) the flow circuit comprises the collector headers and risers, connecting pipes and storage tank. For each component of the flow circuit, the Darcy–Weisbach equation for friction head loss needs to be employed, given by Eq. (5.6).

The Reynolds number, used to find the type of flow as indicated in Section 5.1.1, at the estimated flow rate is calculated using the following correlation for viscosity as a function of temperature in °C:

$$\mu = \frac{0.1}{2.1482\left(T - 8.435 + \sqrt{8087.4 + (T - 8.435)^2}\right) - 120} \tag{11.41}$$

The last term in Eq. (5.6) is included to account for minor friction losses associated with bends, tees, and other restrictions in the piping circuit. It should be noted that although the majority of the pressure drop in the flow circuit occurs across the relatively small diameter collector risers, the minor frictional losses are included to enhance the accuracy of the flow rate estimation. Details of these losses are given in Section 5.1.1 in Chapter 5 and Table 5.2.

All the components of the friction head loss in the flow circuit, at the estimated flow rate, are combined and comparison is made with the previously calculated thermosiphon head. If the thermosiphon head does not balance the frictional losses to within 1% a new guess of the flow rate through the connecting pipes is made by successive substitution. The procedure is repeated with the new estimate of flow rate until the convergence is within 1%.

A new guess of the thermosiphon flow rate can be obtained from (Malkin, 1985):

$$\dot{m}_{new} = \rho A \sqrt{\frac{2gh_T}{\left(\frac{fl}{d} + k\right)_p + \left(\frac{fl}{d} + k\right)_r + \left(\frac{fl}{d} + k\right)_h}} \tag{11.42}$$

where subscript p stands for connecting pipes, r for riser, and h for headers.

Convergence is usually obtained by applying no more than three iterations. The resulting single value for monthly average flow rate is that which balances the thermosiphon head with the frictional losses in the flow circuit. As was pointed out before, the solar fraction is calculated, assuming a fixed flow rate operating in an equivalent active system. As in the standard f-chart method, this procedure is carried out for each month of the year, with a previous month's "equivalent average" flow rate as the initial guess of flow rate for the new month. The fraction of the annual load supplied by solar energy is the sum of the monthly solar energy contributions divided by the annual load as given by Eq. (11.12).

The difference between the TRNSYS simulations and the modified f-chart design method had a reported root mean square error of 2.6% (Malkin, 1985). The procedure is clarified by the following example.

EXAMPLE 11.10

Calculate the solar contribution of a thermosiphon solar water-heating system located in Nicosia, Cyprus for the month of January. The system has the following characteristics.

1. Slope of collector $= 45°$
2. Monthly average solar radiation $= 8960$ kJ/m^2 day (From Appendix 7, Table A7.12)
3. Monthly average ambient temperature $= 12.1$ °C (From Appendix 7, Table A7.12)
4. Monthly average clearness index $= 0.49$ (From Appendix 7, Table A7.12)
5. Number of collector panels $= 2$
6. Collector area per panel $= 1.35$ m^2
7. Collector test $F_R U_L = 21.0$ kJ/h m^2 °C
8. Collector $F_R(\tau\alpha) = 0.79$
9. Collector test flow rate $= 71.5$ kg/h m^2
10. Number of risers per panel $= 10$
11. Riser diameter $= 0.015$ m
12. Combined header length per panel $= 1.9$ m
13. Header diameter $= 0.028$ m
14. Tank-collector connecting pipe length $= 2.5$ m
15. Collector-tank connecting pipe length $= 0.9$ m
16. Connecting pipe diameter $= 0.022$ m
17. Number of bends in connecting pipe $= 2$
18. Connecting pipe heat-loss coefficient $= 10.0$ kJ/h m^2 °C
19. Storage tank volume $= 160$ L
20. Storage tank height $= 1$ m
21. Storage tank diameter $= 0.45$ m
22. Daily load draw off $= 150$ L
23. Mains water temperature $= 18$ °C
24. Auxiliary set temperature $= 60$ °C
25. Height $H_1 = 0.05$ m
26. Height $H_2 = 1.12$ m
27. Height $H_3 = 2.1$ m
28. Height $H_5 = 1.27$ m

Solution

Initially the radiation on the collector surface is required and to save some space the results of Example 11.5 are used. So from Table 11.6 the required value is equal to 14,520 kJ/m^2.

EXAMPLE 11.10—CONT'D

An initial estimate of the equivalent average flow rate is 15 kg/h m² or 40.5 kg/h for a 2.7 m² collector area. The collector performance parameters $F_R U_L$ and $F_R(\tau\alpha)$ are corrected for the assumed flow rate, which is different than the test flow rate by using Eq. (5.4b) using the parameter $F'U_L$ estimated from Eq. (5.3). Thus:

From Eq. (5.3):

$$F'U_L = \frac{-\dot{m}_T c_p}{A_c}\ln\left(1 - \frac{F_R U_L A_c}{\dot{m}_T c_p}\right) = \frac{-71.5 \times 2.7 \times 4.19}{2.7}\ln\left(1 - \frac{21 \times 2.7}{71.5 \times 2.7 \times 4.19}\right)$$

$$= 21.77 \text{ kJ/h m}^2\,{}^{\circ}\text{C}$$

From Eq. (5.4b):

$$r = \frac{\dot{m}_t\left(1 - \exp\left(-\frac{F'U_L A_c}{\dot{m}_t c_p}\right)\right)}{\dot{m}_T\left(1 - \exp\left(-\frac{F'U_L A_c}{\dot{m}_T c_p}\right)\right)} = \frac{40.5\left(1 - \exp\left(-\frac{21.77 \times 2.7}{40.5 \times 4.19}\right)\right)}{2.7 \times 71.5\left(1 - \exp\left(-\frac{21.77 \times 2.7}{71.5 \times 2.7 \times 4.19}\right)\right)} = 0.876$$

and

$$F_R U_L = 0.876 \times 21 = 18.40 \text{ kJ/h m}^2\,{}^{\circ}\text{C}$$

$$F_R(\tau\alpha) = 0.876 \times 0.79 = 0.692$$

Thermal losses from the connecting pipes are estimated from Eqs. (5.64b) and (5.64c):

$$\frac{(\tau\alpha)'}{(\tau\alpha)} = \frac{1}{1 + \dfrac{U_p A_{p,o}}{(\dot{m}c_p)_c}} = \frac{1}{1 + \dfrac{10 \times \pi \times 0.022 \times 0.9}{40.5 \times 4.19}} = 0.996$$

$$\frac{U'_L}{U_L} = \frac{1 - \dfrac{U_p A_{p,i}}{(\dot{m}c_p)_c} + \dfrac{U_p(A_{p,i} + A_{p,o})}{A_c F_R U_L}}{1 + \dfrac{U_p A_{p,o}}{(\dot{m}c_p)_c}} = \frac{1 - \dfrac{10 \times \pi \times 0.022 \times 0.9}{40.5 \times 4.19} + \dfrac{10(\pi \times 0.022 \times 0.9 + \pi \times 0.022 \times 2.5)}{2.7 \times 18.4}}{1 + \dfrac{10 \times \pi \times 0.022 \times 2.5}{40.5 \times 4.19}} = 1.033$$

Therefore:

$$F_R U'_L = 1.033 \times 18.4 = 19.01 \text{ kJ/h m}^2\,{}^{\circ}\text{C}$$

$$F_R(\tau\alpha)' = 0.996 \times 0.692 = 0.689$$

For simplification it is assumed that $F_R(\tau\alpha) = F_R(\overline{\tau\alpha}) = 0.689$

From Eq. (11.21):

$$X_{\text{mix}} = \frac{A_c F_R U'_L (11.6 + 1.18T_w + 3.86T_m - 2.32\overline{T}_a)\Delta t}{L}$$

$$= \frac{2.7 \times 19.01(11.6 + 1.18 \times 60 + 3.86 \times 18 - 2.32 \times 12.1) \times 24}{150 \times 4.19(60 - 18)}$$

$$= 0.047(123.808) = 5.82$$

From Eq. (11.22):

$$Y_{\text{mix}} = \frac{A_c F_R(\overline{\tau\alpha})\overline{H}_t N}{L} = \frac{2.7 \times 0.692 \times 14520}{150 \times 4.19(60 - 18)} = 1.03$$

Continued

EXAMPLE 11.10—CONT'D

It should be noted that in both the above equations we do not multiply by the number of days in a month as the load is estimated on a daily basis.

Storage capacity $= 160/2.7 = 59.3$ L/m^2, which is different from the standard value of 75 L/m^2, so a correction is required estimated from Eq. (11.14): $X_{mix,c} = 5.82 \times (59.3/75)^{-0.25} = 6.17$.

From Eq. (11.13): $f_{mix} = 1.029Y_{mix} - 0.065X_{mix} - 0.245Y^2_{mix} + 0.0018X^2_{mix} + 0.0215Y^3_{mix} = 1.029 \times 1.03 - 0.065 \times 6.17 - 0.245 \times (1.03)^2 + 0.0018 \times (6.17)^2 + 0.0215 \times (1.03)^3 = 0.49$.

To estimate the collector pump operation time a number of parameters are required. From Table 11.10 $\beta_m = L + 29° = 35.15 + 29 = 64.15°$.

From Eq. (11.27): $\overline{K}'_T = \overline{K}_T \cos[0.8(\beta_m - \beta)] = 0.49 \times \cos[0.8 \times (64.15 - 45)] = 0.47$.

From Eq. (11.26b):

$$A = -4.86 \times 10^{-3} + 7.56 \times 10^{-3}\overline{K}'_T - 3.81 \times 10^{-3}\left(\overline{K}'_T\right)^2$$

$$= -4.86 \times 10^{-3} + 7.56 \times 10^{-3} \times 0.47 - 3.81 \times 10^{-3}(0.47)^2 = -0.00215$$

From Eq. (11.26c):

$$B = 5.43 \times 10^{-6} - 1.23 \times 10^{-5}\overline{K}'_T + 7.62 \times 10^{-6}\left(\overline{K}'_T\right)^2$$

$$= 5.43 \times 10^{-6} - 1.23 \times 10^{-5} \times 0.47 + 7.62 \times 10^{-6}(0.47)^2 = 1.332 \times 10^{-6}$$

From Eq. (11.29):

$$\overline{I}_c = \frac{F_R U_L\left(\overline{T}_i - \overline{T}_a\right)}{F_R(\overline{\tau\alpha})} = \frac{19.01(18 - 12.1)}{3.6 \times 0.689} = 45.22 \text{ W/m}^2$$

From Eq. (11.28):

$$\overline{N}_p = -\overline{H}_t\left(A + 2B\overline{I}_c\right) = -14,520\left(-0.00215 + 2 \times 1.332 \times 10^{-6} \times 45.22\right)/3.6 = 8.2 \text{ h}$$

The ratio $\left(\overline{M}_c/M_L\right)$ is equal to $(8.2 \times 40.5)/150 = 2.21$.

From Eq. (11.24):

$$\frac{\Delta X}{\Delta X_{max}} = \frac{1.040\left(\overline{M}_c/M_L\right)}{\left[0.726\left(\overline{M}_c/M_L\right) + 1.564f_{mix} - 2.760f^2_{mix}\right]^2 + 1}$$

$$= \frac{1.040 \times 2.21}{\left[0.726 \times 2.21 + 1.564 \times 0.49 - 2.760 \times (0.49)^2\right]^2 + 1} = 0.59$$

So

$$X_{str} = X_{mix}\left(1 - \frac{\Delta X}{\Delta X_{max}}\right) = 6.17(1 - 0.59) = 2.5297$$

To find $(\overline{\tau\alpha})$ when $F_R = 1$ we solve for $F_{R,high}/F_{R,use}$ at a very high flow rate, say 10,000 kg/h. This gives $r = 1.04$ and $(\overline{\tau\alpha}) = 0.79 \times 1.04 = 0.8216$.

Therefore, from Eq. (11.23):

$$Y_{max} = \frac{A_c(\overline{\tau\alpha})\overline{H}_t N}{L} = \frac{2.7 \times 0.8216 \times 14520}{150 \times 4.19(60 - 18)} = 1.22$$

EXAMPLE 11.10—CONT'D

So from Eq. (11.24):

$$Y_{str} = Y_{mix} + (Y_{max} - Y_{mix})\frac{\Delta X}{\Delta X_{max}} = 1.03 + (1.22 - 1.03) \times 0.59 = 1.1421$$

From Eq. (11.13): $f_{str} = 1.029 Y_{str} - 0.065 X_{str} - 0.245 Y_{str}^2 + 0.0018 X_{str}^2 + 0.0215 Y_{str}^3 = 1.029 \times 1.1421 - 0.065 \times 2.5297 - 0.245 \times (1.1421)^2 + 0.0018 \times (2.5297)^2 + 0.0215 \times (1.1421)^3 = 0.73$.

Then the Phillips stratification coefficient is evaluated:

From Eq. (11.32):

$$M = \frac{A_{cyl}k}{\dot{m}c_p H} = \frac{[\pi(0.45)^2/4] \times 0.6}{40.5 \times 4.19 \times 1} = 0.00056$$

From Eq. (11.33):

$$E = \frac{F_R U_L}{\dot{m}c_p} = \frac{2.7 \times 19.01}{40.5 \times 4.19} = 0.30$$

From Eq. (11.34):

$$K_{str} = \frac{\ln\left(1/1 - E\right)}{E\left(1 + M\,\ln\left(1/1 - E\right)\right)} = \frac{\ln\left(1/1 - 0.3\right)}{0.3\left(1 + 0.00056\,\ln\left(1/1 - 0.3\right)\right)} = 1.19$$

From Eq. (11.30):

$$\frac{\overline{T}_{tank} - T_m}{T_{set} - T_m} = 0.117 f_{str} + 0.356 f_{str}^2 + 0.424 f_{str}^3$$

$$= 0.117 \times (0.73) + 0.356 \times (0.73)^2 + 0.424 \times (0.73)^3 = 0.440$$

Therefore: $\overline{T}_{tank} = T_m + 0.440(T_{set} - T_m) = 18 + 0.440(60 - 18) = 36.5\,°C$

From Eq. (11.36):

$$\overline{T}_i = K_{str}\overline{T}_{tank} + (1 - K_{str})\left(\frac{F_R(\overline{\tau\alpha})}{(F_R U_L)(\overline{N}_p)}\overline{H}_t + \overline{T}_a\right)$$

$$= 1.19 \times 36.5 + (1 - 1.19)\left(\frac{0.692}{19.01 \times 8.2}14{,}520 + 12.1\right) = 28.9\,°C$$

Similarly from Eq. (11.38):

$$\overline{T}_o = \overline{T}_i + \frac{A_c}{\dot{m}c_p \overline{N}_p}\left[\overline{H}_t F_R(\overline{\tau\alpha}) - F_R U_L \overline{N}_p(\overline{T}_i - \overline{T}_a)\right]$$

$$= 28.9 + \frac{2.7}{40.5 \times 4.19 \times 8.2}[14{,}520 \times 0.689 - 19.01 \times 8.2(28.9 - 12.1)] = 43.2\,°C$$

Continued

EXAMPLE 11.10—CONT'D

Then the specific gravity is estimated from these two temperatures using Eq. (11.40) as:

$S_i = 0.995749$ and $S_o = 0.991014$. Using the various values of heights and noting that H_4 is equal to H_5 plus the tank height [$=1.27 + 1$] in Eq. (11.39):

$$h_T = \frac{1}{2}(S_i - S_o)\left[2(H_3 - H_1) - (H_2 - H_1) - \frac{(H_3 - H_5)^2}{(H_4 - H_5)}\right]$$

$$= \frac{1}{2}(0.995749 - 0.991014)\left[2(2.1 - 0.05) - (1.12 - 0.05) - \frac{(2.1 - 1.27)^2}{(2.27 - 1.27)}\right]$$

$$= 0.005543 \text{ m}$$

From Eq. (11.41) using the mean tank temperature: $\mu = 0.000701$ kg/m^2 s.

Hydraulic Considerations.

The specific gravity of the water in the storage tank, obtained using the mean storage tank temperature in Eq. (11.40) is equal to 0.993439. The mean tank temperature is assumed to flow through the connecting pipes. Therefore, in connecting pipes the velocity is equal to:

$$v_c = \frac{\dot{m}_t}{3600 S_t \rho A_c} = \frac{40.5}{3600 \times 0.993439 \times 1000 \times \pi(0.022)^2/4} = 0.0298 \text{ m/s}$$

which gives: $\text{Re} = \frac{S_t \rho v_c D_c}{\mu} = \frac{0.993439 \times 1000 \times 0.0298 \times 0.022}{0.000701} = 929$

Using Eq. (5.7a): $f = 64/929 = 0.069$. Correcting for developing flow in pipes using Eq. (5.7c):

$$f = 1 + \frac{0.038}{\left(\dfrac{L}{d\text{Re}}\right)^{0.964}} = 1 + \frac{0.038}{\left(\dfrac{2.5 + 0.9}{0.022 \times 929}\right)^{0.964}} = 1.2141$$

Therefore $f = 0.069 \times 1.2141 = 0.084$.

Finally, the equivalent pipe length is equal to the actual length of connecting pipes plus the number of bends multiplied by 30 (Table 5.2) multiplied by the pipe diameter $= (2.5 + 0.9) + 2 \times 30 \times 0.022 = 4.72$ m. The pipes are of equal diameter so there is no contraction or expansion loss. There is only a loss due to tank entry which is equal to 1 (Table 5.2), so $k = 1$. Using Eq. (5.6):

$$H_f = \frac{fLv^2}{2dg} + \frac{kv^2}{2g} = \frac{0.084 \times 4.72 \times (0.0298)^2}{2 \times 0.022 \times 9.81} + \frac{1 \times (0.0298)^2}{2 \times 9.81} = 0.000861 \text{ m}$$

Similar calculations for risers give:

Flow rate in risers $= 2.025$ kg/h.

Velocity in risers $= 0.0032$ m/s.

$\text{Re} = 68$

$f = 0.965$ m.

$H_f = 5.147 \times 10^{-5}$ m.

For headers the flow is given by: $\sum_{i=1}^{20} \dot{m}_r/20 = 21.263$ kg/h

Flow rate in risers $= 0.0097$ kg/h.

Velocity in headers $= 0.0096$ m/s.

$\text{Re} = 384.9$

$f = 0.0012$ m.

$H_f = 0.00106$ m.

Therefore the total friction head H_f is equal to: $H_f = 0.000861 + 5.147 \times 10^{-5} + 0.00012 = 0.00103$ m.

EXAMPLE 11.10—CONT'D

By comparing this value with the thermosiphon head estimated before we get a percentage difference of:

$$\% \text{ difference} = \frac{h_T - H_f}{h_T} \times 100 = \frac{0.005543 - 0.00103}{0.005543} \times 100 = 81.4\%$$

As the percentage difference is more than 1% the new flow rate is estimated by Eq. (11.42) which gives 93.8 kg/h.

By repeating this procedure for the new flow rate we get a percentage difference of −29.5%. A new guess of Eq. (11.42) gives a flow rate of 82.4 kg/h which gives a percentage difference of 5.8%. So a third guessing is required which gives a new flow rate equal to 83.7 kg/h which gives:

$f_{str} = 0.66$
$h_T = 0.002671.$
$H_f = 0.002670.$

% difference = 0.04%, which is below 1% so this solution is considered as final. Therefore, the solar contribution at this flow rate is 0.66. This flow rate can also be used as the initial guess for the estimation of the next month's solar contribution if the estimation is done on an annual basis.

If should be noted that this flow rate is equal to 31 kg/h/m² [=83.7 kg/h/2.7 m²] which is about 43% of the test flow rate. Additionally, the number of iterations required depends on how far is the initial guessing of thermosiphonic flow rate from the actual one.

It should be noted that as in earlier examples and due to the large number of calculations required, the use of a spreadsheet program greatly facilitates the calculations, especially at the iteration stage.

11.1.5 General remarks

The *f*-chart design method is used to quickly estimate the long-term performance of solar energy systems of standard configurations. The input data needed are the monthly average radiation and temperature, the monthly load required to heat a building and its service water, and the collector performance parameters obtained from standard collector tests. A number of assumptions are made for the development of the *f*-chart method. The main ones include assumptions that the systems are well built, system configuration and control are close to the ones considered in the development of the method, and the flow rate in the collectors is uniform. If a system under investigation differs considerably from these conditions, then the *f*-chart method cannot give reliable results.

It should be emphasized that the *f*-chart is intended to be used as a design tool for residential space and domestic water-heating systems of standard configuration. In these systems, the minimum temperature at the load is near 20 °C; therefore, energy above this value of temperature is useful. The *f*-chart method cannot be used for the design of systems that require minimum temperatures substantially different from this minimum value. Therefore, it cannot be used for solar air conditioning systems using absorption chillers, for which the minimum load temperature is around 80 °C.

It should also be understood that, because of the nature of the input data used in the *f*-chart method, there are a number of uncertainties in the results obtained. The first uncertainty is related to the meteorological data used, especially when horizontal radiation data are converted into radiation falling on the inclined collector surface, because average data are used, which may differ considerably from the real values of a particular year, and that all days were considered symmetrical about a solar noon. A second uncertainty is related to the fact that solar energy systems are assumed to be well built with well-insulated storage tanks and no leaks in the system, which is not very correct for air systems, all of which leak to some extent, leading to a degraded performance. Additionally, all liquid storage tanks are assumed to be fully mixed, which leads to conservative long-term performance predictions because it gives overestimation of collector inlet temperature. The final uncertainty is related to the building and hot water loads, which strongly depend on variable weather conditions and the habits of the occupants.

Despite these limitations, the *f*-chart method is a handy method that can easily and quickly be used for the design of residential-type solar heating systems. When the main assumptions are fulfilled, quite accurate results are obtained.

11.1.6 *F*-chart program

Although the *f*-chart method is simple in concept, the required calculations are tedious, particularly for the manipulation of radiation data. The use of computers greatly reduces the effort required. Program *f*-chart (Klein and Beckman, 2005), developed by the originators of TRNSYS, is very easy to use and gives predictions very quickly. Again, in this case, the model is accurate only for solar heating systems of a type comparable to that which was assumed in the development of the *f*-chart.

The *f*-chart program is written in the BASIC programming language and can be used to estimate the long-term performance of solar energy systems that have flat-plate, evacuated tube collectors, compound parabolic collectors, and one- or two-axis tracking concentrating collectors. Additionally, the program includes an active–passive storage system and analyzes the performance of a solar energy system in which energy is stored in the building structure rather than a storage unit (treated with methods presented in the following section) and a swimming pool-heating system that provides estimates of the energy losses from the swimming pool. The complete list of solar energy systems that can be handled by the program is as follows:

- Pebble-bed storage space and domestic water heating systems.
- Water-storage space and/or domestic water heating systems.
- Active collection with building storage space-heating systems.
- Direct gain passive systems.
- Collector-storage wall passive systems.
- Pool heating systems.
- General heating systems, such as process heating systems.
- Integral collector-storage domestic water-heating systems.

The program can also perform economic analysis of the systems. The program, however, does not provide the flexibility of detailed simulations and performance investigations, as TRNSYS does.

11.2 Utilizability method

In the previous section, the *f*-chart method is presented. Due to the limitations outlined in Section 11.1.5, the *f*-chart method cannot be used for systems in which the minimum temperature supplied to a load is not near 20 °C. Most of the systems that cannot be simulated with *f*-chart can be modeled with the utilizability method or its enhancements.

The *utilizability method* is a design technique used for the calculation of the long-term thermal collector performance for certain types of systems. Initially suggested by Whillier (1953), the method, referred to as the Φ-*curve method*, is based on the solar radiation statistic, and the necessary calculations have to be done at hourly intervals about solar noon each month. Subsequently, the method was generalized for the time of year and geographic location by Liu and Jordan (1963). Their generalized Φ-curves, generated from daily data, gave the ability to calculate utilizability curves for any location and tilt by knowing only the clearness index, K_T. Afterward, the work by Klein (1978) and Collares-Pereira and Rabl (1979a) eliminated the necessity of hourly calculations. The monthly average daily utilizability, $\overline{\Phi}$ reduced much of the complexity and improved the utility of the method.

11.2.1 Hourly utilizability

The utilizability method is based on the concept that only radiation that is above a critical or threshold intensity is useful. *Utilizability*, Φ, is defined as the fraction of insolation incident on a collector's surface that is above a given threshold or critical value.

We saw in Chapter 3, Section 3.3.4, Eq. (3.61), that a solar collector can give useful heat only if solar radiation is above a critical level. When radiation is incident on the tilted surface of a collector, the utilizable energy for any hour is $(I_t - I_{tc})^+$, where the plus sign indicates that this energy can be only positive or zero. The fraction of the total energy for the hour that is above the critical level is called the *utilizability* for that hour, given by:

$$\Phi_h = \frac{(I_t - I_{tc})^+}{I_t} \tag{11.43}$$

Utilizability can also be defined in terms of rates, using G_t and G_{tc}, but because radiation data are usually available on an hourly basis, the hourly values are preferred and are also in agreement with the basis of the method.

Utilizability for a single hour is not very useful, whereas utilizability for a particular hour of a month having N days, in which the average radiation for the hour is $\overline{I}_t$, is very useful, given by:

$$\Phi = \frac{1}{N} \sum_{1}^{N} \frac{(I_t - I_{tc})^+}{\overline{I}_t} \tag{11.44a}$$

In this case, the average utilizable energy for the month is given by:

$$N\overline{I}_t\Phi \tag{11.44b}$$

Such calculations can be done for all hours of the month, and the results can be added up to get the utilizable energy of the month. Another required parameter is the dimensionless critical radiation level, defined as:

$$X_c = \frac{I_{tc}}{\overline{I}_t} \tag{11.45}$$

For each hour or hour pair, the monthly average hourly radiation incident on the collector is given by:

$$\bar{I}_t = (\bar{H}r - \bar{H}_D r_d)R_B + \bar{H}_D r_d \left[\frac{1 + \cos(\beta)}{2}\right] + \bar{H}r\rho_G \left[\frac{1 - \cos(\beta)}{2}\right] \tag{11.46}$$

Dividing by $\bar{H}$ and using Eq. (2.82a),

$$\bar{I}_t = \bar{K}_T \bar{H}_o \left\{ \left(r - \frac{\bar{H}_D}{\bar{H}}r_d\right)R_B + \frac{\bar{H}_D}{\bar{H}}r_d \left[\frac{1 + \cos(\beta)}{2}\right] + r\rho_G \left[\frac{1 - \cos(\beta)}{2}\right] \right\} \tag{11.47}$$

The ratios r and r_d can be estimated from Eqs. (2.83) and (2.84), respectively.

Liu and Jordan (1963) constructed a set of Φ curves for various values of $\bar{K}_T$. With these curves, it is possible to predict the utilizable energy at a constant critical level by knowing only the long-term average radiation. Later on Clark et al. (1983) developed a simple procedure to estimate the generalized Φ functions, given by:

$$\Phi = \begin{cases} 0 \text{ if } X_c \geq X_m \\ \left(1 - \dfrac{X_c}{X_m}\right)^2 \text{ if } X_m = 2 \\ \text{otherwise,} \\ \left| |g| - \left[g^2 + (1 + 2g)\left(1 - \dfrac{X_c}{X_m}\right)^2\right]^{1/2} \right| \end{cases} \tag{11.48a}$$

where

$$g = \frac{X_m - 1}{2 - X_m} \tag{11.48b}$$

$$X_m = 1.85 + 0.169 \frac{\bar{R}_h}{\bar{k}_T^2} + 0.0696 \frac{\cos(\beta)}{\bar{k}_T^2} - 0.981 \frac{\bar{k}_T}{\cos^2(\delta)} \tag{11.48c}$$

The monthly average hourly clearness index, $\bar{k}_T$, based on Eq. (2.82c), is given by:

$$\bar{k}_T = \frac{\bar{I}}{\bar{I}_o} \tag{11.49}$$

and can be estimated using Eqs. (2.83) and (2.84) as:

$$\bar{k}_T = \frac{\bar{I}}{\bar{I}_o} = \frac{r}{r_d}\bar{K}_T = \frac{r}{r_d}\frac{\bar{H}}{\bar{H}_o} = [\alpha + \beta \cos(h)]\bar{K}_T \tag{11.50}$$

where α and β can be estimated from Eqs. (2.84b) and (2.84c), respectively. If necessary, $\overline{H}_o$ can be estimated from Eq. (2.79) or obtained directly from Table 2.5.

The ratio of monthly average hourly radiation on a tilted surface to that on a horizontal surface, $\overline{R}_h$, is given by:

$$\overline{R}_h = \frac{\overline{I}_t}{\overline{I}} = \frac{\overline{I}_t}{r\overline{H}} \tag{11.51}$$

The Φ curves are used hourly, which means that three to six hourly calculations are required per month if hour pairs are used. For surfaces facing the equator, where hour pairs can be used, the monthly average daily utilizability, $\overline{\Phi}$ presented in the following section can be used and is a more simple way of calculating the useful energy. For surfaces that do not face the equator or for processes that have critical radiation levels that vary consistently during the days of a month, however, the hourly Φ curves need to be used for each hour.

11.2.2 Daily utilizability

As can be understood from the preceding description, a large number of calculations are required to use the Φ curves. For this reason, Klein (1978) developed the monthly average daily utilizability, $\overline{\Phi}$ concept. *Daily utilizability* is defined as the sum over all hours and days of a month of the radiation falling on a titled surface that is above a given threshold or critical value, which is similar to the one used in the Φ concept, divided by the monthly radiation, given by:

$$\overline{\Phi} = \sum_{\text{days}} \sum_{\text{hours}} \frac{(I_t - I_{tc})^+}{N\overline{H}_t} \tag{11.52a}$$

The monthly utilizable energy is then given by the product:

$$N\overline{H}_t\overline{\Phi} \tag{11.52b}$$

The value of $\overline{\Phi}$ for a month depends on the distribution of hourly values of radiation in that month. Klein (1978) assumed that all days are symmetrical about solar noon, and this means that $\overline{\Phi}$ depends on the distribution of daily total radiation, i.e., the relative frequency of occurrence of below-average, average, and above-average daily radiation values. In fact, because of this assumption, any departure from this symmetry within days leads to increased values of $\overline{\Phi}$. This means that the $\overline{\Phi}$ calculated gives conservative results.

Klein developed the correlations of $\overline{\Phi}$ as a function of $\overline{K}_T$, a dimensionless critical radiation level, $\overline{X}_c$, and a geometric factor $\overline{R}/R_n$. The parameter $\overline{R}$ is the monthly ratio of radiation on a tilted surface to that on a horizontal surface, $\overline{H}_t/\overline{H}$, given by Eq. (2.107), and R_n is the ratio for the hour centered at noon of radiation on the tilted surface to that on a horizontal surface for an average day of the month, which is similar to Eq. (2.99) but rewritten for the noon hour in terms of $r_d H_D$ and $r H$ as:

$$R_n = \left(\frac{I_t}{I}\right)_n = \left(1 - \frac{r_{d,n} H_D}{r_n H}\right) R_{B,n} + \left(\frac{r_{d,n} H_D}{r_n H}\right)\left[\frac{1 + \cos(\beta)}{2}\right] + \rho_G\left[\frac{1 - \cos(\beta)}{2}\right] \tag{11.53}$$

where $r_{d,n}$ and r_n are obtained from Eqs. (2.83) and (2.84), respectively, at solar noon ($h = 0°$). It should be noted that R_n is calculated for a day that has a total radiation equal to the monthly average daily total radiation, i.e., a day for which $H = \overline{H}$ and R_n is not the monthly average value of R at noon. The term H_D/H is given from Erbs et al. (1982) as follows:

For $h_{ss} \le 81.4°$,

$$\frac{H_D}{H} = \begin{cases} 1.0 - 0.2727 K_T + 2.4495 K_T^2 - 11.9514 K_T^3 + 9.3879 K_T^4 & \text{for} \quad K_T < 0.715 \\ 0.143 & \text{for} \quad K_T \ge 0.715 \end{cases} \tag{11.54a}$$

For $h_{ss} > 81.4°$,

$$\frac{H_D}{H} = \begin{cases} 1.0 + 0.2832 K_T - 2.5557 K_T^2 + 0.8448 K_T^3 & \text{for} \quad K_T < 0.722 \\ 0.175 & \text{for} \quad K_T \ge 0.722 \end{cases} \tag{11.54b}$$

The monthly average critical radiation level, $\overline{X}_c$, is defined as the ratio of the critical radiation level to the noon radiation level on a day of the month in which the radiation is the same as the monthly average, given by:

$$\overline{X}_c = \frac{I_{tc}}{r_n R_n \overline{H}} \tag{11.55}$$

The procedure followed by Klein (1978) was that, for a given $\overline{K}_T$ a set of days was established that had the correct long-term average distribution of K_T values. The radiation in each of the days in a sequence was divided into hours, and these hourly values of radiation were used to find the total hourly radiation on a tilted surface, I_t. Subsequently, critical radiation levels were subtracted from the I_t values and summed as shown in Eq. (11.52) to get the $\overline{\Phi}$ values. The $\overline{\Phi}$ curves calculated in this manner can be obtained from graphs or the following relation:

$$\overline{\Phi} = \exp\left\{\left[A + B\left(\frac{R_n}{\overline{R}}\right)\right]\left[\overline{X}_c + C\overline{X}_c^2\right]\right\} \tag{11.56a}$$

where

$$A = 2.943 - 9.271\overline{K}_T + 4.031\overline{K}_T^2 \tag{11.56b}$$

$$B = -4.345 + 8.853\overline{K}_T - 3.602\overline{K}_T^2 \tag{11.56c}$$

$$C = -0.170 - 0.306\overline{K}_T + 2.936\overline{K}_T^2 \tag{11.56d}$$

EXAMPLE 11.11

A north-facing surface located in an area that is at 35°S latitude is tilted at 40°. For the month of April, when $\overline{H} = 17.56$ MJ/m², critical radiation is 117 W/m², and $\rho_G = 0.25$, calculate $\overline{\Phi}$ and the utilizable energy.

Solution

For April, the mean day from Table 2.1 is $N = 105$ and $\delta = 9.41°$. From Eq. (2.15), the sunset time $h_{ss} = 83.3°$. From Eqs. (2.84b), (2.84c), and (2.84a), we have:

$$\alpha = 0.409 + 0.5016 \times \sin(h_{ss} - 60) = 0.409 + 0.5016 \times \sin(83.3 - 60) = 0.607$$

$$\beta = 0.6609 - 0.4767 \times \sin(h_{ss} - 60) = 0.6609 - 0.4767 \times \sin(83.3 - 60) = 0.472$$

$$r_n = \frac{\pi}{24}(\alpha + \beta \cos(h))\frac{\cos(h) - \cos(h_{ss})}{\sin(h_{ss}) - \left(\frac{2\pi h_{ss}}{360}\right)\cos(h_{ss})}$$

$$= \frac{\pi}{24}(0.607 + 0.472 \cos(0))\frac{\cos(0) - \cos(83.3)}{\sin(83.3) - \left(\frac{2\pi(83.3)}{360}\right)\cos(83.3)} = 0.152$$

From Eq. (2.83), we have:

$$r_{d,n} = \left(\frac{\pi}{24}\right)\frac{\cos(h) - \cos(h_{ss})}{\sin(h_{ss}) - \left(\frac{2\pi h_{ss}}{360}\right)\cos(h_{ss})} = \left(\frac{\pi}{24}\right)\frac{\cos(0) - \cos(83.3)}{\sin(83.3) - \left[\frac{2\pi(83.3)}{360}\right]\cos(83.3)} = 0.140$$

From Eq. (2.90a), for the Southern Hemisphere (plus sign instead of minus),

$$R_{B,n} = \frac{\sin(L + \beta)\sin(\delta) + \cos(L + \beta)\cos(\delta)\cos(h)}{\sin(L)\sin(\delta) + \cos(L)\cos(\delta)\cos(h)}$$

$$= \frac{\sin(-35 + 40)\sin(9.41) + \cos(-35 + 40)\cos(9.41)\cos(0)}{\sin(-35)\sin(9.41) + \cos(-35)\cos(9.41)\cos(0)} = 1.396$$

From Eq. (2.79) or Table 2.5, $H_o = 24.84$ kJ/m², and from Eq. (2.82a),

$$\overline{K}_T = \frac{17.56}{24.84} = 0.707.$$

For a day in which $H = \overline{H}$, $K_T = 0.707$, and from Eq. (11.54b),

$$\frac{H_D}{H} = 1.0 + 0.2832 K_T - 2.5557 K_T^2 + 0.8448 K_T^3$$

$$= 1.0 + 0.2832 \times 0.707 - 2.5557(0.707)^2 + 0.8448(0.707)^3 = 0.221$$

Then, from Eq. (11.53),

$$R_n = \left(1 - \frac{r_{d,n}H_D}{r_n H}\right)R_{B,n} + \left(\frac{r_{d,n}H_D}{r_n H}\right)\left[\frac{1 + \cos(\beta)}{2}\right] + \rho_G\left[\frac{1 - \cos(\beta)}{2}\right]$$

$$= \left(1 - \frac{0.140 \times 0.221}{0.152}\right)1.396 + \left(\frac{0.140 \times 0.221}{0.152}\right)\left[\frac{1 + \cos(40)}{2}\right]$$

$$+ 0.25\left[\frac{1 - \cos(40)}{2}\right] = 1.321$$

Continued

EXAMPLE 11.11—CONT'D

From Eq. (2.109), for the Southern Hemisphere (plus sign instead of minus),

$$h'_{ss} = \min\{h_{ss}, \cos^{-1}[-\tan(L+\beta)\tan(\delta)]\}$$

$$= \min\{83.3, \cos^{-1}[-\tan(-35+40)\tan(9.41)]\} = \min\{83.3, 90.8\} = 83.3°$$

From Eq. (2.108), for the Southern Hemisphere (plus sign instead of minus),

$$\overline{R}_B = \frac{\cos(L+\beta)\cos(\delta)\sin(h'_{ss}) + (\pi/180)h'_{ss}\sin(L+\beta)\sin(\delta)}{\cos(L)\cos(\delta)\sin(h_{ss}) + (\pi/180)h_{ss}\sin(L)\sin(\delta)}$$

$$= \frac{\cos(-35+40)\cos(9.41)\sin(83.3) + (\pi/180)83.3 \times \sin(-35+40)\sin(9.41)}{\cos(-35)\cos(9.41)\sin(83.3) + (\pi/180)83.3 \times \sin(-35)\sin(9.41)} = 1.496$$

From Eq. (2.105d),

$$\frac{\overline{H}_D}{\overline{H}} = 1.311 - 3.022\overline{K}_T + 3.427\overline{K}_T^2 - 1.821\overline{K}_T^3$$

$$= 1.311 - 3.022 \times 0.707 + 3.427(0.707)^2 - 1.821(0.707)^3 = 0.244$$

From Eq. (2.107),

$$\overline{R} = \frac{\overline{H}_t}{\overline{H}} = \left(1 - \frac{\overline{H}_D}{\overline{H}}\right)\overline{R}_B + \frac{\overline{H}_D}{\overline{H}}\left[\frac{1+\cos(\beta)}{2}\right] + \rho_G\left[\frac{1-\cos(\beta)}{2}\right]$$

$$= (1 - 0.244) \times 1.496 + 0.244\left[\frac{1+\cos(40)}{2}\right] + 0.25\left[\frac{1-\cos(40)}{2}\right] = 1.376$$

Now,

$$\frac{R_n}{\overline{R}} = \frac{1.321}{1.376} = 0.96$$

From Eq. (11.55), the dimensionless average critical radiation level is:

$$\overline{X}_c = \frac{I_{tc}}{r_n R_n \overline{H}} = \frac{117 \times 3600}{0.152 \times 1.321 \times 17.56 \times 10^6} = 0.119$$

From Eq. (11.56):

$$A = 2.943 - 9.271\overline{K}_T + 4.031\overline{K}_T^2 = 2.943 - 9.271 \times 0.707 + 4.031(0.707)^2 = -1.597$$

$$B = -4.345 + 8.853\overline{K}_T - 3.602\overline{K}_T^2 = -4.345 + 8.853 \times 0.707 - 3.602(0.707)^2 = 0.114$$

$$C = -0.170 - 0.306\overline{K}_T + 2.936\overline{K}_T^2 = -0.170 - 0.306 \times 0.707 + 2.936(0.707)^2 = 1.081$$

$$\overline{\Phi} = \exp\left\{\left[A + B\left(\frac{R_n}{\overline{R}}\right)\right]\left[\overline{X}_c + C\overline{X}_c^2\right]\right\}$$

$$= \exp\{[-1.597 + 0.114(0.96)][0.119 + 1.081(0.119)^2]\} = 0.819$$

Finally, the month utilizable energy is:

$$N\overline{H}_t\overline{\Phi} = N\overline{H}\,\overline{R}\,\overline{\Phi} = 30 \times 17.56 \times 1.376 \times 0.819 = 593.7 \text{ MJ/m}^2$$

Both the Φ and the $\overline{\Phi}$ concepts can be applied in a variety of design problems, such as heating systems and passively heated buildings, where the unutilizable energy (excess energy) that cannot be stored in the building mass can be estimated. Examples of these applications are given in the following sections.

11.2.3 Design of active systems with the utilizability method

The method can be developed for an hourly or daily basis. These are treated separately in this section.

Hourly utilizability

Utilizability can also be defined as the fraction of incident solar radiation that can be converted into a useful heat. It is the fraction utilized by a collector having no optical losses and a heat removal factor of unity, i.e., $F_R(\tau\alpha) = 1$, operating at a fixed inlet to ambient temperature difference. It should be noted that the utilizability of this collector is always less than 1, since thermal losses exist in the collector.

The Hottel–Whillier equation (Hottel and Whillier, 1955) relates the rate of useful energy collection by a flat-plate solar collector, Q_u, to the design parameters of the collector and meteorological conditions. This is given by Eq. (3.60) in Chapter 3, Section 3.3.4. This equation can be expressed in terms of the hourly radiation incident on the collector plane, I_t, as:

$$Q_u = A_c F_R [I_t(\tau\alpha) - U_L(T_i - T_a)]^+ \tag{11.57}$$

where

F_R = collector heat removal factor;
A_c = collector area (m^2);
$(\tau\alpha)$ = effective transmittance–absorptance product;
I_t = total radiation incident on the collector surface per unit area (kJ/m^2);
U_L = energy loss coefficient (kJ/m^2 K);
T_i = inlet collector fluid temperature (°C); and
T_a = ambient temperature (°C).

The radiation level must exceed a critical value before useful output is produced. This critical level is found by setting Q_u in Eq. (11.57) equal to 0. This is given in Eq. (3.61), but in terms of the hourly radiation incident on the collector plane, it is given by:

$$I_{tc} = \frac{F_R U_L(T_i - T_a)}{F_R(\tau\alpha)} \tag{11.58}$$

The useful energy gain can thus be written in terms of critical radiation level as:

$$Q_u = A_c F_R(\tau\alpha)(I_t - I_{tc})^+ \tag{11.59}$$

The plus superscript in Eqs. (11.57) and (11.59) and in the following equations indicates that only positive values of I_{tc} are considered. If the critical radiation level is constant for a particular hour of the month having N days, then the monthly average hourly output for this hour is:

$$\overline{Q}_u = \frac{A_c F_R(\tau\alpha)}{N} \sum_N (I_t - I_{tc})^+ \tag{11.60}$$

Because the monthly average radiation for this particular hour is $\overline{I}_t$, the average useful output can be expressed by:

$$\overline{Q}_u = A_c F_R(\tau\alpha)\overline{I}_t \Phi \tag{11.61}$$

where Φ is given by Eq. (11.44). This can be estimated from the generalized Φ curves or Eq. (11.48), given earlier for the dimensionless critical radiation level, X_c, given by Eq. (11.45), which can now be written in terms of the collector parameters, using Eq. (11.58), as:

$$X_c = \frac{I_{tc}}{\bar{I}_t} = \frac{F_R U_L (T_i - T_a)}{F_R(\tau\alpha)_n \frac{(\tau\alpha)}{(\tau\alpha)_n} \bar{I}_t} \tag{11.62}$$

where $(\tau\alpha)/(\tau\alpha)_n$ can be determined for the mean day of the month, shown in Table 2.1, and the appropriate hour angle and can be estimated with the incidence angle modifier constant, b_o, from Eq. (4.25).

With Φ known, the utilizable energy is $\bar{I}_t \Phi$. The main use of hourly utilizability is to estimate the output of processes that have a critical radiation level, X_c, that changes considerably during the day, which can be due to collector inlet temperature variation.

EXAMPLE 11.12

Suppose that a collector system supplies heat to an industrial process. The collector inlet temperature (process return temperature) varies as shown in Table 11.11 but, for a certain hour, is constant during the month. The calculation is done for the month of April, where $\bar{K}_T = 0.63$. The system is located at 35°N latitude and the collector characteristics are $F_R U_L = 5.92$ W/m^2 °C, $F_R(\tau\alpha)_n = 0.82$, tilted at 40°, and the incidence angle modifier constant $b_o = 0.1$. The weather conditions are also given in the table. Calculate the energy output of the collector.

Table 11.11 Collector inlet temperature and weather conditions for Example 11.12.

Hour	T_i (°C)	T_a (°C)	$\bar{I}_t$ (MJ/m^2)
8–9	25	9	1.52
9–10	25	11	2.36
10–11	30	13	3.11
11–12	30	15	3.85
12–13	30	18	3.90
13–14	45	16	3.05
14–15	45	13	2.42
15–16	45	9	1.85

Solution

First, the incidence angle is calculated, from which the incidence angle modifier is estimated. The estimations are done on the half hour; for the hour 8–9, the hour angle is $-52.5°$. From Eq. (2.20),

$$\cos(\theta) = \sin(L - \beta)\sin(\delta) + \cos(L - \beta)\cos(\delta)\cos(h)$$
$$= \sin(35 - 40)\sin(9.41) + \cos(35 - 40)\cos(9.41)\cos(-52.5) = 0.584 \quad \text{or} \quad \theta = 54.3°.$$

From Eq. (4.25),

$$K_\theta = \frac{(\tau\alpha)}{(\tau\alpha)_n} = 1 - b_o\left[\frac{1}{\cos(\theta)} - 1\right] = 1 - 0.1\left[\frac{1}{\cos(54.3)} - 1\right] = 0.929$$

The dimensionless critical radiation level, X_c, is given by Eq. (11.62):

$$X_c = \frac{I_{tc}}{\bar{I}_t} = \frac{F_R U_L (T_i - T_a)}{F_R(\tau\alpha)_n \frac{(\tau\alpha)}{(\tau\alpha)_n} \bar{I}_t} = \frac{5.92(25 - 9) \times 3600}{0.82 \times 0.929 \times 1.52 \times 10^6} = 0.294$$

EXAMPLE 11.12—CONT'D

From Table 2.5, $\overline{H}_o = 35.8$ MJ/m^2. From the input data and Eq. (2.82a),

$$\overline{H} = \overline{K}_T \overline{H}_o = 35.8 \times 0.63 = 22.56 \text{ MJ/m}^2$$

To avoid repeating the same calculations as in previous examples, some values are given directly. Therefore, $h_{ss} = 96.7°$, $\alpha = 0.709$, and $\beta = 0.376$. From Eq. (2.84a),

$$r = \frac{\pi}{24}(\alpha + \beta \cos(h))\frac{\cos(h) - \cos(h_{ss})}{\sin(h_{ss}) - \left(\dfrac{2\pi h_{ss}}{360}\right)\cos(h_{ss})}$$

$$= \frac{\pi}{24}(0.709 + 0.376 \ \cos(54.3))\frac{\cos(-52.5) - \cos(96.7)}{\sin(96.7) - \left(\dfrac{2\pi(96.7)}{360}\right)\cos(96.7)} = 0.075$$

From Eq. (2.83), we have:

$$r_d = \left(\frac{\pi}{24}\right)\frac{\cos(h) - \cos(h_{ss})}{\sin(h_{ss}) - \left(\dfrac{2\pi h_{ss}}{360}\right)\cos(h_{ss})} = \left(\frac{\pi}{24}\right)\frac{\cos(-52.5) - \cos(96.7)}{\sin(96.7) - \left[\dfrac{2\pi(96.7)}{360}\right]\cos(96.7)} = 0.080$$

From Eq. (11.51),

$$\overline{R}_h = \frac{\overline{I}_t}{\overline{I}} = \frac{\overline{I}_t}{r\overline{H}} = \frac{1.52}{0.075 \times 22.56} = 0.898$$

The monthly average hourly clearness index, $\overline{k}_T$, is given by Eq. (11.50):

$$\overline{k}_T = \frac{\overline{I}}{\overline{I}_o} = \frac{r}{r_d}\overline{K}_T = \frac{r}{r_d}\frac{\overline{H}}{\overline{H}_o} = [\alpha + \beta \cos(h)]\overline{K}_T = [0.709 + 0.376 \ \cos(-52.5)] \times 0.63 = 0.591$$

From Eq. (11.48c),

$$X_m = 1.85 + 0.169\frac{\overline{R}_h}{\overline{k}_T^2} + 0.0696\frac{\cos(\beta)}{\overline{k}_T^2} - 0.981\frac{\overline{k}_T}{\cos^2(\delta)}$$

$$= 1.85 + 0.169\frac{0.898}{(0.591)^2} + 0.0696\frac{\cos(40)}{(0.591)^2} - 0.981\frac{0.591}{\cos^2(9.41)} = 1.841$$

From Eq. (11.48b),

$$g = \frac{X_m - 1}{2 - X_m} = \frac{1.841 - 1}{2 - 1.841} = 5.289$$

From Eq. (11.48a),

$$\Phi = \left| |g| - \left[g^2 + (1 + 2g)\left(1 - \frac{X_c}{X_m}\right)^2\right]^{1/2} \right|$$

$$= \left| |5.289| - \left[5.289^2 + (1 + 2 \times 5.289)\left(1 - \frac{0.294}{1.841}\right)^2\right]^{1/2} \right| = 0.723$$

Finally the useful gain (UG) of the collector for that hour is (April has 30 days):

$$F_R(\tau\alpha)_n\frac{(\tau\alpha)}{(\tau\alpha)_n}N\overline{I}_t\Phi = 0.82 \times 0.929 \times 30 \times 1.52 \times 0.723 = 25.11 \text{ MJ/m}^2$$

The results for the other hours are shown in Table 11.12.
The useful gain for the month is equal to 427.6 MJ/m^2.

Continued

EXAMPLE 11.12—CONT'D

Table 11.12 Results for all hours in Example 11.12.

Hour	h (°)	θ (°)	K_θ	X_c	r_d	r	$\bar{R}_h$	$\bar{k}_T$	X_m	g	Φ	UG
8–9	−52.5	54.3	0.929	0.294	0.080	0.075	0.898	0.591	1.841	5.289	0.723	25.11
9–10	−37.5	40.1	0.969	0.159	0.100	0.101	1.036	0.635	1.776	3.464	0.845	47.54
10–11	−22.5	26.7	0.988	0.144	0.114	0.120	1.149	0.666	1.737	2.802	0.859	64.93
11–12	−7.5	16.2	0.996	0.102	0.122	0.132	1.293	0.682	1.747	2.953	0.900	84.90
12–13	7.5	16.2	0.996	0.080	0.122	0.132	1.310	0.682	1.753	3.049	0.921	88.01
13–14	22.5	26.7	0.988	0.250	0.114	0.120	1.127	0.666	1.728	2.676	0.760	56.34
14–15	37.5	40.1	0.969	0.355	0.100	0.101	1.062	0.635	1.787	3.695	0.669	38.59
15–16	52.5	54.3	0.929	0.544	0.080	0.075	1.093	0.591	1.936	14.63	0.525	22.20
											Total =	**427.6**

Although the Φ curves method is a very powerful tool, caution is required to avoid possible misuse. For example, due to finite storage capacity, the critical level of collector inlet temperature for liquid-based domestic solar heating systems varies considerably during the month, so the Φ curves method cannot be applied directly. Exceptions to this rule are air heating systems during winter, where the inlet air temperature to the collector is the return air from the house, and systems with seasonal storage where, due to its size, storage tank temperatures show small variations during the month.

Daily utilizability

As indicated in Section 11.2.2, the use of Φ curves involves a lot of calculations. Klein (1978) and Collares-Pereira and Rabl (1979b;c) simplified the calculations for systems for which a critical radiation level can be used for all hours of the month.

Daily utilizability is defined as the sum for a month over all hours and all days of the radiation on a tilted surface that is above a critical level, divided by the monthly radiation. This is given in Eq. (11.52). The critical level, I_{tc}, is similar to Eq. (11.58), but in this case, the monthly average $(\tau\alpha)$ product must be used and the inlet and ambient temperatures are representative temperatures for the month:

$$I_{tc} = \frac{F_R U_L \left(T_i - \overline{T}_a\right)}{F_R(\tau\alpha)_n \frac{\overline{(\tau\alpha)}}{(\tau\alpha)_n}} \tag{11.63}$$

In Eq. (11.63), the term $\overline{(\tau\alpha)}/(\tau\alpha)_n$ can be estimated with Eq. (11.11). The monthly average critical radiation ratio is the ratio of the critical radiation level, I_{tc}, to the noon radiation level for a day of the month in which the total radiation for the day is the same as the monthly average. In equation form,

$$\overline{X}_c = \frac{I_{tc}}{r_n R_n \overline{H}} = \frac{\dfrac{F_R U_L \left(T_i - \overline{T}_a\right)}{F_R \overline{(\tau\alpha)}}}{r_n R_n \overline{K}_T \overline{H}_o} \tag{11.64}$$

The monthly average daily useful energy gain is given by:

$$\overline{Q}_u = A_c F_R \overline{(\tau\alpha)} \overline{H}_t \overline{\Phi} \tag{11.65}$$

Daily utilizability can be obtained from Eq. (11.56).

It should be noted that, even though monthly average daily utilizability reduces the complexity of the method, calculations can be still quite tedious, especially when monthly average hourly calculations are required.

It is also noticeable that the majority of the aforementioned methods for computing solar energy utilizability have been derived as fits to North American data versus the clearness index, which is the parameter used to indicate the dependence of the climate. Carvalho and Bourges (1985) applied some of these methods to European and African locations and compared results with values obtained from long-term measurements. Results showed that these methods can give acceptable results when the actual monthly average daily irradiation on the considered surface is known.

Examples of this method are given in the next section, where the $\overline{\Phi}$ and *f*-chart methods are combined.

11.3 The $\overline{\Phi}$, *f*-chart method

The utilizability design concept is useful when the collector operates at a known critical radiation level during a specific month. In a practical system, however, the collector is connected to a storage tank, so the monthly sequence of weather and load time distributions cause a fluctuating storage tank temperature and thus a variable critical radiation level. On the other hand, the *f*-chart was developed to overcome the restriction of a constant critical level but is restricted to systems delivering a load near 20 °C.

Klein and Beckman (1979) combined the utilizability concept described in the previous section with the *f*-chart to produce the $\overline{\Phi}$, *f*-chart design method for a closed loop solar energy system, shown in Fig. 11.11. The method is not restricted to loads that are at 20 °C. In this system, the storage tank is assumed to be pressurized or filled with a liquid of high boiling point so that no energy dumping occurs through the relief valve. The auxiliary heater is in parallel with the solar energy system. In these systems, energy supplied to the load must be above a specified minimum useful temperature, T_{min}, and it must be used at a constant thermal efficiency or coefficient of performance so that the load on the solar energy system can be estimated. The return temperature from the load is always at or above T_{min}. Because the performance of a heat pump or a heat engine varies with the temperature level of supplied energy, this design method is not suitable for this kind of application. It is useful, however, in absorption refrigerators, industrial process heating, and space heating systems.

The maximum monthly average daily energy that can be delivered from the system shown in Fig. 11.11 is given by:

$$\sum \overline{Q}_u = A_c F_R(\overline{\tau\alpha})\overline{H}_t\overline{\Phi}_{max} \tag{11.66}$$

This is the same as Eq. (11.65), except that $\overline{\Phi}$ is replaced with $\overline{\Phi}_{max}$, which is the maximum daily utilizability, estimated from the minimum monthly average critical radiation ratio:

$$\overline{X}_{c,min} = \frac{\dfrac{F_R U_L(T_{min} - \overline{T}_a)}{F_R(\overline{\tau\alpha})}}{r_n R_n \overline{K}_T \overline{H}_o} \tag{11.67}$$

FIGURE 11.11

Schematic diagram of a closed-loop solar energy system.

Klein and Beckman (1979) correlated the results of many detailed simulations of the system shown in Fig. 11.11, for various storage size–collector-area ratios, with two dimensionless variables. These variables are similar to the ones used in the *f*-chart but are not the same. Here, the *f*-chart dimensionless parameter Y (plotted on the ordinate of the *f*-chart) is replaced by $\overline{\Phi}_{max}Y$, given by:

$$\overline{\Phi}_{max}Y = \overline{\Phi}_{max}\frac{A_c F_R(\overline{\tau\alpha})N\overline{H}_t}{L} \tag{11.68}$$

And the *f*-chart dimensionless parameter X (plotted on the abscissa of the *f*-chart) is replaced by a modified dimensionless variable, X', given by:

$$X' = \frac{A_c F_R U_L(100)\Delta t}{L} \tag{11.69}$$

In fact, the change in the X dimensionless variable is that the parameter $(100 - \overline{T}_a)$ is replaced with an empirical constant 100.

The $\overline{\Phi}$, *f*-charts can be obtained from actual charts or from the following analytical equation (Klein and Beckman, 1979):

$$f = \overline{\Phi}_{max}Y - 0.015[\exp(3.85f) - 1][1 - \exp(-0.15X')]R_s^{0.76} \tag{11.70}$$

where R_s = ratio of standard storage heat capacity per unit of collector area of 350 kJ/m^2 °C to actual storage capacity, given by (Klein and Beckman, 1979):

$$R_s = \frac{350}{\dfrac{Mc_p}{A_c}} \tag{11.71}$$

where M = actual mass of storage capacity (kg).

Although, in Eq. (11.70), *f* is included on both sides of equation, it is relatively easy to solve for *f* by trial and error. Since the $\overline{\Phi}$, *f*-charts are given for various storage capacities and the user has to interpolate, the use of Eq. (11.70) is preferred, so the actual charts are not included in this book. The $\overline{\Phi}$, *f*-charts are used in the same way as the *f*-charts. The values of $\overline{\Phi}_{max}$ Y, and X' need to be calculated from the long-term radiation data for the particular location and load patterns. As before, *fL* is the average monthly contribution of the solar energy system, and the monthly values can be summed and divided by the total annual load to obtain the annual fraction, F.

EXAMPLE 11.13

An industrial process heat system has a 50 m^2 collector. The system is located at Nicosia, Cyprus (35°N latitude), and the collector characteristics are $F_R U_L = 5.92$ W/m^2 °C, $F_R(\tau\alpha)_n = 0.82$, tilted at 40°, and double glazed. The process requires heat at a rate of 15 kW at a temperature of 70 °C for 10 h each day. Estimate the monthly and annual solar fractions. Additional information is $(\tau\alpha)_n = 0.96$, storage volume = 5000 L. The weather conditions, as obtained from Appendix 7, are given in Table 11.13. The values of the last column are estimated from Eq. (2.82a).

As can be seen, the values of $\overline{H}_o$ are slightly different from those shown in Table 2.5 for 35°N latitude. This is because the actual latitude of Nicosia, Cyprus, is 35.15°N, as shown in Appendix 7.

Continued

EXAMPLE 11.13–CONT'D

Table 11.13 Weather conditions for Example 11.13.

Month	$\overline{H}$ (MJ/m^2)	$\overline{T}_a$ (°C)	$\overline{K}_T$	$\overline{H}_o$ (MJ/m^2)
January	8.96	12.1	0.49	18.29
February	12.38	11.9	0.53	23.36
March	17.39	13.8	0.58	29.98
April	21.53	17.5	0.59	36.49
May	26.06	21.5	0.65	40.09
June	29.20	25.8	0.70	41.71
July	28.55	29.2	0.70	40.79
August	25.49	29.4	0.68	37.49
September	21.17	26.8	0.66	32.08
October	15.34	22.7	0.60	25.57
November	10.33	17.7	0.53	19.49
December	7.92	13.7	0.47	16.85

Table 11.14 Results of radiation coefficients for Example 11.13.

Month	$\overline{R}_B$	$\overline{H}_D/\overline{H}$	$\overline{R}$	r_n	$r_{d,n}$	$R_{B,n}$	H_D/H	R_n
January	1.989	0.40	1.570	0.168	0.156	1.716	0.590	1.283
February	1.624	0.36	1.381	0.156	0.144	1.429	0.505	1.225
March	1.282	0.36	1.162	0.144	0.133	1.258	0.469	1.119
April	1.000	0.35	0.982	0.133	0.123	1.074	0.450	1.018
May	0.827	0.29	0.867	0.126	0.116	0.953	0.336	0.955
June	0.757	0.25	0.812	0.122	0.112	0.929	0.235	0.921
July	0.787	0.25	0.834	0.124	0.114	0.924	0.235	0.939
August	0.921	0.27	0.934	0.130	0.120	1.020	0.276	1.008
September	1.160	0.29	1.103	0.140	0.129	1.180	0.316	1.117
October	1.503	0.34	1.316	0.152	0.141	1.400	0.432	1.216
November	1.885	0.36	1.548	0.164	0.153	1.648	0.505	1.311
December	2.113	0.42	1.620	0.171	0.159	1.797	0.630	1.285

Solution

To simplify the solution, most of the results are given directly in Table 11.14. These concern $\overline{R}_B$, given by Eq. (2.108); $\overline{H}_D/\overline{H}$, given by Eqs. (2.105c) and (2.105d); $\overline{R}$, given by Eq. (2.107); r_n and $r_{d,n}$, given by Eqs. (2.84) and (2.83), respectively, at noon ($h = 0°$); $R_{B,n}$, given by Eq. (2.90a) at noon; H_D/H, given by Eq. (11.54); and R_n, given by Eq. (11.53).

Subsequently, the data for January are presented. First, we need to estimate $(\overline{\tau\alpha})/(\tau\alpha)_n$. For this estimation, we need to know $\overline{S}$ and then apply Eq. (11.10) to find the required parameter. From Eqs. (3.4a) and (3.4b),

EXAMPLE 11.13—CONT'D

$$\theta_{e,D} = 59.68 - 0.1388\beta + 0.001497\beta^2 = 59.68 - 0.1388 \times 40 + 0.001497 \times 40^2 = 57°$$

$$\theta_{e,G} = 90 - 0.5788\beta + 0.002693\beta^2 = 90 - 0.5788 \times 40 + 0.002693 \times 40^2 = 71°$$

From Fig. 3.27, for a double-glazed collector,

$$(\overline{\tau\alpha})_D / (\tau\alpha)_n = 0.87$$

and

$$(\overline{\tau\alpha})_G / (\tau\alpha)_n = 0.57$$

Therefore,

$$(\overline{\tau\alpha})_D = (\tau\alpha)_n \times 0.87 = 0.96 \times 0.87 = 0.835$$

$$(\overline{\tau\alpha})_G = (\tau\alpha)_n \times 0.57 = 0.96 \times 0.57 = 0.547$$

These values are constant for all months. For the beam radiation, we use Figures A3.8(a) and A3.8(b) to find the equivalent angle for each month and Fig. 3.27 to get $(\tau\alpha)/(\tau\alpha)_n$. The 12 angles are 40, 42, 44, 47, 50, 51, 51, 49, 46, 43, 40, and 40, from which 12 values are read from Fig. 3.27 and the corresponding values are given in Table 11.15. The calculations for January are as follows:

$$(\overline{\tau\alpha})_B = (\tau\alpha)_n \frac{(\tau\alpha)}{(\tau\alpha)_n} = 0.96 \times 0.96 = 0.922$$

From the data presented in previous tables,

$$\overline{H}_D = \overline{H}\frac{\overline{H}_D}{\overline{H}} = 8.96 \times 0.40 = 3.58 \text{ MJ/m}^2$$

From Eq. (2.106),

$$\overline{H}_B = \overline{H} - \overline{H}_D = 8.96 - 3.58 = 5.38 \text{ MJ/m}^2$$

and

$$\overline{S} = \overline{H}_B \overline{R}_B (\overline{\tau\alpha})_B + \overline{H}_D (\overline{\tau\alpha})_D \left[\frac{1+\cos(\beta)}{2}\right] + \overline{H}\rho_G (\overline{\tau\alpha})_G \left[\frac{1-\cos(\beta)}{2}\right]$$

$$= 5.38 \times 1.989 \times 0.922 + 3.58 \times 0.835 \left[\frac{1+\cos(40)}{2}\right] + 8.96 \times 0.2 \times 0.547 \left[\frac{1-\cos(40)}{2}\right]$$

$$= 12.62 \text{ MJ/m}^2$$

From Eq. (11.10),

$$(\overline{\tau\alpha}) = \frac{\overline{S}}{\overline{H}_t} = \frac{\overline{S}}{\overline{H}R} = \frac{12.62}{8.96 \times 1.570} = 0.90$$

The results for the other months are shown in Table 11.15.

Continued

EXAMPLE 11.13–CONT'D

Table 11.15 Results of $(\overline{\tau\alpha})/(\tau\alpha)_n$ for other months for Example 11.13.

Month	$(\tau\alpha)/(\tau\alpha)_n$	$(\overline{\tau\alpha})_B$	$\overline{S}$ (MJ/m^2)	$(\overline{\tau\alpha})$	$(\overline{\tau\alpha})/(\tau\alpha)_n$
January	0.96	0.922	12.62	0.90	0.94
February	0.96	0.922	15.31	0.90	0.94
March	0.95	0.912	17.85	0.88	0.92
April	0.93	0.893	18.33	0.87	0.91
May	0.92	0.883	19.42	0.86	0.90
June	0.91	0.874	20.25	0.85	0.89
July	0.91	0.874	20.36	0.86	0.90
August	0.92	0.883	20.53	0.86	0.90
September	0.93	0.893	20.37	0.87	0.91
October	0.95	0.912	17.92	0.89	0.93
November	0.96	0.922	14.36	0.90	0.94
December	0.96	0.922	11.50	0.90	0.94

Now we can proceed with the $\overline{\Phi}$ f-chart method calculations. Again, the estimations for January are shown in detail below. The minimum monthly average critical radiation ratio is given by Eq. (11.67):

$$\overline{X}_{c,min} = \frac{\dfrac{F_R U_L\left(T_{min} - \overline{T}_a\right)}{F_R(\overline{\tau\alpha})}}{r_n R_n \overline{K}_T \overline{H}_o} = \frac{\dfrac{F_R U_L\left(T_{min} - \overline{T}_a\right)}{F_R(\tau\alpha)_n \frac{(\overline{\tau\alpha})}{n(\tau\alpha)_n}}}{r_n R_n \overline{K}_T \overline{H}_o} = \frac{\dfrac{5.92 \times 3600(70 - 12.1)}{0.82 \times 0.94}}{0.168 \times 1.283 \times 0.49 \times 18.29 \times 10^6} = 0.83$$

From Eq. (11.56),

$$A = 2.943 - 9.271\overline{K}_T + 4.031\overline{K}_T^2 = 2.943 - 9.271 \times 0.49 + 4.031(0.49)^2 = -0.6319$$

$$B = -4.345 + 8.853\overline{K}_T - 3.602\overline{K}_T^2 = -4.345 + 8.853 \times 0.49 - 3.602(0.49)^2 = -0.8719$$

$$C = -0.170 - 0.306\overline{K}_T + 2.936\overline{K}_T^2 = -0.170 - 0.306 \times 0.49 + 2.936(0.49)^2 = 0.3850$$

$$\overline{\Phi}_{max} = \exp\left\{\left[A + B\left(\frac{R_n}{\overline{R}}\right)\right]\left[\overline{X}_c + C\overline{X}_c^2\right]\right\}$$

$$= \exp\left\{\left[-0.6319 - 0.8719\left(\frac{1.283}{1.570}\right)\right][0.83 + 0.3850(0.83)^2]\right\} = 0.229$$

The load for January is:

$$15 \times 10 \times 3600 \times 31 = 16.74 \times 10^6 \ kJ = 16.74 \ GJ$$

From Eq. (11.68),

$$\overline{\Phi}_{max}Y = \overline{\Phi}_{max}\frac{A_c F_R(\overline{\tau\alpha})N\overline{H}_t}{L} = \overline{\Phi}_{max}\frac{A_c F_R(\tau\alpha)_n \frac{(\overline{\tau\alpha})}{(\tau\alpha)_n}N\overline{H}R}{L}$$

$$= 0.229\frac{50 \times 0.82 \times 0.94 \times 31 \times 8.96 \times 10^6 \times 1.570}{16.74 \times 10^9} = 0.230$$

EXAMPLE 11.13–CONT'D

From Eq. (11.69),

$$X' = \frac{A_c F_R U_L (100) \Delta t}{L} = \frac{50 \times 5.92 \times 100 \times 24 \times 31 \times 3600}{16.74 \times 10^9} = 4.74$$

The storage parameter, R_s, is estimated with Eq. (11.71):

$$R_s = \frac{350}{\frac{Mc_p}{A_c}} = \frac{350}{\frac{5000 \times 4.19}{50}} = 0.835$$

Finally, f can be calculated from Eq. (11.70). The solar contribution is fL. The calculations for the other months are shown in Table 11.16. The use of a spreadsheet program greatly facilitates calculations.

Table 11.16 Monthly calculations for Example 11.13.

Month	$\overline{X}_{c,min}$	$\overline{\Phi}_{max}$	$\overline{\Phi}_{max}Y$	X'	L (GJ)	f	fL (GJ)
January	0.83	0.229	0.230	4.74	16.74	0.22	3.68
February	0.68	0.274	0.334	4.74	15.12	0.32	4.84
March	0.57	0.315	0.445	4.74	16.74	0.42	7.03
April	0.51	0.355	0.519	4.74	16.20	0.48	7.78
May	0.45	0.390	0.602	4.74	16.74	0.55	9.21
June	0.39	0.445	0.713	4.74	16.20	0.64	10.37
July	0.35	0.494	0.804	4.74	16.74	0.71	11.89
August	0.35	0.497	0.809	4.74	16.74	0.71	11.89
September	0.37	0.478	0.771	4.74	16.20	0.68	11.02
October	0.47	0.398	0.567	4.74	16.74	0.52	8.70
November	0.65	0.300	0.342	4.74	16.20	0.32	5.18
December	0.89	0.222	0.203	4.74	16.74	0.20	3.35
					Total = 197.10		Total = 94.94

The annual fraction is given by Eq. (11.12):

$$F = \frac{\sum f_i L_i}{\sum L_i} = \frac{94.94}{197.10} = 0.482 \quad \text{or} \quad 48.2\%$$

It should be pointed out that the $\overline{\Phi}$ f-chart method overestimates the monthly solar fraction, f. This is due to assumptions that there are no losses from the storage tank and that the heat exchanger is 100% efficient. These assumptions require certain corrections, which follow.

11.3.1 Storage tank-losses correction

The rate of energy lost from the storage tank to the environment, which is at temperature T_{env}, is given by:

$$\dot{Q}_{st} = (UA)_s (T_s - T_{env}) \tag{11.72}$$

The storage tank losses for the month can be obtained by integrating Eq. (11.72), considering that $(UA)_s$ and T_{env} are constant for the month:

$$Q_{st} = (UA)_s (\overline{T}_s - T_{env}) \Delta t \tag{11.73}$$

where $\overline{T}_s$ = monthly average storage tank temperature (°C).

Therefore, the total load on the solar energy system is the actual load required by a process and the storage tank losses. Because the storage tanks are usually well insulted, storage tank losses are small and the tank rarely drops below the minimum temperature. The fraction of the total load supplied by the solar energy system, including storage tank losses, is given by:

$$f_{TL} = \frac{L_s + Q_{st}}{L_u + Q_{st}} \tag{11.74}$$

where

L_s = solar energy supplied to the load (GJ);
L_u = useful load (GJ).

Therefore, after Q_{st} is estimated, f_{TL} can be obtained from the $\overline{\Phi}$ f-charts as usual. The solar fraction f can also be represented by L_s/L_u, i.e., the solar energy supplied to the load divided by the useful load, then Eq. (11.74) becomes:

$$f = f_{TL} \left(1 + \frac{Q_{st}}{L_u} \right) - \frac{Q_{st}}{L_u} \tag{11.75}$$

Storage tank losses can be estimated by considering that the tank remains at T_{min} during the month or by assuming that the average tank temperature is equal to the monthly average collector inlet temperature, $\overline{T}_i$, which can be estimated by the $\overline{\Phi}$ charts. Finally, the average daily utilizability is given by (Klein and Beckman, 1979):

$$\overline{\Phi} = \frac{f_{TL}}{Y} \tag{11.76}$$

For the estimation of the tank losses with Eq. (11.73), Klein and Beckman (1979) recommend the use of the mean of T_{min} and $\overline{T}_i$. The process is iterative, i.e., $\overline{T}_i$ is assumed, from which Q_{st} is estimated. From this, the f_{TL} is estimated with the $\overline{\Phi}$ f-charts; subsequently, $\overline{\Phi}$ is estimated from Eq. (11.76) and $\overline{X}_c$ is obtained from $\overline{\Phi}$ charts, from which $\overline{T}_i$ is estimated from Eq. (11.67). This new value of $\overline{T}_i$ is compared with the initially assumed value and a new iteration is carried out if necessary. Finally, Eq. (11.75) is used to estimate the solar fraction, f.

EXAMPLE 11.14

For the industrial process heat system in Example 11.13, estimate the storage tank losses for the month of June by considering the ambient temperature where the tank is located to be at 18 °C and the tank $(UA)_s = 6.5$ W/°C.

Solution

To solve this problem, we have to assume an average tank temperature. For June, we assume a value of 72 °C. The tank losses are estimated with Eq. (11.73):

$$Q_{st} = (UA)_s \left(\overline{T}_s - T_{env}\right)\Delta t = 6.5(72 - 18) \times 30 \times 24 \times 3600 = 0.91 \text{ GJ}$$

The total load would then be $= 16.20 + 0.91 = 17.11$ GJ. Because the load is indirectly proportional to the dimensionless parameters, the new values are 16.20/17.11 times the values given in Example 11.13. Therefore,

$$\overline{\Phi}_{max} Y = 0.713 \frac{16.20}{17.11} = 0.675$$

and

$$X' = 4.74 \frac{16.20}{17.11} = 4.49$$

From Eq. (11.70), we get $f_{TL} = 0.61$. From Eq. (11.68), we can estimate Y:

$$Y = \frac{A_c F_R(\overline{\tau\alpha})N\overline{H}_t}{L} = \frac{A_c F_R(\overline{\tau\alpha})N\overline{H}R}{L}$$

$$= \frac{50 \times 0.82 \times 0.89 \times 30 \times 29.2 \times 10^6 \times 0.812}{17.11 \times 10^9} = 1.517$$

From Eq. (11.76),

$$\overline{\Phi} = \frac{f_{TL}}{Y} = \frac{0.61}{1.517} = 0.402$$

The $\overline{K}_T$ in June is 0.70; therefore, the coefficients are $A = -1.5715$, $B = 0.0871$, and $C = 1.0544$. Now, from Eq. (11.56a) by trial and error, the new value of $\overline{X}_c = 0.43$ from the original of 0.39.

As in Eq. (11.67) $\overline{X}_c$ is directly proportional to temperature difference, then the original difference of $(70 - 25.8) = 44.2$ °C must be increased by the ratio 0.43/0.39. Therefore,

$$\Delta T \frac{0.43}{0.39} = \overline{T}_i - \overline{T}_a$$

or

$$\overline{T}_i = \Delta T \frac{0.43}{0.39} + \overline{T}_a = 44.2 \frac{0.43}{0.39} + 25.8 = 74.5 \text{ °C}$$

The average tank temperature is then equal to $(74.5 + 70)/2 = 72.3$ °C. This is very near the original assumption, so no iterations are required.

The solar fraction is then obtained from Eq. (11.75):

$$f = f_{TL}\left(1 + \frac{Q_{st}}{L_u}\right) - \frac{Q_{st}}{L_u} = 0.61\left(1 + \frac{0.91}{16.2}\right) - \frac{0.91}{16.2} = 0.59$$

Therefore, the consideration of tank losses reduces the fraction for June from 64% to 59%.

11.3.2 Heat exchanger correction

The heat exchanger increases the storage tank temperature by adding a thermal resistance between the tank and the load. This results in a reduction in the useful energy collection by having higher collector inlet temperatures and an increase in the storage tank losses. The average increase in tank temperature that is necessary to supply the required energy to load is given by (Klein and Beckman, 1979):

$$\Delta T = \frac{fL/\Delta t_L}{\varepsilon_L C_{min}} \tag{11.77}$$

where

Δt_L = number of seconds during a month the load is required (s);
ε_L = effectiveness of load heat exchanger; and
C_{min} = minimum capacitance of the two fluid streams in the heat exchanger (W/°C).

The temperature difference found by Eq. (11.77) is added to the T_{min} to find the monthly average critical radiation from Eq. (11.67).

EXAMPLE 11.15

As in Example 11.14, add the effect of a load heat exchanger to the performance for the month of June for the system in Example 11.13. The effectiveness of the heat exchanger is 0.48 and its capacitance is 3200 W/°C.

Solution

Here we need to assume a storage tank temperature increase of 5 °C due to the action of the load heat exchanger. From Eq. (11.67),

$$\overline{X}_{c,min} = \frac{F_R U_L (T_{min} - \overline{T}_a)}{\dfrac{F_R(\overline{\tau\alpha})}{r_n R_n \overline{K}_T \overline{H}_o}} = \frac{5.92 \times 3600 (70 + 5 - 25.8)}{\dfrac{0.82 \times 0.89}{0.122 \times 0.921 \times 0.70 \times 41.71 \times 10^6}} = 0.44$$

The $\overline{K}_T$ in June is 0.70; therefore, the coefficients are $A = -1.5715$, $B = 0.0871$, and $C = 1.0544$. From Eq. (11.56a) $\overline{\Phi}_{max} = 0.387$. As the use of heat exchanger increases the tank temperature, we need to assume a new tank temperature, as in the previous example. Let us assume a tank temperature of 77 °C. For this temperature, from Eq. (11.73), $Q_s = 0.99$ GJ and the total load is $16.20 + 0.99 = 17.19$ GJ. Therefore, as in previous example,

$$Y = \frac{A_c F_R(\overline{\tau\alpha}) N \overline{H}_t}{L}$$

$$= \frac{A_c F_R(\overline{\tau\alpha}) N \overline{H} \overline{R}}{L} = \frac{50 \times 0.82 \times 0.89 \times 30 \times 29.2 \times 10^6 \times 0.812}{17.19 \times 10^9} = 1.510$$

and

$$\overline{\Phi}_{max} Y = 0.387 \times 1.510 = 0.584$$

From Eq. (11.69),

$$X' = \frac{A_c F_R U_L (100) \Delta t}{L} = \frac{50 \times 5.92 \times 100 \times 24 \times 30 \times 3600}{17.19 \times 10^9} = 4.46$$

EXAMPLE 11.15—CONT'D

From Eq. (11.70), $f_{TL} = 0.54$. Then, we have to check the increase of storage tank temperature assumption. From Eq. (11.76),

$$\overline{\Phi} = \frac{f_{TL}}{Y} = \frac{0.54}{1.510} = 0.358$$

From Eq. (11.56a) by trial and error the new value of $\overline{X}_c = 0.47$ from the original of 0.39. From Eq. (11.67),

$$\overline{T}_i - \overline{T}_a = \frac{F_R(\overline{\tau\alpha})r_n R_n \overline{K}_T \overline{H}_o \overline{X}_c}{F_R U_L} = \frac{0.82 \times 0.89 \times 0.122 \times 0.921 \times 0.70 \times 41.71 \times 10^6 \times 0.47}{5.92 \times 3600} = 52.8\,°C$$

and

$$\overline{T}_i = 52.8 + 25.8 = 78.6\,°C$$

The average tank temperature for the losses is then equal to $(75 + 78.6)/2 = 76.8\,°C$. This is effectively the same as the one originally assumed, so no iteration is required. From Eq. (11.75),

$$f = f_{TL}\left(1 + \frac{Q_{st}}{L_u}\right) - \frac{Q_{st}}{L_u} = 0.54\left(1 + \frac{0.99}{16.2}\right) - \frac{0.99}{16.2} = 0.51$$

Finally, we also need to check the assumption of storage tank temperature increase (5 °C) due to the action of the load heat exchanger. From Eq. (11.77).

$$\Delta T = \frac{fL/\Delta t_L}{\varepsilon_L C_{min}} = \frac{0.51 \times 16.2 \times 10^9/(10 \times 30 \times 3600)}{0.48 \times 3200} = 5\,°C$$

Because this is the same value as the original assumption, no iterations are required and the calculations are complete. Therefore, the solar fraction for June dropped from 64% to 51% due to the presence of the load heat exchanger. This substantial drop in performance is due to the increase of tank temperature and the increased tank losses at the higher temperature.

Klein and Beckman (1979) also performed a validation study to compare the results of the present method with those obtained by the TRNSYS program. Comparisons between the $\overline{\Phi}$ f-chart estimates and TRNSYS calculations were performed for three types of systems: space heating, air conditioning (using a LiBr absorption chiller operated at $T_{min} = 77\,°C$), and process heating applications ($T_{min} = 60\,°C$). The comparisons show that, although there are some particular circumstances in which $\overline{\Phi}$ f-chart will yield inaccurate results, the method can be used to predict the performance of a wide variety of solar energy systems.

11.4 Unutilizability method

Passive solar energy systems are described in Chapter 6, Section 6.2. It is of interest to the designer to be able to estimate the long-term performance of passive systems. In this way, the designer could evaluate how much of the absorbed energy cannot be used because it is available at a time when the loads are satisfied or exceed the capacity of the building to store energy.

The unutilizability method is an extension of the utilizability method that is suitable for direct gain, collector storage walls, and active collection with passive storage (hybrid) systems. These are treated separately in the following sections. The unutilizability (called UU) method, developed by Monsen et al. (1981, 1982), is based on the concept that a passively heated building can be considered a collector with finite heat capacity. As in the case of the *f*- and $\overline{\Phi}$ chart methods, the estimations are carried out on a monthly basis and the result is to give the annual auxiliary energy required. The building thermal load is required for the present method. For this purpose, the methods presented in Chapter 6, Section 6.1, can be used. These vary from the detailed heat balance and transfer function methods to the simple degree day method.

11.4.1 Direct gain systems

The utilization of large glazed areas and massive thermal storage structures in passive heating systems is an effective and simple means of collecting and storing solar energy in buildings. The method of analysis for this type of system is presented by Monsen et al. (1981). The design of such a passive system cannot be based on a fixed design indoor temperature, as in active systems. The monthly energy streams of a direct gain structure are shown in Fig. 11.12. As can be seen, the energy absorbed by the passive system is expressed as:

$$\overline{H}_t N (\overline{\tau\alpha}) A_r = N \overline{S} A_r \tag{11.78}$$

where

A_r = area of the collector (receiving) window (m^2);
$(\overline{\tau\alpha})$ = product of monthly average value of window transmissivity and room absorptivity.

The monthly average energy absorbed $\overline{S}$ is given by Eq. (3.1a) by replacing the hourly direct and diffuse radiation terms with the monthly average terms, as indicated by Eq. (3.1b) in Chapter 3:

$$\overline{S} = \overline{H}_B \overline{R}_B (\overline{\tau\alpha})_B + \overline{H}_D (\overline{\tau\alpha})_D \left[\frac{1 + \cos(\beta)}{2} \right] + \rho_G (\overline{H}_B + \overline{H}_D)(\overline{\tau\alpha})_G \left[\frac{1 - \cos(\beta)}{2} \right] \tag{11.79}$$

The energy lost through the building envelope is shown in Fig. 11.12 as the load *L*. This is estimated by considering that the transmittance of the glazing is 0, given by:

$$L = (UA)_h (T_b - T_a) \tag{11.80}$$

FIGURE 11.12

Monthly energy flows of a direct gain building.

where

$(UA)_h$ = product of the overall heat transfer coefficient and the area of the building structure, including the direct gain windows (W/°C);

T_b = indoor base temperature (°C).

When solar energy is not enough to supply the load, auxiliary energy, Q_{aux}, must be provided. Also, there might be excess absorbed solar energy, above what is required to cover the load, that cannot be stored and must be dumped, denoted by Q_D. Sometimes during a month, sensible heat may be stored or removed from the building structure, provided it has thermal capacitance, called *stored energy*, not shown in Fig. 11.12.

Two limiting cases here need to be investigated separately. In the first, we assume an infinite storage capacity, and in the second, zero storage capacity. In the first case, during a month, all the energy absorbed in excess of the load is stored in the building structure. The infinite capacitance of the building structure implies a constant temperature of the conditioned space. This stored energy is used when required to cover the load, thus it offsets the auxiliary energy, given by the monthly energy balance, as:

$$Q_{aux,i} = \left(L - \overline{H}_t N (\overline{\tau\alpha}) A_r\right)^+ \tag{11.81}$$

The plus superscript in Eq. (11.81) indicates that only positive values are considered. Additionally, no month-to-month carry-over is considered.

For the second limiting case, the building structure has zero storage capacity, and any energy deficits are covered with auxiliary energy, whereas any excess solar energy must be dumped. The temperature of the building is again constant but this time is due to the addition or removal of energy. The rate of energy dumped can be obtained from an instantaneous energy balance, given by:

$$\dot{Q}_{D,z} = [I_t(\tau\alpha)A_r - (UA)_h(T_b - T_a)]^+ \tag{11.82}$$

Similar to the case of solar collectors in an active solar energy system, a critical radiation level can be defined as the level at which the gains are equal to the losses, given by:

$$I_{tc} = \frac{(UA)_h(T_b - T_a)}{(\tau\alpha)A_r} \tag{11.83}$$

Because we have zero storage capacity, any radiation above this critical level is unutilizable and is dumped. Therefore, the dumped energy for the month, Q_D, is given by:

$$Q_{D,z} = A_r(\overline{\tau\alpha}) \int_{month} (I_t - I_{tc})^+ dt \tag{11.84}$$

Over a month, I_{tc} can be considered to be constant, and from Eq. (11.83), its monthly average value is given by:

$$I_{tc} = \frac{(UA)_h(T_b - \overline{T}_a)}{(\overline{\tau\alpha})A_r} \tag{11.85}$$

Energy below I_{tc} is useful, whereas energy above I_{tc} is dumped. Equation (11.84) can be expressed in terms of the monthly average utilizability given by Eq. (11.52), and Q_D may be written as:

$$Q_{D,z} = \overline{H}_t N(\overline{\tau\alpha})A_r \overline{\Phi} \tag{11.86}$$

It is worth noting that, for a passive solar heating system, $\overline{\Phi}$ is a measure of the amount of solar energy that cannot be used to reduce auxiliary energy, and it may be called *unutilizability*.

Using a monthly energy balance, the amount of auxiliary energy required by the zero storage capacity building can be estimated as being equal to the load plus dumped energy minus the absorbed solar energy, given by:

$$Q_{aux,z} = L - \left(1 - \overline{\Phi}\right)\overline{H}_t N(\overline{\tau\alpha})A_r \tag{11.87}$$

Therefore, Eqs. (11.81) and (11.87) give the limits on the amount of auxiliary energy of a real building. Correlations have been developed by Monsen et al. (1981, 1982) in terms of the fraction of the load covered by solar energy. Similar to the active solar energy systems, the solar fraction is equal to $f = 1 - (Q_{aux}/L)$. For these correlations, two dimensionless parameters are specified, X and Y. The X dimensionless parameter is the solar-load ratio, defined as (Monsen et al., 1981):

$$X = \frac{N\overline{S}A_r}{L} = \frac{\overline{H}_t N(\overline{\tau\alpha})A_r}{L} \tag{11.88}$$

For the infinite capacitance system, by dividing all terms of Eq. (11.81) by L, X is equal to the solar fraction, f_i, given by:

$$f_i = X = 1 - \frac{Q_{aux,i}}{L} \tag{11.89}$$

For the zero capacitance case,

$$f_z = 1 - \frac{Q_{aux,z}}{L} \tag{11.90}$$

Replacing $Q_{aux,z}$ from Eq. (11.87), we get:

$$f_z = \left(1 - \overline{\Phi}\right)X \tag{11.91}$$

The Y dimensionless parameter is the monthly ratio of the maximum storage capacity of the building to the solar energy that would be dumped if the building had zero thermal capacitance. It is therefore called the *storage-dump ratio*, given by (Monsen et al., 1981):

$$Y = \frac{C_b \Delta T_b}{\overline{H}_t(\overline{\tau\alpha})A_r \overline{\Phi}} = \frac{N C_b \Delta T_b}{Q_{D,z}} \tag{11.92}$$

where

C_b = effective thermal capacitance, i.e., mass times heat capacity (J/°C);
ΔT_b = the difference of upper and lower temperatures, i.e., the temperature range the building is allowed to float (°C).

The two limiting cases have Y values equal to 0 for the zero-storage capacity building and infinity for the infinite-storage capacity building. The values of the effective thermal capacitance, given by

Barakat and Sander (1982), are equal to 60 kJ/m² °C for light-construction buildings, 153 kJ/m² °C for medium-construction buildings, 415 kJ/m² °C for heavy-construction buildings, and 810 kJ/m² °C for very heavy-construction buildings.

Finally, the correlation of the monthly solar fraction is given in terms of X, Y, and $\overline{\Phi}$ and is given by (Monsen et al., 1981):

$$f = \min\{PX + (1-P)(3.082 - 3.142\overline{\Phi})[1 - \exp(-0.329X)], 1\} \qquad (11.93a)$$

where

$$P = [1 - \exp(-0.294Y)]^{0.652} \qquad (11.93b)$$

The auxiliary energy can be calculated from the solar fraction as:

$$Q_{aux} = (1-f)L \qquad (11.94)$$

EXAMPLE 11.16

A residential building located at 35°N latitude is well insulated and has a direct gain passive solar energy system. Estimate the fraction of heating load supplied by the solar energy system for December and the auxiliary energy required, given the following information:

Window area = 10 m².
Effective heat capacity of the building, C_b = 60 MJ/°C.
Allowable temperature swing = 7 °C.
Low set-point temperature = 18.3 °C.
U value of window without night insulation = 5.23 W/m² °C.
Building UA, excluding direct gain window = 145 W/°C.
Degree days for December, estimated at a base temperature of 18.3 °C = 928 °C-days.
Mean ambient temperature, $\overline{T}_a$ = 11.1 °C.
Monthly average daily total radiation $\overline{H}$ = 9.1 MJ/m².
Monthly average $(\overline{\tau\alpha})$ = 0.76.

Solution
The thermal load needs to be calculated first. The UA for the building, including the direct gain window, is:

$$(UA)_h = 145 + (10 \times 5.23) = 197.3 \text{ W/°C}$$

From Eq. (6.24),

$$L = (UA)_h(DD)_h = 197.3 \times 928 \times 24 \times 3600 = 15.82 \text{ GJ}$$

Because the latitude of the building under consideration is the same as the one used in Example 11.13, $r_n = 0.171$, $r_{dn} = 0.159$.
From Table 2.5, $\overline{H}_o$ = 16.8 MJ/m², which from Eq. (2.82a) gives $\overline{K}_T$ = 0.54. For December, h_{ss} = 72.7°; from Eq. (11.54a), $\overline{H}_D/\overline{H}$ = 0.483; from Eq. (2.105c), $\overline{H}_D/\overline{H}$ = 0.35; and from Eq. (2.108), $\overline{R}_B$ = 2.095. Using $\beta = 90°$ (vertical surface) and assuming a ground reflectance of 0.2, $R_{B,n}$ = 1.603, R_n = 1.208, $\overline{R}$ = 1.637.
From Eq. (11.88),

$$X = \frac{N\overline{S}A_r}{L} = \frac{N\overline{HR}(\overline{\tau\alpha})A_r}{L} = \frac{31 \times 9.1 \times 10^6 \times 1.637 \times 0.76 \times 10}{15.82 \times 10^9} = 0.222$$

Continued

EXAMPLE 11.16–CONT'D

From Eq. (11.85),

$$I_{tc} = \frac{(UA)_h(T_b - \overline{T}_a)}{(\overline{\tau\alpha})A_r} = \frac{197.3(18.3 - 11.1)}{0.76 \times 10} = 186.9 \ W/m^2$$

The next parameter that we need to calculate is $\overline{\Phi}$. From Eq. (11.55),

$$\overline{X}_c = \frac{I_{tc}}{r_n R_n \overline{H}} = \frac{186.9 \times 3600}{0.171 \times 1.208 \times 9.1 \times 10^6} = 0.358$$

From Eq. (11.56),

$$A = 2.943 - 9.271\overline{K}_T + 4.031\overline{K}_T^2 = 2.943 - 9.271 \times 0.54 + 4.031(0.54)^2 = -0.888$$

$$B = -4.345 + 8.853\overline{K}_T - 3.602\overline{K}_T^2 = -4.345 + 8.853 \times 0.54 - 3.602(0.54)^2 = -0.615$$

$$C = -0.170 - 0.306\overline{K}_T + 2.936\overline{K}_T^2 = -0.170 - 0.306 \times 0.54 + 2.936(0.54)^2 = 0.521$$

$$\overline{\Phi} = \exp\left\{ \left[A + B\left(\frac{R_n}{\overline{R}}\right)\right] \left[\overline{X}_c + C\overline{X}_c^2\right] \right\} = \exp\left\{ \left[-0.888 - 0.615\left(\frac{1.208}{1.637}\right)\right] \left[0.358 + 0.521(0.358)^2\right] \right\} = 0.566$$

From Eq. (11.92),

$$Y = \frac{C_b \Delta T_b}{\overline{H}_t(\overline{\tau\alpha})A_r \overline{\Phi}} = \frac{C_b \Delta T_b}{\overline{H}R(\overline{\tau\alpha})A_r \overline{\Phi}} = \frac{60 \times 7}{9.1 \times 1.637 \times 0.76 \times 10 \times 0.566} = 6.55$$

From Eq. (11.93b),

$$P = [1 - \exp(-0.294Y)]^{0.652} = [1 - \exp(-0.294 \times 6.55)]^{0.652} = 0.902$$

From Eq. (11.93a),

$$f = \min\{PX + (1-P)(3.082 - 3.142\overline{\Phi})[1 - \exp(-0.329X)], 1\}$$

$$= \min\left\{ \begin{array}{l} 0.902 \times 0.222 + (1 - 0.902)(3.082 - 3.142 \times 0.566) \\ \times [1 - \exp(-0.329 \times 0.222)], 1 \end{array} \right\} = \min\{0.21, 1\} = 0.21$$

Finally, from Eq. (11.94),

$$Q_{aux} = (1-f)L = (1 - 0.21) \times 15.82 = 12.50 \ GJ$$

11.4.2 Collector storage walls

The thermal analysis of collector storage walls is presented in Section 6.2.1, Chapter 6, where a diagram of the wall and the thermal gains and losses are given. The unutilizability concept, developed by Monsen et al. (1982), can also be applied in this case to determine the auxiliary energy required to cover the energy supplied by the solar energy system. Again here, two limiting cases are investigated: zero and infinite capacitance buildings. For the infinite thermal capacitance case, all net monthly heat gain from the storage wall, Q_g, given by Eq. (6.52), can be used. The monthly energy balance of the infinite capacitance building is given by:

$$Q_{aux,i} = (L_m - Q_g)^+ \tag{11.95}$$

where L_m = monthly energy loss from the building (kJ), given by Eq. (6.45).

For the zero thermal capacitance case, which applies to both the storage wall and the building structure, maximum auxiliary energy is required. The collector storage wall in this case acts as a radiation shield that alters the amplitude but not the time of the solar gains to the building. The monthly energy balance of the zero capacitance building is given by:

$$Q_{\text{aux,z}} = \left(L_m - Q_g + Q_D\right)^+ \tag{11.96}$$

The dumped energy, Q_D, can be determined by integrating $\dot{Q}_D$, the rate at which excess energy must be removed to prevent the room temperature from reaching a value above the high thermostat set temperature. The rate of dumped energy, $\dot{Q}_D$, is the difference between the rate of heat transfer through the collector storage wall into the building and the rate of heat loss from the building structure, given by:

$$\dot{Q}_D = [U_k A_w (T_w - T_R) - (UA)(T_b - T_a)]^+ \tag{11.97}$$

where U_k = overall heat-transfer coefficient of the thermal storage wall including glazing, given by Eq. (6.50) (W/m^2 °C).

For the case of the zero thermal capacitance-collector storage wall, an energy balance gives:

$$I_t(\tau\alpha)A_w = U_o A_w (T_w - T_a) + U_k A_w (T_w - T_R) \tag{11.98}$$

where U_o = overall heat transfer coefficient from the outer wall surface through the glazing to the ambient, without night insulation (W/m^2 °C).

Solving Eq. (11.98) for T_w gives:

$$T_w = \frac{I_t(\tau\alpha) + U_o T_a + U_k T_R}{U_o + U_k} \tag{11.99}$$

Substituting T_w from Eq. (11.99) to Eq. (11.97) gives:

$$\dot{Q}_D = \left[U_k A_w \left(\frac{I_t(\tau\alpha) - U_o(T_R - T_a)}{U_o + U_k} \right) - (UA)(T_b - T_a) \right]^+ \tag{11.100}$$

This equation can be integrated over a month to give Q_D by assuming that $(\tau\alpha)$ and T_a are constant and equal to their mean monthly values $(\overline{\tau\alpha})$ and $\overline{T}_a$:

$$Q_D = \frac{U_k A_w (\overline{\tau\alpha})}{U_o + U_k} \sum (I_t - I_{tc})^+ \tag{11.101}$$

where I_{tc} is the critical radiation level, which makes $\dot{Q}_D$ equal to 0, given by:

$$I_{tc} = \frac{1}{(\overline{\tau\alpha})A_w} \left[(UA)\left(\frac{U_o}{U_k} + 1\right) \frac{T_b - \overline{T}_a}{T_R - \overline{T}_a} + U_o A_w \right] (T_R - \overline{T}_a) \tag{11.102}$$

It should be noted that the summation in Eq. (11.101) is the same as the summation in daily utilizability $\overline{\Phi}$ given by Eq. (11.52); therefore, Eq. (11.101) becomes:

$$Q_D = \frac{U_k A_w N \overline{S} \, \overline{\Phi}}{U_o + U_k} \tag{11.103}$$

The solar fractions corresponding to the limits of performance of the collector storage wall systems, given by Eqs. (11.95) and (11.96), are:

$$f_i = 1 - \frac{Q_{aux,i}}{L_m + L_w} = \frac{L_w + Q_g}{L_m + L_w} \tag{11.104}$$

$$f_z = 1 - \frac{Q_{aux,z}}{L_m + L_w} = f_i - \frac{U_k}{U_o + U_k} \overline{\Phi} X \tag{11.105}$$

where X is the solar-load ratio given by:

$$X = \frac{N \overline{S} A_w}{L_m + L_w} \tag{11.106}$$

Two parameters are then required: the storage capacity of the building, S_b, and of the storage wall, S_w. The building storage capacity for a month is given by (Monsen et al., 1982):

$$S_b = C_b(\Delta T_b)N \tag{11.107}$$

where

C_b = effective building storage capacitance (J/°C);
ΔT_b = allowable temperature swing, the difference between the high and low thermostat settings (°C).

The storage capacity of the wall for the month is given by (Monsen et al., 1982):

$$S_w = c_p w A_w \rho (\Delta T_w)N \tag{11.108}$$

where

c_p = heat capacity of the wall (J/kg °C);
ρ = density of the wall (kg/m^3);
w = thickness of the wall (m); and
ΔT_w = one half of the difference of the monthly average temperatures of the outside and the inside surfaces of the wall (°C).

The heat transfer through the wall into the heated space, Q_g, given in terms of ΔT_w, is:

$$Q_g = \frac{2kA_w}{w}(\Delta T_w)\Delta t N \tag{11.109}$$

Solving Eq. (11.109) in terms of ΔT_w and substituting into Eq. (11.108),

$$S_w = \frac{\rho c_p w^2 Q_g}{2k\Delta t} \tag{11.110}$$

A dimensionless parameter called the *storage-dump ratio* needs to be specified. It is defined as the ratio of a weighted storage capacity of the building and wall to the energy that would be dumped by a building having zero capacitance, given by (Monsen et al., 1982):

$$Y = \frac{S_b + 0.047 S_w}{Q_D} \tag{11.111}$$

The solar fraction here is given by:

$$f = 1 - \frac{Q_{aux}}{L_m + L_w} \tag{11.112}$$

A correlation of the solar fraction, f, was developed from simulations as a function of f_i and Y (Monsen et al., 1982):

$$f = \min\{Pf_i + 0.88(1 - P)[1 - \exp(-1.26f_i)], 1\} \tag{11.113a}$$

where

$$P = [1 - \exp(-0.144Y)]^{0.53} \tag{11.113b}$$

It should be noted that here the X dimensionless parameter is not used in the correlation for the solar fraction. The auxiliary energy required for a month is given by:

$$Q_{aux} = (1 - f)(L_m + L_w) \tag{11.114}$$

The monthly auxiliary energy requirements are then added to obtain the annual auxiliary energy needs of the building.

The steps to follow to estimate the annual performance of the collector storage wall are the following:

1. Estimate the absorbed solar radiation for each month.
2. Estimate the loads L_m and L_w from Eqs. (6.45) and (6.46), respectively, by taking into account internal heat generation, if it exists.
3. Estimate the heat gain across the collector storage wall, Q_g, using Eq. (6.52).
4. Estimate daily utilizability and the energy dump that would occur in a zero capacitance system, Q_D, from Eq. (11.103).
5. Estimate f_i, S_b, and S_w from Eqs. (11.104), (11.107) and (11.110), respectively.
6. Estimate Y from Eq. (11.111).
7. Finally, estimate the monthly fraction, f, and auxiliary energy, Q_{aux}.

EXAMPLE 11.17

The building of Example 11.16 is fitted with a collector storage wall. All the problem data of Example 11.16 apply here with the following additional information about the collector storage wall:

Density = 2200 kg/m^3.
Heat capacity = 910 J/kg °C.
Wall thickness, $w = 0.40$ m.
Loss coefficient from the wall to the ambient, $\overline{U}_o = 4.5$ W/m^2 °C.
Overall heat transfer coefficient of the wall including glazing, $U_w = 2.6$ W/m^2 °C.
Thermal conductivity of the wall, $k = 1.85$ W/m °C.
Estimate the solar fraction for December and the auxiliary energy required for the month.

Continued

EXAMPLE 11.17—CONT'D

Solution

Initially, we need to calculate the loads L_m and L_w. From Eq. (6.45),

$$L_m = (UA)_h(DD)_h = 197.3 \times 928 \times 24 \times 3600 = 15.82 \text{ GJ}$$

As the room temperature is the same as the base temperature, $(DD)_h = (DD)_R$. From Eq. (6.46),

$$L_w = U_w A_w (DD)_R = 2.6 \times 10 \times 928 \times 24 \times 3600 = 2.08 \text{ GJ}$$

From Eq. (6.50),

$$U_k = \frac{h_i k}{w h_i + k} = \frac{8.33 \times 1.85}{0.4 \times 8.33 + 1.85} = 2.97 \text{ W/m}^2 \,°\text{C}$$

(note from Chapter 6, Section 6.2.1, $h_i = 8.33$ W/m^2 °C).
From Example 11.16, $\overline{R} = 1.637$. From Eq. (2.107),

$$\overline{H}_t = \overline{R}\,\overline{H} = 1.637 \times 9.1 = 14.9 \text{ MJ/m}^2$$

From Eq. (6.51),

$$\overline{T}_w = \frac{\overline{H}_t(\overline{\tau\alpha}) + (U_k \overline{T}_R + \overline{U}_o \overline{T}_a)\Delta t}{(U_k + \overline{U}_o)\Delta t}$$

$$= \frac{14.9 \times 10^6 \times 0.76 + (2.97 \times 18.3 + 4.5 \times 11.1) \times 24 \times 3600}{(2.97 + 4.5) \times 24 \times 3600} = 31.5 \,°\text{C}$$

Therefore, the total heat transferred into the room through the storage wall is given by Eq. (6.52):

$$Q_g = U_k A_w (\overline{T}_w - \overline{T}_a)\Delta t N = 2.97 \times 10(31.5 - 18.3) \times 24 \times 3600 \times 31 = 1.05 \text{ GJ}$$

From Eq. (11.102),

$$I_{tc} = \frac{1}{(\overline{\tau\alpha})A_w}\left[(UA)\left(\frac{U_0}{U_k} + 1\right)\frac{T_b - \overline{T}_a}{T_R - \overline{T}_a} + U_o A_w\right](T_R - \overline{T}_a)$$

$$= \frac{1}{0.76 \times 10}\left[145\left(\frac{4.5}{2.97} + 1\right)\frac{18.3 - 11.1}{18.3 - 11.1} + 4.5 \times 10\right](18.3 - 11.1) = 388.1 \text{ W/m}^2$$

From Eq. (11.55),

$$\overline{X}_c = \frac{I_{tc}}{r_n R_n \overline{H}} = \frac{388.1 \times 3600}{0.171 \times 1.208 \times 9.1 \times 10^6} = 0.743$$

From Example 11.16, $A = -0.888$, $B = -0.615$, and $C = 0.521$. From Eq. (11.56),

$$\overline{\Phi} = \exp\left\{\left[A + B\left(\frac{R_n}{\overline{R}}\right)\right]\left[\overline{X}_c + C\overline{X}_c^2\right]\right\}$$

$$= \exp\left\{\left[-0.888 - 0.615\left(\frac{1.208}{1.637}\right)\right]\left[0.743 + 0.521(0.743)^2\right]\right\} = 0.251$$

EXAMPLE 11.17—CONT'D

From Eq. (11.103),

$$Q_D = \frac{U_k A_w N \overline{S} \Phi}{U_o + U_k} = \frac{U_k A_w N H R (\overline{\tau \alpha}) \overline{\Phi}}{U_o + U_k}$$

$$= \frac{2.97 \times 10 \times 31 \times 9.1 \times 10^6 \times 1.637 \times 0.76 \times 0.251}{4.5 + 2.97} = 0.350 \text{ GJ}$$

From Eq. (11.104),

$$f_i = \frac{L_w + Q_g}{L_m + L_w} = \frac{2.08 + 1.05}{15.82 + 2.08} = 0.175$$

From Eq. (11.107),

$$S_b = C_b (\Delta T_b) N = 60 \times 10^6 \times 7 \times 31 = 13.02 \text{ GJ}$$

From Eq. (11.110),

$$S_w = \frac{\rho c_p w^2 Q_g}{2k \Delta t} = \frac{2200 \times 910 \times 0.4^2 \times 1.05 \times 10^9}{2 \times 1.85 \times 24 \times 3600} = 1.052 \text{ GJ}$$

From Eq. (11.111),

$$Y = \frac{S_b + 0.047 S_w}{Q_D} = \frac{13.02 + 0.047 \times 1.052}{0.350} = 37.34$$

From Eq. (11.113b),

$$P = [1 - \exp(-0.144 Y)]^{0.53} = [1 - \exp(-0.144 \times 37.34)]^{0.53} = 0.998$$

From Eq. (11.113a),

$$f = \min\{P f_i + 0.88(1 - P)[1 - \exp(-1.26 f_i)], 1\}$$
$$= \min\{0.998 \times 0.175 + 0.88(1 - 0.998)[1 - \exp(-1.26 \times 0.175)], 1\}$$
$$= \min\{0.175, 1\} = 0.175$$

Finally, from Eq. (11.114),

$$Q_{aux} = (1 - f)(L_m + L_w) = (1 - 0.175)(15.82 + 2.08) = 14.77 \text{ GJ}$$

11.4.3 Active collection with passive storage systems

The third type of system analyzed with the unutilizability method concerns active air or liquid collector systems used to heat a building that utilizes the building structure for storage. The advantages of such a system are: the control of heat collection with the solar collector; the elimination of separate storage, which reduces the cost and complexity of the system; and the relative simplicity of the system. The disadvantages include the large temperature swings of the building, which are inevitable when the building provides the storage, and the limits of the solar energy that can be given to the building in order not to exceed the allowable temperature swing. The method of estimation, developed by Evans and Klein (1984), is similar to that of Monsen et al. (1981) for direct gain passive systems, outlined in Section 11.4.1. In this system, two critical radiation levels are specified: one for the collector system and one for the building. As in previous systems, the limits on performance are required by considering the two extreme cases, i.e., the infinite and zero capacitance buildings. As before, the performance of real buildings is determined on correlations based on simulations.

As indicated in Section 11.2.3, the output of an active collector can be expressed with Eq. (11.65). For a month output, it becomes:

$$\sum Q_u = A_c \overline{H}_t F_R (\overline{\tau\alpha}) N \overline{\Phi}_c = A_c \overline{S} F_R N \overline{\Phi}_c \tag{11.115}$$

where $\overline{\Phi}_c$ = monthly average utilizability associated with solar energy collection.

The critical radiation level used to determine $\overline{\Phi}_c$ is similar to Eq. (11.63), given by:

$$I_{tc,c} = \frac{F_R U_L (\overline{T}_i - \overline{T}_a)}{F_R (\overline{\tau\alpha})} \tag{11.116}$$

where $\overline{T}_i$ = monthly average inlet temperature, the building temperature during collection (°C).

It should be noted that, in both limiting cases of zero and infinite capacitance, T_i is constant. For real cases, this temperature is higher than the minimum building temperature and varies slightly, but this does not affect $\sum Q_u$ too much. For a building with infinite storage capacity, the monthly energy balance is:

$$Q_{aux,i} = \left(L - \sum Q_u\right)^+ \tag{11.117}$$

For the zero capacitance building, energy has to be dumped if solar input exceeds the load. The intensity of radiation incident on the collector that is adequate to meet the building load without dumping is called the *dumping critical radiation level*, given on a monthly average basis by:

$$I_{tc,d} = \frac{(UA)_h (\overline{T}_b - \overline{T}_a) + A_c F_R U_L (\overline{T}_i - \overline{T}_a)}{A_c F_R (\overline{\tau\alpha})} \tag{11.118}$$

where

$(UA)_h$ = overall loss coefficient−area product of the building (W/K);
$\overline{T}_b$ = average building base temperature (°C); and
$\overline{T}_i$ = average building interior temperature (°C).

Therefore, for the zero capacitance building, a radiation level above $I_{tc,c}$ is necessary for collection to take place and $I_{tc,d}$ for the collector to meet the building load without dumping on a monthly basis. Energy greater than $I_{tc,d}$ is dumped, estimated by:

$$Q_D = A_c F_R \overline{S} N \overline{\Phi}_d \tag{11.119}$$

where $\overline{\Phi}_d$ = monthly average utilizability, in fact unutilizability, based on $I_{tc,d}$.

So, for the zero-capacitance building, the energy supplied from the collector system that is useful in meeting the load is the difference between the total energy collected and the energy dumped, given by:

$$Q_{u,b} = \sum Q_u - Q_D = A_c F_R \overline{S} N (\overline{\Phi}_c - \overline{\Phi}_d) \tag{11.120}$$

The monthly auxiliary energy required for the zero-capacitance building is then:

$$Q_{\text{aux,z}} = \left(L - \sum Q_{\text{u}} - Q_{\text{D}} \right)^{+} \tag{11.121}$$

The limits of auxiliary energy requirements are given by Eqs. (11.117) and (11.121), and the auxiliary of a real building with finite capacitance can be obtained by correlations of the solar fraction, f, with two dimensionless coefficients, the solar-load ratio, X, and the storage-dump ratio, Y, given by:

$$X = \frac{A_{\text{c}} F_{\text{R}} \bar{S} N}{L} \tag{11.122}$$

$$Y = \frac{C_{\text{b}} \Delta T_{\text{b}}}{A_{\text{c}} F_{\text{R}} \bar{S} \, \bar{\Phi}_{\text{d}}} = \frac{C_{\text{b}} N \Delta T_{\text{b}}}{Q_{\text{D}}} \tag{11.123}$$

Finally, the correlation of the monthly solar fraction, f, with monthly collection utilizability, $\bar{\Phi}_{\text{c}}$, and the monthly dumping utilizability, $\bar{\Phi}_{\text{d}}$, is given by:

$$f = PX\bar{\Phi}_{\text{c}} + (1 - P)\left(3.082 - 3.142\bar{\Phi}_{\text{u}}\right)\left[1 - \exp(-0.329X)\right] \tag{11.124a}$$

$$P = \left[1 - \exp(-0.294Y)\right]^{0.652} \tag{11.124b}$$

$$\bar{\Phi}_{\text{u}} = 1 - \bar{\Phi}_{\text{c}} + \bar{\Phi}_{\text{d}} \tag{11.124c}$$

The parameter $\bar{\Phi}_{\text{u}}$ is the zero capacitance building unutilizability resulting from the energy loss from the collectors $(1 - \bar{\Phi}_{\text{c}})$ and the energy loss from dumping, $\bar{\Phi}_{\text{d}}$. It should be noted that the correlation for f is very similar to Eq. (11.93) for direct gain systems, for which $I_{\text{tc,c}} = 0$ and $\bar{\Phi}_{\text{c}} = 1$. It then follows that X and Y are the same as in Eqs. (11.88) and (11.92).

EXAMPLE 11.18

The system described in Example 11.16 is heated by a hybrid collector–passive storage system. The solar collector is air type with an area equal to 30 m², $F_{\text{R}}(\tau\alpha)_n = 0.65$, $F_{\text{R}}U_{\text{L}} = 4.95$ W/m² °C, and $(\tau\alpha)_n = 0.91$. In this case, the room temperature is 20 °C. Estimate the solar fraction for December and the auxiliary energy required.

Solution

From Eq. (11.116),

$$I_{\text{tc,c}} = \frac{F_{\text{R}}U_{\text{L}}\left(\bar{T}_{\text{i}} - \bar{T}_{\text{a}}\right)}{F_{\text{R}}(\overline{\tau\alpha})} = \frac{4.95(20 - 11.1)}{0.65\left(\dfrac{0.76}{0.91}\right)} = 81.2 \ \text{W/m}^2$$

From Eq. (11.118),

$$I_{\text{tc,d}} = \frac{(UA)_{\text{h}}\left(\bar{T}_{\text{b}} - \bar{T}_{\text{a}}\right) + A_{\text{c}} F_{\text{R}} U_{\text{L}}\left(\bar{T}_{\text{i}} - \bar{T}_{\text{a}}\right)}{A_{\text{c}} F_{\text{R}}(\overline{\tau\alpha})}$$

$$= \frac{197.3(18.3 - 11.1) + 30 \times 4.95(20 - 11.1)}{30 \times 0.65\left(\dfrac{0.76}{0.91}\right)} = 168.4 \ \text{W/m}^2$$

Continued

EXAMPLE 11.18—CONT'D

From Eq. (11.55),

$$\overline{X}_{c,c} = \frac{I_{tc,c}}{r_n R_n \overline{H}} = \frac{81.2 \times 3600}{0.171 \times 1.208 \times 9.1 \times 10^6} = 0.156$$

From Example 11.16, $A = -0.888$, $B = -0.615$, and $C = 0.521$. From Eq. (11.56),

$$\overline{\Phi}_c = \exp\left\{ \left[A + B\left(\frac{R_n}{\overline{R}}\right) \right] \left[\overline{X}_{c,c} + C\overline{X}_{c,c}^2 \right] \right\}$$

$$= \exp\left\{ \left[-0.888 - 0.615\left(\frac{1.208}{1.637}\right) \right] \left[0.156 + 0.521(0.156)^2 \right] \right\} = 0.797$$

and

$$\overline{X}_{c,d} = \frac{I_{tc,d}}{r_n R_n \overline{H}} = \frac{168.4 \times 3600}{0.171 \times 1.208 \times 9.1 \times 10^6} = 0.323$$

Similarly, $\overline{\Phi}_d = 0.603$.
From Eq. (11.122),

$$X = \frac{A_c F_R \overline{S} N}{L} = \frac{A_c F_R \overline{H} R (\overline{\tau \alpha}) N}{L} = \frac{30 \left(\frac{0.65}{0.91}\right) \times 9.1 \times 10^6 \times 1.637 \times 0.76 \times 31}{15.82 \times 10^9} = 0.475$$

From Eq. (11.123),

$$Y = \frac{C_b \Delta T_b}{A_c F_R \overline{S} \overline{\Phi}_d} = \frac{C_b \Delta T_b}{A_c F_R \overline{H} R (\overline{\tau \alpha}) \overline{\Phi}_d} = \frac{60 \times 10^6 \times 7}{30 \left(\frac{0.65}{0.91}\right) \times 9.1 \times 10^6 \times 1.637 \times 0.76 \times 0.603} = 2.871$$

From Eq. (11.124b),

$$P = [1 - \exp(-0.294Y)]^{0.652} = [1 - \exp(-0.294 \times 2.871)]^{0.652} = 0.693$$

From Eq. (11.124c),

$$\overline{\Phi}_u = 1 - \overline{\Phi}_c + \overline{\Phi}_d = 1 - 0.797 + 0.603 = 0.806$$

From Eq. (11.124a),

$$f = PX\overline{\Phi}_c + (1 - P)(3.082 - 3.142\overline{\Phi}_u)[1 - \exp(-0.329X)]$$
$$= 0.693 \times 0.475 \times 0.797 + (1 - 0.693)(3.082 - 3.142 \times 0.806)$$
$$\times [1 - \exp(-0.329 \times 0.475)] = 0.287$$

Finally, from Eq. (11.94),

$$Q_{aux} = (1 - f)L = (1 - 0.287)15.82 = 11.28 \text{ GJ}$$

11.5 Modeling and simulation of solar energy systems

In this chapter so far, we have seen simple methods that can be used to design active solar energy systems of standard configuration using f-chart and other solar processes with the utilizability methods. Although these methods are proven to be accurate enough and possible to carry out with hand calculations, the most accurate way to estimate the performance of solar processes is with detailed simulation.

The proper sizing of the components of a solar energy system is a complex problem that includes both predictable (collector and other performance characteristics) and unpredictable (weather data) components. The initial step in modeling a system is the derivation of a structure to be used to represent the system. It will become apparent that there is no unique way of representing a given system. Since the way the system is represented often strongly suggests specific modeling approaches, the possibility of using alternative system structures should be left open while the modeling approach selection is being made. The structure that represents the system should not be confused with the real system. The structure will always be an imperfect copy of reality. However, the act of developing a system structure and the structure itself will foster an understanding of the real system. In developing a structure to represent a system, system boundaries consistent with the problem being analyzed are first established. This is accomplished by specifying what items, processes, and effects are internal to the system and what items, processes, and effects are external.

Simplified analysis methods have the advantages of computational speed, low cost, rapid turn-around (which is especially important during iterative design phases), and ease of use by persons with little technical experience. The disadvantages include limited flexibility for design optimization, lack of control over assumptions, and a limited selection of systems that can be analyzed. Therefore, if the system application, configuration, or load characteristics under consideration are significantly non-standard, a detailed computer simulation may be required to achieve accurate results.

Computer modeling of thermal systems presents many advantages, the most important of which are the following:

1. They eliminate the expense of building prototypes.
2. Complex systems are organized in an understandable format.
3. They provide a thorough understanding of the system operation and component interactions.
4. It is possible to optimize the system components.
5. They estimate the amount of energy delivery from the system.
6. They provide temperature variations of the system.
7. They estimate the design variable changes on system performance by using the same weather conditions.

Simulations can provide valuable information on the long-term performance of solar energy systems and the system dynamics. This includes temperature variations, which may reach to values above the degradability limit (e.g., for selective coatings of collector absorber plates) and water boiling, with consequent heat dumping through the relief valve. Usually, the detail or type of output obtained by a program is specified by the user; the more detailed the output required, the more intensive are the calculations, which leads to extended computer time required to obtain the results.

Over the years a number of programs have been developed for the modeling and simulation of solar energy systems. Some of the most popular ones are described briefly in this section. These are the well-known programs TRNSYS, WATSUN, and Polysun. The chapter concludes with a brief description of artificial intelligence techniques used recently for the modeling and performance evaluation of solar and other energy systems.

11.5.1 TRNSYS simulation program

TRNSYS is an acronym for "transient simulation", which is a quasi-steady simulation model. This program, currently in version 18 (Klein et al., 2018), was developed at the University of Wisconsin by the members of the Solar Energy Laboratory and written in the FORTRAN computer language. The first version was developed in 1977 and, so far, has undergone 13 major revisions. The program was originally developed for use in solar energy applications, but the use was extended to include a large variety of thermal and other processes, such as hydrogen production, photovoltaics (PVs), and many more. The program consists of many subroutines that model subsystem components. The mathematical models for the subsystem components are given in terms of their ordinary differential or algebraic equations. With a program such as TRNSYS, which can interconnect system components in any desired manner, solve differential equations, and facilitate information output, the entire problem of system simulation reduces to a problem of identifying all the components that make up the particular system and formulating a general mathematical description of each (Kalogirou, 2004b). The users can also create their own programs, which are no longer required to be recompiled with all other program subroutines but just as a dynamic link library (DLL) file with any FORTRAN compiler and put in a specified directory.

Simulations generally require some components that are not ordinarily considered as part of the system. Such components are utility subroutines and output-producing devices. The TYPE number of a component relates the component to a subroutine that models that component. Each component has a unique TYPE number. The UNIT number is used to identify each component (which can be used more than once). Although two or more system components can have the same TYPE number, each must have a unique UNIT number. Once all the components of the system have been identified and a mathematical description of each component is available, it is necessary to construct an information flow diagram for the system. The purpose of the information flow diagram is to facilitate identification of the components and the flow of information between them. Each component is represented as a box, which requires a number of constant PARAMETERS and time-dependent INPUTS and produces time-dependent OUTPUTS. An information flow diagram shows the manner in which all system components are interconnected. A given OUTPUT may be used as an INPUT to any number of other components. From the flow diagram, a deck file has to be constructed, containing information on all components of the system, the weather data file, and the output format.

Subsystem components in the TRNSYS include solar collectors, differential controllers, pumps, auxiliary heaters, heating and cooling loads, thermostats, pebble bed storage, relief valves, hot-water cylinders, heat pumps, and many more. Some of the main ones are shown in Table 11.17. There are also subroutines for processing radiation data, performing integrations, and handling input and output. Time steps down to 1/1000 h (3.6 s) can be used for reading weather data, which makes the program very flexible with respect to using measured data in simulations. Simulation time steps at a fraction of an hour are also possible.

In addition to the main TRNSYS components, an engineering consulting company specializing in the modeling and analysis of innovative energy systems and buildings, Thermal Energy System Specialists (TESS), developed libraries of components for use in TRNSYS. Currently, the TESS library includes more than 500 TRNSYS components. Each of the component libraries comes with a TRNSYS Model File (*.tmf) to use in the Simulation Studio interface, source code, and an example TRNSYS Project (*.tpf) that demonstrates typical uses of the component models found in that library.

Table 11.17 Main components in the standard library of TRNSYS 18.

Building loads and structures	Hydronics
Energy/degree-hour house	Pump
Roof and attic	Fan
Detailed zone	Pipe
Overhang and wingwall shading	Duct
Thermal storage wall	Various fittings (tee-piece, diverter, tempering valve)
Attached sunspace	Pressure relief valve
Detailed multi-zone building	**Output devices**
Controller components	Printer
Differential controllers	Online plotter
Three-stage room thermostat	Histogram plotter
PID controller	Simulation summary
Microprocessor controller	Economics
Collectors	**Physical phenomena**
Flat-plate collector	Solar radiation processor
Performance map solar collector	Collector array shading
Theoretical flat-plate collector	Psychrometrics
Thermosiphon collector with integral storage	Weather-data generator
Evacuated tube solar collector	Refrigerant properties
Compound parabolic collector	Undisturbed ground-temperature profile
Electrical components	**Thermal storage**
Regulators and inverters	Stratified fluid storage tank
Photovoltaic array	Rock-bed thermal storage
Photovoltaic-thermal collector	Algebraic tank
Wind-energy conversion system	Variable volume tank
Diesel engine generator set	Detailed storage tank
Power conditioning	**Utility components**
Lead-acid battery	Data file reader
Heat exchangers	Time-dependent forcing function
Constant effectiveness heat exchanger	Quantity integrator
Counter flow heat exchanger	Calling Excell
Cross-flow heat exchanger	Calling EES
Parallel-flow heat exchanger	Calling CONTAM
Shell and tube heat exchanger	Calling MATLAB
Waste heat recovery	Calling COMIS
HVAC equipment	Holiday calculator
Auxiliary heater	Input value recall
Dual source heat pump	**Weather-data reading**
Cooling towers	Standard format files
Single-effect hot water-fired absorption chiller	User format files

With the release of the TESS Libraries version 18.0 (numbering system changed to be compatible with TRNSYS version number), over 90 new models with many new examples have been added from the previous version 2.0. According to TRNSYS website (www.trnsys.com/tess-libraries/) TESS Component Libraries for TRNSYS 18 fall into 14 categories as follows:

1. *Application Components.* This is an assortment of scheduling and setpoint applications that use the TRNSYS Simulation Studio plug-in feature and are useful for creating daily, weekly, monthly schedules, normalized occupancy, lighting, or equipment schedules and setpoints for thermostats.

2. *Cogeneration (CHP) Components.* This is also known as the Combined Heat and Power (CHP) Library and features many steam system components: pumps, valves, superheaters, desuperheaters, turbines, etc. These components can be used to simulate different cogeneration or trigeneration systems at different scales.

3. *Controller Components.* This includes numerous controllers, which can be used in practically any TRNSYS simulation and range from a simple thermostat controller to a complex multi-stage differential controller with minimum on and off times and from a tempering valve controller to an outside air-reset controller.

4. *Electrical Components.* This includes components for solar PV and combined solar thermal (PV/T) modeling in TRNSYS and also building integrated PV models, a PV array shading component, an equipment outage component, and lighting controls.

5. *Geothermal Heat Pump (GHP) Components.* This includes not only extensive ground heat exchanger models (horizontal multi-layer ground heat exchanger and the vertical heat exchanger), but also buried single and double pipes, and various heat pump models.

6. *Ground Coupling Components.* This includes components to compute the energy transfer between an object (building slabs, basements, buried thermal storage tanks, etc.) and the surrounding ground.

7. *High-Temperature Solar Components.* Many of the models in the standard TRNSYS and TESS Solar Collector Libraries assume that the specific heat of the collector's working fluid is constant. This assumption, while appropriate for low-temperature applications, is not appropriate for high-temperature collectors or for the heat transfer fluids that they typically employ. The TESS High-Temperature Solar Library contains components that have temperature-dependent thermophysical fluid properties, such as a parabolic trough collector, valves, pumps, expansion tanks, and pipes.

8. *HVAC Equipment Components.* This library contains more than 90 different components for modeling anything related to heating, ventilation, and air conditioning of buildings as well as residential, commercial, and industrial HVAC components.

9. *Hydronic Components.* This contains components for a variety of fans, pumps, valves, plenums, duct and pipe components that are essential for working with fluid loops in a TRNSYS simulation.

10. *Loads and Structures Components.* This library contains alternatives to the standard TRNSYS building model components and includes a synthetic building (load-curve generator), a simple multi-zone building and methods for imposing pre-calculated building loads (generated not only by TRNSYS but also by other software) onto a TRNSYS system or plant simulation.

11. *Optimization Components.* This is also known as TRNOPT, which is a tool that couples the TRNSYS simulation with GenOpt program for the minimization of a cost or error function

(see Section 11.6.2). TRNOPT may also be used to calibrate simulation results to the data from the actual system.

12. *Solar Collector Components.* This contains 18 different solar collector components varying from different theoretical and test-based collectors to collectors with different glazings.

13. *Storage Tank Components.* This contains in addition to the standard vertical cylindrical storage tank, spherical, rectangular and horizontal cylindrical tank models. It includes also a wrap-around heat exchanger tank, aquastats, a heat pump water heater and a zip water heater.

14. *Utility Components.* This is a collection of useful components for TRNSYS simulations including random number generators, water draw profiles, an "event triggered" printer, a wind speed calculator, an economics routine, building infiltration models, and many more.

Another interesting application developed for TRNSYS is the library for Solar Thermal Electric Components (STEC). This library includes the necessary components, which can be used for modeling and simulation of Concentrating Solar Power (CSP) systems and comprises models for thermodynamic properties, Brayton and Rankine Cycles, solar thermal electric and storage sub-libraries.

Model validation studies have been conducted to determine the degree to which the TRNSYS program serves as a valid simulation program for a physical system. The use of TRNSYS for the modeling of a thermosiphon solar water heater was also validated by the author and found to be accurate within 4.7% (Kalogirou and Papamarcou, 2000).

TRNSYS used to be a non-user-friendly program; the latest versions of the program (version 16, 17 and 18), however, operate in a graphic interface environment called the *simulation studio*. In this environment, icons of ready-made components are dragged and dropped from a list and connected together according to the real system configuration, in a way similar to the way piping and control wires connect the components in a real system. Each icon represents the detailed program of each component of the system and requires a set of inputs (from other components or data files) and a set of constant parameters, which are specified by the user. Each component has its own set of output parameters, which can be saved in a file, plotted, or used as input in other components. Thus, once all the components of the system are identified, they are dragged and dropped in the working project area and connected together to form the model of the system to be simulated. By double-clicking on each icon, the parameters and the inputs can be easily specified in ready-made tables. Additionally, by double-clicking on the connecting lines, the user can specify which outputs of one component are inputs to another. The project area also contains a weather processing component, printers, and plotters through which the output data are viewed or saved to data files. The model diagram of a thermosiphon solar water heating system is shown in Fig. 11.13.

More details about the TRNSYS program can be found in the program manual (Klein et al., 2018) and the paper by Beckman (1998). Numerous applications of the program are mentioned in the literature. Some typical examples are for the modeling of a thermosiphon system (Kalogirou and Papamarcou, 2000; Kalogirou, 2009), modeling and performance evaluation of solar DHW systems (Oishi and Noguchi, 2000), investigation of the effect of load profile (Jordan and Vajen, 2000), modeling of industrial process heat applications (Kalogirou, 2003a; Benz et al., 1999; Schweiger et al., 2000, Ktistis et al., 2022), and modeling and simulation of a lithium bromide absorption system (Florides et al., 2002). As an example, the results of modeling a thermosiphon system are given (Kalogirou, 2009). The system model diagram is shown in Fig. 11.13 and its specifications in

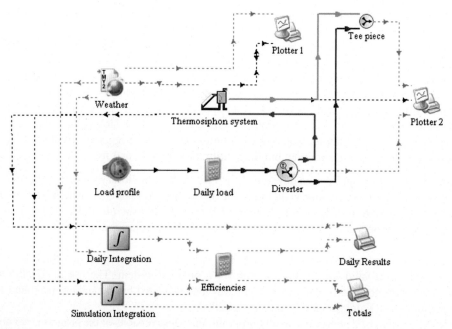

FIGURE 11.13

Model diagram of a thermosiphon solar water-heating system in simulation studio.

Table 11.18 Specifications of the thermosiphon solar water system.

Parameter	Value
Collector area (m^2)	2.7 (two panels)
Collector slope (°)	40
Storage capacity (L)	150
Auxiliary capacity (kW)	3
Heat exchanger	Internal
Heat exchanger area (m^2)	3.6
Hot water demand (L)	120 (four persons)

Table 11.18. The system monthly energy flows are shown in Fig. 11.14, which includes the total radiation incident on the collector (Q_{ins}), the useful energy supplied from the collectors (Q_u), the hot-water energy requirements (Q_{load}), the auxiliary energy demand (Q_{aux}), the heat losses from the storage tank (Q_{env}), and the solar fraction.

The system is simulated using the typical meteorological year of Nicosia, Cyprus. As it can be seen from the curve of total radiation incident on collector (Q_{ins}), the maximum value occurs in the month

FIGURE 11.14

Energy flows of the thermosiphon solar water heater.

of August (1.88 GJ). The useful energy supplied from the collectors (Q_u) is maximized in the month of April (0.62 GJ). It can be also seen from Fig. 11.14 that there is a reduction in the incident solar radiation and consequently the useful energy collected during the month of May. This is a characteristic of the climatic conditions of Nicosia and is due to the development of clouds as a result of excessive heating of the ground and thus excessive convection, especially in the afternoon hours.

From the curve of the energy lost from storage tank (Q_{env}), it can be seen that, during the summer months, the energy lost from storage to surroundings is maximized. This is true because, in these months, the temperature in the storage tank is higher and consequently more energy is lost.

Referring to the curve of hot water load (Q_{load}), there is a decrease of hot-water load demand during the summer months. This is attributed to the fact that, during the summer months, the total incidence of solar radiation is higher, which results to higher temperatures in the cold water storage tank (usually located on top of the solar collector). Consequently, the hot water demand from the hot-water storage tank during these months is reduced.

The variation in the annual solar fraction is also shown in Fig. 11.14. The solar fraction, f, is a measure of the fractional energy savings relative to that used for a conventional system. As can be seen, the solar fraction is lower in the winter months and higher, reaching 100%, in the summer months. The annual solar fraction is determined to be 79% (Kalogirou, 2009).

TRNSYS demo

A fully operational demo version of the program is available on the TRNSYS website (http://trnsus. com/demo/index.html). The demo version includes over 40 examples and allows users to create a simulation with up to five components and buildings with up to two thermal zones. It should be noted that the demo version expires annually in September and users can download a new demo each year so as to have access to the latest features. The examples include:

1. Buildings
2. Solar Domestic Heating Water (SDHW) system
3. Photovoltaics
4. PVT systems
5. Stand-alone power systems
   And many more.

Assignments that can be performed using these demo examples are included in the problems section of this and other chapters. The students are expected to explore these examples to understand how the various components are connected and how the parameters, input, and outputs are used, how the time step variables are integrated to give monthly and annual outputs, and generally how to set up and run a simulation of complex energy systems. The program also includes a number of weather files the user can specify. If a particular area is missing and you have the weather data file, you can add it manually in the /weather folder of the program. For example, for Europe, weather data files in TMY2 format are available for Uccle, Zurich, Praha, Stuttgart, Madrid, Nice, London, Rome, Stockholm, and Ankara.

Running the program is not easy for new users, but by reading the documentation files that come with the demo and by trying the various demos, this can be done quite easily. Numerous YouTube videos demonstrate the use of this program. A few simple instructions to use the program are as follows:

1. To edit the parameters click its icon. For example, to change the collector area double click the collector icon and change the value of parameter 2.
2. The collector slope is specified in the weather processor (Zurich).
3. To run a program, you can click the "running person" figure on the LHS of the screen.
4. To view results, you need to open the external files. Click from the main menu at the top of the screen, Calculate/Open/External Files/***.totals (to see the annual total results).
5. For the whole year simulation. Click the "gear" sign on the LHS menu and change the start time to January 1 and the stop time to December 31.

11.5.2 WATSUN simulation program

WATSUN simulates active solar energy systems and was developed originally by the Watsun Simulation Laboratory of the University of Waterloo in Canada in the early 1970 and 1980s (WATSUN, 1996). The program fills the gap between simpler spreadsheet-based tools used for quick assessments and the more complete, full simulation programs that provide more flexibility but are harder to use. The complete list of systems that can be simulated by the original program is as follows:

- Domestic hot water system with or without storage and heat exchanger.
- Domestic hot water system with stratified storage tank.
- Phase change system, for boiling water.
- Sun switch system, stratified tank with heater.
- Swimming pool-heating system (indoor or outdoor).
- Industrial process heat system, reclaim before collector.
- Industrial process heat system, reclaim after collector.
- Industrial process heat system, reclaim before collector with storage.
- Industrial process heat system, reclaim after collector with storage.
- Variable volume tank-based system.
- Space heating system for a one-room building.

Recently, Natural Resources Canada (NRCan) developed a new version of the program, WATSUN (2009) (NRCan, 2009). This is also used for the design and simulation of active solar energy systems and is provided free from the NRCan website (NRCan, 2009). The two programs share the same name, focus on the hourly simulation of solar energy systems, and use similar equations for modeling some components; however, the new program was redeveloped from scratch, in C++, using object-oriented techniques.

The program currently models two kinds of systems: solar water-heating systems without storage and solar water-heating systems with storage. The second actually covers a multitude of system configurations, in which the heat exchanger can be omitted, the auxiliary tank-heater can be replaced with an inline heater, and the preheat tank can be either fully mixed or stratified. Simple entry forms are used, where the main parameters of the system (collector size and performance equation, tank size, etc.) can be entered easily. The program simulates the interactions between the system and its environment on an hourly basis. This can sometimes, however, break down into sub-hourly time steps, when required by the numerical solver, usually when on–off controllers change state. It is a ready-made program that the user can learn and operate easily. It combines collection, storage, and load information provided by the user with hourly weather data for a specific location and calculates the system state every hour. WATSUN provides information necessary for long-term performance calculations. The program models each component in the system, such as the collector, pipes, and tanks, individually and provides globally convergent methods to calculate their state.

WATSUN uses weather data, consisting of hourly values for solar radiation on the horizontal plane, dry bulb ambient temperature, and in the case of unglazed collectors, wind speed. At the moment, WATSUN TMY files and comma or blank delimited ASCII files are recognized by the program.

The WATSUN simulation interacts with the outside world through a series of files. A file is a collection of information, labeled and placed in a specific location. Files are used by the program to input and output information. One input file, called the *simulation data file*, is defined by the user. The simulation program then produces three output files: a listing file, an hourly data file, and a monthly data file.

The system is an assembly of collection devices, storage devices, and load devices that the user wants to assess. The system is defined in the simulation data file. The file is made up of data blocks that contain groups of related parameters. The simulation data file controls the simulation. The parameters in this file specify the simulation period, weather data, and output options. The simulation data file also contains information about the physical characteristics of the collector, the storage device(s), the heat exchangers, and the load.

The outputs of the program include a summary of the simulation as well as a file containing simulation results summed by month. The monthly energy balances of the system include solar gains, energy delivered, auxiliary energy, and parasitic gains from pumps. This file can be readily imported into spreadsheet programs for further analysis and plotting graphs. The program also gives the option to output data on an hourly or even sub-hourly basis, which gives the user the option to analyze the result of the simulation in greater detail and facilitates comparison with monitored data, when these are available.

Another use of the program is the simulation of active solar energy systems for which monitored data are available. This can be done either for validation purposes or to identify areas of improvement in the way the system works. For this purpose, WATSUN allows the user to enter monitored data from a separate file, called the *alternate input file*. Monitored climatic data, energy collected, and many other data can be read from the alternate input file and override the values normally used by the program. The program can also print out strategic variables (such as collector temperature or the temperature of water delivered to the load) on an hourly basis for comparison with monitored values.

The program was validated against the TRNSYS program using several test cases. Program-to-program comparisons with TRNSYS were very favorable; differences in predictions of yearly energy delivered were less than 1.2% in all configurations tested (Thevenard, 2008).

11.5.3 Polysun simulation program

The Polysun program provides dynamic annual simulations of solar thermal systems and helps to optimize them (Polysun, 2008). The program is user-friendly and the graphical user interface permits comfortable and clear input of all system parameters. All aspects of the simulation are based on physical models that work without empirical correlation terms. The basic systems that can be simulated include:

- Domestic hot water.
- Space heating.
- Swimming pools.
- Process heating.
- Cooling.

The input of the required data is very easy and done in a ready-made graphical environment like the one shown in Fig. 11.15. The input of the various parameters for each component of the system can be done by double-clicking on each component. Such templates are available for all types of systems that can be modeled with Polysun, and there is a template editor for users who want to create their own, tailored to the requirements of specific products. The modern and appealing graphical user interface makes it easy and fast to access the software. The convenient modular unit construction system allows the combination and parameterization of different system components through simple menu prompts.

Polysun is now in version 5.10 and makes the design of solar thermal systems simple and professional. An earlier version of Polysun was validated by Gantner (2000) and found to be accurate to within 5%−10%. The characteristics of the various components of the system can be obtained from ready-made catalogs, which include a large variety of market available components, but the user can add also the characteristics of a component, such as a collector, not included in the catalogs. The components included in catalogs comprise storage tanks, solar collectors, pipes, boilers, pumps, heat exchangers, heat pumps, buildings, swimming pools, PV modules and inverters, and many more. The program also features simple analysis and evaluation of simulations through graphics and reports. Worldwide meteorological data for 6300 locations are available, and new locations can be individually defined. There is also provision to specify the temperature of cold water and of the storage room. All features of the program are given in English, Spanish, Portuguese, French, Italian, Czech, and German.

Storage tanks can be specified with up to 10 connection ports, up to six internal heat exchangers, up to three internal heaters, and an internal tank and coil heat exchanger. The output of the program includes solar fraction, energy values (on the loop and component levels), temperatures, flow rate, and status for all components as a curve diagram, economic analysis, and summary of the most relevant values as a PDF file.

FIGURE 11.15

Polysun graphical environment.

The simulation algorithm provides dynamic simulation, including variable time steps, flow rate calculation including consideration of pressure drop, and material properties depending on temperature.

11.5.4 System advisor model

The System Advisor Model (SAM) is a performance and financial model designed to facilitate decision-making for persons involved in the renewable energy sector as engineers, developers, or researchers. Program SAM makes performance predictions and estimates the cost of energy for grid-connected power projects based on installation and operating costs and system design parameters that the users specify as inputs to the model. The program can be downloaded for free from the National Renewable Energy Laboratory (NREL) Website (https://sam.nrel.gov/). A short description of the program and its capabilities are given in Appendix 10. A full description can be obtained from the SAM help file, which is like a user manual and the program website. Many helpful videos are also available on the program website and YouTube.

Program SAM was originally called the "Solar Advisor Model" and was developed by the NREL in collaboration with Sandia National Laboratories in 2005. The first public version was released in 2007, enabling solar energy professionals to analyze photovoltaic systems and CSP parabolic trough systems in the same modeling platform using consistent financial assumptions. In 2010, the name of the program was changed to "System Advisor Model" to better represent the addition of nonsolar technologies. In relevant chapters of the book, some assignments are given.

11.6 Artificial intelligence in solar energy systems

Artificial intelligence (AI) systems are widely accepted as a technology that offers an alternative way to tackle complex and ill-defined problems. These systems can learn from examples, are fault tolerant in the sense that they are able to handle noisy and incomplete data, are able to deal with non-linear problems, and once trained, can perform prediction and generalization at high speed (Rumelhart et al., 1986). They have been used in diverse applications in control, robotics, pattern recognition, forecasting, medicine, power systems, manufacturing, optimization, signal processing, and the social/psychological sciences. They are particularly useful in system modeling, such as in implementing complex mappings and system identification. AI systems comprise areas such as artificial neural networks, genetic algorithms, fuzzy logic, and various hybrid systems, which combine two or more techniques.

Artificial neural networks (ANNs) mimic somewhat the learning process of a human brain. ANNs are collections of small, individually interconnected processing units. Information is passed between these units along interconnections. An incoming connection has two values associated with it: an input value and a weight. The output of the unit is a function of the summed value. ANNs, though implemented on computers, are not programmed to perform specific tasks. Instead, they are trained with respect to data sets until they learn the patterns used as inputs. Once they are trained, new patterns may be presented to them for prediction or classification. ANNs can automatically learn to recognize patterns in data from real systems or physical models, computer programs, or other sources. They can handle many inputs and produce answers in a form suitable for designers.

Genetic algorithms (GAs) are inspired by the way living organisms adapt to the harsh realities of life in a hostile world, i.e., by evolution and inheritance. In the process the algorithm imitates the evolution of a population by selecting only fit individuals for reproduction. Therefore, a genetic algorithm is an optimum search technique based on the concepts of natural selection and survival of the fittest. It works with a fixed-size population of possible solutions to a problem, called *individuals*, which evolve over time. A genetic algorithm utilizes three principal genetic operators: selection, crossover, and mutation.

Fuzzy logic is used mainly in control engineering. It is based on fuzzy logic reasoning, which employs linguistic rules in the form of if-then statements. Fuzzy logic and fuzzy control feature a relative simplification of a control methodology description. This allows the application of a "human language" to describe the problems and their fuzzy solutions. In many control applications, the model of the system is unknown or the input parameters are highly variable and unstable. In such cases, fuzzy controllers can be applied. These are more robust and cheaper than conventional PID controllers. It is also easier to understand and modify fuzzy controller rules, which not only use a human operator's strategy but are expressed in natural linguistic terms.

Hybrid systems combine more than one of these technologies, either as part of an integrated method of problem solution or to perform a particular task, followed by a second technique, which performs some other task. For example, neuro-fuzzy controllers use neural networks and fuzzy logic for the same task, i.e., to control a process; whereas in another hybrid system, a neural network may be used to derive some parameters and a genetic algorithm might be used subsequently to find an optimum solution to a problem.

For the estimation of the flow of energy and the performance of solar energy systems, analytic computer codes are often used. The algorithms employed are usually complicated, involving the solution of complex differential equations. These programs usually require a great deal of computer power and need a considerable amount of time to give accurate predictions. Instead of complex rules and mathematical routines, artificial intelligence systems are able to learn the key information patterns within a multi-dimensional information domain. Data from solar energy systems, being inherently noisy, are good candidate problems to be handled with artificial intelligence techniques.

The major objective of this section is to illustrate how artificial intelligence techniques might play an important role in the modeling and prediction of the performance and control of solar processes. The aim of this material is to enable the reader to understand how artificial intelligence systems can be set up. Various examples of solar energy systems are given as references so that interested readers can find more details. The results presented in these examples are testimony to the potential of artificial intelligence as a design tool in many areas of solar engineering.

Two techniques not included in this book, which are both parts of AI techniques and can handle problems involving energy and renewable energy systems, are machine learning which is a subset of AI, and deep learning, which is a subset of machine learning. According to the literature, machine learning can be used in the case of very complex problems for which traditional techniques cannot find a solution or are difficult to apply, and it is suitable for non-stable environments, as it can easily adapt to new data. In machine learning, there is no complex relationship between inputs and outputs, so it is suitable for problems that require many long lists of rules.

In some cases, deep learning is used to improve the results of machine learning, which implies that models based on deep learning are more accurate. Recently, deep learning has gained popularity due to its accuracy in classification and prediction when trained with an enormous amount of data, and thus, it is applied to complex problems that machine learning is not able to solve. It should be noted that when a small amount of data is available, machine learning algorithms are more suitable than deep learning ones. In addition, deep learning algorithms can use raw data directly and automatically extract the relevant features from them. Generally, there are no specific rules to select the most suitable algorithm for a given application, as each algorithm has its strengths and weaknesses, and the selection of the suitable algorithm depends mainly on factors such as the nature of the problem, the amount of available data, the algorithm implementation and complexity, and finally the expected accuracy and generalization capability. More details on these techniques can be seen in Mellit and Kalogirou (2021) for applications in the diagnosis and remote sensing of solar photovoltaic systems.

11.6.1 Artificial neural networks

The concept of ANN analysis was conceived more than 50 years ago, but only in the last 30 years has applications software been developed to handle practical problems. The purpose of this section is to present a brief overview of how neural networks operate and describe the basic features of some of the mostly used neural network architectures. A review of applications of ANNs in solar energy systems is also included.

ANNs are good for some tasks but lacking in some others. Specifically, they are good for tasks involving incomplete data sets, fuzzy or incomplete information, and highly complex and ill-defined problems, where humans usually decide on an intuitional basis. They can learn from examples and are able to deal with non-linear problems. Furthermore, they exhibit robustness and fault tolerance. The tasks that ANNs cannot handle effectively are those requiring high accuracy and precision, as in logic and arithmetic. ANNs have been applied successfully in a number of application areas. Some of the most important ones are (Kalogirou, 2003b):

- *Function approximation.* The mapping of a multiple input to a single output is established. Unlike most statistical techniques, this can be done with adaptive model-free estimation of parameters.
- *Pattern association and pattern recognition.* This is a problem of pattern classification. ANNs can be effectively used to solve difficult problems in this field—for instance, in sound, image, or video recognition. This task can be made even without an a priori definition of the pattern. In such cases, the network learns to identify totally new patterns.
- *Associative memories.* This is the problem of recalling a pattern when given only a subset clue. In such applications, the network structures used are usually complicated, composed of many interacting dynamical neurons.
- *Generation of new meaningful patterns.* This general field of application is relatively new. Some claims are made that suitable neuronal structures can exhibit rudimentary elements of creativity.

ANNs have been applied successfully in various fields of mathematics, engineering, medicine, economics, meteorology, psychology, neurology, and many others. Some of the most important ones are in pattern, sound, and speech recognition; the analysis of electromyographs and other medical signatures; the identification of military targets; and the identification of explosives in passenger suitcases. They have also being used in forecasting weather and market trends, the prediction of mineral exploration sites, prediction of electrical and thermal loads, adaptive and robotic control, and many others. Neural networks are also used for process control because they can build predictive models of the process from multi-dimensional data routinely collected from sensors.

Neural networks obviate the need to use complex, mathematically explicit formulas, computer models, and impractical and costly physical models. Some of the characteristics that support the success of ANNs and distinguish them from the conventional computational techniques are (Nannariello and Frike, 2001):

- The direct manner in which ANNs acquire information and knowledge about a given problem domain (learning interesting and possibly non-linear relationships) through the "training" phase.
- The ability to work with numerical or analog data that would be difficult to deal with by other means because of the form of the data or because there are many variables.
- Its "black-box" approach, in which the user requires no sophisticated mathematical knowledge.
- The compact form in which the acquired information and knowledge is stored within the trained network and the ease with which it can be accessed and used.
- The ability of solutions provided to be robust, even in the presence of "noise" in the input data.
- The high degree of accuracy reported when ANNs are used to generalize over a set of previously unseen data (not used in the "training" process) from the problem domain.

While neural networks can be used to solve complex problems, they do suffer from a number of shortcomings. The most important of them are:

- The need for data used to train neural networks to contain information that, ideally, is spread evenly throughout the entire range of the system.
- The limited theory to assist in the design of neural networks.

- The lack of guarantee of finding an acceptable solution to a problem.
- The limited opportunities to rationalize the solutions provided.

The following sections briefly explain how the artificial neuron is visualized from a biological one and the steps required to set up a neural network. Additionally, the characteristics of some of the most used neural network architectures are described.

Biological and artificial neurons

A biological neuron is shown in Fig. 11.16. In the brain, coded information flows (using electro-chemical media, the so-called neurotransmitters) from the synapses toward the axon. The axon of each neuron transmits information to a number of other neurons. The neuron receives information at the synapses from a large number of other neurons. It is estimated that each neuron may receive stimuli from as many as 10,000 other neurons. Groups of neurons are organized into subsystems, and the integration of these subsystems forms the brain. It is estimated that the human brain has around 100 billion interconnected neurons.

Figure 11.17 shows a highly simplified model of an artificial neuron, which may be used to stimulate some important aspects of the real biological neuron. An ANN is a group of interconnected artificial neurons, interacting with one another in a concerted manner. In such a system, excitation is applied to the input of the network. Following some suitable operation, it results in a desired output. At the synapses, there is an accumulation of some potential, which in the case of the artificial neurons, is modeled as a connection weight. These weights are continuously modified, based on suitable learning rules.

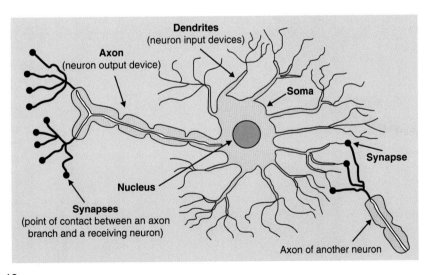

FIGURE 11.16

A schematic of a biological neuron.

FIGURE 11.17

A simplified model of an artificial neuron.

Artificial neural network principles

According to Haykin (1994), a neural network is a massively parallel distributed processor that has a natural propensity for storing experiential knowledge and making it available for use. It resembles the human brain in two respects:

- Knowledge is acquired by the network through a learning process.
- Interneuron connection strengths, known as *synaptic weights*, are used to store the knowledge.

ANN models may be used as an alternative method in engineering analysis and predictions. ANNs mimic somewhat the learning process of a human brain. They operate like a "black-box" model, requiring no detailed information about the system. Instead, they learn the relationship between the input parameters and the controlled and uncontrolled variables by studying previously recorded data, similar to the way a non-linear regression might perform. Another advantage of using ANNs is their ability to handle large, complex systems with many interrelated parameters. They seem to simply ignore excess input parameters that are of minimal significance and concentrate instead on the more important inputs.

A schematic diagram of a typical multi-layer, feed-forward neural network architecture is shown in Fig. 11.18. The network usually consists of an input layer, some hidden layers, and an output layer. In its simple form, each single neuron is connected to other neurons of a previous layer through adaptable synaptic weights. Knowledge is usually stored as a set of connection weights (presumably corresponding to synapse efficacy in biological neural systems). Training is the process of modifying the

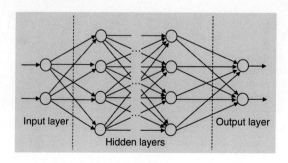

FIGURE 11.18

Schematic diagram of a multi-layer feed-forward neural network.

connection weights in some orderly fashion, using a suitable learning method. The network uses a learning mode, in which an input is presented to the network along with the desired output and the weights are adjusted so that the network attempts to produce the desired output. The weights after training contain meaningful information, whereas before training they are random and have no meaning.

Figure 11.19 illustrates how information is processed through a single node. The node receives the weighted activation of other nodes through its incoming connections. First, these are added up (summation). The result is then passed through an activation function; the outcome is the activation of the node. For each of the outgoing connections, this activation value is multiplied by the specific weight and transferred to the next node.

A training set is a group of matched input and output patterns used for training the network, usually by suitable adaptation of the synaptic weights. The outputs are the dependent variables that the network produces for the corresponding input. It is important that all the information the network needs to learn is supplied to the network as a data set. When each pattern is read, the network uses the input data to produce an output, which is then compared to the training pattern, i.e., the correct or desired output. If there is a difference, the connection weights (usually but not always) are altered in such a direction that the error is decreased. After the network has run through all the input patterns, if the error is still greater than the maximum desired tolerance, the ANN runs again through all the input patterns repeatedly until all the errors are within the required tolerance. When the training reaches a satisfactory level, the network holds the weights constant and the trained network can be used to make decisions, identify patterns, or define associations in new input data sets not used to train it.

By *learning*, we mean that the system adapts (usually by changing suitable controllable parameters) in a specified manner so that some parts of the system suggest a meaningful behavior, projected as output. The controllable parameters have different names, such as *synaptic weights*, *synaptic efficacies*, *free parameters*, and others.

The classical view of learning is well interpreted and documented in approximation theories. In these, learning may be interpreted as finding a suitable hypersurface that fits known input—output data points in such a manner that the mapping is acceptably accurate. Such a mapping is usually accomplished by employing simple non-linear functions that are used to compose the required function (Pogio and Girosi, 1990).

A more general approach to learning is adopted by Haykin (1994), in which learning is a process by which the free parameters of a neural network are adapted through a continuing process of simulation by the environment in which the network is embedded. The type of learning is determined by the manner in which the parameter changes take place.

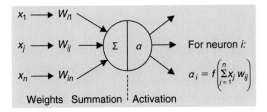

Weights Summation Activation

FIGURE 11.19

Information processing in a neural network unit.

Generally, learning is achieved through any change in any characteristic of a network so that meaningful results are achieved, meaning that a desired objective is met with a satisfactory degree of success. Thus, learning could be achieved through synaptic weight modification, network structure modifications, appropriate choice of activation functions and other ways.

The objective is usually quantified by a suitable criterion or cost function. It is usually a process of minimizing an error function or maximizing a benefit function. In this respect, learning resembles optimization. That is why a genetic algorithm, which is an optimum search technique (see Section 11.6.2), can also be employed to train ANNs.

Several algorithms are commonly used to achieve the minimum error in the shortest time. There are also many alternative forms of neural networking systems and, indeed, many different ways in which they may be applied to a given problem. The suitability of an appropriate paradigm and strategy for an application depends very much on the type of problem to be solved.

The most popular learning algorithms are back-propagation (BP) and its variants (Barr and Feigenbaum, 1981; Werbos, 1974). The *back-propagation algorithm* is one of the most powerful learning algorithms in neural networks. Back-propagation training is a gradient descent algorithm. It tries to improve the performance of the neural network by reducing the total error by changing the weights along its gradient. The error is expressed by the root mean square value (RMS), which can be calculated by:

$$E = \frac{1}{2}\left[\sum_p \sum_i \left|t_{pi} - o_{pi}\right|^2\right]^{1/2} \tag{11.125}$$

where E is the RMS error, t is the network output (target), and o is the desired output vectors over all patterns, p. An error of zero would indicate that all the output patterns computed by the ANN perfectly match the expected values and the network is well trained. In brief, back-propagation training is performed by initially assigning random values to the weight terms (w_{ij})[1] in all nodes. Each time a training pattern is presented to the ANN, the activation for each node, α_{pi}, is computed. After the output of the layer is computed the error term, δ_{pi}, for each node is computed backward through the network. This error term is the product of the error function, E, and the derivative of the activation function and, hence, is a measure of the change in the network output produced by an incremental change in the node weight values. For the output layer nodes and the case of the logistic-sigmoid activation, the error term is computed as:

$$\delta_{pi} = \left(t_{pi} - \alpha_{pi}\right)\alpha_{pi}\left(1 - \alpha_{pi}\right) \tag{11.126}$$

For a node in a hidden layer,

$$\delta_{pi} = \alpha_{pi}\left(1 - \alpha_{pi}\right)\sum_k \delta_{pk}w_{kj} \tag{11.127}$$

In this expression, the k subscript indicates a summation over all nodes in the downstream layer (the layer in the direction of the output layer). The j subscript indicates the weight position in each

---

[1]The j subscript refers to a summation of all nodes in the previous layer of nodes, and the i subscript refers to the node position in the present layer.

node. Finally, the δ and α terms for each node are used to compute an incremental change to each weight term via:

$$\Delta w_{ij} = \varepsilon \left(\delta_{pi} \alpha_{pj} \right) + m w_{ij}(\text{old}) \tag{11.128}$$

The term ε, referred to as the *learning rate*, determines the size of the weight adjustments during each training iteration. The term m is called the *momentum factor*. It is applied to the weight change used in the previous training iteration, $w_{ij}(\text{old})$. Both of these constant terms are specified at the start of the training cycle and determine the speed and stability of the network. The training of all patterns of a training data set is called an *epoch*.

Network parameter selection

Though most scholars are concerned with the techniques to define ANN architecture, practitioners want to apply the ANN architecture to the model and obtain quick results. The term *neural network architecture* refers to the arrangement of neurons into layers and the connection patterns between layers, activation functions, and learning methods. The neural network model and the architecture of a neural network determine how a network transforms its input into an output. This transformation is, in fact, a computation. Often, the success depends on a clear understanding of the problem, regardless of the network architecture. However, in determining which neural network architecture provides the best prediction, it is necessary to build a good model. It is essential to be able to identify the most important variables in a process and generate best-fit models. How to identify and define the best model is very controversial.

Despite the differences between traditional approaches and neural networks, both methods require preparing the model. The classical approach is based on the precise definition of the problem domain as well as the identification of a mathematical function or functions to describe it. It is, however, very difficult to identify an accurate mathematical function when the system is non-linear and parameters vary with time due to several factors. The control program often lacks the capability to adapt to the parameter changes. Neural networks are used to learn the behavior of the system and subsequently to simulate and predict its behavior. In defining the neural network model, first the process and the process control constraints have to be understood and identified. Then, the model is defined and validated.

When using a neural network for prediction, the following steps are crucial. First, a neural network needs to be built to model the behavior of the process, and the values of the output are predicted based on the model. Second, based on the neural network model obtained in the first phase, the output of the model is simulated using different scenarios. Third, the control variables are modified to control and optimize the output.

When building the neural network model, the process has to be identified with respect to the input and output variables that characterize it. The inputs include measurements of the physical dimensions, measurements of the variables specific to the environment or equipment, and controlled variables modified by the operator. Variables that have no effect on the variation of the measured output are discarded. These are estimated by the contribution factors of the various input parameters. These factors indicate the contribution of each input parameter to the learning of the neural network and are usually estimated by the network, depending on the software employed.

The selection of training data plays a vital role in the performance and convergence of the neural network model. An analysis of historical data for identification of variables that are important to the process is important. Plotting graphs to check whether the various variables reflect what is known about the process from operating experience and for discovery of errors in data is very helpful.

All input and output values are usually scaled individually such that the overall variance in the data set is minimized. Therefore, the input and output values are normalized. This is necessary because it leads to faster learning. The scaling used is either in the range -1 to 1 or in the range $0-1$, depending on the type of data and the activation function used.

The basic operation that has to be followed to successfully handle a problem with ANNs is to select the appropriate architecture and the suitable learning rate, momentum, number of neurons in each hidden layer, and the activation function. The procedure for finding the best architecture and the other network parameters is laborious and time-consuming, but as experience is gathered, some parameters can be predicted easily, tremendously shortening the time required.

The first step is to collect the required data and prepare them in a spreadsheet format with various columns representing the input and output parameters. If a large number of sequences or patterns are available in the input data file, to avoid a long training time, a smaller training file may be created, containing as much as possible representative samples of the whole problem domain, in order to select the required parameters and to use the complete data set for the final training.

Three types of data files are required: a training data file, a test data file, and a validation data file. The former and the last should contain representative samples of all the cases the network is required to handle, whereas the test file may contain about 10% of the cases contained in the training file.

During training, the network is tested against the test file to determine accuracy, and training should be stopped when the mean average error remains unchanged for a number of epochs. This is done in order to avoid over-training, in which case, the network learns the training patterns perfectly but is unable to make predictions when an unknown training set is presented to it.

In back-propagation networks, the number of hidden neurons determines how well a problem can be learned. If too many are used, the network will tend to memorize the problem and not generalize well later. If too few are used, the network will generalize well but may not have enough "power" to learn the patterns well. Getting the right number of hidden neurons is a matter of trial and error, since there is no science to it. In general, the number of hidden neurons (N) may be estimated by applying the following empirical formula (Ward Systems Group, Inc., 1996):

$$N = \frac{I + O}{2} + \sqrt{P_i} \tag{11.129}$$

where

I = number of input parameters;
O = number of output parameters; and
P_i = number of training patterns available.

The most important parameter to select in a neural network is the type of architecture. A number of architectures can be used in solar engineering problems. A short description of the most important ones is given in this section: back-propagation (BP), general regression neural networks (GRNNs), and the group method of data handling (GMDH). These are described briefly in the next sections.

Back-propagation architecture

Architectures in the back-propagation category include standard networks, recurrent, feed-forward with multiple hidden slabs, and jump connection networks. Back-propagation networks are known for their ability to generalize well on a wide variety of problems. They are a supervised type of network, i.e., trained with both inputs and outputs. Back-propagation networks are used in a large number of working applications, since they tend to generalize well.

The first category of neural network architectures is the one where each layer is connected to the immediately previous layer (see Fig. 11.18). Generally, three layers (input, hidden, and output) are sufficient for the majority of problems to be handled. A three-layer back-propagation network with standard connections is suitable for almost all problems. One, two, or three hidden layers can be used, however, depending on the problem characteristics. The use of more than five layers in total generally offers no benefits and should be avoided.

The next category of architecture is the recurrent network with dampened feedback from either the input, hidden, or output layer. It holds the contents of one of the layers as it existed when the previous pattern was trained. In this way, the network sees the previous knowledge it had about previous inputs. This extra slab is sometimes called the network's *long-term memory*. The long-term memory re-members the input, output, or hidden layer that contains features detected in the raw data of previous patterns. Recurrent neural networks are particularly suitable for prediction of sequences, so they are excellent for time-series data. A back-propagation network with standard connections, as just described, responds to a given input pattern with exactly the same output pattern every time the input pattern is presented. A recurrent network may respond to the same input pattern differently at different times, depending on the patterns that had been presented as inputs just previously. Thus, the sequence of the patterns is as important as the input pattern itself. Recurrent networks are trained as the standard back-propagation networks, except that patterns must always be presented in the same order. The difference in structure is that an extra slab in the input layer is connected to the hidden layer, just like the other input slab. This extra slab holds the contents of one of the layers (input, output, or hidden) as it existed when the previous pattern was trained.

The third category is the feed-forward network with multiple hidden slabs. These network architectures are very powerful in detecting different features of the input vectors when different activation functions are given to the hidden slabs. This architecture has been used in a number of engineering problems for modeling and prediction with very good results (see the later section, "ANN Applications in Solar Energy Systems"). This is a feed-forward architecture with three hidden slabs, as shown in Fig. 11.20. The information processing at each node is performed by combining all input numerical information from upstream nodes in a weighted average of the form:

$$\beta_i = \sum_j w_{ij}\alpha_{pj} + b_1 \tag{11.130}$$

where

$\alpha_{(pi)}$ = activation for each node;
b_1 = a constant term referred to as the bias.

The final nodal output is computed via the activation function. This architecture has different activation functions in each slab. Referring to Fig. 11.20, the input slab activation function is linear, i.e., $\alpha_{(pi)} = \beta_i$ (where β_i is the weighted average obtained by combining all input numerical information from upstream nodes), while the activations used in the other slabs are as follows.

Gaussian for slab 2,

$$\alpha_{(pi)} = e^{-\beta_i^2} \tag{11.131}$$

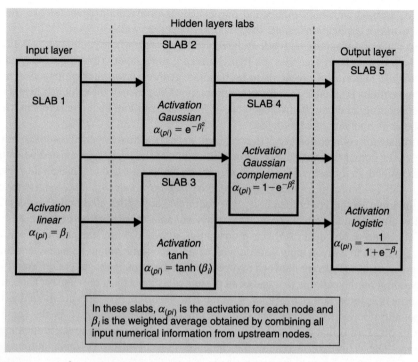

FIGURE 11.20

Feed-forward architecture with multiple hidden slabs.

Tanh for slab 3,

$$\alpha_{(pi)} = \tanh(\beta_i) \tag{11.132}$$

Gaussian complement for slab 4,

$$\alpha_{(pi)} = 1 - e^{-\beta_i^2} \tag{11.133}$$

Logistic for the output slab,

$$\alpha_{(pi)} = \frac{1}{1 + e^{-\beta_i}} \tag{11.134}$$

Different activation functions are applied to hidden layer slabs to detect different features in a pattern processed through a network. The number of hidden neurons in the hidden layers may also be calculated with Eq. (11.129). However, an increased number of hidden neurons may be used to get more "degrees of freedom" and allow the network to store more complex patterns. This is usually done when the input data are highly non-linear. It is recommended in this architecture to use Gaussian function on one hidden slab to detect features in the middle range of the data and Gaussian complement in another hidden slab to detect features from the upper and lower extremes of the data. Combining the two feature sets in the output layer may lead to a better prediction.

General regression neural network architecture

Another type of architecture is general regression neural networks (GRNNs), which are known for their ability to train quickly on sparse data sets. In numerous tests, it was found that a GRNN responds much better than back-propagation to many types of problems, although this is not a rule. It is especially useful for continuous function approximation. A GRNN can have multi-dimensional input, and it will fit multi-dimensional surfaces through data. GRNNs work by measuring how far a given sample pattern is from patterns in the training set in N-dimensional space, where N is the number of inputs in the problem. The Euclidean distance is usually adopted.

A GRNN is a four-layer feed-forward neural network based on the non-linear regression theory, consisting of the input layer, the pattern layer, the summation layer, and the output layer (see Fig. 11.21). There are no training parameters, such as learning rate and momentum, as in back-propagation networks, but a smoothing factor is applied after the network is trained. The smoothing factor determines how tightly the network matches its predictions to the data in the training patterns. Although the neurons in the first three layers are fully connected, each output neuron is connected only to some processing units in the summation layer. The summation layer has two types of processing units: summation units and a single division unit. The number of summation units is always the same as the number of the GRNN output units. The division unit only sums the weighted activations of the pattern units of the hidden layer, without using any activation function.

Each GRNN output unit is connected only to its corresponding summation unit and the division unit (there are no weights in these connections). The function of the output units consists of a simple division of the signal coming from the summation unit by the signal coming from the division unit. The summation and output layers together basically perform a normalization of the output vector, making a GRNN much less sensitive to the proper choice of the number of pattern units. More details on GRNNs can be found in Tsoukalas and Uhrig (1997) and Ripley (1996).

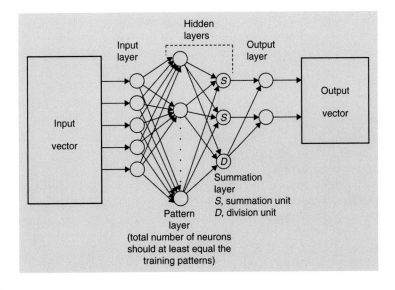

FIGURE 11.21

General regression neural network architecture.

For GRNN networks, the number of neurons in the hidden pattern layer is usually equal to the number of patterns in the training set because the hidden layer consists of one neuron for each pattern in the training set. This number can be made larger if one wants to add more patterns, but it cannot be made smaller.

The training of the GRNN is quite different from the training used in other neural networks. It is completed after presentation of each input−output vector pair from the training data set to the GRNN input layer only once.

The GRNN may be trained using a genetic algorithm (see Section 11.6.2). The genetic algorithm is used to find the appropriate individual smoothing factors for each input as well as an overall smoothing factor. Genetic algorithms use a "fitness" measure to determine which individuals in the population survive and reproduce. Therefore, survival of the fittest causes good solutions to progress. A genetic algorithm works by selective breeding of a population of "individuals", each of which could be a potential solution to the problem. In this case, a potential solution is a set of smoothing factors, and the genetic algorithm seeks to breed an individual that minimizes the mean squared error of the test set, which can be calculated by:

$$E = \frac{1}{p} \sum_p \left(t_p - o_p \right)^2 \tag{11.135}$$

where

E = mean squared error;
t = network output (target); and
o = desired output vectors over all patterns (p) of the test set.

The larger the breeding pool size, the greater is its potential to produce a better individual. However, the networks produced by every individual must be applied to the test set on every reproductive cycle, so larger breeding pools take longer time. After testing all the individuals in the pool, a new "generation" of individuals is produced for testing. Unlike the back-propagation algorithm, which propagates the error through the network many times, seeking a lower mean squared error between the network's output and the actual output or answer, GRNN training patterns are presented to the network only once.

The input smoothing factor is an adjustment used to modify the overall smoothing to provide a new value for each input. At the end of training, the individual smoothing factors may be used as a sensitivity analysis tool; the larger the factor for a given input, the more important that input is to the model, at least as far as the test set is concerned. Inputs with low smoothing factors are candidates for removal for a later trial.

Individual smoothing factors are unique to each network. The numbers are relative to each other within a given network, and they cannot be used to compare inputs from different networks.

If the number of input, output, or hidden neurons is changed, however, the network must be retrained. This may occur when more training patterns are added, because GRNN networks require one hidden neuron for each training pattern.

Group method of data handling neural network architecture

One type of neural network that is very suitable for modeling is the group method of data handling (GMDH) neural network. The GMDH technique was invented by A.G. Ivakhenko, from the Institute of Cybernetics, Ukrainian Academy of Sciences (Ivakhenko, 1968, 1971), but enhanced by others (Farlow, 1984). This technique is also known as *polynomial networks*. Ivakhenko developed the

GMDH technique to build more accurate predictive models of fish populations in rivers and oceans. The GMDH technique worked well for modeling fisheries and many other modeling applications (Hecht-Nielsen, 1991). The GMDH is a feature-based mapping network.

The GMDH technique works by building successive layers, with links that are simple polynomial terms. These polynomial terms are created by using linear and non-linear regression. The initial layer is simply the input layer. The first layer created is made by computing regressions of the input variables, from which the best ones are chosen. The second layer is created by computing regressions of the values in the first layer, along with the input variables. Only the best, called *survivors*, are chosen by the algorithm. This process continues until the network stops getting better, according to a pre-specified selection criterion. More details on the GMDH technique can be found in the book by Hecht-Nielsen (1991).

The resulting network can be represented as a complex polynomial description of the model in the form of a mathematical equation. The complexity of the resulting polynomial depends on the variability of the training data. In some respects, GMDH is very much like using regression analysis but far more powerful. The GMDH network can build very complex models while avoiding overfitting problems. Additionally, an advantage of the GMDH technique is that it recognizes the best variables as it trains and, for problems with many variables, the ones with low contribution can be discarded.

The central idea behind the GMDH technique is that it is trying to build a function (called a *polynomial model*) that behaves as closely as possible to the way the predicted and actual values of the output would. For many end users, it may be more convenient to have a model that is able to make predictions using polynomial formulas that are widely understood than a normal neural network, which operates like a "black-box" model. The most common approach to solving such models is to use regression analysis. The first step is to decide the type of polynomial that regression should find. For example, a good idea is to choose, as terms of the polynomial, powers of input variables along with their covariants and trivariants, such as:

$$\left\{x_1, x_2, x_3, \ldots, x_1^2, x_2^2, x_3^2, \ldots, x_1 x_2, x_1 x_3, \ldots, x_{n-1} x_n, x_1 x_2 x_3, \ldots\right\} \tag{11.136}$$

The next step is to construct a linear combination of all the polynomial terms with variable coefficients. The algorithm determines the values of these coefficients by minimizing the squared sum of differences between sample outputs and model predictions, over all samples.

The main problem when utilizing regression is how to choose the set of polynomial terms correctly. In addition, decisions need to be made on the degree of the polynomial. For example, decisions have to be made on how complex the terms should be or whether the model should evaluate terms such as x^{10}, or maybe limit consideration to terms such as x^4 and lower. The GMDH technique works better than regression by answering these questions before trying all possible combinations.

The decision about the quality of each model must be made using some numeric criterion. The simplest criterion (a form of which is also used in linear regression analysis) is the sum, over all samples, of the squared differences between the actual output (y_a) and the model's prediction (y_p) divided by the sum of the squared actual output. This is called the *normalized mean squared error* (NMSE). In equation form,

$$\text{NMSE} = \frac{\sum_{i=1}^{N} \left(y_a - y_p\right)^2}{\sum_{i=1}^{N} y_a^2} \tag{11.137}$$

However, if only the NMSE is used on real data, the NMSE value gets smaller and smaller as long as extra terms are added to the model. This is because the more complex the model, the more exact it is. This is always true if NMSE is used alone, which determines the quality of the model by evaluating the same information already used to build the model. This results in an "over-complex" model or model overfit, which means the model does not generalize well because it pays too much attention to noise in the training data. This is similar to over-training other neural networks.

To avoid this danger, a more powerful criterion is needed, based on information other than that which was used to build the evaluated model. There are several ways to define such criteria. For example, the squared sum of differences between the known output and model prediction over some other set of experimental data (a test set) may be used. Another way to avoid overfitting is to introduce a penalty for model complexity. This is called the *predicted squared error criterion.*

Theoretical considerations show that increasing model complexity should be stopped when the selection criterion reaches a minimum value. This minimum value is a measure of model reliability.

The method of searching for the best model based on testing all possible models is usually called the *combinatorial GMDH algorithm.* To reduce computation time, the number of polynomial terms used to build the models to be evaluated should be reduced. To do so, a one-stage procedure of model selection should be changed to a multi-layer procedure. In this, the first two input variables are initially taken and combined into a simple set of polynomial terms. For example, if the first two input variables are x_1 and x_2, the set of polynomial terms would be $\{c, x_1, x_2, x_1 \times x_2\}$, where (c) represents the constant term. Subsequently, all possible models made from these terms are checked and the best is chosen; any one of the evaluated models is a candidate for survival.

Then, another pair of input variables is taken and the operation is repeated, resulting in another candidate for survival, with its own value of the evaluation criterion. By repeating the same procedure for each possible pair of n input variables, $n(n-1)/2$ candidates for survival are generated, each with its own value of the evaluation criterion.

Subsequently, these values are compared, and several candidates for survival that give the best approximation of the output variable are chosen. Usually a predefined number of the best candidates are selected for survival and are stored in the first layer of the network and preserved for the next layer. The candidates selected are called *survivors.*

The layer of survivors is used for inputs in building the next layer in the network. The original network inputs used in the first layer may also be chosen as inputs to the new layer. Therefore, the next layer is built with polynomials of this broadened set of inputs. It should be noted that, since some inputs are already polynomials, the next layer may contain very complex polynomials.

The layer building of the GMDH procedure continues as long as the evaluation criteria continue to diminish. Each time a new layer is built the GMDH algorithm checks whether the new evaluation criterion is lower than the previous one and, if this is so, continues training; otherwise, it stops training.

ANN applications in solar energy systems

Artificial neural networks have been used by the author in the field of solar energy, for modeling the heat-up response of a solar steam generating plant (Kalogirou et al., 1998), the estimation of a parabolic trough collector intercept factor (Kalogirou et al., 1996), the estimation of a parabolic trough collector local concentration ratio (Kalogirou, 1996a), the design of a solar steam-generation system (Kalogirou, 1996b), the performance prediction of a thermosiphon solar water heater (Kalogirou et al., 1999a), modeling solar domestic water-heating systems (Kalogirou et al., 1999b), the long-term performance prediction of forced circulation solar domestic water-heating systems (Kalogirou, 2000), and the thermosiphon solar domestic water-heating system's long-term performance prediction (Kalogirou and Panteliou, 2000). A review of these models, together with other applications in the field of renewable energy, is given in an article by Kalogirou (2001). In most of those models, the multiple hidden layer architecture shown in Fig. 11.20 was used. The errors reported are well within acceptable limits, which clearly suggests that ANNs can be used for modeling and prediction in other fields of solar energy engineering. What is required is to have a set of data (preferably experimental) representing the past history of a system so that a suitable neural network can be trained to learn the dependence of expected output on the input parameters.

11.6.2 Genetic algorithms

The genetic algorithm (GA) is a model of machine learning that derives its behavior from a representation of the processes of evolution in nature. This is done by the creation, within a machine or computer, of a population of individuals represented by chromosomes. Essentially, these are a set of character strings that are analogous to the chromosomes in the DNA of human beings. The individuals in the population then go through a process of evolution.

It should be noted that evolution as occurring in nature or elsewhere is not a purposive or directed process, i.e., no evidence supports the assertion that the goal of evolution is to produce humankind. Indeed, the processes of nature seem to end with different individuals competing for resources in the environment. Some are better than others; those that are better are more likely to survive and propagate their genetic material.

In nature, the encoding for the genetic information is done in a way that admits asexual reproduction and typically results in offspring that are genetically identical to the parent. Sexual reproduction allows the creation of genetically radically different offspring that are still of the same general species.

In an oversimplified consideration, at the molecular level, what happens is that a pair of chromosomes bump into one another, exchange chunks of genetic information, and drift apart. This is the recombination operation, which in GAs is generally referred to as *crossover* because of the way that genetic material crosses over from one chromosome to another.

The crossover operation happens in an environment where the selection of who gets to mate is a function of the fitness of the individual, i.e., how good the individual is at competing in its environment. Some GAs use a simple function of the fitness measure to select individuals (probabilistically) to undergo genetic operations, such as crossover or asexual reproduction, i.e., the propagation of genetic material remains unaltered. This is a fitness proportionate selection. Other implementations use a model in which certain randomly selected individuals in a subgroup compete and the fittest is selected. This is called *tournament selection*. The two processes that most contribute to evolution are crossover and fitness-based selection/reproduction. Mutation also plays a role in this process.

GAs are used in a number of application areas. An example of this would be multi-dimensional optimization problems, in which the character string of the chromosome can be used to encode the values for the different parameters being optimized.

Therefore, in practice, this genetic model of computation can be implemented by having arrays of bits or characters to represent the chromosomes. Simple bit manipulation operations allow the implementation of crossover, mutation, and other operations.

When the GA is executed, it is usually done in a manner that involves the following cycle. Evaluate the fitness of all of the individuals in the population. Create a new population by performing operations such as crossover, fitness-proportionate reproduction, and mutation on the individuals whose fitness has just been measured. Discard the old population and iterate using the new population. One iteration of this loop is referred to as a *generation*. The structure of the standard GA is shown in Fig. 11.22 (Zalzala and Fleming, 1997).

With reference to Fig. 11.22, in each generation, individuals are selected for reproduction according to their performance with respect to the fitness function. In essence, selection gives a higher chance of survival to better individuals. Subsequently, genetic operations are applied to form new and possibly better offspring. The algorithm is terminated either after a certain number of generations or when the optimal solution has been found. More details on genetic algorithms can be found in Goldberg (1989), Davis (1991), and Michalewicz (1996).

The first generation (generation 0) of this process operates on a population of randomly generated individuals. From there on, the genetic operations, in concert with the fitness measure, operate to improve the population.

During each step in the reproduction process, the individuals in the current generation are evaluated by a fitness function value, which is a measure of how well the individual solves the problem. Then, each individual is reproduced in proportion to its fitness. The higher the fitness, the higher is its chance to participate in mating (crossover) and produce an offspring. A small number of newborn offspring undergo the action of the mutation operator. After many generations, only those individuals who have the best genetics (from the point of view of the fitness function) survive. The individuals that emerge from this "survival-of-the-fittest" process are the ones that represent the optimal solution to the problem specified by the fitness function and the constraints.

Genetic algorithms are suitable for finding the optimum solution in problems where a fitness function is present. Genetic algorithms use a "fitness" measure to determine which individuals in the population survive and reproduce. Thus, survival of the fittest causes good solutions to progress. A genetic algorithm works by selective breeding of a population of "individuals", each of which could be a potential solution to the problem. The genetic algorithm seeks to breed an individual that either maximizes, minimizes, or is focused on a particular solution to a problem.

Genetic algorithm

```
Begin (1)
    t = 0 [start with an initial time]
    Initialize population, P(t) [initialize a usually random population of individuals]
    Evaluate fitness of population P(t) [evaluate fitness of all individuals in
                                        population]
    While (Generations < Total number) do begin (2)
        t = t + 1 [increase the time counter]
        Select Population P(t) out of Population P(t − 1) [select sub population for
                                                          offspring production]
        Apply crossover on population P(t)
        Apply mutation on population P(t)
        Evaluate fitness of population P(t) [evaluate new fitness of population]
    end (2)
end (1)
```

FIGURE 11.22

The structure of a standard genetic algorithm.

The larger the breeding pool size, the greater the potential for producing a better individual. However, since the fitness value produced by every individual must be compared with all other fitness values of all other individuals on every reproductive cycle, larger breeding pools take longer time. After testing all the individuals in the pool, a new "generation" of individuals is produced for testing.

During the setting up of the genetic algorithm, the user has to specify the adjustable chromosomes, i.e., the parameters that would be modified during evolution to obtain the maximum value of the fitness function. Additionally, the user has to specify the ranges of these values, called *constraints*.

A genetic algorithm is not gradient-based and uses an implicitly parallel sampling of the solutions space. The population approach and multiple sampling means that it is less subject to becoming trapped in local minima than traditional direct approaches and can navigate a large solution space with a highly efficient number of samples. Although not guaranteed to provide the globally optimum solution, GAs have been shown to be highly efficient at reaching a very near optimum solution in a computationally efficient manner.

The genetic algorithm is usually stopped after best fitness remains unchanged for a number of generations or when the optimum solution is reached.

An example of using GAs in this book is given in Chapter 3, Example 3.2, where the two glass temperatures are varied to get the same Q_t/A_c value from Eqs. (3.15), (3.17), and (3.22). In this case, the values of T_{g1} and T_{g2} are the adjustable chromosomes and the fitness function is the sum of the absolute difference between each Q_t/A_c value from the mean Q_t/A_c value (obtained from the aforementioned three equations). In this problem, the fitness function should be 0, so all Q_t/A_c values are equal, which is the objective. Other applications of GAs in solar energy are given in the next section.

GA applications in solar energy systems

Genetic algorithms were used by the author in a number of optimization problems: the optimal design of flat-plate solar collectors (Kalogirou, 2003c), predicting the optimal sizing coefficient of PV supply systems (Mellit and Kalogirou, 2006a), and the optimum selection of the fenestration openings in buildings (Kalogirou, 2007). They have also been used to optimize solar energy systems, in combination with TRNSYS and ANNs (Kalogirou, 2004a). In this, the system is modeled using the TRNSYS computer program and the climatic conditions of Cyprus. An ANN was trained, using the results of a small number of TRNSYS simulations, to learn the correlation of collector area and storage tank size on the auxiliary energy required by the system, from which the life cycle savings can be estimated. Subsequently, a genetic algorithm was employed to estimate the optimum size of these two parameters, for maximizing the life-cycle savings; thus, the design time is reduced substantially. As an example, the optimization of an industrial process heat system employing flat-plate collectors is presented (Kalogirou, 2004a). The optimum solutions obtained from the present methodology give increased life cycle savings of 4.9% and 3.1% when subsidized and non-subsidized fuel prices are used, respectively, as compared to solutions obtained by the traditional trial and error method. The present method greatly reduces the time required by design engineers to find the optimum solution and, in many cases, reaches a solution that could not be easily obtained from simple modeling programs or by trial and error, which in most cases depends on the intuition of the engineer.

GENOPT and TRNOPT programs

When simulation models are used to simulate and design a system, it is usually not easy to determine the parameter values that lead to optimal system performance. This is sometimes due to time constraints, since it is time-consuming for a user to change the input values, run the simulation, interpret the new results, and guess how to change the input for the next trial. Sometimes time is not a problem, but due to the complexity of the system analyzed, the user is just not capable of understanding the non-linear interactions of the various parameters. However, using genetic algorithms, it is possible to do automatic single- or multi-parameter optimization with search techniques that require only little effort. GenOpt is a generic optimization program developed for such system optimization. It was designed by the Lawrence Berkeley National Laboratory and is available free of charge (GenOpt, 2011). GenOpt is used for finding the values of user-selected design parameters that minimize a so-called objective function, such as annual energy use, peak electrical demand, or predicted percentage of dissatisfied people (PPD value), leading to best operation of a given system. The objective function is calculated by an external simulation program, such as TRNSYS (Wetter, 2001). GenOpt can also identify unknown parameters in a data fitting process. GenOpt allows coupling of any simulation program (e.g., TRNSYS) with text-based input—output (I/O) by simply modifying a configuration file, without requiring code modification. Further, it has an open interface for easily adding custom minimization algorithms to its library. This allows the use of GenOpt as an environment for the development of optimization algorithms (Wetter, 2004).

Another tool that can be used is TRNopt, which is an interface program that allows TRNSYS users to quickly and easily utilize the GenOpt optimization tool to optimize combinations of continuous and discrete variables. GenOpt actually controls the simulation and the user sets up the optimization beforehand, using the TRNopt pre-processor program.

11.6.3 Fuzzy logic

Fuzzy logic is a logical system, which is an extension of multi-valued logic. Additionally, fuzzy logic is almost synonymous with the theory of fuzzy sets, a theory that relates to classes of objects without sharp boundaries in which membership is a matter of degree. Fuzzy logic is all about the relative importance of precision, i.e., how important it is to be exactly right when a rough answer will work. Fuzzy inference systems have been successfully applied in fields such as automatic control, data classification, decision analysis, expert systems, and computer vision. Fuzzy logic is a convenient way to map an input space to an output space—as for example, according to hot-water temperature required, to adjust the valve to the right setting, or according to the steam outlet temperature required, to adjust the fuel flow in a boiler. From these two examples, it can be understood that fuzzy logic mainly has to do with the design of controllers.

Conventional control is based on the derivation of a mathematical model of the plant from which a mathematical model of a controller can be obtained. When a mathematical model cannot be created, there is no way to develop a controller through classical control. Other limitations of conventional control are (Reznik, 1997):

- *Plant non-linearity.* Non-linear models are computationally intensive and have complex stability problems.
- *Plant uncertainty.* Accurate models cannot be created due to uncertainty and lack of perfect knowledge.
- *Multi-variables, multi-loops, and environmental constraints.* Multi-variable and multi-loop systems have complex constraints and dependencies.
- *Uncertainty in measurements due to noise.*

- *Temporal behavior.* Plants, controllers, environments, and their constraints vary with time. Additionally, time delays are difficult to model.

The advantages of fuzzy control are (Reznik, 1997):

- Fuzzy controllers are more robust than PID controllers, as they can cover a much wider range of operating conditions and operate with noise and disturbances of different natures.
- Their development is cheaper than that of a model-based or other controller to do the same thing.
- They are customizable, since it is easier to understand and modify their rules, which are expressed in natural linguistic terms.
- It is easy to learn how these controllers operate and how to design and apply them in an application.
- They can model non-linear functions of arbitrary complexity.
- They can be built on top of the experience of experts.
- They can be blended with conventional control techniques.

Fuzzy control should not be used when conventional control theory yields a satisfactory result and an adequate and solvable mathematical model already exists or can easily be created.

Fuzzy logic was initially developed in 1965 in the United States by Professor Lofti Zadeh (1973). In fact, Zadeh's theory not only offered a theoretical basis for fuzzy control but established a bridge connecting artificial intelligence to control engineering. Fuzzy logic has emerged as a tool for controlling industrial processes, as well as household and entertainment electronics, diagnosis systems, and other expert systems. Fuzzy logic is basically a multi-valued logic that allows intermediate values to be defined between conventional evaluations such as yes—no, true—false, black—white, large—small, etc. Notions such as "rather warm" or "pretty cold" can be formulated mathematically and processed in computers. Thus, an attempt is made to apply a more humanlike way of thinking to the programming of computers.

A fuzzy-controller design process contains the same steps as any other design process. One needs initially to choose the structure and parameters of a fuzzy controller, test a model or the controller itself, and change the structure and/or parameters based on the test results (Reznik, 1997). A basic requirement for implementing fuzzy control is the availability of a control expert who provides the necessary knowledge for the control problem (Nie and Linkens, 1995). More details on fuzzy control and practical applications can be found in the works by Zadeh (1973), Mamdani (1974, 1977), and Sugeno (1985).

The linguistic description of the dynamic characteristics of a controlled process can be interpreted as a fuzzy model of the process. In addition to the knowledge of a human expert, a set of fuzzy control rules can be derived by using experimental knowledge. A fuzzy controller avoids rigorous mathematical models and, consequently, is more robust than a classical approach in cases that cannot, or only with great difficulty, be precisely modeled mathematically. Fuzzy rules describe in linguistic terms a quantitative relationship between two or more variables. Processing the fuzzy rules provides a mechanism for using them to compute the response to a given fuzzy controller input.

The basis of a fuzzy or any fuzzy rule system is the inference engine responsible for the inputs' fuzzification, fuzzy processing, and defuzzification of the output. A schematic of the inference engine is shown in Fig. 11.23. *Fuzzification* means that the actual inputs are fuzzified and fuzzy inputs are obtained. *Fuzzy processing* means that the inputs are processed according to the rules set and produces

FIGURE 11.23

Operation of a fuzzy controller.

fuzzy outputs. *Defuzzification* means producing a crisp real value for fuzzy output, which is also the controller output.

The fuzzy logic controller's goal is to achieve satisfactory control of a process. Based on the input parameters, the operation of the controller (output) can be determined. The typical design scheme of a fuzzy logic controller is shown in Fig. 11.24 (Zadeh, 1973). The design of such a controller contains the following steps:

1. Define the inputs and the control variables.
2. Define the condition interface. Inputs are expressed as fuzzy sets.
3. Design the rule base.
4. Design the computational unit. Many ready-made programs are available for this purpose.
5. Determine the rules for defuzzification, i.e., to transform fuzzy control output to crisp control action.

Membership functions

A membership function is a curve that defines how each point in the input space is mapped to a membership value, or degree of membership, between 0 and 1. In the literature, the input space is

FIGURE 11.24

Basic configuration of fuzzy logic controller.

sometimes referred to as the *universe of discourse*. The only condition a membership function must really satisfy is that it must vary between 0 and 1. Additionally, it is possible, in a fuzzy set, to have a partial membership, such as "the weather is rather hot". The function itself can be an arbitrary curve whose shape can be defined as a function that suits the particular problem from the point of view of simplicity, convenience, speed, and efficiency.

Based on signals usually obtained from sensors and common knowledge, membership functions for the input and output variables need to be defined. The inputs are described in terms of linguistic variables as, for example, very high, high, okay, low, and very low, as shown in Fig. 11.25. It should be noted that, depending on the problem, different sensors could be used showing different parameters such as distance, angle, resistance, slope, etc.

The output can be adjusted in a similar way, according to some membership functions—for example, the ones presented in Fig. 11.26. In both cases, membership curves other than the triangular can be used, such as trapezoidal, quadratic, Gaussian (exponential), cos-function, and many others.

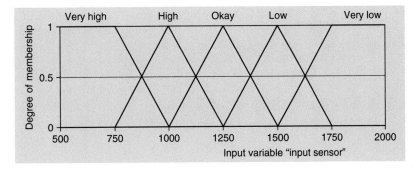

FIGURE 11.25

Membership functions for linguistic variables describing an input sensor.

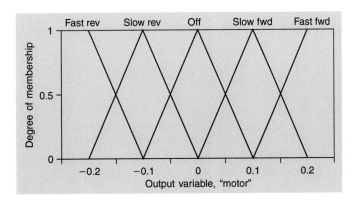

FIGURE 11.26

Membership functions for linguistic variables describing motor operation.

Logical operations

The most important thing to realize about fuzzy logical reasoning is that it is a superset of standard Boolean logic, i.e., if the fuzzy values are kept at their extremes of 1 (completely true) and 0 (completely false), standard logical operations hold. In fuzzy logic, however, the truth of any statement is a matter of degree. The input values can be real numbers between 0 and 1. It should be noted that the results of the statement A AND B, where A and B are limited to the range (0, 1) can be resolved by using min (A, B). Similarly, an OR operation can be replaced with the max function so that A OR B becomes equivalent to max (A, B), and the operation NOT A is equivalent to the operation 1 − A. Given these three functions, any construction can be resolved using fuzzy sets and the fuzzy logical operations AND, OR, and NOT. An example of the operations on fuzzy sets is shown in Fig. 11.27.

In Fig. 11.27, only one particular correspondence between two-valued and multi-valued logical operations for AND, OR, and NOT is defined. This correspondence is by no means unique. In more general terms, what are known as the fuzzy intersection or conjunction (AND), fuzzy union or disjunction (OR), and fuzzy complement (NOT) can be defined.

The intersection of two fuzzy sets, A and B, is specified in general by a binary mapping, T, which aggregates two membership functions as:

$$\mu_{A \cap B}(x) = T[\mu_A(x), \mu_B(x)] \tag{11.138}$$

The binary operator, T, may represent the multiplication of $\mu_A(x)$ and $\mu_B(x)$. These fuzzy intersection operators are usually refined as T norm (triangular norm) operators. Similarly, in fuzzy intersection, the fuzzy union operator is specified in general by a binary mapping, S, as:

$$\mu_{A \cup B}(x) = S[\mu_A(x), \mu_B(x)] \tag{11.139}$$

The binary operator, S, may represent the addition of $\mu_A(x)$ and $\mu_B(x)$. These fuzzy union operators are usually referred to as T conorm (or S norm) operators.

If-then rules

Fuzzy sets and fuzzy operators are the subjects and verbs of fuzzy logic. While the differential equations are the language of conventional control, *if-then* rules, which determine the way a process is controlled, are the language of fuzzy control. Fuzzy rules serve to describe the quantitative relationship between variables in linguistic terms. These if-then rule statements are used to formulate the conditional statements that comprise fuzzy logic. Several rule bases of different complexity can be developed, such as:

IF Sensor 1 is Very Low AND Sensor 2 is Very Low THEN Motor is Fast Reverse.
IF Sensor 1 is High AND Sensor 2 is Low THEN Motor is Slow Reverse.
IF Sensor 1 is Okay AND Sensor 2 is Okay THEN Motor Off.
IF Sensor 1 is Low AND Sensor 2 is High THEN Motor is Slow Forward.
IF Sensor 1 is Very Low AND Sensor 2 is Very High THEN Motor is Fast Forward.

In general form, a single fuzzy IF-THEN rule is of the form:

$$\text{IF } x \text{ is A and } y \text{ is B THEN } z \text{ is C} \tag{11.140}$$

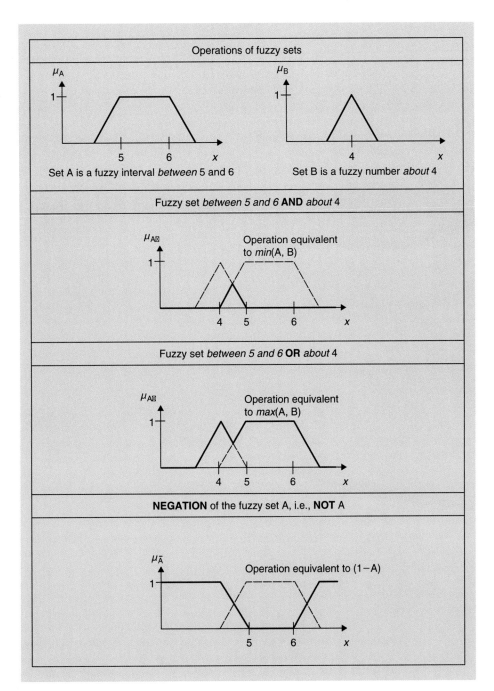

FIGURE 11.27

Operations on fuzzy sets.

where A, B, and C are linguistic values defined by fuzzy sets on the ranges (universe of discourse) X, Y, and Z, respectively. In if-then rules, the term following the IF statement is called the *premise* or antecedent, and the term following THEN is called the *consequent*.

It should be noted that A and B are represented as a number between 0 and 1, and so the antecedent is an interpretation that returns a single number between 0 and 1. On the other hand, C is represented as a fuzzy set, so the consequent is an assignment that assigns the entire fuzzy set C to the output variable z. In the if-then rule, the word *is* gets used in two entirely different ways, depending on whether it appears in the antecedent or the consequent. In general, the input to an if-then rule is the current value of an input variable, in Eq. (11.140), x and y, and the output is an entirely fuzzy set, in Eq. (11.140), z. This will later be defuzzified, assigning one value to the output.

Interpreting an if-then rule involves two distinct parts:

1. Evaluate the antecedent, which involves fuzzifying the input and applying any necessary fuzzy operators.
2. Apply that result to the consequent, known as *implication*.

In the case of two-valued or binary logic, if-then rules present little difficulty. If the premise is true, then the conclusion is true. In the case of a fuzzy statement, if the antecedent is true to some degree of membership, then the consequent is also true to that same degree; that is,

In binary logic, $p \rightarrow q$ (p and q are either both true or both false)
In fuzzy logic, $0.5p \rightarrow 0.5q$ (partial antecedents provide partial implication)

It should be noted that both the antecedent and the consequent parts of a rule can have multiple components. For example, the antecedent part can be:

if temperature is high and sun is shining and pressure is falling, then ...

In this case, all parts of the antecedent are calculated simultaneously and resolved to a single number using the logical operators described previously. The consequent of a rule can also have multiple parts, for example,

if temperature is very hot, then boiler valve is shut and public mains water valve is open

In this case, all consequents are affected equally by the result of the antecedent. The consequent specifies a fuzzy set assigned to the output. The implication function then modifies that fuzzy set to the degree specified by the antecedent. The most common way to modify the output set is truncation using the min function.

In general, interpreting if-then fuzzy rules is a three-part process:

1. Fuzzify inputs. All fuzzy statements in the antecedent are resolved to a degree of membership between 0 and 1.
2. Apply a fuzzy operator to multiple part antecedents. If there are multiple parts to the antecedent, apply fuzzy logic operators and resolve the antecedent to a single number between 0 and 1.
3. Apply the implication method. The degree of support for the entire rule is used to shape the output fuzzy set. The consequent of a fuzzy rule assigns an entire fuzzy set to the output. This fuzzy set is represented by a membership function that is chosen to indicate the quantities of the consequent. If the antecedent is only partially true, then the output fuzzy set is truncated according to the implication method.

Fuzzy inference system

Fuzzy inference is a method that interprets the values in the input vector and, based on some sets of rules, assigns values to the output vector. In fuzzy logic, the truth of any statement becomes a matter of a degree.

Fuzzy inference is the process of formulating the mapping from a given input to an output using fuzzy logic. The mapping then provides a basis from which decisions can be made or patterns discerned. The process of fuzzy inference involves all of the pieces described so far, i.e., membership functions, fuzzy logic operators, and if-then rules. Two main types of fuzzy inference systems can be implemented: Mamdani-type (1977) and Sugeno-type (1985). These two types of inference systems vary somewhat in the way outputs are determined.

Mamdani-type inference expects the output membership functions to be fuzzy sets. After the aggregation process, there is a fuzzy set for each output variable, which needs defuzzification. It is possible, and sometimes more efficient, to use a single spike as the output membership function rather than a distributed fuzzy set. This, sometimes called a *singleton output membership function*, can be considered a pre-defuzzified fuzzy set. It enhances the efficiency of the defuzzification process because it greatly simplifies the computation required by the more general Mamdani method, which finds the centroid of a two-dimensional function. Instead of integrating across the two-dimensional function to find the centroid, the weighted average of a few data points can be used.

The Sugeno method of fuzzy inference is similar to the Mamdani method in many respects. The first two parts of the fuzzy inference process, fuzzifying the inputs and applying the fuzzy operator, are exactly the same. The main difference between Mamdani-type and Sugeno-type fuzzy inferences is that the output membership functions are only linear or constant for the Sugeno-type fuzzy inference. A typical fuzzy rule in a first-order Sugeno fuzzy model has the form:

$$\text{If } x \text{ is A and } y \text{ is B, then } z = px + qy + r \tag{11.141}$$

where A and B are fuzzy sets in the antecedent, while p, q, and r are all constants.

Higher-order Sugeno fuzzy models are possible, but they introduce significant complexity with little obvious merit. Because of the linear dependence of each rule on the system's input variables, the Sugeno method is ideal for acting as an interpolating supervisor of multiple linear controllers that are to be applied, respectively, to different operating conditions of a dynamic non-linear system. A Sugeno fuzzy inference system is extremely well suited to the task of smoothly interpolating the linear gains that would be applied across the input space, i.e., it is a natural and efficient gain scheduler. Similarly, a Sugeno system is suitable for modeling non-linear systems by interpolating multiple linear models.

Fuzzy systems applications in solar energy systems

The applications of fuzzy systems in solar applications are much fewer. They concern the design of a fuzzy single-axis tracking mechanism controller (Kalogirou, 2002) and a neuro-fuzzy-based model for PV power supply system (Mellit and Kalogirou, 2006b). In fact, the membership functions shown in Figs. 11.25 and 11.26 and the rule basis, given previously, are from the first application, whereas the latter is a hybrid system described in the next section.

11.6.4 Hybrid systems

Hybrid systems are systems that combine two or more artificial intelligence techniques to perform a task. The classical hybrid system is the neuro-fuzzy control, whereas other types combine genetic algorithms and fuzzy control or artificial neural networks and genetic algorithms as part of an integrated problem solution or to perform specific, separate tasks of the same problem. Since most of these techniques are problem specific, more details are given here for the first category.

A fuzzy system possesses great power in representing linguistic and structured knowledge using fuzzy sets and performing fuzzy reasoning and fuzzy logic in a qualitative manner. Also, it usually relies on domain experts to provide the necessary knowledge for a specific problem. Neural networks, on the other hand, are particularly effective at representing non-linear mappings in computational fashion. They are "constructed" through training procedures presented to them as samples. Additionally, although the behavior of fuzzy systems can be understood easily due to their logical structure and step by step inference procedures, a neural network generally acts as a "black box", without providing explicit explanation facilities. The possibility of integrating the two technologies was considered into a new kind of system, called *neuro-fuzzy control*, where several strengths of both systems are utilized and combined appropriately.

More specifically, *neuro-fuzzy control* means (Nie and Linkens, 1995):

1. The controller has a structure resulting from a combination of fuzzy systems and ANNs.
2. The resulting control system consists of fuzzy systems and neural networks as independent components performing different tasks.
3. The design methodologies for constructing respective controllers are hybrid ones coming from ideas in fuzzy and neural control.

In this case, a trained neural network can be viewed as a means of knowledge representation. Instead of representing knowledge using if-then localized associations as in fuzzy systems, a neural network stores knowledge through its structure and, more specifically, its connection weights and local processing units, in a distributed or localized manner. Many commercial software (such as Matlab) include routines for neuro-fuzzy modeling.

The basic structure of a fuzzy inference system is described in Section 11.6.3. This is a model that maps the input membership functions, input membership function to rules, rules to a set of output characteristics, output characteristics to output membership functions, and the output membership function to a single-valued output or decision associated with the output. Thus, the membership functions are fixed. In this way, fuzzy inference can be applied to modeling systems whose rule structure is essentially predetermined by the user's interpretation of the characteristics of the variables in the model.

In some modeling situations, the shape of the membership functions cannot be determined by just looking at the data. Instead of arbitrarily choosing the parameters associated with a given membership function, these parameters could be chosen to tailor the membership functions to the input–output data in order to account for these types of variations in the data values. If fuzzy inference is applied to a system for which a past history of input–output data is available, these can be used to determine the membership functions. Using a given input–output data set, a fuzzy inference system can be constructed, whose membership function parameters are tuned or adjusted using a neural network. This is called a *neuro-fuzzy system*.

The basic idea behind a neuro-fuzzy technique is to provide a method for the fuzzy modeling procedure to learn information about a data set, in order to compute the membership function parameters that best allow the associated fuzzy inference system to track the given input–output data. A neural network, which maps inputs through input membership functions and associated parameters, then through output membership functions and associated parameters to outputs, can be used to interpret the input–output map. The parameters associated with the membership functions will change through a learning process. Generally, the procedure followed is similar to any neural network technique described in Section 11.6.1.

It should be noted that this type of modeling works well if the data presented to a neuro-fuzzy system for training and estimating the membership function parameters is representative of the features of the data that the trained fuzzy inference system is intended to model. However, this is not always the case, and data are collected using noisy measurements or training data cannot be representative of all features of the data that will be presented to the model. For this purpose, model validation can be used, as in any neural network system. Model validation is the process by which the input vectors from input–output data sets that the neuro-fuzzy system has not seen before are presented to the trained system to check how well the model predicts the corresponding data set output values.

11.7 Limitations of simulations

Simulations are powerful tools for solar energy systems design, offering a number of advantages, as outlined in the previous sections. However, there are limits to their use. For example, it is easy to make mistakes, such as assuming wrong constants and neglect important factors. As with other engineering calculations, a high level of skill and scientific judgment is required to produce correct, useful results (Kalogirou, 2004b).

It is possible to model a system to a high degree of accuracy to extract the required information. In practice, however, it may be difficult to represent in detail some of the phenomena taking place in real systems. Additionally, physical world problems, such as plugged pipes, leaks, poor system installation, scale on heat exchanger surfaces, problematic operation of controllers, and poor insulation of collectors and other equipment, cannot be easily modeled or accounted for. Moreover, simulation programs deal only with the thermal behavior of the processes, but mechanical and hydraulic considerations can also affect the thermal performance of solar energy systems. An exception to this is the use of artificial intelligence systems when data from real systems are used, where possible problems are embedded into the data used for training the systems.

It should be noted that there is no substitute for carefully executed experiments. Additionally, a combination of system simulation and physical experiments can lead to better understanding of how processes work and thus to better systems. These can reveal whether or not theory is adequate and where difficulties are present in the design and/or operation of the systems. As a conclusion, simulations are powerful tools for the modeling, design, prediction of performance, and research and development of solar energy systems. They must, however, be used very carefully.

No study of solar energy systems is complete unless an economic analysis is carried out. For this purpose, a life cycle analysis is usually performed, as explained in the following chapter.

Exercises

11.1 A house located at 45°N latitude with $UA = 156$ W/°C has a solar energy system which includes 30 m^2 collectors and a 2250 L storage tank. The collector heat-exchanger parameters, obtained from standard collector tests, are $F'_R(\tau\alpha)_n = 0.80$ and $F'_R U_L = 4.25$ W/m^2 °C. The load heat exchanger has $Z = 2.5$, the radiation on the collector surface $\overline{H}_t = 13.5$ MJ/m^2 and

$(\overline{\tau\alpha})/(\tau\alpha)_n = 0.94$. The domestic water-heating load is 1.9 GJ per month. For the month of January, estimate the solar fraction and contribution. The average ambient temperature of the location in January is 3 °C and the degree days are 730.

11.2 If, in Exercise 11.1, $Z = 0.75$ and the storage tank size is halved, what is the new value of solar fraction and contribution?

11.3 A space heating system is located in Boulder, Colorado. Estimate the monthly and annual solar fraction and contribution if the following characteristics apply:

Collector area = 40 m².
Collector $F'_R(\tau\alpha)_n = 0.78$.
Collector $F'_R U_L = 4.21$ W/m² °C.
$(\overline{\tau\alpha})/(\tau\alpha)_n = 0.96$.
Collector slope = 45°
Storage volume = 150 L/m².
Load heat exchanger $Z = 2$ (standard size).
Building $UA = 250$ W/K.
Hot water load = 2.45 GJ per month (constant).
Ground reflectance = 0.2.

11.4 A space heating system is located in an area where $\overline{H}_t = 13.5$ MJ/m², $\overline{T}_a = -2\,°C$, and the degree days are 550. The system uses an air heating system of standard configuration, with standard air flow rate and storage size. Estimate the solar fraction and contribution in January if the hot water load is 1.95 GJ and the system has the following characteristics:

Building $UA = 325$ W/°C.
Collector area = 35 m².
Collector $F'_R(\tau\alpha)_n = 0.78$.
Collector $F'_R U_L = 3.45$ W/m² °C.
$(\overline{\tau\alpha})/(\tau\alpha)_n = 0.94$.

11.5 In Exercise 11.4, what collector area is required to cover 50% of the load?

11.6 A house with $UA = 350$ W/K has an air space-heating system and is located in Springfield, Illinois. The collectors are double glazed, inclined 50°. Estimate the monthly and annual solar fraction and contribution of the system, which uses air collectors that have the following characteristics:

Collector area = 50 m².
Collector $F'_R(\tau\alpha)_n = 0.65$.
Collector $F'_R U_L = 5.45$ W/m² °C.
Air flow rate = 15 L/m² s.
Storage capacity = 0.2 m³/m² of rocks.
Hot water load = 1.95 GJ per month (constant).
Ground reflectance = 0.2.

11.7 Calculate the solar contribution of a thermosiphon solar water-heating system located in a place where local latitude is 40°, for the month of March. The system has the following characteristics:

1. Slope of collector = 45°
2. Monthly average solar radiation on collector surface = 15,900 kJ/m² day

  3. Monthly average ambient temperature $= 9.1\ °C$
  4. Monthly average clearness index $= 0.53$
  5. Number of collector panels $= 2$
  6. Collector area per panel $= 1.5\ m^2$
  7. Collector test $F_R U_L = 19.0\ kJ/h\ m^2\ °C$
  8. Collector $F_R(\tau\alpha) = 0.82$
  9. Collector test flow rate $= 71.5\ kg/h\ m^2$
 10. Number of risers per panel $= 8$
 11. Riser diameter $= 0.012\ m$
 12. Combined header length per panel $= 2\ m$
 13. Header diameter $= 0.028\ m$
 14. Tank-collector connecting pipe length $= 2.1\ m$
 15. Collector-tank connecting pipe length $= 1.1\ m$
 16. Connecting pipe diameter $= 0.028\ m$
 17. Number of bends in connecting pipe $= 2$
 18. Connecting pipe heat loss coefficient $= 11.1\ kJ/h\ m^2\ °C$
 19. Storage tank volume $= 170\ L$
 20. Storage tank height $= 1\ m$
 21. Storage tank diameter $= 0.465\ m$
 22. Daily load draw off $= 160\ L$
 23. Mains water temperature $= 16\ °C$
 24. Auxiliary set temperature $= 60\ °C$
 25. Height $H_1 = 0.07\ m$
 26. Height $H_2 = 1.09\ m$
 27. Height $H_3 = 2.2\ m$
 28. Height $H_5 = 1.31\ m$

11.8 A collector is used in an application located at 40°N latitude and has a slope of 45°. If, for January, the average daily radiation on horizontal is $11.9\ MJ/m^2$ and the critical radiation level for the collector is $156\ W/m^2$, estimate the daily utilizability and the utilizable energy for the month. Ground reflectance $= 0.2$.

11.9 Using the Φ method, estimate the total energy collection in March of a collector located at 35°N latitude that has the following characteristics:
Collector $F_R(\tau\alpha) = 0.81$ (constant value).
Collector $F_R U_L = 5.05\ W/m^2\ °C$.
Collector inclination $= 40°$.
Ground reflectance $= 0.3$.
$\overline{K}_T = 0.55$.
$\overline{H}_o = 29.6\ MJ/m^2$.
$\overline{T}_a = 1\ °C$.
$T_i = 45\ °C$.

11.10 Repeat Exercise 11.9 using the $\overline{\overline{\Phi}}$ method.

11.11 A collector system supplies heat to an industrial process. The collector inlet temperature (process return temperature) varies as shown in the following table but, for a certain hour, is constant during the month. The calculation is done for the month of March, when $\overline{K}_T = 0.55$ and $\rho_G = 0.2$. The system is located at 35°N latitude and the collector characteristics are $F_R U_L = 5.44$ W/m^2 °C, $F_R(\tau\alpha)_n = 0.79$, tilted 40°, and the incidence angle modifier constant $b_o = 0.1$. The weather conditions are also given in the table. Calculate the energy output of the collector.

Hour	T_i (°C)	T_a (°C)	$\overline{I}_t$ (MJ/m^2)
8–9	45	−2	1.48
9–10	45	0	2.13
10–11	60	2	3.05
11–12	60	5	3.67
12–13	60	7	3.85
13–14	75	8	2.95
14–15	75	6	2.32
15–16	75	3	1.80

11.12 The collector system of Exercise 11.11 is located in Albuquerque, New Mexico, and has an area of 60 m^2, $(\tau\alpha)_n = 0.96$, and a storage tank volume of 4000 L. If the process requires heat at a rate of 12 kW at a temperature of 80 °C for 8 h/day, estimate the monthly and annual solar fractions.

11.13 For Exercise 11.12, estimate, for the month of July, the storage tank losses by considering an environmental temperature of 20 °C and $(UA)_s = 4.5$ W/°C. Estimate also the effect of the load on the heat exchanger if the heat exchanger has effectiveness $= 0.52$ and its capacitance is 4000 W/°C.

11.14 A building located in Albuquerque, New Mexico (35°N latitude), has a 12.5 m^2 south-facing window. The UA of the building is 325 W/°C and its thermal capacitance is 18.9 MJ/°C. The window is double glazed and has $U = 3.25$ W/m^2 °C. The room is maintained at 18.3 °C and the allowable temperature swing is equal to 6 °C. For the month of January, if $\rho_G = 0.2$ and monthly average $(\overline{\tau\alpha}) = 0.75$, estimate the auxiliary energy required.

11.15 A building located at 35°N latitude has an active collection-passive storage system. The building has $UA = 500$ W/°C, a thermal capacitance of 21.7 MJ/m^2, the indoor temperature is kept at 20 °C, and the average temperature and degree days for January are 8.9 °C and 875 °C-days, respectively. The allowable temperature swing is 5 °C and $\overline{K}_T = 0.63$. Estimate the auxiliary energy required if the system uses air collectors inclined at 45° with the following characteristics:
Collector area = 50 m^2.
Collector $F_R(\tau\alpha)_n = 0.65$.
Collector $F_R U_L = 5.45$ W/m^2 °C.
$(\overline{\tau\alpha})/(\tau\alpha)_n = 0.85$.

Assignments

1. Design of a solar thermal system

Student name: _____ Number of problem: _____

Design a solar thermal system with an f-chart method that will provide space heating and DHW, which will be installed in one of the cities shown in Appendix 7. Consider the city's mains water monthly temperature to be the same as the mean monthly ambient temperature of the city chosen.

The solar thermal collectors face south and use a mixture of water/glycol ($c_p = 3840$ J/kg°C) as the collector heat transfer fluid, while the thermal storage fluid is water ($c_p = 4180$ J/kg°C). The flow rate of the collector circuit is 0.015 kg/s-m^2. The flow from the hot-water storage tank to the load heat exchanger is 1.5 kg/s, whereas the load heat exchanger's effectiveness is 0.8.

The air flow rate of the space-heating system is 1.1 L/s-$(UA)_L$. Other characteristics of the system are:

1. Collector inclination (β): _____°
2. Building UA: _____ W/°C
3. Number of persons for DHW use: _____
4. How water use temperature (T_w): _____°C
5. How water demand: _____ l/person-day
6. Collector $F_R U_L$: _____ W/m^2-°C
7. Collector $F_R(\tau\alpha)_n$: _____
8. Heat exchanger UA: _____ W/°C
9. Storage capacity: _____ l/m^2
10. Parameter $(\tau\alpha)_n$: _____
11. Ground reflectivity (ρ_G): _____

Estimate the monthly and annual solar contribution of the system for a collector area of 10 m^2.

Problem options for various students

Student	Slope	$(UA)_B$	Persons	T_w	l/P-day	$F_R U_L$	$F_R(\tau\alpha)_n$	Storage (L/m^2)	$(\tau\alpha)_n$	ρ_G
1	30	320	14	45	10	4.4	0.7	50	0.96	0.2
2	32	330	13	50	12	4.6	0.72	55	0.95	0.22
3	34	340	12	55	14	4.8	0.74	60	0.94	0.24
4	36	350	12	60	16	5	0.76	65	0.93	0.26
5	38	360	13	60	18	5.2	0.78	70	0.92	0.28
6	40	370	14	55	20	5.4	0.8	80	0.93	0.3
7	42	380	14	50	18	5.6	0.82	85	0.94	0.28
8	44	390	13	45	16	5.8	0.8	90	0.95	0.26
9	46	400	12	50	14	6	0.78	95	0.96	0.24

Continued

10	48	410	14	55	12	6.2	0.76	100	0.97	0.22
11	30	420	13	60	10	6.4	0.74	105	0.96	0.2
12	32	430	12	55	12	6.6	0.72	110	0.95	0.18
13	34	420	14	60	14	4.4	0.76	115	0.94	0.2
14	36	410	13	45	16	4.6	0.74	120	0.93	0.22
15	38	400	12	55	18	4.8	0.72	125	0.92	0.24

2. Simulation of a solar domestic water-heating system

Using TRNSYS V.18 demo, downloaded as explained in Section 11.5.1, and example SDHW simulate a domestic solar water-heating system with the following characteristics:

Collector parameters
Collector in series $= 2$
    Collector area $= 10$ m^2
    Working fluid $=$ water with $c_p = 4.19$ kJ/kg°C
    Efficiency mode $= 1$
    Testing flow rate $= 40$ kg/hr-m^2
    Intercept efficiency $= 0.82$.
    Efficiency slope $= 15$ kJ/hr-m^{2}K
    Efficiency curvature $= 0.05$ kJ/hr-m^{2}K^2
    Optical mode $= 2$
    First-order IAM $= 0.1$
    Second-order IAM $= 0$

Storage tank parameters
Tank volume $= 0.8$ m^3 (800 L)
    Tank height $= 2$ m.

Pump characteristics
Flow rate $= 400$ kg/h.
    Rated power $= 600$ kJ/h.

The system is to be simulated for the city of Nice, France, for the whole year at a time step of 1 h (changed by selecting the small gear on the LHS icons). For this purpose, the example SDWH can be used and change the parameters of the various components according to the above list.

Report the annual values of the radiation falling on the collector surface, the useful energy collected, the auxiliary energy required, the mean collector efficiency, and the annual solar contribution.

3. Simulation of a simple building with daylight access

Using TRNSYS V.18 demo, downloaded as explained in Section 11.5.1, and example 3D_Building/ 6_Step_Add_Daylight/Building-Step6, simulate a simple building having access to daylight, with the following characteristics:

Period: May 1st - May 31st

City: Madrid, Spain.

When the simulation completes, do not exit the online plotter and use various options at the bottom of the plotter to explore the data produced and mainly evaluate the effect of daylighting.

References

Barakat, S.A., Sander, D.M., 1982. Building Research Note No. 184. Division of Building Research, National Research Council of Canada.

Barr, A., Feigenbaum, E.A., 1981. The Handbook of Artificial Intelligence, vol. 1. Morgan Kaufmann, Los Altos, CA.

Beckman, W.A., 1998. Modern computing methods in solar energy analysis. In: Proceedings of EuroSun '98 on CD-ROM. Portoroz, Slovenia.

Beckman, W.A., Klein, S.A., Duffie, J.A., 1977. Solar Heating Design by the F-Chart Method. Wiley-Interscience, New York.

Benz, N., Gut, M., Belkircher, T., Russ, W., 1999. Solar process heat with non-concentrating collectors for food industry. In: Proceedings of ISES Solar World Congress on CD-ROM. Jerusalem, Israel.

Carvalho, M.J., Bourges, B., 1985. Application of utilizability computation methods to Europe and Africa. In: Intersol 85, Proceedings of the 9th Biennial Congress ISES, vol. 4, pp. 2439–2448.

Clark, D.R., Klein, S.A., Beckman, W.A., 1983. Algorithm for evaluating the hourly radiation utilizability function. ASME J. Sol. Energy Eng. 105, 281–287.

Close, D.J., 1962. The performance of solar water heaters with natural circulation. Sol. Energy 6, 33–40.

Copsey, A.B., 1984. Modification of the F-Chart Method for Solar Domestic Hot Water Systems with Stratified Storage (MS Thesis). University of Wisconsin.

Collares-Pereira, M., Rabl, A., 1979a. Derivation of method for predicting long term average energy delivery of solar collectors. Sol. Energy 23 (3), 223–233.

Collares-Pereira, M., Rabl, A., 1979b. The average distribution of solar radiation-correlations between diffuse and hemispherical and between daily and hourly insolation values. Sol. Energy 22 (2), 155–164.

Collares-Pereira, M., Rabl, A., 1979c. Simple procedure for predicting long term average performance of non-concentrating and concentrating solar collectors. Sol. Energy 23 (3), 235–253.

Davis, L., 1991. Handbook of Genetic Algorithms. Van Nostrand, New York.

Erbs, D.G., Klein, S.A., Duffie, J.A., 1982. Estimation of diffuse radiation fraction for hourly, daily and monthly average global radiation. Sol. Energy 28 (4), 293–302.

Evans, B.L., Klein, S.A., 1984. A design method of active collection-passive storage space heating systems. In: Proceedings of ASME Meeting. Las Vegas, NV.

Evans, D.L., Rule, T.T., Wood, B.D., 1982. A new look at long term collector performance and utilisability. Sol. Energy 28, 13–23.

Fanney, A.H., Klein, S.A., 1983. Performance of solar domestic hot water systems at the National Bureau of Standards. J. Sol. Energy Eng. 105, 311–321.

Farlow, S.J. (Ed.), 1984. Self-organizing Methods in Modeling. Marcel Dekker, New York.

Florides, G., Kalogirou, S., Tassou, S., Wrobel, L., 2002. Modeling and simulation of an absorption solar cooling system for Cyprus. Sol. Energy 72 (1), 43–51.

Gantner, M., 2000. Dynamische Simulation Thermischer Solaranlagen (Diploma Thesis). Hochschule für Technik Rapperswil (HSR), Switzerland.

GenOpt, 2011. A Generic Optimization Program. Available from: http://simulationresearch.lbl.gov/GO/index.html.

Goldberg, D.E., 1989. Genetic Algorithms in Search Optimization and Machine Learning. Addison-Wesley, Reading, MA.

Haykin, S., 1994. Neural Networks: A Comprehensive Foundation. Macmillan, New York.

Hecht-Nielsen, R., 1991. Neurocomputing. Addison-Wesley, Reading, MA.

Hottel, H.C., Whillier, A., 1955. Evaluation of flat plate collector performance. In: Transactions of the Conference on the Use of Solar Energy. Part I, vol. 2. University of Arizona Press, p. 74.

Ivakhenko, A.G., 1968. The group method of data handling—a rival of stochastic approximation. Sov. Autom. Control 1, 43–55 (in Russian).

Ivakhenko, A.G., 1971. Polynomial theory of complex systems. IEEE Trans. Syst. Man Cybern. SMC-12 364–378.

Jordan, U., Vajen, K., 2000. Influence of the DHW load profile on the fractional energy savings: a case study of a solar combi-system with TRNSYS simulations. In: Proceedings of EuroSun 2000 on CD-ROM. Copenhagen, Denmark.

Kalogirou, S.A., 1996a. Artificial neural networks for estimating the local concentration ratio of parabolic trough collectors. In: Proceedings of the EuroSun'96 Conference, vol. 1. Freiburg, Germany, pp. 470–475.

Kalogirou, S.A., 1996b. Design of a solar low temperature steam generation system. In: Proceedings of the EuroSun'96 Conference, vol. 1. Freiburg, Germany, pp. 224–229.

Kalogirou, S.A., 2000. Long-term performance prediction of forced circulation solar domestic water heating systems using artificial neural networks. Appl. Energy 66 (1), 63–74.

Kalogirou, S.A., 2001. Artificial neural networks in renewable energy systems: a review. Renew. Sustain. Energy Rev. 5 (4), 373–401.

Kalogirou, S.A., 2002. Design of a fuzzy single-axis sun tracking controller. Int. J. Renew. Energy Eng. 4 (2), 451–458.

Kalogirou, S.A., 2003a. The potential of solar industrial process heat applications. Appl. Energy 76 (4), 337–361.

Kalogirou, S.A., 2003b. Artificial intelligence for the modelling and control of combustion processes: a review. Prog. Energy Combust. Sci. 29 (6), 515–566.

Kalogirou, S.A., 2003c. Use of genetic algorithms for the optimal design of flat plate solar collectors. In: Proceedings of the ISES 2003 Solar World Congress on CD-ROM. Goteborg, Sweden.

Kalogirou, S.A., 2004a. Optimisation of solar systems using artificial neural networks and genetic algorithms. Appl. Energy 77 (4), 383–405.

Kalogirou, S.A., 2004b. Solar thermal collectors and applications. Prog. Energy Combust. Sci. 30 (3), 231–295.

Kalogirou, S.A., 2007. Use of genetic algorithms for the optimum selection of the fenestration openings in buildings. In: Proceedings of the 2nd PALENC Conference and 28th AIVC Conference on Building Low Energy Cooling and Advanced Ventilation Technologies in the 21st Century. Crete Island, Greece, September 2007, pp. 483–486.

Kalogirou, S.A., 2009. Thermal performance, economic and environmental life cycle analysis of thermosiphon solar water heaters. Sol. Energy 38 (1), 39–48.

Kalogirou, S.A., Panteliou, S., 2000. Thermosiphon solar domestic water heating systems long-term performance prediction using artificial neural networks. Sol. Energy 69 (2), 163–174.

Kalogirou, S.A., Papamarcou, C., 2000. Modeling of a thermosiphon solar water heating system and simple model validation. Renew. Energy 21 (3–4), 471–493.

Kalogirou, S.A., Neocleous, C., Schizas, C., 1996. A comparative study of methods for estimating intercept factor of parabolic trough collectors. In: Proceedings of the Engineering Applications of Neural Networks (EANN'96) Conference. London, pp. 5–8.

Kalogirou, S.A., Neocleous, C., Schizas, C., 1998. Artificial neural networks for modelling the starting-up of a solar steam generator. Appl. Energy 60 (2), 89–100.

Kalogirou, S.A., Panteliou, S., Dentsoras, A., 1999a. Artificial neural networks used for the performance prediction of a thermosiphon solar water heater. Renew. Energy 18 (1), 87–99.

Kalogirou, S.A., Panteliou, S., Dentsoras, A., 1999b. Modelling of solar domestic water heating systems using artificial neural networks. Sol. Energy 65 (6), 335–342.

Klein, S.A., 1976. A Design Procedure for Solar Heating Systems (Ph.D. thesis). Chemical Engineering, University of Wisconsin, Madison.

Klein, S.A., 1978. Calculation of flat-plate collector utilizability. Sol. Energy 21 (5), 393–402.

Klein, S.A., et al., 2018. TRNSYS Version 18 Program Manual. Solar Energy Laboratory, University of Wisconsin, Madison et al.

Klein, S.A., Beckman, W.A., 1979. A general design method for closed-loop solar energy systems. Sol. Energy 22 (3), 269–282.

Klein, S.A., Beckman, W.A., 2005. F-chart Users Manual. University of Wisconsin, Madison.

Klein, S.A., Beckman, W.A., Duffie, J.A., 1976. A design procedure for solar heating systems. Sol. Energy 18 (2), 113–127.

Klein, S.A., Beckman, W.A., Duffie, J.A., 1977. A design procedure for solar air heating systems. Sol. Energy 19 (6), 509–512.

Ktistis, P., Agathkleous, R., Kalogirou, S.A., 2022. A design tool for a Parabolic Trough Collector system for Industrial Process Heat based on dynamic simulation. Renew. Energy 183, 502–514.

Liu, B.Y.H., Jordan, R.C., 1963. The long-term average performance of flat-plate solar energy collectors. Sol. Energy 7 (2), 53–74.

Malkin, M.P., 1985. Design of Thermosiphon Solar Domestic Hot Water Systems (MS thesis). University of Wisconsin.

Malkin, M.P., Klein, S.A., Duffie, J.A., Copsey, A.B., 1987. A design method for thermosiphon solar domestic hot water systems. Trans. ASME 109, 150–155.

Mamdani, E.H., 1974. Application of fuzzy algorithms for control of simple dynamic plant. IEE Proc 121, 1585–1588.

Mamdani, E.H., 1977. Applications of fuzzy set theory to control systems: a survey. In: Gupta, M.M., et al. (Eds.), Fuzzy Automata and Decision Process. North-Holland, Amsterdam, pp. 77–88.

Mellit, A., Kalogirou, S.A., 2006a. Application of neural networks and genetic algorithms for predicting the optimal sizing coefficient of photovoltaic supply systems. In: Proceedings of the IX World Renewable Energy Congress on CD-ROM. Florence, Italy.

Mellit, A., Kalogirou, S.A., 2006b. Neuro-fuzzy based modeling for photovoltaic power supply (PVPS) system. In: Proceedings of the IEEE First International Power and Energy Conference, vol. 1, pp. 88–93. Malaysia.

Mellit, A., Kalogirou, S.A., 2021. Artificial intelligence and internet of things to improve efficacy of diagnosis and remote sensing of solar photovoltaic systems: challenges, recommendations and future directions. Renew. Sustain. Energy Rev. 143 (2021), 110889.

Michalewicz, Z., 1996. Genetic Algorithms + Data Structures = Evolution Programs, third ed. Springer, Berlin.

Mitchell, J.C., Theilacker, J.C., Klein, S.A., 1981. Calculation of monthly average collector operating time and parasitic energy requirements. Sol. Energy 26, 555–558.

Monsen, W.A., Klein, S.A., Beckman, W.A., 1981. Prediction of direct gain solar heating system performance. Sol. Energy 27 (2), 143–147.

Monsen, W.A., Klein, S.A., Beckman, W.A., 1982. The unutilizability design method for collector-storage walls. Sol. Energy 29 (5), 421−429.

Nannariello, J., Frike, F.R., 2001. Introduction to neural network analysis and its applications to building services engineering. Build. Serv. Eng. Res. Technol. 22 (1), 58−68.

Nie, J., Linkens, D.A., 1995. Fuzzy-Neural Control: Principles, Algorithms and Applications. Prentice Hall, Englewood Cliffs, NJ.

NRCan, 2009. Natural Resources Canada, WATSUN. Program available for free download from: http://canmetenergy.nrcan.gc.ca/software-tools/1546.

Oishi, M., Noguchi, T., 2000. The evaluation procedure on performance of SDHW system by TRNSYS simulation for a yearly performance prediction. In: Proceedings of EuroSun 2000 on CD-ROM. Copenhagen, Denmark.

Phillips, W.F., Dave, R.N., 1982. Effects of stratification on the performance of liquid-based solar heating systems. Sol. Energy 29, 111−120.

Pogio, T., Girosi, F., 1990. Networks for approximation and learning. Proc. IEEE 78, 1481−1497.

Polysun, 2008. User Manual for Polysun 4. Vela Solaris AG, Rapperswil.

Reznik, L., 1997. Fuzzy Controllers. Newnes, Oxford.

Ripley, B.D., 1996. Pattern Recognition and Neural Networks. Cambridge University Press, Cambridge, UK.

Rumelhart, D.E., Hinton, G.E., Williams, R.J., 1986. Learning internal representations by error propagation chapter 8. In: Parallel Distributed Processing: Explorations in the Microstructure of Cognition, vol. 1. MIT Press, Cambridge, MA.

Schweiger, H., Mendes, J.F., Benz, N., Hennecke, K., Prieto, G., Gusi, M., Goncalves, H., 2000. The potential of solar heat in industrial processes. A state of the art review for Spain and Portugal. In: Proceedings of EuroSun 2000 Copenhagen, Denmark on CD-ROM.

Sugeno, M., 1985. Industrial Applications of Fuzzy Control. North-Holland, Amsterdam.

Thevenard, D., 2008. A simple tool for the simulation of active solar systems: WATSUN reborn. In: Proceedings of the Third Solar Buildings Conference. Fredericton N.B, pp. 189−196.

Tsoukalas, L.H., Uhrig, R.E., 1997. Fuzzy and Neural Approaches in Engineering. John Wiley & Sons, New York.

Ward Systems Group, Inc., 1996. Neuroshell-2 Program Manual. Frederick, MD.

WATSUN, 1996. WATSUN Users Manual and Program Documentation. 3V.13.3. University of Waterloo. WATSUN Simulation Laboratory, Waterloo, Canada.

Werbos, P.J., 1974. Beyond Regression: New Tools for Prediction and Analysis in the Behavioral Science (Ph.D. thesis). Harvard University, Cambridge, MA.

Wetter, M., 2001. GenOpt: a generic optimization program. In: Proceedings of IBPSA's Building Simulation Conference. Rio de Janeiro, Brazil.

Wetter, M., 2004. GenOpt: A Generic Optimization Program. User Manual. Lawrence Berkeley Laboratory, Berkeley, CA.

Whillier, A., 1953. Solar Energy Collection and its Utilization for House Heating (Ph.D. thesis). Mechanical Engineering, M.I.T., Cambridge, MA.

Zadeh, L.A., 1973. Outline of a new approach to the analysis of complex systems and decision processes. IEEE Trans. Syst. Man Cybern. SMC-3 28−44.

Zalzala, A., Fleming, P., 1997. Genetic Algorithms in Engineering Systems. The Institution of Electrical Engineers, London.

Solar Economic Analysis

Although the resource of a solar energy system—that is, the solar irradiation—is free, the equipment required to collect it and convert it to useful form (heat or electricity) has a cost. Therefore, solar energy systems are generally characterized by high initial cost and low operating costs. To decide to employ a solar energy system, the cost of collectors, other required equipment, and conventional fuel required as backup must be lower than the cost of other conventional energy sources to perform the same task. Thus, the economic problem is to compare an initial known investment with estimated future operating costs, including both the cost to run and maintain the solar energy system and auxiliary energy used as backup. Other factors that need to be considered include the interest paid on money borrowed, taxes if any, insurance cost if any, and resale of equipment at the end of its life.

In previous chapters, various solar energy components and systems are discussed and various methods to determine the long-term thermal performance of the solar systems are presented. It is very important to be able to analyze and assess the economic viability of these systems, in order to convince prospective clients to install a solar energy system. The objective of the economic analysis is to find the right size of a system for a particular application that gives the lowest combination of solar and auxiliary energy costs.

Since the availability of solar energy is intermittent and unpredictable, it is generally not cost-effective to provide 100% of the energy requirements of a thermal system with solar energy year round. This is because, when the system satisfies fully the requirements under the worst operating conditions, it will be greatly oversized during the rest of the year, requiring the dumping of thermal energy, which is not cost-effective. Usually, the other way around is effective, i.e., size the system to satisfy 100% the thermal energy requirements under the system's best operating conditions, usually during summertime, and use auxiliary energy for the rest of the year to back up the solar energy system. The actual size is usually decided by following an economic analysis, as described in this chapter. The best use of solar energy is in conjunction with the type and cost of the conventional fuel used as backup. The target is to design a solar energy system that operates at full or nearly full capacity most of the time and uses auxiliary energy for the rest of the year, although the total percentage of annual demand covered is less than 100%. It can easily be proved that such a system is much more economical than a larger system satisfying fully the thermal load year round. The auxiliary system can also cover the load in extreme weather conditions, thus increasing the reliability of the solar energy system. The annual load factor covered by solar energy compared to the total annual thermal load is called the *solar fraction*, F. It is defined as the ratio of the useful solar energy supplied to the system to the energy needed to heat the water or the building space if no solar energy is used. In other words, F is a measure of the fractional energy savings relative to that used for a conventional energy system. This is expressed as a percentage, given by an equation similar to Eq. (11.1):

Solar Energy Engineering. https://doi.org/10.1016/B978-0-323-99350-0.00004-1

$$F = \frac{L - L_{AUX}}{L} \tag{12.1}$$

where

L = annual energy required by the load (GJ);
L_{AUX} = annual energy required by the auxiliary (GJ).

12.1 Life cycle analysis

The right proportion of solar to auxiliary energy is determined by economic analysis. There are various types of such analysis—some simple and others more complicated, based on thermoeconomics.

The economic analysis of solar energy systems is carried out to determine the least cost of meeting the energy needs, considering both solar and non-solar alternatives. The method employed in this book for the economic analysis is called *life cycle analysis*. This method takes into account the time value of money and allows detailed consideration of the complete range of costs. It also includes inflation when estimating future expenses. In the examples given in this chapter, both dollars ($) and euros (€) are used. The actual monetary value used, however, is not important to the actual method, and life cycle analysis can be used for any monetary system.

Several criteria can be used to evaluate and optimize solar energy systems. The definitions of the most important ones are as follows:

1. *Life cycle cost* (LCC) is the sum of all costs associated with an energy delivery system over its lifetime in today's money, taking into account the time value of money. LCC can also be estimated for a selected period of time. The idea of LCC is to bring back costs that are anticipated in the future to present-day costs by discounting, i.e., by calculating how much would have to be invested at a market discount rate. The market discount rate is the rate of return of the best alternative investment, i.e., putting the money (to be invested) in a bank at the highest possible interest rate.
2. *Life cycle savings* (LCS), for a solar plus auxiliary energy system, is defined as the difference between the LCC of a conventional fuel-only system and the LCC of the solar plus auxiliary system. This is equivalent to the total present worth of the gains from the solar energy system compared to the fuel-only system (Beckman et al., 1977).
3. *Payback time* is defined in many different ways, but the most common one is the time needed for the cumulative fuel savings to become equal to the total initial investment, i.e., it is the time required to get back the money spent to erect the solar energy system from the fuel savings incurred because of the use of the system. This time can be obtained with and without discounting the fuel savings. Other definitions of *payback time* are the time required for the annual solar savings to become positive and the time required for the cumulative solar savings to become zero.
4. *Return on investment* (ROI) is defined as the market discount rate that results in zero LCS, i.e., the discount rate that makes the present worth of the solar and non-solar alternatives equal.

All the software programs described in Chapter 11 have routines for the economic analysis of the modeled systems. The economic analysis of solar energy systems can also be performed with a spreadsheet program. Spreadsheet programs are especially suitable for economic analyses because their general format is a table with cells that can contain values or formulas and they incorporate many built-in functions. An economic analysis is carried out for every year for which various economic parameters are calculated in different columns. For example, the ROI can easily be obtained using different values of the market discount rate until the LCS become 0 by trial and error. A detailed description of the method of economic analysis of solar energy systems using spreadsheets was given by Kalogirou (1996).

12.1.1 Life cycle costing

Life cycle analysis, in fact, reflects the benefits accumulated by the use of solar energy against the fuel savings incurred. Compared to conventional fossil fuel systems, solar energy systems have relatively high initial cost and low operating cost, whereas the opposite is true for conventional systems. Therefore, in a naive selection, based on the initial cost alone, the solar energy system would have no chance to be selected. As will be proved in this chapter, this is not the case when a life cycle analysis is employed, because it considers all costs incurred during the life of the solar energy system. Additionally, one should consider that, as the resource becomes scarce, oil prices will rise, and the higher the fuel cost replaced by the solar energy system, the better are the economic factors, such as the LCS and the payback times. The detrimental effects of the use of conventional fuel on the environment, as outlined in Chapter 1, should not be underestimated. An analysis of the environmental benefits of solar water heating systems was given by Kalogirou (2004).

In a life cycle analysis, both the initial cost and the annual operating costs are considered for the entire life of the solar energy system. These include the initial purchase cost of the system, operating costs for fuel and electricity required for the pumps, interest charges on money borrowed, maintenance costs, and taxes paid, if applicable. There is also a salvage value, which is returned at the end of the life of the system, when the components are sold as scrap metal for recycling.

The initial purchase cost should include the actual equipment cost, designer fee, transportation cost, labor cost to install the system, cost of brackets and any other modifications required to install the system, the value of space required to install the system if this is not installed on the roof of the building, and the profit of the installer. The actual equipment cost includes the solar collectors, storage tank, pumps or fans, piping or ducting, insulation, heat exchangers, and controls.

As the solar energy system size increases, it produces more energy but costs more. It would, therefore, be required to determine the optimum size of the solar energy system that has the maximum LCS or the quickest payback time. The problem of finding the lowest-cost system is a multi-variable one, in which all the components of the system and the system configuration have some effect on its thermal performance and cost. In practice, the total load that needs to be covered is known or given. For example, for a hot water system, it is the hot water demand multiplied by the temperature difference between the supply make-up water and the hot water delivery temperature; in a space heating application, it is the total thermal load. Therefore, the problem is to find the size of the solar energy system with the other parameters, such as the storage capacity, to be fixed relative to the collector area. Additionally, solar energy systems are much more sensitive to the size of the collector array area than to any other component of the system, such as storage. Therefore, to simplify the analysis, the collector size is considered as the optimization parameter for a given load and system characteristics, with the other parameters selected in accordance with the collector size. The total cost of the solar equipment, C_s, is given by the sum of two terms: one is proportional to the collector area, A_c, called *area-*

dependent costs, C_A, and the other is independent of the collector area, called *area-independent costs*, C_I, given by:

$$C_s = C_A A_c + C_I \qquad (12.2)$$

In Eq. (12.2), the area-dependent cost, C_A, includes not only costs related to the purchase and installation of the collector system, such as collector panels, brackets, and piping, but also other costs that depend on the size of the collector system, such as portion of the cost for storage and portion of the solar pump cost. Area-independent cost includes the cost of components not related to the collector area, such as the cost of controls and electrical installation. It should be noted that, if any subsidies apply, these should be subtracted from the total system cost, since this is not really an expenditure. For example, if a 40% subsidy on the initial system cost applies in a country, then if the total system cost is €10,000, the real expenditure required is $0.6 \times 10,000 = €6000$, because the buyer receives from the government $0.4 \times €10,000 = €4000$. Because the subsidy applies at the beginning of the system's life, it is not subject to the effects of time on money and is a present worth in a life savings analysis.

Operating costs, C_o, include maintenance, parasitic, and fuel costs. Maintenance costs are usually considered to be a percentage of the initial investment and are assumed to increase at a certain rate per year of the system operation to account for the system aging. For stationary collectors, maintenance can be assumed to be 1% and, for tracking collectors, 2% of the initial investment, inflated by 0.5% and 1%, respectively, per year of system operation (Kalogirou, 2003). Parasitic costs accounts for the energy required (electricity) to drive the solar pump or blower, depending on the type of system.

Fuel savings are obtained by subtracting the annual cost of the conventional fuel used for the auxiliary energy from the fuel needs of a fuel-only system. The integrated cost of the auxiliary energy use for the first year, that is, solar backup, is given by the formula:

$$C_{AUX} = C_{FA} \int_0^t L_{AUX} dt \qquad (12.3)$$

The integrated cost of the total load for the first year, that is, the cost of conventional fuel without solar energy, is:

$$C_L = C_{FL} \int_0^t L dt \qquad (12.4)$$

where C_{FA} and C_{FL} = cost rates (in \$/GJ) for auxiliary energy and conventional fuel, respectively.

If the same type of fuel is used by both systems, $C_{FA} = C_{FL}$. Both values are equal to the product of fuel calorific value and heater efficiency. Equation (12.4), in fact, gives the fuel costs for the conventional (non-solar) system. In Eq. (12.3), instead of L_{AUX}, the total thermal load, L, can be used, multiplied by $(1 - F)$.

In equation form, the annual cost for both the solar and conventional backup systems to cover the energy need of the thermal load is given by:

$$\text{Annual cost} = \text{mortgage payment} + \text{fuel cost} + \text{maintenance cost} + \text{parasitic energy cost}$$
$$+ \text{property taxes} + \text{insurance cost} - \text{income tax savings} \quad (12.5)$$

It should be noted that not all parameters apply to all possible systems. Only those that apply in each case should be used. For example, if the solar energy system is paid completely with available funds, then no annual mortgage payment is required. This applies to other factors, such as property taxes, insurance costs, and income tax savings, which are different in each country or not applied at all and concerning the insurance may be different for each consumer. Because the rules of tax savings differ from country to country or even in some cases, as in the United States, from state to state and these rules change continuously, this chapter cannot go into detail on this matter. As part of the design process, the designer must adapt to the rules followed in the area where the solar energy system is to be installed. For the United States, the income tax savings depend on whether the system is non-income producing, such as home systems, or income producing, such as an industrial process heating system. The appropriate equations are as follows (Duffie and Beckman, 2006):

For non-income-producing systems,

$$\text{Income tax savings} = \text{effective tax rate} \times (\text{interest payment} + \text{property tax}). \quad (12.6)$$

For income-producing systems,

$$\text{Income tax savings} = \text{effective tax rate} \times \begin{cases} \text{interest payment} \\ +\text{property tax} \\ +\text{fuel expense} \\ +\text{maintenance} \\ +\text{insurance} \\ +\text{parasitic energy} \\ -\text{depreciation} \end{cases} \quad (12.7)$$

State income taxes are deductible from income for federal tax purposes. In cases where the federal taxes are not deductible from the state tax, the effective tax is estimated by Duffie and Beckman (2006):

$$\text{Effective tax rate} = \text{federal tax rate} + \text{state tax rate} - (\text{federal tax rate} \times \text{state tax rate}) \quad (12.8)$$

According to the definition of the LCS given in a previous section, solar savings can also be obtained by the difference between the cost of conventional and solar energy systems:

$$\text{Solar savings} = \text{cost of conventional energy} - \text{cost of solar energy} \quad (12.9)$$

It should be noted that, if savings are negative, then they are deficits (expenses) instead of savings. According to Beckman et al. (1977), in solar savings, costs which are common to both systems are not evaluated. For example, a storage tank is usually installed in both solar and non-solar energy systems, so if the storage tank or other equipment in the two systems is of a different size, the difference in their costs is included as an increment to the solar energy system cost. Therefore, in this concept, it is necessary to consider only the incremental or extra cost in installing the solar energy system, as given by:

Solar savings = fuel savings − extra mortgage payment − extra maintenance − extra insurance

− extra parasitic energy cost − extra property tax + income tax savings

$$(12.10)$$

Concerning the last term of Eq. (12.10), equations similar to Eq. (12.6) and Eq. (12.7) can be written by adding the word *extra* to the various terms.

12.2 Time value of money

As was indicated before, the usual approach in solar process economics is to use a life cycle cost method, which takes into consideration all future expenses and compares the future costs with today's costs. Such a comparison is done by discounting all costs expected in the future to the common basis of present value or present worth, i.e., it is required to find the amount of money that needs to be invested today in order to have funds available to cover the future expenses.

Table 12.1 Present worth, PW_n.

Year (n)	Market discount rate (d)								
	2%	4%	6%	8%	10%	12%	15%	20%	25%
1	0.9804	0.9615	0.9434	0.9259	0.9091	0.8929	0.8696	0.8333	0.8000
2	0.9612	0.9246	0.8900	0.8573	0.8264	0.7972	0.7561	0.6944	0.6400
3	0.9423	0.8890	0.8396	0.7938	0.7513	0.7118	0.6575	0.5787	0.5120
4	0.9238	0.8548	0.7921	0.7350	0.6830	0.6355	0.5718	0.4823	0.4096
5	0.9057	0.8219	0.7473	0.6806	0.6209	0.5674	0.4972	0.4019	0.3277
6	0.8880	0.7903	0.7050	0.6302	0.5645	0.5066	0.4323	0.3349	0.2621
7	0.8706	0.7599	0.6651	0.5835	0.5132	0.4523	0.3759	0.2791	0.2097
8	0.8535	0.7307	0.6274	0.5403	0.4665	0.4039	0.3269	0.2326	0.1678
9	0.8368	0.7026	0.5919	0.5002	0.4241	0.3606	0.2843	0.1938	0.1342
10	0.8203	0.6756	0.5584	0.4632	0.3855	0.3220	0.2472	0.1615	0.1074
11	0.8043	0.6496	0.5268	0.4289	0.3505	0.2875	0.2149	0.1346	0.0859
12	0.7885	0.6246	0.4970	0.3971	0.3186	0.2567	0.1869	0.1122	0.0687
13	0.7730	0.6006	0.4688	0.3677	0.2897	0.2292	0.1625	0.0935	0.0550
14	0.7579	0.5775	0.4423	0.3405	0.2633	0.2046	0.1413	0.0779	0.0440
15	0.7430	0.5553	0.4173	0.3152	0.2394	0.1827	0.1229	0.0649	0.0352
16	0.7284	0.5339	0.3936	0.2919	0.2176	0.1631	0.1069	0.0541	0.0281
17	0.7142	0.5134	0.3714	0.2703	0.1978	0.1456	0.0929	0.0451	0.0225
18	0.7002	0.4936	0.3503	0.2502	0.1799	0.1300	0.0808	0.0376	0.0180
19	0.6864	0.4746	0.3305	0.2317	0.1635	0.1161	0.0703	0.0313	0.0144
20	0.6730	0.4564	0.3118	0.2145	0.1486	0.1037	0.0611	0.0261	0.0115
25	0.6095	0.3751	0.2330	0.1460	0.0923	0.0588	0.0304	0.0105	0.0038
30	0.5521	0.3083	0.1741	0.0994	0.0573	0.0334	0.0151	0.0042	0.0012
40	0.4529	0.2083	0.0972	0.0460	0.0221	0.0107	0.0037	0.0007	0.0001
50	0.3715	0.1407	0.0543	0.0213	0.0085	0.0035	0.0009	0.0001	—

It must be noted that a sum of money at hand today is worth more than the same sum in the future. Therefore, a sum of money or cash flow in the future must be discounted and worth less than its present-day value. A cash flow (F) occurring (n) years from now can be reduced to its present value (P) by:

$$P = \frac{F}{(1 + d)^n} \tag{12.11}$$

where d = market discount rate (%).

Therefore, an expense anticipated to be €100 in 6 years is equivalent to an obligation of €70.50 today at a market discount rate of 6%. To have €100 available in 6 years, it would be necessary to make an investment of €70.50 today at an annual rate of return of 6%.

Equation (12.11) shows that a present worth of a given amount of money is discounted in the future by the factor $(1 + d)^{-n}$ for each year in the future. Therefore, the fraction $(1 + d)^{-n}$ can be used to estimate the present worth of a monetary unit (euro, dollar, etc.), at any year (n), PW_n, given by:

$$PW_n = \frac{1}{(1 + d)^n} \tag{12.12}$$

The present worth can be estimated from Eq. (12.12) or obtained directly from Table 12.1. The present worth can be multiplied by any cash flow in the future at time (n) to give its present value. Its use in a life cycle analysis allows all calculations to be made at present by discounting costs and savings incurred during the life of the system.

EXAMPLE 12.1

You are about to receive €500 over 3 years and there are two options. The first is to receive €100 during year 1, €150 in year 2, and €250 in year 3. The second is to receive nothing in year 1, €200 in year 2, and €300 in year 3. If the interest rate is 8%, which option is more beneficial?

Solution

From Table 12.1 or Eq. (12.12) the present worth for the various years are:

$$\text{Year } 1 = 0.9259$$
$$\text{Year } 2 = 0.8573$$
$$\text{Year } 3 = 0.7938$$

Table 12.2 Present worth of each option in Example 12.1.

Year (n)	PW_n	Annual benefit (€)		Present worth (€)	
		Option 1	Option 2	Option 1	Option 2
1	0.9259	100	0	92.59	0
2	0.8573	150	200	128.60	171.46
3	0.7938	250	300	198.45	238.14
Totals		500	500	419.64	409.60

The present worth of each option is obtained by multiplying PW_n by the annual amount received, as shown in Table 12.2. Therefore, it is better to receive less money sooner than more money later.

Therefore, a discount rate of 8% means that, for an investor, the value of money is worth 8% less in 1 year, or €100 this year has the same value as €108 in 1 year's time.

Similarly, the amount of money needed to purchase an item increases because the value of money decreases. Thus, an expense (C), when inflated at a rate (i) per time period, equals (C) at the end of the first time period, equals $C(1 + i)$ at the end of the second time period, equals $C(1 + i)^2$ at the end of the third time period, and so on. Therefore, with an annual inflation rate (i), a purchase cost (C) at the end of year (n) becomes a future cost (F) according to:

$$F = C(1 + i)^{n-1} \tag{12.13}$$

Thus, a cost that will be €1000 at the end of the first time period will be $1000(1 + 0.07)^5 = €1402.6$ at the end of six time periods at an inflation rate of 7%.

12.3 Description of the life cycle analysis method

In life cycle cost analysis, all anticipated costs are discounted to their present worth and the LCC is the addition of all present worth values. The cash flow for each year can be calculated, and the LCC can be found by discounting each annual cash flow to its present value and finding the sum of these discounted cash flows. Life cycle costing requires that all costs are projected into the future and the results obtained from such an analysis depend extensively on the predictions of these future costs.

In general, the present worth (or discounted cost) of an investment or cost (C) at the end of year (n) at a discount rate of (d) and interest rate of (i) is obtained by combining Eqs. (12.11) and (12.13):

$$PW_n = \frac{C(1 + i)^{n-1}}{(1 + d)^n} \tag{12.14}$$

Equation (12.14) gives the present worth of a future cost or expenditure at the end of (n) years when the cost or expenditure at the end of first year is (C). This equation is useful for estimating the present worth of any one payment of a series of inflating payments. Therefore, in a series of annual payments, if the first payment is €1000, due to inflation, say, at a rate of 5%, the sixth payment will be €1276.28, which is worth only €761 today at a discount rate of 9%. This is obtained by Eq. (12.14):

$$PW_6 = \frac{1000(1 + 0.05)^5}{(1 + 0.09)^6} = €761$$

Equation (12.14) gives the present worth of a single future payment. Summing up all the present worth values of (n) future payments results in the total present worth (TPW), given by:

$$TPW = C\left[\sum_{j=1}^{n} \frac{(1 + i)^{j-1}}{(1 + d)^j} \right] = C[PWF(n, i, d)] \tag{12.15}$$

where PWF(n, i, d) = present worth factor, given by:

$$PWF(n, i, d) = \sum_{j=1}^{n} \frac{(1+i)^{j-1}}{(1+d)^{j}} \tag{12.16}$$

The solution of Eq. (12.16) is as follows.
If $i = d$,

$$PWF(n, i, d) = \frac{n}{1+i} \tag{12.17}$$

If $i \neq d$,

$$PWF(n, i, d) = \frac{1}{d-i}\left[1 - \left(\frac{1+i}{1+d}\right)^{n}\right] \tag{12.18}$$

Equation (12.14) can easily be incorporated into a spreadsheet with the parameters (d) and (i) entered into separate cells and referencing them in the formulas as absolute cells. In this way, a change in either (d) or (i) causes automatic recalculation of the spreadsheet. If the PWF(n, i, d) is multiplied by the first of a series of payments made at the end of each year, the result is the sum of (n) payments discounted to the present with a market discount rate (d). The factor PWF(n, i, d) can be obtained with Eq. (12.17) or Eq. (12.18), depending on the values of (i) and (d), or from the tables in Appendix 8, which tabulate PWF(n, i, d) for the most usual range of parameters.

EXAMPLE 12.2

If the first payment is $600, find the present worth of a series of 10 payments, which are expected to inflate at a rate of 6% per year, and the market discount rate is 9%.

Solution

From Eq. (12.18),

$$PWF(10, 0.06, 0.09) = \frac{1}{0.09 - 0.06}\left[1 - \left(\frac{1+0.06}{1+0.09}\right)^{10}\right] = 8.1176$$

Therefore, the present worth is $600 \times 8.1176 = \$4870.56$.

A mortgage payment is the annual value of money required to cover the funds borrowed at the beginning to install the system. This includes payment of interest and principal. An estimation of the annual mortgage payment can be found by dividing the amount borrowed by the present worth factor (PWF). The PWF is estimated by using the inflation rate equal to 0 (equal payments) and with the market discount rate equal to the mortgage interest rate. The PWF can be obtained from tables (see Appendix 8) or calculated by the following equation, which is obtained from Eq. (12.18):

$$PWF(n_L, 0, d_m) = \frac{1}{d_m}\left[1 - \left(\frac{1}{1+d_m}\right)^{n_L}\right] \tag{12.19}$$

728 Chapter 12 Solar Economic Analysis

where

d_m = mortgage interest rate (%);
n_L = number of years of equal installments for the loan.

Therefore, if the mortgage principal is M, the periodic payment is:

$$\text{Periodic payment} = \frac{M}{\text{PWF}(n_L, 0, d_m)} \qquad (12.20)$$

EXAMPLE 12.3

The initial cost of a solar energy system is $12,500. If this amount is paid with a 20% down payment and the balance is borrowed at a 9% interest rate for 10 years, calculate the annual payments and interest charges for a market discount rate of 7%. Also estimate the present worth of the annual interest payments.

Solution

The actual amount borrowed is $10,000 (=12,500 × 0.8), which is the total present worth of all mortgage payments. The annual mortgage payment is estimated with Eq. (12.20):

$$\text{PWF}(n_L, 0, d_m) = \frac{1}{d_m}\left[1 - \left(\frac{1}{1+d_m}\right)^{n_L}\right] = \frac{1}{0.09}\left[1 - \left(\frac{1}{1+0.09}\right)^{10}\right] = 6.4177$$

Therefore, annual mortgage payment = 10,000/6.4177 = $1558.20.

The annual mortgage payment includes a principal payment and interest charges. Year after year, as the principal remaining on the loan reduces, the interest charge decreases accordingly. The estimation needs to be carried out for every year.

For year 1,

$$\text{Interest payment} = 10,000 \times 0.09 = \$900$$

$$\text{Principal payment} = 1558.20 - 900 = \$658.20$$

$$\text{Principal remaining at the end of year } 1 = 10,000 - 658.20 = \$9341.80$$

$$\text{Present worth of interest payment, from Eq. (12.11)} = \frac{900}{(1+0.07)^1} = \$841.12$$

For year 2,

$$\text{Interest payment} = 9341.80 \times 0.09 = \$840.76$$

$$\text{Principal payment} = 1558.20 - 840.76 = \$717.44$$

$$\text{Principal remaining at the end of year } 2 = 9341.80 - 717.44 = \$8624.36$$

$$\text{Present worth of interest payment, from Eq. (12.11)} = \frac{840.76}{(1+0.07)^2} = \$734.35$$

These calculations are repeated for all other years. The results are shown in Table 12.3.

EXAMPLE 12.3–CONT'D

Table 12.3 Calculations for remaining years in Example 12.3.

Year	Mortgage payment ($)	Interest payment ($)	Principal payment ($)	Principal remaining ($)	PW of interest payment ($)
1	1,558.20	900.00	658.20	9,341.80	841.12
2	1,558.20	840.76	717.44	8,624.36	734.35
3	1,558.20	776.19	782.01	7,842.35	633.60
4	1,558.20	705.81	852.39	6,989.96	538.46
5	1,558.20	629.10	929.10	6,060.86	448.54
6	1,558.20	545.48	1,012.72	5,048.14	363.48
7	1,558.20	454.33	1,103.87	3,944.27	282.93
8	1,558.20	354.98	1,203.22	2,741.05	206.60
9	1,558.20	246.69	1,311.51	1,429.54	134.19
10	1,558.20	128.66	1,429.54	0.00	65.40
Total	$15,582	$5,582	$10,000		$4,248.66

As can be understood from Example 12.3, the calculations can be carried out very easily with the help of a spreadsheet program. Alternatively, the total present worth of all payments can be calculated from the following equation:

$$PW_i = M \left[\frac{PWF(n_{min}, 0, d)}{PWF(n_L, 0, d_m)} + PWF(n_{min}, d_m, d) \left(d_m - \frac{1}{PWF(n_L, 0, d_m)} \right) \right] \tag{12.21}$$

where n_{min} = the minimum of n_L and period of economic analysis.

It should be noted that the period of economic analysis may not coincide with the term of mortgage; for example, the economic analysis may be performed for 20 years, which is the usual life of solar water heating systems, but the loan is to be paid in the first 10 years.

EXAMPLE 12.4

Calculate the total present worth of interest paid (PW_i) in Example 12.3.

Solution

The various PWF values may be obtained from the tables of Appendix 8, as follows:

$$PWF(n_{min}, 0, d) = PWF(10, 0, 0.07) = 7.0236$$

$$PWF(n_L, 0, d_m) = PWF(10, 0, 0.09) = 6.4177$$

$$PWF(n_{min}, d_m, d) = PWF(10, 0.09, 0.07) = 10.172$$

Using Eq. (12.21),

$$PW_i = M \left[\frac{PWF(n_{min}, 0, d)}{PWF(n_L, 0, d_m)} + PWF(n_{min}, d_m, d) \left(d_m - \frac{1}{PWF(n_L, 0, d_m)} \right) \right]$$

$$= 10,000 \left[\frac{7.0236}{6.4177} + 10.172 \left(0.09 - \frac{1}{6.4177} \right) \right] = \$4248.99$$

This is effectively the same answer as the one obtained before.

Life cycle analysis is performed annually and the following are evaluated according to Eq. (12.10) to find the solar savings:

- Fuel savings.
- Extra mortgage payment.
- Extra maintenance cost.
- Extra insurance cost.
- Extra parasitic cost.
- Extra property tax.
- Extra tax savings.

As indicated before, not all these costs may be present in every case, depending on the laws and conditions in each country or region. Additionally, as already indicated, the word *extra* appearing in some of the items assumes that the associated cost is also present for a fuel-only system and, therefore, only the extra part of the cost incurred for the installation of the solar energy system should be included. The inflation, over the period of economic analysis, of the fuel savings is estimated by using Eq. (12.13) with (i) equal to the fuel inflation rate. The parasitic cost is the energy required to power auxiliary items, such as the pump, fan, and controllers. This cost also increases at an inflation rate over the period of economic analysis using Eq. (12.13) with (i) equal to the annual increase in electricity price.

Solar savings for each year are the sums of the items shown above, as shown in Eq. (12.10). Actually, the savings are positive and the costs are negative. Finally, the present worth of each year's solar savings is determined using Eqs. (12.11) and (12.12). The results are estimated for each year. These annual values are then added up to obtain the LCS, according to the equation:

$$PW_{LCS} = \sum_{j=1}^{n} \frac{\text{Solar energy savings}}{(1 + d)^j} \tag{12.22}$$

To understand the method better, various aspects of the economic analysis are examined separately and mainly through examples. In this way, the basic ideas of life cycle analysis are clarified. It should be noted that the costs specified in the various examples that follow are arbitrary and have no significance. Additionally, these costs vary widely according to the type and size of system, location where the solar energy system is installed, laws and other conditions of the country or region, international fuel prices, and international material prices, such as for copper and steel, which affect the cost of the solar equipment.

12.3.1 Fuel cost of non-solar energy system examples

The first example is about the fuel cost of a non-solar or conventional energy system. It examines the time value of an inflating fuel cost.

EXAMPLE 12.5

Calculate the cost of fuel of a conventional (non-solar) energy system for 15 years if the total annual load is 114.9 GJ and the fuel rate is $17.2/GJ, the market discount rate is 7%, and the fuel inflation rate is 4% per year.

Solution

The first-year fuel cost is obtained from Eq. (12.4) as:

$$C_L = C_{FL} \int_0^t L dt = 17.2 \times 114.9 = \$1976.30$$

(because the total annual load is given, the integral is equal to 114.9 GJ).

The fuel costs in various years are shown in Table 12.4. Each year's cost is estimated with Eq. (12.13) or from the previous year's cost multiplied by $(1 + i)$. Each value for the present worth is estimated by the corresponding value of the fuel cost using Eq. (12.11).

Table 12.4 Fuel costs in various years for Example 12.5.

Year	Fuel cost ($)	PW of fuel cost ($)
1	1,976.30	1,847.01
2	2,055.35	1,795.22
3	2,137.57	1,744.89
4	2,223.07	1,695.97
5	2,311.99	1,648.42
6	2,404.47	1,602.20
7	2,500.65	1,557.28
8	2,600.68	1,513.62
9	2,704.70	1,471.18
10	2,812.89	1,429.93
11	2,925.41	1,389.84
12	3,042.42	1,350.87
13	3,164.12	1,313.00
14	3,290.68	1,276.18
15	3,422.31	1,240.40
Total PW of fuel cost		$22,876

Continued

EXAMPLE 12.5–CONT'D

An alternative method is to estimate PWF(n, i, d) from Eq. (12.18) or Appendix 8 and multiply the value with the first year's fuel cost as follows.

From Eq. (12.18),

$$PWF(n, i, d) = \frac{1}{d - i}\left[1 - \left(\frac{1 + i}{1 + d}\right)^n\right]$$

or

$$PWF(15, 0.04, 0.07) = \frac{1}{0.07 - 0.04}\left[1 - \left(\frac{1.04}{1.07}\right)^{15}\right] = 11.5752$$

Present worth of fuel cost $= 1976.3 \times 11.5752 = \$22,876$

As can be seen, this is a much quicker method, especially if the calculations are done manually, and the same result is obtained but the intermediate values cannot be seen.

Although in the previous example, a fixed fuel inflation rate is used for all years, this may vary with time. In the case of a spreadsheet calculation, this can be easily accommodated by having a separate column representing the fuel inflation rate for each year and using this rate in each annual estimation accordingly. So, in this case, either the same value for all years or different values for each year can be used without difficulty. These estimations can also be performed with the help of the PWF, as shown in Example 12.5, but as the number of different rates considered increases, the complexity of the estimation increases, because the PWF needs to be calculated for every time period the rate changes, as shown in the following example.

EXAMPLE 12.6

Calculate the present worth of a fuel cost over 10 years if the first year's fuel cost is €1400 and inflates at 8% for 4 years and 6% for the rest of the years. The market discount rate is 7% per year.

Solution

The problem can be solved by considering two sets of payments at the two inflation rates. The first set of five payments has a first payment that is €1400 and inflates at 8%. Therefore, from Eq. (12.18),

$$PWF(n, i, d) = \frac{1}{d - i}\left[1 - \left(\frac{1 + i}{1 + d}\right)^n\right]$$

or

$$PWF(5, 0.08, 0.07) = \frac{1}{0.07 - 0.08}\left[1 - \left(\frac{1.08}{1.07}\right)^{5}\right] = 4.7611$$

Note that 5 years are used, the first plus four, according to the problem definition. Thus, the present worth of the first set is $1400 \times 4.7611 = €6665.54$.

EXAMPLE 12.6—CONT'D

The second set starts at the beginning of the sixth year and, for this period, $i = 6\%$ per year. To find the initial payment for this set, €1400 is inflated four times by 8% and one time by 6%. Therefore, Initial payment for the second set is $1400(1.08)^4(1.06) = €2018.97$.

As before, for the second series of payments,

$$\text{PWF}(n, i, d) = \frac{1}{d - i}\left[1 - \left(\frac{1+i}{1+d}\right)^n\right]$$

or

$$\text{PWF}(5, 0.06, 0.07) = \frac{1}{0.07 - 0.06}\left[1 - \left(\frac{1.06}{1.07}\right)^5\right] = 4.5864$$

The second set of payments needs to be discounted to the present worth by:

$$\text{PW} = \frac{4.5864 \times 2018.97}{(1.07)^5} = €6602.11$$

So the answer is the sum of the present worth of the two sets of payments: $6665.54 + 6602.11 = €13,267.65$.

12.3.2 Hot-water system example

The example in this section considers a complete solar water heating system. Although different solar energy systems have different characteristics, the way of handling the problems is the same.

EXAMPLE 12.7

A combined solar and auxiliary energy system is used to meet the same load as in Example 12.5. The total cost of the system to cover 65% of the load (solar fraction) is $20,000. The owner will pay a down payment of 20% and the rest will be paid over a 20-year period at an interest rate of 7%. Fuel costs are expected to rise at 9% per year. The life of the system is considered to be 20 years, and at the end of this period, the system will be sold for 30% of its original value. In the first year, the extra maintenance, insurance, and parasitic energy costs are $120 and the extra property tax is $300; both are expected to increase by 5% per year. The general market discount rate is 8%. The extra property taxes and interest on the mortgage are deducted from the income tax, which is at a fixed rate of 30%. Find the present worth of the solar savings.

Solution

The values estimated for the various costs and savings for the entire life of the system are shown in Table 12.5. Year zero includes only the down payment, whereas the values for the first year are as given by the problem definition. In the table, the savings are positive and the expenses (or payments) are negative. The down payment is equal to $0.2 \times 20,000 = \$4000$. The annual mortgage payment is found from Eq. (12.20):

$$\text{Periodic payment} = \frac{M}{\text{PWF}(n_L, 0, d_m)} = \frac{20,000 \times 0.8}{\text{PWF}(20, 0, 0.07)} = \frac{16,000}{10.594} = \$1510.29$$

Continued

EXAMPLE 12.7−CONT'D

Table 12.5 Estimated costs and savings for the system in Example 12.7.

1	2	3	4	5	6	7	8
Year	Extra mortgage payment ($)	Fuel savings ($)	Extra insurance, maintenance, and parasitic cost ($)	Extra property tax ($)	Income tax savings ($)	Solar savings ($)	PW of solar savings ($)
0	–	–	–	–	–	−4000.00	−4000.00
1	−1510.29	1284.70	−120.00	−300.00	426.00	−219.59	−203.32
2	−1510.29	1400.32	−126.00	−315.00	422.30	−128.66	−110.31
3	−1510.29	1526.35	−132.30	−330.75	418.26	−28.73	−22.80
4	−1510.29	1663.72	−138.92	−347.29	413.84	81.07	59.59
5	−1510.29	1813.46	−145.86	−364.65	409.01	201.66	137.25
6	−1510.29	1976.67	−153.15	−382.88	403.73	334.08	210.52
7	−1510.29	2154.57	−160.81	−402.03	397.98	479.42	279.74
8	−1510.29	2348.48	−168.85	−422.13	391.71	638.92	345.19
9	−1510.29	2559.85	−177.29	−443.24	384.88	813.91	407.16
10	−1510.29	2790.23	−186.16	−465.40	377.45	1005.83	465.89
11	−1510.29	3041.35	−195.47	−488.67	369.36	1216.29	521.64
12	−1510.29	3315.07	−205.24	−513.10	360.57	1447.01	574.63
13	−1510.29	3613.43	−215.50	−538.76	351.01	1699.89	625.05
14	−1510.29	3938.64	−226.28	−565.69	340.63	1977.01	673.10
15	−1510.29	4293.12	−237.59	−593.98	329.37	2280.63	718.95
16	−1510.29	4679.50	−249.47	−623.68	317.14	2613.20	762.77
17	−1510.29	5100.65	−261.94	−654.86	303.89	2977.44	804.71
18	−1510.29	5559.71	−275.04	−687.61	289.51	3376.29	844.91
19	−1510.29	6060.08	−288.79	−721.99	273.94	3812.95	883.51
20	−1510.29	6605.49	−303.23	−758.09	257.07	4290.95	920.62
20						6000.00	1287.29
Total present worth of solar savings							$6186.07

The first year fuel savings is 114.9 GJ × 0.65 = 74.69 GJ. According to Example 12.5, this corresponds to $1284.70.

$$\text{The interest paid for the first year} = 16,000 \times 0.07 = \$1120$$

$$\text{The principal payment} = 1510.29 - 1120 = \$390.29$$

$$\text{Principal balance} = 16,000 - 390.29 = \$15,609.71$$

$$\text{Tax savings} = 0.3(1120 + 300) = \$426$$

The annual solar savings is the sum of the values in columns 2 to 6. These are then brought to a present worth value using the market discount rate of 8%. Year 20 is repeated twice so as to include the resale value of 20,000 × 0.3 = $6000. This is a positive value as it is a saving.

EXAMPLE 12.7—CONT'D

The sum of all the values in the last column is the total present worth of the savings of the solar energy system as compared to a fuel-only system. These are the savings the owner would have by installing and operating the solar energy system instead of buying fuel for a conventional system.

As can be understood from the analysis, a supplementary table is required with analysis of the money borrowed (to find tax savings) and cumulative solar savings, as in Table 12.6. The last column is required in the estimation of payback time (see Section 12.3.4).

Table 12.6 Supplementary table for Example 12.7.

Year	Interest paid ($)	Principal payment ($)	Principal balance ($)	Cumulative solar savings ($)
0	0	0	16,000.00	−4,000.0
1	1,120.00	390.29	15,609.71	−4,203.3
2	1,092.68	417.61	15,192.10	−4,313.6
3	1,063.45	446.84	14,745.26	−4,336.4
4	1,032.17	478.12	14,267.14	−4,276.8
5	998.70	511.59	13,755.55	−4,139.6
6	962.89	547.40	13,208.15	−3,929.1
7	924.57	585.72	12,622.43	−3,649.3
8	883.57	626.72	11,995.71	−3,304.1
9	839.70	670.59	11,325.12	−2,897.0
10	792.76	717.53	10,607.59	−2,431.1
11	742.53	767.76	9,839.84	−1,909.5
12	688.79	821.50	9,018.34	−1,334.8
13	631.28	879.01	8,139.33	−709.8
14	569.75	940.54	7,198.80	−36.7
15	503.92	1,006.37	6,192.42	682.3
16	433.47	1,076.82	5,115.60	1,445.0
17	358.09	1,152.20	3,963.41	2,249.7
18	277.44	1,232.85	2,730.56	3,094.7
19	191.14	1,319.15	1,411.41	3,978.2
20	98.80	1,411.49	0	4,898.8
20				6,186.1

Another way of solving this problem is to carry out separate life cycle analyses for the solar and non-solar energy systems. In this case, the total present worth of the solar savings would be obtained by subtracting the total present worth of the two systems. It should be noted, however, that, in this case, equipment common to both systems needs to be considered in the analysis, so more information than that given is required.

12.3.3 Hot-water system optimization example

When a solar energy system is designed, the engineer seeks to find a solution that gives the maximum LCS of the installation. Such savings represent the money that the user/owner will save because of the use of a solar energy system instead of buying fuel. To find the optimum size system that gives the maximum LCS, various sizes are analyzed economically. When the present values of all future costs are estimated for each of the alternative systems under consideration, including solar and non-solar options, the system that yields the lowest LCC or the maximum LCS is the most cost-effective.

As an example, a graph of LCS against the collector area is shown in Fig. 12.1. For this graph, all other parameters except the collector area are kept constant. As can be seen, the LCS start at a negative value for a collector area equal to 0 (or of very small area), representing the total value of money required for fuel for a non-solar energy system, and as solar collectors are added to the system, the LCS reach a maximum and then drop. An increase in collector area beyond the maximum point gives lower life cycle savings (than the maximum value), as the bigger initial expenditure required for the solar energy system cannot replace the cost of the fuel saved. It even gives negative LCS for large areas in multiplication of the optimum value, which represent the money lost by the owner in erecting and operating the solar energy system instead of buying the fuel.

In previous examples, the annual fraction of load met by the solar energy system and the collector area, and thus the solar energy system cost, were given. In the following example, the relationship between the solar fraction, F, and collector area, obtained from thermal performance calculations, are given instead, so the objective is to find the area (system size) that gives the highest LCS.

FIGURE 12.1

Variation of collector area with life cycle savings.

EXAMPLE 12.8

If, in Example 12.7, the area-dependent costs are $250/m^2 and the area-independent cost is $1250 and all other parameters are kept constant, find the optimum area of the solar energy system that maximizes the LCS. From a thermal analysis of the solar process, the relationship of the collector area and solar fraction are as given in Table 12.7.

Table 12.7 Relationship of collector area and solar fraction in Example 12.8.

Area (m^2)	Annual solar fraction (F)
0.01	0
25	0.35
50	0.55
75	0.65
100	0.72
125	0.77
150	0.81

Solution

To solve this problem, the life cycle savings method needs to be applied for each collector area. A spreadsheet calculation of this can very easily be done by changing the collector area and the first year fuel savings, which can be estimated from the annual solar fraction. The complete results are shown in Table 12.8.

Table 12.8 First-year fuel and solar savings for Example 12.8.

Area (m^2)	Annual solar fraction (F)	Installed cost ($)	First-year fuel savings ($)	Solar savings ($)
0.01	0	1250	0	−5,680.7
25	0.35	7,500	691.8	3,609.9
50	0.55	11,250	1,087.1	6,898.9
75	0.65	20,000	1,284.7	6,186.1
100	0.72	26,250	1,423.1	4,275.0
125	0.77	32,500	1,521.9	1,562.3
150	0.81	38,750	1,600.9	−1,551.2

The life cycle solar savings are plotted against the collector area in Fig. 12.2. As can be seen, the maximum occurs at $A_c = 60$ m^2, where the LCS = $7013.70.

Continued

EXAMPLE 12.8–CONT'D

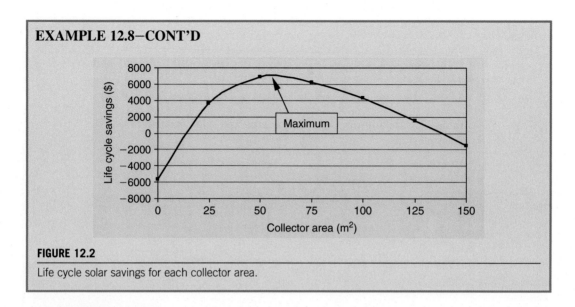

FIGURE 12.2

Life cycle solar savings for each collector area.

12.3.4 Payback time

The *payback time* is defined in many ways. Three of the most useful ones are shown in Section 12.1. As indicated, the most common one is the time needed for the cumulative fuel savings to equal to the total initial investment, i.e., it is the time required to get back the money spent to erect the solar energy system from the fuel savings incurred because of the use of the system. This time can be obtained with or without discounting the fuel savings.

Not discounting fuel savings

Initially, we consider that the fuel savings are not discounted. The fuel saved in a year (j) is given by:

$$\text{Fuel saved in year } j = FLC_{F1}(1 + i_F)^{j-1} \tag{12.23}$$

where

F = solar fraction, obtained from Eq. (12.1);

L = load (GJ);

C_{F1} = first year unit energy cost delivered from fuel (like parameters C_{FA} and C_{FL}, it is the product of fuel heating value and heater efficiency) ($/GJ); and

i_F = fuel inflation rate.

It should be noted that the product FL represents the energy saved because of the use of solar energy. Summing the fuel saved in year (j) over the payback time (n_P) and equating to the initial system cost, (C_s), given by Eq. (12.2), gives:

$$\sum_{j=1}^{n_p} FLC_{F1}(1 + i_F)^{j-1} = C_s \tag{12.24}$$

The summation gives:

$$C_s = \frac{FLC_{F1}[(1 + i_F)^{n_P} - 1]}{i_F} \quad (12.25)$$

Solving Eq. (12.25) for the payback time, n_P, gives:

$$n_P = \frac{\ln\left(\dfrac{C_s i_F}{FLC_{F1}} + 1\right)}{\ln(1 + i_F)} \quad (12.26)$$

Another way to determine the payback time is to use the PWF values tabulated in Appendix 8. Here, the sum of the fuel savings is given by the multiplication of the first year's saving, FLC_{F1} and PWF at zero discount rate. In equation form, this is given by:

$$FLC_{F1} \times PWF(n_P, i_F, 0) = C_s \quad (12.27)$$

Therefore, the payback time can be found by interpolation from the tables in Appendix 8, for which, PWF $= C_s/FLC_{F1}$.

EXAMPLE 12.9

Find the undiscounted payback time of a solar energy system that covers 63% of an annual load of 185 GJ and costs €15,100. The first-year fuel cost rate is €9.00/GJ and inflates at 9% per year.

Solution

Using Eq. (12.26),

$$n_P = \frac{\ln\left(\dfrac{C_s i_F}{FLC_{F1}} + 1\right)}{\ln(1 + i_F)} = \frac{\ln\left(\dfrac{15,100 \times 0.09}{0.63 \times 185 \times 9} + 1\right)}{\ln(1 + 0.09)} = 9.6 \text{ years}$$

The same result can be obtained from tables of Appendix 8. In this case, PWF $= C_s/FLC_{F1} = 15,100/(0.63 \times 185 \times 9) = 14.395$. This is very close to the first value ($d = 0$) of the column for $i = 9\%$ for $n = 10$ (PWF $= 15.193$). Interpolation can be used to get a more correct answer, but the exact result can be obtained with more accuracy with Eq. (12.26).

Discounting fuel savings

The procedures followed to equate discounted fuel costs to initial investment are similar. For discounted fuel costs, Eq. (12.27) becomes:

$$FLC_{F1} \times PWF(n_P, i_F, d) = C_s \quad (12.28)$$

Similarly, the payback time is given by the following:
If $i_F \neq d$,

$$n_P = \frac{\ln\left[\dfrac{C_s(i_F - d)}{FLC_{F1}} + 1\right]}{\ln\left(\dfrac{1 + i_F}{1 + d}\right)} \quad (12.29)$$

If $i_F = d$,

$$n_P = \frac{C_s(1 + i_F)}{FLC_{F1}}$$ (12.30)

EXAMPLE 12.10

Repeat Example 12.9 with fuel costs discounted at a rate of 7%.

Solution

Using Eq. (12.29),

$$n_P = \frac{\ln\left[\dfrac{C_s(i_F - d)}{FLC_{F1}} + 1\right]}{\ln\left(\dfrac{1 + i_F}{1 + d}\right)} = \frac{\ln\left(\dfrac{15,100(0.09 - 0.07)}{0.63 \times 185 \times 9} + 1\right)}{\ln\left(\dfrac{1.09}{1.07}\right)} = 13.7 \text{ years}$$

Other definitions of *payback time* are the time required for the annual solar savings to become positive and the time required for the cumulative solar savings to become 0. These can be determined by life cycle analysis. Using the results of Example 12.7, the payback time is.

1. The time required for the annual solar savings to become positive ≈ 4 years;
2. The time required for the cumulative solar savings to become zero ≈ 15 years.

12.4 The P_1, P_2 method

Another way of viewing the calculations of Example 12.7 is to obtain the present worth of each column and sum them to obtain the present worth of solar savings, using appropriate signs for each column. Therefore, the LCS of a solar energy system over a conventional system is expressed as the difference between a reduction in the fuel costs and an increase in expenses incurred as a result of the additional investment required for the solar energy system, given by:

$$LCS = P_1 C_{F1} FL - P_2 C_s$$ (12.31)

where

P_1 = ratio of life cycle fuel cost savings to first-year fuel savings;
P_2 = ratio of life cycle expenditure incurred from the additional investment to the initial investment.

The economic parameter P_1 is given by:

$$P_1 = (1 - Ct_e)\text{PWF}(n_e, i_F, d)$$ (12.32)

where

t_e = effective income tax rate;
C = flag indicating whether the system is commercial or non-commercial,

$$C = \begin{Bmatrix} 1 \rightarrow \text{commercial} \\ 0 \rightarrow \text{non-commercial} \end{Bmatrix} \qquad (12.33)$$

For example, in the United States, the effective tax rate is given by Eq. (12.8).

The economic parameter P_2 includes seven terms:

1. Down payment, $P_{2,1} = D$

2. Life cycle cost of the mortgage principal and interest,

$$P_{2,2} = (1 - D)\frac{\text{PWF}(n_{\min}, 0, d)}{\text{PWF}(n_L, 0, d_m)}$$

3. Income tax deductions of the interest,

$$P_{2,3} = (1 - D)t_e \left\{ \text{PWF}(n_{\min}, d_m, d) \left[d_m - \frac{1}{\text{PWF}(n_L, 0, d_m)} \right] + \frac{\text{PWF}(n_{\min}, 0, d)}{\text{PWF}(n_L, 0, d_m)} \right\}$$

4. Maintenance, insurance and parasitic energy costs,

$$P_{2,4} = (1 - Ct_e)M_1\text{PWF}(n_e, i, d)$$

5. Net property tax costs,

$$P_{2,5} = t_p(1 - t_e)V_1\text{PWF}(n_e, i, d)$$

6. Straight line depreciation tax deduction,

$$P_{2,6} = \frac{Ct_e}{n_d}\text{PWF}(n'_{\min}, 0, d)$$

7. Present worth of resale value,

$$P_{2,7} = \frac{R}{(1 + d)^{n_e}}$$

And P_2 is given by:

$$P_2 = P_{2,1} + P_{2,2} - P_{2,3} + P_{2,4} + P_{2,5} - P_{2,6} - P_{2,7} \qquad (12.34)$$

where

D = ratio of down payment to initial total investment;
M_1 = ratio of first year miscellaneous costs (maintenance, insurance, and parasitic energy costs) to the initial investment;
V_1 = ratio of assessed value of the solar energy system in the first year to the initial investment;
t_p = property tax, based on assessed value;

n_e = term of economic analysis;

n'_{min} = years over which depreciation deductions contribute to the analysis (usually the minimum of n_e and n_d, the depreciation lifetime in years); and

R = ratio of resale value at the end of its life to the initial investment.

It should be noted that, as before, not all these costs may be present, according to the country or region laws and regulations. Additionally, the contributions of loan payments to the analysis depend on n_L and n_e. If $n_L \leq n_e$, all n_L payments will contribute. If, on the other hand, $n_L \geq n_e$, only n_e payments will be made during the period of analysis. Accounting for loan payments after n_e depends on the reasoning for choosing the particular n_e. If n_e is the period over which the discounted cash flow is estimated without consideration for the costs occurring outside this period, $n_{min} = n_e$. If n_e is the expected operating life of the system and all payments are expected to be made as scheduled, $n_{min} = n_L$. If n_e is chosen as the time of sale of the system, the remaining loan principal at n_e would be repaid at that time and the life cycle mortgage cost would consist of the present worth of n_e load payments plus the principal balance at n_e. The principal balance should then be deduced from the resale value.

EXAMPLE 12.11

Repeat Example 12.7 using the P_1, P_2 method.

Solution

As noted in Example 12.7, the system is not income producing; therefore, $C = 0$. The ratio P_1 is calculated with Eq. (12.32):

$$P_1 = \text{PWF}(n, i_F, d) = \text{PWF}(20, 0.09, 0.08) = 20.242$$

The various terms of parameter P_2 are as follows:

$$P_{2,1} = D = 0.2$$

$$P_{2,2} = (1-D)\frac{\text{PWF}(n_{min}, 0, d)}{\text{PWF}(n_L, 0, d_m)} = (1-0.2)\frac{\text{PWF}(20, 0, 0.08)}{\text{PWF}(20, 0, 0.07)} = 0.8\frac{9.8181}{10.594} = 0.741$$

$$P_{2,3} = (1-D)t_e\left\{\text{PWF}(n_{min}, d_m, d)\left[d_m - \frac{1}{\text{PWF}(n_L, 0, d_m)}\right] + \frac{\text{PWF}(n_{min}, 0, d)}{\text{PWF}(n_L, 0, d_m)}\right\}$$

$$= (1-0.2) \times 0.3 \left\{ \begin{array}{l} \text{PWF}(20, 0.07, 0.08)\left[0.07 - \dfrac{1}{\text{PWF}(20, 0, 0.07)}\right] \\[2mm] + \dfrac{\text{PWF}(20, 0, 0.08)}{\text{PWF}(20, 0, 0.07)} \end{array} \right\}$$

$$= 0.8 \times 0.3\left[16.977\left(0.07 - \frac{1}{10.594}\right) + \frac{9.8181}{10.594}\right] = 0.123$$

$$P_{2,4} = (1 - Ct_e)M_1\text{PWF}(n_e, i, d) = (120/20,000)\text{PWF}(20, 0.05, 0.08) = 0.006 \times 14.358 = 0.0861$$

$$P_{2,5} = t_p(1 - t_e)V_1\text{PWF}(n_e, i, d)$$

$$= (300/20,000)(1 - 0.3) \times 1 \times \text{PWF}(20, 0.05, 0.08) = 0.015 \times 0.7 \times 14.358 = 0.151$$

EXAMPLE 12.11—CONT'D

$$P_{2,6} = \frac{Ct_e}{n_d} \text{PWF}(n'_{\min}, 0, d) = 0$$

$$P_{2,7} = \frac{R}{(1+d)^{n_e}} = \frac{0.3}{1.08^{20}} = 0.064$$

Finally, from Eq. (12.34),

$$P_2 = P_{2,1} + P_{2,2} - P_{2,3} + P_{2,4} + P_{2,5} - P_{2,6} - P_{2,7}$$
$$= 0.2 + 0.741 - 0.123 + 0.0861 + 0.151 - 0.064 = 0.9911$$

From Eq. (12.31),

$$\text{LCS} = P_1 C_{F1} FL - P_2 C_s = 20.242 \times 17.2 \times 0.65 \times 114.9 - 0.9911 \times 20,000 = \$6180.50$$

This is effectively the same answer as the one obtained in Example 12.7.

As can be seen from this example, the P_1, P_2 method is quick and easy to carry out manually.

As also can be seen, Eqs. (12.32) and (12.34) include only PWF values and ratios of payments to initial investment of the system and do not include as inputs the collector area and solar fraction. Therefore, as P_1 and P_2 are independent of A_c and F, systems in which the primary design variable is the collector area, A_c, can be optimized using Eq. (12.31). This is analyzed in the following section.

12.4.1 Optimization using P_1, P_2 method

As we have already seen in solar energy system design, the collector area is considered as the primary parameter for a given load and system configuration. The collector area is also the optimization parameter, i.e., the designer seeks to find the collector area that gives the highest LCS. A method for the economic optimization was given in Section 12.3.3, in which LCS are plotted against the collector area, A_c, to find the area that maximizes savings. The optimization procedure can be simplified if the LCS can be expressed mathematically in terms of the collector area. Therefore, the optimum is obtained when:

$$\frac{\partial(\text{LCS})}{\partial A_c} = 0 \tag{12.35}$$

or, by using Eq. (12.31) for LCS and Eq. (12.2) for C_s,

$$P_1 C_{F1} L \frac{\partial F}{\partial A_c} - P_2 C_A = 0 \tag{12.36}$$

Rearranging, the maximum savings are obtained when the relationship between the collector area and solar fraction satisfies the following relation:

$$\frac{\partial F}{\partial A_c} = \frac{P_2 C_A}{P_1 C_{F1} L} \tag{12.37}$$

According to Eq. (12.37), the optimum collector area occurs where the slope of the F versus A_c curve is $P_2 C_A / P_1 C_{F1} L$. This condition is shown in Fig. 12.3.

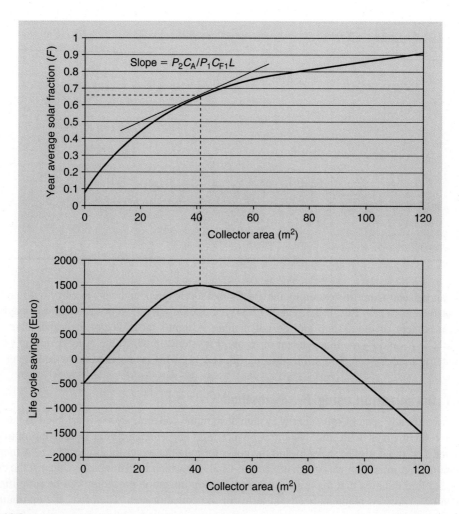

FIGURE 12.3

Optimum collector area determination from the slope of the F versus A_c curve.

EXAMPLE 12.12

For a residential liquid-based solar space heating system, the following information is given:

Annual heating load = 161 GJ.

First-year fuel cost rate, C_{F1} = $8.34/GJ.

Area-dependent cost = $210/m².

Area-independent cost = $1150.

Market discount rate, d = 8%.

Mortgage interest rate, d_m = 6%.

General inflation rate, i = 5%.

Fuel inflation rate, i_F = 9%.

Term of economic analysis = 20 years

Term of mortgage loan = 10 years

EXAMPLE 12.12−CONT'D

Down payment = 20%.
Ratio of first year miscellaneous costs to the initial investment, $M_1 = 0.01$.
Ratio of assessed value of the system in the first year to the initial investment, $V_1 = 1$.
Ratio of resale value = 0.3.
Property tax, $t_p = 2\%$.
Effective income tax, $t_e = 35\%$.
In addition, the solar fraction to the collector area varies as shown in Table 12.9.
Determine the optimum collector area that maximizes the LCS and the LCS.

Table 12.9 Solar fraction to collector area in Example 12.12.

Area (m²)	Annual solar fraction (F)
0	0
20	0.29
50	0.53
80	0.68
100	0.72

Solution

As the system is residential $C = 0$. The PWFs can be estimated from Eq. (12.18) or the tables of Appendix 8 as follows:

$$\text{PWF}(n_e, i_F, d) = \text{PWF}(20, 0.09, 0.08) = 20.242$$

$$\text{PWF}(n_{min}, 0, d) = \text{PWF}(10, 0, 0.08) = 6.7101$$

$$\text{PWF}(n_L, 0, d_m) = \text{PWF}(10, 0, 0.06) = 7.3601$$

$$\text{PWF}(n_{min}, d_m, d) = \text{PWF}(10, 0.06, 0.08) = 8.5246$$

$$\text{PWF}(n_e, i, d) = \text{PWF}(20, 0.05, 0.08) = 14.358$$

From Eq. (12.32),

$$P_1 = (1 - Ct_e)\text{PWF}(n, i_F, d) = 20.242$$

The various terms of parameter P_2 are as follows:

$$P_{2,1} = D = 0.2$$

$$P_{2,2} = (1 - D)\frac{\text{PWF}(n_{min}, 0, d)}{\text{PWF}(n_L, 0, d_m)} = (1 - 0.2)\frac{6.7101}{7.3601} = 0.729$$

$$P_{2,3} = (1 - D)t_e \left[\text{PWF}(n_{min}, d_m, d)\left(d_m - \frac{1}{\text{PWF}(n_L, 0, d_m)}\right) + \frac{\text{PWF}(n_{min}, 0, d)}{\text{PWF}(n_L, 0, d_m)}\right]$$

$$= (1 - 0.2) \times 0.3\left[8.5246\left(0.06 - \frac{1}{7.3601}\right) + \frac{6.7101}{7.3601}\right] = 0.064$$

$$P_{2,4} = (1 - Ct_e)M_1\text{PWF}(n_e, i, d) = 0.01 \times 14.358 = 0.1436$$

$$P_{2,5} = t_p(1 - t_e)V_1\text{PWF}(n_e, i, d) = 0.02(1 - 0.35) \times 1 \times 14.358 = 0.187$$

Continued

EXAMPLE 12.12—CONT'D

$$P_{2,6} = \frac{Ct_e}{n_d} \text{PWF}(n'_{min}, 0, d) = 0$$

$$P_{2,7} = \frac{R}{(1+d)^{n_e}} = \frac{0.3}{1.08^{20}} = 0.064$$

Finally, from Eq. (12.34),

$$P_2 = P_{2,1} + P_{2,2} - P_{2,3} + P_{2,4} + P_{2,5} - P_{2,6} - P_{2,7}$$

$$= 0.2 + 0.729 - 0.064 + 0.1436 + 0.187 - 0.064 = 1.1316$$

From Eq. (12.37),

$$\frac{\partial F}{\partial A_c} = \frac{P_2 C_A}{P_1 C_{F1} L} = \frac{1.1316 \times 210}{20.242 \times 8.34 \times 161} = 0.00874$$

The data of Table 12.9 can be plotted, and the optimum value of A_c can be found from the slope, which must be equal to 0.00874. This can be done graphically or using the trend line of the spreadsheet, as shown in Fig. 12.4, and equating the

FIGURE 12.4

Table 12.9 plotted using the spreadsheet trend line.

derivative of the curve equation with the slope (0.00874). By doing so, the only unknown in the derivative of the trend line is the collector area, and either by solving the second-order equation or through trial and error, the area can be found that gives the required value of the slope (0.00874).

As can be seen, the optimum solution occurs at about a collector area of 30 m^2 and $F = 0.39$ (The exact solution gives $A_c = 28.7$ m^2 and $F = 0.374$ but 30 m^2 is selected as the closer round figure). For this size, the total cost of the solar energy system is obtained from Eq. (12.2) as:

$$C_s = 210 \times 30 + 1150 = \$7450$$

The life cycle savings are obtained from Eq. (12.31) as:

$$\text{LCS} = P_1 C_{F1} F L - P_2 C_s = 20.242 \times 8.34 \times 161 \times 0.39 - 1.1316 \times 7450 = \$2170$$

12.5 Uncertainties in economic analysis

Due to the nature of the economic analysis, i.e., predicting the way various costs will occur during the life of a solar energy system, a number of uncertainties are involved in the method. The person responsible for the economic analysis of a solar energy system must consider a number of economic parameters and how these will develop in the years to come. A usual technique is to find how these parameters evolved in previous years and expect the same behavior to be reflected in future years. These two periods are usually equal to the expected life of the system. Additionally, the prediction of future energy costs is difficult because international oil prices change according to the quantity supplied by the oil-producing countries. Therefore, it is desirable to be able to determine the effect of uncertainties on the results obtained from an economic analysis.

For a given set of economic conditions, the change in LCS resulting from a change in a particular parameter, say, Δx_j, can be obtained from:

$$\Delta \text{LCS} = \frac{\partial \text{LCS}}{\partial x_i} \Delta x_i = \frac{\partial}{\partial x_i} [P_1 C_{F1} LF - P_2 (C_A A_c + C_1)] \Delta x_i \tag{12.38}$$

When uncertainties exist in more than one parameter, the maximum possible uncertainty is given by:

$$\Delta \text{LCS} = \sum_{i=1}^{n} \left| \frac{\partial \text{LCS}}{\partial x_i} \right| \Delta x_i \tag{12.39}$$

Therefore, the most probable uncertainty in the LCS can be written as:

$$\Delta \text{LCS}_p = \sqrt{\sum_{i=1}^{n} \left(\frac{\partial \text{LCS}}{\partial x_i} \Delta x_i \right)^2} \tag{12.40}$$

From Eq. (12.38),

$$\frac{\partial \text{LCS}}{\partial x_i} = \frac{\partial (P_1 C_{F1} LF)}{\partial x_i} - \frac{\partial [P_2 (C_A A_c + C_1)]}{\partial x_i} \tag{12.41}$$

The partial derivatives of the ratios P_1 and P_2 can be obtained using Eqs. (12.32) and (12.34) for the most crucial parameters, as follows:

For the fuel inflation rate,

$$\frac{\partial P_1}{\partial i_F} = (1 - Ct_e) \frac{\partial \text{PWF}(n_e, i_F, d)}{\partial i_F} \tag{12.42}$$

For the general inflation rate,

$$\frac{\partial P_2}{\partial i} = [(1 - Ct_e) M_1 + (1 - t_e) t_p V_1] \frac{\partial \text{PWF}(n_e, i, d)}{\partial i} \tag{12.43}$$

For the resale value,

$$\frac{\partial P_2}{\partial R} = \frac{1 - Ct_{\mathrm{e}}}{(1+d)^{n_{\mathrm{e}}}} \tag{12.44}$$

The partial derivative of the LCS with respect to the solar fraction is:

$$\frac{\partial \mathrm{LCS}}{\partial F} = P_1 C_{\mathrm{F1}} L \tag{12.45}$$

The partial derivatives of all parameters can be seen in Brandemuehl and Beckman (1979).

Finally, it is necessary to know the partial derivatives of the PWF values. Using Eqs. (12.17) and (12.18), the following equations can be obtained, as given by Duffie and Beckman (2006):

If $i = d$,

$$\frac{\partial \mathrm{PWF}(n, i, d)}{\partial n} = \frac{1}{1+i} = \frac{1}{1+d} \tag{12.46}$$

$$\frac{\partial \mathrm{PWF}(n, i, d)}{\partial i} = \frac{n(n-1)}{2(1+i)^2} \tag{12.47}$$

$$\frac{\partial \mathrm{PWF}(n, i, d)}{\partial d} = -\frac{n(n+1)}{2(1+d)^2} \tag{12.48}$$

If $i \neq d$,

$$\frac{\partial \mathrm{PWF}(n, i, d)}{\partial n} = -\frac{1}{d-i}\left(\frac{1+i}{1+d}\right)^n \ln\left(\frac{1+i}{1+d}\right) \tag{12.49}$$

$$\frac{\partial \mathrm{PWF}(n, i, d)}{\partial i} = \frac{1}{d-i}\left[\mathrm{PWF}(n, i, d) - \frac{n}{1+i}\left(\frac{1+i}{1+d}\right)^n\right] \tag{12.50}$$

$$\frac{\partial \mathrm{PWF}(n, i, d)}{\partial d} = \frac{1}{d-i}\left[\frac{n}{1+d}\left(\frac{1+i}{1+d}\right)^n - \mathrm{PWF}(n, i, d)\right] \tag{12.51}$$

It should be noted that, in order to estimate the uncertainties of more than one variable, the same procedure can be used to determine the appropriate terms in Eqs. (12.39) and (12.40). A much easier way to determine uncertainties is to use a spreadsheet for the calculations. In this case, the change of one or more parameters in appropriate cells (e.g., cells containing i, d, i_F, etc.) causes automatic recalculation of the spreadsheet, and the new value of the LCS is obtained immediately.

EXAMPLE 12.13

If, in a domestic solar energy system economic analysis, the fuel inflation rate is taken as 8%, find the uncertainty in the LCS if the fuel inflation rate is ±2%. The solar energy system replaces 65% of the annual load, the first year fuel cost is $950, the initial installation cost is $8,500, $n_{\mathrm{e}} = 20$ years, $d = 6\%$, $P_1 = 21.137$, and $P_2 = 1.076$.

Solution
The LCS of the system without uncertainty is obtained from Eq. (12.31):

$$\mathrm{LCS} = P_1 C_{\mathrm{F1}} F L - P_2 C_{\mathrm{s}} = 21.137 \times 950 \times 0.65 - 1.076 \times 8500 = \$3906$$

EXAMPLE 12.13—CONT'D

The fuel inflation rate affects only P_1. Therefore, from Eq. (12.42), for $C = 0$,

$$\frac{\partial P_1}{\partial i_F} = \frac{\partial PWF(n_e, i_F, d)}{\partial i_F}$$

From Eq. (12.50),

$$\frac{\partial PWF(n, i, d)}{\partial i} = \frac{1}{d - i}\left[PWF(n, i, d) - \frac{n}{1 + i}\left(\frac{1 + i}{1 + d}\right)^n\right]$$

$$= \frac{1}{0.06 - 0.08}\left[PWF(20, 0.08, 0.06) - \frac{20}{1.08}\left(\frac{1.08}{1.06}\right)^{20}\right] = 212.4$$

The uncertainty in the LCS can be obtained from Eqs. (12.38)–(12.41), all of which give the same result, as only one variable is considered. Therefore, from Eq. (12.38),

$$\Delta LCS = \frac{\partial LCS}{\partial i_F}\Delta i_F = C_{F1}LF\frac{\partial P_1}{\partial i_F}\Delta i_F = 950 \times 0.65 \times 212.4 \times 0.02 = \$2623$$

Therefore, the uncertainty in the LCS is almost equal to the two thirds of the originally estimated LCS, and this is for just 2% in uncertainty in fuel inflation rate.

Exercises

12.1 What is the present worth of $500 in 10 years if the market discount rate is 6%?

12.2 Estimate the present worth of a €1000 payment in 7 years for a market discount rate of 8% and inflation rate of 4%.

12.3 If the initial cost of a solar energy system is €7500, the mortgage term is 12 years, and the interest rate is 9%, find the annual payment.

12.4 The initial cost of a solar energy system is €14,000. If this amount is paid with a 30% down payment and the balance is borrowed at 8% interest for 12 years, calculate the annual payments and interest charges for a market discount rate of 6%. Also estimate the present worth of the annual payments.

12.5 Calculate the total present worth of the interest paid (PW_i) in Exercise 12.4.

12.6 Calculate the present worth of a 10-period series of payments, the first of which is $980, payable at the end of the first period, inflating by 6% per period, and the discount rate is 8%.

12.7 Calculate the cost of fuel of a conventional (non-solar) energy system for 12 years, if the total annual load is 152 GJ and the fuel price is $14/GJ, the market discount rate is 8%, and the fuel inflation rate is 5% per year.

12.8 Calculate the present worth of a fuel cost over 12 years, if the first year's fuel cost is €1050 and inflates at 7% for 4 years and 5% for the rest of the years. The market discount rate is 9% per year.

12.9 The area-dependent cost of a solar energy system is $175/m^2 and the area-independent cost is $3350. The down payment of the initial expenditure is 25% and the rest is paid in equal installments over 20 years at an interest rate of 7%. The backup fuel costs $12/GJ and its price inflates at 6% per year. Find the optimum system, if the combination of collector area and load covered is as shown in the following table and the total annual load is 980 GJ. The life of the system is considered to be 20 years, and at the end of this period, the system will be sold at 25% of its original value. In the first year, the extra maintenance, insurance, and parasitic energy costs are equal to 1% of the initial investment and the extra property tax is 1.5% of the initial investment, both expected to increase by 3% per year. The general market discount rate is 8%. The extra property taxes and interest on mortgage are deducted from the income tax, which is at a fixed rate of 30%.

Collector area (m^2)	Energy covered (GJ)
0	0
100	315
200	515
300	653
400	760
500	843

12.10 Tabulate the annual cash flows of a solar energy system that has an initial cost of $7000 for a 12-year analysis, where the following economic parameters apply:
Down payment = 20%.
Interest rate = 9%.
First-year fuel saving = $1250.
Market discount rate = 8%.
Fuel inflation rate = 8%.
Maintenance and parasitic cost = 0.5% increasing at 1% per year.
Resale value = 25%.
Also estimate the life cycle savings and the return on investment.

12.11 Repeat Exercise 12.10 using the P_1, P_2 method.

12.12 Find the undiscounted and discounted payback times of a solar energy system that covers 73% of an annual load of 166 GJ and costs €13,300. The first year fuel cost rate is €11.00/GJ and inflates at 8% per year and fuel costs are discounted at a rate of 6%.

12.13 For a residential solar space heating system, the following information is given:
Annual heating load = 182 GJ.
First-year fuel cost rate, C_{F1} = $9/GJ.
Area-dependent cost = $190/m^2.
Area-independent cost = $1200.
Market discount rate, d = 8%.
Mortgage interest rate, d_m = 7%.
General inflation rate, i = 6%.
Fuel inflation rate, i_F = 7%.

Term of economic analysis $= 12$ years
Term of mortgage load $= 8$ years
Down payment $= 15\%$.
Ratio of first-year miscellaneous costs to the initial investment, $M_1 = 0.01$.
Ratio of assessed value of the system in the first year to the initial investment, $V_1 = 1$.
Ratio of resale value $= 0.4$.
Property tax, $t_p = 2\%$.
Effective income tax, $t_e = 40\%$.

   In addition, the solar fraction to the collector area varies as follows. Determine the optimum collector area that maximizes the LCS and the LCS.

Area (m^2)	Annual solar fraction (F)
0	0
10	0.32
20	0.53
30	0.66
40	0.73

12.14 Estimate the uncertainty in the LCS of Exercise 12.13, if the uncertainty in general inflation rate is $\pm 3\%$.

Assignment #1

As an assignment, the student is required to construct a spreadsheet program that can be used to carry out most of the exercises in this chapter. Some of them are given at the companion site.

Assignment #2: optimized design of a solar thermal system

Student name: _____ Number of problem: _____
   Extend Assignment #1 of Chapter 11 to optimize the size of the solar thermal system by finding the optimum net aperture area of the solar collectors that provides maximum LCS. This applies to the same exercise according to the student number and the same city considered in the previous assignment.
   The purchase of the system is made by paying 20% of the total cost of the system as a down payment, and the rest is paid using a loan with a 20-year duration, according to the data below. Some common parameters are:

1. The cost of maintenance $= 1\%$ of the total cost of the system, increased by 0.5% per year.
2. The increase in auxiliary energy $= 1\%$ per year.
3. Total system life $= 20$ years.

   Other characteristics of the system are:

1. Area independent cost: _____ €
2. Area dependent cost: _____ €/m^2

3. Market discount rate (d): ____%
4. Interest rate: ____ %
5. Auxiliary energy cost: ____ €/GJ
6. System resale value: ____ %

First find the LCS for the system for a net collector area of 10 m² as in Chapter 11, Assignment 1, and then the collector area that maximizes the LCS of the system.

Problem options for various students:

Student number	Area independent cost (€)	Area dependent cost (€/m²)	Market discount rate (d) (%)	Interest rate (%)	Auxiliary energy cost (€/GJ)	Resale value (%)
1	500	300	4	4	28	20
2	600	280	4.2	5	30	22
3	700	260	4.4	4	32	24
4	800	240	4.6	5	30	16
5	900	220	4.8	4	28	28
6	1000	200	5	5	30	30
7	900	220	4.8	4	32	28
8	800	240	4.6	5	30	26
9	700	260	4.4	4	28	24
10	600	280	4.2	5	30	22
11	500	300	4	4	32	20
12	600	280	4.2	5	30	22
13	700	260	4.4	4	32	24
14	800	240	4.6	5	28	16
15	900	220	4.8	4	30	28

References

Beckman, W.A., Klein, S.A., Duffie, J.A., 1977. Solar Heating Design by the F-Chart Method. Wiley-Interscience, New York.
Brandemuehl, M.J., Beckman, W.A., 1979. Economic evaluation and optimization of solar heating systems. Sol. Energy 23 (1), 1−10.
Duffie, J.A., Beckman, W.A., 2006. Solar Engineering of Thermal Processes, third ed. John Wiley & Sons, New York.
Kalogirou, S.A., 1996. Economic analysis of solar energy systems using spreadsheets. In: Proceedings of the World Renewable Energy Congress IV, Denver, Colorado, vol. 2, pp. 1303−1307.
Kalogirou, S.A., 2003. The potential of solar industrial process heat applications. Appl. Energy 76 (4), 337−361.
Kalogirou, S.A., 2004. Environmental benefits of domestic solar energy systems. Energy Convers. Manage. 45 (18−19), 3075−3092.

Wind Energy Systems

Winds are movements of air masses in the atmosphere and are an indirect action of solar radiation falling on earth, as they are generated primarily by temperature differences within the air layer due to differential solar heating. Thus, wind energy can be considered to be a form of solar energy, a renewable energy resource. A brief historical introduction into wind energy and the systems used to harness it are given in Section 1.6.1 of Chapter 1. In this chapter, more details are given mainly about the wind characteristics and how these are evaluated in order to decide whether a site is suitable for wind exploitation, the aerodynamic analysis of wind turbines, based on the axial momentum theory, economics of wind energy and wind energy exploitation problems.

As we have seen in earlier chapters, the energy per unit surface area received from the sun depends on the latitude of a location, therefore different parts of the world receive different amounts of energy. This creates temperature differences and hence pressure gradients which, together with the Coriolis force associated with the rotation of the Earth, which deflects the air, except exactly on the equator, and the centripetal force, induce the movements of the air masses known as the "*gradient wind*".

The economies of wind power strongly depend on wind characteristics and mainly the magnitude of wind speed and its duration. Therefore, the first thing usually considered to examine the suitability of a site for wind energy exploitation is the wind characteristics analyzed in the following section.

13.1 Wind characteristics

As was seen above, the wind is caused by differences in pressure. Generally, the two major driving forces that drive large-scale winds are the different heating between the equator and the poles and the effect of the Earth's rotation. Closer to the Earth's surface, friction causes the wind to be slower.

Theoretically, at the Earth's surface, the wind blows from high-pressure areas to low-pressure areas. However, at the intermediate and higher latitudes, its direction is modified by the earth's rotation. In the northern hemisphere, the wind rotates counter-clockwise around cyclonic areas and clockwise around anticyclonic areas. In the southern hemisphere, these wind directions are reversed. On a terrestrial scale regular pressure systems produce the main winds, called *dominant winds*. Atmospheric circulation may be represented as shown in Fig. 13.1, which is useful in identifying the most important global wind characteristics.

As can be seen, three somewhat individualized cells exist: a tropical cell, also called *Hadley cell* because of George Hadley who originally described it; a temperate cell, also called *Ferrel cell* because of William Ferrel; and a polar cell. These turn one against the other like the gears in a gear transmission system. Additionally, the north and south tropical cells are divided from one another by the equatorial plane which is a low-pressure area, also called *intertropical convergence zone*. Moreover, the tropical

Solar Energy Engineering. https://doi.org/10.1016/B978-0-323-99350-0.00015-6
753

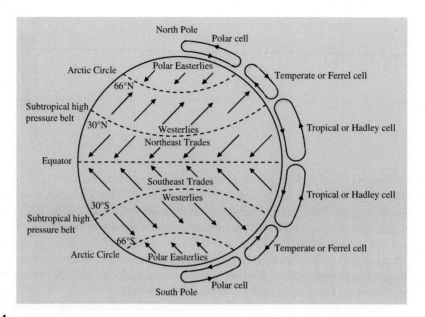

FIGURE 13.1

General world-wind circulations.

cell is divided from the temperate cell at about 30 °N and 30 °S latitudes by the subtropical high-pressure zones and the temperate cell is divided from the polar cell at about the Arctic Circle. The air from the subtropical high-pressure zone moves back toward the equator. This moving air is deflected by the Coriolis Effect to create the Northeast Trades (right deflection) and Southeast Trades (left deflection). The surface air moving toward the poles from the subtropical high zone is also deflected by Coriolis acceleration producing the Westerlies. Between the latitudes of 30 ° and 66 ° North and South, upper air winds blow generally toward the poles. Once again, Coriolis force deflects this wind to cause it to flow west to east forming the polar jet stream at roughly 66 ° North and South. On the Earth's surface at about 66 ° North and South latitude, the subtropical Westerlies collide with cold air traveling from the poles.

As can also be seen from Fig. 13.1 the cyclones near the equator are much thicker than the ones near the poles. The situation depicted in Fig. 13.1 is often altered because of unequal heating of the oceans' and continents' surface, existence of vegetation and seasonal variations that deform and modify the high-and low-pressure zones. There are additionally atmospheric disturbances created by masses of cold air that move occasionally from the poles toward the equator.

As a general guide, usually the most favorable areas for wind energy production are situated on sites near the seashores or in the sea.

13.1.1 Wind speed profiles

When dealing with wind energy we are interested in the wind speed and distribution up to about a height of 150 m from the land surface. The shape and roughness of the terrain over which the wind is blowing will impose a frictional effect on the wind speed near the surface. This is affected by both the height and the spacing of the roughness elements found on the surface of the earth, expressed by a parameter called, the *surface roughness length*, z_0. This is defined as the height above ground in meters at which the wind speed is theoretically equal to zero. Typical values of z_0 are: 0.0051 m for beach, ice, snow landscape and oceans, 0.032 m for low grass, airports and empty crop areas, 0.103 m for high-grass and low-crops areas, 0.254 m for tall row crops and low-woods areas and 0.505 for forests areas.

Because of the surface roughness effects the wind speed is lower closer to the ground. It is found that wind velocity increases with height above the ground according to the power law profile given by:

$$\frac{V_1}{V_2} = \left(\frac{z_1}{z_2}\right)^a \tag{13.1}$$

where

V_1 = wind velocity at some reference height z_1;
V_2 = wind velocity at height z_2; and
a = Hellman exponent.

The constant Hellman exponent (a) depends on the nature of the surface, stability of the air, temperature, time of the day, season, and surface roughness. It can be obtained from Lysen (1983):

$$a = 0.096[\ln(z_0)] + 0.016[\ln(z_0)]^2 + 0.24 \tag{13.2}$$

Equation (13.1) can also be used to correct the data recorded at a certain height to the data that apply to the required height. This is particularly convenient when examining the suitability of a site as weather masts for wind measurement have usually a height of 10 m, whereas wind data are required at the hub height of the wind turbines which nowadays is more than 100 m. Another way to calculate the Hellman exponent (a) is to perform measurements at the same time at two heights, say 10 and 50 m, and fit the data to obtain (a) from Eq. (13.1). Subsequently, use Eq. (13.1) to estimate the wind speed at other heights, say 150 m, that data have not being recorded, using the value of (a) calculated before. If detailed information does not exist, then for stable flow, a generally accepted value for (a) is 1/7 or 0.143. This procedure is explained better in the following example.

EXAMPLE 13.1

The following data of wind speed were measured at a particular site at the heights indicated in Table 13.1. Determine the best fit of the Hellman exponent and use the calculated value to predict the speed at a height of 150 m from the ground.

Solution

The pairs of data shown in Table 13.1 are plotted in Fig. 13.2. Using Excel function "linear trendline" and setting the line passing through 0, the equation shown is obtained. As can be seen, the data are fitted well with an R^2 value equal to 0.9952.

Table 13.1 Wind speed data at two different heights in m/s for Example 13.1.

10 m	50 m
6	7.4
7.8	9.9
5.2	6.6
6.1	7.8
8.2	10.3
9.1	11.3
9.8	12.5
5.8	7.2

Continued

EXAMPLE 13.1—CONT'D

FIGURE 13.2

Plot of the data for Example 13.1.

The equation on Fig. 13.2 is equivalent to $V_{50} = SV_{10}$, where S is the slope. So the slope is equal to 1.2558. Therefore, from Eq. (13.1):

$$S = \frac{V_{50}}{V_{10}} = \left(\frac{50}{10}\right)^a = 5^a$$

Taking the natural logarithm of both sides: $a = \ln(S)/\ln(5) = \ln(1.2558)/\ln(5) = 0.1415 \approx 0.142$.
As can be seen, the value obtained is very near the general value of 0.143.
To obtain the data for the 150 m height we use again Eq. (13.1) as:

$$V_{150} = V_{50}\left(\frac{150}{50}\right)^{0.142} \quad \text{which for the first reading is equal to: } V_{150} = 7.4\left(\frac{150}{50}\right)^{0.142} = 8.6 \text{ m/s}.$$

The complete set of results is shown in Table 13.2.

Table 13.2 Results for wind speed in m/s at 150 m for Example 13.1.

50 m	150 m
7.4	8.6
9.9	11.6
6.6	7.8
7.8	9.0
10.3	11.9
11.3	13.2
12.5	14.6
7.2	8.3

It should be noted that usually a lot more data than the ones used here are available, recorded over a long period of time to have a reliable prediction.

By assuming that wind speed grows logarithmically with the height, the wind speed at a certain height can also be estimated as a function of the height z and the roughness length z_0 from (de Jongh and Rijs, 2004):

$$V(z) = V_r \frac{\ln\left(\dfrac{z}{z_0}\right)}{\ln\left(\dfrac{z_r}{z_0}\right)} \tag{13.3}$$

where V_r is the reference speed at a reference height z_r, usually 10 m.

13.1.2 Wind speed variation with time

Knowledge of the behavior (variation and duration) of the wind is required so as to be able to understand and predict the performance of wind turbines. This usually varies with time and depends on the parameters listed above but mainly on the climate and topography of the region and the surface condition of the terrain around the site of interest. Figure 13.3 shows typical recordings of wind speed with an anemometer at three heights. As can be seen, near the ground turbulent fluctuations are random in nature and cannot be predicted with deterministic equations, therefore, usually statistical techniques are employed.

Time-varying wind speeds or the uncertainty about the magnitude of wind speed at the next minute or next hour is a great challenge for wind turbine designers. For example, a designer would like to know if a sudden strong wind or wind gust might happen which could change rapidly the wind turbine power production and would necessitate the use of the control system to adjust the speed of the rotor to optimize the turbine power output. Moreover, the designer would like to know the distribution of periods with low wind speed, which is valuable for the sizing of appropriate storage devices (Nfaoui, 2012).

FIGURE 13.3

Wind speed at three heights.

FIGURE 13.4

Frequency data for wind speed.

Additionally, not only the distribution of the wind speeds over a day or a year is important but also the number of hours per month or per year during which the given wind speeds occurred, i.e., the frequency distribution of the wind speed. A typical frequency distribution is usually presented as a histogram as shown in Fig. 13.4, for a selected year for a particular site. In this figure the maximum value corresponds to the most frequent wind speed and this occurs for the value of 6 m/s.

It is also important to know the number of hours over a time period that a wind turbine will not operate. In this case, it is necessary to add the number of hours in all intervals above a given wind speed for the whole examination period. In this way the duration distribution graph can be obtained, as the one shown in Fig. 13.5 for the data shown in Fig. 13.4. The flatter the duration curve, i.e., the longer one specific wind speed persists, the more constant the wind regime is, or, the steeper the duration curve is, the more irregular the wind regime is (Nfaoui, 2012).

In many calculations the standard deviation of wind velocity is required, given by:

$$\sigma_{\mathrm{w}} = \sqrt{\frac{1}{N-1} \sum_{i=1}^{N} \left(V_i - \overline{V}\right)^2} \tag{13.4}$$

FIGURE 13.5

Duration distribution histogram.

where the average velocity is given by:

$$\overline{V} = \frac{1}{N} \sum_{i=1}^{N} (V_i)$$

(13.5)

13.1.3 Statistical representation of wind speed

As was seen above, the real wind data are variable with time and usually a large number of these data are used to evaluate the suitability of a site for wind energy exploitation. Therefore, to be useable, the wind data must be reduced without losing any information, usually done using statistical techniques. What is required is to use mathematical functions that approach the velocity frequency data in a histogram as closely as possible, as shown in Fig. 13.6 for the data in Fig. 13.4. For this purpose the Weibull distribution function is usually used, as it gives a good match with the experimental data.

Weibull distribution is a continuous probability distribution function, named after Waloddi Weibull who originally developed it (Weibull, 1951). Weibull distribution is a tool to represent the frequency distribution of wind speed in a compact form. In practice, the power output of a wind turbine is estimated over an average year, based on the power curve of the turbine (see Section 13.3.2) and the predicted wind speed distribution at the hub height.

For this purpose the two parameter Weibull distribution is used, expressed mathematically as:

$$f(u) = \frac{k}{C} \left(\frac{u}{C}\right)^{k-1} \exp\left(-\left(\frac{u}{C}\right)^k\right)$$

(13.6)

where

$f(u)$ = frequency of occurrence of wind speed u;
C = Scale parameter (m/s); and
k = shape parameter ($-$).

FIGURE 13.6

Power density function for the data shown in Fig. 13.4.

The influence on the shape of the $f(u)$ curve for different values of the shape parameter (k) is shown in Fig. 13.7. Two special cases of the Weibull distribution are for $k = 1$, the exponential distribution, and $k = 2$, the Rayleigh distribution. In wind speed calculations the Rayleigh distribution is preferred. The two parameters can be obtained from:

$$k = \left[\left(\frac{\sum\limits_{i=1}^{N} n_i V_i^k \ln(V_i)}{\sum\limits_{i=1}^{N} n_i V_i^k} \right) - \left(\frac{\sum\limits_{i=1}^{N} n_i V_i}{N} \right) \right]^{-1} \tag{13.7a}$$

$$C = \left[\frac{\sum\limits_{i=1}^{N} n_i V_i^k}{N} \right]^{\frac{1}{k}} \tag{13.7b}$$

where N is the total number of observations of non-zero wind speed.

The cumulative distribution function $F(u)$ gives the probability of wind speed exceeding the value of u, given by:

$$F(u) = 1 - \exp\left(-\left(\frac{u}{C}\right)^k \right) \tag{13.8}$$

Solving the Weibull function for various cases we get:

$$\text{General solution}: \quad \overline{u^m} = C^m \Gamma\left(1 + \frac{m}{k}\right) \tag{13.9a}$$

$$\text{Mean value}: \quad \overline{u} = C\Gamma\left(1 + \frac{1}{k}\right) \tag{13.9b}$$

$$\text{Mean square}: \quad \overline{u^2} = C^2 \Gamma\left(1 + \frac{2}{k}\right) \tag{13.9c}$$

FIGURE 13.7

The shape of the weibull distribution function for different values of the shape parameter, k.

$$\text{Mean cube}: \quad \overline{u^3} = C^3\Gamma\left(1 + \frac{3}{k}\right) \tag{13.9d}$$

$$\text{Variance}: \quad C^2\left[\Gamma\left(1 + \frac{2}{k}\right) - \Gamma^2\left(1 + \frac{1}{k}\right)\right] \tag{13.9e}$$

where

$\Gamma(x) =$ gamma function.

Gamma function can be obtained from Table 13.3 or Excel using the function GAMMALN which gives the natural logarithm of gamma function.

One property of gamma function which is very useful is the following:

$$\Gamma(1 + n) = n\Gamma(n) \tag{13.10}$$

Table 13.3 Gamma function.

n	$\Gamma(n)$	n	$\Gamma(n)$	n	$\Gamma(n)$	n	$\Gamma(n)$
1.00	1.00000	1.25	0.90640	1.50	0.88623	1.75	0.91906
1.01	0.99433	1.26	0.90440	1.51	0.88659	1.76	0.92137
1.02	0.98884	1.27	0.90250	1.52	0.88704	1.77	0.92376
1.03	0.98355	1.28	0.90072	1.53	0.88757	1.78	0.92623
1.04	0.97844	1.29	0.89904	1.54	0.88818	1.79	0.92877
1.05	0.97350	1.30	0.89747	1.55	0.88887	1.80	0.93138
1.06	0.96874	1.31	0.89600	1.56	0.88964	1.81	0.93408
1.07	0.96415	1.32	0.89464	1.57	0.89049	1.82	0.93685
1.08	0.95973	1.33	0.89338	1.58	0.89142	1.83	0.93969
1.09	0.95546	1.34	0.89222	1.59	0.89243	1.84	0.94261
1.10	0.95135	1.35	0.89115	1.60	0.89352	1.85	0.94561
1.11	0.94740	1.36	0.89018	1.61	0.89468	1.86	0.94869
1.12	0.94359	1.37	0.88931	1.62	0.89592	1.87	0.95184
1.13	0.93993	1.38	0.88854	1.63	0.89724	1.88	0.95507
1.14	0.93642	1.39	0.88785	1.64	0.89864	1.89	0.95838
1.15	0.93304	1.40	0.88726	1.65	0.90012	1.90	0.96177
1.16	0.92980	1.41	0.88676	1.66	0.90167	1.91	0.96523
1.17	0.92670	1.42	0.88636	1.67	0.90330	1.92	0.96877
1.18	0.92373	1.43	0.88604	1.68	0.90500	1.93	0.97240
1.19	0.92089	1.44	0.88581	1.69	0.90678	1.94	0.97610
1.20	0.91817	1.45	0.88566	1.70	0.90864	1.95	0.97988
1.21	0.91558	1.46	0.88560	1.71	0.91057	1.96	0.98374
1.22	0.91311	1.47	0.88563	1.72	0.91258	1.97	0.98768
1.23	0.91075	1.48	0.88575	1.73	0.91467	1.98	0.99171
1.24	0.90852	1.49	0.88595	1.74	0.91683	1.99	0.99581
						2.00	1.00000

13.1.4 Wind resources

The most convenient quantity for the evaluation of the wind potential of a site is the power of the wind. At a particular instant the mass flow rate of air $\dot{m}$ with density ρ passing from an area A and kinetic energy $V^2/2$ gives wind power, P_w, equal to:

$$P_w = \frac{1}{2}\dot{m}V^2 = \frac{1}{2}\rho A V^3 \tag{13.11}$$

The power density, E, which is equal to the wind power divided by the area, is given by:

$$E = \frac{P_w}{A} = \frac{1}{2}\rho V^3 \tag{13.12}$$

Thus, the available wind power density is proportional to the cube of the wind speed, so using Eq. (13.9d), we get:

$$E = \frac{1}{2}\rho C^3 \Gamma\left(1 + \frac{3}{k}\right) \tag{13.13}$$

Therefore, from the above equations it can be concluded that the wind power is proportional to the density of the air, which means that at higher altitudes a wind turbine will produce less energy for the same wind speed. Additionally, wind energy is proportional to the swept area of the wind turbine and thus the square of the rotor diameter and most importantly to the cube of the wind speed, which means that 10% more speed will give 33% more energy.

To get the mean wind power however, it is not correct to use the cube of the average wind speed, but we need to consider the wind speed distribution using:

$$\overline{P_w} = \frac{A}{2N}\sum_{i=1}^{N}\rho_i V_i^3 \tag{13.14}$$

Therefore, the correct estimation of the mean wind power presumes knowledge of the wind speed distribution in the time period N. So, with a time series which includes the wind speed and density, the mean wind power can be estimated. The mean wind power can also be combined with the characteristics of the wind turbine for the calculation of the wind energy produced for any time period. In this case the power output is:

$$P = \frac{1}{2}\rho V^3 A_R \eta \tag{13.15}$$

where

A_R = rotor area of the wind turbine (m^2);
η = combined power coefficient.

The combined power coefficient is the product of the mechanical efficiency η_m, the electrical efficiency η_e and the aerodynamic efficiency C_P. All three efficiencies depend on the wind speed and the wind power. It has been shown by Betz law that the maximum possible value of aerodynamic efficiency is 59% (see Section 13.2), so even if there are no mechanical or electrical losses the maximum wind energy that can be converted into electrical energy by the wind turbine is 59%.

The power production for a particular wind turbine is determined by the power curve of a wind turbine (see Section 13.3.2). By combining the turbine power curve with the wind speed distribution at hub height, the mean annual energy production, AEP, can be estimated from:

$$\text{AEP} = N_0 \int_{V_{\text{start}}}^{V_{\text{stop}}} P(u)f(u)du \tag{13.16}$$

where

$P(u)$ = power curve function;
$f(u)$ = wind speed probability function, given by Eq. (13.6);
V_{start} = cut-in wind speed;
V_{stop} = cut-out wind speed; and
N_0 = number of hours in a year (8760).

The variance, which is the square of the standard deviation, can be obtained by solving Eq. (13.9b) for C and applying it in Eq. (13.9e). After some manipulations:

$$\sigma^2 = \overline{V}^2 \left[\frac{\Gamma\left(1 + \dfrac{2}{k}\right)}{\Gamma^2\left(1 + \dfrac{1}{k}\right)} - 1 \right] \tag{13.17}$$

Therefore, once the average and standard deviation of the wind velocity are known, parameters C and k can be found by simultaneous solution of Eqs. (13.9b) and (13.17).

The power of the wind can be estimated by Eq. (13.11) or the power density can be estimated by Eq. (13.12). By using the combined power coefficient the power of a wind turbine can be estimated by Eq. (13.15). Using the Weibull distribution the mean wind power that can be obtained from a wind turbine is given by:

$$\overline{P} = \eta E A_R = \eta \overline{\rho} A_R \frac{\overline{u^3}}{2} = \frac{1}{2} \eta \overline{\rho} A_R C^3 \Gamma\left(1 + \frac{3}{k}\right) \tag{13.18}$$

EXAMPLE 13.2

In an area where a wind turbine is installed the mean density of the air is 1.23 kg/m³. The wind turbine has a diameter equal to 36 m and its hub height is 80 m. The area is clear from any obstacles, the wind flow is stable and the wind turbine has a combined efficiency equal to 53.21%. At the height of 10 m the mean wind speed is 4.23 m/s and the standard deviation is 2.21 m/s. How much is the mean power of the wind turbine?

Solution

The Weibull parameters at 80 m can be found by using the standard deviation at 10 m. As no other data are given the generally accepted value for the Hellman exponent for stable flow is used which is equal to 1/7 (0.143).
So from Eq. (13.1):

$$\frac{V_1}{V_2} = \left(\frac{z_1}{z_2}\right)^a \Rightarrow \frac{V_2}{V_1} = \left(\frac{z_2}{z_1}\right)^a = \left(\frac{80}{10}\right)^{0.143} = 1.346$$

Therefore, the mean wind at the height of 80 m is: $V_{80} = 4.23 \times 1.346 = 5.69$ m/s.
And the standard deviation is $\sigma_{80} = 2.21 \times 1.346 = 2.97$ m/s.

Continued

EXAMPLE 13.2—CONT'D

The two Weibull parameters C and k can be found by simultaneous solution of Eqs. (13.9b) and (13.17). This can be done using program EES (Engineering Equation Solver). For this purpose the limited student version (does not allow copy, paste and save) can be used, obtained from www.fchart.com by writing the following commands:

u = C*gamma_(1 + 1/k)
sigma^2 = u^2*((gamma_(1 + (2/k)/(Gamma_(1 + (1/k))^2) − 1))
sigma = 2.97
u = 5.69.

They can also be estimated in Excel by using the function⇒ "goal seek". In fact the parameter C can be estimated by the mean wind speed and a random value of the parameter k and the difference between σ^2, calculated from Eq. (13.17) and the same random value of k, and the square of the σ value given should be zero (using goal seek). In this way k can be found (which gives a difference near zero) and at the same time C is estimated, which is also based on the value of k.

Using either of these procedures the values obtained are: $C = 6.421$ and $k = 2.003$.

Now using Eq. (13.18):

$$\overline{P} = \frac{1}{2}\eta\overline{\rho}A_R C^3 \Gamma\left(1 + \frac{3}{k}\right) = \frac{1}{2} \times 0.5321 \times 1.23 \times \frac{\pi \times (36)^2}{4} \times (6.421)^3 \times \Gamma\left(1 + \frac{3}{2.003}\right)$$

$$= 117 \text{ kW}$$

It should be noted that as $3/2.003 = 1.4977$, the last term of the above equation is equal to 2.4977 which cannot be evaluated directly from Table 13.3. So Eq. (13.10) should be used as:

$$\Gamma(1 + 1.4977) = 1.4977 \times \Gamma(1.4977) = 1.4977 \times 0.88623 = 1.3273$$

13.1.5 Wind resource atlases

Measurements for wind energy applications and resource assessments based on modeling tools have been carried out in many countries of the world. The Danish National Laboratory, RISO, developed a user-friendly topographical wind flow model called WAsP, which allows relatively inexpert users access to a very powerful tool of computational fluid dynamics. This tool can produce reliable results for local flows and has been used extensively by the wind energy community. Using this tool, wind atlases of the world have been produced by applying the WAsP program in more than 100 countries and territories around the world for national, regional and local studies. More information can be obtained from www.windatlas.dk. The world atlas showing wind speeds in m/s at 10 m above ground level is presented in Fig. 13.8.

13.1.6 Detailed study of wind speed

Wind is a highly variable energy resource and its speed varies stochastically both time wise and space wise. In order to control the dynamics of a wind system to make the required adjustments to a predicted change in wind speed, short-term wind speed forecasting is required. This is also very important for the integration of wind produced electrical energy into the grid. For this purpose a lot of methods have been used including statistical methods, Markov chain models, physical models, such as numerical weather forecast and others. Lately artificial neural networks (see Chapter 11), have been employed

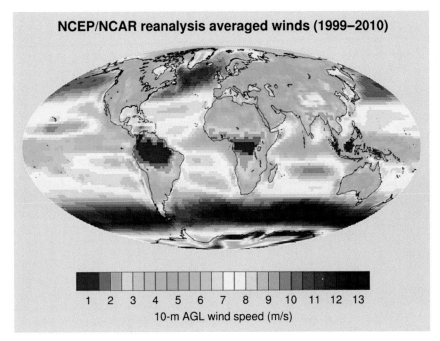

FIGURE 13.8

World wind atlas from RISO produced using the WAsP tool showing wind speed at 10 m above ground level (AGL).

With permission from Department of wind energy at the technical University of Denmark (DTU).

with success. These however are not suitable for hand calculations, so are not described in this book. The interested reader can find more information in other dedicated publications.

13.2 One-dimensional model for wind turbines

The wind turbine is the device used to convert wind speed into useful power. Details of the characteristics of wind turbines are given in Section 13.3. In this section the one-dimensional wind turbine model, which is based on the axial momentum theory, is presented, based on the analysis of Betz performed in 1926. The assumptions Betz made to develop this theory are the following:

- incompressible medium;
- no frictional drag (inviscid flow);
- axisymmetric flow;
- ideal rotor having an infinite number of blades;
- homogenous flow;
- uniform thrust over the rotor area;

- non-rotating wake region; and
- static pressure far before and far behind the rotor is equal to the undisturbed ambient static pressure.

The basics of axial momentum theory are still considered the backbone in rotor design of wind turbines. A wind rotor can extract power from the wind because it slows down the wind, i.e., the wind speed behind the rotor is lower than in front of the rotor. This fact is shown schematically in Fig. 13.9, which will be used to explain the model.

The momentum theory consists basically of control volume integrals for conservation of mass, axial and angular momentum balances and energy conservation (Sorensen, 2012). From this theory the Betz limit, which is the maximum power a wind turbine can produce, is derived.

The simple axial momentum theory is originally developed by Rankine and Froude. Based on Fig. 13.9, consider an axial flow of air with speed V_∞ passing through a turbine of rotor area A_R with constant axial load (thrust) T. Denoting by V_R the axial velocity in the rotor plane, by V_∞ the undisturbed axial velocity of the air and by ρ the density of air, in the wake region the air regains its undisturbed pressure value, $P_w = P_\infty$ (Sorensen, 2012). The model considers the stream tube that encloses the rotor disk, as shown in Fig. 13.9, and the two areas of importance are denoted as A_∞ and A_w being the cross-sectional areas of the air flow far upstream and far downstream of the rotor, respectively. The equation of continuity requires that the rate of mass flow $\dot{m}$ is constant in each cross-section. Thus,

$$\dot{m} = \rho V_\infty A_\infty = \rho V_R A_R = \rho V_w A_w \tag{13.19}$$

Axial momentum balance for the stream tube considered, results in the following equation for thrust:

$$T = \dot{m}(V_\infty - V_w) = \rho V_R A_R (V_\infty - V_w) \tag{13.20}$$

The pressure difference can be estimated using Bernoulli's equation. Applying the Bernoulli equation in front and behind the rotor, the total pressure head of the air in the stream will be decreased by:

$$\Delta P = \frac{1}{2}\rho\left(V_\infty^2 - V_w^2\right) \tag{13.21}$$

Actually, the pressure drop takes place across the rotor and represents the thrust:

$$T = A_R \Delta P \tag{13.22}$$

Therefore, substituting Eqs. (13.20) and (13.21) into Eq. (13.22), gives:

$$V_R = \frac{1}{2}(V_\infty + V_w) \tag{13.23}$$

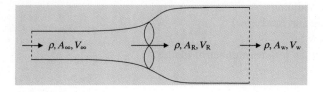

FIGURE 13.9

Control volume of one-dimensional wind turbine model.

Now, a new parameter is introduced, the axial interference factor, as follows:

$$\alpha = \frac{V_\infty - V_R}{V_\infty} \tag{13.24}$$

So, by using Eqs. (13.23) and (13.24) and performing various substitutions, we get:

$$V_R = (1 - \alpha) V_\infty \tag{13.25a}$$

$$V_w = (1 - 2\alpha) V_\infty \tag{13.25b}$$

From Eq. (13.25b) it is clear that the maximum value the axial interference coefficient α, can take is 0.5 otherwise the wake velocity would be negative which is impossible.

By substituting Eq. (13.25b) into Eq. (13.20), we get the following expression for the thrust:

$$T = 2\rho A_R V_\infty^2 \alpha (1 - \alpha) \tag{13.26}$$

Similarly, the power extraction is equal to $V_R T$, thus using Eqs. (13.25a) and (13.26) we get:

$$P = 2\rho A_R V_\infty^3 \alpha (1 - \alpha)^2 \tag{13.27}$$

Finally, we introduce the dimensionless thrust and power coefficients as:

$$C_T \equiv \frac{T}{\frac{1}{2} \rho A_R V_\infty^2} \tag{13.28a}$$

$$C_P \equiv \frac{P}{\frac{1}{2} \rho A_R V_\infty^3} \tag{13.28b}$$

Applying Eqs. (13.26) and (13.27) into Eq. (13.28) we get:

$$C_T = 4\alpha (1 - \alpha) \tag{13.29a}$$

$$C_P = 4\alpha (1 - \alpha)^2 \tag{13.29b}$$

The maximum value of the power coefficient can be obtained by differentiating Eq. (13.29b) with respect to the axial interference factor, α, and setting the derivative to zero. Thus:

$$\frac{dC_P}{d\alpha} = \frac{d(4\alpha - 8\alpha^2 + 4\alpha^3)}{d\alpha} = 4 - 16\alpha + 12\alpha^2 = 0 \tag{13.30}$$

Equation (13.30) is a second-order equation and has two solutions; $\alpha = 1$ and $\alpha = 1/3$. Applying $\alpha = 1$ gives the minimum value as $C_P = 0$. Therefore, the maximum value occurs at $\alpha = 1/3$, which when applied to Eq. (13.29b), gives:

$$C_{P,max} = 4 \times \frac{1}{3} \times \left(1 - \frac{1}{3}\right)^2 = \frac{16}{27} = 0.593 \tag{13.31}$$

This value is called the Betz limit and states that the maximum power that can be obtained from a wind turbine is equal to 59.3%. This value does not include the losses due to rotation of the wake and therefore it represents a conservative upper maximum.

As was said before a wind rotor can extract power from the wind because it slows down the wind, i.e., the wind speed behind the rotor is lower than in front of the rotor. Too much slowing down causes the air to flow around the wind rotor area instead of through the area and as is shown above the maximum power extraction is reached when the wind velocity in the wake of the rotor is 1/3 of the undisturbed wind velocity V_∞.

As can be seen from Eq. (13.29), both C_T and C_P are related only to the interference factor, α, so by eliminating α, the following relationship can be obtained between the two coefficients:

$$C_P = \frac{C_T}{2}(1 \pm \sqrt{1 - C_T}) \tag{13.32}$$

Therefore, by knowing any one of the two coefficients of an ideal wind turbine, the other can be estimated. Using a spreadsheet Figs. 13.10 and 13.11 were produced which are more handy to be used. It should be noted that although the $\pm$ sign is retained in Eq. (13.32), because of the solution of the second-order equation, only the positive values give a correct result.

FIGURE 13.10

Relationship between the power coefficient, C_P, and the thrust coefficient, C_T.

FIGURE 13.11

Relationship between the power coefficient, C_P, the thrust coefficient, C_T and the axial interference factor, α.

The factor 16/27 is called the Betz coefficient and represents the maximum fraction that an ideal wind rotor under the given conditions can extract from the flow. The power output of real wind turbines can be estimated using the power coefficient of a real turbine, $C_{P,real}$, from:

$$P_{real} = P_{Betz} \frac{C_{P,real}}{C_{P,Betz}}$$ (13.33)

EXAMPLE 13.3

In an area where a large wind turbine is installed, the speed of the wind is 9.6 m/s, the pressure is 1 atm and the temperature is 15 °C. The wind turbine rotor diameter is 118 m. Estimate the power density of the wind, the maximum power of the wind turbine, the thrust exerted on the turbine when working at maximum power and the maximum power output of a real wind turbine which has a power coefficient equal to 0.45.

Solution

From Appendix 5, Table A5.1 by interpolation the density of the air at 15 °C is 1.234 kg/m³. Therefore, the wind power density can be obtained from Eq. (13.12):

$$E = \frac{P_w}{A} = \frac{1}{2}\rho V^3 = \frac{1}{2} \times 1.234 \times (9.6)^3 = 546 \text{ W/m}^2$$

The maximum wind turbine power is estimated at the Betz limit, so from Eq. (13.15) using only $C_{P,Betz}$ for η:

$$P_{Betz} = \frac{1}{2}\rho V^3 A_R C_{P,Betz} = \frac{1}{2} \times 1.234 \times (9.6)^3 \times \frac{\pi(118)^2}{4} \times 0.593 = 3.54 \text{ MW}$$

The thrust exerted on the turbine, which represents the force required to keep the turbine in place can be obtained from Eq. (13.26) using $\alpha = 1/3$ as the estimation is done at the maximum power:

$$T = 2\rho A_R V_\infty^2 \alpha(1-\alpha) = 2 \times 1.234 \times \frac{\pi(118)^2}{4} \times (9.6)^2 \times \frac{1}{3}\left(1-\frac{1}{3}\right) = 553 \text{ kN}$$

It should be noted that the value of P_{Betz} can also be obtained from Eq. (13.27) using $\alpha = 1/3$. Finally, the maximum power of a real wind turbine can be estimated from Eq. (13.33):

$$P_{real} = P_{Betz} \frac{C_{P,real}}{C_{P,Betz}} = 3.54 \frac{0.45}{0.593} = 2.69 \text{ MW}$$

The detailed aerodynamic analysis of wind turbines is beyond the scope of this book. For this purpose, various numerically based aerodynamic rotor models have been developed in the past, ranging from simple lifting line-wake models to full-blown Navier–Stokes-based CFD models (Sorensen, 2012). These however are mostly computer-based models and are not suitable for hand calculations, so they are not included in this book. The following section gives technical details of wind turbines.

13.3 Wind turbines

Wind turbines are considered a matured technology and nowadays are provided commercially in a wide range of capacities varying from 400 W to 7.5 MW. As was seen above the theoretical maximum aerodynamic conversion efficiency of wind turbines, from wind to mechanical power is 59%. However, the fundamental properties of even the most efficient of modern aerofoil sections, used for blades of large and medium size wind turbines, limit the peak achievable efficiency to around 48%. In practice the need to economize on blade costs tends to lead to the construction of slender bladed, fast running wind turbines with peak efficiencies a little below the optimum, say 45%. The average, year-round efficiency of most turbines is about half this figure. This is due to the need to shut down the wind turbine in low or high winds and to limit the power once the rated level is reached. Further, a reduction of the average efficiency is caused by generator (and gearbox) losses and by the fact that the machine does not always operate in its optimum working point (Beurskens and Garrad, 1996).

In spite of its maturity, new control strategies and improved energy storage systems may increase the production of wind turbines. A picture of how the wind turbine diameter and power have increased in the last 30 years is shown in Fig. 13.12. The largest units today can produce 7.5 MW and have a rotor diameter of 170 m. It is anticipated that in the next few years machines with diameters of 250 m capable of producing 20 MW will be produced. These are possible today after the recent developments in carbon fiber-reinforced blade production, which allows the manufacture of long and strong blades with small weight.

As can be seen from Eq. (13.11) and other equations, the power of the wind is proportional to the rotor area. Therefore, the size of the rotor determines the power output of a wind turbine and the larger the diameter the more energy it can capture from a flowing wind. So according to Fig. 13.12, a 600 kW wind turbine will have a rotor diameter of 60 m whereas a 5 MW machine uses a diameter of 124 m.

13.3.1 Types of wind turbines

The two main concepts of wind turbines are the horizontal axis wind turbines (HAWTs) and the vertical axis wind turbines (VAWTs). All particular models of wind turbines generally belong to one of these categories. The world market is dominated however by the HAWT. Various designs have been proposed in the past in an attempt to reduce cost and increase efficiency.

The main types of wind turbines are shown in Fig. 13.13. These are the ones that have some potential as many other types have been proposed with limited success. Today the basic system of HAWT starts with a large rotor comprising two, three or four blades mounted on a horizontal shaft at the top of a tower. There is also a single-bladed machine, not shown in Fig. 13.13a, which offers the

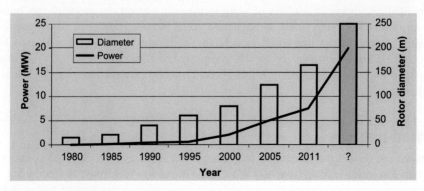

FIGURE 13.12

Evolution of wind turbine rotor diameter and power from 1980.

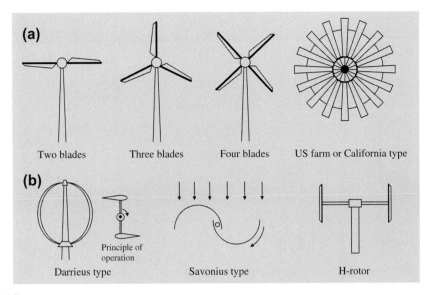

FIGURE 13.13

Main types of wind turbines. (a) Horizontal axis wind turbines. (b) Vertical axis wind turbines.

lowest cost and the lowest weight solution but must offset the counterweight. The most feasible design offering the lowest cost is the two-bladed machine, the drawback however of both one- and two-bladed machines is the high level of noise generated. Four-blade machines offer a good rotor balance but are heavier and less cost-effective (Sorensen, 2009). Nowadays most of the turbines are manufactured with three blades which is a compromise between the various types as they offer smoother rotational operation, lower noise and not a very high cost. The last type of HAWT shown is the farm or California-type wind turbine, used with success some years ago for water pumping.

The alternative is the VAWTs. In this case the weight is supported by a ground level bearing and both the gearbox and the generator can be at ground level, which makes maintenance easier compared to HAWT. Another advantage of VAWT is that it operates with wind blowing from any direction and unlike with HAWTs, no pointing mechanism is required to align with wind direction. Several designs of VAWT have been proposed, the most notable are the Darrieus, the Savonius and the H-rotor machines shown in Fig. 13.13b. Vertical-axis machines are good for pumping water and other high-torque low-speed applications and are not usually connected to the electric power grid.

The Darrieus type, first proposed in 1931 by the French engineer G.J.M. Darrieus, comprises two thin curved blades attached to a vertical shaft. The blades have an aerofoil shape as shown in the detail showing the principle of operation. It is also possible to have three or four blades.

The Savonius wind turbine is one of the simplest turbines, and was invented by the Finnish engineer S.J. Savonius in 1922. It is a drag-type device, consisting of two or three scoops. In the sectional view shown in Fig. 13.13b, a two-scoop machine would look like an "S". The differential drag in the two scoops causes the Savonius turbine to spin. Instead of adding more scoops in the same plane it is preferable to have many two-scoop arrangements located one on top of the other at different orientations. An alternative to this design is to have a long helical scoop, which also gives smooth torque

FIGURE 13.14

Combination of Darrieus and Savonius vertical wind turbines.

output. This design is often preferred in applications in buildings because it can be used also as a decorative element.

Because Savonius turbines are drag-type devices, they extract much less of the wind's power. For this reason some designs propose the combination of Darrieus and Savonius turbines as shown in Fig. 13.14.

Finally, the H-rotor design shown in Fig. 13.13b can be constructed with three or four blades; the three-blade design is preferred. The blades of this system are also of the aerofoil shape but they are straight instead of curved as in the Darrieus design.

Another type of VAWT, not shown in Fig. 13.13b, uses semi-circular cups instead of scoops. Usually three such cups are used and this type of system is mainly used in anemometers, which is the device used to measure wind speed.

Other classifications of wind turbines are possible and these can be according to:

- Direction of rotation—clockwise rotation (seen from the wind direction) dominates the market.
- Rotor control by stall-controlled (or fixed blades) or variable pitch blades.
- Electrical generators which may be of the induction type or of the DC generators with converter type.
- Transmission system which may include a gearbox in connection with induction generators, direct-drive operation with no gearbox in connection with multi-pole induction or synchronous generators or variable-speed generators.

A schematic diagram of the equipment found on the weather tight compartment at the top of a tower of an HAWT, called *nacelle*, is shown in Fig. 13.15. The slow rotation of the shaft of an HAWT is normally increased with a gearbox and passed to the generator. Gearbox and generator are attached directly to the turbine shaft though a braking system as shown. The electricity produced from the generator is taken with cables through the tower to a substation and eventually to the grid. The rotor, gearbox and generator are mounted on a platform which is able to rotate or yaw about a vertical axis so that the rotor shaft is perpendicular to the wind direction.

FIGURE 13.15

Schematic diagram of equipment contained in a nacelle.

Stalled-controlled wind turbines have their rotor blades fixed at a certain angle on the hub. This is a simple design but it has the drawbacks of low efficiency at low wind speed, no assisted start and variations in the maximum steady-state speed which may be caused because of the variations in wind speed, air density and grid frequency. On the contrary the blades of pitch-controlled turbines can be controlled to have their blade cord parallel to the wind direction. This solves the disadvantages of the stall-controlled machines but adds extra complexity and cost.

Modern wind turbines rotate at speeds between 5 and 20 rpm whereas generators operate between 800 and 3000 rpm. Therefore, there is a need for a speed-up gearbox to increase the speed to the required level. Such a gearbox must be heavy duty because it has to withstand not only the rotational speed from the rotor but also the bending moment created because of the variability of the wind with height and the force created by the generator trying to synchronize its speed with the grid frequency. Gearboxes require greasing and frequent maintenance that has to be done at the hub level and is the most likely component of the system that can create problems or fail. For this reason direct-drive systems have been developed in which the rotor is connected directly to the generator, as indicated in the classification of wind turbines above. Another design gaining ground nowadays is the use of a variable-speed generator. The variation in the speed of variable speed systems is about 10%−20% and this reduces drastically the mechanical stresses on the gearbox and increases the overall efficiency of a wind turbine.

Towers used in wind turbines have been constructed in a variety of styles. Originally the farm type wind turbines employed the lattice metal tower but this is replaced by cylindrical steel or concrete designs. The cylindrical metal design dominates the market today. The diameter of the tower depends on the rotor diameter and tower height, both depending on the nominal capacity of the turbine and the need to locate the rotor at high level to avoid the turbulent air near the ground and the higher available wind speed. Towers are fixed on a large diameter underground concrete base, which will require a substantial excavation during the construction phase. Towers house also the wires transmitting the generated electricity, the control and isolation systems as well as an elevator and a stair for the maintenance personnel to visit the nacelle. Photographs inside such a tower are shown in Fig. 13.16.

FIGURE 13.16

Photos inside a wind turbine tower showing the transmission wires and the stair (left), isolation and control system and elevator (right).

13.3.2 Power characteristics of wind turbines

The characteristics of a wind turbine are given in terms of the power curve. As can be seen from Eq. (13.16) the mean annual energy production, AEP, depends on the power curve function $P(u)$. A power curve gives the relation between wind speed and power output of a wind turbine. Figure 13.17 shows on the same graph the power curve and the efficiency of a wind turbine. The efficiency typically reaches a maximum value at a wind speed of 7−9 m/s.

On the same graph the Betz limit is also drawn to show the differences of actual efficiency to the ideal one. The nominal or rated power is usually obtained at wind speeds greater than 12 m/s. The cut-in speed is the speed of wind required to set the turbine in motion whereas the cut-out speed is the wind speed at which the turbine stops for safety reasons.

FIGURE 13.17

Power and efficiency characteristics of a wind turbine.

13.3.3 Offshore wind turbines

Generally it is more expensive to build a wind park offshore but the associated cost is offset by the higher annual yield as at these locations there is higher average wind speed. The global offshore wind energy resource is estimated to be about 37,500 TWh (Leutz et al., 2002). Although today there are solutions to install wind turbines in sites where the water is deep, the most feasible solutions is for sites where the water is relatively shallow and are at a distance of about 25 km from the shoreline so as the turbines are not visible. However distances as close as 15 km present moderate visual effects.

The wind in offshore locations is generally less disturbed because the surface of the wind tends to be smoother and there are no obstacles to create turbulence in the wind. This is not the case in rough seas with big waves. Seas with persistent roughness should be avoided for wind exploration.

The major barrier in offshore wind energy exploitation is the cost. Depending on the distance from the shoreline and the depth of seawater the additional cost can be between 40% and 100% compared to onshore installations. This includes increased installation cost as well as the higher grid connection cost. Additionally, there is an increase in the maintenance cost, which in addition to the normal standard cost requires boat transportation to each wind turbine. These costs are usually alleviated by the higher wind speeds available and the stability of the resource, which lead to higher electrical energy output from the wind, which may give more than 50% higher annual production compared to onshore turbines of the same capacity. To be feasible however, bigger size turbines need to be used so as to use a smaller number of them for the same energy output.

A number of methods have been used to install wind turbines in the sea. In shallow waters the normal concrete foundation can be used. For higher depths, single pillar or monopile support, tripod support or a floating support can be used. Details of these are beyond the purpose of this book.

Finally, a future energy offshore system could combine offshore wind systems with wave systems. Such a combination offers interesting possibilities which could smooth the fluctuations in the produced electrical energy but no such systems have been constructed yet.

13.3.4 Wind parks

Wind parks or wind farms, as they are also known, cluster together a number of wind turbines. A photo of one such park is shown in Fig. 13.18. Usually, large wind parks are installed in low population areas or offshore. Although this creates extra costs to transfer the electrical energy to the populated areas, it reduces the reaction of the public who oppose the development of wind energy. Grid connection costs are reduced drastically by combining a number of large wind parks located in the same area. The connection of the wind turbines of a wind park to the grid will normally require a substation to raise the voltage to the value of the distribution grid voltage which is usually very high to reduce transmission losses. So in the case of combining a number of large wind parks located in the same area together, a lower voltage substation may be required for each wind park and a common large one to connect to the high-voltage grid transmission system.

Wind turbines in a wind park are spaced so as to have a distance between them of 5−10 rotor diameter so as to avoid wind interference effects. This means that a large area will be required to install many wind turbines although the actual area that the turbines will occupy will be very small.

FIGURE 13.18

Photo of a wind park.

13.4 Economic issues

The economics of wind energy systems depend strongly on the available wind speed, because as was seen previously the wind power is directly proportional to the third power of wind speed. Therefore, by doubling the wind speed eight times more power is obtained. For example a 1.5 MW wind turbine installed at a site with mean wind speed of 5.5 m/s will generate about 1000 MWh/year. At a site with average wind speed of 8.5 m/s the turbine will generate 4500 MWh/year and if the wind speed is 10.5 m/s it will produce 8000 MWh/year (Sorensen, 2009). Therefore, it is clear from this discussion that the selection of a suitable site is very important for the economics of a wind energy exploitation project. Other factors that are important are the distance from the existing electricity grid, the availability of roads and generally infrastructure at a location, the type of terrain, which is related to the turbulence created in the wind and determines the height of the wind turbines and the existence of natural habitat in the area. With the modern wind turbines, the minimum wind speeds which are considered economically exploitable are 5–5.5 m/s for onshore and 6.5 m/s for offshore sites. The latter is bigger due to the higher installation cost (Sorensen, 2009).

Wind energy exploitation, especially in the case of large wind parks, requires a large financial investment. Therefore, they require a serious planning and careful study and preparation of the site. The most important study required is the confirmation of the available wind speed to a potential site. For this purpose short- and long-term wind speed measurements will be required to verify the wind regime with on-site measurements of at least a year.

Costs of wind turbine installations usually include:

- Turbines installation cost which includes foundations, transportation and cranes;
- Road construction cost;
- Underground cabling within the wind park;
- Substation cost;
- Cost of control and monitoring center;
- Testing and commissioning cost; and
- Administrative, financing and legal costs.

Most of these costs are not dependent on the size of the park so larger wind parks comprising many wind turbines benefit from economies of scale. The normal lifetime of wind turbines is 20−25 years.

In many countries wind power is already competitive with fossil and nuclear power when external/social costs are included. The often-perceived disadvantage that wind (and solar) energy is an intermittent (stochastically varying) source, not representing any capacity credit, making the resource of uncertain value for large-scale electricity production, is not true. Utility studies have shown that wind energy does represent a certain capacity credit, although the factor is 2−3 times lower than the value for nuclear and fossil fuel-fired plants. Thus, wind energy replaces fossil fuels and saves capacity of other generating plants.

Wind energy technology has progressed significantly over the last 2 decades, driving down capital costs to as low as €1000 to €1350 per kW for onshore installations and €1500 to €2000 per kW for offshore installations, depending on the particular site conditions. At this level of capital costs, wind power is already economical at locations with fairly good wind resources. Therefore, the average annual growth in worldwide wind energy capacity was over 30% for the last 5 years. The total worldwide installed capacity reached a level of 59 GW in 2005 whereas by the end of 2021 it reached 845 GW with 102 GW added in 2021. More details on these issues are given in Chapter 1, Section 1.3.6.

The world's total theoretical potential for onshore wind power is around 55 TW with a practical potential of at least 2 TW, which is about two-thirds of the entire present worldwide generating capacity. The offshore wind-energy potential is even larger.

Wind energy can compete with energy from other sources (coal, oil and nuclear) only under favorable wind and grid conditions. Further decrease of cost will extend the market potential for wind turbine systems considerably. Decrease of the cost of wind energy can be achieved by reducing the relative investment cost, introducing reliability design methods, and exploiting the best available wind sites.

13.5 Wind-energy exploitation problems

The growth of the installed wind power is hampered by a number of barriers. These are public acceptance, land requirements, visual impact, audible noise, telecommunication interference, and various impacts on natural habitat and wild life.

Although wind energy systems operate cost-effectively exploiting a renewable energy resource, it is not considered beneficial by all. Many lobbies exist in countries with high penetration of wind energy systems, trying to resist further development of onshore wind parks. It is a fact that wind energy turbines present some problems as shown above; the most serious are the visibility and noise problems. Wind parks cover a large area and it is not possible to hide them. Large wind turbines are visible over long distances. The concept that wind energy is very good "but not in my back yard" is the main argument of various opposition groups.

Noise is in fact not a problem at distances of more than 500 m from a wind turbine and noise from modern wind turbines is much reduced compared to earlier machines. The noise is low-frequency and is generated by the movement of the blades in the air and from the gearbox and the generator. The blade noise is the most serious of these. There is also some noise, dust and other disturbance presented during the installation of the wind turbines but this is temporary.

From the ecology point of view there are some problems related to possible loss of habitat for birds, reduction of breeding and feeding areas and problems related to fatal collisions on the wind turbine blades although this is not as serious as fatal collisions with cars and buildings. This is a problem only if a wind park is located at an area which is a passage of migratory birds, so these areas should be avoided.

Finally, under certain circumstances wind turbines can cause electromagnetic interference affecting television signal transmission and telecommunications. These can be mitigated with simple remedial measures and by selecting sites which are at a safe distance from relaying stations.

Most of the above problems can be solved by offshore wind turbine installation, although these plants also create some problems as they interfere with fisheries and shipping and affect marine life. During construction there are also some temporary problems concerning the disruption of the seabed. Offshore wind parks also interfere with radar signals although modern radar systems are less affected by such interferences.

Exercises

13.1 The following data of wind speed in m/s were measured at a particular site at the heights indicated in the table. Determine the best fit of the Hellman exponent and use the calculated value to predict the speed at a height of 120 m from the ground.

10 m	40 m
8.0	9.8
9.7	11.9
7.6	9.4
7.0	8.9
7.2	8.8
9.1	11.9
9.8	12.1
6.8	8.6

13.2 In an area near an airport what is the value of the Hellman coefficient?

13.3 In an offshore area the wind speed at a reference height of 10 m is 6.9 m/s. What is the speed at a height of 100 m?

13.4 For average wind speed of 7 m/s and standard deviation of 2.2 m/s estimate the two Weibull parameters C and k.

13.5 In an area where a wind turbine is installed, the mean density of the air is 1.225 kg/m³. The wind turbine has a diameter equal to 46 m and its hub height is 100 m. The area is clear from any obstacles, the wind flow is stable and the wind turbine has a combined efficiency of 53.29%. At the height of 10 m the mean wind speed is 5.25 m/s and the standard deviation is 2.75 m/s. How much is the mean power of the wind turbine?

13.6 If the axial velocity in the rotor plane is 5.5 m/s and the undisturbed wind velocity is 7.5 m/s what is the value of the thrust and power coefficients?

13.7 In an area where a large wind turbine is installed, the speed of the wind is 12.2 m/s, the pressure is 1 atm and the temperature is 20 °C. The wind turbine rotor diameter is 125 m. Estimate the power density of the wind, the maximum power of the wind turbine, the thrust exerted on the turbine when working at maximum power and the maximum power output of a real wind turbine which has a power coefficient equal to 0.48.

Assignment

With the help of the System Advisor Model (SAM), details of which are given in Appendix 10, design a 50 MW wind energy system. This can be done quite easily by running the Wind Wizard. This can be accessed by choosing "Quick start for new users" at the bottom left corner of the welcome screen of SAM and then selecting "Wind Wizard," which will guide you through the steps to select the area of installation, model of a wind turbine, and other parameters. Alternatively, a normal procedure can be followed starting from Wind and then selecting the type of financial model to use. You can choose different types of financial models, various types of wind turbines, and incentives. For actual installation and operating costs of each combination, find the most cost-effective combination of equipment setup. For parameters not specified above, you can use realistic figures (or the program default values), which need to be listed in the report produced.

References

Beurskens, J., Garrad, A., 1996. Wind energy. In: The Proceedings of EuroSun 96 Conference, Freiburg, Germany, vol. 4, pp. 1373−1388.

de Jongh, J.A., Rijs, R.P.P. (Eds.), 2004. Wind Resources. Arrakis, p. 34.

Leutz, R., Ackerman, T., Suzuki, A., Akisawa, A., Kashiwagi, T., 2002. Technical offshore wind energy potential around the globe. In: Proceedings of the European Wind Energy Conference, Copenhagen, Denmark.

Lysen, E.H., 1983. Introduction to Wind Energy. CWD 82-1, Amersfoort, Netherlands.

Nfaoui, H., 2012. Wind energy potential. In: Sayigh, A. (Ed.), Comprehensive Renewable Energy. Chapter 2.04, Major Reference Works, vol. 2. Elsevier, pp. 73−92.

Sorensen, B. (Ed.), 2009. Renewable Energy Focus Handbook. Academic Press, Elsevier, ISBN 978-0-12-374705-1, pp. 435−444. Wind power (Chapter 9.1).

Sorensen, J.N., 2012. Aerodynamic analysis of wind turbines. In: Sayigh, A. (Ed.), Comprehensive Renewable Energy. Chapter 2.08, Major Reference Works, vol. 2. Elsevier, pp. 225−240.

Weibull, W., 1951. A statistical distribution function of wide applicability. J. Appl. Mech. 18 (3), 293−297.

Appendix 1: Nomenclature

In this book, all the symbols used are explained when they first occur. This appendix includes a general list to consult in case of doubt. It should be noted, however, that only the general or most used symbols are included in the following list and not variations, which are usually denoted with subscripts. The appendix initially presents the nomenclature for radiation and then the general list of symbols. Finally, a list of abbreviations is provided.

Radiation nomenclature

G	Irradiance (W/m^2)
H	Irradiation for a day (J/m^2)
I	Irradiation for an hour (J/m^2)
R	Radiation tilt factor

Subscripts

B	Beam
c	Critical
D	Diffuse
G	Ground reflected
n	Normal
t	Radiation on tilted plane

Symbols

A	Aperture area (m^2)
a	Hellman coefficient
A_a	Absorber area (m^2)
A_c	Total collector aperture area (m^2)
A_f	Collector geometric factor
A_r	Receiver area (m^2)
A_R	Rotor area (m^2)
b_0	Incidence angle modifier constant-first order
b_1	Incidence angle modifier constant-second order
C	Collector concentration ratio $= A_a/A_r$; capacitance rate; speed of light; useful accumulator capacity (Wh); mass concentration (kg of salts/kg of water); scale parameter
c_0	Intercept efficiency $= F_R(\tau\alpha)$
c_1	First-order coefficient of the collector efficiency (W/m^2 °C)
c_2	Second-order coefficient of the collector efficiency (W/m^2 °C^2)
C_A	Cost per unit of collector area or area-dependent cost ($/m^2)
C_b	Bond conductance (W/m °C)
C_{F1}	Cost of fuel during first year of operation ($)
C_I	Area-independent cost ($)
C_P	Power coefficient
c_p	Specific heat at constant pressure (J/kg °C)

C_s	Total cost of solar equipment ($)
C_T	Thrust coefficient
d	Market discount rate (%); interest rate (%); pipe diameter (m)
D	Riser tube outside diameter (m); energy demand (J)
d^*	Universal nonrandom error parameter due to receiver mislocation and reflector profile errors, $d^* = d_r/D$
D_i	Tube inside diameter (m)
d_m	Mortgage interest rate (%)
D_o	Tube outside diameter (m)
d_r	Displacement of receiver from focus (m)
E	Emissive power (W/m^2 μm); exergy (W); voltage source (V); root mean square error; power density (W/m^2); collector effectiveness
E_{in}	Exergy in (W)
E_{out}	Exergy out (W)
E_p	Photon energy (J)
E_{PV}	Energy delivered from a PV array (J)
F	Annual fraction by solar energy; view factor; fin efficiency; cash flow ($); fraction of time night insulation is used; fraction of window shaded
f	Focal distance (m); monthly solar fraction; friction factor; fraction of steam in turbine bleed
F'	Collector efficiency factor
F''	Flow factor
F_R	Heat removal factor
F'_R	Collector heat exchanger efficiency factor
f_{TL}	Fraction of total load supplied by solar, including tank losses
g	Acceleration due to gravity (m/s)
G_{on}	Extraterrestrial radiation measured on a normal surface (W/m^2)
Gr	Grashof number
G_{sc}	Solar constant = 1366.1 W/m^2
H	Enthalpy (J)
h	Hour angle (degrees); Specific enthalpy (kJ/kg); Planck's constant = 6.625×10^{-34} (J s); shape parameter
h_c	Convection heat-transfer coefficient (W/m^2 °C)
h_{fi}	Heat-transfer coefficient inside absorber tube or duct (W/m^2 °C)
H_o	Total extraterrestrial radiation on a horizontal surface over a day (J/m^2)
h_p	Height of a parabola (m)
H_p	Latus rectum of a parabola (m) = opening of the parabola at focal point
h_r	Radiation heat-transfer coefficient (W/m^2 °C)
h_{ss}	Sunset hour angle (degrees)
H_{ss}	Sunset time
h_T	Thermosiphon head (m)
$\overline{H}_t$	Monthly average daily radiation incident on the collector surface per unit area (J/m^2)
H_t	Tower height (m)
h_v	Volumetric heat-transfer coefficient (W/m^3 °C)
h_w	Wind heat-transfer coefficient (W/m^2 °C)
i	Inflation rate (%)
I	Irreversibility (W); current (A)
i_F	Fuel inflation rate (%)
I_o	Dark saturation current (A)
I_{ph}	Photocurrent (A)

I_{sc}	Short-circuit current (A)
J	Radiosity (W/m^2); mass flux (kg/m^2 s)
K	Extinction coefficient
k	Thermal conductivity (W/m °C); friction head (m)
k_o	Intercept efficiency $= F_R\eta_o$
k_1	First-order coefficient of the collector efficiency (W/m^2 °C) $= c_1/C$
k_2	Second-order coefficient of the collector efficiency (W/m^2 °C^2) $= c_2/C$
K_{str}	Stratification coefficient
K_T	Daily clearness index
k_T	Hourly clearness index
K_θ	Incidence angle modifier
L	Half distance between two consecutive riser pipes (m) $= (W - D)/2$; collector length (m); latitude (degrees); annual energy required by the load (J); thickness of glass cover (m); pipe length (m); mean daily electrical energy consumption (Wh)
L_{AUX}	Annual energy required by the auxiliary (J)
L_m	Monthly load (J)
L_S	Load covered by solar energy (J)
m	Air mass
M	Mass flow number; mass of storage capacity (kg); molar mass (kg/mole); mixing number
$\dot{m}$	Mass flow rate of fluid (kg/s)
N	Days in month; number of moles; number of bends in serpentine collector
n	Number of years; refraction index; number of reflections
N_g	Number of glass covers
N_p	Average daily collector operating time (h)
n_p	Payback time (years); number of photons
N_s	Entropy generation number
Nu	Nusselt number
P	Profile angle (degrees); power (W); pressure (Pa)
P_1	Ratio of life cycle fuel savings to first year fuel energy cost
P_2	Ratio of cost owed to initial cost
P_a	Partial water vapor pressure of air (Pa)
P_h	Collector header pressure drop (Pa)
Pr	Prandtl number
P_s	Saturation water-vapor pressure at air temperature, t_a (Pa)
P_w	Saturation water-vapor pressure at water temperature, t_w (Pa), wind power (W)
PW_n	Present worth after n years
Q	Energy (J); rate (energy per unit time) of heat transfer (W)
q	Energy per unit time per unit length or area (W/m or W/m^2); heat loss (J/m^2 day)
q^*	Irradiation per unit of collector area (W/m^2)
Q^*	Solar radiation incident on collector (W)
Q_{aux}	Auxiliary energy (J)
Q_D	Dump energy (J)
q'_{fin}	Useful energy conducted per unit fin length (J/m)
Q_g	Net monthly heat gain from storage wall (J)
Q_l	Load or demand energy (J); rate of energy removed from storage tank (W)
Q_o	Rate of heat loss to ambient (W)
q_o^*	Radiation falling on the receiver (W/m^2)
Q_s	Radiation emitted by the sun (W/m^2)

Q_S	Solar energy delivered (J)
Q_{tl}	Rate of energy loss from storage tank (W)
q'_{tube}	Useful energy conducted per unit tube length (J/m)
Q_u	Useful energy collected (J); rate of collected energy delivered to storage tank (W); rate of useful energy delivered by the collector (W)
q'_u	Useful energy gain per unit length (J/m)
r	Ratio of total radiation in an hour to total in a day, Parabolic reflector radius (m); water recovery rate
R	Receiver radius (m); ratio of total radiation on tilted plane to horizontal plane; thermal resistance (m² °C/W), membrane salts rejection rate
Ra	Rayleigh number
r_d	Ratio of diffuse radiation in an hour to diffuse in a day
Re	Reynolds number
R_s	Ratio of standard storage heat capacity per unit area to actual = 350 kJ/m² °C
R_{total}	Total thermal resistance (m² °C/W)
S	Absorbed solar radiation per unit area (J/m²); Seebeck coefficient
s	Specific entropy (J/kg °C)
S_{gen}	Generated entropy (J/°C)
S_M	Membrane active area (m²)
T	Absolute temperature (K)
t	Time
T^*	Apparent sun temperature as an exergy source, ~4500 K
T_a	Ambient temperature (°C)
$\overline{T}_a$	Monthly average ambient temperature (°C)
T_{av}	Average collector fluid temperature (°C)
T_b	Local base temperature (°C)
T_C	Absolute temperature of the PV cell (K), cold reservoir temperature (K)
T_c	Cover temperature (°C)
T_f	Local fluid temperature (°C)
T_{fi}	Temperature of the fluid entering the collector (°C)
T_H	Temperature of hot reservoir (K)
T_i	Collector inlet temperature (°C)
T_m	Public mains water temperature (°C)
T_o	Ambient temperature (K); temperature of the fluid leaving the collector (°C)
T_{oi}	Collector outlet initial water temperature (°C)
T_{ot}	Collector outlet water temperature after time t (°C)
T_p	Average temperature of the absorbing surface (°C); stagnation temperature (°C)
T_r	Temperature of the absorber (°C); receiver temperature (°C)
T_{ref}	Empirically derived reference temperature = 100 °C
T_s	Apparent black body temperature of the sun, ~6000 K
T_w	Minimum acceptable hot-water temperature (°C)
U	Overall heat-transfer coefficient (W/m² °C)
U_b	Bottom heat-loss coefficient (W/m² °C)
U_e	Edges heat loss coefficient (W/m² °C)
U_L	Solar collector overall heat-loss coefficient (W/m² °C)
U_o	Heat-transfer coefficient from fluid to ambient air (W/m² °C)
U_r	Receiver-ambient heat-transfer coefficient based on A_r (W/m² °C)
U_t	Top heat-loss coefficient (W/m² °C)
v	Fluid velocity (m/s)

V	Wind velocity (m/s); volumetric consumption (l); voltage (V)
V_∞	Undisturbed wind velocity (m/s)
V_L	Load flow rate from storage tank (kg/s)
V_{oc}	Open-circuit voltage (V)
V_R	Axial velocity in the rotor plane (m/s); Return-flow rate to storage tank (kg/s)
V_w	Air velocity in the wake region (m/s)
W	Distance between riser tubes (m); net work output (J/kg)
W_a	Collector aperture (m)
X	Dimensionless collector loss ratio
x	Mass concentration of LiBr in solution; mole fraction
X'	Modified dimensionless collector loss ratio
X_c	Corrected value of X; dimensionless critical radiation level
Y	Dimensionless absorbed energy ratio
Y_c	Corrected value of Y
Z	Dimensionless load heat exchanger parameter
z	Solar azimuth angle (degrees)
z_0	Surface roughness length (m)
z_r	Reference height (m)
Z_s	Surface azimuth angle (degrees)

Greek

α_α	Absorber absorptance
α	Fraction of solar energy reaching surface that is absorbed (absorptivity); solar altitude angle; thermal diffusivity (m^2/s); interference factor (degrees)
β	Collector slope (degrees); misalignment angle error (degrees)
β'	Volumetric coefficient of expansion (1/K)
$\beta*$	Universal nonrandom error parameter due to angular errors, $\beta* = \beta C$
γ	Collector intercept factor; average bond thickness (m); correction factor for diffuse radiation (CPC collectors)
δ	Absorber (fin) thickness (m); declination (degrees)
Δ	Expansion or contraction of a collector array (mm)
ΔT	Temperature difference $= T_i - T_a$
Δx	Elemental fin or riser tube distance (m)
ε	Specific exergy (J/kg)
ε_g	Emissivity of glass covers
ε_p	Absorber plate emittance
λ	Wavelength (m)
η	Efficiency; combined power coefficient
η_0	Collector optical efficiency
η_p	Plant efficiency
η_t	Tower efficiency
ν	Frequency (per second); kinetic viscosity (m^2/s)
π	Transmembrane osmotic pressure
ρ	Density (kg/m^3); reflectance; reflectivity
ρ_m	Mirror reflectance
θ	Dimensionless temperature $= T/T_0$; angle of incidence (degrees)
θ_c	Acceptance half-angle for CPC collectors (degrees)

θ_e	Effective incidence angle (degrees)
θ_m	Collector half acceptance angle (degrees)
σ	Stefan–Boltzmann constant = 5.67×10^{-8} W/m^2 K^4
σ^*	Universal random error parameter, $\sigma^* = \sigma C$
σ_{sun}	Standard deviation of the energy distribution of the sun's rays at normal incidence
σ_{slope}	Standard deviation of the distribution of local slope errors at normal incidence
σ_{mirror}	Standard deviation of the variation in diffusivity of the reflective material at normal incidence
τ	Transmittance
τ_α	Absorber transmittance
$(\tau\alpha)$	Transmittance–absorptance product
$(\overline{\tau\alpha})$	Monthly average transmittance–absorptance product
$(\tau\alpha)_B$	Transmittance–absorptance product for estimating incidence-angle modifier for beam radiation
$(\tau\alpha)_D$	Transmittance–absorptance product for estimating incidence-angle modifier for sky (diffuse) radiation
$(\tau\alpha)_G$	Transmittance–absorptance product for estimating incidence-angle modifier for ground-reflected radiation
Φ	Zenith angle (degrees); utilizability
φ	Parabolic angle (degrees), i.e., the angle between the axis and the reflected beam at the focus of the parabola
φ_r	Collector rim angle (degrees)

Note: A horizontal bar over the symbols indicates monthly average values.

Abbreviations

AEP	Annual energy production
AFC	Alkaline fuel cell
AFP	Advance flat plate
ANN	Artificial neural network
AST	Apparent solar time
BIPV	Building integrated PV
BP	Backpropagation
CLFR	Compact linear Fresnel reflector
COP	Coefficient of performance
CPC	Compound parabolic collector
CPV	Concentrating photovoltaic
CSP	Concentrating solar power
CTC	Cylindrical trough collector
CTF	Conduction transfer function
DAS	Data acquisition system
DD	Degree days
DS	Daylight saving
ED	Electrodialysis
ER	Energy recovery
ET	Equation of time
ETC	Evacuated tube collector

E−W	East−west
FF	Fill factor, fouling factor
FPC	Flat-plate collector
GA	Genetic algorithm
GMDH	Group method of data handling
GRNN	General regression neural network
HFC	Heliostat field collector
HVAC	Heating, ventilating, and air conditioning
ICPC	Integrated compound parabolic collector
ISCCS	Integrated solar combined-cycle system
LCC	Life cycle cost
LCR	Local concentration ratio
LCS	Life cycle savings
LFR	Linear Fresnel reflector
LL	Local longitude
LOP	Loss of load probability
LST	Local standard time
LTV	Long tube vertical
MCFC	Molten carbonate fuel cell
MEB	Multiple-effect boiling
MES	Multiple-effect stack
MPP	Maximum power point
MPPT	Maximum power-point tracker
MSF	Multistage flash
MVC	Mechanical vapor compression
NIST	National Institute of Standards and Technology
NOCT	Nominal operating cell temperature
NTU	Number of transfer units
N−S	North−south
PAFC	Phosphoric acid fuel cell
PDR	Parabolic dish reflector
PEFC	Polymer electrolyte fuel cell
PEMFC	Proton exchange membrane fuel cell
PTC	Parabolic trough collector
PV	Photovoltaic
PWF	Present worth factor
RES	Renewable energy systems
RH	Relative humidity
RMS	Root mean square
RMSD	Root mean standard deviation
RO	Reverse osmosis
ROI	Return on investment
RTF	Room transfer function
SC	Shading coefficient
SEGS	Solar electric generating systems

SHGC	Solar heat gain coefficient
SL	Standard longitude
SOC	State of charge
SOFC	Solid oxide fuel cell
SRC	Standard rating conditions
TCF	Temperature correction factor
TDS	Total dissolved solids
TFM	Transfer function method
TI	Transparent insulation
TPV	Thermophotovoltaic
TVC	Thermal vapor compression
VC	Vapor compression

Appendix 2: Definitions

This appendix presents the definitions of various terms used in solar engineering. A valuable source of definitions is the ISO 9488 standard. The previous version of the standard gave the solar energy vocabulary in three languages; in the new version (year 2020) only English is used.

Absorber Component of a solar energy collector that collects and retains as much of the radiation from the sun as possible. A heat transfer fluid flows through the absorber or conduits attached to the absorber.

Absorber area The maximum projected area of the absorber. The surface area of the absorber which is designed to absorb solar radiation.

Absorptance The ratio of absorbed solar radiation to that of incident radiation. Absorptivity is the property of absorbing radiation, possessed by all materials to varying extents.

Absorption air conditioning Achieves a cooling effect through the absorption–desorption process without the requirement of large shaft work input.

Altazimuth tracker Sun following device which uses the solar elevation angle and the azimuth angle of the sun as the coordinates of movement.

Air mass Measure of the length of the path though the earth's atmosphere at sea level traversed by light rays from a celestial body, expressed with reference to the path length along the vertical.

Aperture The opening through which non-concentrated solar energy is admitted. Aperture area is the maximum projected area through which non-concentrated solar radiation enters the collector.

Auxiliary system A system that provides a backup to the solar energy system during cloudy periods or nighttime.

Azimuth angle The angle between the north–south line at a given location and the projection of the sun–earth line in the horizontal plane.

Battery An electrical energy storage system using reversible chemical reactions.

Beam radiation Radiation incident on a given plane and originating from a small solid angle centered on the sun's disk.

Brayton cycle A heat engine that uses the thermodynamic cycle used in jet (combustion turbine) engines.

Cadmium sulfide (Cds) A yellow-orange chemical compound produced from cadmium metal. As a semiconductor, Cds is always n-type.

Capital cost The cost of equipment, construction, land, and other items required to construct a facility.

Cavity receiver A receiver in the form of a cavity where the solar radiation enters through one or more openings (apertures) and is absorbed on the internal heat-absorbing surfaces.

Collector cover Transparent material/s that cover the absorber to reduce heat losses and provide weather protection without having a direct contact with the absorber.

Collector efficiency (η) The ratio of the thermal power removed by the heat transfer fluid, to the product of the collector area and the solar irradiance falling on the collector.

Collector efficiency factor The ratio of the energy delivered by a solar collector to the energy that would be delivered if the entire absorber were at the average fluid temperature in the collector.

Collector flow factor The ratio of the energy delivered by a solar collector to the energy that would be delivered if the average fluid temperature in the collector were equal to the fluid inlet temperature.

Collector tilt angle The angle at which the collector-aperture plane is tilted from the horizontal plane.

Concentration ratio The ratio of aperture to receiver area of a solar energy collector.

Concentrating collector A solar collector that uses reflectors or lenses to redirect and concentrate the solar radiation passing through the aperture onto an absorber.

Cover plate Transparent material used to cover a collector—absorber plate so that the solar energy is trapped by the "greenhouse effect".

CPC collector Compound parabolic concentrator, a non-imaging collector consisting of two parabolas one facing the other.

Declination The angle subtended between the earth—sun line and the plane of the equator (north positive).

Differential thermostat Control of one or more actuators by means of a temperature difference between two temperature sensors, using hysteresis between on and off.

Diffuse radiation Radiation from the sun scattered by the atmosphere that falls on a plane of a given orientation.

Direct radiation Radiation from the sun received from a narrow solid-angle measured from a point on the earth's surface.

Direct system A solar heating system in which the heated water to be consumed by the user passes directly through the collector.

Drainback system Solar thermal system in which as part of the normal working cycle, the heat transfer fluid is drained from the solar collector into a storage device when the pump is turned off, and refills the collector when the pump is turned on.

Draindown system Direct solar heating system in which the water can be drained from the collector and run to waste, usually to prevent freezing.

Efficiency The ratio of the measure of a desired effect to the measure of the input effect, both expressed in the same units.

Emittance (ε) The ratio of radiation emitted by a real surface to the radiation emitted by a perfect radiator (blackbody) at the same temperature.

Equatorial tracker Sun-following device having an axis of rotation parallel to the Earth's axis.

Evacuated collector Collector in which the space between the absorber and the cover is evacuated.

Evacuated tube collector A collector employing a glass tube with an evacuated space between the tube and the absorber.

Evaporator A heat exchanger in which a fluid undergoes a liquid to vapor-phase change.

Extraterrestrial radiation Solar radiation received on a surface at the limit of the earth's atmosphere.

Flat-plate collector A stationary collector that can collect both direct and diffuse radiations and can heat water or air. The two basic designs of water flat-plate collectors are the header and riser type and the serpentine type. The absorbing surface is essentially planar.

Forced-circulation system System which utilizes a pump or a fan to circulate the heat transfer fluid through the collector.

Fresnel collector A line-focus solar thermal collector that uses reflectors composed of at least two longitudinal segments with parallel axes to concentrate the solar radiation onto a fixed receiver.

Geometric factor A measure of the effective reduction of the aperture area of a concentrator due to abnormal incidence effects.

Glazing Glass, plastic, or other transparent material covering the collector absorber surface.

Global radiation Hemispherical solar radiation received by a horizontal plane, i.e., the total of beam and diffuse radiation.

Greenhouse effect A heat transfer effect where heat loss from the surfaces is controlled by suppressing the convection loss, frequently incorrectly attributed to the suppression of radiation from an enclosure.

Gross collector area Maximum projected area of the complete collector, excluding any integral means of mounting and connecting fluid pipework.

Heat exchanger Device used to transfer heat between two fluid streams without mixing them.

Heat pipe A passive heat exchanger employing the principles of evaporation and condensation to transfer heat at high levels of effectiveness.

Heat pump A device that transfers heat from a relatively low-temperature reservoir to one at a higher temperature by the input of shaft work.

Heat removal factor The ratio of the energy delivered by a solar energy collector to the energy that would be delivered if the entire absorber were at the fluid inlet temperature.

Heliostat A device to direct sunlight toward a fixed target.

Hole A vacant electron state in a valence band—behaves like a positively charged electron.

Hour angle The angle between the sun projection on the equatorial plane at a given time and the sun projection on the same plane at solar noon.

Incident angle The angle between the sun's rays and a line normal to the irradiated surface.

Indirect system A solar heating system in which a heat transfer fluid other than the water to be consumed is circulated through the collector and, with a heat exchanger, transfers its heat to the water to be consumed.

Insolation A term applying specifically to solar energy irradiation (J/m^2).

Integrated collector storage (ICS) A solar heating system in which the solar energy collector also functions as the storage device.

Intercept factor The ratio of the energy intercepted by the receiver to the energy reflected by the focusing device.

Irradiance (G) Quotient of the radiant flux incident on a surface and the area of that surface, or the rate at which radiant energy is incident on a surface per unit area of that surface (W/m^2).

Irradiation (H) The incident energy per unit area on a surface found by integration of irradiance over a specified time (usually, an hour or a day) (J/m^2).

Latitude The angular distance north (+) or south (−) of the equator, measured in degrees.

Line focus collector A concentrating collector that concentrates solar radiation in one plane, producing a linear focus.

Local solar time System of astronomical time in which the sun always crosses the true north–south meridian at 12 noon. This system of time differs from local clock time according to the longitude, time zone, and equation of time.

n-type Semiconductor doped with impurities so as to have free electrons in the conduction band.

Non-imaging collector Concentrating collector that concentrates solar radiation onto a relatively small receiver without creating an image of the sun on the receiver.

Open-circuit voltage Photovoltaic voltage developed on an open circuit, which is the maximum available at a given irradiance.

Optical efficiency (η_o) The ratio of the energy absorbed by the receiver to the energy incident on the concentrator's aperture when the collector inlet temperature is near ambient temperature. It is the maximum efficiency a collector can have. This is similar to the peak collector efficiency which applies to flat plate collectors.

p-type Semiconductor doped with impurities so as to have vacancies (holes) in the valence band.

Parabola Curve formed by the locus of a point moving in a plane so that its distances from a fixed point (focus) and a fixed straight line (directrix) are equal.

Parabolic dish reflector Paraboloidal dish, dual axis tracking, solar thermal concentrator that focuses radiant energy onto an attached point focus receiver or engine-receiver unit.

Parabolic trough collector A paraboloidal trough (line focus collector), single-axis tracking, solar thermal concentrator that focuses radiant energy onto an attached linear focus receiver.

Parasitic energy Energy, usually electricity, consumed by pumps, fans, and controls in a solar heating system.

Passive system System using the sun's energy without mechanical systems' support.

Payback period Length of time required to recover the investment in a project by benefits accruing from the investment.

Peak collector efficiency (or zero loss efficiency) Efficiency of the collector when the mean fluid temperature is equal to the ambient temperature. This applies mainly to non-concentrating collectors.

Photovoltaic effect The generation of an electromotive force when radiant energy falls on the boundary between certain dissimilar substances in close contact.

Photovoltaic thermal (PV/T) collector A hybrid device used to convert solar radiation to thermal and electrical energy.

p–n junction Junction of dissimilar semiconductor materials, where electrons move from one type to another under specific conditions.

Point focus collector A concentrating collector that focuses solar radiation on a point.

Present value The value of a future cash flow discounted to the present.

Pyranometer Radiometer designed for measuring the solar irradiance on a plane receiver surface.

Pyrheliometer Radiometer using a collimated detector for measuring the direct solar irradiance under normal incidence.

Radiation The emission or transfer of energy in the form of electromagnetic wave.

Radiosity The rate at which radiant energy leaves a surface per unit area by combined emission, reflection, and transmission (W/m^2).

Rankine cycle A closed-loop heat engine cycle using various components, including a working fluid pumped under pressure to a boiler where heat is added, expanded in a turbine where work is generated, and condensed in a condenser that rejects low-grade heat to the environment.

Receiver Part of a concentrating collector to which the solar radiation is finally directed or redirected, consisting of an absorber and any glazing cover through which radiation must pass.

Reflectance (ρ) The ratio of radiation reflected from a surface to that incident on the surface. Reflectivity is the property of reflecting radiation, possessed by all materials to varying extents, called the *albedo* in atmospheric references.

Selective surface A surface whose optical properties of reflectance, absorptance, transmittance, and emittance are wavelength-dependent.

Sky temperature Equivalent temperature of a blackbody radiating, on a horizontal surface, the same longwave radiation as the atmosphere.

Silicon cells Photovoltaic cells made principally of silicon, which is a semiconductor.

Solar altitude angle The angle between the line joining the center of the solar disc to the point of observation at any given instant and the horizontal plane through that point of observation.

Solar collector A device designed to absorb solar radiation and transfer the thermal energy so produced to a fluid passing through it.

Solar combisystem Solar heating system providing both hot water and space heating.

Solar contribution Energy supplied by the solar part of a system.

Solar constant The intensity of solar radiation outside the earth's atmosphere, at the average earth—sun distance (149.5×10^6 km), on a surface perpendicular to the sun's rays.

Solar distillation Process in which the sun's energy is utilized for the purification of sea, brackish, or poor quality water. The greenhouse effect is utilized to trap heat to evaporate the liquid. The vapor so formed then condenses on the cover plate and can be collected for use.

Solar energy Energy, in the form of electromagnetic energy, emitted from the sun and generated by means of a fusion reaction within the sun.

Solar fraction (f or F) Energy supplied by the solar energy system divided by the total system load, i.e., the part of the load covered by the solar energy system.

Solar heating system System using solar energy for the delivery of thermal energy.

Solar-only system Solar heating system without any other heat source.

Solar ponds Ponds of stratified water that collect and retain heat. Convection normally present in ponds is suppressed by imposing a stable density gradient of dissolved salts.

Solar preheat system Solar heating system to preheat a fluid prior to its entry into any other type of heater.

Solar radiation Radiant energy received from the sun both directly as beam component, diffusely by scattering from the sky and reflection from the ground.

Solar noon Local time of day when the sun crosses the observer's meridian.

Solar simulator A device equipped with an artificial source of radiant energy simulating solar radiation.

Solar time Time based on the apparent angular motion of the sun across the sky.

Stagnation The status of a collector or system when no heat is being removed by a heat transfer fluid.

Steady state Status of a collector when the rate of heat removal plus heat losses is equal to the solar flux.

Sun path diagram Diagram of solar altitude versus solar azimuth, showing the position of the sun as a function of time for various dates of the year.

Thermal stratification The state of a fluid in which the special distribution of temperature only depends on the vertical direction. Thermal stratification is commonly observed in liquid stores in which the temperature increases from bottom to the top, due to density gradients.

Thermosiphon system The convective circulation of fluid occurring in a closed system wherein less-dense warm-fluid rises, displaced by denser cooler fluid in the same fluid loop.

Tracking collector Solar collector that moves in one or two axes to follow the apparent motion of the sun during the day.

Tracking system The motors, gears, actuators, and controls necessary to maintain a device (usually a concentrator) in a focus position, orientated with respect to the sun.

Transmittance (τ) The ratio of the radiant energy transmitted by a given material to the radiant energy incident on a surface of that material, depends on the angle of incidence.

Wind and infrared sensitive collector (WISC) Solar thermal collector that is particularly sensitive to the wind and/or infrared radiation. This refers to unglazed collectors.

Unglazed collector A solar collector with no cover over the absorber.

Zenith angle Angular distance of the sun from the vertical.

Appendix 3: Sun Diagrams

This appendix contains useful diagrams (Figs. A3.1–A3.8) for the quick reading of various design parameters. Initially, however, Table A3.1 shows the approximate dates for various values of solar declination, which are useful for approximations using sun path diagrams.

Table A3.1 Approximate dates (day number, N) for various values of solar declination.

Declination	Approximate dates
+23° 27	June 21 (172)
+20°	May 20 (140), July 23 (204)
+15°	May 2 (121), August 11 (223)
+10°	April 17 (107), August 26 (238)
+5°	April 3 (93), September 8 (251)
0°	March 21 (80), September 21 (263)
−5°	March 9 (68), October 3 (276)
−10°	February 23 (55), October 17 (290)
−15°	February 10 (41), November 1 (304)
−20°	January 21 (21), November 19 (323)
−23° 27	December 21 (355)

FIGURE A3.1

Sunset and sunrise angles in various months for several latitudes.

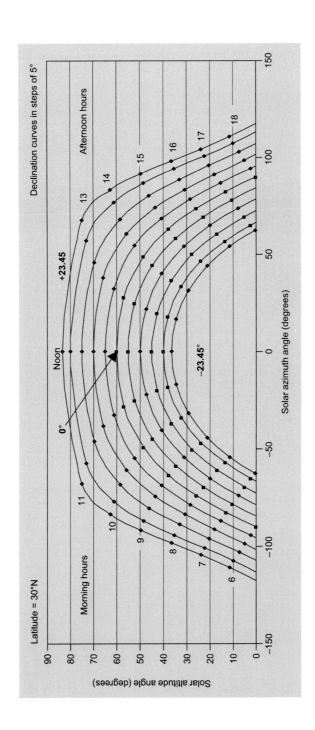

FIGURE A3.2

Sun path diagram for 30°N latitude.

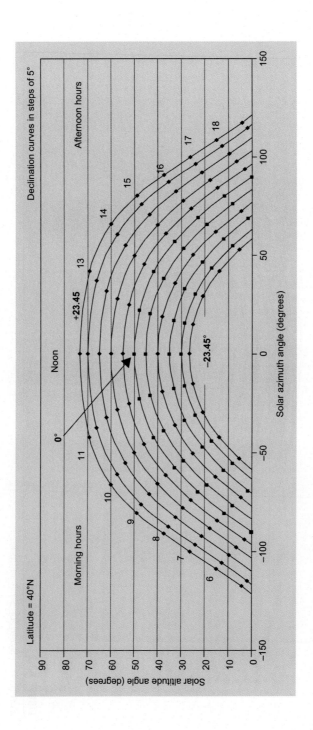

FIGURE A3.3

Sun path diagram for 40°N latitude.

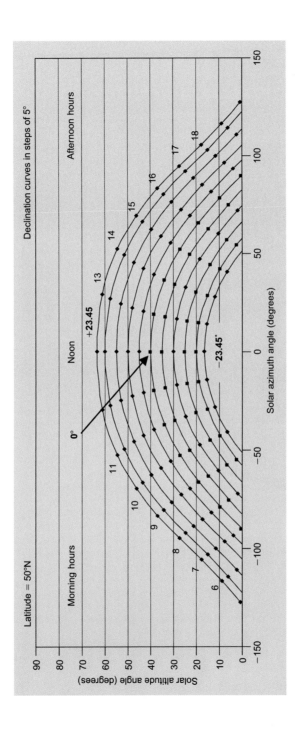

FIGURE A3.4

Sun path diagram for 50°N latitude.

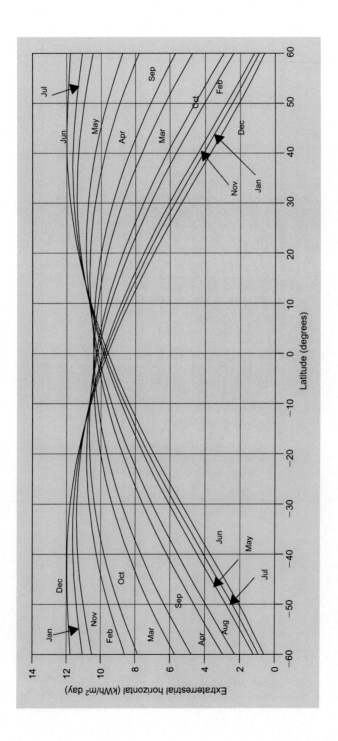

FIGURE A3.5

Monthly average daily extraterrestrial insolation on a horizontal surface (kWh/m² day).

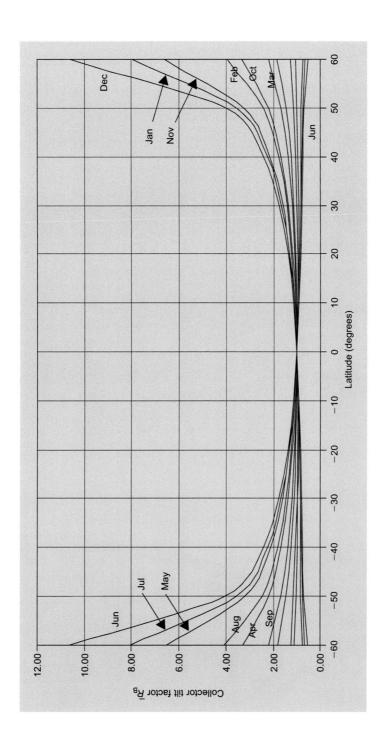

FIGURE A3.6

Values of $\overline{R}_B$ for a collector slope equal to the latitude.

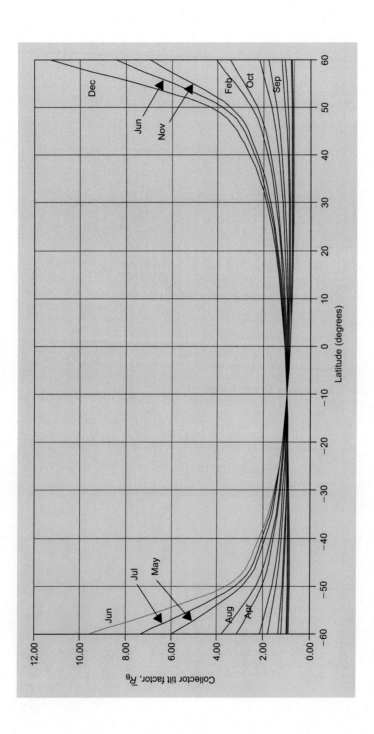

FIGURE A3.7

Values of $\overline{R}_B$ for a collector slope equal to the latitude $+10°$.

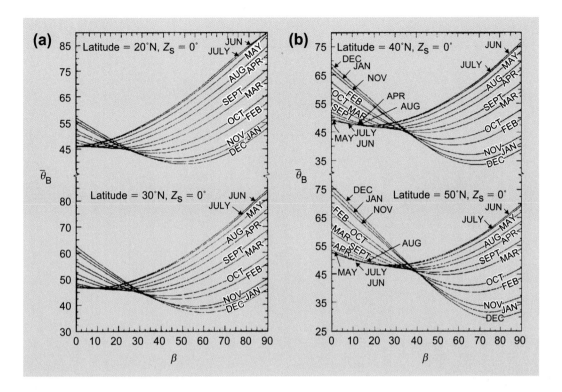

FIGURE A3.8

Monthly average beam-incidence angle for various surface locations and orientations for the Northern Hemisphere (*reprinted from* Klein, 1979, *with permission from Elsevier*). *Note:* For the Southern Hemisphere, interchange January with July, February with August, March with September, April with October, May with November, June with December, July with January, August with February, September with March, October with April, November with May, and December with June.

FIGURE A3.8 Cont'd.

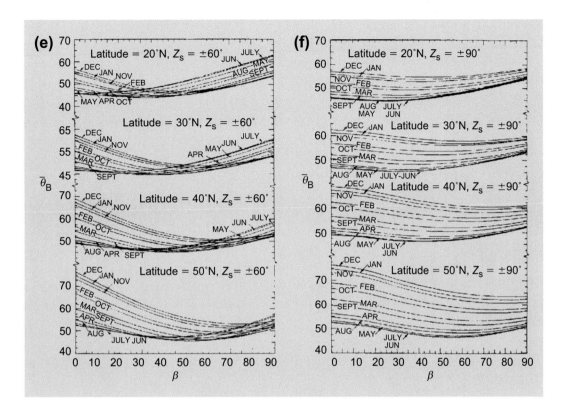

FIGURE A3.8 Cont'd.

Reference

Klein, S.A., 1979. Calculation of monthly-average transmittance–absorptance product. Sol. Energy 23 (6), 547–551.

Appendix 4: Terrestrial Spectral Irradiance

The absorbance, reflectance, and transmittance of terrestrial solar energy are important factors in solar thermal system performance, photovoltaic system performance, materials studies, biomass studies, and solar simulation activities. These optical properties are normally functions of wavelength, which requires that the spectral distribution of the solar flux is known before the solar-weighted property can be calculated. To compare the performance of competitive products, a reference standard solar spectral irradiance distribution is desirable. Table A4.1, given in this appendix, is constructed from data contained in ISO 9845-1:1992, Reference Solar Spectral Irradiance at the Ground at Different Receiving Conditions, Part 1, Direct Normal and Hemispherical Solar Irradiance for Air Mass 1.5 (see Fig. A4.1).

Table A4.1 Direct normal solar irradiance at air mass 1.5.

λ (μm)	E (W/m² μm)	λ (μm)	E (W/m² μm)	λ (μm)	E (W/m² μm)
0.3050	3.4	0.7100	1002.4	1.3500	30.1
0.3100	15.8	0.7180	816.9	1.3950	1.4
0.3150	41.1	0.7240	842.8	1.4425	51.6
0.3200	71.2	0.7400	971.0	1.4625	97.0
0.3250	100.2	0.7525	956.3	1.4770	97.3
0.3300	152.4	0.7575	942.2	1.4970	167.1
0.3350	155.6	0.7625	524.8	1.5200	239.3
0.3400	179.4	0.7675	830.7	1.5390	248.8
0.3450	186.7	0.7800	908.9	1.5580	249.3
0.3500	212.0	0.8000	873.4	1.5780	222.3
0.3600	240.5	0.8160	712.0	1.5920	227.3
0.3700	324.0	0.8230	660.2	1.6100	210.5
0.3800	362.4	0.8315	765.5	1.6300	224.7
0.3900	381.7	0.8400	799.8	1.6460	215.9
0.4000	556.0	0.8600	815.2	1.6780	202.8
0.4100	656.3	0.8800	778.3	1.7400	158.2
0.4200	690.8	0.9050	630.4	1.8000	28.6
0.4300	641.9	0.9150	565.2	1.8600	1.8
0.4400	798.5	0.9250	586.4	1.9200	1.1
0.4500	956.6	0.9300	348.1	1.9600	19.7
0.4600	990.0	0.9370	224.2	1.9850	84.9
0.4700	998.0	0.9480	271.4	2.0050	25.0
0.4800	1046.1	0.9650	451.2	2.0350	92.5
0.4900	1005.1	0.9500	549.7	2.0650	56.3
0.5000	1026.7	0.9935	630.1	2.1000	82.7
0.5100	1066.7	1.0400	582.9	2.1480	76.5
0.5200	1011.5	1.0700	539.7	2.1980	66.4
0.5300	1084.9	1.1000	366.2	2.2700	65.0
0.5400	1082.4	1.1200	98.1	2.3600	57.6
0.5500	1102.2	1.1300	169.5	2.4500	19.8
0.5700	1087.4	1.1370	118.7	2.4940	17.0
0.5900	1024.3	1.1610	301.9	2.5370	3.0
0.6100	1088.8	1.1800	406.8	2.9410	4.0
0.6300	1062.1	1.2000	375.2	2.9730	7.0
0.6500	1061.7	1.2350	423.6	3.0050	6.0
0.6700	1046.2	1.2900	365.7	3.0560	3.0
0.6900	859.2	1.3200	223.4	3.1320	5.0

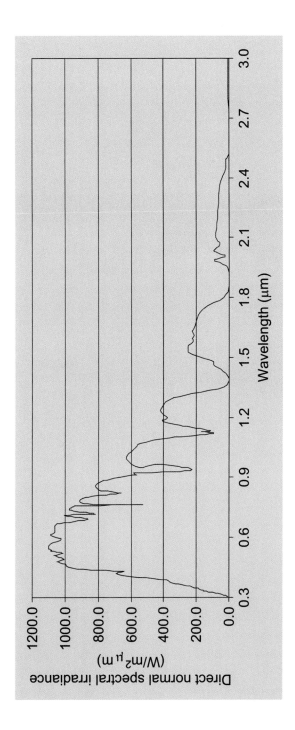

FIGURE A4.1

Direct normal solar irradiance at air mass 1.5.

Appendix 5: Thermo-physical Properties of Materials and Other Useful Information

This appendix includes various tables giving the thermo-physical properties of various materials.

Table A5.1 Physical properties of air at atmospheric pressure.

T (K)	ρ (kg/m^3)	c_p (kJ/kg °C)	μ (kg/m s) $\times 10^{-5}$	ν (m^2/s) $\times 10^{-6}$	k (W/m °C)	α (m^2/s) $\times 10^{-4}$	Pr
100	3.6010	1.0266	0.692	1.923	0.00925	0.0250	0.770
150	2.3675	1.0099	1.028	4.343	0.01374	0.0575	0.753
200	1.7684	1.0061	1.329	7.490	0.01809	0.1017	0.739
250	1.4128	1.0053	1.488	9.490	0.02227	0.1316	0.722
300	1.1774	1.0057	1.983	16.84	0.02624	0.2216	0.708
350	0.9980	1.0090	2.075	20.76	0.03003	0.2983	0.697
400	0.8826	1.0140	2.286	25.90	0.03365	0.3760	0.689
450	0.7833	1.0207	2.484	31.71	0.03707	0.4222	0.683
500	0.7048	1.0295	2.671	37.90	0.04038	0.5564	0.680
550	0.6423	1.0392	2.848	44.34	0.04360	0.6532	0.680
600	0.5879	1.0551	3.018	51.34	0.04659	0.7512	0.680
650	0.5430	1.0635	3.177	58.51	0.04953	0.8578	0.682
700	0.5030	1.0752	3.332	66.25	0.05230	0.9672	0.684
750	0.4709	1.0856	3.481	73.91	0.05509	1.0774	0.686
800	0.4405	1.0978	3.625	82.29	0.05779	1.1951	0.689
850	0.4149	1.1095	3.765	90.75	0.06028	1.3097	0.692
900	0.3925	1.1212	3.899	99.30	0.06279	1.4271	0.696
950	0.3716	1.1321	4.023	108.2	0.06225	1.5510	0.699
1000	0.3524	1.1417	4.152	117.8	0.06752	1.6779	0.702

T, temperature; ρ, density; c_p, specific heat capacity; μ, viscosity; $\nu = \mu/\rho$, kinetic viscosity; k, thermal conductivity; $\alpha = k/c_p\rho$, heat (thermal) diffusivity; Pr, Prandtl number.

Table A5.2 Physical properties of saturated liquid water.

T (°C)	ρ (kg/m³)	c_p (kJ/kg °C)	v (m²/s) $\times 10^{-6}$	k (W/m °C)	α (m²/s) $\times 10^{-7}$	Pr	β (K^{-1}) $\times 10^{-3}$
0	1002.28	4.2178	1.788	0.552	1.308	13.6	
20	1000.52	4.1818	1.006	0.597	1.430	7.02	0.18
40	994.59	4.1784	0.658	0.628	1.512	4.34	
60	985.46	4.1843	0.478	0.651	1.554	3.02	
80	974.08	4.1964	0.364	0.668	1.636	2.22	
100	960.63	4.2161	0.294	0.680	1.680	1.74	
120	945.25	4.250	0.247	0.685	1.708	1.446	
140	928.27	4.283	0.214	0.684	1.724	1.241	
160	909.69	4.342	0.190	0.680	1.729	1.099	
180	889.03	4.417	0.173	0.675	1.724	1.004	
200	866.76	4.505	0.160	0.665	1.706	0.937	
220	842.41	4.610	0.150	0.652	1.680	0.891	
240	815.66	4.756	0.143	0.635	1.639	0.871	
260	785.87	4.949	0.137	0.611	1.577	0.874	
280	752.55	5.208	0.135	0.580	1.481	0.910	
300	714.26	5.728	0.135	0.540	1.324	1.019	

T, temperature; ρ, density; c_p, specific heat capacity; $v = \mu/\rho$, kinetic viscosity; k, thermal conductivity; $\alpha = k/c_p\rho$, heat (thermal) diffusivity; Pr, Prandtl number; β, coefficient of volumetric expansion of fluid.

Table A5.3 Properties of materials.

Material	Specific heat (kJ/kg K)	Density (kg/m³)	Thermal conductivity (W/m K)	Volumetric specific heat (kJ/m³ K)
Aluminum	0.896	2700	211	2420
Concrete	0.92	2240	1.73	2060
Copper	0.383	8795	385	3370
Fiberglass	0.71–0.96	5–30	0.0519	4–30
Glass	0.82	2515	1.05	2060
Polyurethane	1.6	24	0.0245	38
Rock pebbles	0.88	1600	1.8	1410
Steel	0.48	7850	47.6	3770
Water	4.19	1000	0.596	4190

Table A5.4 Thermal properties of building and insulation materials.

Material	Density (kg/m³)	Thermal conductivity (W/m K)
Granite	2,500–2,700	2.80
Crystalline rock	2,800	3.50
Basaltic rock	2,700–3,000	3.50
Marble	2800	3.50
Limestone, soft	1,800	1.10
Limestone, hard	2,200	1.70
Limestone, very hard	2,600	2.30
Plaster (cement + sand)	–	1.39
Gypsum board	700,900	0.210.25
Concrete, medium density	2,000	1.35
Concrete, high density	2,400	2.00
Hollow brick	1,000	0.25
Solid brick	1,600	0.70
Fiber wool	50	0.041
Expanded polystyrene	Min. 20	0.041
Extruded polystyrene	>20	0.030
Polyurethane	>30	0.025
Air	1.23	0.025
Argon	1.70	0.017
Krypton	3.56	0.0090
Xenon	5.58	0.0054

Table A5.5 Thermal resistance of stagnant air and surface resistance (m^2 K/W).

Thickness of air (mm)	Heat-Flow direction		
	Sideways	Up	Down
5	0.11	0.11	0.11
7	0.13	0.13	0.13
10	0.15	0.15	0.15
15	0.17	0.16	0.17
25	0.18	0.16	0.19
50	0.18	0.16	0.21
100	0.18	0.16	0.22
300	0.18	0.16	0.23
Surface resistance			
Internal surface	0.12	0.11	0.16
External surface	0.044 (applies to all directions)		

Table A5.6 Decimal multiples.

Multiple	Prefix	Symbol
10^{24}	yotta	Y
10^{21}	zeta	Z
10^{18}	exa	E
10^{15}	peta	P
10^{12}	tera	T
10^{9}	giga	G
10^{6}	mega	M
10^{3}	kilo	k
10^{2}	hecto	h
10^{1}	deca	da
10^{-1}	deci	d
10^{-2}	centi	c
10^{-3}	milli	m
10^{-6}	micro	μ
10^{-9}	nano	n
10^{-12}	pico	p
10^{-15}	femto	f
10^{-18}	atto	a
10^{-21}	zepto	z
10^{-24}	yocto	y

Table A5.7 Curve fits for saturated water and steam.

Range	Relation	Correlation
Saturated water and steam, temperature in °C as input		
$T = 1-100°C$	$h_f = 4.1861(T) + 0.0836$	$R^2 = 1.0$
	$h_g = -0.0012(T^2) + 1.8791(T) + 2500.5$	$R^2 = 1.0$
	$s_f = -2.052 \times 10^{-5}(T^2) + 1.507 \times 10^{-2}(T) + 2.199 \times 10^{-3}$	$R^2 = 1.0$
	$s_g = 7.402 \times 10^{-5}(T^2) - 2.515 \times 10^{-2}(T) + 9.144$	$R^2 = 0.9999$
Saturated water and steam pressure in bars as input		
$P = 0.01-1$ bar	$h_f = -15772.4(P^6) + 52298.1(P^5) - 67823.6(P^4)$ $+ 43693.9(P^3) - 14854.1(P^2) + 2850.04(P) + 21.704$	$R^2 = 0.9981$
	$h_g = -6939.53(P^6) + 22965.64(P^5) - 29720.13(P^4)$ $+ 19105.32(P^3) - 6481.2(P^2) + 1232.74(P) + 2510.81$	$R^2 = 0.9978$
	$s_f = -55.76(P^6) + 184.508(P^5) - 238.5798(P^4)$ $+ 153.024(P^3) - 51.591(P^2) + 9.6043(P) + 0.0869$	$R^2 = 0.9973$
	$s_g = 92.086(P^6) - 304.24(P^5) + 392.33(P^4) - 250.3(P^3)$ $+ 83.356(P^2) - 14.841(P) + 8.9909$	$R^2 = 0.9955$
$P = 1.1-150$ bar	$h_f = -3.016 \times 10^{-10}(P^6) + 2.416 \times 10^{-7}(P^5) - 7.429 \times 10^{-5}(P^4)$ $+ 0.011(P^3) - 0.85596(P^2) + 37.0458(P) + 442.404$	$R^2 = 0.9984$
	$h_g = -3.48 \times 10^{-10}(P^6) + 2.261 \times 10^{-7}(P^5) - 5.6965 \times 10^{-5}(P^4)$ $+ 6.9969 \times 10^{-3}(P^3) - 0.441(P^2) + 12.458(P) + 2685.153$	$R^2 = 0.9961$
	$s_f = -9.656 \times 10^{-12}(P^6) + 4.743 \times 10^{-9}(P^5) - 9.073 \times 10^{-7}(P^4)$ $+ 8.565 \times 10^{-5}(P^3) - 4.213 \times 10^{-3}(P^2) + 0.1148(P) + 1.3207$	$R^2 = 0.9976$
	$s_g = 9.946 \times 10^{-12}(P^6) - 4.8593 \times 10^{-9}(P^5) + 9.225 \times 10^{-7}(P^4)$ $- 8.602 \times 10^{-5}(P^3) + 4.13 \times 10^{-3}(P^2) - 0.1058(P) + 7.3187$	$R^2 = 0.9955$

Table A5.8 Relations for the estimation of the Nusselt number used for the estimation of the convective heat transfer coefficient, $h_{c,p-g}$.

Authors	Relation	Constraints
Randall et al. (1979)	$Nu = 0.118\left[Ra_L \cos^2(\beta - 45)\right]^{0.29}$	$2.8 \times 10^3 < Ra_L < 2.2 \times 10^5$ $45° < \beta < 90°$
Yin et al. (1978)	$Nu = 0.210 * Gr^{0.269}\left(\dfrac{H}{L}\right)^{-0.131}$	$1.5 \times 10^3 < Gr < 7 \times 10^6$ $\beta = 90°$
Buchberg et al. (1976)	$Nu = 1 + 1.446\left(1 - \dfrac{1708}{Ra_L\cos(\beta)}\right)^+$	$1708 < Ra_L\cos(\beta) < 5900$
	$Nu = 0.229\left(Ra_L\cos(\beta)\right)^{0.252}$	$5900 < Ra_L\cos(\beta) < 9.2 \times 10^4$
	$Nu = 0.157\left(Ra_L\cos(\beta)\right)^{0.285}$	$9.2 \times 10^4 < Ra_L\cos(\beta) < 10^6$ For all three cases: $0° < \beta < 60°$

β, *collector slope in degrees from horizontal;* H, *enclosure height (m);* L, *enclosure width (m);* RA$_L$, *Rayleigh number with L as the characteristic length.*

Table A5.9 Relations for the estimation of the sky temperature (T_{sky}).

Author	Cloud condition	Relation
Tang et al. (2004)	Clear sky	$T_{sky} = T_a(0.754 + 0.0044T_{db})^{0.25}$
Chen et al. (1995)	Clear sky	$T_{sky} = T_a(0.736 + 0.00577T_{db})^{0.25}$
Melchor (1982)	Clear sky	$T_{sky} = T_a(0.56 + 0.08P_v^{0.5})^{0.25}$
Aubinet (1994)	Cloudy sky	$T_{sky} = 94 + 12.6\ln(P_v) - 13K_T + 0.341T_a$

P_v, *vapor pressure (Pa)*; T_a, *ambient temperature (K)*; T_{dp}, *Dew point temperature (^oC)*.

Table A5.10 Relations for estimating the heat transfer coefficient with convection from the glass cover to the ambient air.

Authors	Relation	Constraints
McAdams (1954)	$h_w = 5.7 + 3.8V$ $h_w = 6.47V^{0.78}$	For $V < 5$ m/s For $V > 5$ m/s
Watmuff et al. (1977)	$h_w = 2.3 + 3V$	For $0 < V < 7$ m/s
Sparrow et al. (1979)	$h_w = \dfrac{0.86\rho c_p V}{Pr^{2/3}\,Re^{1/2}}$	For $0 < V < 7$ m/s
Kumar et al. (1997)	$h_w = 10.03 + 4.687V$	For $0 < V < 4$ m/s
Sharples and Charlesworth (1998)	$h_w = 8.3 + 2.2V$ $h_w = 6.5 + 3.3V$	For $0.8 < V < 6.5$ m/s, $\theta=0°$ For $0.8 < V < 6.5$ m/s, $\theta=90°$

V, *wind velocity (m/s)*; θ, *wind direction*.

Table A5.11 Physical properties of argon.

T (°C)	ρ (kg/m³)	c_p (kJ/kg K)	μ (kg/m s) $\times 10^{-5}$	ν (m²/s) $\times 10^{-6}$	k (W/m °C)	α (m²/s) $\times 10^{-4}$	Pr
−20	1.925	0.5222	1.969	10.229	0.01544	0.1536	0.6659
0	1.784	0.5219	2.102	11.783	0.01648	0.1770	0.6654
20	1.662	0.5216	2.231	13.424	0.01750	0.2019	0.6650
40	1.555	0.5214	2.356	15.151	0.01848	0.2279	0.6648
60	1.462	0.5213	2.479	16.956	0.01945	0.2552	0.6645
80	1.379	0.5212	2.599	18.847	0.02039	0.2837	0.6644
100	1.305	0.5211	2.717	20.820	0.02131	0.3134	0.6642
120	1.238	0.5210	2.832	22.876	0.02221	0.3443	0.6641
140	1.178	0.5209	2.944	24.992	0.02309	0.3763	0.6641
160	1.124	0.5208	3.054	27.171	0.02396	0.4093	0.6640
180	1.074	0.5208	3.162	29.441	0.02480	0.4434	0.6640
200	1.029	0.5207	3.268	31.759	0.02564	0.4785	0.6639
220	0.9870	0.5207	3.373	34.174	0.02645	0.5147	0.6639
240	0.9485	0.5207	3.475	36.637	0.02725	0.5517	0.6639
260	0.9129	0.5206	3.575	39.161	0.02804	0.5900	0.6639
280	0.8799	0.5206	3.674	41.755	0.02881	0.6289	0.6639
300	0.8492	0.5206	3.771	44.407	0.02957	0.6689	0.6640
320	0.8205	0.5206	3.867	47.130	0.03032	0.7098	0.6640
340	0.7938	0.5206	3.961	49.899	0.03106	0.7516	0.6640
360	0.7687	0.5205	4.054	52.738	0.03178	0.7943	0.6640
380	0.7451	0.5205	4.146	55.644	0.03250	0.8380	0.6641
400	0.7230	0.5205	4.236	58.589	0.03320	0.8822	0.6641
420	0.7021	0.5205	4.325	61.601	0.03390	0.9276	0.6642

T, temperature; ρ, density; c_p, specific heat capacity; μ, viscosity; $\nu = \mu/\rho$, kinetic viscosity; k, thermal conductivity; $\alpha = k/c_p\rho$, heat (thermal) diffusivity; Pr, Prandtl number.

Table A5.12 Physical properties of krypton.

T (°C)	ρ (kg/m³)	c_p (kJ/kg K)	μ (kg/m s) $\times 10^{-5}$	ν (m²/s) $\times 10^{-6}$	k (W/m °C)	α (m²/s) $\times 10^{-4}$	Pr
−20	4.048	0.2489	2.143	5.294	0.008011	0.0795	0.6684
0	3.749	0.2495	2.298	6.130	0.008584	0.0918	0.6679
20	3.491	0.2492	2.450	7.018	0.009148	0.1052	0.6675
40	3.267	0.2490	2.599	7.955	0.009703	0.1193	0.6672
60	3.070	0.2489	2.746	8.945	0.01025	0.1341	0.6670
80	2.895	0.2488	2.890	9.983	0.01078	0.1497	0.6668
100	2.739	0.2487	3.032	11.070	0.01131	0.1660	0.6667
120	2.600	0.2486	3.171	12.196	0.01183	0.1830	0.6666
140	2.473	0.2485	3.307	13.372	0.01233	0.2006	0.6665
160	2.359	0.2485	3.441	14.587	0.01283	0.2189	0.6665
180	2.255	0.2484	3.572	15.840	0.01332	0.2378	0.6664
200	2.159	0.2484	3.701	17.142	0.01379	0.2571	0.6664
220	2.071	0.2484	3.827	18.479	0.01426	0.2772	0.6664
240	1.990	0.2483	3.950	19.849	0.01472	0.2979	0.6664
260	1.916	0.2483	4.071	21.247	0.01517	0.3189	0.6664
280	1.846	0.2483	4.189	22.692	0.01561	0.3406	0.6664
300	1.782	0.2483	4.305	24.158	0.01604	0.3625	0.6664
320	1.722	0.2482	4.418	25.656	0.01646	0.3851	0.6664
340	1.666	0.2482	4.528	27.179	0.01687	0.4080	0.6664
360	1.613	0.2482	4.636	28.741	0.01727	0.4314	0.6664
380	1.563	0.2482	4.741	30.333	0.01766	0.4552	0.6664
400	1.517	0.2482	4.844	31.931	0.01804	0.4791	0.6664
420	1.473	0.2482	4.944	33.564	0.01841	0.5036	0.6665

T, temperature; ρ, density; c_p, specific heat capacity; μ, viscosity; $\nu = \mu/\rho$, kinetic viscosity; k, thermal conductivity; $\alpha = k/c_p\rho$, heat (thermal) diffusivity; Pr, Prandtl number.

Table A5.13 Physical properties of xenon.

T (°C)	ρ (kg/m³)	c_p (kJ/kg K)	μ (kg/m s) × 10^{-5}	ν (m²/s) × 10^{-6}	k (W/m °C)	α (m²/s) × 10^{-4}	Pr
−20	6.374	0.1612	1.996	3.131	0.004855	0.0473	0.6630
0	5.898	0.1606	2.136	3.622	0.005182	0.0547	0.6619
20	5.488	0.1601	2.275	4.145	0.005507	0.0627	0.6614
40	5.133	0.1598	2.412	4.699	0.005830	0.0711	0.6611
60	4.821	0.1595	2.548	5.285	0.006150	0.0800	0.6610
80	4.545	0.1593	2.683	5.903	0.006468	0.0893	0.6609
100	4.299	0.1592	2.816	6.550	0.006784	0.0991	0.6608
120	4.079	0.1591	2.948	7.227	0.007098	0.1094	0.6608
140	3.880	0.159	3.079	7.936	0.007409	0.1201	0.6607
160	3.700	0.1589	3.209	8.673	0.007718	0.1313	0.6607
180	3.536	0.1588	3.337	9.437	0.008024	0.1429	0.6606
200	3.386	0.1588	3.464	10.230	0.008328	0.1549	0.6605
220	3.248	0.1587	3.590	11.053	0.008630	0.1674	0.6604
240	3.121	0.1587	3.715	11.903	0.008929	0.1803	0.6602
260	3.003	0.1587	3.838	12.781	0.009226	0.1936	0.6601
280	2.894	0.1586	3.960	13.683	0.009520	0.2074	0.6599
300	2.793	0.1586	4.081	14.612	0.009812	0.2215	0.6597
320	2.699	0.1586	4.201	15.565	0.01010	0.2359	0.6594
340	2.611	0.1586	4.319	16.542	0.01039	0.2509	0.6592
360	2.528	0.1585	4.436	17.547	0.01067	0.2663	0.6589
380	2.450	0.1585	4.551	18.576	0.01095	0.2820	0.6587
400	2.377	0.1585	4.666	19.630	0.01123	0.2981	0.6584
420	2.309	0.1585	4.779	20.697	0.01151	0.3145	0.6580

T, temperature; ρ, density; c_p, specific heat capacity; μ, viscosity; ν = μ/ρ, kinetic viscosity; k, thermal conductivity; α = $k/c_p\rho$, heat (thermal) diffusivity; Pr, Prandtl number.

Table A5.14 Physical properties of hellion.

T (°C)	ρ (kg/m³)	c_p (kJ/kg K)	μ (kg/m s) × 10⁻⁵	ν (m²/s) × 10⁻⁶	k (W/m °C)	α (m²/s) × 10⁻⁴	Pr
−20	0.1926	5.193	1.775	92.160	0.1387	1.3868	0.6645
0	0.1785	5.193	1.869	104.706	0.1462	1.5772	0.6640
20	0.1663	5.193	1.962	117.980	0.1535	1.7775	0.6637
40	0.1557	5.193	2.052	131.792	0.1607	1.9875	0.6633
60	0.1464	5.193	2.141	146.243	0.1677	2.2058	0.6631
80	0.1381	5.193	2.229	161.405	0.1746	2.4346	0.6629
100	0.1307	5.193	2.315	177.123	0.1814	2.6727	0.6628
120	0.1240	5.193	2.400	193.548	0.1881	2.9211	0.6627
140	0.1180	5.193	2.484	210.508	0.1947	3.1774	0.6626
160	0.1126	5.193	2.567	227.975	0.2012	3.4409	0.6626
180	0.1076	5.193	2.649	246.190	0.2076	3.7153	0.6626
200	0.1031	5.193	2.729	264.694	0.2139	3.9952	0.6625
220	0.09889	5.193	2.809	284.053	0.2202	4.2879	0.6626
240	0.09503	5.193	2.888	303.904	0.2264	4.5877	0.6626
260	0.09147	5.193	2.966	324.259	0.2325	4.8947	0.6626
280	0.08816	5.193	3.043	345.168	0.2385	5.2095	0.6627
300	0.08509	5.193	3.120	366.671	0.2445	5.5333	0.6627
320	0.08222	5.193	3.196	388.713	0.2504	5.8646	0.6628
340	0.07954	5.193	3.271	411.240	0.2562	6.2026	0.6629
360	0.07703	5.193	3.345	434.246	0.262	6.5497	0.6629
380	0.07467	5.193	3.419	457.881	0.2678	6.9063	0.6630
400	0.07245	5.193	3.492	481.988	0.2735	7.2694	0.6631
420	0.07036	5.193	3.565	506.680	0.2791	7.6386	0.6632

T, temperature; ρ, density; c_p, specific heat capacity; μ, viscosity; $\nu = \mu/\rho$, kinetic viscosity; k, thermal conductivity; $\alpha = k/c_p\rho$, heat (thermal) diffusivity; Pr, Prandtl number.

Table A5.15 Thermophysical properties of Cu–water nanofluid.

T (K)	ρ (kg/m³)	k (W/m °C)	c_p (kJ/kg K)
100	9012	482	0.252
200	8976	413	0.356
300	8933	401	0.385
400	8888	393	0.397
600	8791	379	0.417
800	8689	366	0.433
1000	8580	352	0.451
1200	8462	339	0.480

T, temperature; ρ, density; k, thermal conductivity; c_p, specific heat capacity.

Table A5.16 Thermophysical properties of Al$_2$O$_3$–water nanofluid.

T (K)	ρ (kg/m³)	k (W/m °C)	c_p (kJ/kg K)
100	3987	55.2	0.415
200	3978	45.6	0.590
300	3970	36.0	0.765
400	3962	26.4	0.940
500	3953	21.1	1.025
600	3944	15.8	1.110
700	3934	13.1	1.145
800	3925	10.4	1.180
900	3914	9.125	1.203
1000	3904	7.850	1.225
1100	3894	6.575	1.248
1200	3883	5.300	1.270

T, temperature; ρ, density; k, thermal conductivity; c_p, specific heat capacity.

References

Aubinet, M., 1994. Longwave sky radiation parametrizations. Sol. Energy 53 (2), 147–154.

Buchberg, H., Catton, I., Edwards, D., 1976. Natural convection in enclosed spaces—a review of application to solar energy collection. J. Heat Trans. 98 (2), 182.

Chen, B., Clark, D., Maloney, J., Mei, W., Kasher, J., 1995. Measurement of night sky emissivity in determining radiant cooling from cool storage roofs and roof ponds. In: Proceedings of the 20th National Passive Solar Conference, 20, pp. 310–313. Minneapolis, MN, US.

Kumar, S., Sharma, V., Kandpal, T., Mullick, S., 1997. Wind induced heat losses from outer cover of solar collectors. Renew. Energy 10 (4), 613–616.

McAdams, W., 1954. Heat Transmission, 3rd ed. McGraw-Hill, New York-London.

Melchor, C.V., 1982. New formula for the equivalent night sky emissivity. (model A). Solar Energy 28 (6), 489–498.

Randall, K., Mitchell, J., El-Wakil, M., 1979. Natural convection heat transfer characteristics of flat plate enclosures. J. Heat Trans. 101 (1), 120–125.

Sparrow, E., Ramsey, J., Mass, E., 1979. Effect of finite width on heat transfer and fluid flow about an inclined rectangular plate. J. Heat Trans. 101 (2), 199.

Sharples, S., Charlesworth, P., 1998. Full-scale measurements of wind-induced convective heat transfer from a roof-mounted flat plate solar collector. Sol. Energy 62 (2), 69–77.

Tang, R., Etzion, Y., Meir, I.A., 2004. Estimates of clear night sky emissivity in the Negev Highlands, Israel. Ener. Conver. Manage. 45 (11), 1831–1843.

Yin, S., Wung, T., Chen, K., 1978. Natural convection in an air layer enclosed within rectangular cavities. Int. J. Heat and Mass Trans. 21 (3), 307–315.

Watmuff, J., Charters, W., Proctor, D., 1977. Solar and wind induced external coefficients for solar collectors. Revue Int. d'Heliotech. 2 (56).

Appendix 6: Equations for the Curves of Figs. 3.38–3.40

When greater accuracy is required, the following equations can be used to represent the curves plotted in Figs. 3.38–3.40. The various symbols used are presented in Fig. A6.1. The subscript T is for the truncated compound parabolic collector (CPC) design.

The following equations apply for a full and truncated (subscript T) CPCs (Welford and Winston, 1978):

$$f = \alpha'[1 + \sin(\theta_c)] \tag{A6.1}$$

$$\alpha = \frac{\alpha'}{\sin(\theta_c)} \tag{A6.2}$$

$$h = \frac{f \cos(\theta_c)}{\sin^2(\theta_c)} \tag{A6.3}$$

$$\alpha_T = \frac{f \sin(\Phi_T - \theta_c)}{\sin^2\left(\dfrac{\Phi_T}{2}\right)} - \alpha' \tag{A6.4}$$

$$h_T = \frac{f \cos(\Phi_T - \theta_c)}{\sin^2\left(\dfrac{\Phi_T}{2}\right)} \tag{A6.5}$$

For a truncated CPC,

$$C = \frac{\alpha_T}{\alpha'} \tag{A6.6}$$

For a full CPC,

$$C = \frac{\alpha}{\alpha'} \tag{A6.7}$$

By replacing α from Eq. (A6.2),

$$C = \frac{1}{\sin(\theta_c)} \tag{A6.8}$$

The reflector area per unit depth of a truncated CPC, A_{RT}, is given by:

$$\frac{A_{RT}}{2\alpha_T} = \frac{f}{2}\left[\frac{\cos(\Phi/2)}{\sin^2(\Phi/2)} + \ln \cot\left(\frac{\Phi}{4}\right)\right]\Bigg|_{\theta_c + \pi/2}^{\Phi_T} \tag{A6.9}$$

For Eq. (A6.9), if $\Phi_T = 2\theta_c$, $A_{RT} = A_R$.

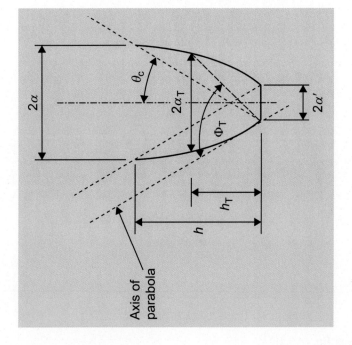

FIGURE A6.1

A truncated CPC. Its height-to-aperture ratio is about one half the full height of a CPC.

The average number of reflections, n, is given by (Rabl, 1976):

$$n = \max\left\{ C\frac{A_{RT}}{4\alpha_T} - \frac{x^2 - \cos^2(\theta)}{2[1 + \sin(\theta)]}, 1 - \frac{1}{C} \right\} \tag{A6.10}$$

where

$$x = \left[\frac{1 + \sin(\theta)}{\cos(\theta)}\right]\left\{ -\sin(\theta) + \left[1 + \frac{h_T}{h}\cot^2(\theta)\right]^{1/2} \right\} \tag{A6.11}$$

References

Rabl, A., 1976. Optical and thermal properties of compound parabolic concentrators. Sol. Energy 18 (6), 497–511.

Welford, W.T., Winston, R., 1978. The Optics of Non-imaging Concentrators. Academic Press, New York.

Appendix 7: Meteorological Data

This appendix lists meteorological data for various locations. Because this information can typically be obtained over the Internet, data for only a few selected locations are presented—mostly used in examples and problems of the book. The data presented for US locations are from http://www.nrel.gov/rredc, except the monthly average clearness index, which is calculated from Eq. (2.82a), and the estimation of extraterrestrial horizontal radiation given by Eq. (2.79) for an average day in Table 2.1. For other locations, the NASA Internet site, http://eosweb.larc.nasa.gov/cgi-bin/sse/grid.cgi?email= (requires free registration), can be used by entering the longitude and latitude of each location found from www.infoplease.com/atlas/latitude-longitude.html. For the degree days presented, the base temperature for both cooling and heating is 18.3 °C for US locations and 18 °C for all others.

The data recorded are the following:

$\overline{H}$ = Monthly average radiation on a horizontal surface (MJ/m^2).
$\overline{K}_T$ = Monthly average clearness index.
$\overline{T}_a$ = Monthly average ambient temperature (°C).
HDD = Heating degree days (°C-days).
CDD = Cooling degree days (°C-days).

The data reported are for the following locations.
In the United States: Albuquerque, NM; Boulder, CO; Las Vegas, NV; Los Angeles, CA; Madison, WI; Phoenix, AZ; San Antonio, TX; Springfield, IL.
In Europe: Almeria, ES; Athens, GR; London, UK; Nicosia, CY; Rome, IT.
In the rest of the world: Adelaide, AU; Montreal, CA; New Delhi, IN; Pretoria, SA; Rio de Janeiro, BR.

United States

Table A7.1 Albuquerque, NM: latitude (N), 35.05°, longitude (W), 106.62°.

	Jan	Feb	Mar	Apr	May	Jun	Jul	Aug	Sep	Oct	Nov	Dec
$\overline{H}$	11.52	15.12	19.44	24.48	27.72	29.16	27.00	24.84	21.24	16.92	12.60	10.44
$\overline{K}_T$	0.63	0.65	0.66	0.68	0.69	0.70	0.66	0.67	0.67	0.67	0.65	0.62
$\overline{T}_a$	1.2	4.4	8.3	12.9	17.9	23.4	25.8	24.4	20.3	13.9	6.8	1.8
HDD	531	389	312	167	49	0	0	0	10	144	345	512
CDD	0	0	0	4	36	155	233	188	70	6	0	0

Table A7.2 Boulder, CO: latitude (N), 40.02°; longitude (W), 105.25°.

	Jan	Feb	Mar	Apr	May	Jun	Jul	Aug	Sep	Oct	Nov	Dec
$\overline{H}$	8.64	11.88	15.84	20.16	22.32	24.84	24.12	21.60	18.00	13.68	9.36	7.56
$\overline{K}_T$	0.57	0.58	0.58	0.58	0.56	0.60	0.59	0.59	0.60	0.61	0.57	0.55
$\overline{T}_a$	−1.3	0.8	3.9	9.0	14.0	19.4	23.1	21.9	16.8	10.8	3.9	−0.6
HDD	608	492	448	280	141	39	0	0	80	238	433	586
CDD	0	0	0	0	6	71	148	113	35	4	0	0

Table A7.3 Las Vegas, NV: latitude (N), 36.08°; longitude (W), 115.17°.

	Jan	Feb	Mar	Apr	May	Jun	Jul	Aug	Sep	Oct	Nov	Dec
$\overline{H}$	10.80	14.40	19.44	24.84	28.08	30.24	28.44	25.92	22.32	16.92	12.24	10.08
$\overline{K}_T$	0.61	0.63	0.67	0.70	0.70	0.73	0.70	0.70	0.71	0.69	0.65	0.62
$\overline{T}_a$	7.5	10.6	13.5	17.8	23.3	29.4	32.8	31.5	26.9	20.2	12.8	7.6
HDD	336	216	162	79	8	0	0	0	0	34	169	332
CDD	0	0	12	64	163	332	449	408	258	91	0	0

Table A7.4 Los Angeles, CA: latitude (N), 33.93°; longitude (W), 118.4°.

	Jan	Feb	Mar	Apr	May	Jun	Jul	Aug	Sep	Oct	Nov	Dec
$\overline{H}$	10.08	12.96	17.28	21.96	23.04	23.76	25.56	23.40	19.08	15.12	11.52	9.36
$\overline{K}_T$	0.53	0.54	0.58	0.61	0.58	0.57	0.63	0.62	0.59	0.59	0.57	0.54
$\overline{T}_a$	13.8	14.2	14.4	15.6	17.1	18.7	20.6	21.4	21.1	19.3	16.4	13.8
HDD	143	119	124	88	53	30	5	3	12	18	71	143
CDD	0	4	4	6	14	42	76	98	94	49	14	4

Table A7.5 Madison, WI: latitude (N), 43.13°; longitude (W), 89.33°.

	Jan	Feb	Mar	Apr	May	Jun	Jul	Aug	Sep	Oct	Nov	Dec
$\overline{H}$	6.84	10.08	13.32	16.92	20.88	23.04	22.32	19.44	14.76	10.08	6.12	5.40
$\overline{K}_T$	0.52	0.54	0.51	0.50	0.53	0.55	0.55	0.54	0.51	0.48	0.42	0.46
$\overline{T}_a$	−8.9	−6.3	0.2	7.4	13.6	19.0	21.7	20.2	15.4	9.4	1.9	−5.7
HDD	844	691	563	327	163	38	7	21	93	277	493	746
CDD	0	0	0	0	17	58	110	78	7	0	0	0

Table A7.6 Phoenix, AZ: latitude (N), 33.43°; longitude (W), 112.02°.

	Jan	Feb	Mar	Apr	May	Jun	Jul	Aug	Sep	Oct	Nov	Dec
$\overline{H}$	11.52	15.48	19.80	25.56	28.80	30.24	27.36	25.56	21.96	17.64	12.96	10.80
$\overline{K}_T$	0.60	0.64	0.65	0.71	0.72	0.73	0.67	0.68	0.68	0.68	0.64	0.61
$\overline{T}_a$	12.0	14.3	16.8	21.1	26.0	31.2	34.2	33.1	29.8	23.6	16.6	12.3
HDD	201	126	101	42	4	0	0	0	0	9	74	192
CDD	4	12	53	123	242	387	491	457	343	173	23	4

Table A7.7 San Antonio, TX: latitude (N), 29.53°; longitude (W), 98.47°.

	Jan	Feb	Mar	Apr	May	Jun	Jul	Aug	Sep	Oct	Nov	Dec
$\overline{H}$	11.16	14.04	17.28	19.80	21.60	24.12	24.84	23.04	19.44	16.20	12.24	10.44
$\overline{K}_T$	0.52	0.54	0.54	0.54	0.54	0.59	0.61	0.61	0.58	0.58	0.54	0.52
$\overline{T}_a$	9.6	11.9	16.5	20.7	24.2	27.9	29.4	29.4	26.3	21.2	15.8	11.2
HDD	274	184	93	18	0	0	0	0	0	17	100	227
CDD	4	6	36	89	181	287	344	343	238	106	23	7

Table A7.8 Springfield, IL: latitude (N), 39.83°; longitude (W), 89.67°.

	Jan	Feb	Mar	Apr	May	Jun	Jul	Aug	Sep	Oct	Nov	Dec
$\overline{H}$	7.56	10.44	13.32	18.00	21.60	23.40	23.04	20.52	16.56	12.24	7.92	6.12
$\overline{K}_T$	0.49	0.51	0.48	0.52	0.54	0.56	0.57	0.56	0.55	0.54	0.48	0.44
$\overline{T}_a$	−4.3	−1.8	4.9	11.8	17.5	22.7	24.7	23.2	19.6	13.1	6.1	−1.3
HDD	703	564	417	201	92	4	0	4	24	174	368	608
CDD	0	0	0	4	67	136	198	154	63	12	0	0

Europe

Table A7.9 Almeria, ES: latitude (N), 36.83°; longitude (W), 2.45°.

	Jan	Feb	Mar	Apr	May	Jun	Jul	Aug	Sep	Oct	Nov	Dec
$\overline{H}$	9.83	12.89	17.35	22.03	24.48	27.40	27.54	24.52	19.44	14.08	10.26	8.57
$\overline{K}_T$	0.56	0.56	0.58	0.61	0.61	0.65	0.67	0.66	0.61	0.56	0.54	0.53
$\overline{T}_a$	11.0	11.8	13.7	15.8	18.7	22.5	25.1	25.5	22.8	19.0	15.0	12.2
HDD	210	168	128	70	20	0	0	0	0	10	88	172
CDD	0	0	1	5	41	133	221	237	147	45	3	0

Table A7.10 Athens, GR: latitude (N), 37.98°; longitude (E), 23.73°.

	Jan	Feb	Mar	Apr	May	Jun	Jul	Aug	Sep	Oct	Nov	Dec
$\overline{H}$	7.70	10.37	14.40	19.33	23.15	26.86	26.50	23.83	18.76	12.38	7.85	6.23
$\overline{K}_T$	0.45	0.46	0.49	0.54	0.57	0.64	0.65	0.64	0.60	0.51	0.42	0.40
$\overline{T}_a$	10.2	10.1	12.2	16.1	21.1	25.7	28.1	27.9	24.5	20.1	15.2	11.5
HDD	234	218	179	71	6	0	0	0	0	13	87	195
CDD	0	0	0	10	95	225	308	305	195	81	9	0

Table A7.11 London, UK: latitude (N), 51.50°; longitude, 0.00°.

	Jan	Feb	Mar	Apr	May	Jun	Jul	Aug	Sep	Oct	Nov	Dec
$\overline{H}$	2.95	5.26	8.82	13.39	16.96	17.89	17.93	15.62	10.55	6.44	3.56	2.23
$\overline{K}_T$	0.35	0.37	0.39	0.42	0.44	0.43	0.45	0.46	0.41	0.38	0.36	0.32
$\overline{T}_a$	4.1	4.3	6.6	8.8	12.8	16.2	18.8	18.9	15.7	11.9	7.4	4.9
HDD	429	381	348	273	163	73	22	22	76	183	316	405
CDD	0	0	0	0	2	15	44	50	9	1	0	0

Table A7.12 Nicosia, CY: latitude (N), 35.15°; longitude (E), 33.27°.

	Jan	Feb	Mar	Apr	May	Jun	Jul	Aug	Sep	Oct	Nov	Dec
$\overline{H}$	8.96	12.38	17.39	21.53	26.06	29.20	28.55	25.49	21.17	15.34	10.33	7.92
$\overline{K}_T$	0.49	0.53	0.58	0.59	0.65	0.70	0.70	0.68	0.66	0.60	0.53	0.47
$\overline{T}_a$	12.1	11.9	13.8	17.5	21.5	25.8	29.2	29.4	26.8	22.7	17.7	13.7
HDD	175	171	131	42	3	0	0	0	0	1	36	128
CDD	0	0	1	26	112	234	348	353	263	146	29	0

Table A7.13 Rome, IT: latitude (N), 41.45°; longitude (E), 12.27°.

	Jan	Feb	Mar	Apr	May	Jun	Jul	Aug	Sep	Oct	Nov	Dec
$\overline{H}$	7.13	10.51	15.55	19.73	24.41	27.50	27.61	24.16	18.29	12.24	7.60	6.08
$\overline{K}_T$	0.49	0.52	0.56	0.57	0.61	0.65	0.68	0.66	0.61	0.55	0.47	0.47
$\overline{T}_a$	9.6	9.5	11.2	13.1	17.6	21.4	24.7	25.1	21.8	18.6	14.2	10.9
HDD	247	233	204	146	37	2	0	0	0	17	108	207
CDD	0	0	0	0	23	99	202	221	118	42	2	0

Rest of the world

Table A7.14 Adelaide, AU: latitude (S), 34.92°; longitude (E), 138.60°.

	Jan	Feb	Mar	Apr	May	Jun	Jul	Aug	Sep	Oct	Nov	Dec
$\overline{H}$	24.66	22.36	17.96	13.61	9.94	8.32	9.22	11.92	15.73	19.69	22.75	24.19
$\overline{K}_T$	0.57	0.57	0.55	0.54	0.52	0.51	0.53	0.53	0.54	0.54	0.54	0.54
$\overline{T}_a$	22.8	23.0	20.5	17.4	13.9	11.2	10.1	10.9	13.3	16.0	19.3	21.5
HDD	2	2	13	50	124	190	228	206	142	85	33	10
CDD	152	147	90	32	6	0	0	0	7	24	73	118

Table A7.15 Montreal, CA: latitude (N), 45.50°; longitude (W), 73.58°.

	Jan	Feb	Mar	Apr	May	Jun	Jul	Aug	Sep	Oct	Nov	Dec
$\overline{H}$	5.69	9.07	13.03	16.06	18.32	20.20	19.87	17.68	13.57	8.57	5.22	4.61
$\overline{K}_T$	0.47	0.51	0.51	0.48	0.46	0.48	0.49	0.50	0.48	0.42	0.38	0.44
$\overline{T}_a$	−11.2	−9.6	−4.2	4.7	12.6	18.5	21.0	19.9	15.1	7.7	0.7	−7.1
HDD	912	788	689	397	178	43	7	17	103	317	519	783
CDD	0	0	0	0	7	53	97	80	23	0	0	0

Table A7.16 New Delhi, IN: latitude (N), 28.60°; longitude (E), 77.20°.

	Jan	Feb	Mar	Apr	May	Jun	Jul	Aug	Sep	Oct	Nov	Dec
$\overline{H}$	13.68	16.85	20.88	22.68	23.11	21.85	18.79	17.28	18.18	17.39	15.05	12.67
$\overline{K}_T$	0.61	0.62	0.64	0.60	0.57	0.53	0.46	0.45	0.53	0.60	0.64	0.60
$\overline{T}_a$	13.3	16.6	22.6	28.0	31.1	31.7	29.2	28.0	26.7	23.7	19.3	14.7
HDD	129	48	2	0	0	0	0	0	0	0	5	79
CDD	2	19	149	295	399	405	346	311	269	190	62	4

Table A7.17 Pretoria, SA: latitude (S), 24.70°; longitude (E), 28.23°.

	Jan	Feb	Mar	Apr	May	Jun	Jul	Aug	Sep	Oct	Nov	Dec
$\overline{H}$	23.76	22.10	20.16	17.89	16.60	15.23	16.52	18.68	21.96	22.57	23.04	23.40
$\overline{K}_T$	0.55	0.55	0.57	0.60	0.68	0.69	0.72	0.69	0.67	0.59	0.55	0.54
$\overline{T}_a$	23.2	23.1	22.1	19.4	16.0	12.6	12.4	15.5	19.4	21.3	22.0	22.5
HDD	0	0	0	8	59	148	161	80	18	6	2	0
CDD	163	145	130	56	10	0	0	9	62	109	122	142

Table A7.18 Rio de Janeiro, BR: latitude (S), 22.90°; longitude (W), 43.23°.

	Jan	Feb	Mar	Apr	May	Jun	Jul	Aug	Sep	Oct	Nov	Dec
$\overline{H}$	18.76	19.48	17.14	15.48	13.18	13.14	13.18	15.55	15.05	17.06	17.89	18.04
$\overline{K}_T$	0.44	0.48	0.48	0.51	0.52	0.57	0.55	0.55	0.45	0.44	0.43	0.42
$\overline{T}_a$	24.6	24.7	23.7	22.5	20.6	19.6	19.4	20.5	21.3	22.3	22.8	23.6
HDD	0	0	0	0	2	8	17	10	6	3	1	0
CDD	212	196	189	148	98	74	72	98	110	142	153	187

Appendix 8: Present Worth Factors

In all tables in this appendix, the columns represent interest rates (%) and rows market discount rates (%).

Table A8.1 $n = 5$.

d	0	1	2	3	4	5	6	7	8	9	10
0	5.0000	5.1010	5.2040	5.3091	5.4163	5.5256	5.6371	5.7507	5.8666	5.9847	6.1051
1	4.8534	4.9505	5.0495	5.1505	5.2534	5.3585	5.4655	5.5747	5.6859	5.7993	5.9149
2	4.7135	4.8068	4.9020	4.9990	5.0980	5.1989	5.3018	5.4067	5.5136	5.6226	5.7336
3	4.5797	4.6695	4.7610	4.8544	4.9495	5.0466	5.1455	5.2463	5.3491	5.4538	5.5606
4	4.4518	4.5382	4.6263	4.7161	4.8077	4.9010	4.9962	5.0932	5.1920	5.2927	5.3954
5	4.3295	4.4127	4.4975	4.5839	4.6721	4.7619	4.8535	4.9468	5.0419	5.1388	5.2375
6	4.2124	4.2925	4.3742	4.4574	4.5423	4.6288	4.7170	4.8068	4.8984	4.9916	5.0867
7	4.1002	4.1774	4.2561	4.3363	4.4181	4.5014	4.5864	4.6729	4.7611	4.8509	4.9424
8	3.9927	4.0671	4.1430	4.2204	4.2992	4.3795	4.4613	4.5447	4.6296	4.7162	4.8043
9	3.8897	3.9614	4.0346	4.1092	4.1852	4.2626	4.3415	4.4219	4.5038	4.5872	4.6721
10	3.7908	3.8601	3.9307	4.0026	4.0759	4.1506	4.2267	4.3042	4.3831	4.4636	4.5455
11	3.6959	3.7628	3.8309	3.9003	3.9711	4.0432	4.1166	4.1913	4.2675	4.3451	4.4241
12	3.6048	3.6694	3.7351	3.8022	3.8705	3.9401	4.0109	4.0831	4.1566	4.2314	4.3077
13	3.5172	3.5796	3.6432	3.7079	3.7739	3.8411	3.9095	3.9792	4.0502	4.1224	4.1960
14	3.4331	3.4934	3.5548	3.6174	3.6811	3.7460	3.8121	3.8794	3.9480	4.0177	4.0888
15	3.3522	3.4104	3.4698	3.5303	3.5919	3.6546	3.7185	3.7835	3.8498	3.9172	3.9858
16	3.2743	3.3307	3.3881	3.4466	3.5061	3.5668	3.6285	3.6914	3.7554	3.8206	3.8869
17	3.1993	3.2539	3.3094	3.3660	3.4236	3.4823	3.5420	3.6028	3.6647	3.7277	3.7918
18	3.1272	3.1800	3.2337	3.2885	3.3442	3.4010	3.4587	3.5176	3.5774	3.6384	3.7004
19	3.0576	3.1087	3.1608	3.2138	3.2677	3.3227	3.3786	3.4355	3.4934	3.5524	3.6124
20	2.9906	3.0401	3.0905	3.1418	3.1941	3.2473	3.3014	3.3565	3.4126	3.4697	3.5277

i

Table A8.2 $n = 10$.

d	i										
	0	1	2	3	4	5	6	7	8	9	10
0	10.000	10.462	10.950	11.464	12.006	12.578	13.181	13.816	14.487	15.193	15.937
1	9.4713	9.9010	10.354	10.832	11.335	11.865	12.425	13.014	13.635	14.289	14.979
2	8.9826	9.3825	9.8039	10.248	10.716	11.209	11.728	12.275	12.851	13.458	14.097
3	8.5302	8.9029	9.2954	9.7087	10.144	10.603	11.085	11.594	12.129	12.692	13.286
4	8.1109	8.4586	8.8246	9.2098	9.6154	10.042	10.492	10.965	11.462	11.986	12.537
5	7.7217	8.0464	8.3881	8.7476	9.1258	9.5238	9.9425	10.383	10.846	11.334	11.847
6	7.3601	7.6637	7.9830	8.3188	8.6720	9.0434	9.4340	9.8447	10.277	10.731	11.208
7	7.0236	7.3078	7.6065	7.9205	8.2506	8.5976	8.9624	9.3458	9.7488	10.172	10.618
8	6.7101	6.9764	7.2562	7.5501	7.8590	8.1836	8.5246	8.8828	9.2593	9.6547	10.070
9	6.4177	6.6674	6.9298	7.2053	7.4946	7.7984	8.1176	8.4527	8.8047	9.1743	9.5625
10	6.1446	6.3791	6.6253	6.8837	7.1550	7.4398	7.7388	8.0526	8.3820	8.7279	9.0909
11	5.8892	6.1097	6.3410	6.5837	6.8383	7.1055	7.3858	7.6800	7.9887	8.3126	8.6524
12	5.6502	5.8576	6.0752	6.3033	6.5425	6.7934	7.0566	7.3326	7.6221	7.9257	8.2442
13	5.4262	5.6216	5.8263	6.0410	6.2660	6.5018	6.7491	7.0083	7.2801	7.5651	7.8638
14	5.2161	5.4003	5.5932	5.7953	6.0071	6.2291	6.4616	6.7053	6.9607	7.2284	7.5089
15	5.0188	5.1925	5.3745	5.5650	5.7646	5.9736	6.1926	6.4219	6.6621	6.9137	7.1773
16	4.8332	4.9973	5.1691	5.3489	5.5371	5.7341	5.9404	6.1564	6.3826	6.6194	6.8674
17	4.6586	4.8137	4.9760	5.1458	5.3235	5.5094	5.7040	5.9076	6.1207	6.3437	6.5772
18	4.4941	4.6409	4.7943	4.9548	5.1227	5.2983	5.4819	5.6741	5.8751	6.0853	6.3053
19	4.3389	4.4779	4.6232	4.7750	4.9338	5.0997	5.2733	5.4547	5.6444	5.8429	6.0504
20	4.1925	4.3242	4.4618	4.6056	4.7558	4.9128	5.0769	5.2484	5.4277	5.6151	5.8110

Table A8.3 $n = 15$.

d	0	1	2	3	4	5	6	7	8	9	10
0	15.000	16.097	17.293	18.599	20.024	21.579	23.276	25.129	27.152	29.361	31.772
1	13.865	14.851	15.926	17.098	18.375	19.767	21.285	22.942	24.748	26.718	28.867
2	12.849	13.738	14.706	15.759	16.906	18.156	19.517	21.000	22.616	24.377	26.297
3	11.938	12.741	13.614	14.563	15.596	16.719	17.942	19.273	20.722	22.300	24.017
4	11.118	11.845	12.634	13.492	14.423	15.436	16.536	17.733	19.035	20.451	21.991
5	10.380	11.039	11.754	12.530	13.372	14.286	15.279	16.357	17.529	18.802	20.187
6	9.7122	10.311	10.960	11.664	12.426	13.254	14.151	15.125	16.182	17.329	18.575
7	9.1079	9.6535	10.244	10.883	11.575	12.325	13.138	14.019	14.974	16.010	17.134
8	8.5595	9.0573	9.5954	10.177	10.807	11.488	12.225	13.024	13.889	14.826	15.842
9	8.0607	8.5159	9.0073	9.5380	10.111	10.731	11.402	12.127	12.912	13.761	14.681
10	7.6061	8.0230	8.4726	8.9576	9.4811	10.046	10.657	11.317	12.030	12.802	13.636
11	7.1909	7.5735	7.9856	8.4297	8.9085	9.4249	9.9822	10.584	11.233	11.935	12.694
12	6.8109	7.1627	7.5411	7.9485	8.3872	8.8598	9.3693	9.9187	10.511	11.151	11.842
13	6.4624	6.7864	7.1346	7.5090	7.9116	8.3450	8.8116	9.3143	9.8560	10.440	11.070
14	6.1422	6.4412	6.7621	7.1067	7.4769	7.8750	8.3031	8.7638	9.2598	9.7940	10.370
15	5.8474	6.1237	6.4200	6.7378	7.0789	7.4451	7.8386	8.2616	8.7165	9.2060	9.7328
16	5.5755	5.8313	6.1053	6.3989	6.7136	7.0512	7.4135	7.8025	8.2205	8.6697	9.1527
17	5.3242	5.5615	5.8153	6.0869	6.3778	6.6895	7.0236	7.3820	7.7667	8.1796	8.6233
18	5.0916	5.3120	5.5475	5.7992	6.0685	6.3567	6.6654	6.9962	7.3507	7.7310	8.1392
19	4.8759	5.0809	5.2998	5.5335	5.7832	6.0501	6.3357	6.6414	6.9688	7.3196	7.6957
20	4.6755	4.8666	5.0703	5.2875	5.5194	5.7671	6.0318	6.3148	6.6176	6.9417	7.2887

i

Table A8.4 $n = 20$.

d	i = 0	1	2	3	4	5	6	7	8	9	10
0	20.000	22.019	24.297	26.870	29.778	33.066	36.786	40.995	45.762	51.160	57.275
1	18.046	19.802	21.780	24.009	26.524	29.362	32.568	36.190	40.284	44.913	50.150
2	16.351	17.885	19.608	21.546	23.728	26.186	28.958	32.084	35.612	39.594	44.093
3	14.877	16.221	17.727	19.417	21.317	23.453	25.857	28.564	31.613	35.050	38.926
4	13.590	14.771	16.092	17.571	19.231	21.093	23.185	25.536	28.180	31.156	34.506
5	12.462	13.503	14.665	15.965	17.419	19.048	20.874	22.922	25.222	27.806	30.710
6	11.470	12.391	13.417	14.562	15.840	17.269	18.868	20.659	22.665	24.916	27.442
7	10.594	11.411	12.320	13.332	14.459	15.717	17.122	18.692	20.448	22.414	24.617
8	9.8181	10.546	11.353	12.250	13.247	14.358	15.596	16.977	18.519	20.242	22.169
9	9.1285	9.7785	10.498	11.296	12.181	13.164	14.258	15.476	16.834	18.349	20.039
10	8.5136	9.0959	9.7390	10.450	11.238	12.112	13.082	14.160	15.359	16.694	18.182
11	7.9633	8.4866	9.0632	9.6998	10.403	11.182	12.044	13.001	14.063	15.243	16.556
12	7.4694	7.9410	8.4596	9.0307	9.6607	10.356	11.125	11.977	12.920	13.967	15.129
13	7.0248	7.4509	7.9186	8.4326	8.9983	9.6218	10.310	11.070	11.910	12.841	13.872
14	6.6231	7.0094	7.4323	7.8962	8.4057	8.9660	9.5830	10.263	11.014	11.844	12.762
15	6.2593	6.6103	6.9939	7.4137	7.8738	8.3788	8.9338	9.5445	10.217	10.959	11.779
16	5.9288	6.2487	6.5975	6.9784	7.3951	7.8514	8.3520	8.9017	9.5062	10.172	10.905
17	5.6278	5.9199	6.2379	6.5845	6.9628	7.3764	7.8291	8.3252	8.8697	9.4680	10.126
18	5.3527	5.6203	5.9110	6.2271	6.5715	6.9472	7.3577	7.8067	8.2985	8.8379	9.4301
19	5.1009	5.3465	5.6128	5.9019	6.2162	6.5584	6.9316	7.3389	7.7843	8.2718	8.8061
20	4.8696	5.0956	5.3402	5.6052	5.8928	6.2053	6.5453	6.9159	7.3202	7.7619	8.2452

Table A8.5 $n = 25$.

d	|	0	1	2	3	4	5	6	7	8	9	10
0		25.000	28.243	32.030	36.459	41.646	47.727	54.865	63.249	73.106	84.701	98.347
1		22.023	24.752	27.929	31.633	35.958	41.014	46.933	53.869	62.003	71.550	82.762
2		19.523	21.832	24.510	27.622	31.245	35.470	40.401	46.164	52.906	60.800	70.051
3		17.413	19.375	21.644	24.272	27.322	30.867	34.994	39.804	45.417	51.974	59.639
4		15.622	17.298	19.229	21.459	24.038	27.028	30.498	34.531	39.224	44.693	51.071
5		14.094	15.532	17.184	19.085	21.277	23.810	26.740	30.137	34.079	38.660	43.990
6		12.783	14.024	15.444	17.072	18.943	21.098	23.585	26.458	29.784	33.639	38.112
7		11.654	12.729	13.954	15.356	16.961	18.803	20.923	23.364	26.183	29.440	33.210
8		10.675	11.611	12.674	13.885	15.269	16.851	18.666	20.750	23.148	25.912	29.103
9		9.8226	10.641	11.568	12.620	13.817	15.182	16.743	18.530	20.580	22.936	25.648
10		9.0770	9.7960	10.607	11.525	12.566	13.749	15.097	16.636	18.396	20.412	22.727
11		8.4217	9.0560	9.7693	10.574	11.482	12.512	13.682	15.012	16.530	18.264	20.248
12		7.8431	8.4051	9.0349	9.7426	10.540	11.440	12.459	13.615	14.929	16.425	18.133
13		7.3300	7.8300	8.3884	9.0138	9.7159	10.506	11.398	12.406	13.548	14.846	16.322
14		6.8729	7.3195	7.8167	8.3716	8.9926	9.6892	10.473	11.356	12.353	13.483	14.764
15		6.4641	6.8646	7.3089	7.8033	8.3547	8.9713	9.6625	10.439	11.314	12.301	13.417
16		6.0971	6.4575	6.8562	7.2983	7.7898	8.3377	8.9500	9.6357	10.406	11.272	12.249
17		5.7662	6.0918	6.4508	6.8476	7.2875	7.7763	8.3207	8.9286	9.6090	10.372	11.230
18		5.4669	5.7620	6.0864	6.4439	6.8390	7.2766	7.7626	8.3036	8.9072	9.5822	10.339
19		5.1951	5.4635	5.7576	6.0809	6.4370	6.8303	7.2657	7.7489	8.2864	8.8857	9.5555
20		4.9476	5.1924	5.4600	5.7532	6.0753	6.4300	6.8215	7.2547	7.7351	8.2692	8.8642

Table A8.6 $n = 30$.

d	|	0	1	2	3	4	5	6	7	8	9	10
							i					
0		30.000	34.785	40.568	47.575	56.085	66.439	79.058	94.461	113.283	136.308	164.494
1		25.808	29.703	34.389	40.042	46.878	55.164	65.225	77.462	92.367	110.545	132.735
2		22.396	25.589	29.412	34.002	39.529	46.201	54.270	64.050	75.922	90.353	107.916
3		19.600	22.235	25.374	29.126	33.624	39.029	45.541	53.404	62.914	74.435	88.413
4		17.292	19.481	22.076	25.163	28.846	33.254	38.541	44.900	52.563	61.813	73.000
5		15.372	17.203	19.363	21.919	24.955	28.571	32.891	38.065	44.276	51.746	60.748
6		13.765	15.307	17.116	19.246	21.765	24.751	28.302	32.537	37.601	43.668	50.953
7		12.409	13.716	15.241	17.028	19.131	21.612	24.549	28.037	32.190	37.147	43.076
8		11.258	12.372	13.667	15.176	16.942	19.017	21.461	24.351	27.778	31.851	36.704
9		10.274	11.230	12.335	13.618	15.111	16.856	18.904	21.313	24.157	27.523	31.518
10		9.4269	10.253	11.202	12.299	13.569	15.046	16.771	18.792	21.166	23.965	27.273
11		8.6938	9.4112	10.232	11.175	12.262	13.520	14.982	16.687	18.681	21.022	23.776
12		8.0552	8.6819	9.3954	10.211	11.147	12.225	13.472	14.918	16.603	18.572	20.879
13		7.4957	8.0462	8.6699	9.3795	10.190	11.119	12.188	13.423	14.855	16.520	18.464
14		7.0027	7.4888	8.0371	8.6578	9.3634	10.169	11.091	12.151	13.375	14.792	16.438
15		6.5660	6.9975	7.4819	8.0278	8.6456	9.3473	10.147	11.063	12.115	13.327	14.729
16		6.1772	6.5620	6.9921	7.4748	8.0185	8.6332	9.3310	10.126	11.035	12.078	13.279
17		5.8294	6.1742	6.5579	6.9868	7.4677	8.0091	8.6208	9.3146	10.104	11.007	12.041
18		5.5168	5.8271	6.1710	6.5538	6.9813	7.4604	7.9995	8.6082	9.2981	10.083	10.979
19		5.2347	5.5150	5.8247	6.1679	6.5496	6.9757	7.4531	7.9898	8.5956	9.2816	10.061
20		4.9789	5.2333	5.5132	5.8222	6.1646	6.5453	6.9700	7.4456	7.9801	8.5828	9.2649

Appendix 9: Excel Functions—Goal Seek and Solver

Two very useful functions built into Excel that can help to find solutions or optimize results obtained from an exercise are *Goal Seek* and *Solver*. These are available under the *Data* → *What-If Analysis* menu. To work with these functions, basic knowledge of Excel is required.

A9.1 Function *Goal Seek*

Goal Seek determines the value a user would need to enter into a single cell to produce a desired result in a dependent cell. This is particularly important in explicit equations, which require a trial-and-error procedure because the unknown quantity is present on both sides of the equation. In such a case, *Goal Seek* attempts to find the value of the unknown quantity that equalizes the left-hand and right-hand sides of the equation.

Generally, with *Goal Seek*, Excel determines what value in an input cell produces a desired result in a formula cell. In most cases used in this book, *Goal Seek* atttempts to find the value for a cell that makes another cell, containing an equation, equal to zero. This equation may be entered in a problem just to equate the two sides of the explicit equation and find the value of the unknown that gives a zero result so as to make the two sides equal.

An example of *Goal Seek* is presented here to give the reader an idea of what this tool can do. Suppose you want to purchase a house, but according to your salary, you can afford to pay only 1000 Euro per month as a mortgage payment. The requirements of the bank which will lend you the money are that it can give you a 30-year (360 months) loan at an interest rate of 4.5% based on a 25% down payment. So the problem is determining the maximum value of the property you can purchase based on these constraints. Using an approximate purchase value of 300,000 Euro and the other inputs, as shown in Fig. A9.1, the monthly payment is 1140 Euro, which is more than the target value of 1000 Euro. The monthly payment is obtained from the Build-in Excel function *PMT(rate,nper,pv)*, where *nper* = loan period in months and *pv* = purchase value. Thus, for the current problem, this is equal using PMT(B5/ 12,B4,-B8).

To find the required monthly payment, you choose *Data* → *What-If Analysis* → *Goal Seek*. The *Goal Seek* dialog box then appears, as shown in Fig. A9.2. The logic is that you want to *Set cell:* B9 to 1000 *By changing cell:* B2. These values are entered in the various dialog boxes, as shown, by either typing them into the cells or pointing with the mouse. Subsequently, OK is clicked to start the *Goal Seek* process. It should be noted that the $ sign before the column letter and row number indicates an absolute (i.e., unchanging) cell address.

In milliseconds, Excel displays the *Goal Seek Status* window, as shown in Fig. A9.3, which shows the target value and the current value that Excel has calculated. As shown, an exact value is calculated, and cell B2 now shows the value of 263,148 Euro, which is the purchase value, with the constraints indicated above, that results in a monthly loan payment of exactly 1000 euro. By clicking *OK*, this solution is accepted. If *Cancel* is pressed, the worksheet is restored to the original values before *Goal Seek* was chosen.

	A	B
1	**Inputs:**	
2	Purchase Price =	€300,000
3	Down payment =	25%
4	Loan term (months) =	360
5	Interest rate =	4.50%
6		
7	**Results:**	
8	Loan Amount =	€225,000
9	Monthly Payment =	€1,140
10		

FIGURE A9.1

Goal Seek, mortgage payment example.

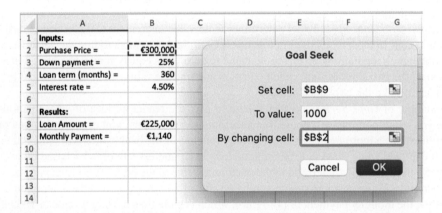

FIGURE A9.2

The *Goal Seek* window.

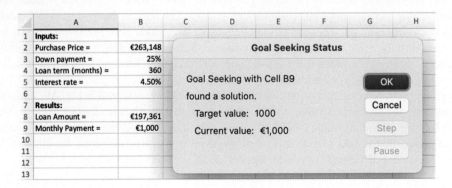

FIGURE A9.3

The *Goal Seek Status* window.

It is possible, according to the problem, that Excel cannot find a solution that the user is seeking simply because such a solution does not exist. In such a case, the **Goal Seeking Status** box informs the user of this fact.

In cases where Excel cannot provide a solution, and the user is sure that such a solution exists, the following options can be investigated (Aleksander and Kusleika, 2019):

1. Change the value of the input **By changing cell** in the **Goal Seek** dialog box (Fig. A9.2) to a value that is closer to the expected solution and press **OK**. If this value is not known, different values can be used, as in a trial-and-error procedure.
2. Adjust the **Maximum iterations** setting in the **Formulas** tab of Excel **Options** (**Excel Preferences** in Mac). Increasing this number may require more time to solve the problem, but many more possible solutions are evaluated.
3. Check the logic of the problem to make sure that the cell containing the formula depends on the changing cell specified.

Examples of using **Goal Seek** in this book can be seen in the problems for Chapter 3 (for example, Problem 3.1) related to the requirement to assume the glazing temperature to get an equal amount of heat flow from the plate to the glass cover and from the glass cover to the ambient. Other examples appear in Chapters 9, 11, 12, and 13.

A9.2 Function *Solver*

Goal Seek is a very useful tool, but it is only suitable for a single variable or one adjustable cell and returns only a single solution. The **Solver** tool becomes more useful with more variables, as the user can specify multiple adjustable cells and specify constraints on the values that the adjustable cells can have. **Solver** can generate a solution that maximizes or minimizes a cell and can generate multiple solutions to a problem. The **Solver** tool determines the values that you need to enter in multiple input cells to produce the result the user wants. Furthermore, because the user can specify additional constraints to the problem, a significant problem-solving ability is available. However, **Solver** can be much more complicated than **Goal Seek**.

Solver can typically be used in cases where a target cell depends on other cells and formulas which can be maximized, minimized, or set equal to a specified value; the target cell depends on a group of cells, called *changing cells*, which the **Solver** can modify to affect the target cell; and the solution must obey to certain limitations, called *constraints*.

Therefore, after a worksheet is set up, the **Solver** can be used to adjust the changing cells to produce the result that the user wants in the target cell, which at the same time meets all user-specified constraints. This is better explained by a few examples.

It should be noted that if **Solver** is not available in the **Data** menu, the user must install it, which can be done with a simple process as follows.

On a Windows PC:

1. Choose **Files** → **Options**
2. Select **Add-Ins** tab
3. At the bottom of the dialog box, select **Excel Add-Ins** from the **Manager** drop-down list and then click **Go**, after which the Add-ins dialog box appears
4. Check the box next to the **Solver Add-In** and then click **OK.**

On a Mac:

1. Open *Tools* on the top drop-down menu
2. Select *Excel Add-Ins*
3. From the box, check *Solver Add-In* and then click *OK.*

Example 1: Solving simultaneous linear equations

Suppose that you have the following set of three linear equations with three variables:

$$4x + 2y - z = 0$$
$$3x - 2y + 3z = 2$$
$$-4x - 3y + 2z = 1$$

So what is required is to find values for x, y, and z that satisfy all three equations. Figure A9.4 shows the worksheet that can be set up to solve this problem. This has three named cells all initialized to a value of 1:

x: B7
y: B8
z: B9

The three equations are entered in cells D7:D9 as follows:

D7: $= (4*B7) + (2*B8) - (B9)$
D8: $= (3*B7) - (2*B8) + (3*B9)$
D9: $= -(4*B7) - (3*B8) + (2*B9)$

These formulas use the values in cells B7:B9 for x, y, and z, respectively, to solve the three equations, which do not solve the problem because a value of 1 is used for all. The desired values are

	A	B	C	D	E
1	Example 1: Solving three simultaneous linear equations				
2	4x+2y-z=0				
3	3x-2y+3z=2				
4	-4x-3y+2z=1				
5					
6	Variable	Value		Formula	Desired value
7	x	1	Equation 1:	5	0
8	y	1	Equation 2:	4	2
9	z	1	Equation 3:	-5	1
10					

FIGURE A9.4

Solving three simultaneous linear equations with a *Solver* example.

shown in cells E7:E9, where constants on the RHS of the three equations are used. So a correct solution would be to find the values of x, y, and z that return values for cells D7:D9 equal to those in cells E7:E9.

Therefore, the **Solver** will need to adjust the values in x, y and z, i.e., changing the values in cells B7:B9 subject to the following constraints:

D7 = E7
D8 = E8
D9 = E9

The **Solver** window is shown in Fig. A9.5. The **Set Objective** could be any one of the three equations in cells D7 to D9. Additionally, for this example, it is not important which one of the three options—**Max**, **Min**, or **Value of**—is set, as the solution depends on the three constraints.

FIGURE A9.5

The window **Solver Parameters** for Example 1.

To run the *Solver*, press *Solve* at the bottom of the box. After a few seconds, the screen shown in Fig. A9.6 is obtained. And if the solution is satisfactory, i.e., the values in cells D7 = E7, D8 = E8, and D9 = E9, which indicates that the problem is solved satisfactorily, *OK* is clicked. As shown in Fig. A9.7, this is achieved by having x = 2, y = −7, and z=−6.

FIGURE A9.6

The window *Solver Results*.

	A	B	C	D	E
1	Example 1: Solving three simultaneous linear equations				
2	4x+2y-z=0				
3	3x-2y+3z=2				
4	-4x-3y+2z=1				
5					
6	Variable	Value		Formula	Desired value
7	x	2	Equation 1:	0	0
8	y	-7	Equation 2:	2	2
9	z	-6	Equation 3:	1	1
10					

FIGURE A9.7

The solved worksheet—Example 1.

Example 2: Solving the double-glazed flat plate collector example (Example 3.2)

Here the unknown parameters are the temperatures of the two glasses, T_{g1} and T_{g2}. These need to be guessed, and the solution obtained is correct if the three heat flows from the absorbing plate to glass 1, from glass 1 to glass 2, and from glass 2 to ambient are equal. Figure A9.8 shows the worksheet used to

	A	B	C	D	E	F	G	H	I	J	K	L
1	Example 3-2	Solver Example 2										
2	Inputs:											
3	Parameter	Symbol	Value	Units								
4	Collector slope=	β	35	Deg								
5	Collector area=	A_c	2	m								
6	Glass thickness=	L_g	0.004	m								
7	Plate thickness=	L_p	0.0005	m								
8	Space between glasses=	S_g	0.02	-								
9	Space between place-glass=	S_{p-g}	0.04	-								
10	Insulation thickness=	X_{ins}	0.05	m								
11	Insulation conductivity=	K_{ins}	0.05	W/m-K								
12	Plate temperaure=	T_p	353	K								
13	Ambient Temperature=	T_a	288	K								
14	Glass emissivity=	ε_g	0.88	-								
15	Plate emissivity	ε_p	0.1	-								
16	Wind Velocity=	V	2.5	m/s								
17	Collector length=	L	2	m								
18	Sigma=	σ	5.67E-08	W/m^2-K^4								
19	Gravity acceleration=	g	9.81	m/s^2								
20		**Assume**				**Air properties** (values should be within the mean temperatures, cells C30 and C31)						
21		T_{g1} =	300.00	K		Temp. (K)	v	Pr	k	μ	C_p	
22		T_{g2} =	310.00	K		300	1.68E-05	0.708	0.02624	1.983	1.0057	
23						350	2.08E-05	0.697	0.03003	2.075	1.009	
24	**Calculations**											
25	Parameter	Value	Units									
26	$h_{r,p-g2}$ =	0.819	W/m^2-K									
27	$h_{r,g2-g1}$ =	5.057	W/m^2-K									
28	$h_{r,g1-a}$ =	5.074	W/m^2-K									
29	Properties:		Mean Temp. (K)	v	Pr	k						
30	Mean temp between Tp and Tg2:	331.50	1.932E-05	0.701	0.0286							
31	Mean temp between Tg2 and Tg1:	305.00	1.720E-05	0.707	0.0266							
32												
33	Ra_{rp-g2}	152961										
34	Ra_{rg2-g1}	6148.4										
35												
36	$h_{c,p-g2}$ =	2.997	W/m^2-K									
37	$h_{c,g2-g1}$ =	2.181	W/m^2-K									
38	$h_{c,g1-a}$ =	11.294	W/m^2-K									
39												
40	$Q1/A_c$ =	164.1	W/m^2									
41	$Q2/A_c$ =	72.4	W/m^2	U_t =	2.168	W/m^2-K						
42	$Q3/A_c$ =	196.4	W/m^2									
43	AVERAGE =>	144.29	Differences=>	19.8	71.9	52.1	143.8					
44			Desired values=	0	0	0						
45			USE Solver to find values of Tg2 and Tg1 which will make differences = zero									

FIGURE A9.8

The worksheet for Example 2.

solve the problem with the two values of the glass covers guessed at 300 and 310 K, respectively (in cells C21 and C22). As shown by the heat flow results in cells B40: B42, these guesses are not correct, as the three heat flows should be the same. The differences in cells D43:F43 should also be zero. These cells contain a formula giving the difference between each heat flow and the average value of the three heat flows contained in cell B43.

The **Solver Parameters** window set up for this example is shown in Fig. A9.9. As before, the **Set objective** can be cell D43 (used), E43, or F43. The changing cells contain the two guessed values of T_{g1} and T_{g2} in cells C21:C22. The constraints are D43 = D44, E43 = E44, and F43 = F44, representing the result of the differences mentioned above and the desired values set to zero in all cases in cells D44: F44.

The solved problem is shown in Fig. A9.10. As can be seen, the two correct temperatures are $T_{g1} = 296.80$ K and $T_{g2} = 314.81$ K. These values make all heat flows equal to 143.3 W/m^2 and all the differences equal to zero. The final value of U_t obtained in this problem is 2.205 $W/m^2 K$.

FIGURE A9.9

The **Solver Parameters** window for Example 2.

	A	B	C	D	E	F	G	H	I	J	K	L
1	Example 3-2	Solver Example 2										
2	Inputs:											
3	Parameter	Symbol	Value	Units								
4	Collector slope=	β	35	Deg								
5	Collector area=	A_c	2	m								
6	Glass thickness=	L_g	0.004	m								
7	Plate thickness=	L_p	0.0005	m								
8	Space between glasses=	S_g	0.02	-								
9	Space between place-glass=	S_{p-g}	0.04	-								
10	Insulation thickness=	X_{ins}	0.05	m								
11	Insulation conductivity=	K_{ins}	0.05	W/m-K								
12	Plate temperaure=	T_p	353	K								
13	Ambient Temperature=	T_a	288	K								
14	Glass emissivity=	ε_g	0.88	-								
15	Plate emissivity	ε_p	0.1	-								
16	Wind Velocity=	V	2.5	m/s								
17	Collector length=	L	2	m								
18	Sigma=	σ	5.67E-08	$W/m^2\text{-}K^4$								
19	Gravity acceleration=	g	9.81	m/s^2								
20		**Assume**				Air properties (values should be within the mean temperatures, cells C30 and C31)						
21		$T_{g1} =$	296.80	K		Temp. (K)	v	Pr	k	µ	C_p	
22		$T_{g2} =$	314.81	K		300	1.68E-05	0.708	0.02624	1.983	1.0057	
23						350	2.08E-05	0.697	0.03003	2.075	1.009	
24	**Calculations**											
25	Parameter	Value	Units									
26	$h_{r,p-g2} =$	0.836	$W/m^2\text{-}K$									
27	$h_{r,g2-g1} =$	5.101	$W/m^2\text{-}K$									
28	$h_{r,g1-a} =$	4.991	$W/m^2\text{-}K$									
29	Properties:		Mean Temp. (K)	v	Pr	k						
30	Mean temp between Tp and Tg2:		333.91	1.951E-05	0.701	0.0288						
31	Mean temp between Tg2 and Tg1:		305.81	1.726E-05	0.707	0.0267						
32												
33	$Ra_{,p-g2}$	132118										
34	$Ra_{,g2-g1}$	10960.2										
35												
36	$h_{c,p-g2} =$	2.917	$W/m^2\text{-}K$									
37	$h_{c,g2-g1} =$	2.855	$W/m^2\text{-}K$									
38	$h_{c,g1-a} =$	11.294	$W/m^2\text{-}K$									
39												
40	$Q1/A_c =$	**143.3**	W/m^2									
41	$Q2/A_c =$	**143.3**	W/m^2	$U_t =$	2.205	$W/m^2\text{-}K$						
42	$Q3/A_c =$	**143.3**	W/m^2									
43	AVERAGE =>	143.30	Differences=>	0.0	0.0	0.0	0.0					
44			Desired values=	0	0	0						
45			USE Solver to find values of Tg2 and Tg1 which will make differences = zero									

FIGURE A9.10

The final worksheet for Example 2.

A9.2.1 Solver options

Solver is generally a complex tool, and what is presented here is just to indicate its use for the purpose of solving the problems in this book. Readers interested in learning more may visit the website of the company that developed the tool, *Frontline Systems,* at www.solver.com, for more examples and other helpful information. Several *Solver* Options can be adjusted by the user by pressing *Options* in the

Solver Parameters window, as shown in Fig. A9.5 or Fig. A9.9. By doing so, the window shown in Fig. A9.11 is obtained.

From these settings, many aspects of the solution process can be controlled, as well as load and save model specifications in a worksheet range. Usually, the user wants to save to a model using more than one set of *Solver* parameters with a worksheet. This is done automatically by Excel, which saves the parameters in the current worksheet. If additional models are saved, Excel stores the information as formulae corresponding to the specifications. The last cell in the saved range is an array formula that holds the options settings.

Sometimes it happens to get a report that the *Solver* cannot find a solution, although you know that a solution exists. In such a case, one or more of the *Solver Options* shown in Fig. A9.11 can be changed before trying again. The various *Solver* options are (Aleksander and Kusleika, 2019):

Constraint Precision: This value specifies how close the Cell Reference and Constraint formulae must be to satisfy a constraint. Lower precision leads to shorter run times.

FIGURE A9.11

The *Solver* - *Options* window.

Use Automatic Scaling: This is ticked when the problem involves large differences in magnitude, for example, when the user attempts to maximize a percentage by varying cells that are very large.

Show Iteration Results: If this is selected, *Solver* is instructed to pause the results after each iteration.

Ignore Integer Constraints: When this box is checked, *Solver* ignores constraints that specify that a particular cell must be an integer. This option allows *Solver* to find a solution that cannot be obtained otherwise.

Max Time: If time constraints exist, this setting specifies the time in seconds that the *Solver* will spend to find a solution to a problem. If an exceeded time limit is reported, this setting can be increased.

Iterations: This entry indicates the maximum number of trial solutions the *Solver* will perform.

Max Subproblems: This is used in complex problems to specify the maximum number of sub-problems that may be explored by the Evolutionary algorithm.

Max Feasible Solutions: This feature is also used in complex problems to specify the maximum number of feasible solutions that may be explored by the Evolutionary algorithm.

Reference

Aleksander, M., Kusleika, D., 2019. Excel® 2019 Bible. Wiley.

Appendix 10: System Advisor Model

Section 11.5.4 provides an introduction to the System Advisor Model. Further details are given here. The System Advisor Model (SAM) Version 2020.11.29 is a performance and financial model designed to facilitate decision-making for persons working in renewable energy as engineers, developers, or researchers. As mentioned in Chapter 11, SAM makes performance predictions and estimates the energy cost for grid-connected power projects based on installation and operating costs and system design parameters that users specify as model inputs. The program can be downloaded free from the National Renewable Energy Laboratory (NREL) website (https://sam.nrel.gov/). The following is a short description of the program and its capabilities; a full description can be obtained from the SAM help file, which is like a user manual on which the following description is based, and from the program website.

The first step in creating a SAM file is to choose a performance and financial model for the project. SAM automatically fills the input variables with a set of default values for the type of project chosen. It is the responsibility of the user/designer to review and modify all the input data according to the analysis to be performed. Subsequently, the user must provide information about a project's location, the type of equipment employed, the cost of installing and operating the system, and financial and incentive (subsidy) parameters.

Each performance model in SAM represents a part of the system, and each financial model represents a project's financial structure. As in all models, the user must specify input data to describe the performance characteristics of physical equipment in the system and the various costs involved. SAM also requires a weather file describing the renewable energy resource and weather conditions for the project location. Various such files are available in the program database, can be downloaded from the internet, or can be created by the user with available data. Additionally, SAM includes several databases of performance data and coefficients for system components such as wind turbines, parabolic trough receivers and collectors, photovoltaic modules and inverters, and biopower combustion systems. For all input variables, the user can use the default values provided or change them according to the project.

After running the simulation, SAM displays the results in tables and graphs, showing the net present value (NPV) of the project, the first-year annual production, detailed annual cash flows, hourly performance data, and other metrics. All graphs and tables can be exported in various formats to be included in reports and analyzed further with spreadsheet programs like Excel.

A10.1 Performance models included in the program

SAM's performance models carry out hour-by-hour calculations of a power system's electric output, generating a set of 8760 hourly values that represent the system's electricity production over a single year. Some performance models also support sub-hourly simulations. The current version of the SAM includes models for the following types of systems:
- Photovoltaic systems (flat-plate and concentrating)
- Parabolic trough concentrating solar power (CSP)
- Power tower concentrating solar power (molten salt)
- Linear Fresnel concentrating solar power (heat transfer fluid and direct steam)

- Industrial process heat models for parabolic troughs and linear collectors
- Conventional fossil-fuel thermal
- Solar water heating for residential or commercial buildings
- Wind power systems
- Marine energy
- Geothermal power
- Biomass combustion

The user can compare different kinds of projects by creating more than one case in a file. For example, one could compare the savings of a residential rooftop solar water heater to those of a photovoltaic system. Or, for a large utility-scale project, one could compare the power purchase price that would be required to make a wind, a photovoltaic, or a CSP project profitable at a given location. However, it should be noted that SAM does not model hybrid power systems; for example, a single project that combines wind turbines and photovoltaic modules cannot be modeled.

A10.2 Financial models

The financial models calculate financial metrics for various kinds of power projects based on the cash flows of a project for the analysis period that the user specifies. The financial models use the electrical output of the system calculated by the performance model to calculate the series of annual cash flows. SAM includes financial models for the following types of projects, all based on USA rules and conditions:

- Residential (retail electricity rates)
- Commercial (retail rates)
- Third-party ownership (retail rates and power purchase agreement)
- Power purchase agreement (PPA) (utility-scale or power generation project)
- Merchant plant

A10.2.1 Residential and commercial projects

Residential and commercial projects are considered to be financed through either a loan or cash payment and recover investment costs by selling electricity through either a net metering or time-of-use pricing agreement. For these projects, the following financial metrics are reported in addition to a detailed annual cash flow:

- Levelized cost of energy (LCOE)
- Electricity bill with and without the power system
- After-tax NPV
- Payback period

A10.2.2 Power purchase agreement (PPA) projects

PPA projects are assumed to sell electricity through a power purchase agreement at a fixed price with optional annual escalation and time-of-delivery (TOD) factors. For these projects, SAM calculates a detailed annual cash flow and the following metrics:

- Levelized cost of energy (LCOE)
- PPA price (electricity sales price)

- Internal rate of return
- Net present value (NPV)
- Debt percent or debt service coverage ratio

SAM can either calculate the internal rate of return based on a power price or calculate the power price based on the rate of return, both specified by the user.

A10.2.3 Project cash flow

SAM calculates financial metrics from project cash flows representing the value of energy savings for projects using retail electricity rates and the value of revenue from electricity sales for projects selling electricity under a power purchase agreement. For PPA partnership models, SAM calculates cash flows from the project perspective and from the perspective of each partner. The project's annual cash flows include:

- Value of electricity sales (or savings) and incentive payments
- Installation costs
- Operating, maintenance, and equipment replacement costs
- Debt principal and interest payments
- Tax benefits and liabilities (accounting for any tax credits for which the project is eligible)
- Incentive payments
- Project and partner's internal rate of return requirements (for PPA projects)

A10.2.4 Incentives

The financial model can account for a wide range of incentive payments and tax credits:
- Investment based incentives (IBI)
- Capacity-based incentives (CBI)
- Production-based incentives (PBI)
- Investment tax credits (ITC)
- Production tax credits (PTC)
- Depreciation (modified accelerated depreciation schedule [MACRS], straight-line, custom, bonus, etc.)

A10.3 Performance models short descriptions

SAM performance models run hourly or subhourly simulations to calculate the electrical power output of a system. The sum of these values over a year gives the total annual output used by the financial model to calculate the project's annual cash flows and other financial metrics.

It should be noted that the solar water heating models calculate the thermal output of the system by assuming that it replaces the electricity that would normally heat water in a conventional electric water heating system.

Moreover, because of the nature of the technology, the geothermal model calculates system performance over a period of years rather than hours. SAM's detailed photovoltaic, PVWatts, solar water heating, and physical trough models can perform sub-hourly simulations, but to do so, sub-hourly weather data are required. The following is a short description of the various models.

A10.3.1 **Photovoltaic systems**

SAM models grid-connected photovoltaic systems consisting of a photovoltaic array and inverter. The array can consist of flat-plate or concentrating photovoltaic (CPV) modules with one axis, two axes, or no tracking in three different models as follows.

Detailed photovoltaic model

This model calculates the electrical output of a grid-connected photovoltaic system using module and inverter models. It models the effect of temperature on module performance and has options for calculating shading and other losses of the system. The model also includes a system sizing calculator to enable the user to determine the number of modules and inverters required by the system.

PVWatts model

The PVWatts model can simulate a grid-connected photovoltaic system using a few basic inputs to describe the nameplate capacity of the system, the array orientation and mounting type, and system losses. Three types of modules are considered by PVWatts, which makes internal assumptions about the characteristics of the module and the inverter. The PVWatts model can be used for a preliminary project analysis before information about the type of equipment planned to be used in the system is available.

High-concentration photovoltaic model

This model is appropriate for grid-connected photovoltaic systems with high-concentration photovoltaic modules (HCPV). The concentrating photovoltaic model also uses separate models to represent the module and inverter.

A10.3.2 **Energy storage**

Detailed photovoltaic–battery model

This model couples a grid-connected photovoltaic system to a battery bank, which can be connected to either the DC side or the AC side of the system. The model must be used with a financial model as a behind-the-meter application for Residential, Commercial, and Third-Party Ownership financial models and a front-of-the-meter application for PPA or Merchant Plant financial models.

PVWatts–battery model

The PVWatts–battery model is a simplified implementation of the PV–battery model for behind-the-meter applications with Residential, Commercial, or Third-Party Ownership financial models.

Generic system–battery model

The detailed PV-Battery model couples a grid-connected power system to a battery bank, which can be connected to either the DC or the AC side of the system. The model must be used with a financial model as a behind-the-meter application for Residential, Commercial, and Third-Party Ownership financial models and a front-of-the-meter application for PPA or Merchant Plant financial models.

A10.3.3 **Concentrating solar power**

The concentrated solar power (CSP) models concern grid-connected thermal power plants which generate steam to drive an electric power generation plant or industrial process heat. It includes eight different models.

Parabolic trough (physical model)

This model calculates the electricity delivered to the grid by a parabolic trough solar plant that provides thermal energy to a power block for electricity generation, with an optional thermal energy storage system. This physical model characterizes many parabolic trough collector system components from the first principles of heat transfer and thermodynamics.

Parabolic trough (empirical model)

This empirical model is used for the same type of parabolic trough collector system as the physical trough model but employs a set of curve-fit equations derived from regression analysis of data measured from the SEGS plants in California, USA. Therefore, it is limited to modeling systems comprising components for which there are measured data.

Power tower molten salt model

The power tower molten salt models a system called a central receiver system, which consists of a heliostat field, a tower and receiver, a power block, and an optional storage system. Heliostats focus direct normal solar radiation onto a receiver at the top of the tower, where molten salt is heated and pumped to the power block. Subsequently, the power block generates steam that drives a conventional steam turbine and generator to convert the thermal energy to electricity.

Linear Fresnel molten salt model

This model uses a system comprising a field of Fresnel reflectors that focus sunlight on an absorber in the focal plane located above the reflector. The absorber circulates a heat transfer fluid that transfers heat to a power block. The system may also include thermal storage.

Linear Fresnel direct steam model

The linear Fresnel direct steam model is a system model consisting of a field of Fresnel reflectors. Steam is used as the heat transfer fluid throughout the system. This system also may not include thermal storage.

Concentrating solar power generic model

The CSP model allows the user to model a system consisting of a solar field, a power block with a conventional steam turbine, and an optional thermal energy storage system. The solar field is represented using a set of optical efficiency values for different sun angles. It can be used for any solar technology that uses solar energy to generate steam for electric power generation.

   This family of models also includes two industrial process heat systems models as follows.

Parabolic trough—heat model

This model uses a version of the CSP physical parabolic trough model with no power cycle modified for process heat applications.

Linear Fresnel direct steam—heat model

This model is for a linear collector using direct steam with no power cycle that is modified for process heat applications.

A10.3.4 Marine energy

Marine energy systems convert ocean wave or tidal energy to electricity. It includes two models as follows.

Wave energy model

This model is for a system that uses a wave energy converter to convert the energy of ocean waves into electricity.

Tidal energy

This model is for a system that uses a tidal energy converter to convert the energy of ocean tides into electricity.

A10.3.5 Wind energy

The wind energy model is of a single wind turbine or a project with two or more wind turbines that produce electrical power to be sold to the grid.

A10.3.6 Fuel cell—photovoltaic—battery

This combines the PVWatts, battery, and fuel cell models for systems with a combined photovoltaic array and fuel cell and an optional battery bank.

A10.3.7 Geothermal

It should be noted that the geothermal power model is for electricity-generating systems only and does not include ground source heat pumps or geo-exchange systems. This model considers using ground heat to drive a steam-electric power generation plant. SAM analyzes the plant's performance over its lifetime, considering that changes in the resource and electrical output occur monthly over several years rather than over hours for 1 year, as in solar and other technologies modeled in SAM.

A10.3.8 Solar water heating

This model considers a two-tank glycol system with an auxiliary electric heater and storage tank for residential or commercial applications. The model allows the user to modify the location (weather data file), hot water load profiles, and performance characteristics of the collector, heat exchanger, and storage tanks.

A10.3.9 **Biomass combustion**

This system considers the burning of biomass, with or without supplementary coal, in a combustion system to generate steam that drives an electric power generation plant.

A10.3.10 **Generic system**

This is used to model a basic conventional power plant and enables the user to compare a renewable energy project to a basic conventional power plant using consistent financial assumptions. This model allows the user to characterize the plant's performance by specifying either the nameplate capacity and capacity factor value, assuming that the plant generates power at a constant rate over the year, or by specifying the hourly or sub-hourly generation profile, assuming that the plant generates power according to the generation profile specified.

A10.4 **Financial models**

SAM's financial models calculate a project's cash flow over a user-specified analysis period. The cash flow includes the value of electricity generated by the system, incentives, the cost of installation, operation and maintenance, taxes, and debt. It should be noted that SAM is designed to calculate the value of electricity generated by a system. The economic metrics reported by the program are based on units of electrical energy rather than thermal energy. As already indicated, the solar water heating model calculates the value of the electricity saved by the system, assuming that heat from the system displaces the heat that would have been generated by a conventional electric water heater without the solar system.

The financial models can represent two main types of projects:
- Residential and commercial projects that buy and sell electricity at retail rates and displace power purchases from the grid;
- PPA projects that sell electricity at wholesale rates to meet internal rate of return requirements.

A10.4.1 **Residential and commercial owner**

In SAM, residential and commercial projects are considered to buy and sell power at retail rates. They may be financed through either a loan or without debt, i.e., full cash payment. These projects recover investment costs by selling electricity at rates established by the electricity service provider. SAM calculates metrics for these projects at the project level, assuming that a single entity develops, owns, and operates the project.

For residential and commercial projects, SAM calculates the project LCOE, which represents the cost of installing and operating the system, including debt and tax costs and accounting for incentives. The model also calculates the net present value (NPV) of the after-tax cash flow and a payback period representing the number of years required for the cumulative after-tax cash flow to cover the initial equity investment in the project.

Commercial projects may qualify for tax deductions under MACRS as described in the United States tax code. SAM provides options for specifying custom depreciation schedules in addition to the MACRS mid-quarter and half-year schedules on the depreciation page.

Residential and commercial projects are typically smaller than 500 kW, although SAM does not restrict system sizes, so it is possible to model any size system using either the residential or commercial financial model.

SAM's Electricity Rates page provides a range of options for specifying the retail electricity rate structure for a project. The rate structure may include any of the following:
- Flat buy and sell rates (with or without net metering)
- Time-of-use energy charges
- Monthly demand charges (either fixed or time-of-use)
- Tiered rates
- Fixed monthly charges

For projects with demand charges and tiered rates, SAM requires electric load data, which are specified on the Electric Load page.

A10.4.2 Third-party ownership

In a third-party ownership project, a photovoltaic system is installed on a residential or commercial property. The property owner, or customer, makes an agreement with a third party who installs, operates, and owns the system. The system reduces the customer's electricity bill, and the customer makes payments to the third-party owner for the system, either through a lease agreement or a power purchase agreement.

A10.4.3 Power purchase agreement

Power purchase agreement (PPA) projects sell all of the electricity generated by the power system at a price negotiated through a PPA to meet a set of equity returns requirements. A PPA project may involve one or two parties. SAM provides options for calculating a power purchase price given a target internal rate of return or for calculating the rate of return given a power purchase price.

The power price can be made to vary annually over the analysis period using either a fixed annual escalation rate or by assigning a separate price to each year. The price can be made to vary diurnally using time-of-delivery (TOD) factors defined either using hour-by-month schedules or hourly (or sub-hourly) time-series factors.

A10.4.4 Single owner

In the Single-Owner model, the project owner builds, owns, and operates the power system and has sufficient tax liability to fully utilize all tax benefits. The owner may be either the original developer or a third-party tax investor that purchases the project from the developer.

A10.4.5 Partnership flip with and without debt

The partnership flip with debt (Leveraged Partnership Flip) and partnership flip without debt (All Equity Partnership Flip) models are for two-party projects that involve equity investments by a project developer and a third-party tax investor.

The investor has sufficient tax liability from its other business operations to fully utilize tax credits and depreciation deductions in years when the project generates benefits. The project sets up a limited liability entity, and once the project begins generating and selling electricity, all of the project's net cash flows and tax benefits are passed through this entity to its owners.

The project allocates a majority of the cash and tax benefits to the investor when the project begins operation and until the "flip year" when the tax investor receives a pre-negotiated after-tax internal rate of return or "target IRR." Once the target IRR is reached, a majority of the cash and any remaining tax benefits are allocated to the developer.

A10.4.6 Sale-leaseback

The Sale-Leaseback model is another two-party structure that involves a tax investor purchasing 100% of the project from the developer and then leasing it back to the developer. This structure differs from the partnership flip structures in that the tax investor and the developer do not share the project cash and tax benefits (or liability). Instead, each party has its own separate cash flow and taxable income.

The purchase price paid by the tax investor is equal to the total project cost, less a lease payment and the value of working capital reserve accounts. The developer typically funds the reserve accounts to ensure it has some financial exposure. The tax investor receives lease payments from the developer and any ownership-related incentives such as tax credits, incentive payments, and depreciation tax deductions. The developer operates the project and keeps any excess cash flow from operations after payment of all operating expenses and lease payments. This provides the developer with an incentive to operate the project as efficiently as possible.

A10.4.7 Merchant plant

The Merchant Plant model is similar to the Single-Owner model described above, except that power generated by the system is sold at market prices that can vary on an hourly (or sub-hourly), daily, or monthly basis over the project life. Revenue can be from energy production, ancillary services, capacity payments, or a combination of the three.

A10.4.8 Levelized cost of energy calculator

The LCOE calculator uses a simple method to calculate the LCOE of a project. The installation cost, operating costs, and a fixed charge rate as inputs are provided by the user, and the model calculates the LCOE based on the annual energy generated by the system. The calculator can also calculate the fixed charge rate when the basic financial parameters are provided by the user.

A10.5 Solar and wind resource files

As already indicated, SAM comes with a small set of sample weather files for the solar and wind performance models. The user can use the download tools on the Location and Resource page to download solar resource data and from the Wind Resource input page to download wind data.

Index

Note: 'Page numbers followed by "*f*" indicate figures and "*t*" indicate tables.'

Printed in the United States
by Baker & Taylor Publisher Services